Senior Biology 2

Student Workbook

Senior *Biology 2*

Student Workbook

Previous annual editions 2002-2009
Ninth Edition 2011

FOURTH PRINTING with corrections

ISBN 978-1-877462-60-3

Published by **BIOZONE International Ltd**

Printed by REPLIKA PRESS PVT LTD using paper produced from renewable and waste materials

About the Writing Team

Tracey Greenwood joined the staff of Biozone at the beginning of 1993. She has a Ph.D in biology, specializing in lake ecology, and taught undergraduate and graduate biology at the University of Waikato for four years.

Kent Pryor has a BSc from Massey University majoring in zoology and ecology. He was a secondary school teacher in biology and chemistry for 9 years before joining Biozone as an author in 2009.

Richard Allan has had 11 years experience teaching senior biology at Hillcrest High School in Hamilton, New Zealand. He attained a Masters degree in biology at Waikato University, New Zealand.

Purchases of this workbook may be made direct from the publisher:

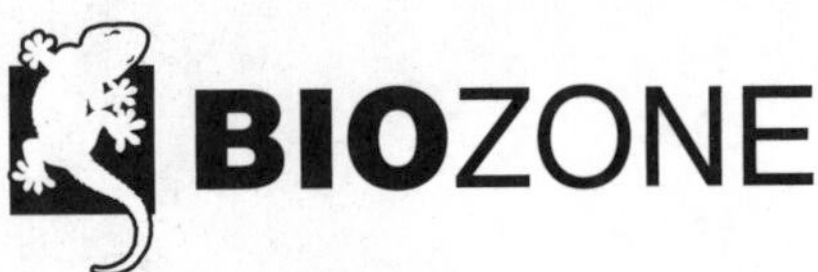

www.thebiozone.com

NORTH & SOUTH AMERICA, AFRICA:

BIOZONE International Ltd.
P.O. Box 13-034, Hamilton 3251, **New Zealand**
Telephone: +64 7-856-8104
FREE Fax: 1-800-717-8751 (USA-Canada)
FAX: +64 7-856-9243
E-mail: sales@biozone.co.nz

UNITED KINGDOM:

BIOZONE Learning Media (UK) Ltd.
Bretby Business Park, Ashby Road, Bretby, Burton upon Trent, DE15 0YZ, **UK**
Telephone: +44 1283-553-257
FAX: +44 1283-553-258
E-mail: sales@biozone.co.uk

ASIA & AUSTRALIA:

BIOZONE Learning Media Australia
P.O. Box 2841, Burleigh BC, QLD 4220, **Australia**
Telephone: +61 7-5535-4896
FAX: +61 7-5508-2432
E-mail: sales@biozone.com.au

Preface to the 2011 Edition

This is the ninth edition of Biozone's **Student Workbook** for students in biology programs at grades 11 and 12 or equivalent. This title, and its companion volume, are particularly well suited to students taking International Baccalaureate (IB) Biology, Advanced Placement (AP) Biology, or Honors Biology. Biozone's aim with this release is to build on the successful features of previous editions, while specifically focusing on scientific literacy and learning within relevant contexts. The workbook is a substantial revision and marks an important shift in several respects from earlier versions of this product.

- Content reorganization. In this edition we have brought together the content for classification, animal and plant form and function, and ecology into one volume, and shifted the chapters on evolution into Senior Biology 1. This reorganization of content follows the schedule of the AP biology program and provides a more cohesive coverage of material in these areas. Extension material is provided on the Teacher Resource CD-ROM (for separate purchase).
- International Baccalaureate students and teachers will find that, although core content has shifted between volumes (Senior Biology 1 and 2), it should be easier to locate material. Activities suitable for HL-only are indicated in the 'Contents', as is material that is not required under the IB scheme (there is a limited amount of this). **IB Options (A-H) are provided as complete units on the IB Options CD-ROM** (for separate purchase). Options C-E are also adequately covered within the workbooks for those making those option choices.
- A contextual approach. We encourage students to become thinkers through the application of their knowledge in appropriate contexts. Many chapters are prefaced with an account examining a 'biological story' related to the theme of the chapter. This approach provides a context for the material to follow and an opportunity to focus on comprehension and the synthesis of ideas. Throughout the workbook, there are many examples of applying knowledge within context. Examples among many include the search for blood substitutes, renal dialysis, contraception, cancer treatment using monoclonal antibodies, and exercise physiology.
- Concept maps introduce each main part of the workbook, integrating the content across chapters to encourage linking of ideas.
- An easy-to-use chapter introduction comprising succinct learning objectives, a list of key terms, and a short summary of key concepts.
- An emphasis on acquiring skills in scientific literacy. Each chapter includes a comprehension and/or literacy activity, and the appendix (a new feature) includes references for works cited throughout the text.
- *Web links* and *Related Activities* support the material provided on each activity page.

A Note to the Teacher

This workbook is a student-centered resource, and benefits students by facilitating independent learning and critical thinking. This workbook is just that; a place for your answers notes, asides, and corrections. It is **not a textbook** and annual revisions are our commitment to providing a current, flexible, and engaging resource. The low price is a reflection of this commitment. Please **do not photocopy** the activities. If you think it is worth using, then we recommend that the students themselves own this resource and keep it for their own use. I thank you for your support.
Richard Allan

Acknowledgements

We would like to thank those who have contributed to this edition • Dr. John Craig for permission to use his material on the behavior of swamphen • Mary McDougall, Sue FitzGerald and Gwen Gilbert for their efficient handling of the office • TechPool Studios, for their clipart collection of human anatomy: Copyright ©1994, TechPool Studios Corp. USA (some of these images were modified by R. Allan and T. Greenwood) • Totem Graphics, for their clipart collection • Corel Corporation, for vector clipart from the Corel MEGAGALLERY collection • 3D artwork created using Poser IV, Curious Labs and Bryce.

Photo Credits

Royalty free images, purchased by Biozone International Ltd, are used throughout this workbook and have been obtained from the following sources: **Corel** Corporation from various titles in their Professional Photos CD-ROM collection; **IMSI** (International Microcomputer Software Inc.) images from IMSI's MasterClips® and MasterPhotosTM Collection, 1895 Francisco Blvd. East, San Rafael, CA 94901-5506, USA; ©1996 **Digital Stock**, Medicine and Health Care collection; **©Hemera** Technologies Inc, 1997-2001; © 2005 JupiterImages Corporation www.clipart.com; ©1994., **©Digital Vision**; Gazelle Technologies Inc.; **PhotoDisc®**, Inc. USA, www.photodisc.com • 3D modeling software, Poser IV (Curious Labs) and Bryce.

The writing team would like to thank the following individuals and institutions who kindly provided photographs: • The late Dr. M. Soper, for his photograph of the waxeye feeding chicks • Stephen Moore, for his photo of a hydrophyte, *Myriophyllum* and for his photos of stream invertebrates • Dr Roger Wagner, Dept of Biological Sciences, University of Delaware, for the LS of a capillary • Dan Butler for his photograph of a finger injury • PASCO for their photographs of probeware (available for students of biology in the USA) • The three-spined stickleback image, which was originally prepared by Ellen Edmonson as part of the 1927-1940 New York Biological Survey. Permission for use granted by the New York State Department of Environmental Conservation. • Wellington Harrier Athletics Club • Helen Hall for the picture of her late husband Richard Hall • Janice Windsor, for photographs taken 'on safari' in East Africa • LJ Grauke, USDA-ARS Pecan Breeding & Genetics, for his photograph of budscales in shagbark hickory • Simon Pollard, for his photograph of the naked mole rat • UC Regents David campus • D. Fankhauser-University of Cincinnati, Clermont College • Bruce Wetzel and Harry Schaefer, National Cancer Institute • Charles Goldberg, UCSD School of Medicine • Ian Smith • Rogan Colbourne • Karen Nichols for the photograph of the digger wasp

Contributors identified by coded credits are as follows: **BF**: Brian Finerran (Uni. of Canterbury), **BH**: Brendan Hicks (Uni. of Waikato), **BOB**: Barry O'Brien (Uni. of Waikato), **CDC**: Centers for Disease Control and Prevention, Atlanta, USA, **COD**: Colin O'Donnell, **DEQ**: Dept of Environment Queensland Ltd., **DH**: Don Horne, **DOC**: Dept of Conservation (NZ), **DNRI**: Dept of Natural Resources, Illinois, **DS**: Digital Stock; **EII**: Education Interactive Imaging, **EW**: Environment Waikato, **GW**: Graham Walker, **IF**: I. Flux (DoC), **IS**: Ian Smith, **JB-BU**: Jason Biggerstaff, Brandeis University, **JDG**: John Green (Uni. of Waikato), **NASA**: National Aeronautics and Space Administration, **NOAA**: National Oceanic and Atmospheric Administration www.photolib.noaa.gov **NYSDEC**: New York State Dept of Environmental Conservation, **PH**: Phil Herrity, **RA**: Richard Allan, **RCN**: Ralph Cocklin, **TG**: Tracey Greenwood, **WBS**: Warwick Silvester, University of Waikato, **WMU**: Waikato Microscope Unit.

We also acknowledge the photographers that have made their images available through **Wikimedia Commons** under Creative Commons Licences 2.5. or 3.0: • Jon Mollivan • Psychonaught • Hans Hillewaert • WmPearl • US Fish and Wildlife • Mike Baird • Vlmastra • Cereal Research Centre, AAFC • New York State Dept of Environmental Conservation • Gina Mikel • Andreas Trepte • Velela

Special thanks to all the partners of the Biozone

Cover Photographs

Main photograph: The Eurasian lynx (*Lynx lynx*) ranges from central and northern Europe across Asia. It was considered locally extinct in some parts of its former range but resettlement programs begun in the 1970s, together with a ban on hunting, have been successful in reestablishing breeding populations in the wild.

PHOTO: ©Jasper Doest/Foto Natura/Minden Pictures

Background photograph: Autumn leaves, Image ©2005 JupiterImages Corporation www.clipart.com

Contents

CODES: Δ Upgraded ☆ New activity * Not for IB-SL ** Not for IB (or extension) † IB Option

CONTENTS *(continued)*

CODES: Δ Upgraded ☆ New activity * Not for IB-SL ** Not for IB (or extension) † IB Option

CONTENTS *(continued)*

CODES: Δ Upgraded ☆ New activity * Not for IB-SL ** Not for IB (or extension) † IB Option

Getting The Most From This Resource

This workbook is designed as a resource to increase your understanding and enjoyment of biology. While this workbook meets the needs of most general biology courses, it also provides specific keyed objectives for the **International Baccalaureate** (IB) and **Advanced Placement** (AP) courses. Consult the Syllabus Guides on pages 6-8 of this workbook to establish where material for your syllabus is covered. It is hoped that this workbook will reinforce and extend the ideas developed by your teacher. It must be emphasized that this workbook is **not a textbook**. It is designed to complement the biology textbooks provided for your course. Each topic in the workbook includes the following useful features:

Features of the Concept Map

Each major section of the workbook has central theme:
Part 1: Diversity of organisms
Part 2: The structure and function of animals
Part 3: The structure and function of plants
Part 4: Ecology and the biosphere.
The themes in Senior Biology 1 and 2 also encapsulate the recurring themes (I-VIII) of the AP scheme.

Encouraging Key Competencies

Thinking - bringing ideas together
Relating to others - communicating
Using language, symbols, and text
Managing self - independence
Participating and contributing

Each section of the workbook emphasizes skills and knowledge to be gained.

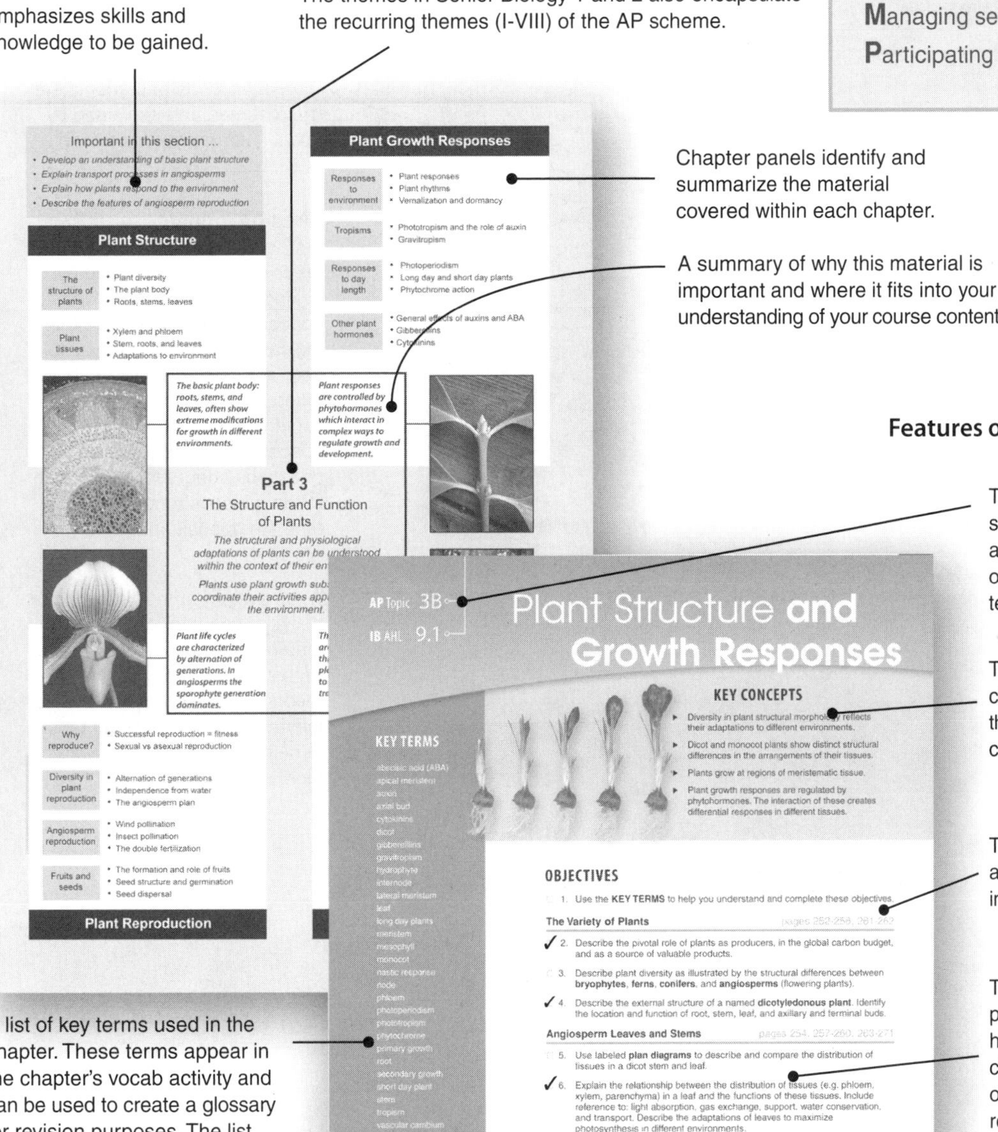

Chapter panels identify and summarize the material covered within each chapter.

A summary of why this material is important and where it fits into your understanding of your course content.

Features of the Chapter Topic Page

The part of the AP or IB (SL/HL) scheme to which this chapter applies. For other courses, objectives can be assigned at the teacher's discretion.

The important key ideas in this chapter. You should have a thorough understanding of the concepts summarized here.

The page numbers for the activities covering the material in this subsection of objectives.

The objectives provide a point by point summary of what you should have achieved by the end of the chapter. An equivalent set of objectives, for teacher's-only reference, is provided in the Teacher's Handbook. These provide extra explanatory detail and examples.

You can use the check boxes to mark objectives to be completed (a **dot** to be done; a **tick** when completed).

A list of key terms used in the chapter. These terms appear in the chapter's vocab activity and can be used to create a glossary for revision purposes. The list represents the minimum literacy requirement for the chapter.

Periodicals of interest are identified by title on a tab on the activity page to which they are relevant. The full citation appears in the **Appendix** on the page indicated.

The Weblinks on many of the activities can be accessed through the web links page at: *www.thebiozone.com/weblink/SB2-2603.html* See page 5 for more details.

Extra resources for this chapter are available on the Teacher Resource CD-ROM (for separate purchase).

Using the Activities

The activities make up most of the content of this book. Your teacher may use the activity pages to introduce a topic for the first time, or you may use them to revise ideas already covered by other means. They are excellent for use in the classroom, as homework exercises and topic revision, and for self-directed study and personal reference.

Perforations allow easy removal so that pages can be submitted for grading or kept in a separate folder of related work.

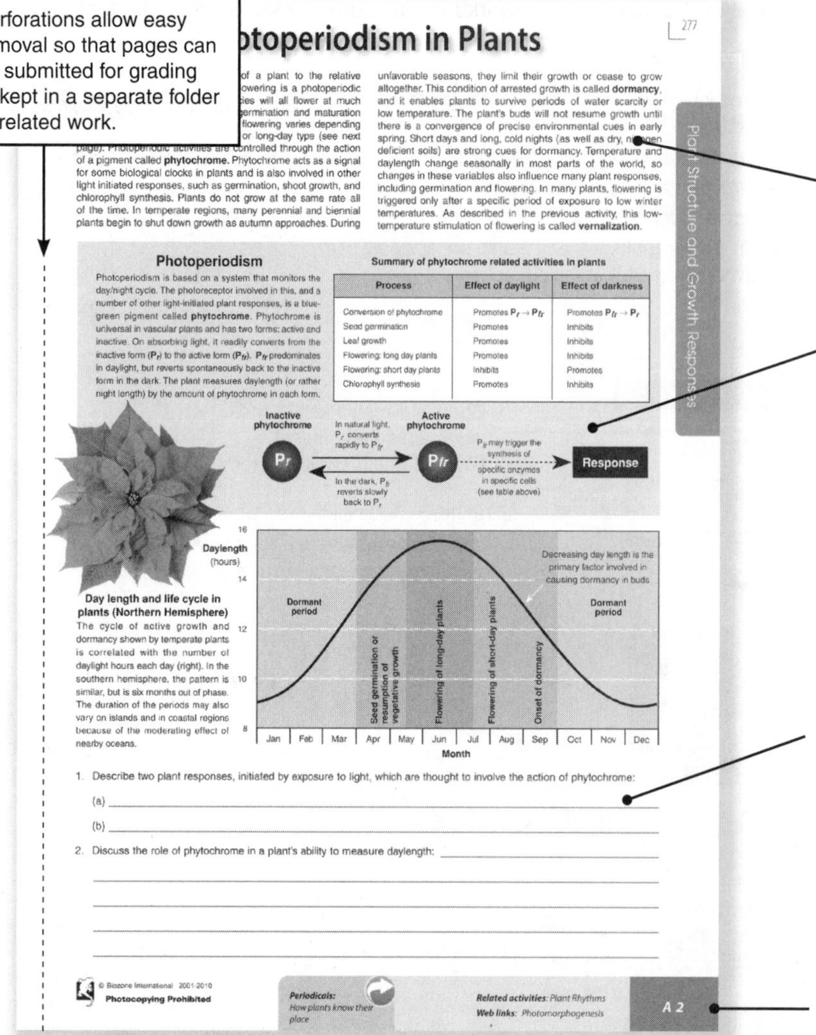

Introductory paragraph: The introductory paragraph provides essential background and provides the focus of the page. Note words that appear in bold, as they are 'key words' worthy of including in a glossary of terms for the topic.

Easy to understand diagrams: The main ideas of the topic are represented and explained by clear, informative diagrams.

Write-on format: Your understanding of the main ideas of the topic is tested by asking questions and providing spaces for your answers. Where indicated by the space available, your answers should be concise. Questions requiring more explanation or discussion are spaced accordingly. Answer the questions adequately according to the questioning term used (see the opposite page).

A tab system at the base of each activity page is a relatively new feature in Biozone's workbooks. With it, we have tagged valuable resources to the activity to which they apply. Use the guide below to help you use the tab system most effectively.

Using page tabs more effectively

Periodicals: *How plants know their place*

Related activities: *Plant Rhythms*

Web links: *Photomorphogenesis*

A 2

Students (and teachers) who would like to know more about this topic area are encouraged to locate the periodical cited on the Periodicals tab.
Articles of interest directly relevant to the topic content are cited. The full citation appears in the Appendix as indicated at the beginning of the topic chapter.

Related activities

Other activities in the workbook cover related topics or may help answer the questions on the page. In most cases, extra information for activities that are coded R can be found on the pages indicated here.

Web links

This citation indicates a valuable video clip or animation that can be accessed from the web links page specifically for this workbook. www.thebiozone.com/weblink/SB2-2603.html

INTERPRETING THE ACTIVITY CODING SYSTEM

Type of Activity

- **D** = includes some data handling or interpretation
- **P** = includes a paper practical
- **R** = *may* require extra reading (e.g. text or other activity)
- **A** = includes application of knowledge to solve a problem
- **E** = extension material

Level of Activity

- **1** = generally simpler, including mostly describe questions
- **2** = more challenging, including explain questions
- **3** = challenging content and/or questions, including discuss

Explanation of Terms

Questions come in a variety of forms. Whether you are studying for an exam, or writing an essay, it is important to understand exactly what the question is asking. A question has two parts to it: one part of the question will provide you with information, the second part of the question will provide you with instructions as to how to answer the question. Following these instructions is most important.

Commonly used Terms in Biology

The following terms are frequently used when asking questions in examinations and assessments. Most of these are listed in the IB syllabus document as action verbs indicating the depth of treatment required for a given statement. Students should have a clear understanding of each of the following terms and use this understanding to answer questions appropriately.

Account for: Provide a satisfactory explanation or reason for an observation.

Analyze: Interpret data to reach stated conclusions.

Annotate: Add **brief** notes to a diagram, drawing or graph.

Apply: Use an idea, equation, principle, theory, or law in a new situation.

Appreciate: To understand the meaning or relevance of a particular situation.

Calculate: Find an answer using mathematical methods. Show the working unless instructed not to.

Compare: Give an account of similarities and differences between two or more items, referring to both (or all) of them throughout. Comparisons can be given using a table. Comparisons generally ask for similarities more than differences (see contrast).

Construct: Represent or develop in graphical form.

Contrast: Show differences. Set in opposition.

Deduce: Reach a conclusion from information given.

Define: Give the precise meaning of a word or phrase as concisely as possible.

Derive: Manipulate a mathematical equation to give a new equation or result.

Describe: Give an account, including all the relevant information.

Design: Produce a plan, object, simulation or model.

Determine: Find the only possible answer.

Discuss: Give an account including, where possible, a range of arguments, assessments of the relative importance of various factors, or comparison of alternative hypotheses.

Distinguish: Give the difference(s) between two or more different items.

Often students in examinations know the material but fail to follow instructions and, as a consequence, do not answer the question appropriately. Examiners often use certain key words to introduce questions. Look out for them and be absolutely clear as to what they mean. Below is a list of commonly used terms that you will come across and a brief explanation of each.

Draw: Represent by means of pencil lines. Add labels unless told not to do so.

Estimate: Find an approximate value for an unknown quantity, based on the information provided and application of scientific knowledge.

Evaluate: Assess the implications and limitations.

Explain: Give a clear account including causes, reasons, or mechanisms.

Identify: Find an answer from a number of possibilities.

Illustrate: Give concrete examples. Explain clearly by using comparisons or examples.

Interpret: Comment upon, give examples, describe relationships. Describe, then evaluate.

List: Give a sequence of names or other brief answers with no elaboration. Each one should be clearly distinguishable from the others.

Measure: Find a value for a quantity.

Outline: Give a brief account or summary. Include essential information only.

Predict: Give an expected result.

Solve: Obtain an answer using algebraic and/or numerical methods.

State: Give a specific name, value, or other answer. No supporting argument or calculation is necessary.

Suggest: Propose a hypothesis or other possible explanation.

Summarize: Give a brief, condensed account. Include conclusions and avoid unnecessary details.

In Conclusion

Students should familiarize themselves with this list of terms and, where necessary throughout the course, they should refer back to them when answering questions. The list of terms mentioned above is not exhaustive and students should compare this list with past examination papers and essays etc. and add any new terms (and their meaning) to the list above. The aim is to become familiar with interpreting the question and answering it appropriately.

Resources Information

Your set textbook should be a starting point for information about the content of your course. There are also many other resources available, including journals, magazines, supplementary texts, dictionaries, computer software, and the internet. Your teacher will have some prescribed resources for your use, but a few of the readily available periodicals are listed here for quick reference. The titles of relevant articles are listed with the activity to which they relate and are cited in the appendix. Please note that listing any product in this workbook does not, in any way, denote Biozone's endorsement of that product and Biozone does not have any business affiliation with the publishers listed herein.

Supplementary Texts

Supplementary texts are those that cover a specific topic or range of topics, rather than an entire course. All titles are available in North America unless indicated by (§). For further details or to make purchases, link to the publisher via: **www.thebiozone.com > Resources > Supplementary > International**

Barnard, C., F. Gilbert, & P. McGregor, 2007
Asking Questions in Biology: Key Skills for Practical Assessments & Project Work, 256 pp.
Publisher: Benjamin Cummings
ISBN: 978-0132224352
Comments: *Covers many aspects of design, analysis and presentation of practical work in senior level biology.*

Clegg, C.J., 2003
Green Plants: The Inside Story, 96 pp.
Publisher: Hodder Murray
ISBN: 0 7195 7553 2
The emphasis in this text is on flowering plants. Topics include leaf, stem, and root structure in relation to function, reproduction, economic botany, and sensitivity and adaptation.

Adds, J., E. Larkcom & R. Miller, 2004.
Exchange and Transport, Energy and Ecosystems, revised edition 240 pp.
Publisher: Nelson Thornes
ISBN: 0-7487-7487-4
Covers exchange processes, transport systems, adaptation, and sexual reproduction. Practical activities are included in several of the chapters.

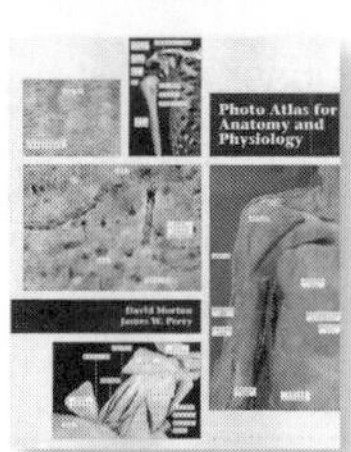

Morton, D. & J.W. Perry, 1998
Photo Atlas for Anatomy & Physiology, 160 pp.
Publisher: Brooks Cole.
ISBN: 0-534-51716-1
Comments: *An excellent photographic guide to lab work. Also available: PhotoAtlas for Botany (1998) ISBN: 0-534-52938-0 and Photo Atlas for Biology (1995) ISBN: 0-534-23556-5*

Periodicals, Magazines and Journals

Details of the periodicals referenced in this workbook are listed below. For enquiries and further details regarding subscriptions, link to the relevant publisher via Biozone's resources hub or by going to: **www.thebiozone.com > Resources > Journals**

Biological Sciences Review (Biol. Sci. Rev.)
An excellent quarterly publication for teachers and students. The content is current and the language is accessible. Subscriptions available from Philip Allan Publishers, Market Place, Deddington, Oxfordshire OX 15 OSE.
Tel. 01869 338652 **Fax**: 01869 338803
E-mail: sales@philipallan.co.uk

New Scientist: *Published weekly and found in many libraries. It often summarizes the findings published in other journals. Articles range from news releases to features.*
Subscription enquiries:
Tel. (UK and international): +44 (0)1444 475636. (US & Canada) 1 888 822 3242.
E-mail: ns.subs@qss-uk.com

Scientific American: *A monthly magazine containing mostly specialist feature articles. Articles range in level of reading difficulty and assumed knowledge.*
Subscription enquiries:
Tel. (US & Canada) 800-333-1199.
Tel. (outside North America): 515-247-7631
Web: www.sciam.com

The American Biology Teacher: *The official, peer-reviewed journal of the National Association of Biology Teachers. Published nine times a year and containing information and activities relevant to the teaching of biology in the US and elsewhere.* Enquiries: NABT, 12030 Sunrise Valley Drive, #110, Reston, VA 20191-3409
Web: www.nabt.org

Biology Dictionaries

Access to a good biology dictionary is of great value when dealing with the technical terms used in biology. Below are some biology dictionaries that you may wish to locate or purchase. They can usually be obtained directly from the publisher or they are all available (at the time of printing) from www.amazon.com. For further details of text content, or to make purchases, link to the relevant publisher via Biozone's resources hub or by typing: **www.thebiozone.com > Resources > Dictionaries**

Allaby, M. (ed).
A Dictionary of Zoology 2 reissue ed., 2003, 608 pp. Oxford University Press.
ISBN: 019860758X
Wide coverage of terms in animal behavior, ecology, physiology, genetics, cytology, evolution, and zoogeography. Full taxonomic coverage of most phyla.

Hale, W.G. **Collins: Dictionary of Biology** 4 ed. 2005, 528 pp. Collins.
ISBN: 0-00-720734-4.
Updated to take in the latest developments in biology and now internet-linked. (§ This latest edition is currently available only in the UK. The earlier edition, ISBN: 0-00-714709-0, is available though amazon.com in North America).

Henderson, E. Lawrence. **Henderson's Dictionary of Biological Terms**, 2008, 776 pp. Benjamin Cummings. **ISBN**: 978-0321505798
This edition has been updated, rewritten for clarity, and reorganised for ease of use. An essential reference and the dictionary of choice for many.

Making www.thebiozone.com Work For You

The current internet address (URL) for the web site is displayed here. You can type a new address directly into this space.

Use Google to search for web sites of interest. The more precise your search words are, the better the list of results. EXAMPLE: If you type in "biotechnology", your search will return an overwhelmingly large number of sites, many of which will not be useful to you. Be more specific, e.g. "biotechnology medicine DNA uses".

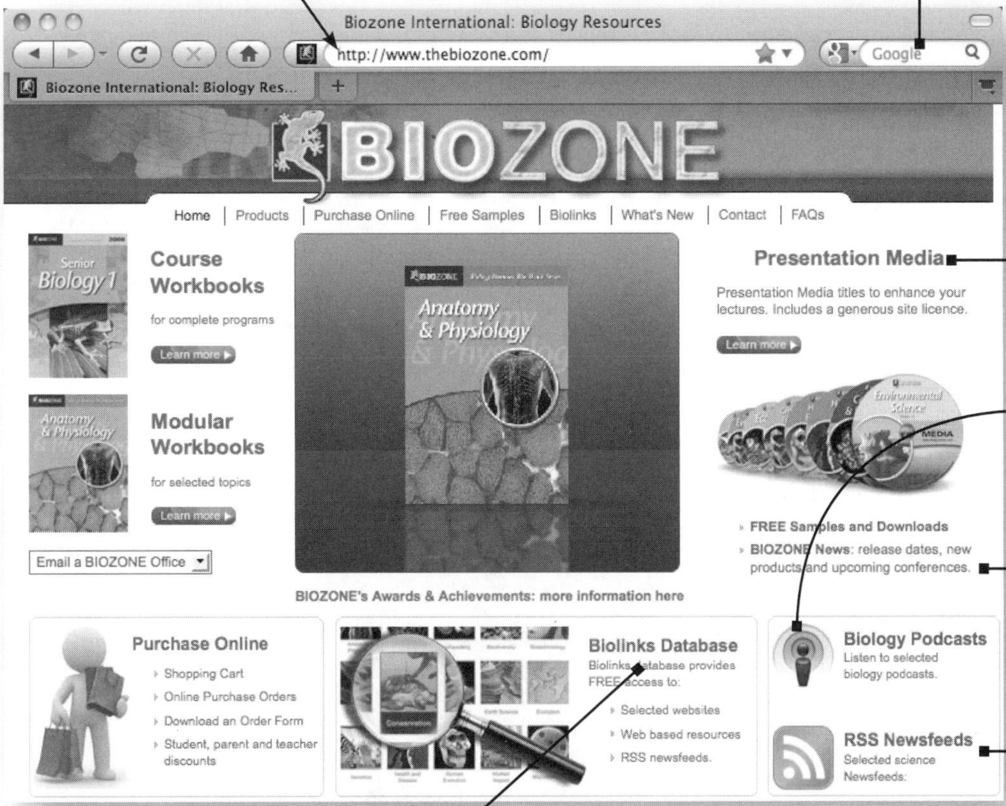

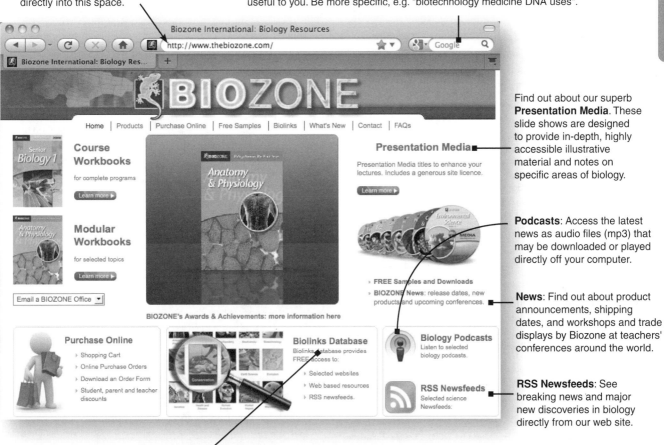

Find out about our superb **Presentation Media**. These slide shows are designed to provide in-depth, highly accessible illustrative material and notes on specific areas of biology.

Podcasts: Access the latest news as audio files (mp3) that may be downloaded or played directly off your computer.

News: Find out about product announcements, shipping dates, and workshops and trade displays by Biozone at teachers' conferences around the world.

RSS Newsfeeds: See breaking news and major new discoveries in biology directly from our web site.

Access the **BioLinks** database of web sites related to each major area of biology. It's a great way to quickly find out more on topics of interest.

Weblinks: www.thebiozone.com/weblink/SB2-2603.html

BOOKMARK WEBLINKS BY TYPING IN THE ADDRESS: IT IS NOT ACCESSIBLE DIRECTLY FROM BIOZONE'S WEBSITE

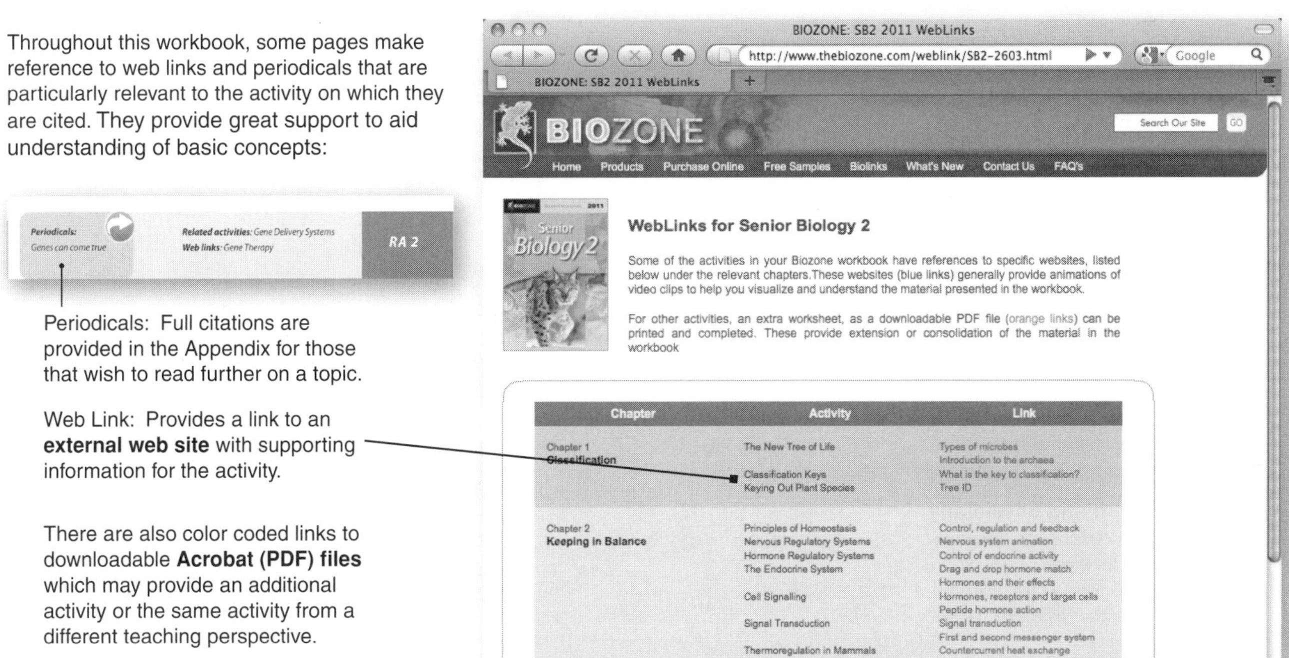

Throughout this workbook, some pages make reference to web links and periodicals that are particularly relevant to the activity on which they are cited. They provide great support to aid understanding of basic concepts:

Periodicals: Full citations are provided in the Appendix for those that wish to read further on a topic.

Web Link: Provides a link to an **external web site** with supporting information for the activity.

There are also color coded links to downloadable **Acrobat (PDF) files** which may provide an additional activity or the same activity from a different teaching perspective.

International Baccalaureate Course

The International Baccalaureate (IB) biology course is divided into three sections: core, additional higher level material, and option material. All **IB candidates** must complete the **core** topics. Higher level students are also required to undertake Additional Higher Level **(AHL)** material as part of the core. Options fall into three categories (see the following page): those specific to standard level students **(OPT-SL)**, one only specific to higher level students **(OPT-HL)** and those offered to both **(OPT-SL/HL)**. All candidates are required to study two options. All candidates must also carry out **practical work** and must participate in **the group 4 project**. In the guide below, we have indicated where the relevant material can be found: SB1 for Senior Biology 1 and SB2 for Senior Biology 2.

Topic			See workbook
CORE: *(All students)*			
1	**Statistical analysis**		
1.1	Mean and SD, t-test, correlation.	SB1	Skills in Biology
	For this CORE topic also see the TRC: Spreadsheets and Statistics		
2	**Cells**		
2.1	Cell theory. Cell and organelle sizes. Surface area to volume ratio. Emergent properties. Cell specialization and differentiation. Stem cells.	SB1	Cell Structure, Processes in Cells
2.2	Prokaryotic cells: ultrastructure & function.	SB1	Cell Structure
2.3	Eukaryotic cells: ultrastructure & function. Prokaryotic vs eukaryotic cells. Plant vs animal cells. Extracellular components.	SB1	Cell Structure, Processes in Cells
2.4	Membrane structure. Active and passive transport. Diffusion and osmosis.	SB1	Processes in Cells
2.5	Cell division and the origins of cancer.	SB1	Processes in Cells
	For extension on this topic also see the TRC: The Cell Theory		
3	**The chemistry of life**		
3.1	Elements of life. The properties and importance of water.	SB1	The Chemistry of Life
3.2	Structure and function of carbohydrates, lipids, and proteins.	SB1	The Chemistry of Life
3.3	Nucleotides and the structure of DNA.	SB1	The Chemistry of Life
3.4	Semi-conservative DNA replication.	SB1	Molecular Genetics
3.5	RNA and DNA structure. The genetic code. Transcription. Translation.	SB1	Molecular Genetics
3.6	Enzyme structure and function.	SB1	The Chemistry of Life
3.7	Cellular respiration and ATP production.	SB1	Cellular Energetics
3.8	Biochemistry of photosynthesis. Factors affecting photosynthetic rates.	SB1	Cellular Energetics
4	**Genetics**		
4.1	Eukaryote chromosomes. Genomes. Gene mutations and consequences.	SB1	Chromosomes & Meiosis
4.2	Meiosis and non-disjunction. Karyotyping and pre-natal diagnosis.	SB1	Chromosomes & Meiosis
4.3	Theoretical genetics: alleles and single gene inheritance, sex linkage, pedigrees.	SB1	Heredity
4.4	Genetic engineering and biotechnology: PCR, gel electrophoresis, DNA profiling. HGP. Transformation. GMOs. Cloning.	SB1	Nucleic Acid Technology
	For extension on this topic also see the TRC: Engineering Solutions		
5	**Ecology and evolution**		
5.1	Ecosystems. Food chains and webs. Trophic levels. Ecological pyramids. The role of decomposers in recycling nutrients.	SB2	Habitat & Distribution, Community Ecology
5.2	The greenhouse effect. The carbon cycle. Precautionary principle. Global warming.	SB2	Ecosystem Ecology & Human Impact
5.3	Factors influencing population size. Population growth.	SB2	Population Ecology
5.4	Genetic variation. Sexual reproduction as a source of variation in species.	SB1	Chromosomes & Meiosis,
	Evidence for evolution: natural selection. Evolution in response to environmental change.	SB1	The Origin & Evolution of Life, Speciation
5.5	Classification. Binomial nomenclature. Features of plant & animal phyla. Keys.	SB2	Classification
	For extension on this topic also see the TRC: Classification of Life		
6	**Human health and physiology**		
6.1	Role of enzymes in digestion. Structure and function of the digestive system.	SB2	Eating to Live
6.2	Structure and function of the heart. The control of heart activity. Blood & vessels.	SB2	Life Blood
6.3	Pathogens and their transmission. Antibiotics. Role of skin as a barrier to infection. Role of phagocytic leukocytes. Antigens & antibody production. HIV/AIDS.	SB2	Defending Against Disease
6.4	Gas exchange. Ventilation systems. Control of breathing.	SB2	Breath of Life
6.5	Principles of homeostasis. Control of body temperature and blood glucose. Diabetes. Role of the nervous and endocrine systems in homeostasis.	SB2	Keeping in Balance, Responding to the Environment
6.6	Human reproduction and the role of hormones. Reproductive technologies and ethical issues.	SB2	The Next Generation
COMPULSORY: AHL Topics *(HL students only)*			
7	**Nucleic acids and proteins**		
7.1	DNA structure, exons & introns (junk DNA)	SB1	Molecular Genetics
7.2	DNA replication, including the role of enzymes and Okazaki fragments.	SB1	Molecular Genetics
7.3	DNA alignment, transcription. The removal of introns to form mature mRNA.	SB1	Molecular Genetics
7.4	The structure of tRNA and ribosomes. The process of translation. Peptide bonds.	SB1	Molecular Genetics, The Chemistry of Life
7.5	Protein structure and function.	SB1	The Chemistry of Life
7.6	Enzymes: induced fit model. Inhibition. Allostery in the control of metabolism.	SB1	The Chemistry of Life
	For extension on this topic see the TRC: The Meselson-Stahl Experiment		
8	**Cell respiration and photosynthesis**		
8.1	Structure and function of mitochondria. Biochemistry of cellular respiration.	SB1	Cellular Energetics
8.2	Chloroplasts, the biochemistry and control of photosynthesis, chemiosmosis.	SB1	Cellular Energetics
9	**Plant science**		
9.1	Structure and growth of a dicot plant. Function and distribution of tissues in leaves. Dicots vs monocots. Plant modifications. Auxins.	SB2	Plant Structure & Growth Responses
9.2	Support in terrestrial plants. Transport in angiosperms: ion movement through soil, active ion uptake by roots, transpiration, translocation. Abscisic acid. Xerophytes.	SB2	Plant Support & Transport
9.3	Dicot flowers. Pollination and fertilization. Seeds: structure, germination, dispersal. Flowering and phytochrome.	SB2	Plant Reproduction
10	**Genetics**		
10.1	Meiosis, and the process of crossing over. Mendel's law of independent assortment.	SB1	Chromosomes & Meiosis
10.2	Dihybrid crosses. Types of chromosomes.	SB1	Heredity
10.3	Polygenic inheritance.	SB1	Heredity
	For extension on this topic also see the TRC: Chromosome Mapping		
11	**Human health and physiology**		
11.1	Blood clotting. Clonal selection. Acquired immunity. Antibodies and monoclonal antibodies. Vaccination.	SB2	Defending Against Disease
11.2	Nerves, muscles, bones and movement. Joints. Skeletal muscle and contraction.	SB2	Muscles & Movement
11.3	Excretion. Structure and function of the human kidney. Urine production. Diabetes.	SB2	Regulating Fluids and Removing Wastes
11.4	Testis and ovarian structure. Spermatogenesis and oogenesis. Fertilization and embryonic development. The placenta. Birth. Role of hormones.	SB2	The Next Generation

International Baccalaureate Course *continued*

Topic		See workbook

OPTIONS: OPT - SL *(SL students only)*

A Human nutrition and health

A.1 Diet and malnutrition. Deficiency & supplements. PKU.
A.2 Energy content of food types. BMI. Obesity and anorexia. Appetite control.
A.3 Special diet issues; breastfeeding vs bottle-feeding, type II diabetes, cholesterol.

● *Provided as a separate complete unit on the IB OPTIONS CD-ROM*

B Physiology of exercise

B.1 Locomotion in animals. Roles of nerves, muscles, and bones in movement. Joints. Skeletal muscle and contraction.
B.2 Training and the pulmonary system.
B.3 Training and the cardiovascular system.
B.4 Respiration and exercise intensity. Roles of myoglobin and adrenaline. Oxygen debt and lactate in muscle fatigue.
B.5 Exercise induced injuries and treatment.

● *Provided as a separate complete unit on the IB OPTIONS CD-ROM*

C Cells and energy

	Topic		See workbook
C.1	Protein structure and function. Fibrous and globular proteins.	SB1	The Chemistry of Life
C.2	Enzymes: induced fit model. Inhibition. Allostery in the control of metabolism.	SB1	The Chemistry of Life
C.3	Biochemistry of cellular respiration.	SB1	Cellular Energetics
C.4	The biochemistry of photosynthesis including chemiosmosis. Action and absorption spectra. Limiting factors.	SB1	Cellular Energetics

● *Provided as a separate complete unit on the IB OPTIONS CD-ROM*

OPTIONS: OPT - SL/HL *(SL and HL students)*

D Evolution

	Topic		See workbook
D.1	Prebiotic experiments. Comets. Protobionts and prokaryotes. Endosymbiotic theory.	SB1	The Origin & Evolution of Life
D.2	Species, gene pools, speciation. Types and pace of evolution. Transient vs balanced polymorphism.	SB1	Speciation, Patterns of Evolution
D.3	Fossil dating. Primate features. Hominid features. Diet and brain size correlation. Genetic and cultural evolution.	●	The Evolution of Humans (**TRC**)
	D.4-D.5 is extension for HL only		
D.4	The Hardy-Weinberg principle.	SB1	Speciation
D.5	Biochemical evidence for evolution. Biochemical variations indicating phylogenetic relationships.	SB1	The Origin and Evolution of Life
	Classification. Cladistics and cladograms	SB2	Classification

● *Provided as a separate complete unit on the IB OPTIONS CD-ROM*

E Neurobiology and behavior

	Topic		See workbook
E.1	Stimuli, responses and reflexes in the context of animal behavior. Animal responses and natural selection.	SB2	Nerves, Muscles & Movement, Animal Behavior
E.2	Sensory receptors. Structure and function of the human eye and ear.	SB2	Nerves, Muscles & Movement
E.3	Innate vs learned behavior and its role in survival. Learned behavior and birdsong.	SB2	Animal Behavior
E.4	Presynaptic neurons at synapses. Examples of excitatory and inhibitory psychoactive drugs. Effects of drugs on synaptic transmission. Causes of addiction.	SB2	Aspects covered in Nerves, Muscles & Movement
	E.5-E.6 is extension for HL only		
E.5	Structure and function of the human brain. ANS control. Pupil reflex and its use in testing for death. Hormones as painkillers.	SB2	Aspects covered in Nerves, Muscles & Movement
E.6	Social behavior and organization. The role of altruism in sociality. Foraging behavior. Mate selection. Rhythmical behavior.	SB2	Animal Behavior

● *Provided as a separate complete unit on the IB OPTIONS CD-ROM*

F Microbes and Biotechnology

F.1 Classification. Diversity of Archaea and Eubacteria. Diversity of viruses. Diversity of microscopic eukaryotes.
F.2 Roles of microbes in ecosystems. Details of the nitrogen cycle including the role of bacteria. Sewage treatment. Biofuels.
F.3 Reverse transcription. Somatic vs germline, gene therapy. Viral vectors.
F.4 Microbes involved in food production of beer, wine, bread, and soy sauce. Food preservation. Food poisoning.

F.5-F.6 is extension for HL only

F.5 Metabolism of microbes. Modes of nutrition. Cyanobacterium. Bioremediation.
F.6 Pathogens and disease: influenza virus, malaria, bacterial infections. Controlling microbes. Epidemiology. Prion hypothesis.

● *Provided as a separate complete unit on the IB OPTIONS CD-ROM*

G Ecology and conservation

G.1 Factors affecting plant and animal distribution. Sampling. Ecological niche and the competitive exclusion principle. Species interactions. Measuring biomass.
G.2 Trophic levels. Ecological pyramids. Primary vs secondary succession. Biome vs biosphere. Plant productivity (includes calculating gross and net production, and biomass).
G.3 Conservation of biodiversity. Diversity index. Human impact on ecosystems: alien species. Biological control. Effect of CFCs on ozone layer. UV radiation absorption.

G.4-G.5 is extension for HL only

G.4 Monitoring environmental change. Biodiversity. Endangered species. Conservation Strategies. Extinction.
G.5 *r*-strategies and K-strategies.Mark-and-recapture sampling. Fisheries conservation.

● *Provided as a separate complete unit on the IB OPTIONS CD-ROM*

OPTION: OPT - HL *(HL students only)*

H Further human physiology

H.1 Hormones and their modes of action. Hypothalamus and pituitary gland. Control of ADH secretion.
H.2 Digestion and digestive juices. Stomach ulcers and stomach cancers. Role of bile.
H.3 Structure of villi. Absorption of nutrients and transport of digested food.
H.4 The structure and function of the liver (including role in nutrient processing and detoxification). Liver damage from alcohol.
H.5 The cardiac cycle and control of heart rhythm. Atherosclerosis, coronary thrombosis and coronary heart disease.
H.6 Gas exchange: oxygen dissociation curves and the Bohr shift. Ventilation rate and exercise. Breathing at high altitude. Causes and effects of asthma.

● *Provided as a separate complete unit on the IB OPTIONS CD-ROM*

Practical Work *(All students)*

Practical work consists of short and long term investigations, and an interdisciplinary project (The Group 4 project). Also see the "Guide to Practical Work" on the last page of this introductory section.

Short and long term investigations

Investigations should reflect the breadth and depth of the subjects taught at each level, and include a spread of content material from the core, options, and AHL material, where relevant.

The Group 4 project

All candidates must participate in the group 4 project. In this project it is intended that students analyze a topic or problem suitable for investigation in each of the science disciplines offered by the school (not just in biology). This project emphasizes the processes involved in scientific investigations rather than the products of an investigation.

Advanced Placement Course

The Advanced Placement (AP) biology course is designed to be equivalent to a college introductory biology course. It is to be taken by students after successful completion of first courses in high school biology and chemistry. In the guide below, we have indicated where the relevant material can be found: SB1 for Senior Biology 1 and SB2 for Senior Biology 2. Because of the general nature of the AP curriculum document, the detail given here is based on Biozone's interpretation of the scheme.

Topic I: Molecules and Cells

A Chemistry of life

	Topic		See workbook
1	The chemical & physical properties of water. The importance of water to life.	SB1	The Chemistry of Life
2	The role of carbon. Structure and function of carbohydrates, lipids, nucleic acids, and proteins. The synthesis and breakdown of macromolecules.	SB1	The Chemistry of Life, Molecular Genetics, Processes in Cells
3	The laws of thermodynamics and their relationship to biochemical processes. Free energy changes.	SB1	The Chemistry of Life
4	The action of enzymes and their role in the regulation of metabolism. Enzyme specificity. Factors affecting enzyme activity. Applications of enzymes.	SB1	The Chemistry of Life

● *For extension on this topic also see the TRC: Industrial Microbiology*

B Cells

	Topic		See workbook
1	Comparison of prokaryotic and eukaryotic cells, including evolutionary relationships.	SB1 SB1	Cell Structure, The Origin & Evolution of Life

● *For extension on this topic also see the TRC: The Cell Theory*

	Topic		See workbook
2	Membrane structure: fluid mosaic model. Active and passive transport.	SB1	Processes in Cells
3	Structure and function of organelles. Comparison of plant and animal cells. Cell size and surface area: volume ratio. Organization of cell function.	SB1	Cell Structure, Processes in Cells
4	Mitosis and the cell cycle. Mechanisms of cytokinesis. Cancer (tumour formation) as the result of uncontrolled cell division.	SB1	Processes in Cells

C Cellular energetics

	Topic		See workbook
1	Nature and role of ATP. Anabolic and catabolic processes. Chemiosmosis.	SB1	Cellular Energetics
2	Structure and function of mitochondria. Biochemistry of cellular respiration, including the role of oxygen in energy yielding pathways. Anaerobic systems.	SB1	Cellular Energetics
3	Structure and function of chloroplasts. Biochemistry of photosynthesis. Adaptations for photosynthesis in different environments.	SB1	Cellular Energetics

● *For extension on this topic also see the TRC: Events in Biochemistry*

Topic II: Heredity and Evolution

A Heredity

	Topic		See workbook
1	The importance of meiosis in heredity. Gametogenesis. Similarities and differences between gametogenesis in animals and plants.	SB1 SB1	Chromosomes & Meiosis Processes in Cells
2	Structure of eukaryotic chromosomes. Heredity of genetic information.	SB1	Chromosomes & Meiosis
3	Mendel's laws. Inheritance patterns.	SB1	Heredity

● *For extension on this topic also see the TRC: Chromsome Mapping*

B Molecular genetics

	Topic		See workbook
1	RNA and DNA structure and function. Eukaryotic and prokaryotic genomes.	SB1	Molecular Genetics

For extension on this topic see the TRC: The Meselsohn-Stahl Experiment

	Topic		See workbook
2	Gene expression in prokaryotes and eukaryotes. The *Lac* operon model.	SB1	Molecular Genetics
3	Causes of mutations. Gene mutations (e.g. sickle cell disease). Chromosomal mutations (e.g. Down syndrome).	SB1	Mutation
4	Viral structure and replication.	SB1	Molecular Genetics
5	Nucleic acid technology and applications. legal and ethical issues.	SB1	Nucleic Acid Technology

● *For extension on this topic also see the TRC: Engineering Solutions*

C Evolutionary biology

	Topic		See workbook
1	The origins of life on Earth. Prebiotic experiments. Origins of prokaryotic cells. Endosymbiotic theory.	SB1	The Origin & Evolution of Life
2	Evidence for evolution. Dating of fossils.	SB1	The Origin & Evolution of Life

● *For extension on this topic also see the TRC: Dating the Past*

	Topic		See workbook
3	The species concept. Mechanisms of evolution: natural selection, speciation, macroevolution.	SB1	Speciation, Patterns of Evolution

● *For extension on this topic also see the TRC: A Case Study in Evolution*

Topic III: Organisms and Populations

A Diversity of organisms

	Topic		See workbook
1	Evolutionary patterns: major body plans of plants and animals.	SB2	Classification
2	Diversity of life: representative members from the five kingdoms Monera (=Prokaryotae), Fungi, Protista (=Protoctista), Animalia and Plantae.	SB2	Classification
3	Phylogenetic classification. Binomial nomenclature. Five kingdom classification. Use of dichotomous keys.	SB2	Classification
4	Evolutionary relationships: genetic and morphological characters. Phylogenies.	SB2	Classification

● *For extension on this topic also see the TRC: Practical Classification*

B Structure and function of plants and animals

	Topic		See workbook
1	Plant and animal reproduction and development (includes humans). Adaptive significance of reproductive features and their regulation.	SB2	The Next Generation, Plant Reproduction

● *For extension material see the TRC: Mammalian Patterns of Reproduction*

	Topic		See workbook
2	Organization of cells, tissues & organs.		
	The structure and function of animal and plant organ systems. Adaptive features that have contributed to the success of plants and animals in occupying particular terrestrial niches.	SB2	PART 2 (Animals), chapters as required. PART 3 (Plants), chapters as required
3	Plant and animal responses to environmental cues. The role of hormones in these responses.	SB2	Respondng to the Environment, Plant Structure and Growth Responses

● *For extension material see the TRC: Migratory Navigation in Birds*

C Ecology

	Topic		See workbook
1	Factors influencing population size. Population growth curves.	SB2	Population Ecology
2	Abiotic and biotic factors: effects on community structure and ecosystem function. Trophic levels: energy flows through ecosystems and relationship to trophic structure. Nutrient cycles.	SB2	Habitat and Distribution, Community Ecology, Ecosystem Ecology and Human Impact

● *For extension material see the TRC: Production and Trophic Efficiency*

	Topic		See workbook
3	Human influence on biogeochemical cycles: (e.g. use of fertilizers).	SB2	Ecosystem Ecology and Human Impact

● *For extension material see the TRC: Sustainable Futures*

Practical Work

Integrated practicals as appropriate
"Guide to Practical Work" in this introductory section.
Senior Biology 1: Skills in Biology (for reference throughout the course).
● TRC: *Spreadsheets and Statistics*

Guide to Practical Work

A practical or laboratory component is an essential part of any biology course, especially at senior level. It is through your practical sessions that you are challenged to carry out experiments drawn from many areas within modern biology. Both AP and IB courses have a strong practical component, aimed at providing a framework for your laboratory experience. Well executed laboratory and field sessions will help you to understand problems, observe accurately, make hypotheses, design and implement controlled experiments, collect and analyze data, think analytically, and communicate your findings in an appropriate way using tables and graphs. The outline below provides some guidelines for AP and IB students undertaking their practical work. Be sure to follow required safety procedures at all times during practical work.

International Baccalaureate Practical Work

The practical work carried out by IB biology students should reflect the depth and breadth of the subject syllabus, although there may not be an investigation for every syllabus topic. All candidates must participate in the group 4 project, and the internal assessment (IA) requirements should be met via a spread of content from the core, options and, where relevant, AHL material. A wide range of IA investigations is possible: short laboratory practicals and longer term practicals or projects, computer simulations, data gathering and analysis exercises, and general laboratory and field work.

Suitable material, or background preparation, for this component can be found in this workbook and its companion title, Senior Biology 2.

College Board's AP® Biology Lab Topics

Each of the 12 set laboratory sessions in the AP course is designed to complement a particular topic area within the course. The basic structure of the lab course is outlined below:

LAB 1: **Diffusion and osmosis**
Overview: To investigate diffusion and osmosis in dialysis tubing. To investigate the effect of solute concentration on water potential (ψ) in plant tissues.
Aims: An understanding of passive transport mechanisms in cells, and an understanding of the concept of water potential, solute potential, and pressure potential, and how these are measured.

LAB 2: **Enzyme catalysis**
Overview: To investigate the conversion of hydrogen peroxide to water and oxygen gas by catalase.
Aims: An understanding of the effects of environmental factors on the rate of enzyme catalyzed reactions.

LAB 3: **Mitosis and meiosis**
Overview: To use prepared slides of onion root tips to study plant mitosis. To simulate the phases of meiosis by using chromosome models.
Aims: Recognition of stages in mitosis in plant cells and calculation of relative duration of cell cycle stages. An understanding of chromosome activity during meiosis and an ability to calculate map distances for genes.

LAB 4: **Plant pigments and photosynthesis**
Overview: To separate plant pigments using chromatography. To measure photosynthetic rate in chloroplasts.
Aims: An understanding of Rf values. An understanding of the techniques used to determine photosynthetic rates. An ability to explain variations in photosynthetic rate under different environmental conditions.

LAB 5: **Cell(ular) respiration**
Overview: To investigate oxygen consumption during germination (including the effect of temperature).
Aims: An understanding of how cell respiration rates can be calculated from experimental data. An understanding of the relationship between gas production and respiration rate, and the effect of temperature on this.

LAB 6: **Molecular biology**
Overview: To investigate the basic principles of molecular biology through the transformation of *E.coli* cells. To investigate the use of restriction digestion and gel electrophoresis.
Aims: An understanding of the role of plasmids as vectors, and the use of gel electrophoresis to separate DNA fragments of varying size. An ability to design appropriate experimental procedures and use multiple experimental controls.

LAB 7: **Genetics of organisms**
Overview: Use *Drosophila* to perform genetic crosses. To collect and analyze the data from these crosses.
Aims: An understanding of the independent assortment of two genes and an ability to determine if genes are autosomal or sex linked from the analysis of the results of multigeneration genetic crosses.

LAB 8: **Population genetics and evolution**
Overview: To learn about the Hardy-Weinberg law of genetic equilibrium and study the relationship between evolution and changes in allele frequency.
Aims: An ability to calculate allele and genotype frequencies using the Hardy-Weinberg formula. An understanding of natural selection and other causes of microevolution.

LAB 9: **Transpiration**
Overview: To investigate transpiration in plants under controlled conditions. To examine the organization of plant stems and leaves as they relate to this.
Aims: An understanding of the effects of environmental variables on transpiration rates. An understanding of the relationship between the structure and function of the tissues involved.

LAB 10: **Physiology of the circulatory system**
Overview: To measure (human) blood pressure and pulse rate under different conditions. To analyze these variables and relate them to an index of fitness. To investigate the effect of temperature on heart rate in *Daphnia*.
Aims: An understanding of blood pressure and pulse rate, and their measurement and significance with respect to fitness. An understanding of the relationship between heart rate and temperature in a poikilotherm.

LAB 11: **Animal behavior**
Overview: To investigate responses in pillbugs (woodlice). To investigate mating behavior in fruit flies.
Aims: To understand and describe aspects of animal behavior. To understand the adaptiveness of appropriate behaviors.

LAB 12: **Dissolved oxygen & aquatic primary productivity**
Overview: To measure & analyze dissolved oxygen concentration in water samples. To measure and analyze the primary productivity of natural waters or lab cultures.
Aims: An understanding of primary productivity and its measurement. To use a controlled experiment to investigate the effect of changing light intensity on primary productivity.

Important in this section ...

- *Recall the role of evolution in giving rise to diversity*
- *Explain the principles of a classification system*
- *Use characteristic features to describe taxa*
- *Explain the use of cladistics in modern taxonomy*

The Evolution of Diversity (SB1)

RECALL ***Evolutionary processes are responsible for the diversity of life on Earth.***

RECALL ***The classification of organisms should reflect their evolutionary history***

RECALL ***Modern classification schemes are based on shared derived characters***

Recognizing Diversity

Characteristic features
- What is a characteristic feature?
- The significance of shared derived characters (synapomorphies)

Animal body plans and symmetry
- Bilateral vs radial symmetry
- Protostomes and deuterostomes

Features of specific taxa
- Archaea
- Eubacteria
- Fungi
- Protista
- Plantae
- Animalia

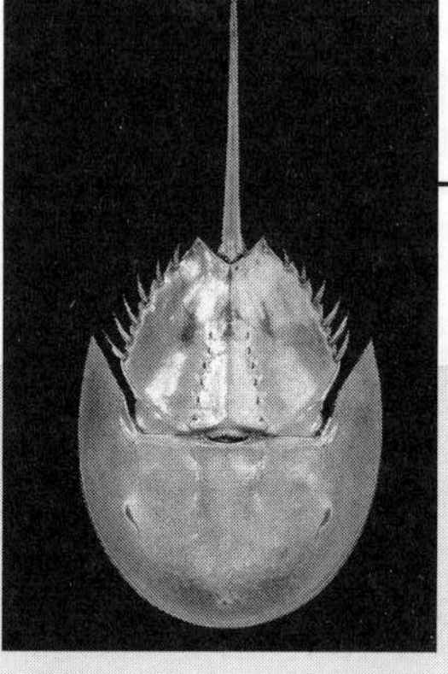

Diversity on Earth has arisen through evolution, so classification should reflect this history.

The members of a taxon share derived features that set them apart from other such taxonomic groups.

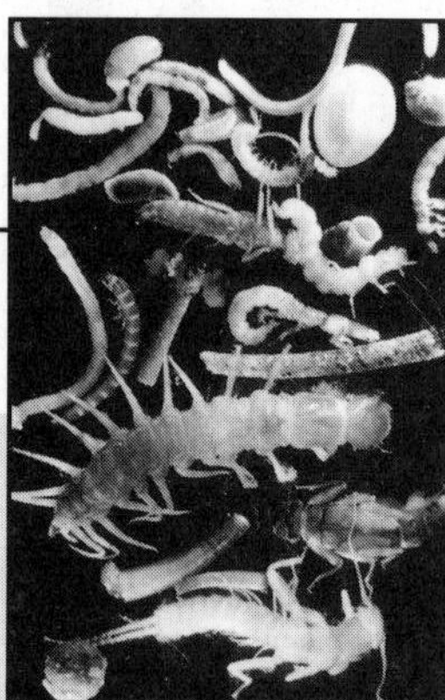

Part 1

The Diversity of Organisms

Classification provides a way to appreciate and quantify the Earth's biodiversity.

Classification schemes should reflect phylogeny of organisms, but these schemes are not always the most familiar to us.

Binomial nomenclature provides an unambiguous way to identify organisms.

New classification schemes emphasize monophyletic groups rather than morphological similarity

An unambiguous naming system is essential when studying organisms.

New classification systems
- 6 Kingdoms vs 3 Domains
- The influence of molecular genetics: the rise of modern cladistics

Phylogenetic systematics
- Shared derived characters and common ancestry
- The rule of parsimony
- Constructing basic cladograms

Note: cladistic analyses ignores sensible, clearly defined paraphyletic groups such as reptiles, so in practice, these older taxonomic groupings persist.

Phylogenetic Systematics

Classification systems
- Binomial nomenclature
- Three Domains
- Six and five Kingdom classification
  - Kingdom
  - Phylum/Division
  - Class
  - Order
  - Family
  - Genus
  - Species

Dichotomous keys
- The principle of dichotomous keys
- Using keys to identify real organisms

The Practicalities of Taxonomy

AP Topic 3A

IB SL/HL 5.5

Classification

KEY CONCEPTS

- Classification enables us to recognize and quantify the biological diversity on Earth.
- Organisms are assigned to taxonomic categories based on shared derived characteristics.
- Organisms are identified using binomial nomenclature: genus and species.
- Dichotomous classification keys can be used to identify unknown organisms.

KEY TERMS

Animalia
Archaebacteria (Archaea)
binomial nomenclature
class
cladistics
cladogram
classification key
common name
dichotomous
distinguishing feature
Eubacteria
Eukarya
Eukaryotae
family
Fungi
genus
kingdom
Monera
monophyletic
morphology
order
paraphyletic
phylogeny
phylogenetic systematics
phylum
Plantae
Prokaryotae
Protista
shared derived character (=synapomorphy)
species
taxon (pl. taxa)
taxonomic category

OBJECTIVES

☐ 1. Use the **KEY TERMS** to help you understand and complete these objectives.

Biodiversity

☐ 2. Define biodiversity and explain the importance of classification in recognizing, appreciating, and conserving the biodiversity on Earth.

Classification of Life

pages 12-30

☐ 3. Describe the principles and importance of biological classification.

☐ 4. Describe the use of **cladistic analyses** to produce monophyletic phylogenies based on **shared derived characters** (synapomorphies). Describe how cladistic analyses differ from traditional systematics and explain how and why the phylogenetic trees (**cladograms**) produced may differ.

☐ 5. Describe the **distinguishing features** of each kingdom in the five kingdom classification system.

☐ 6. Know that there are other classification systems based on phylogenetic relationships determined in more recent times:
- the six kingdom system, which recognizes Archaebacteria and Eubacteria.
- the three domain system, which recognizes Archaea, Eubacteria, and Eukarya.

☐ 7. Explain how organisms are assigned to different taxonomic categories on the basis of their shared derived characteristics.

☐ 8. Recognize at least seven major **taxonomic categories**: kingdom, phylum, class, order, family, genus, and species.

☐ 9. Explain how **binomial nomenclature** is used to classify organisms.

☐ 10. Recall what is meant by a distinguishing feature.

☐ 11. Recognize and describe examples of morphological differences within the same species.

☐ 12 Explain the principles by which **dichotomous classification keys** are used to identify organisms.

☐ 13. Use simple dichotomous keys to recognize and classify some common organisms (e.g. insects, legumes, fruits).

Periodicals:
listings for this chapter are on page 388

Weblinks:
www.thebiozone.com/weblink/SB2-2603.html

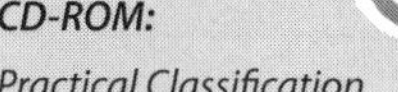

Teacher Resource CD-ROM:
Practical Classification

The New Tree of Life

Taxonomy is the science of classification and, like all science, it is a dynamic field, subject to change in the light of new information. With the advent of DNA sequencing technology, scientists began to analyze the genomes of many bacteria. In 1996, the results of a scientific collaboration examining DNA evidence confirmed the proposal that life comprises three major evolutionary lineages (domains) and not two as was the convention. The recognized lineages were the **Bacteria** (formerly Eubacteria), the **Eukarya**, and the **Archaea** (formerly the Archaebacteria). The new classification reflects the fact that there are very large differences between the Archaea and the Eubacteria. All three domains probably had a distant common ancestor.

A Five (or Six) Kingdom World (right)

The diagram (right) represents the **five kingdom system** of classification commonly represented in many biology texts. It recognizes two basic cell types: prokaryote and eukaryote. Superkingdom Prokaryotae includes all bacteria and cyanobacteria (the Kingdom Monera). Superkingdom Eukaryotae includes includes protists, fungi, plants, and animals. More recently, based on 16S ribosomal RNA sequence comparisons, Carl Woese divided the prokaryotes into two kingdoms, the Eubacteria and Archaebacteria. This **six-kingdom system** is also commonly recognized in texts.

A New View of the World (below)

In 1996, scientists deciphered the full DNA sequence of an unusual bacterium called *Methanococcus jannaschii.* An **extremophile**, this methane-producing archaebacterium lives at 85°C; a temperature lethal for most bacteria as well as eukaryotes. The DNA sequence confirmed that life consists of three major evolutionary lineages, not the two that have been routinely described. Only 44% of this archaebacterium's genes resemble those in bacteria or eukaryotes, or both.

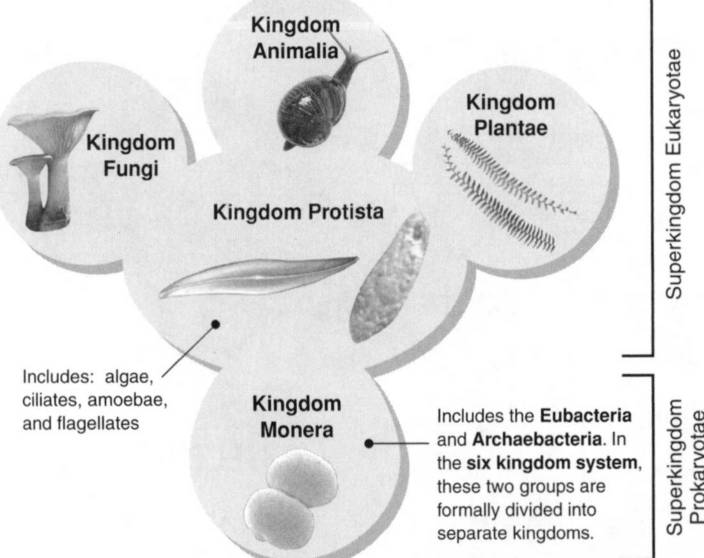

Domain Bacteria

Lack a distinct nucleus and cell organelles. Generally prefer less extreme environments than Archaea. Includes well-known pathogens, many harmless and beneficial species, and the cyanobacteria (photosynthetic bacteria containing the pigments chlorophyll *a* and phycocyanin).

Domain Archaea

Closely resemble eubacteria in many ways but cell wall composition and aspects of metabolism are very different. Live in extreme environments similar to those on primeval Earth. They may utilize sulfur, methane, or halogens (chlorine, fluorine), and many tolerate extremes of temperature, salinity, or pH.

Domain Eukarya

Complex cell structure with organelles and nucleus. This group contains four of the kingdoms classified under the more traditional system. Note that Kingdom Protista is separated into distinct groups: e.g. amoebae, ciliates, flagellates.

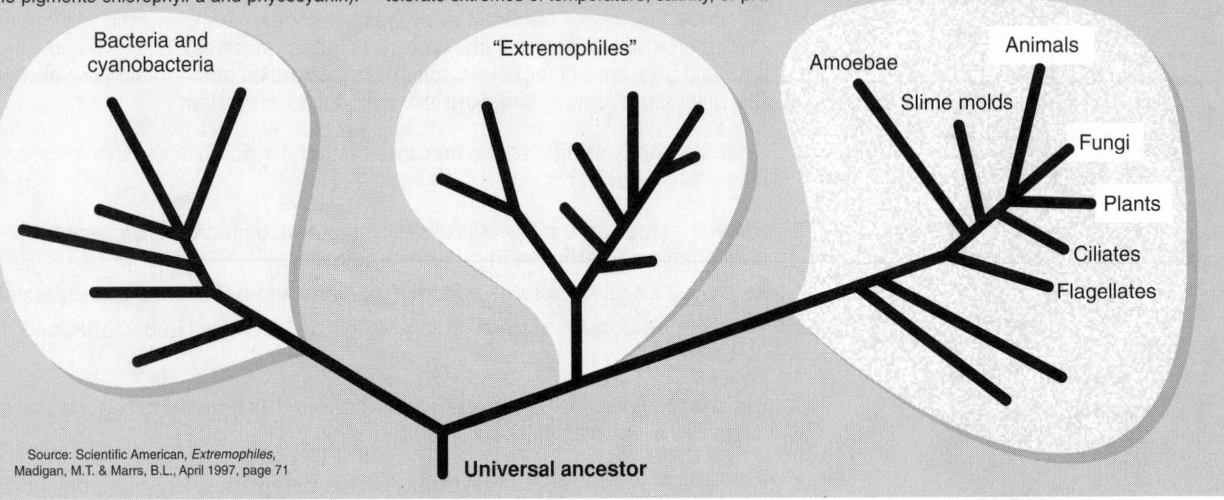

Source: Scientific American, *Extremophiles*, Madigan, M.T. & Marrs, B.L., April 1997, page 71

1. Explain why some scientists have recommended that the conventional classification of life be revised so that the Archaea, Bacteria and Eukarya are three separate domains:

2. Describe one feature of the three domain system that is very different from the five kingdom classification:

3. Describe one way in which the three domain system and the six kingdom classification are alike:

Phylogenetic Systematics

The aim of classification is to organize species in a way that most accurately reflects their evolutionary history (**phylogeny**). Each successive group in the taxonomic hierarchy should represent finer and finer branching from a common ancestor. Traditional classification systems emphasize morphological similarities in order to group species into genera and other higher level taxa. In contrast, **cladistic analysis** relies on **shared derived characteristics** (**synapomorphies**), and emphasizes features that are the result of shared ancestry (homologies), rather than convergent evolution. Technology has assisted taxonomy by providing biochemical evidence for the relatedness of species. Traditional and cladistic schemes do not necessarily conflict, but there have been reclassifications of some taxa (notably the primates, but also the reptiles, dinosaurs, and birds). Popular classifications will probably continue to reflect similarities and differences in appearance, rather than a strict evolutionary history. In this respect, they are a compromise between phylogeny and the need for a convenient filing system for species diversity.

Constructing a Simple Cladogram

A table listing the features for comparison allows us to identify where we should make branches in the **cladogram**. An outgroup (one which is known to have no or little relationship to the other organisms) is used as a basis for comparison.

Comparative features	Taxa: Jawless fish (outgroup)	Bony fish	Amphibians	Lizards	Birds	Mammals
Vertebral column	✔	✔	✔	✔	✔	✔
Jaws	✘	✔	✔	✔	✔	✔
Four supporting limbs	✘	✘	✔	✔	✔	✔
Amniotic egg	✘	✘	✘	✔	✔	✔
Diapsid skull	✘	✘	✘	✔	✔	✘
Feathers	✘	✘	✘	✘	✔	✘
Hair	✘	✘	✘	✘	✘	✔

The table above lists features shared by selected taxa. The outgroup (jawless fish) shares just one feature (vertebral column), so it gives a reference for comparison and the first branch of the cladogram (tree).

As the number of taxa in the table increases, the number of possible trees that could be drawn increases exponentially. To determine the most likely relationships, the rule of **parsimony** is used. This assumes that the tree with the least number of evolutionary events is most likely to show the correct evolutionary relationship.

Three possible cladograms are shown on the right. The top cladogram requires six events while the other two require seven events. Applying the rule of parsimony, the top cladogram must be taken as correct.

Parsimony can lead to some confusion. Some evolutionary events have occurred multiple times. An example is the evolution of the four chambered heart, which occurred separately in both birds and mammals. The use of fossil evidence and DNA analysis can help to solve problems like this.

Possible Cladograms

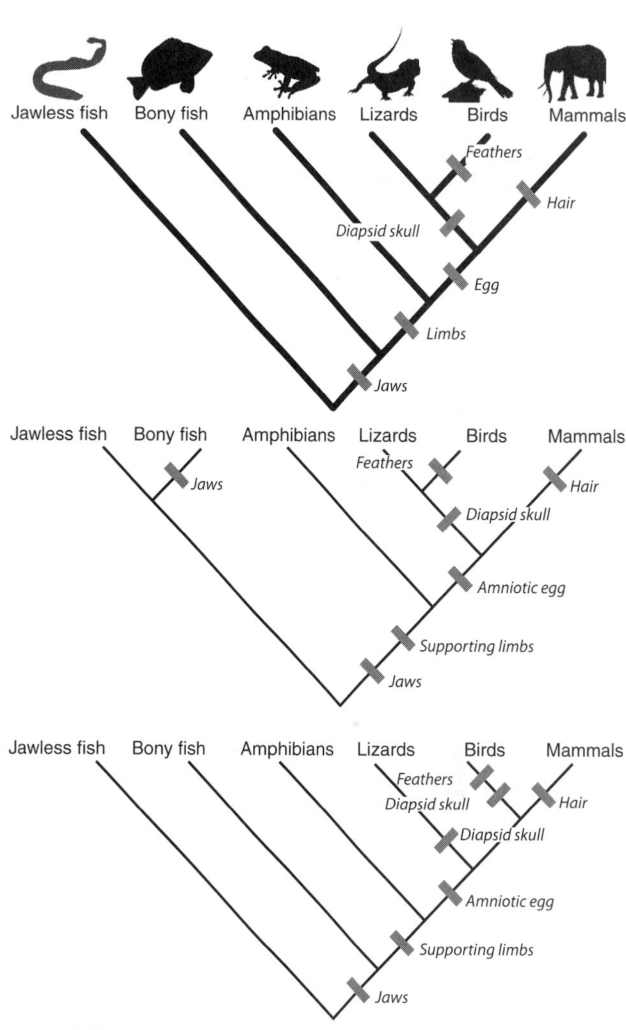

Using DNA Data

DNA analysis has allowed scientists to confirm many phylogenies and refute or redraw others. In a similar way to morphological differences, DNA sequences can be tabulated and analyzed. The ancestry of whales has been in debate since Darwin. The radically different morphologies of whales and other mammals makes it difficult work out the correct phylogenetic tree. However recently discovered fossil ankle bones, as well as DNA studies, show whales are more closely related to hippopotami than to any other mammal. Coupled with molecular clocks, DNA data can also give the time between each split in the lineage.

The DNA sequences on the right show part of a the nucleotide subset 141-200 and some of the matching nucleotides used to draw the cladogram. Although whales were once thought most closely related to pigs, based on the DNA analysis the most parsimonious tree disputes this.

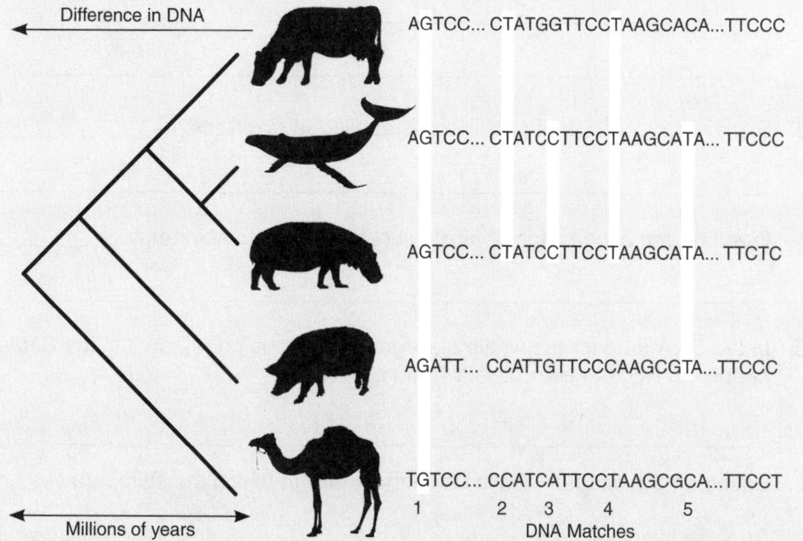

Related activities: *Classification System, The New Tree of Life*

A Classical Taxonomic View

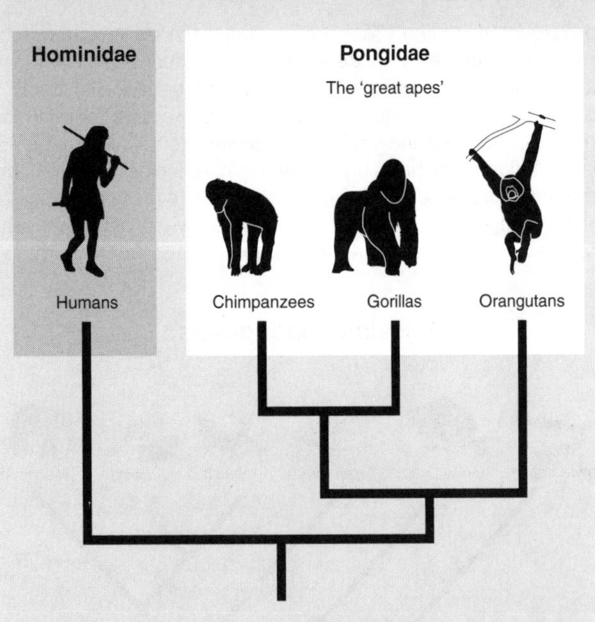

On the basis of overall anatomical similarity (e.g. bones and limb length, teeth, musculature), apes are grouped into a family (Pongidae) that is separate from humans and their immediate ancestors (Hominidae). The family Pongidae (the great apes) is not monophyletic (of one phylogeny), because it stems from an ancestor that also gave rise to a species in another family (i.e. humans). This traditional classification scheme is now at odds with schemes derived after considering genetic evidence.

A Cladistic View

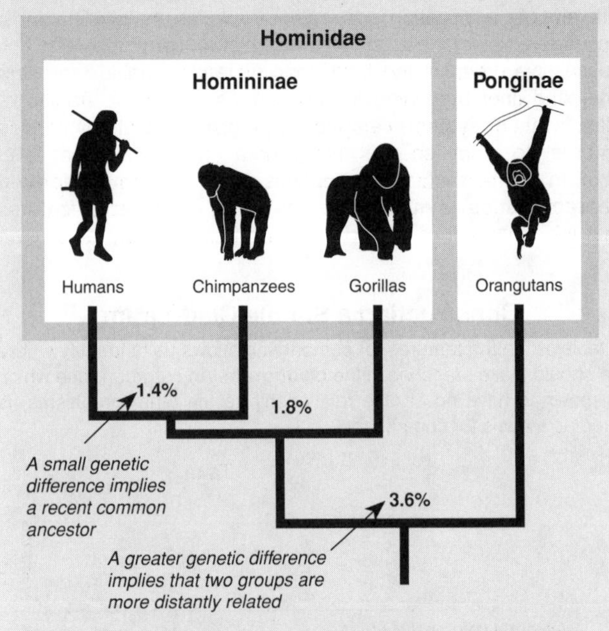

Based on the evidence of genetic differences (% values above), chimpanzees and gorillas are more closely related to humans than to orangutans, and chimpanzees are more closely related to humans than they are to gorillas. Under this scheme there is no true family of great apes. The family Hominidae includes two subfamilies: Ponginae and Homininae (humans, chimpanzees, and gorillas). This classification is monophyletic: the Hominidae includes all the species that arise from a common ancestor.

1. Briefly explain the benefits of classification schemes based on:

   (a) Morphological characters: ______________________

   (b) Relatedness in time (from biochemical evidence): ______________________

2. Explain the difference between a shared characteristic and a shared derived characteristic: ______________________

3. Explain how the rule of parsimony is applied to cladistics: ______________________

4. Describe the contribution of biochemical evidence to taxonomy: ______________________

5. In the DNA data for the whale cladogram (previous page) identify the DNA match that shows a mutation event must have happened twice in evolutionary history.

   ____

6. Based on the diagram above, state the family to which the chimpanzees belong under:

   (a) A traditional scheme: ______________ (b) A cladistic scheme: ______________

Features of Taxonomic Groups

In order to distinguish organisms, it is desirable to classify and name them (a science known as **taxonomy**). An effective classification system requires features that are distinctive to a particular group of organisms. Revised classification systems, recognizing three domains (rather than five or six kingdoms) are now recognized as better representations of the true diversity of life. However, for the purposes of describing the groups with which we are most familiar, the five kingdom system (used here) is still appropriate. The distinguishing features of some major **taxa** are provided in the following pages by means of diagrams and brief summaries. Note that most animals show **bilateral symmetry** (body divisible into two halves that are mirror images). **Radial symmetry** (body divisible into equal halves through various planes) is a characteristic of cnidarians and ctenophores.

SUPERKINGDOM: PROKARYOTAE (Bacteria)

- Also known as monerans or prokaryotes.
- Two major bacterial lineages are recognized: the primitive **Archaebacteria** and the more advanced **Eubacteria**.
- All have a prokaryotic cell structure: they lack the nuclei and chromosomes of eukaryotic cells, and have smaller (70S) ribosomes.
- Have a tendency to spread genetic elements across species barriers by sexual conjugation, viral transduction and other processes.
- Can reproduce rapidly by binary fission in the absence of sex.
- Have evolved a wider variety of metabolism types than eukaryotes.
- Bacteria grow and divide or aggregate into filaments or colonies of various shapes.
- They are taxonomically identified by their appearance (form) and through biochemical differences.

Species diversity: 10 000 + Bacteria are rather difficult to classify to the species level because of their relatively rampant genetic exchange, and because their reproduction is usually asexual.

Eubacteria

- Also known as 'true bacteria', they probably evolved from the more ancient Archaebacteria.
- Distinguished from Archaebacteria by differences in cell wall composition, nucleotide structure, and ribosome shape.
- Diverse group includes most bacteria.
- The **gram stain** provides the basis for distinguishing two broad groups of bacteria. It relies on the presence of peptidoglycan (unique to bacteria) in the cell wall. The stain is easily washed from the thin peptidoglycan layer of gram negative walls but is retained by the thick peptidoglycan layer of gram positive cells, staining them a dark violet colour.

Gram-Positive Bacteria

The walls of gram positive bacteria consist of many layers of peptidoglycan forming a thick, single-layered structure that holds the gram stain.

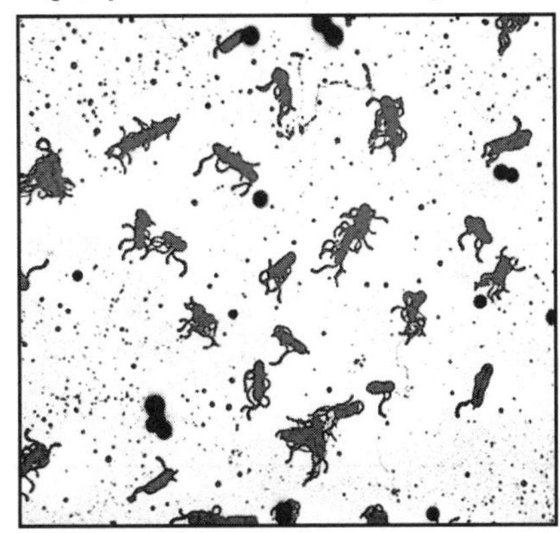

Bacillus alvei: a gram positive, flagellated bacterium. Note how the cells appear dark.

Gram-Negative Bacteria

The cell walls of gram negative bacteria contain only a small proportion of peptidoglycan, so the dark violet stain is not retained by the organisms.

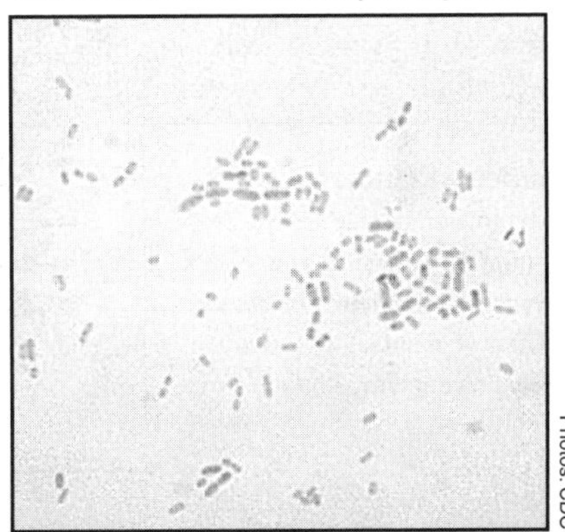

Alcaligenes odorans: a gram negative bacterium. Note how the cells appear pale.

Photos: CDC

SUPERKINGDOM EUKARYOTAE
Kingdom: FUNGI

- Heterotrophic.
- Rigid cell wall made of chitin.
- Vary from single celled to large multicellular organisms.
- Mostly saprotrophic (i.e. feeding on dead or decaying material).
- Terrestrial and immobile.

Examples:
Mushrooms/toadstools, yeasts, truffles, morels, molds, and lichens.

Species diversity: 80 000 +

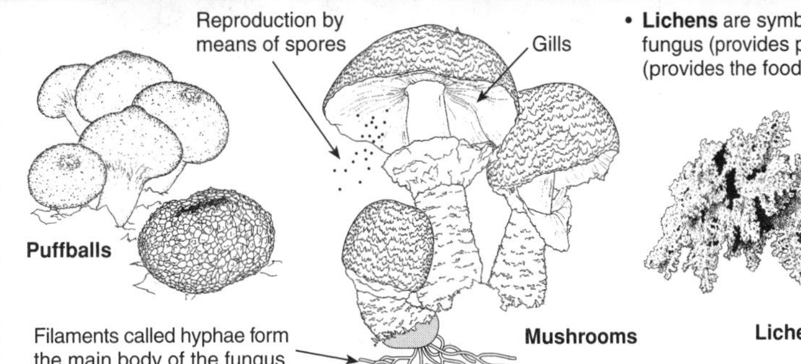

- **Lichens** are symbiotic associations of a fungus (provides protection) and an alga (provides the food).

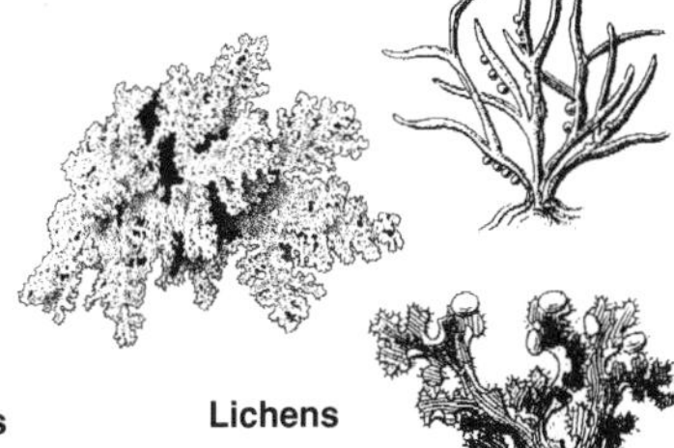

Kingdom: PROTISTA

- A diverse group of organisms that do not fit easily into other taxonomic groups.
- Unicellular or simple multicellular.
- Widespread in moist or aquatic environments.

Examples of algae: green, brown, and red algae, dinoflagellates, diatoms.

Examples of protozoa: amoebas, foraminiferans, radiolarians, ciliates.

Species diversity: 55 000 +

Algae 'plant-like' protists

- Autotrophic (photosynthesis)
- Characterized by the type of chlorophyll present

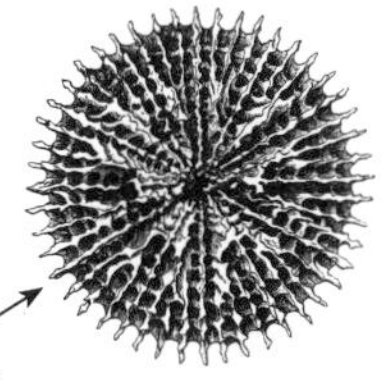

Protozoa 'animal-like' protists

- Heterotrophic nutrition and feed via ingestion
- Most are microscopic (5 µm-250 µm)

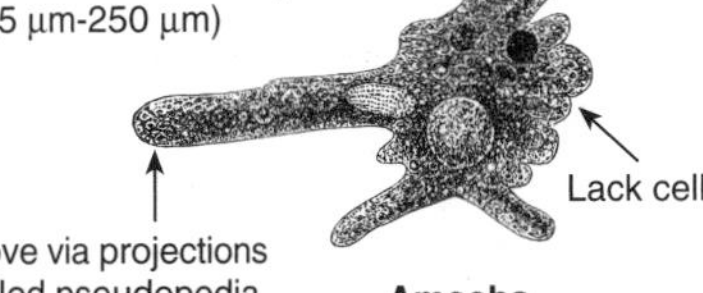

Kingdom: PLANTAE

- Multicellular organisms (the majority are photosynthetic and contain chlorophyll).
- Cell walls made of cellulose; Food is stored as starch.
- Subdivided into two major divisions based on tissue structure: ***Bryophytes*** (non-vascular) and ***Tracheophytes*** (vascular) plants.

Non-Vascular Plants:

- Non vascular, lacking transport tissues (no xylem or phloem).
- They are small and restricted to moist, terrestrial environments.
- Do not possess 'true' roots, stems or leaves

Phylum Bryophyta: Mosses, liverworts, and hornworts.

Species diversity: 18 600 +

Phylum: Bryophyta

Sexual reproductive structures

Flattened thallus (leaf like structure)

Liverworts

Sporophyte: reproduce by spores

Rhizoids anchor the plant into the ground

Mosses

Vascular Plants:

- Vascular: possess transport tissues.
- Possess true roots, stems, and leaves, as well as stomata.
- Reproduce via spores, not seeds.
- Clearly defined *alternation of sporophyte and gametophyte generations.*

Seedless Plants:

Spore producing plants, includes:

Phylum Filicinophyta: Ferns

Phylum Sphenophyta: Horsetails

Phylum Lycophyta: Club mosses

Species diversity: 13 000 +

Phylum: Lycophyta

Leaves

Club moss

Phylum: Sphenophyta

Leaves

Horsetail

Phylum: Filicinophyta

Reproduce via spores on the underside of leaf

Large dividing leaves called fronds

Rhizome

Adventitious roots

Fern

Seed Plants:

Also called Spermatophyta. Produce seeds housing an embryo. Includes:

Gymnosperms

- Lack enclosed chambers in which seeds develop.
- Produce seeds in cones which are exposed to the environment.

Phylum Cycadophyta: Cycads

Phylum Ginkgophyta: Ginkgoes

Phylum Coniferophyta: Conifers

Species diversity: 730 +

Phylum: Cycadophyta | Phylum: Ginkophyta | Phylum: Coniferophyta

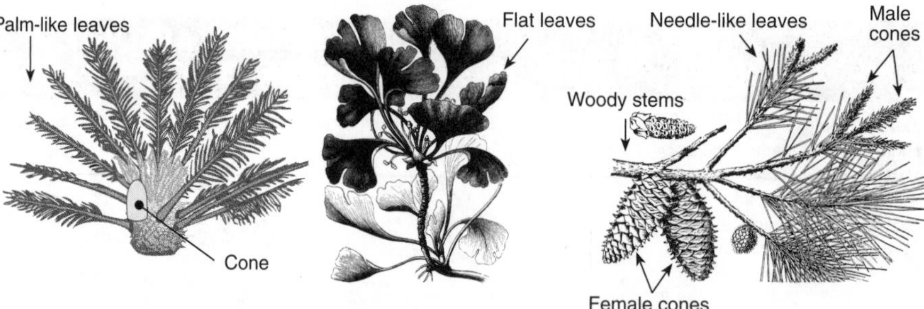

Cycad | Ginkgo | Conifer

Angiosperms

Phylum: Angiospermophyta

- Seeds in specialized reproductive structures called flowers.
- Female reproductive ovary develops into a fruit.
- Pollination usually via wind or animals.

Species diversity: 260 000 +

The phylum Angiospermophyta may be subdivided into two classes:

Class *Monocotyledoneae* (Monocots)

Class *Dicotyledoneae* (Dicots)

Angiosperms: **Monocotyledons**

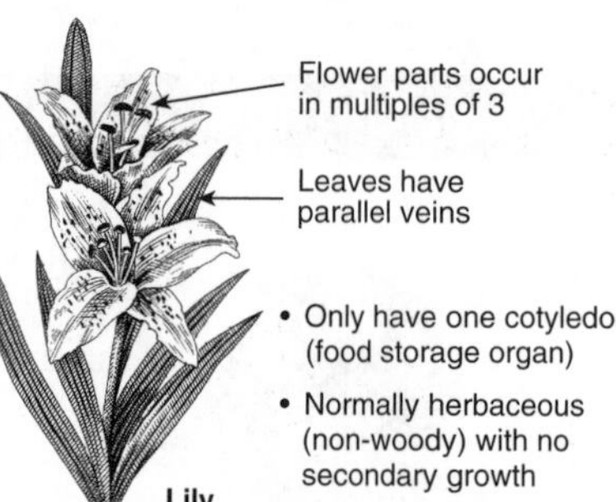

- Only have one cotyledon (food storage organ)
- Normally herbaceous (non-woody) with no secondary growth

Examples: cereals, lilies, daffodils, palms, grasses.

Angiosperms: **Dicotyledons**

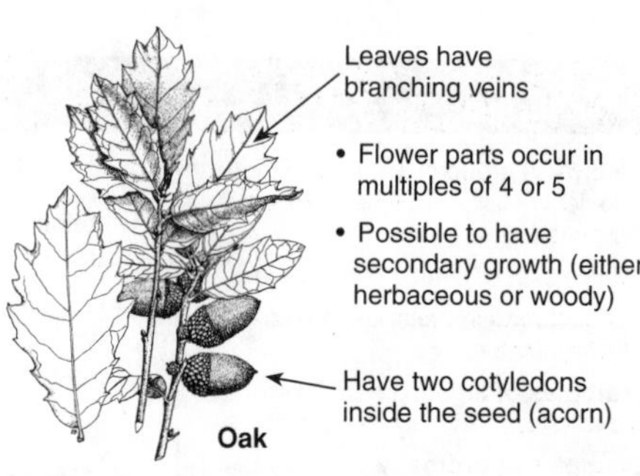

- Flower parts occur in multiples of 4 or 5
- Possible to have secondary growth (either herbaceous or woody)

Examples: many annual plants, trees and shrubs.

Kingdom: ANIMALIA

- Over 800 000 species described in 33 existing phyla.
- Multicellular, heterotrophic organisms.
- Animal cells lack cell walls.
- Further subdivided into various major phyla on the basis of body symmetry, type of body cavity, and external and internal structures.

Phylum: Rotifera

- A diverse group of small organisms with sessile, colonial, and planktonic forms.
- Most freshwater, a few marine.
- Typically reproduce via cyclic parthenogenesis.
- Characterized by a wheel of cilia on the head used for feeding and locomotion, a large muscular pharynx (mastax) with jaw like trophi, and a foot with sticky toes.

Species diversity: 1500 +

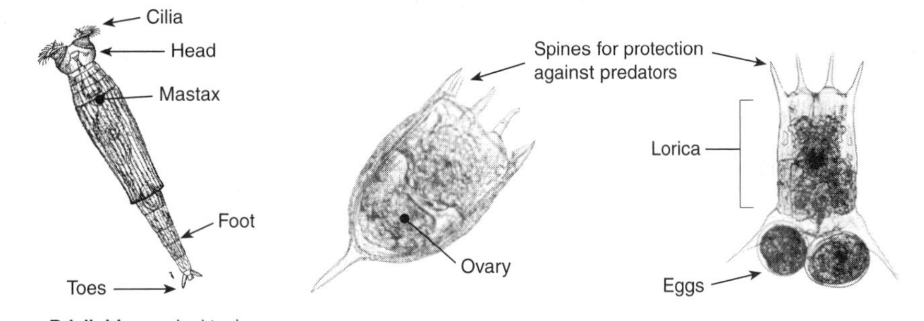

Bdelloid: non planktonic, creeping rotifer

Planktonic forms swim using their crown of cilia

Phylum: Porifera

- Lack organs.
- All are aquatic (mostly marine).
- Asexual reproduction by budding.
- Lack a nervous system.

Examples: sponges.

Species diversity: 8000 +

- Capable of regeneration (the replacement of lost parts)
- Possess spicules (needle-like internal structures) for support and protection

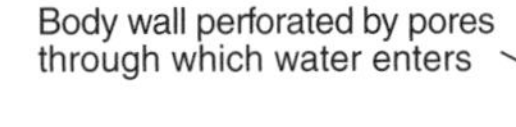

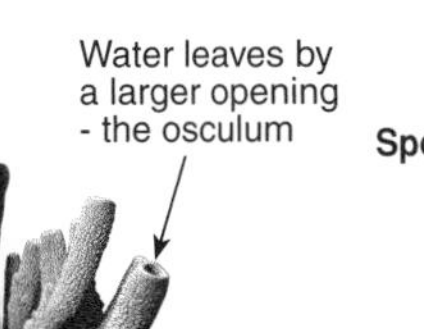

Tube sponge

Sponge

Phylum: Cnidaria

- Two basic body forms:

  Medusa: umbrella shaped and free swimming by pulsating bell.

  Polyp: cylindrical, some are sedentary, others can glide, or somersault or use tentacles as legs.
- Some species have a life cycle that alternates between a polyp stage and a medusa stage.
- All are aquatic (most are marine).

Examples: Jellyfish, sea anemones, hydras, and corals.

Species diversity: 11 000 +

Some have air-filled floats

Nematocysts (stinging cells)

Jellyfish (Portuguese man-o-war)

Colonial polyps

Single opening acts as mouth and anus

Polyps stick to seabed

Sea anemone

Polyps may aggregate in colonies

Brain coral

Contraction of the bell propels the free swimming medusa

Phylum: Platyhelminthes

- Unsegmented body.
- Flattened body shape.
- Mouth, but no anus.
- Many are parasitic.

Examples: Tapeworms, planarians, flukes.

Species diversity: 20 000 +

Liver fluke

Hooks

Detail of head (scolex)

Tapeworm

Planarian

Phylum: Nematoda

- Tiny, unsegmented roundworms.
- Many are plant/animal parasites

Examples: Hookworms, stomach worms, lung worms, filarial worms

Species diversity: 80 000 - 1 million

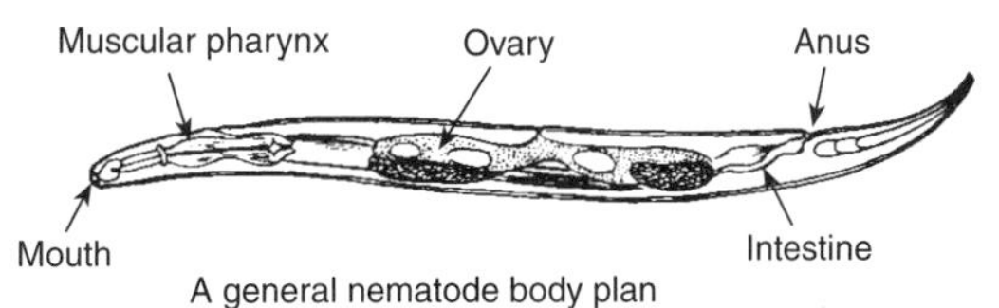

A general nematode body plan

A roundworm parasite

Phylum: Annelida

- Cylindrical, segmented body with chaetae (bristles).
- Move using hydrostatic skeleton and/or parapodia (appendages).

Examples: Earthworms, leeches, polychaetes (including tubeworms).

Species diversity: 15 000 +

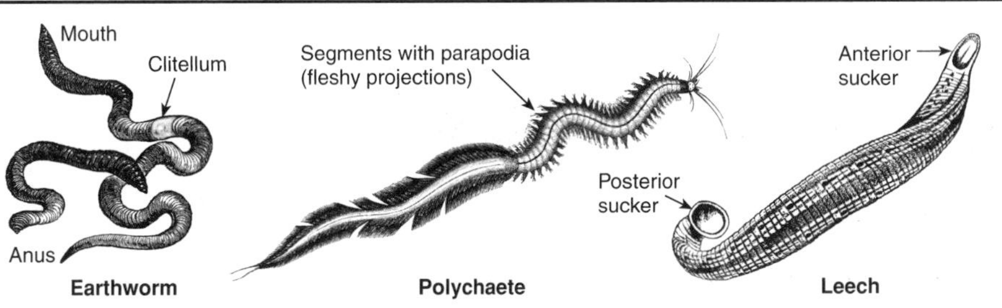

Earthworm

Polychaete

Leech

Kingdom: ANIMALIA *(continued)*

Phylum: Mollusca

- Soft bodied and unsegmented.
- Body comprises head, muscular foot, and visceral mass (organs).
- Most have radula (rasping tongue).
- Aquatic and terrestrial species.
- Aquatic species possess gills.

Examples: Snails, mussels, squid.

Species diversity: 110 000 +

Class: Bivalvia

Radula lost in bivalves
Mantle secretes shell
Two shells hinged together
Scallop

Class: Gastropoda

Mantle secretes shell
Muscular foot for locomotion
Tentacles with eyes
Head
Land snail

Class: Cephalopoda

Well developed eyes
Squid
Foot divided into tentacles

Phylum: Arthropoda

- Exoskeleton made of chitin.
- Grow in stages after moulting.
- Jointed appendages.
- Segmented bodies.
- Heart found on dorsal side of body.
- Open circulation system.
- Most have compound eyes.

Species diversity: 1 million +
Make up 75% of all living animals.

Arthropods are subdivided into the following classes:

Class: Crustacea (crustaceans)
- Mainly marine.
- Exoskeleton impregnated with mineral salts.
- Gills often present.
- Includes: Lobsters, crabs, barnacles, prawns, shrimps, isopods, amphipods
- **Species diversity**: 35 000 +

Class: Arachnida (chelicerates)
- Almost all are terrestrial.
- 2 body parts: cephalothorax and abdomen (except horseshoe crabs).
- Includes: spiders, scorpions, ticks, mites, horseshoe crabs.
- **Species diversity**: 57 000 +

Class: Insecta (insects)
- Mostly terrestrial.
- Most are capable of flight.
- 3 body parts: head, thorax, abdomen.
- Include: Locusts, dragonflies, cockroaches, butterflies, bees, ants, beetles, bugs, flies, and more
- **Species diversity**: 800 000 +

Myriapoda (=many legs)
Class: Diplopoda (millipedes)
- Terrestrial.
- Have a rounded body.
- Eat dead or living plants.
- **Species diversity**: 2000 +

Class: Chilopoda (centipedes)
- Terrestrial.
- Have a flattened body.
- Poison claws for catching prey.
- Feed on insects, worms, and snails.
- **Species diversity**: 7000 +

Class: Crustacea

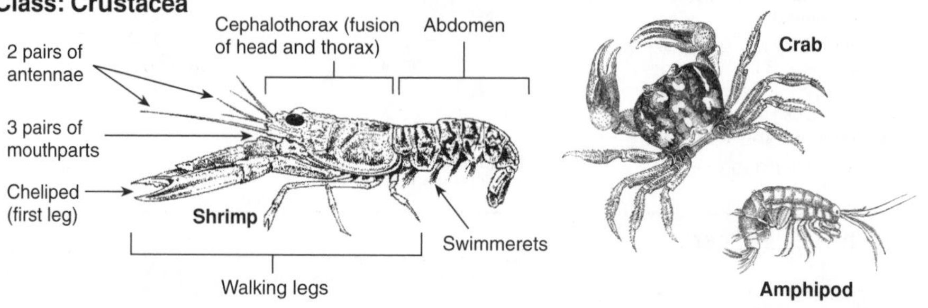

Class: Arachnida

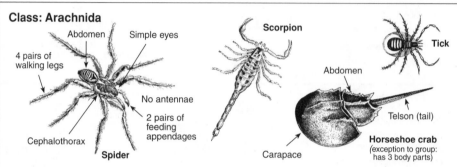

Abdomen
Simple eyes
4 pairs of walking legs
No antennae
2 pairs of feeding appendages
Cephalothorax
Spider
Scorpion
Tick
Abdomen
Telson (tail)
Carapace
Horseshoe crab (exception to group: has 3 body parts)

Class: Insecta

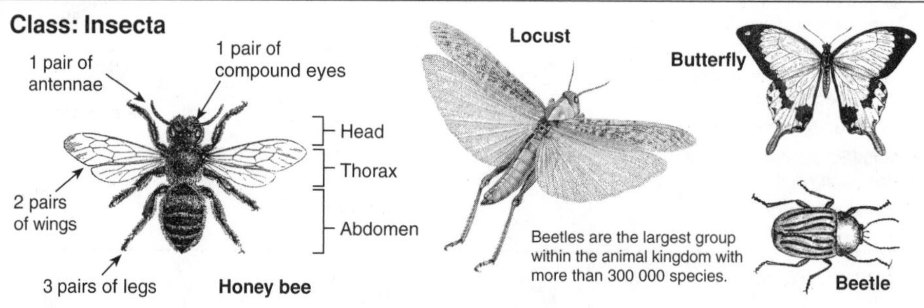

Class: Diplopoda

Body with many similar segments
Clearly defined head
1 pair of antennae
Each segment has 2 pairs of legs
1 pair of mouthparts

Class: Chilopoda

Body with many similar segments
1 pair of large antennae
Clearly defined head
1 pair of mouthparts
Each segment has 1 pair of legs

Phylum: Echinodermata

- Rigid body wall, internal skeleton made of calcareous plates.
- Many possess spines.
- Ventral mouth, dorsal anus.
- External fertilization.
- Unsegmented, marine organisms.
- Tube feet for locomotion.
- Water vascular system.

Examples: Starfish, brittlestars, feather stars, sea urchins, sea lilies.

Species diversity: 6000 +

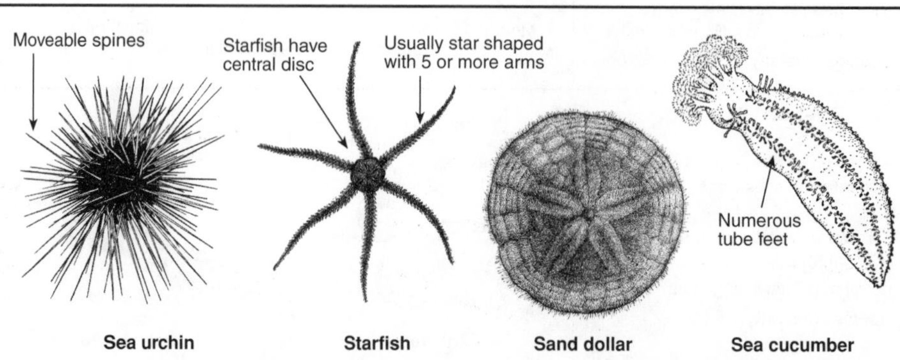

Kingdom: ANIMALIA *(continued)*

Phylum: Chordata

- Dorsal notochord (flexible, supporting rod) present at some stage in the life history.
- Post-anal tail present at some stage in their development.
- Dorsal, tubular nerve cord.
- Pharyngeal slits present.
- Circulation system closed in most.
- Heart positioned on ventral side.

Species diversity: 48 000 +

- A very diverse group with several sub-phyla:
  - Urochordata (sea squirts, salps)
  - Cephalochordata (lancelet)
  - Craniata (vertebrates)

Sub-Phylum Craniata (vertebrates)

- Internal skeleton of cartilage or bone.
- Well developed nervous system.
- Vertebral column replaces notochord.
- Two pairs of appendages (fins or limbs) attached to girdles.

Further subdivided into:

Class: Chondrichthyes (cartilaginous fish)

- Skeleton of cartilage (not bone).
- No swim bladder.
- All aquatic (mostly marine).
- Include: Sharks, rays, and skates.

Species diversity: 850 +

Class: Osteichthyes (bony fish)

- Swim bladder present.
- All aquatic (marine and fresh water).

Species diversity: 21 000 +

Class: Amphibia (amphibians)

- Lungs in adult, juveniles may have gills (retained in some adults).
- Gas exchange also through skin.
- Aquatic and terrestrial (limited to damp environments).
- Include: Frogs, toads, salamanders, and newts.

Species diversity: 3900 +

Class Reptilia (reptiles)

- Ectotherms with no larval stages.
- Teeth are all the same type.
- Eggs with soft leathery shell.
- Mostly terrestrial.
- Include: Snakes, lizards, crocodiles, turtles, and tortoises.

Species diversity: 7000 +

Class: Aves (birds)

- Terrestrial endotherms.
- Eggs with hard, calcareous shell.
- Strong, light skeleton.
- High metabolic rate.
- Gas exchange assisted by air sacs.

Species diversity: 8600 +

Class: Mammalia (mammals)

- Endotherms with hair or fur.
- Mammary glands produce milk.
- Glandular skin with hair or fur.
- External ear present.
- Teeth are of different types.
- Diaphragm between thorax/abdomen.

Species diversity: 4500 +

Subdivided into three subclasses: *Monotremes, marsupials, placentals.*

Class: Chondrichthyes (cartilaginous fish)

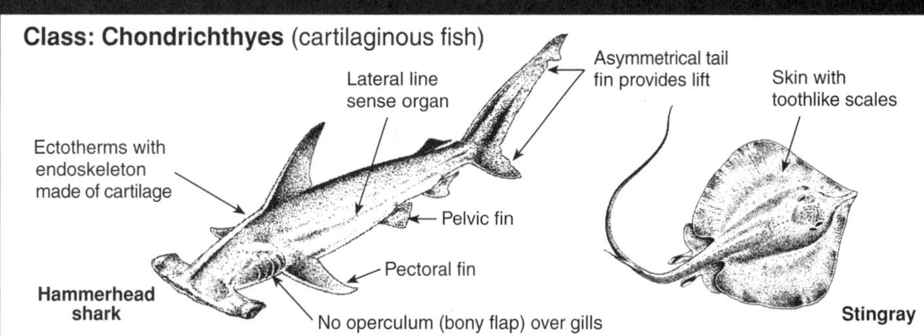

Class: Osteichthyes (bony fish)

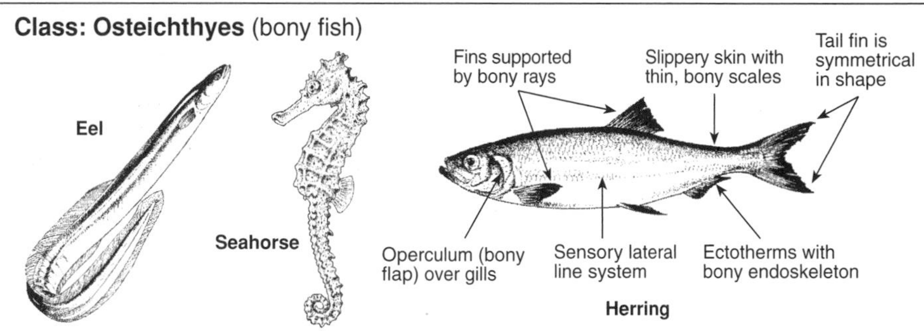

Class: Amphibia

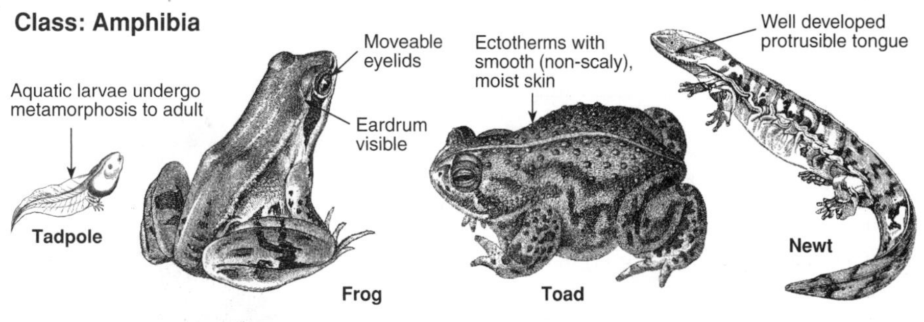

Class: Reptilia

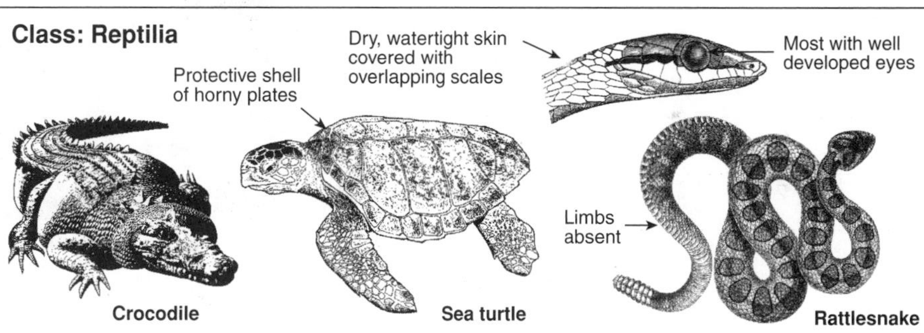

Class: Aves

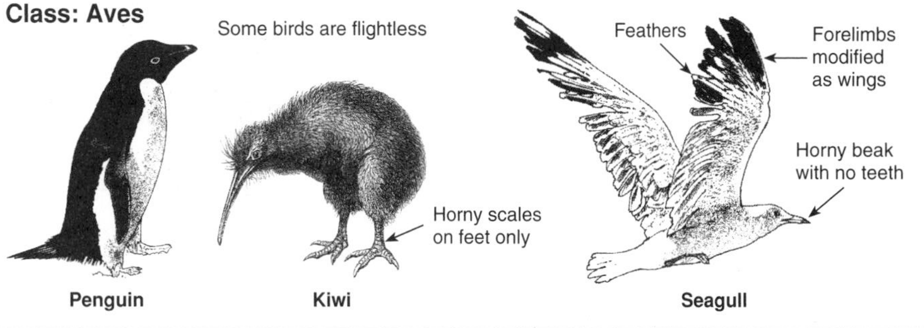

Class: Mammalia

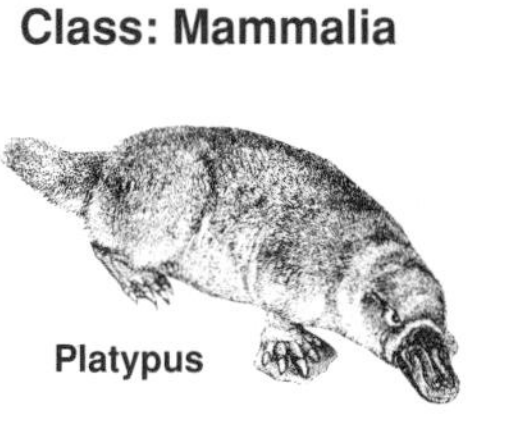

Monotremes
Egg laying mammals

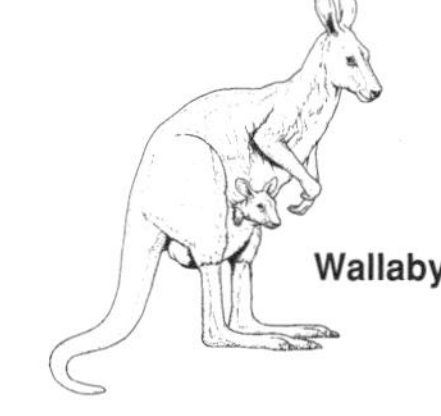

Marsupials
Give birth to live, very immature young which then develop in a pouch

Placentals
Have a placenta and give birth to live, well developed young

Classification

Features of the Five Kingdoms

The classification of organisms into taxonomic groups is based on how biologists believe they are related in an evolutionary sense. Organisms in a taxonomic group share features which set them apart from other groups. By identifying these **distinguishing features**, it is possible to develop an understanding of the evolutionary history of the group. The focus of this activity is to summarize the distinguishing features of each of the five kingdoms in the five kingdom classification system.

1. Distinguishing features of Kingdom **Prokaryotae**:

2. Distinguishing features of Kingdom **Protista**:

3. Distinguishing features of Kingdom **Fungi**:

4. Distinguishing features of Kingdom **Plantae**:

5. Distinguishing features of Kingdom **Animalia**:

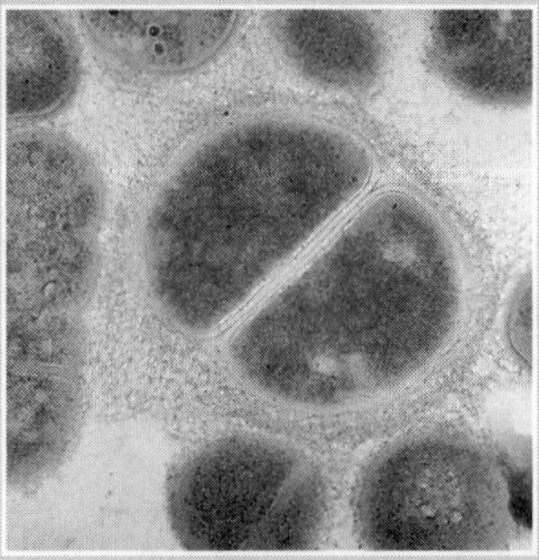

EII

***Staphylococcus* dividing**

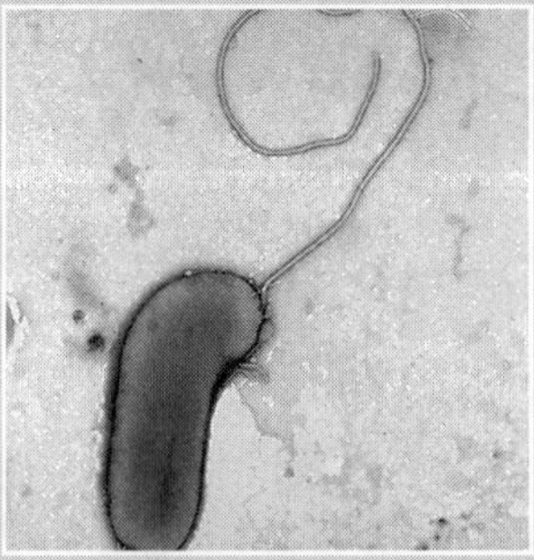

EII

Helicobacter pylori

CDC

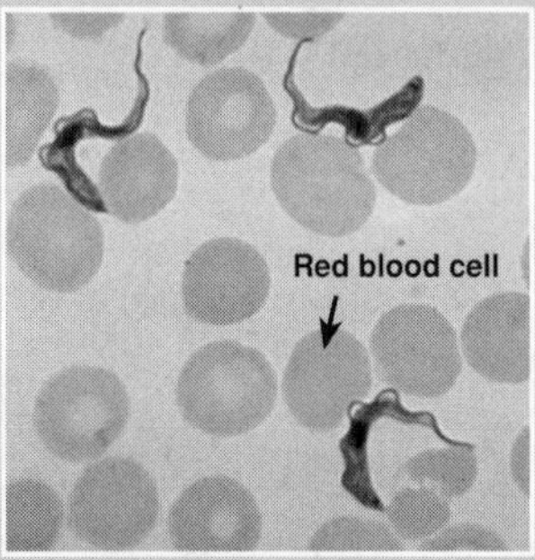

Red blood cell

***Trypanosoma* parasite**

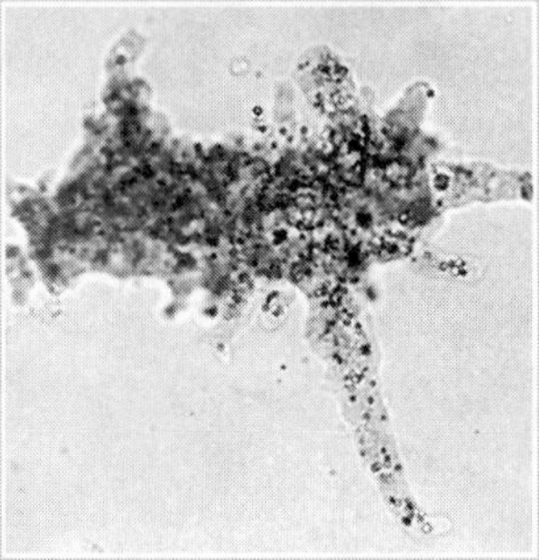

RCN

Amoeba

Mushrooms

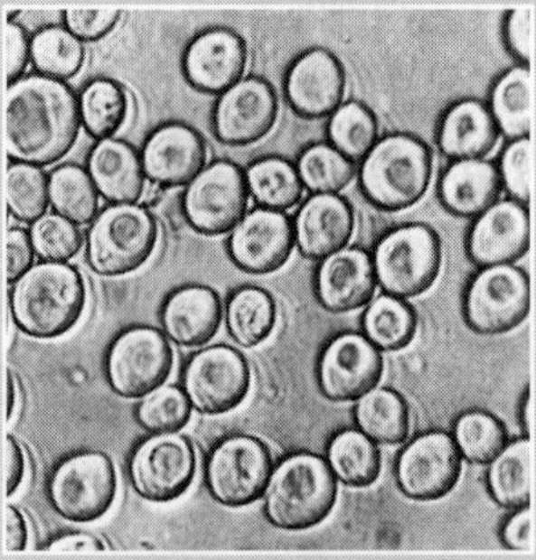

Yeast cells in solution

Maple seeds

Pea plants

RA

Cicada moulting

Gibbon

Classification System

The classification of organisms is designed to reflect how they are related to each other. The fundamental unit of classification of living things is the **species** (see Senior Biology 1*: The Species Concept* for a discussion of what we understand by a species). Its members are so alike genetically that they can interbreed. This genetic similarity also means that they are almost identical in their physical and other characteristics. Species are classified further into larger, more comprehensive categories (higher taxa). It must be emphasized that all such higher classifications are human inventions to suit a particular purpose.

1. The table below shows part of the classification for humans using the seven major levels of classification. For this question, use the example of the classification of the Ethiopian hedgehog, on the next page, as a guide.

   (a) Complete the list of the classification levels on the left hand side of the table below:

	Classification level	Human classification
1.		
2.		
3.		
4.		
5.	Family	Hominidae
6.		
7.		

   (b) The name of the Family that humans belong to has already been entered into the space provided. Complete the classification for humans (*Homo sapiens*) on the table above.

2. Describe the two-part scientific naming system (called the **binomial system**) which is used to name organisms:

3. Give two reasons why the classification of organisms is important:

   (a)

   (b)

4. Traditionally, the classification of organisms has been based largely on similarities in physical appearance. More recently, new methods involving biochemical comparisons have been used to provide new insights into how species are related. Describe an example of a biochemical method for comparing how species are related:

5. As an example of physical features being used to classify organisms, mammals have been divided into three major sub-classes: monotremes, marsupials, and placentals. Describe the main physical feature distinguishing each of these taxa:

   (a) Monotreme:

   (b) Marsupial:

   (c) Placental:

Periodicals:
A passion for order

***Related activities**: Phylogenetic Systematics, Classification Keys*

Classification of the Ethiopian Hedgehog

Below is the classification for the **Ethiopian hedgehog**. Only one of each group is subdivided in this chart showing the levels that can be used in classifying an organism. Not all possible subdivisions have been shown here. For example, it is possible to indicate such categories as **super-class** and **sub-family**. The only natural category is the **species**, often separated into geographical **races**, or **sub-species**, which generally differ in appearance.

Kingdom: **Animalia**
Animals; one of five kingdoms

Phylum: **Chordata**
Animals with a notochord (supporting rod of cells along the upper surface)
tunicates, salps, lancelets, and vertebrates

23 other phyla

Sub-phylum: **Vertebrata**
Animals with backbones
fish, amphibians, reptiles, birds, mammals

Class: **Mammalia**
Animals that suckle their young on milk from mammary glands
placentals, marsupials, monotremes

Sub-class: **Eutheria or Placentals**
Mammals whose young develop for some time in the female's reproductive tract gaining nourishment from a placenta
placental mammals

Order: **Insectivora**
Insect eating mammals
An order of over 300 species of primitive, small mammals that feed mainly on insects and other small invertebrates.

17 other orders

Sub-order: **Erinaceomorpha**
The hedgehog-type insectivores. One of the three suborders of insectivores. The other suborders include the tenrec-like insectivores (*tenrecs and golden moles*) and the shrew-like insectivores (*shrews, moles, desmans, and solenodons*).

Family: **Erinaceidae**
The only family within this suborder. Comprises two subfamilies: the true or spiny hedgehogs and the moonrats (gymnures). Representatives in the family include the common European hedgehog, desert hedgehog, and the moonrats.

Genus: ***Paraechinus***
One of eight genera in this family. The genus *Paraechinus* includes three species which are distinguishable by a wide and prominent naked area on the scalp.

7 other genera

Species: ***aethiopicus***
The Ethiopian hedgehog inhabits arid coastal areas. Their diet consists mainly of insects, but includes small vertebrates and the eggs of ground nesting birds.

3 other species

The order *Insectivora* was first introduced to group together shrews, moles, and hedgehogs. It was later extended to include tenrecs, golden moles, desmans, tree shrews, and elephant shrews and the taxonomy of the group became very confused. Recent reclassification of the elephant shrews and tree shrews into their own separate orders has made the Insectivora a more cohesive group taxonomically.

Ethiopian hedgehog
Paraechinus aethiopicus

Classification Keys

Classification systems provide biologists with a way in which to identify species. They also indicate how closely related, in an evolutionary sense, each species is to others. An organism's classification should include a clear, unambiguous **description**, an accurate **diagram**, and its unique name, denoted by the **genus** and **species**. Classification keys are used to identify an organism and assign it to the correct species (assuming that the organism has already been formally classified and is included in the key). Typically, keys are **dichotomous** and involve a series of linked steps. At each step, a choice is made between two features; each alternative leads to another question until an identification is made. If the organism cannot be identified, it may be a new species or the key may need revision. Two examples of **dichotomous keys** are provided here. The first (below) describes features for identifying the larvae of various genera within the order Trichoptera (caddisflies). From this key you should be able to assign a generic name to each of the caddisfly larvae pictured. The key on the next page identifies aquatic insect orders.

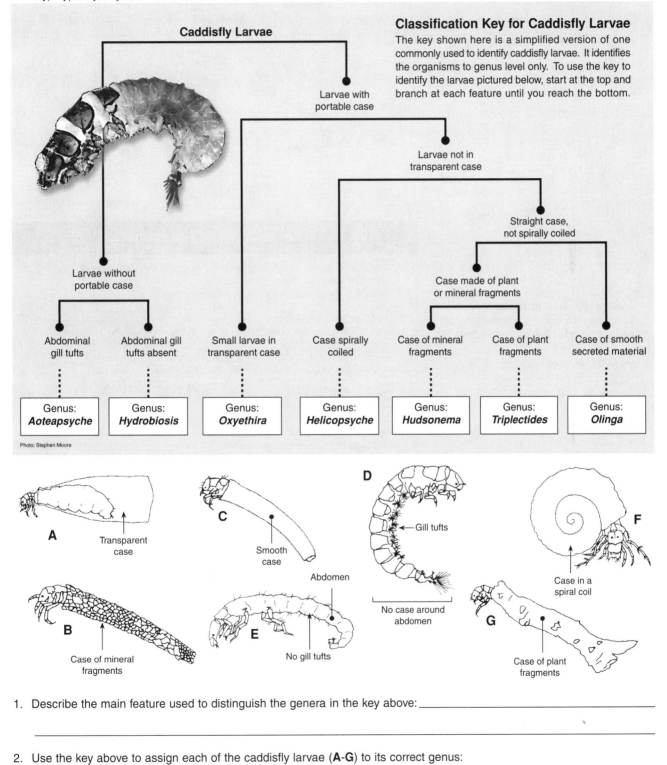

Classification Key for Caddisfly Larvae

The key shown here is a simplified version of one commonly used to identify caddisfly larvae. It identifies the organisms to genus level only. To use the key to identify the larvae pictured below, start at the top and branch at each feature until you reach the bottom.

Caddisfly Larvae

- Larvae without portable case
  - Abdominal gill tufts → Genus: ***Aoteapsyche***
  - Abdominal gill tufts absent → Genus: ***Hydrobiosis***
- Larvae with portable case
  - Small larvae in transparent case → Genus: ***Oxyethira***
  - Larvae not in transparent case
    - Case spirally coiled → Genus: ***Helicopsyche***
    - Straight case, not spirally coiled
      - Case made of plant or mineral fragments
        - Case of mineral fragments → Genus: ***Hudsonema***
        - Case of plant fragments → Genus: ***Triplectides***
      - Case of smooth secreted material → Genus: ***Olinga***

Photo: Stephen Moore

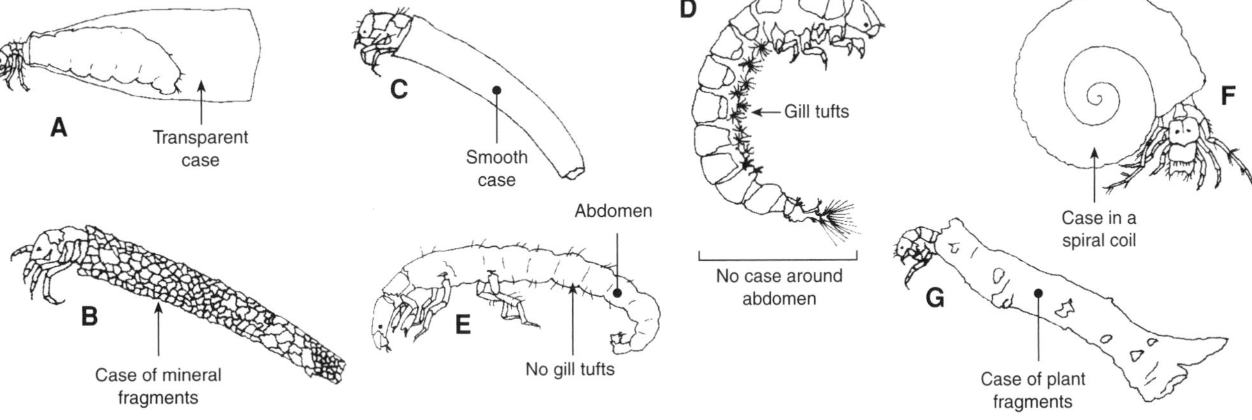

1. Describe the main feature used to distinguish the genera in the key above: ______________________

____

2. Use the key above to assign each of the caddisfly larvae (**A**-**G**) to its correct genus:

A: ____________ D: ____________ G: ____________

B: ____________ E: ____________

C: ____________ F: ____________

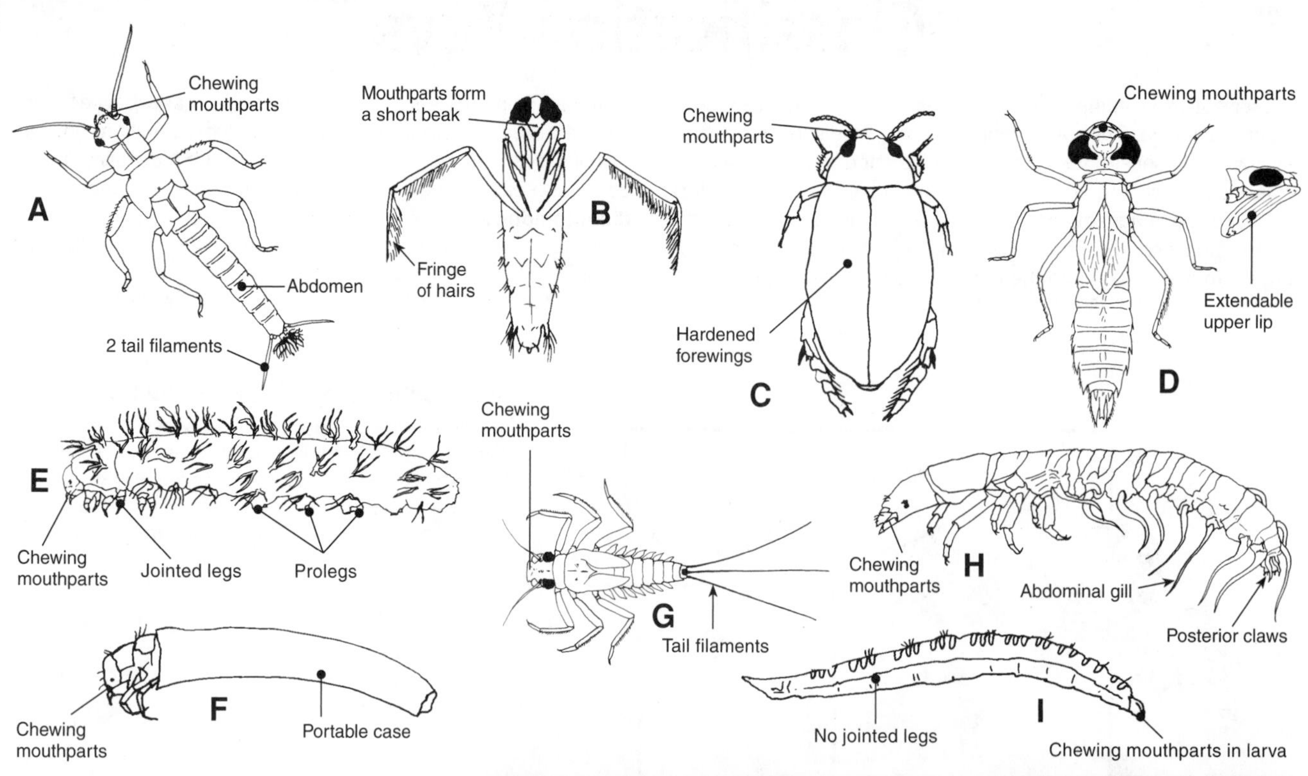

3. Use the simplified key to identify each of the orders (by order or common name) of aquatic insects (**A**-**I**) pictured above:

(a) Order of insect A: ______________________

(b) Order of insect B: ______________________

(c) Order of insect C: ______________________

(d) Order of insect D: ______________________

(e) Order of insect E: ______________________

(f) Order of insect F: ______________________

(g) Order of insect G: ______________________

(h) Order of insect H: ______________________

(i) Order of insect I: ______________________

	Key to Orders of Aquatic Insects	
1	Insects with chewing mouthparts; forewings are hardened and meet along the midline of the body when at rest (they may cover the entire abdomen or be reduced in length).	**Coleoptera** (beetles)
	Mouthparts piercing or sucking and form a pointed cone With chewing mouthparts, but without hardened forewings	*Go to 2* *Go to 3*
2	Mouthparts form a short, pointed beak; legs fringed for swimming or long and spaced for suspension on water.	**Hemiptera** (bugs)
	Mouthparts do not form a beak; legs (if present) not fringed or long, or spaced apart.	*Go to 3*
3	Prominent upper lip (labium) extendable, forming a food capturing structure longer than the head.	**Odonata** (dragonflies & damselflies)
	Without a prominent, extendable labium	*Go to 4*
4	Abdomen terminating in three tail filaments which may be long and thin, or with fringes of hairs.	**Ephemeroptera** (mayflies)
	Without three tail filaments	*Go to 5*
5	Abdomen terminating in two tail filaments	**Plecoptera** (stoneflies)
	Without long tail filaments	*Go to 6*
6	With three pairs of jointed legs on thorax	*Go to 7*
	Without jointed, thoracic legs (although non-segmented prolegs or false legs may be present).	**Diptera** (true flies)
7	Abdomen with pairs of non-segmented prolegs bearing rows of fine hooks.	**Lepidoptera** (moths and butterflies)
	Without pairs of abdominal prolegs	*Go to 8*
8	With eight pairs of finger-like abdominal gills; abdomen with two pairs of posterior claws.	**Megaloptera** (dobsonflies)
	Either, without paired, abdominal gills, or, if such gills are present, without posterior claws.	*Go to 9*
9	Abdomen with a pair of posterior prolegs bearing claws with subsidiary hooks; sometimes a portable case.	**Trichoptera** (caddisflies)

Keying Out Plant Species

Dichotomous keys are a useful tool in biology and can enable identification to the species level provided the characteristics chosen are appropriate for separating species. Keys are extensively used by botanists as they are quick and easy to use in the field, although they sometimes rely on the presence of particular plant parts such as fruits or flowers. Some also require some specialist knowledge of plant biology. The following simple activity requires you to identify five species of the genus *Acer* from illustrations of the leaves. It provides valuable practice in using characteristic features to identify plants to species level.

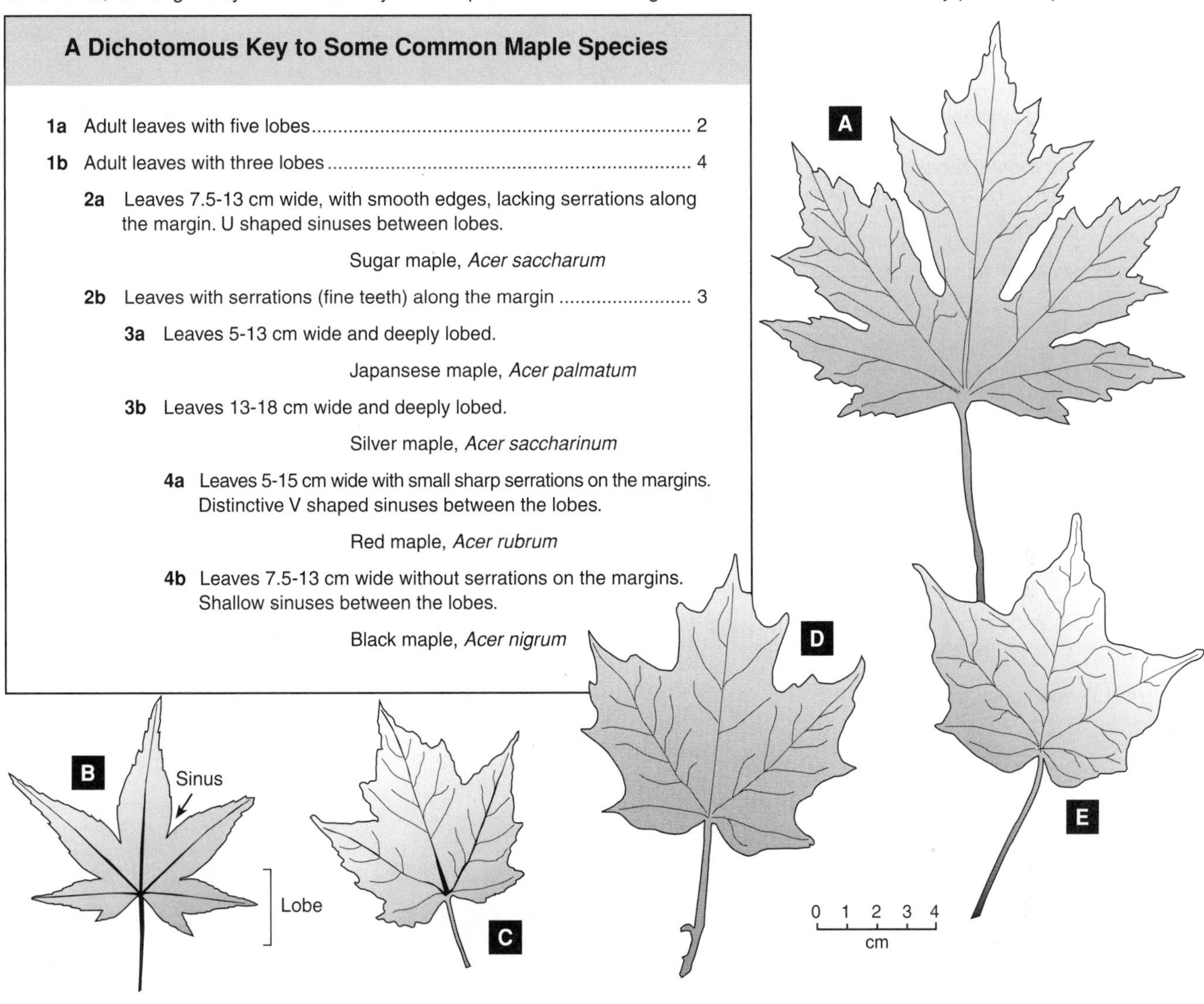

A Dichotomous Key to Some Common Maple Species

1a Adult leaves with five lobes 2

1b Adult leaves with three lobes 4

2a Leaves 7.5-13 cm wide, with smooth edges, lacking serrations along the margin. U shaped sinuses between lobes.

Sugar maple, *Acer saccharum*

2b Leaves with serrations (fine teeth) along the margin 3

3a Leaves 5-13 cm wide and deeply lobed.

Japansese maple, *Acer palmatum*

3b Leaves 13-18 cm wide and deeply lobed.

Silver maple, *Acer saccharinum*

4a Leaves 5-15 cm wide with small sharp serrations on the margins. Distinctive V shaped sinuses between the lobes.

Red maple, *Acer rubrum*

4b Leaves 7.5-13 cm wide without serrations on the margins. Shallow sinuses between the lobes.

Black maple, *Acer nigrum*

1. Use the dichotomous key to the common species of *Acer* to identify the species illustrated by the leaves (drawn to scale). Begin at the top of the key and make a choice as to which of the illustrations best fits the description:

   (a) Species A: ______________________

   (b) Species B: ______________________

   (c) Species C: ______________________

   (d) Species D: ______________________

   (e) Species E: ______________________

2. Identify a feature that could be used to identify maple species when leaves are absent: ______________________

3. Suggest why it is usually necessary to consider a number of different features in order to classify plants to species level:

   ____

   ____

4. When identifying a plant, suggest what you should be sure of before using a key to classify it to species level:

   ____

   ____

Features of Microbial Groups

A microorganism (or microbe) is literally a microscopic organism. The term is usually reserved for the organisms studied in microbiology: bacteria, fungi, microscopic protistans, and viruses. The first three of these represent three of the five kingdoms for which you described distinguishing features in an earlier activity (viruses are non-cellular and therefore not included in the five-kingdom classification). Most microbial taxa, but particularly the fungi, also have macroscopic representatives. The distinction between a macrofungus and a microfungus is an artificial but convenient one. Unlike microfungi, which are made conspicuous by the diseases or decay they cause, macrofungi are most likely to be observed with the naked eye. The microfungi include yeasts and pathogenic species. Macrofungi, e.g. mushrooms, toadstools, and lichens, are illustrated in *Features of Macrofungi and Plants*.

1. Describe aspects of each of the following for the bacteria and cyanobacteria (Kingdom Prokaryotae):

   (a) Environmental range: ______________________

   ____

   (b) Ecological role: ______________________

   ____

2. Identify an example within the bacteria of the following:

   (a) Photosynthetic: ______________________

   (b) Pathogen: ______________________

   (c) Decomposer: ______________________

   (d) Nitrogen fixer: ______________________

3. Describe aspects of each of the following for the microscopic protistans (Kingdom Protista):

   (a) Environmental range: ______________________

   ____

   (b) Ecological role: ______________________

   ____

4. Identify an example within the protists of the following:

   (a) Photosynthetic: ______________________

   (b) Pathogen: ______________________

   (c) Biological indicator: ______________________

5. Describe aspects of each of the following for the microfungi (Kingdom Fungi):

   (a) Environmental range: ______________________

   ____

   (b) Ecological role: ______________________

   ____

6. Identify examples within the microfungi of the following:

   (a) Animal pathogen: ______________________

   ____

   (b) Plant pathogens: ______________________

   ____

   ____

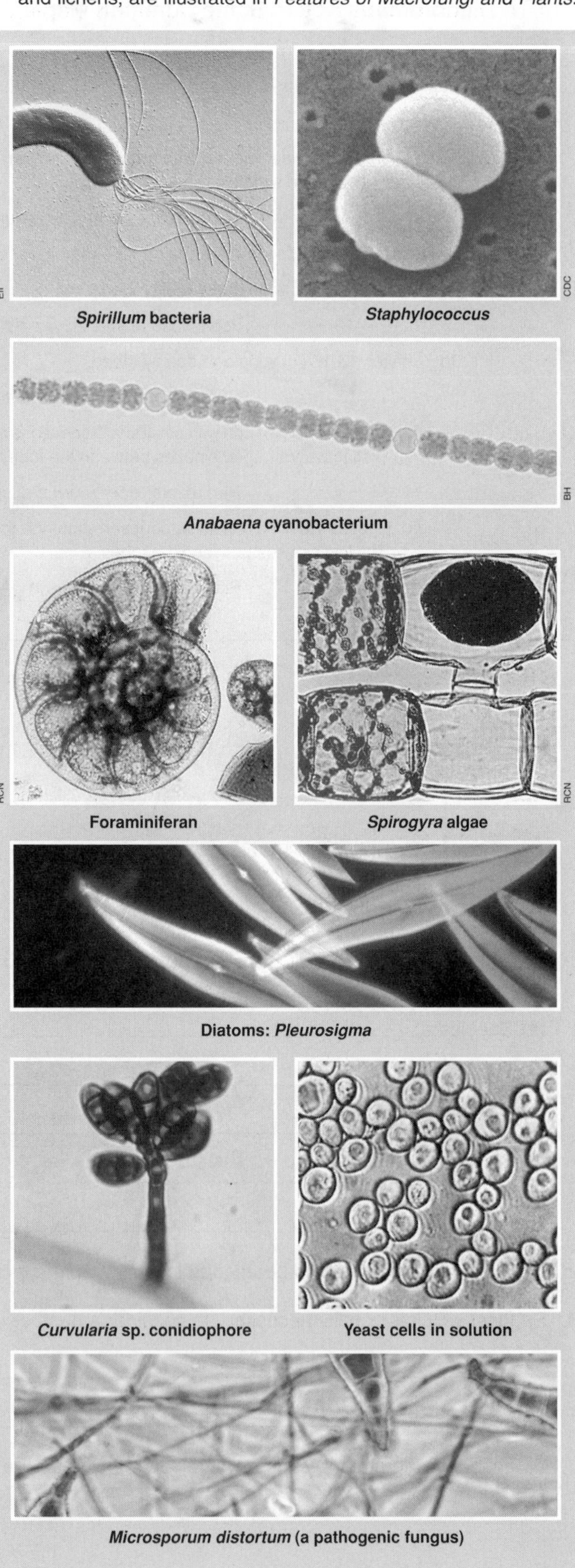

Spirillum bacteria

Staphylococcus

Anabaena cyanobacterium

Foraminiferan

Spirogyra algae

Diatoms: *Pleurosigma*

Curvularia sp. conidiophore

Yeast cells in solution

Microsporum distortum (a pathogenic fungus)

Related activities: *Features of Taxonomic Groups, Bacterial Cells*

Features of Macrofungi and Plants

Although plants and fungi are some of the most familiar organisms in our environment, their classification has not always been straightforward. We know now that the plant kingdom is monophyletic, meaning that it is derived from a common ancestor. The variety we see in plant taxa today is a result of their enormous diversification from the first plants. Although the fungi were once grouped together with the plants, they are unique organisms that differ from other eukaryotes in their mode of nutrition, structural organization, growth, and reproduction. The focus of this activity is to summarize the features of the fungal kingdom (**macrofungi**), the major divisions of the plant kingdom, and the two classes of flowering plants (angiosperms).

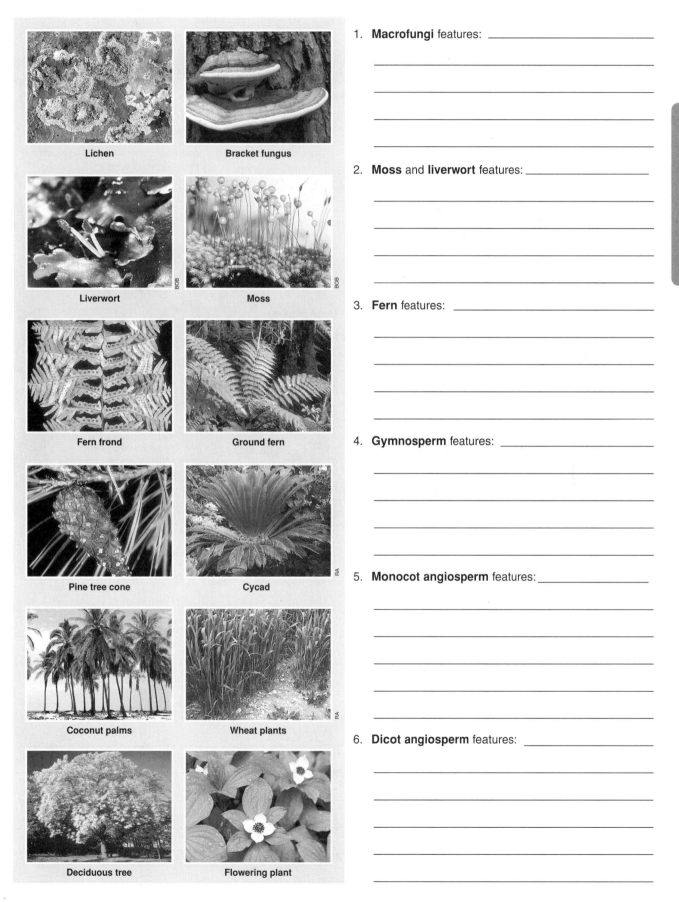

Lichen

Bracket fungus

Liverwort

Moss

Fern frond

Ground fern

Pine tree cone

Cycad

Coconut palms

Wheat plants

Deciduous tree

Flowering plant

1. **Macrofungi** features: ____________________

2. **Moss** and **liverwort** features: ____________________

3. **Fern** features: ____________________

4. **Gymnosperm** features: ____________________

5. **Monocot angiosperm** features: ____________________

6. **Dicot angiosperm** features: ____________________

Animal Symmetry

All the Eumetazoa (almost all animals but not sponges) show some kind of body symmetry, indicating symmetry is a fundamental aspect of animal architecture. By recognizing it, we divide animals into two broad categories. Radially symmetrical animals (the Branch Radiata) have body parts radiating from a central axis, whereas bilaterally symmetrical animals (the Branch Bilateria) have mirror-image left-right sides and an anterior and posterior end. Bilateral symmetry is the most common within the animal kingdom and is considered to be an adaptation to the motile lifestyle that most animals have. Radial symmetry in all life stages occurs in only two phyla (the cnidarians and ctenophores). Some animals (such as the echinoderms) show radial symmetry as adults, but this is secondary (their larvae show bilateral symmetry).

Body Symmetry

Mollusc

Bilateral Symmetry

Most animals show **bilateral symmetry**, where the body can be divided evenly through only one plane. The division goes through the midline of the body to produce roughly equivalent, mirror image halves.

Sand dollar

Radial Symmetry

Coelenterates (e.g. sea anemones) and adult echinoderms (e.g. sand dollar, above), show **radial symmetry**, where similar body parts are arranged around a central body axis. Radial symmetry is thought to have evolved originally as an adaptation to a **sedentary** lifestyle. Some of the organisms that show radial symmetry today are sedentary (e.g. coral and sea anemones) but many others are quite active (e.g. most echinoderms and jellyfish).

Surfaces in Bilateral Animals

Bilateral symmetry is considered to be an adaptation to motility (movement). To locate or describe body structures it is useful to know the basic terms: dorsal, ventral, lateral, anterior, and posterior.

Anterior (front): The front end of an animal often has a head with sense organs for detecting stimuli encountered as the animal moves through its environment.

Dorsal (top): The "back" surface of an animal (in vertebrates, the surface where the vertebral column is located).

Lateral (side): A structure is lateral if it is to one side of the body e.g. the ears are lateral to the nose in humans.

Ventral (bottom): The undersurface or belly side of an animal.

Posterior (rear): The posterior is the rear of the animal. It may help in locomotion (with a tail) or it may just follow behind.

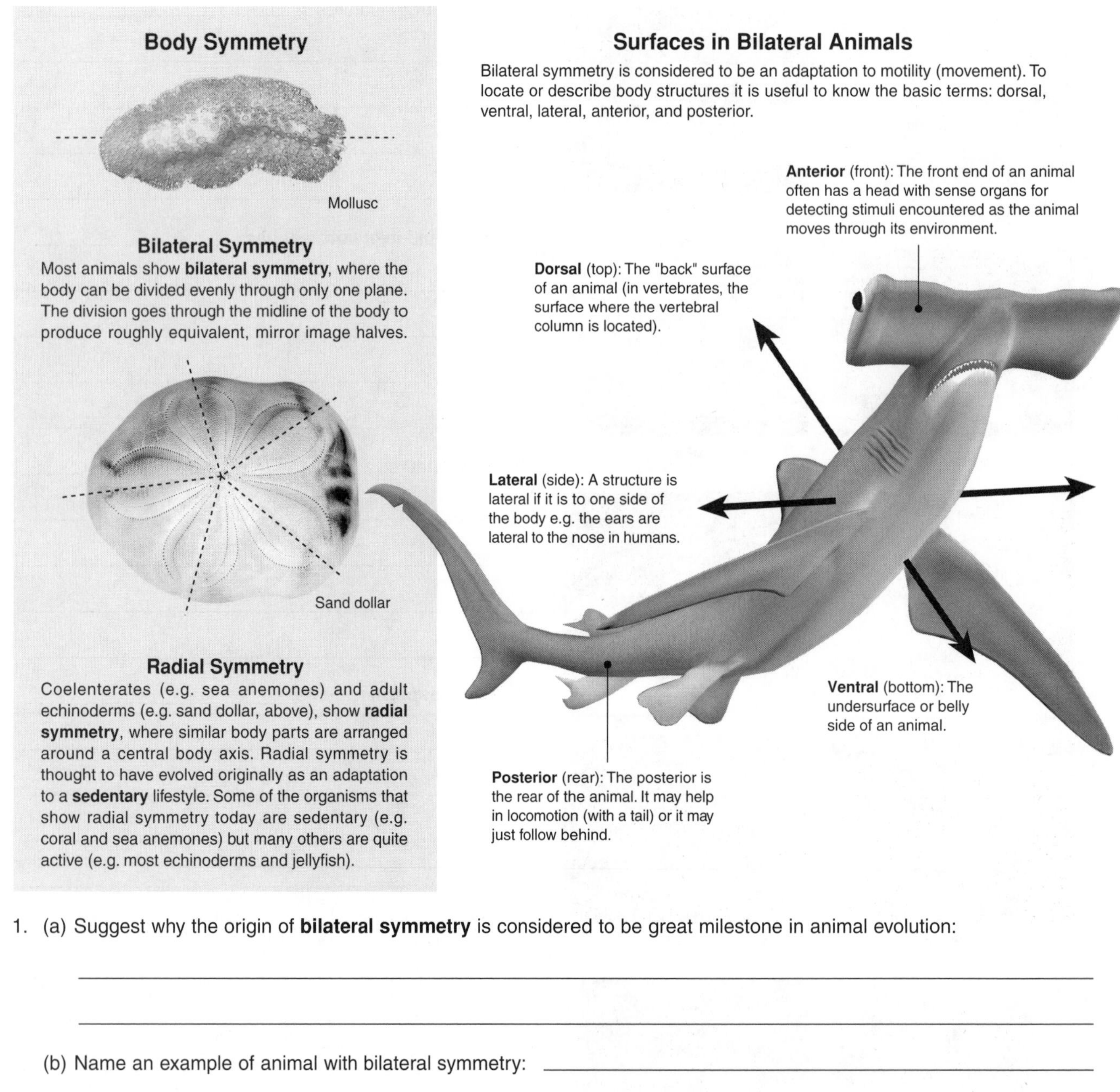

1. (a) Suggest why the origin of **bilateral symmetry** is considered to be great milestone in animal evolution:

(b) Name an example of animal with bilateral symmetry: _____

2. (a) Explain why **radial symmetry** is considered to have evolved in response to a sedentary lifestyle:

(b) Suggest where the sense organs of a radially symmetrical animal would be located: _____

(c) Name an example of an animal with radial symmetry: _____

3. Sponges are animals but not Eumetazoans and show no symmetry. What could this suggest about Kingdom Animalia?

Related activities: Phylogenetic Systematics, Features of Taxonomic Groups

Features of Animal Taxa

The animal kingdom is classified into about 35 major **phyla**. Representatives of the more familiar taxa are illustrated below: **cnidarians** (includes jellyfish, sea anemones, and corals), **annelids** (segmented worms), **arthropods** (insects, crustaceans, spiders, scorpions, centipedes and millipedes), **molluscs** (snails, bivalve shellfish, squid and octopus), **echinoderms** (starfish and sea urchins), **vertebrates** from the phylum **chordates** (fish, amphibians, reptiles, birds, and mammals). The **arthropods** and the **vertebrates** have been represented in more detail, giving the **classes** for each of these **phyla**. This activity asks you to describe the **distinguishing features** of each of the taxa represented below.

Sea anemones

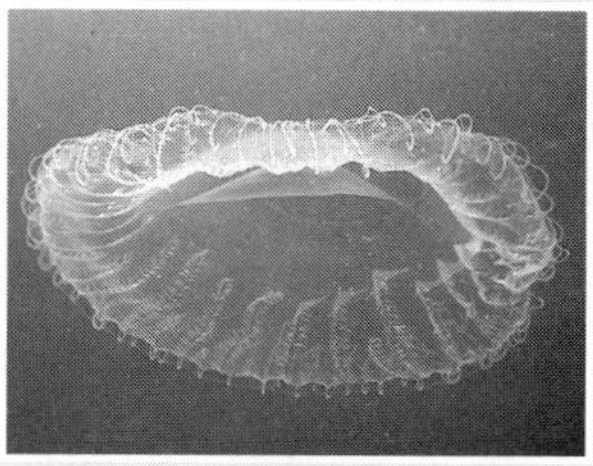

Jellyfish

Tubeworms

Earthworm

BOB

Long-horned beetle

Butterfly

Crab

Woodlouse

Scorpion

Spider

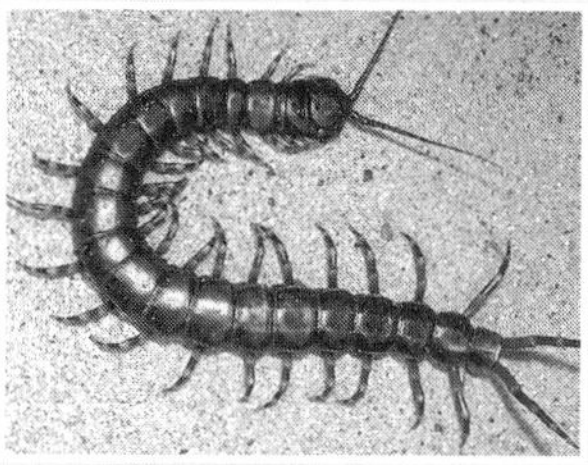

Centipede

Millipede

1. **Cnidarian** features: ______

2. **Annelid** features: ______

3. **Insect** features: ______

4. **Crustacean** features: ______

5. **Arachnid** features: ______

6. **Myriapod** (class Chilopoda and Diplopoda) features: ______

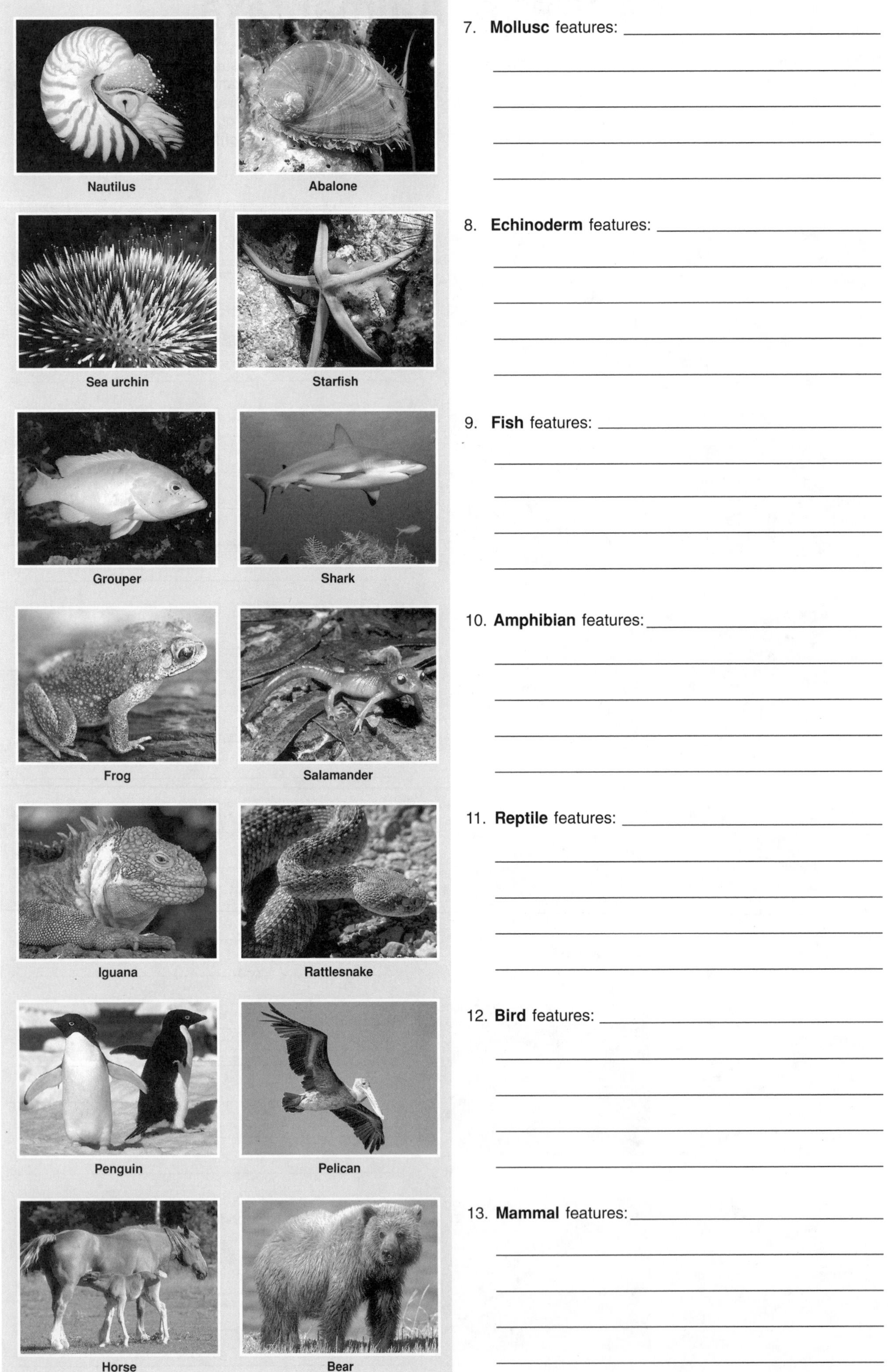

Nautilus

Abalone

Sea urchin

Starfish

Grouper

Shark

Frog

Salamander

Iguana

Rattlesnake

Penguin

Pelican

Horse

Bear

7. **Mollusc** features: ____________________

____

____

____

____

8. **Echinoderm** features: ____________________

____

____

____

____

9. **Fish** features: ____________________

____

____

____

____

10. **Amphibian** features: ____________________

____

____

____

____

11. **Reptile** features: ____________________

____

____

____

____

12. **Bird** features: ____________________

____

____

____

____

13. **Mammal** features: ____________________

____

____

____

____

KEY TERMS Crossword

Complete the crossword below, which will test your understanding of key terms in this chapter and their meanings

Clues Across

2. A phylogenetic tree based on shared derived characters.
3. A taxonomic category ranking below a phylum or division and above an order.
5. A unit of classification – a group of related genera.
7. Group of multicellular organisms that use chlorophyll to harness light to produce carbohydrates. .
8. Group of organisms with cells containing organelles. DNA is present as chromosomes.
10. The largest, least specific unit of classification within a kingdom.
13. The smallest, most precise unit of classification; a closely related group of organisms able to interbreed.
14. Kingdom of multicellular, heterotrophic organisms. Cells do not possess cell walls.
15. Phylum whose name literally means "jointed foot".
16. Kingdom including organisms with cells that lack organelles and have a single circular chromosome.
17. A unit of classification. A group of related species.
18. Diverse group of eukaryotes, either unicellular or multicellular but lacking specialized tissues.
19. Another term for a shared derived character.
21. Historically, the highest rank in biological taxonomy.
22. A group of organisms that possess a chitinous cell wall, feed using hyphae, and reproduce by spores.

Clues Down

1. Sub phylum of the kingdom Animalia. Possess backbones, normally made of bone, that protect a dorsal nerve cord.
2. Primitive group of animals with jelly-like bodies and a gastrovascular cavity.
4. Arthropod with eight legs. Includes spiders and scorpions.
6. The study of form. How an organism physically appears
8. A large and diverse group of bacteria, lacking in photosynthetic pigments. Possess simple cells that may be spherical or rod shaped.
9. Ancient line of bacteria distinct from the eubacteria. It includes a number of extremophiles.
11. Phylum characterized by the presence of a fleshy, muscular foot and mantle. Most have shells, although this may be internal in some.
12. Key that gives two options at each step of the identification process.
20. Unit of classification used to group together related families.

Important in this section ...

- *Develop your understanding of the structure and function of animal body systems*
- *Explain the relationship between adaptation and environment with reference to specific examples*

Eating to Live

Feeding and nutrition	• Parasitic nutrition • Holozoic nutrition • Methods of feeding
Digestion and absorption	• Adaptations to diet (guts and dentition) • Extracellular digestion • Maximizing nutrient uptake
Human digestion	• Gut structure and function • Nutrient absorption and transport

The adaptations of teeth and guts reflect the diverse mechanisms by which animals procure their food.

Regulating Fluids, Removing Wastes

Osmo-regulation	• Fluid balance in water and on land • Water budgets
Sources of wastes	• Waste ions and carbon dioxide • Protein metabolism and nitrogen
The principles of excretion	• The toxicity of nitrogenous wastes • Ultrafiltration and secretion
Excretory organs	• Excretory organs in invertebrates • Gills as excretory organs • Kidneys
Human excretion	• Urine formation in the nephron • Regulating urine output • Renal dialysis and kidney transplant

Metabolism creates nitrogenous wastes, which must be excreted in a form that is appropriate in the animal's environment.

Keeping in Balance

Animals maintain a steady state in spite of external fluctuations.

This state of homeostasis is maintained through negative feedback mechanisms operating through the interplay of nervous and hormonal systems.

Insects, birds, fish, and mammals have evolved different systems for the exchange of gases with the environment

The Breath of Life

Gas exchange systems; who needs one ?	• Cellular respiration • Features of respiratory surfaces • The constraints of environment • Gas exchange efficiencies
Diversity in gas exchange systems	• Tracheal systems in insects • Parabronchial lungs in birds • Gills in fish • Alveolar lungs in mammals
Human respiratory physiology	• Lung structure and function • Control of respiratory activity • The effects of altitude

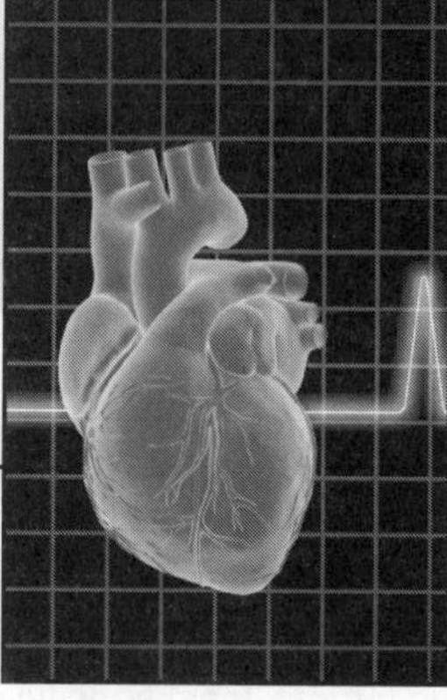

Circulatory systems have various roles in different animal taxa and this is reflected in their structure.

Life Blood

Internal transport systems; who needs one ?	• The constraints of size • Features of internal transport systems • The constraints of environment • Transport efficiencies
Diversity in internal transport systems	• Open circulatory systems • Closed circulatory systems • Blood vessels • Hearts and pumping efficiencies
Human circulatory physiology	• Heart structure and function • Control of heart activity • The effects of exercise • Hemostasis and blood substitutes

Defending Against Disease

Infection and disease
- The nature of pathogens - HIV
- Transmission of disease
- Self and non-self

The first line of defense
- The role of intact skin
- Tears, mucus, saliva
- Mucous membranes

The second line of defense
- Antimicrobial substances
- Inflammation and fever
- Phagocytosis

The third line of defense
- Immune responses: good and bad
- Immunological memory
- Monoclonal antibodies

Muscles and Movement

Support systems
- Animal skeletons
- The constraints of environment

Animal movement
- Swimming: reducing drag
- Flight: achieving lift
- Running: limb structure & stride length

The human musculo-skeletal system
- The human skeleton
- Muscles, tendons, and ligaments
- The mechanics of movement - joints
- Types of movement

Muscle function
- The sliding filament theory
- Muscle innervation
- Fatigue and oxygen debt

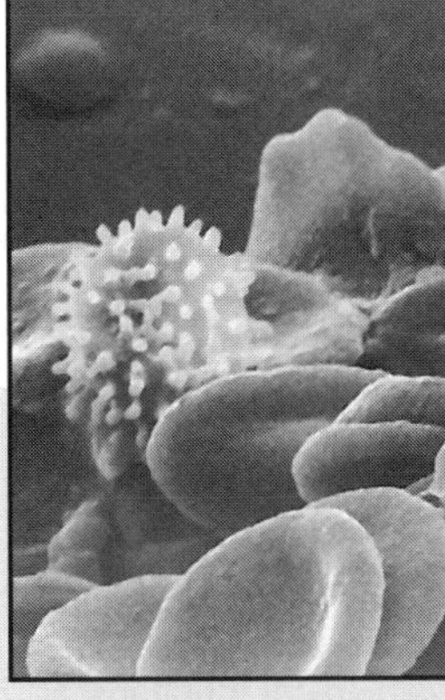

Defenses against foreign material include barriers, phagocytosis, and the immune response. Some pathogens evade these defenses.

Animal support structures can be internal or external. Muscular contraction against a skeletal framework produces movement.

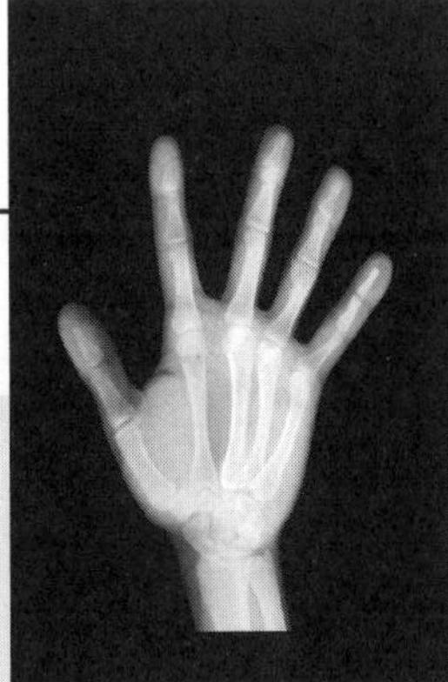

Part 2

The Structure and Function of Animals

The structural and physiological adaptations of animals can be understood within the context of the environment in which they live.

Systems function to maintain an internal steady state environment.

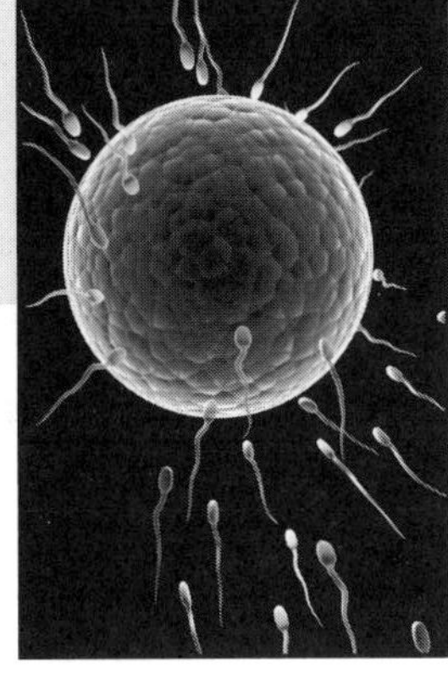

The reproductive strategies of animals reflect their evolutionary history and the constraints of environment.

Animal responses to stimuli, from simple reflexes to learning, are adaptive and appropriate in the given environment

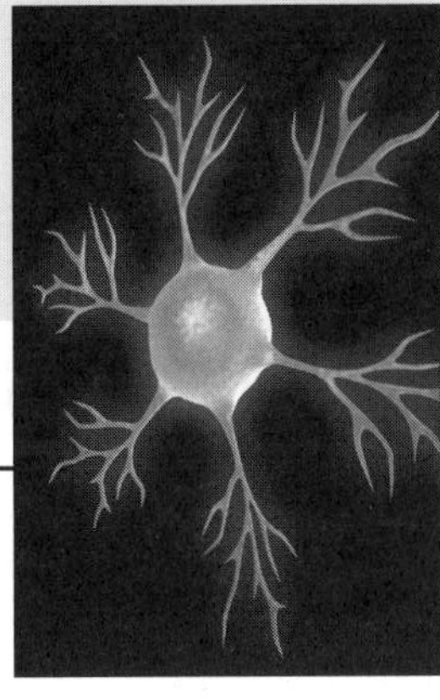

The Next Generation

Why reproduce?
- Successful reproduction = fitness
- Sexual vs asexual reproduction

Reproductive strategies
- The constraints of environment
- Egg laying vs vivpary
- Mammalian patterns of reproduction

Human reproduction and development
- Male & female reproductive systems
- The menstrual cycle
- Pregnancy, birth, and lactation
- Growth and development
- Reproductive technologies

Responding to the Environment

Nervous systems
- Invertebrate nervous systems
- The move towards cephalization
- The vertebrate plan

The human nervous system
- Central and peripheral NS
- Neuron structure and function
- Synaptic transmission

Sensory reception
- The basis of sensory perception
- Receptors as biological transducers
- Vision as a case study

Behavior
- Communication and sociality
- Timing and behavioral rhythms
- Navigation and homing
- The adaptive value of learning

AP Topic 3B

IB SL/HL 6.5

Keeping in Balance

KEY TERMS

autocrine signaling
countercurrent heat exchange
cyclic AMP
diabetes mellitus
endocrine gland
endocrine signaling
endocrine system
glucagon
homeostasis
hormone
humoral
hyperthermia
hypothalamus
hypothermia
insulin
islets of Langerhans
ligand
negative feedback
nervous system
neural
paracrine signaling
pituitary gland
positive feedback
second messenger
signal transduction
thermoregulation

Periodicals:
listings for this chapter are on page 388

Weblinks:
www.thebiozone.com/weblink/SB2-2603.html

Teacher Resource CD-ROM:
Type 1 Diabetes

KEY CONCEPTS

- Homeostasis is maintained using hormonal and nervous mechanisms via negative feedback.
- Cells respond to messages through signal transduction pathways.
- Thermoregulation enables maintenance of an optimum body temperature for metabolism.
- The regulation of blood glucose relies on the interplay of two antagonistic hormones.

OBJECTIVES

□ 1. Use the **KEY TERMS** to help you understand and complete these objectives.

Principles of Homeostasis — pages 35-38

□ 2. Explain the role of homeostasis in providing independence from the fluctuating external environment.

□ 3. Explain how **negative feedback** stabilizes systems against excessive change. Using examples, explain the role of **receptors**, **effectors**, and negative feedback in homeostasis. Recognize **positive feedback** as a destabilizing mechanism with a role in certain physiological functions.

Nerves and Hormones in Mammals — pages 39-46

□ 4. Describe the general structure of the nervous system, relating the structure to the way in which animals receive stimuli and generate a response.

□ 5. Describe the general organization of the endocrine system and explain the role of hormones in maintaining homeostasis. Explain how hormones act and why they have wide-ranging physiological effects.

□ 6. Describe the role of the **hypothalamus** and **pituitary gland** in homeostasis.

□ 7. Contrast the speed of nervous and endocrine responses. Explain the role of feedback mechanisms in regulating nervous and endocrine activity.

□ 8. Describe types of **cell signaling** and identify the involvement of signaling molecules in a **signal transduction pathway**.

Thermoregulation — pages 49-55

□ 9. Describe the mechanisms for thermoregulation in an **ectotherm** (e.g. a reptile) and in an **endothermic homeotherm** (e.g. a placental mammal).

□ 10. Explain how body temperature is regulated in a mammal. Include reference to the role of the **hypothalamus** and the **autonomic nervous system**. Describe disruptions to normal thermoregulatory controls, e.g. **hypothermia**.

Regulating Blood Glucose — pages 47-48, 56, 82

□ 11. Describe the structure of the liver and its role in regulating blood nutrient levels and in nutrient storage. Include reference to its central role in fat, carbohydrate, and protein metabolism, and in breakdown of red blood cells.

□ 12. Explain how blood glucose level is regulated in humans, including reference to the role of negative feedback mechanisms and the hormones **insulin** and **glucagon**. Recall the role of the liver in glucose-glycogen conversions.

□ 13. Describe and explain disorders of glucose metabolism, e.g. type 1 and/or type 2 diabetes mellitus.

Principles of Homeostasis

Homeostasis the relative physiological constancy of the body, despite external fluctuations. Homeostasis of the internal environment is an essential feature of complex animals and it is the job of the body's **organ systems** to maintain it, even as they make necessary exchanges with the environment. Homeostatic control systems have three functional components: a receptor to detect change, a control centre, and an effector to direct an appropriate response. In **negative feedback** systems, movement away from a steady state triggers a mechanism to counteract further change in that direction. Using negative feedback systems, the body counteracts disturbances and restores the steady state. **Positive feedback** is also used in physiological systems, but to a lesser extent since positive feedback leads to the response escalating in the same direction.

Organ systems maintain a constant internal environment that provides for the needs of all the body's cells, making it possible for animals to move through different and often highly variable external environments. This representation of a mammal shows how organ systems permit exchanges with the environment. The exchange surfaces of organ systems are usually internal, but may be connected to the environment via openings on the body surface.

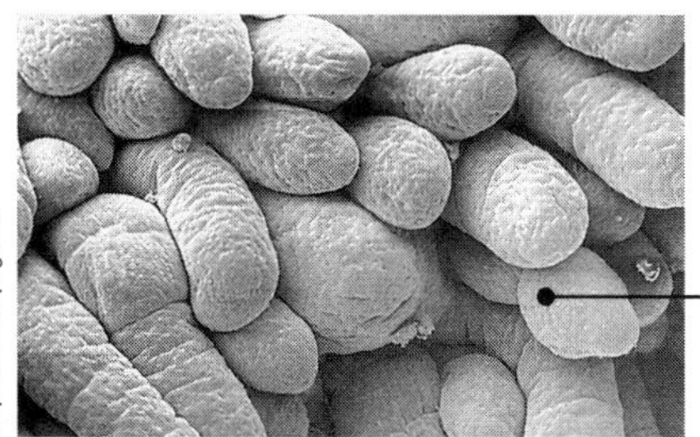

The finger-like villi of the small intestine greatly expand the surface area for nutrient absorption.

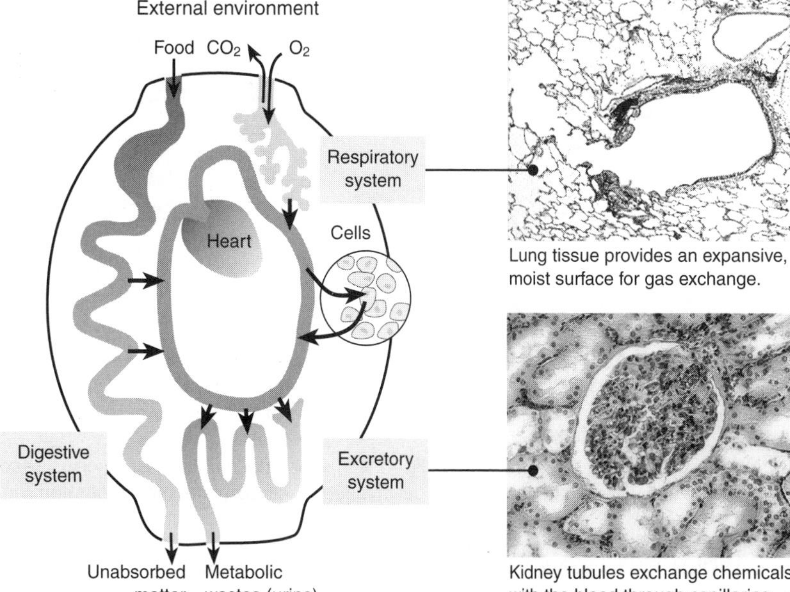

Lung tissue provides an expansive, moist surface for gas exchange.

Kidney tubules exchange chemicals with the blood through capillaries.

All photos this page: EII

Negative Feedback and Control Systems

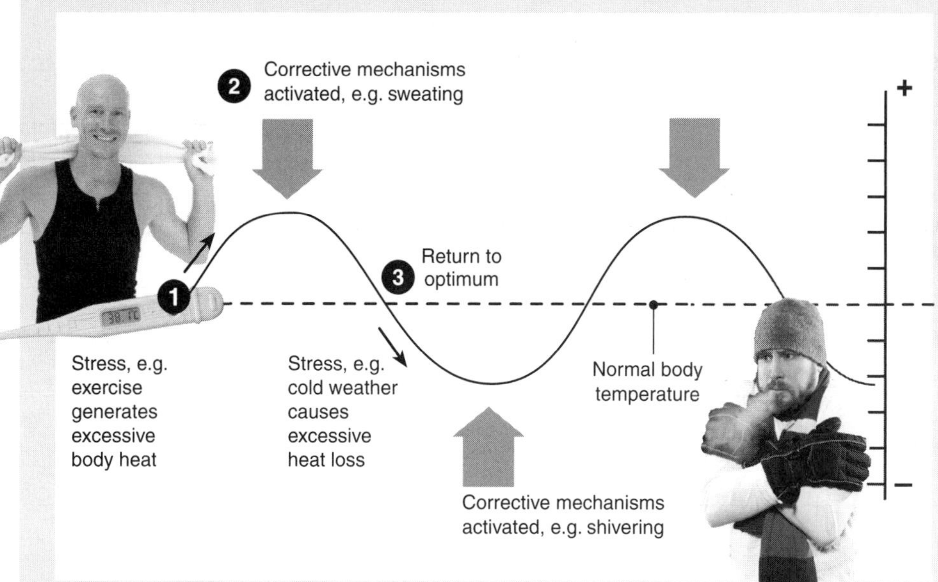

1. A stressor, e.g. exercise, takes the internal environment away from optimum.
2. Stress is detected by receptors and corrective mechanisms (e.g. sweating or shivering) are activated.
3. Corrective mechanisms act to restore optimum conditions.

Negative feedback acts to counteract departures from steady state. The diagram shows how stress is counteracted in the case of body temperature.

1. Describe the three main components of a regulatory control system in the human body:

2. Explain how negative feedback mechanisms maintain homeostasis in a variable environment:

Fever, Positive Feedback and Response Escalation

Positive feedback causes large deviations from the original levels
Normal temperature cycle (fluctuations around a set point)
Normal body temperature 36.2 to 37.2°C
Pathogen enters body
Pathogen detected. Body temperature begins to rise
Fever peaks and body temperature then begins to fall
+
–

1. Body temperature fluctuates on a normal, regular basis around a narrow set point.
2. Pathogen enters the body.
3. The body detects the pathogen and macrophages attack it. Macrophages release interleukins which stimulate the hypothalamus to increase prostaglandin production and reset the body's thermostat to a higher 'fever' level by shivering (the chill phase).
4. The fever breaks when the infection subsides. Levels of circulating interleukins (and other fever-associated chemicals) fall, and the body's thermostat is reset to normal. This ends the positive feedback escalation and normal controls resume. If the infection persists, the escalation may continue, and the fever may intensify. Body temperatures in excess of 43°C are often fatal or result in brain damage.

Unlike negative feedback, positive feedback will push physiological levels out of the normal range. Positive feedback is inherently unstable, but it has a specific biological purpose. It will cause an escalation in response in order to achieve a particular result.

Normally, a positive feedback loop is ended when the natural resolution is reached. Labour is another example of where positive feedback plays an important role.

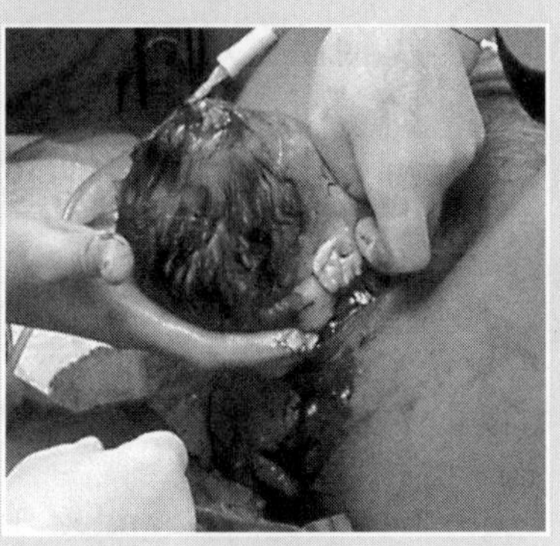

LABOUR: Positive feedback: oxytocin and uterine contraction. Result: delivery of infant. During childbirth (far right), the release of oxytocin intensifies the contractions of the uterus so that labour proceeds to its conclusion. The birth itself restores the system by removing the initiating stimulus.

3. (a) Explain the biological purpose of positive feedback loops and describe an example:

   (b) Explain why positive feedback is inherently unstable (compare with negative feedback):

   (c) Explain how a positive feedback loop is normally stopped:

   (d) Describe a situation in which this might not happen and the result:

4. Explain the value in having separate negative feedback mechanisms to control departures in different directions:

Maintaining Homeostasis

The various organ systems of the body act to maintain homeostasis through a combination of hormonal and nervous mechanisms. In everyday life, the body must regulate respiratory gases, protect itself against agents of disease (pathogens), maintain fluid and salt balance, regulate energy and nutrient supply, and maintain a constant body temperature. All these must be coordinated and appropriate responses made to incoming stimuli. In addition, the body must be able to repair itself when injured and be capable of reproducing (leaving offspring).

Regulating Respiratory Gases

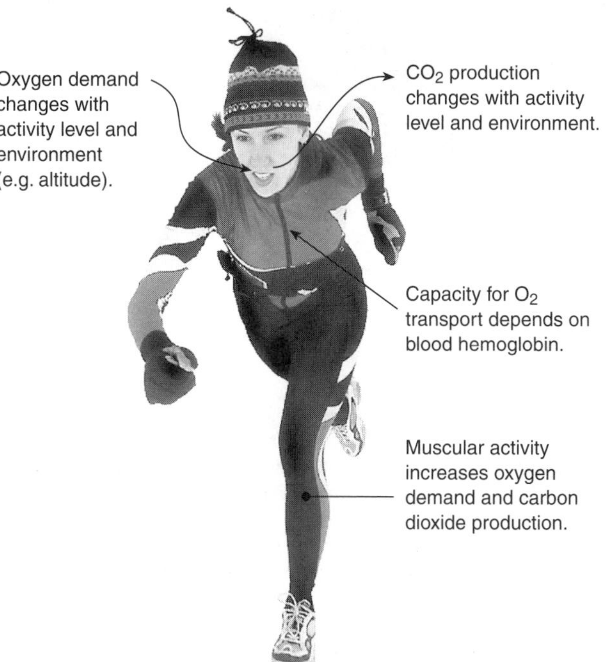

Oxygen must be delivered to all cells and carbon dioxide (a waste product of cellular respiration) must be removed. **Breathing** brings in oxygen and expels CO_2, and the cardiovascular and lymphatic systems circulate these respiratory gases (the oxygen mostly bound to hemoglobin). The rate of breathing is varied according to oxygen demands (as detected by CO_2 levels in the blood).

Coping with Pathogens

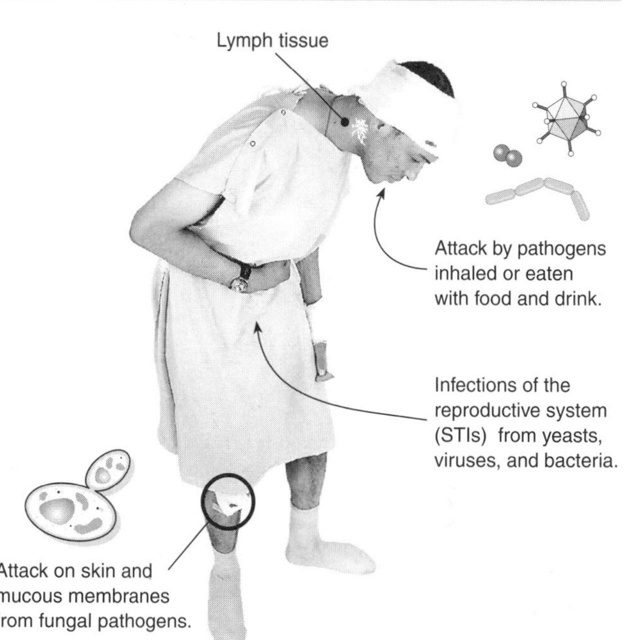

Lymph tissue
Attack by pathogens inhaled or eaten with food and drink.
Infections of the reproductive system (STIs) from yeasts, viruses, and bacteria.
Attack on skin and mucous membranes from fungal pathogens.

All of us are under constant attack from pathogens (disease causing organisms). The body has a number of mechanisms that help to prevent the entry of pathogens and limit the damage they cause if they do enter the body. The skin, the digestive system, and the immune system are all involved in the body's defense, while the cardiovascular and lymphatic systems circulate the cells and antimicrobial substances involved.

Maintaining Nutrient Supply

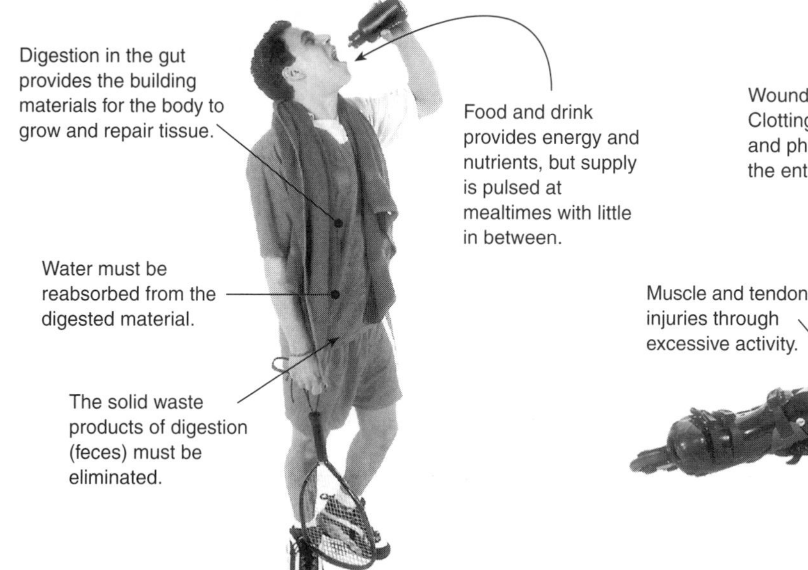

Digestion in the gut provides the building materials for the body to grow and repair tissue.
Food and drink provides energy and nutrients, but supply is pulsed at mealtimes with little in between.
Water must be reabsorbed from the digested material.
The solid waste products of digestion (feces) must be eliminated.

Food and drink must be taken in to maintain the body's energy supplies. The digestive system makes these nutrients available, and the cardiovascular system distributes them throughout the body. Food intake is regulated largely through nervous mechanisms while hormones control the regulation of cellular uptake of glucose.

Repairing Injuries

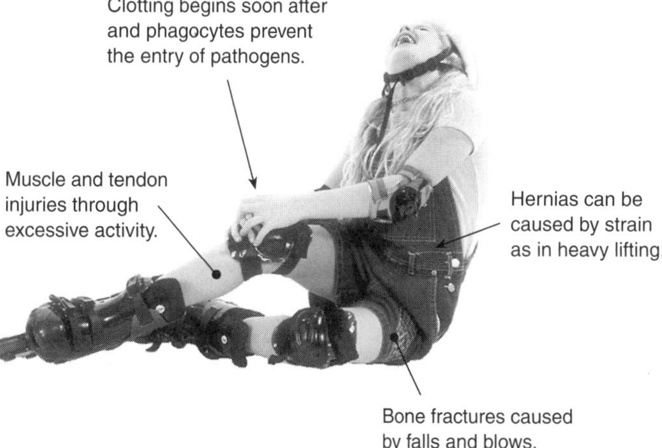

Wounds result in bleeding. Clotting begins soon after and phagocytes prevent the entry of pathogens.
Muscle and tendon injuries through excessive activity.
Hernias can be caused by strain as in heavy lifting.
Bone fractures caused by falls and blows.

Damage to body tissues triggers the **inflammatory response** and white blood cells move to the injury site. The inflammatory response is started (and ended) by chemical signals (e.g. from histamine and prostaglandins) released when tissue is damaged. The cardiovascular and lymphatic systems distribute the cells and molecules involved.

Related activities: *The Body's Defenses, Nervous Regulatory Systems, Hormonal Regulatory Systems, Exercise & Blood Flow*

Maintaining Fluid and Ion Balance

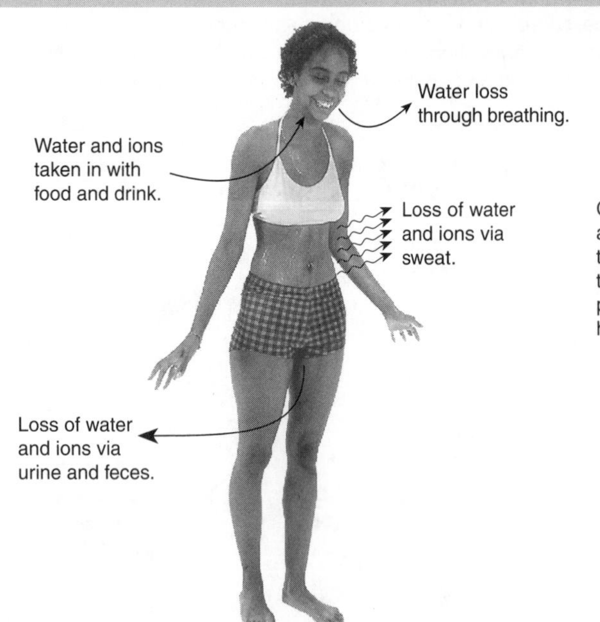

Water loss
through breathing.
Water and ions
taken in with
food and drink.
Loss of water
and ions via
sweat.
Loss of water
and ions via
urine and feces.

Fluid and electrolyte balance in the body is maintained by the kidneys (although the skin is also important). Osmoreceptors monitor blood volume and bring about the release of regulatory hormones; the kidneys regulate reabsorption of water and sodium from blood in response to levels of the hormones ADH and aldosterone. The cardiovascular and lymphatic system distribute fluids around the body.

Coordinating Responses

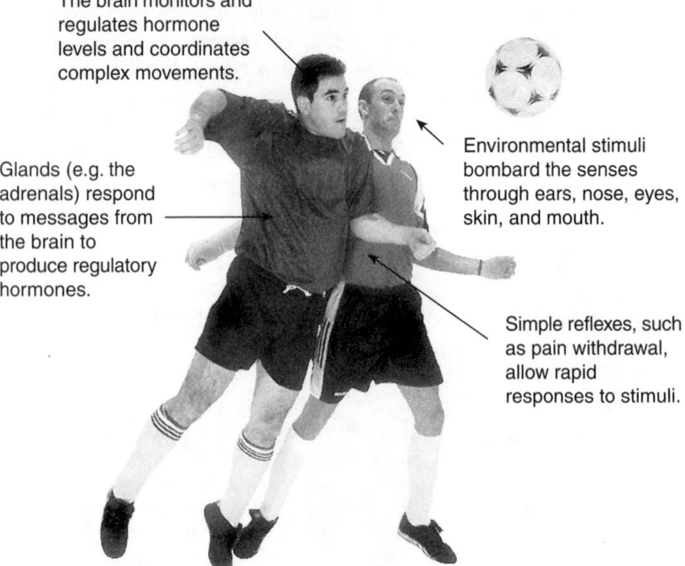

The body is constantly bombarded by stimuli from the environment. The brain sorts these stimuli into those that require a response and those that do not. Responses are coordinated via nervous or hormonal controls. Simple nervous responses (reflexes) act quickly. Hormones, which are distributed by the cardiovascular and lymphatic systems, take longer to produce a response and the response is more prolonged.

1. Describe two mechanisms that operate to restore homeostasis after infection by a pathogen:

   (a) ____

   ____

   (b) ____

   ____

2. Describe two mechanisms by which responses to stimuli are brought about and coordinated:

   (a) ____

   ____

   (b) ____

   ____

3. Explain two ways in which water and ion balance are maintained. Name the organ(s) and any hormones involved:

   (a) ____

   ____

   (b) ____

   ____

4. Explain two ways in which the body regulates its respiratory gases during exercise:

   (a) ____

   ____

   (b) ____

   ____

Nervous Regulatory Systems

An essential feature of living organisms is their ability to coordinate their activities. In multicellular animals, such as mammals, detecting and responding to environmental change, and regulating the internal environment (homeostasis) is brought about by two coordinating systems: the nervous and endocrine systems. Although structurally these two systems are quite different, they frequently interact to coordinate behavior and physiology. The nervous system contains cells called neurons (or nerve cells). Neurons are specialized to transmit information in the form of electrochemical impulses (action potentials). The nervous system is a signalling network with branches carrying information directly to and from specific target tissues. Impulses can be transmitted over considerable distances and the response is very precise and rapid. Whilst it is extraordinarily complex, comprising millions of neural connections, its basic plan (below) is quite simple. Further detail on nervous system structure and function is provided in the topic: *Responding to the Environment*.

Coordination by the Nervous System

The vertebrate nervous system consists of the central nervous system (brain and spinal cord), and the nerves and receptors outside it (peripheral nervous system). Sensory input to receptors comes via stimuli. Information about the effect of a response is provided by feedback mechanisms so that the system can be readjusted. The basic organization of the nervous system can be simplified into a few key components: the sensory receptors, a central nervous system processing point, and the effectors which bring about the response (below):

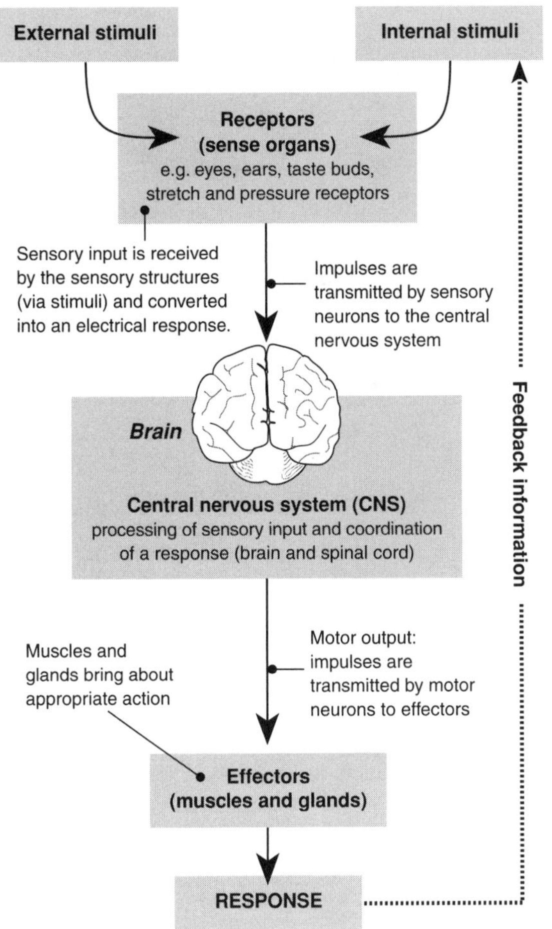
External stimuli
Internal stimuli
Receptors (sense organs) e.g. eyes, ears, taste buds, stretch and pressure receptors
Sensory input is received by the sensory structures (via stimuli) and converted into an electrical response.
Impulses are transmitted by sensory neurons to the central nervous system
Brain
Central nervous system (CNS) processing of sensory input and coordination of a response (brain and spinal cord)
Feedback information
Muscles and glands bring about appropriate action
Motor output: impulses are transmitted by motor neurons to effectors
Effectors (muscles and glands)
RESPONSE

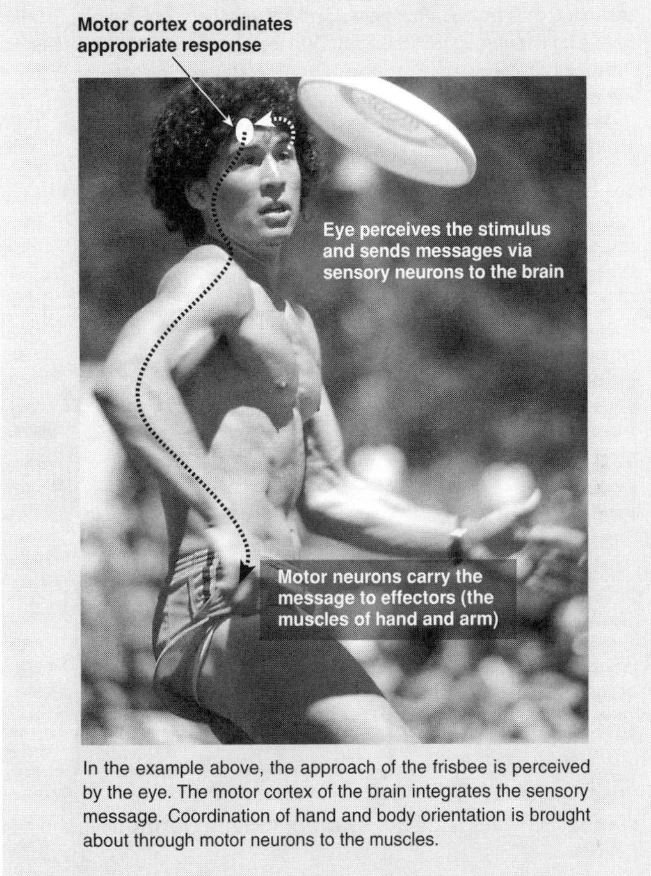
Motor cortex coordinates appropriate response
Eye perceives the stimulus and sends messages via sensory neurons to the brain
Motor neurons carry the message to effectors (the muscles of hand and arm)

In the example above, the approach of the frisbee is perceived by the eye. The motor cortex of the brain integrates the sensory message. Coordination of hand and body orientation is brought about through motor neurons to the muscles.

Comparison of nervous and hormonal control

	Nervous control	Hormonal control
Communication	Impulses across synapses	Hormones in the blood
Speed	Very rapid (within a few milliseconds)	Relatively slow (over minutes, hours, or longer)
Duration	Short term and reversible	Longer lasting effects
Target pathway	Specific (through nerves) to specific cells	Hormones broadcast to target cells everywhere
Action	Causes glands to secrete or muscles to contract	Causes changes in metabolic activity

1. Identify the three basic components of a nervous system and explain how they function to maintain homeostasis:

2. Describe two differences between nervous control and endocrine (hormonal) control of body systems:

(a)

(b)

Hormonal Regulatory Systems

The endocrine system regulates the body's processes by releasing chemical messengers (hormones) into the bloodstream. Hormones are potent chemical regulators: they are produced in minute quantities yet can have a large effect on metabolism. The endocrine system comprises endocrine cells (organized into endocrine glands), and the hormones they produce. Unlike exocrine glands (e.g. sweat and salivary glands), endocrine glands are ductless glands, secreting hormones directly into the bloodstream rather than through a duct or tube. Some organs (e.g. the pancreas) have both endocrine and exocrine regions, but these are structurally and functionally distinct. The basis of hormonal control and the role of negative feedback mechanisms in regulating hormone levels are described below.

The Mechanism of Hormone Action

Endocrine cells produce hormones and secrete them into the bloodstream where they are distributed throughout the body. Although hormones are broadcast throughout the body, they affect only specific target cells. These target cells have receptors on the plasma membrane which recognize and bind the hormone (see inset, below right). The binding of hormone and receptor triggers the response in the target cell. Cells are unresponsive to a hormone if they do not have the appropriate receptors.

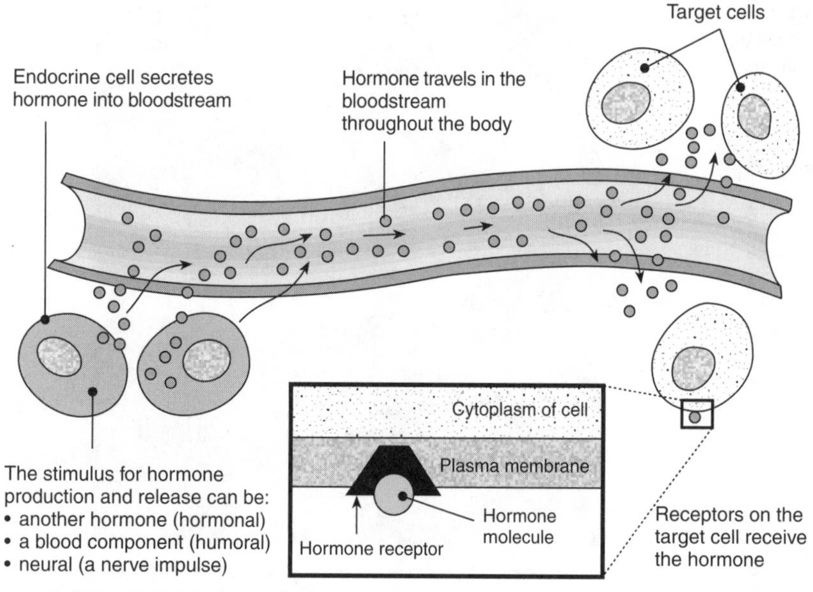

The stimulus for hormone production and release can be:
- another hormone (hormonal)
- a blood component (humoral)
- neural (a nerve impulse)

Antagonistic Hormones

Insulin secretion

Blood glucose rises: insulin is released

Raises blood glucose level

Lowers blood glucose level

Blood glucose falls: glucagon is released

Glucagon secretion

The effects of one hormone are often counteracted by an opposing hormone. Feedback mechanisms adjust the balance of the two hormones to maintain a physiological function. Example: insulin decreases blood glucose and glucagon raises it.

1. (a) Explain what is meant by **antagonistic hormones** and describe an example of how two such hormones operate:

   Example:

   (b) Explain the role of feedback mechanisms in adjusting hormone levels (explain using an example if this is helpful):

2. Explain how a hormone can bring about a response in target cells even though all cells may receive the hormone:

3. Explain why hormonal control differs from nervous system control with respect to the following:

   (a) The speed of hormonal responses is slower:

   (b) Hormonal responses are generally longer lasting:

Related activities: *Nervous Regulatory Systems, Control of Blood Glucose*
Web links: *Control of Endocrine Activity*

The Endocrine System

The nervous and endocrine systems interact to regulate the body's activities. The endocrine system comprises **endocrine glands** and their **hormones**. Endocrine glands, are **ductless glands** (compare with **exocrine glands** which deliver secretions via a duct) and are distributed throughout the body, frequently associated with the organs of other body systems. Under appropriate stimulation they secrete **hormones**, which are carried in the blood to exert a specific effect on target cells, before being broken down and excreted. Hormones may be amino acids, peptides, proteins (often modified), fatty acids, or steroids. Some basic features of the human endocrine system are explained below. The hypothalamus is part of the brain and, although it is not strictly an endocrine gland, it contains neurosecretory cells, and links the nervous and endocrine systems. The hypothalamus, together with the pituitary, adrenal, and thyroid glands, form a central axis of endocrine control that regulates much of the body's metabolic activity.

Hypothalamus
Coordinates nervous and endocrine systems. Secretes releasing hormones, which regulate the hormones of the anterior pituitary. Produces oxytocin and ADH, which are released from the posterior pituitary.

Pineal
This small gland in the brain secretes melatonin, which regulates the sleep-wake cycle. Melatonin secretion follows a circadian rhythm and coodinates reproductive hormones too.

Thyroid gland
Secretes thyroxine, an iodine containing hormone needed for normal growth and development. Thyroxine stimulates metabolism and growth via protein synthesis.

Pancreatic islets
Specialised α and β endocrine cells in the pancreas produce glucagon and insulin. Together, these control blood sugar levels.

Ovaries (in females)
The ovaries produce oestrogen and progesterone.
These hormones control and maintain female characteristics, stimulate the menstrual cycle, maintain pregnancy, and prepare the mammary glands for lactation.

Pituitary gland
The pituitary is located below the hypothalamus. It secretes at least nine hormones that regulate the activities of other endocrine glands.

Parathyroid glands
On the surface of the thyroid, they secrete PTH (parathyroid hormone), which regulates blood calcium levels and promotes the release of calcium from bone. High levels of calcium in the blood inhibit PTH secretion.

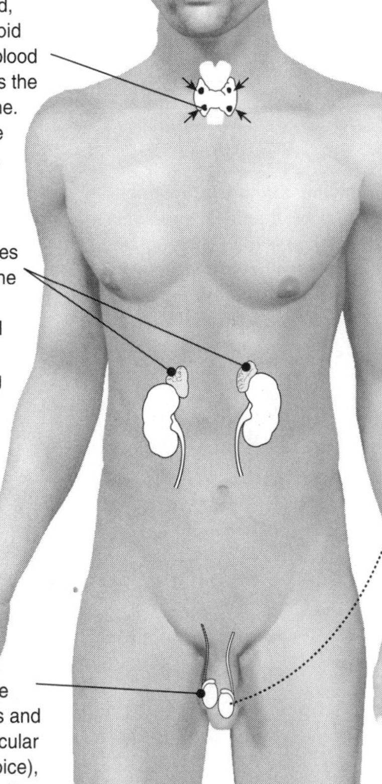

Adrenal glands
The adrenal medulla produces adrenaline and noradrenaline responsible for the fight or flight response. The adrenal cortex produces various steroid hormones, including cortisol (response to stress) and aldosterone (sodium regulation).

Testes (in males)
The testes of males produce testosterone, which controls and maintains "maleness" (muscular development and deeper voice), and promotes sperm production.

1. Explain how a hormone is different from a neurotransmitter: ______________________________

____

2. Using ruled lines, connect each of the following endocrine glands with its correct role in the body

(a) Pituitary gland	The hormone from this gland regulates the levels of calcium in the blood
(b) Ovaries	Master gland secreting at least nine hormones, including growth hormone and TSH
(c) Pineal gland	Produces hormones involved in the regulation of metabolic rate
(d) Parathyroid glands	Secretes melatonin to regulate sleep patterns and cycles of reproductive hormones
(e) Thyroid	Produce estrogen and progesterone in response to hormones from the pituitary

3. Review the three types of stimuli for hormone release and describe a specific example of each:

(a) Hormonal stimulus: ______________________________

(b) Humoral stimulus: ______________________________

(c) Neural stimulus: ______________________________

Cell Signaling

Cells use **signals** (chemical messengers) to gather information about, and respond to, changes in their cellular environment and for communication between cells. The signaling and response process is called the **signal transduction pathway**, and often involves a number of enzymes and molecules in a **signal cascade** which causes a large response in the target cell. Cell signaling pathways are categorized primarily on the distance over which the signal molecule travels to reach its target cell, and generally fall into three categories. The **endocrine** pathway involves the transport of hormones over large distances through the circulatory system. During **paracrine** signaling, the signal travels an intermediate distance to act upon neighboring cells. **Autocrine** signaling involves a cell producing and reacting to its own signal. These three pathways are illustrated below.

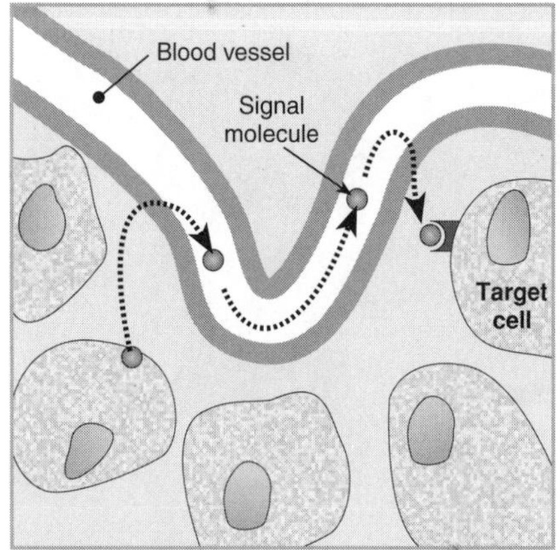

Endocrine signaling: Hormone signals are released by ductless endocrine glands and carried long distances through the body by the circulatory system to the target cells. Examples include sex hormones, growth factors, and neurohormones such as dopamine.

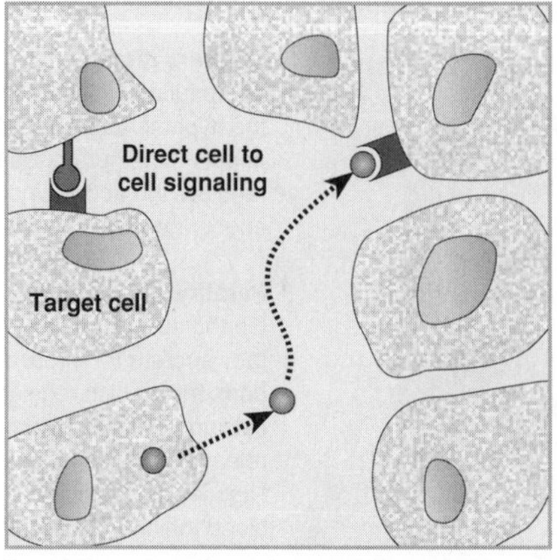

Paracrine signaling: Signals released from a cell act upon target cells within the immediate vicinity. The chemical messenger can be transferred through the extracellular fluid (e.g. at synapses) or directly between cells, which is important during embryonic development.

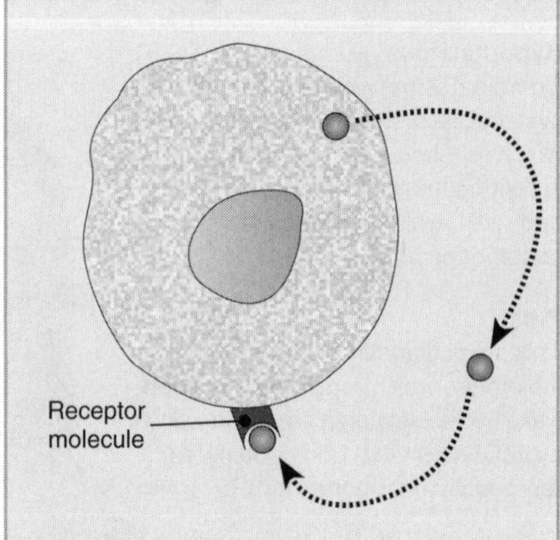

Autocrine signaling: Cells produce and react to their own signals. In vertebrates, when a foreign antibody enters the body, some T-cells (lymphocytes) produce a growth factor to stimulate their own production. The increased number of T-cells helps to fight the infection.

Signaling Receptors and Signaling Molecules

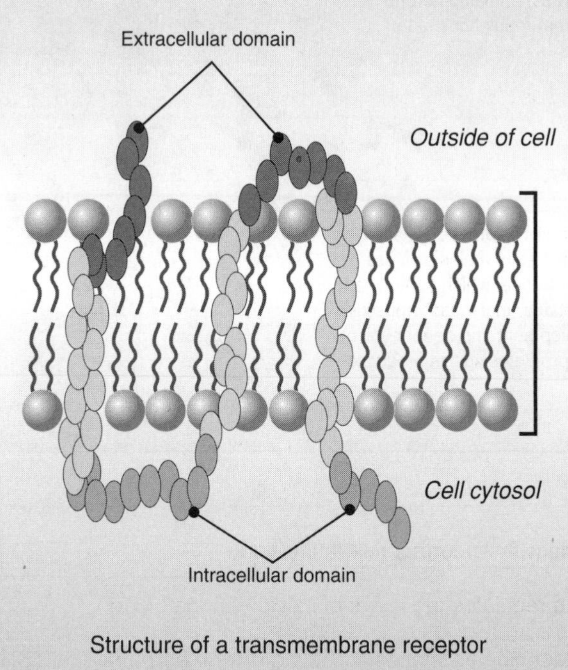

Structure of a transmembrane receptor

Insulin like growth factor 1 (IGF-1)

Progesterone

Examples of cell signaling molecules

The binding sites of cell receptors are very specific; they only bind certain **ligands** (signal molecules). This stops them from reacting to every signal bombarding the cell. Receptors fall into two main categories :

- **Cytoplasmic receptors** Cytoplasmic receptors, located within the cell cytoplasm, bind ligands which are able to cross the plasma membrane unaided.
- **Transmembrane receptors** These span the cell membrane and bind ligands which cannot cross the plasma membrane on their own. They have an extra-cellular domain outside the cell, and an intracellular domain within the cell cytosol. Ion channels, protein kinases and G-protein linked receptors are examples of transmembrane receptors (see diagram on left).

1. Briefly describe the three types of cell signaling:

   (a) ______________________________

   (b) ______________________________

   (c) ______________________________

2. Identify the components that all three cell signaling types have in common: ______________________________

____

____

Related activities: *Hormonal Regulatory Systems, Signal Transduction*
Web links: *Hormones, Receptors, and Target Cells, Peptide Hormone Action*

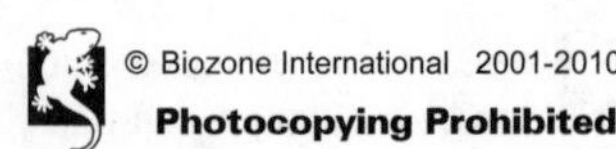

Signal Transduction

Once released, a hormone is carried in the blood to affect specific target cells. Water soluble hormones are carried free in the blood, whilst steroid and thyroid hormones are carried bound to plasma proteins. Target cells have receptors to bind the hormone, initiating a cascade of reactions which results in a specific target cell response (e.g. protein synthesis, change in membrane permeability, enzyme activation, or secretion). **Peptide hormones** operate by interacting with transmembrane receptors and activating a second messenger system (e.g. cyclic AMP). **Steroid hormones** enter the cell to interact directly with intracellular cytoplasmic receptors. Once the target cell responds, the response is recognized by the hormone-producing cell through a feedback signal and the hormone is degraded.

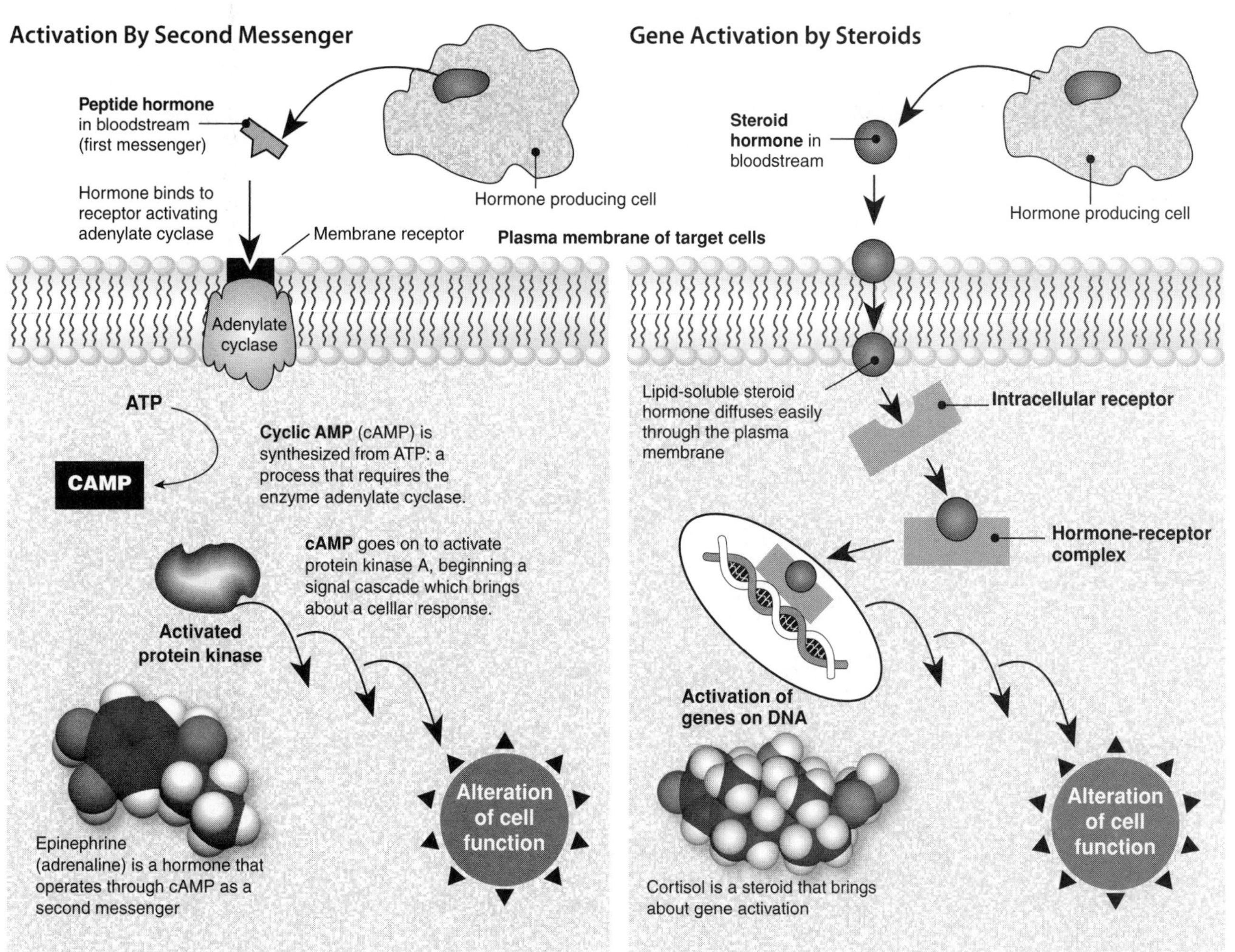
Activation By Second Messenger
Peptide hormone in bloodstream (first messenger)
Hormone producing cell
Hormone binds to receptor activating adenylate cyclase
Membrane receptor
Plasma membrane of target cells
Adenylate cyclase
ATP
CAMP
Cyclic AMP (cAMP) is synthesized from ATP: a process that requires the enzyme adenylate cyclase.
cAMP goes on to activate protein kinase A, beginning a signal cascade which brings about a celllar response.
Activated protein kinase
Alteration of cell function
Epinephrine (adrenaline) is a hormone that operates through cAMP as a second messenger
Gene Activation by Steroids
Steroid hormone in bloodstream
Hormone producing cell
Lipid-soluble steroid hormone diffuses easily through the plasma membrane
Intracellular receptor
Hormone-receptor complex
Activation of genes on DNA
Alteration of cell function
Cortisol is a steroid that brings about gene activation

Cyclic AMP is a **second messenger** linking the hormone to the cellular response. Cellular concentration of cAMP increases markedly once a hormone binds and the cascade of enzyme-driven reactions is initiated.

Steroid hormones alter cellular function through direct activation of genes. Once inside the target cell, steroids bind to intracellular receptor sites, creating hormone-receptor complexes that activate specific genes.

1. Describe the two mechanisms by which a hormone can bring about a cellular response:

   (a) ____

   ____

   (b) ____

   ____

2. State in what way these two mechanisms are alike: ____

   ____

   ____

3. Explain how a very small amount of hormone is able to exert a disproportionately large effect on a target cell:

   ____

   ____

   ____

Keeping in Balance

Related activities: *Hormonal Regulatory Systems, Cell Signaling*
Web links: *Signal Transduction, First and Second Messenger System*

The Hypothalamus and Pituitary

The **hypothalamus** is located below the thalamus, just above the brain stem and the pituitary gland, with which it has a close structural and functional relationship. Information comes to the hypothalamus through sensory pathways from sensory receptors. On the basis of this information, the hypothalamus controls and integrates many basic physiological activities (e.g. temperature regulation, food and fluid intake, and sleep), including the reflex activity of the **autonomic nervous system**.

One of the most important functions of the hypothalamus is to link the nervous system to the endocrine system (via the pituitary). The hypothalamus contains **neurosecretory cells**. These are specialized secretory neurons, which are both nerve cells and endocrine cells. They produce hormones (usually peptides) in the cell body, which are packaged into droplets and transported along the axon. At the axon terminal, the **neurohormone** is released into the blood in response to nerve impulses.

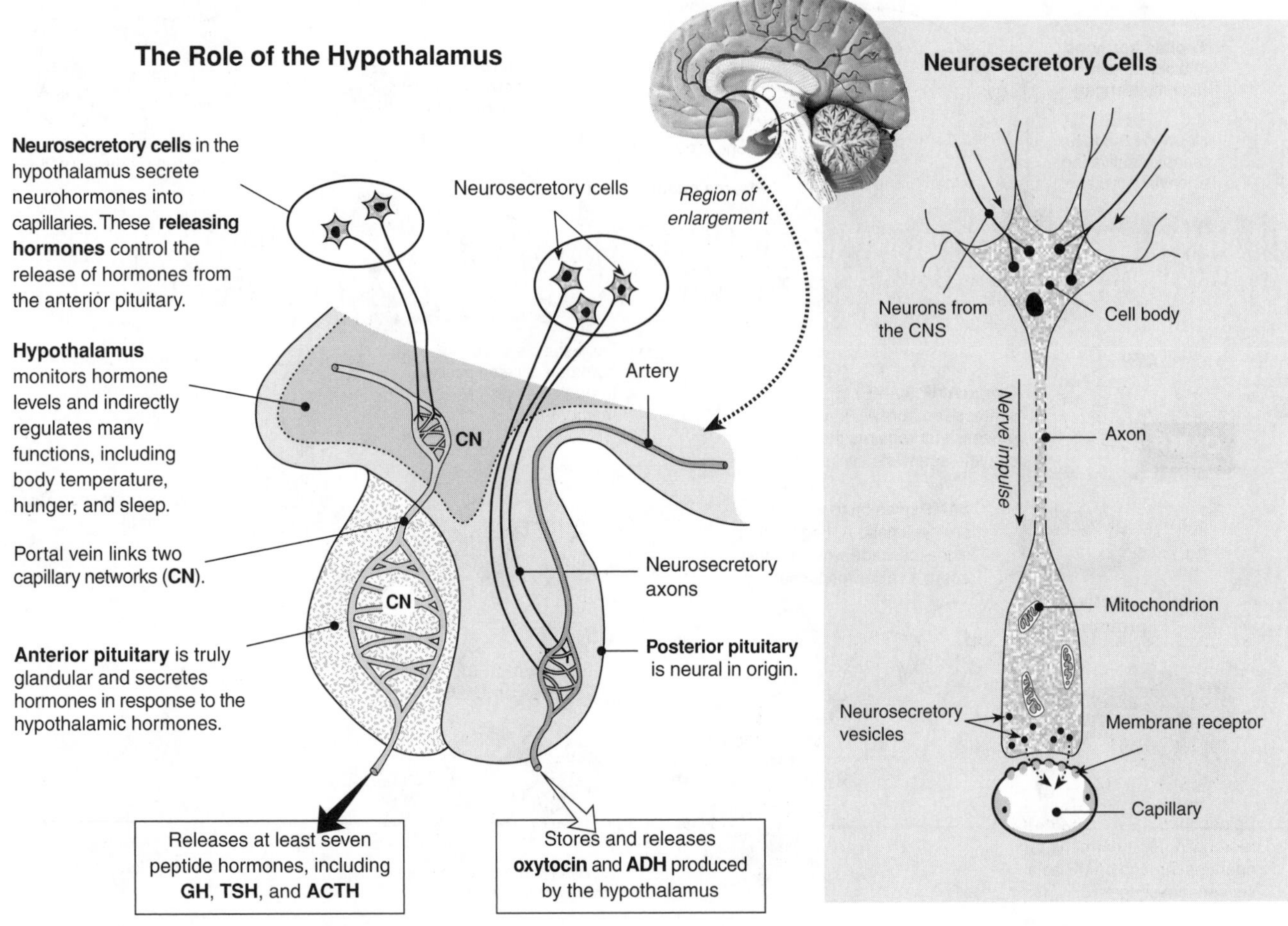
The Role of the Hypothalamus
Neurosecretory cells in the hypothalamus secrete neurohormones into capillaries. These releasing hormones control the release of hormones from the anterior pituitary.
Hypothalamus monitors hormone levels and indirectly regulates many functions, including body temperature, hunger, and sleep.
Portal vein links two capillary networks (CN).
Anterior pituitary is truly glandular and secretes hormones in response to the hypothalamic hormones.
Neurosecretory cells
Region of enlargement
Artery
CN
CN
Neurosecretory axons
Posterior pituitary is neural in origin.
Releases at least seven peptide hormones, including GH, TSH, and ACTH
Stores and releases oxytocin and ADH produced by the hypothalamus
Neurosecretory Cells
Neurons from the CNS
Cell body
Nerve impulse
Axon
Mitochondrion
Neurosecretory vesicles
Membrane receptor
Capillary

1. (a) Explain how the anterior and posterior pituitary differ with respect to their relationship to the hypothalamus:

   (b) Explain how these differences relate to the nature of the hormonal secretions for each region:

2. Describe the role of the neurohormones released by the hypothalamus:

3. Explain why the adrenal and thyroid glands atrophy if the pituitary gland ceases to function:

4. Although the anterior pituitary is often called the master gland, the hypothalamus could also claim that title. Explain:

Related activities: *Hormones of the Pituitary*

Hormones of the Pituitary

The **hypothalamus** is located at the base of the brain, just above the pituitary gland. Information comes to the hypothalamus through sensory pathways from the sense organs. On the basis of this information, the hypothalamus controls and integrates many basic physiological activities (e.g. temperature regulation, food and fluid intake, and sleep), including the reflex activity of the **autonomic nervous system**. The pituitary gland comprises two regions: the **posterior pituitary** and the **anterior pituitary**.

The **posterior pituitary** is neural (nervous) in origin and is essentially an extension of the hypothalamus. Its neurosecretory cells release oxytocin and ADH directly into the bloodstream in response to nerve impulses. The **anterior pituitary** is connected to the hypothalamus by blood vessels and receives releasing and inhibiting factors from the hypothlamus via a capillary network. These releasing factors regulate the secretion of the anterior pituitary's hormones.

ANTERIOR PITUITARY

The anterior pituitary releases at least seven **peptide hormones** (below) into the blood from simple secretory cells. The release of these hormones is regulated by releasing and inhibiting hormones from the hypothalamus.

POSTERIOR PITUITARY

The posterior pituitary develops as an extension of the hypothalamus. The release of its two hormones, oxytocin and antidiuretic hormone, occurs directly as a result of nervous input to the hypothalamus.

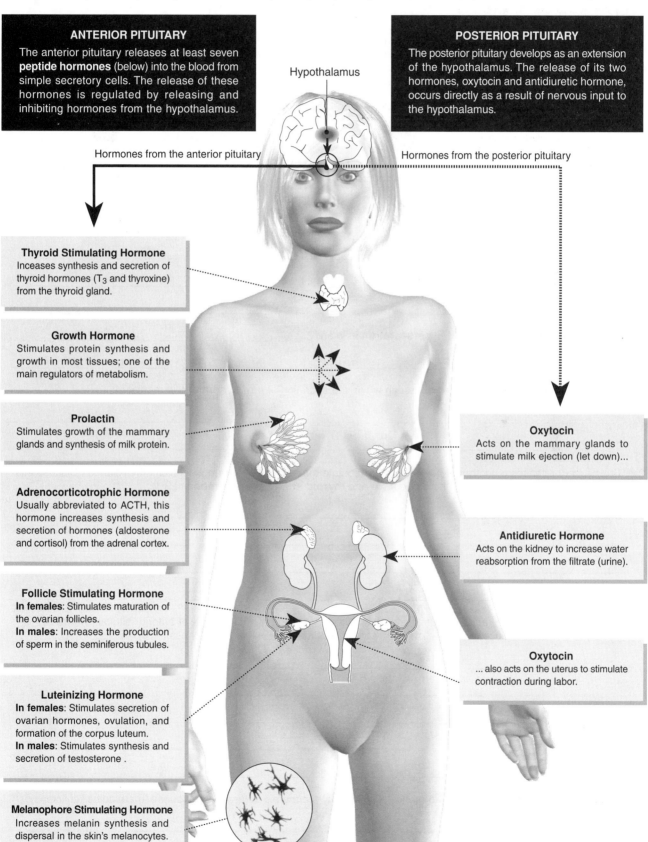

Hypothalamus

Hormones from the anterior pituitary

Hormones from the posterior pituitary

Thyroid Stimulating Hormone
Inceases synthesis and secretion of thyroid hormones (T_3 and thyroxine) from the thyroid gland.

Growth Hormone
Stimulates protein synthesis and growth in most tissues; one of the main regulators of metabolism.

Prolactin
Stimulates growth of the mammary glands and synthesis of milk protein.

Adrenocorticotrophic Hormone
Usually abbreviated to ACTH, this hormone increases synthesis and secretion of hormones (aldosterone and cortisol) from the adrenal cortex.

Follicle Stimulating Hormone
In females: Stimulates maturation of the ovarian follicles.
In males: Increases the production of sperm in the seminiferous tubules.

Luteinizing Hormone
In females: Stimulates secretion of ovarian hormones, ovulation, and formation of the corpus luteum.
In males: Stimulates synthesis and secretion of testosterone .

Melanophore Stimulating Hormone
Increases melanin synthesis and dispersal in the skin's melanocytes.

Oxytocin
Acts on the mammary glands to stimulate milk ejection (let down)...

Antidiuretic Hormone
Acts on the kidney to increase water reabsorption from the filtrate (urine).

Oxytocin
... also acts on the uterus to stimulate contraction during labor.

Keeping in Balance

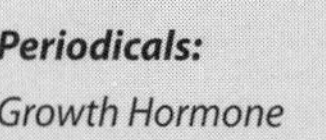

Effects of Growth Hormone

Growth hormone (GH) is released in response to GHRH (growth hormone releasing hormone) from the hypothalamus. GH acts both directly and indirectly to affect metabolic activities associated with growth.

GH directly stimulates metabolism of fat, but its major role is to stimulate the liver and other tissues to secrete IGF-I (Insulin-like Growth Factor) and through this stimulate bone and muscle growth. GH secretion is regulated is via negative feedback:

High levels of IGF-1

- *suppress secretion of GHRH*
- *stimulate release of somatostatin from the hypothalamus. Somatostatin suppresses GH secretion (not shown).*

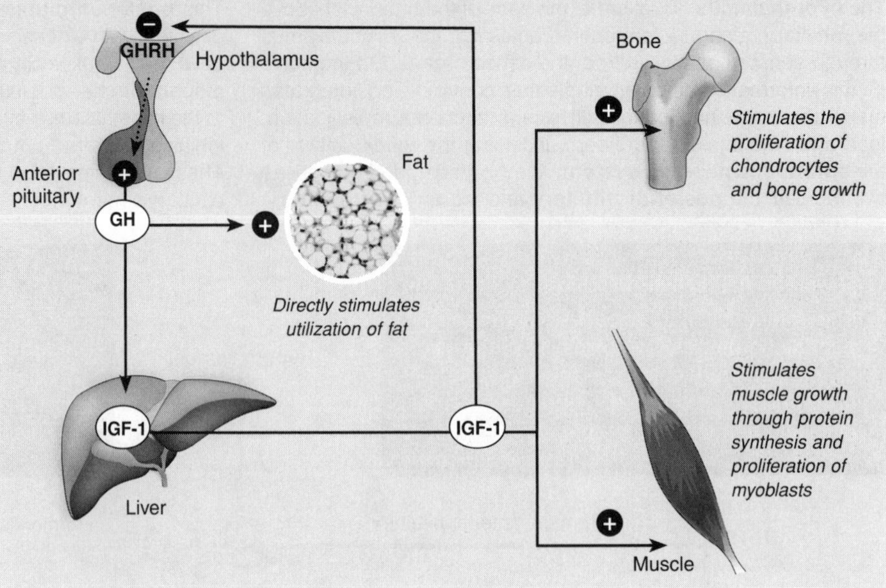

1. (a) Describe the metabolic effects of growth hormone: ______________________

   (b) Predict the effect of chronic **GH deficiency** of GH in infancy: ______________________

   (c) Predict the effect of chronic **GH hypersecretion** in infancy: ______________________

   (d) Describe the two main mechanisms through which the secretion of growth hormone is regulated:

2. "The pituitary releases a number of hormones that regulate the secretion of hormones from other glands". Discuss this statement with reference to growth hormone (GH) and **thyroid stimulating hormone** (TSH):

3. Using the example of TSH and its target tissue (the thyroid), explain how the release of anterior pituitary hormones is regulated. Include reference to the role of negative feedback mechanisms in this process:

4. Iodine is needed to produce thyroid hormones. Explain why the thyroid enlarges in response to an iodine deficiency:

The Liver's Homeostatic Role

The liver, located just below the diaphragm and making up 3-5% of body weight, is the largest homeostatic organ. It performs a vast number of functions including production of bile, storage and processing of nutrients, and detoxification of poisons and metabolic wastes. The liver has a **unique double blood supply** and up to 20% of the total blood volume flows through it at any one time. This rich vascularization makes it the central organ for regulating activities associated with the blood and circulatory system. In spite of the complexity of its function, the liver tissue and the liver cells themselves are structurally relatively simple. Features of liver structure and function are outlined below. The histology of the liver in relation to its role is described on the next page.

Homeostatic Functions of the Liver

The liver is one of the largest and most complex organs in the body. It has a central role as an organ of homeostasis and performs many functions, particularly in relation to the regulation of blood composition. General functions of the liver are outlined below. Briefly summarized, the liver:

1. Secretes bile, important in emulsifying fats in digestion.
2. Metabolizes amino acids, fats, and carbohydrates (below).
3. Synthesizes glucose from non-carbohydrate sources when glycogen stores are exhausted (gluconeogenesis).
4. Stores iron, copper, and some vitamins (A, D, E, K, B_{12}).
5. Converts unwanted amino acids to urea (urea cycle).
6. Manufactures heparin and plasma proteins (e.g. albumin).
7. Detoxifies poisons or turns them into less harmful forms.
8. Some liver cells phagocytose worn-out blood cells.
9. Synthesizes cholesterol from acetyl coenzyme A.

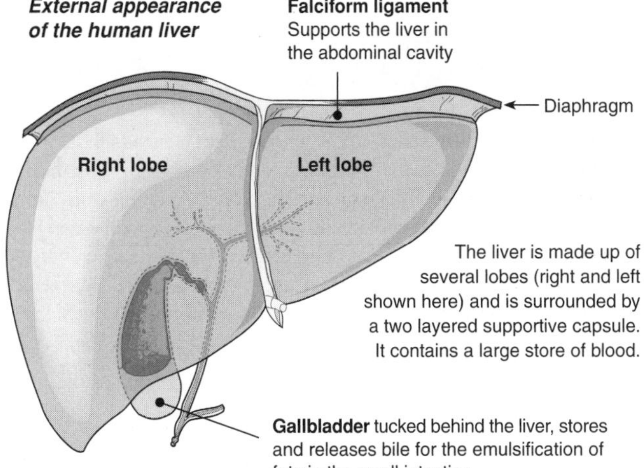

External appearance of the human liver
Falciform ligament Supports the liver in the abdominal cavity
Diaphragm
Right lobe
Left lobe
The liver is made up of several lobes (right and left shown here) and is surrounded by a two layered supportive capsule. It contains a large store of blood.
Gallbladder tucked behind the liver, stores and releases bile for the emulsification of fats in the small intestine

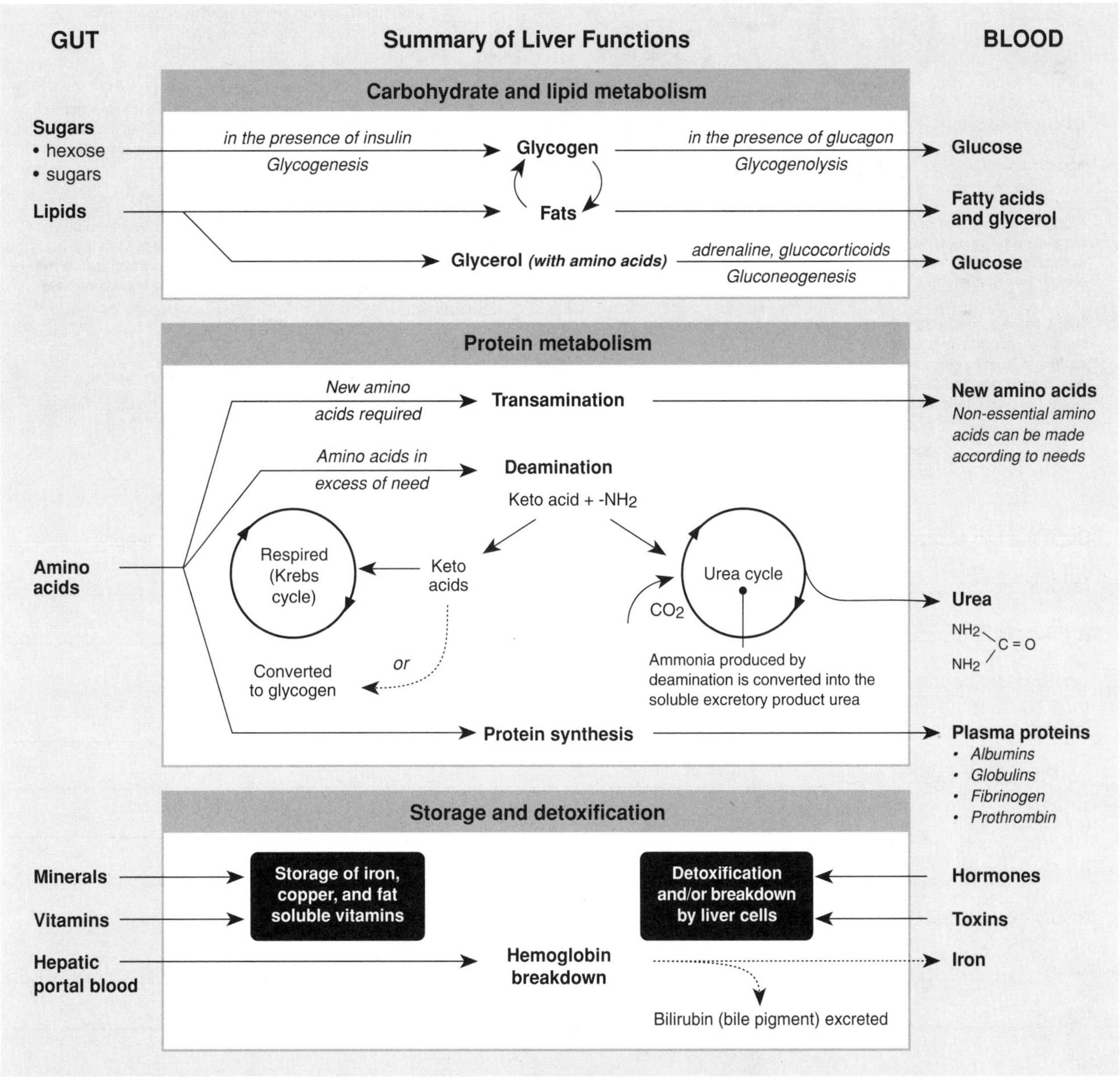

Summary of Liver Functions

GUT → BLOOD

Carbohydrate and lipid metabolism

- **Sugars** (• hexose • sugars) → *in the presence of insulin* / *Glycogenesis* → **Glycogen** → *in the presence of glucagon* / *Glycogenolysis* → **Glucose**
- Glycogen ⇄ **Fats**
- **Lipids** → **Fats** → **Fatty acids and glycerol**
- Lipids → **Glycerol** ***(with amino acids)*** → *adrenaline, glucocorticoids* / *Gluconeogenesis* → **Glucose**

Protein metabolism

- **Amino acids** → *New amino acids required* → **Transamination** → **New amino acids** (*Non-essential amino acids can be made according to needs*)
- Amino acids → *Amino acids in excess of need* → **Deamination** → Keto acid + $-NH_2$
  - Keto acids → Respired (Krebs cycle) *or* Converted to glycogen
  - $-NH_2$ + CO_2 → Urea cycle → **Urea** (NH_2, C = O, NH_2)
  - Ammonia produced by deamination is converted into the soluble excretory product urea
- Amino acids → **Protein synthesis** → **Plasma proteins**
  - *Albumins*
  - *Globulins*
  - *Fibrinogen*
  - *Prothrombin*

Storage and detoxification

- **Minerals**, **Vitamins** → **Storage of iron, copper, and fat soluble vitamins**
- **Hormones**, **Toxins** → **Detoxification and/or breakdown by liver cells**
- **Hepatic portal blood** → **Hemoglobin breakdown** → **Iron**; Bilirubin (bile pigment) excreted

Periodicals: *Metabolic powerhouse, The liver in health & disease*

Related activities: *The Digestive Role of the Liver*
Web links: *In and Out of Cells*

The Internal Structure of the Liver

Radiating cords of hepatocytes

CV

Blood flows towards the central vein

Schematic illustration of the arrangement of lobules and portal tracts in liver tissue

Bile ductule

Branch of hepatic portal vein

Branch of hepatic artery

Portal tract (triad)

Bile flow

Central vein

Fibrous connective tissue capsule surrounds lobules

Lobule

The connective tissue capsule covering the liver branches through the tissue, dividing it into functional units called **lobules**. A lobule consists of rows (**cords**) of **hepatocytes** (liver cells) arranged in a radial pattern around a central vein. Between the cords are blood spaces called **sinusoids** and small channels through which the bile flows (the **bile canaliculi**). Between the lobules are branches of the hepatic artery, hepatic portal vein, and bile duct. These form a **portal tract** (triad). Lymphatic vessels and nerves are also found in this area (not shown). **The photograph above** shows the histology of a liver lobule in a human, with the central vein, cords of liver cells, and sinusoids (dark spaces).

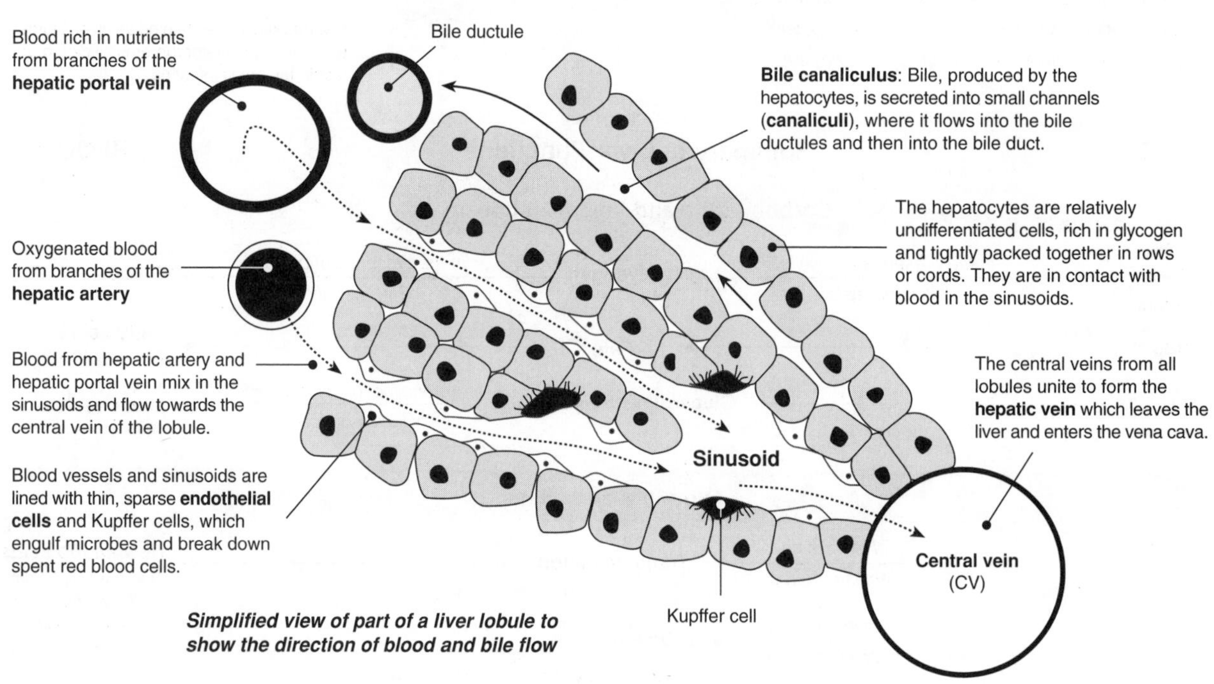

Blood rich in nutrients from branches of the **hepatic portal vein**
Bile ductule
Bile canaliculus: Bile, produced by the hepatocytes, is secreted into small channels (**canaliculi**), where it flows into the bile ductules and then into the bile duct.
Oxygenated blood from branches of the **hepatic artery**
The hepatocytes are relatively undifferentiated cells, rich in glycogen and tightly packed together in rows or cords. They are in contact with blood in the sinusoids.
Blood from hepatic artery and hepatic portal vein mix in the sinusoids and flow towards the central vein of the lobule.
The central veins from all lobules unite to form the **hepatic vein** which leaves the liver and enters the vena cava.
Sinusoid
Blood vessels and sinusoids are lined with thin, sparse **endothelial cells** and Kupffer cells, which engulf microbes and break down spent red blood cells.
Central vein (CV)
Kupffer cell

Simplified view of part of a liver lobule to show the direction of blood and bile flow

1. State the two sources of blood supply to the liver, describing the primary physiological purpose of each supply:

   (a) Supply 1: __________ Purpose: __________

   (b) Supply 2: __________ Purpose: __________

2. Briefly describe the role of the following structures in liver tissue:

   (a) Bile canaliculi:

   (b) Phagocytic Kupffer cells:

   (c) Central vein:

   (d) Sinusoids:

3. Briefly explain three important aspects of **either** protein metabolism **or** carbohydrate metabolism in the liver:

   (a)

   (b)

   (c)

The Skin's Role in Homeostasis

The skin and its associated structures provide physical and chemical protection for most parts of all other body systems. It has a critical role in thermoregulation and in the absorption of sunlight and synthesis of a vitamin D precursor. The skin is a dry membrane, made up of an outer **epidermis** and underlying **dermis**. The subcutaneous tissue beneath the dermis is not part of the skin, but it does anchor the skin to underlying organs, thereby insulating and protecting them. The homeostatic interactions of the skin with other body systems are described below (highlighted panels).

Endocrine system

- Estrogens help to maintain skin hydration.
- Androgens activate the sebaceous glands and help to regulate the growth of hair.
- Skin pigmentation may change when hormones fluctuate, e.g. during pregnancy.

Respiratory system

- Provides oxygen to the cells of the skin via gas exchange with the blood.
- Removes carbon dioxide (gaseous metabolic waste) from the cells of the skin via gas exchange with the blood.

Cardiovascular system

- Blood vessels transport O_2 and nutrients to the skin and remove wastes (via the blood).
- The skin prevents fluid loss and acts as a reservoir for blood.
- Dilation and constriction of blood vessels is important thermoregulatory mechanism.
- The blood supplies substances required for functioning of the skin's glands.

Digestive system

- Skin synthesizes vitamin D, which is required for absorption of calcium from the gut.
- Digestive system provides nutrients for growth, repair, and maintenance of the skin (delivered via the cardiovascular system).

Skeletal system

- Skin absorbs ultraviolet light and produces a vitamin D precursor. Vitamin D is involved in calcium and phosphorus metabolism, and is needed for normal calcium absorption and deposition of calcium salts in bone.

Nervous system

- Many sensory organs and simple receptors are located in the skin.
- The nervous system regulates blood vessel dilation and sweat gland secretion.
- CNS interprets sensory information from the skin's sensory receptors.
- Nervous stimulation causes erection of hair (thermoregulatory response).

Lymphatic system and immunity

- Tissue fluid bathes and nourishes skin cells. Lymphatic vessels collect and return tissue fluid to the general circulation.

Urinary system

- The skeleton protects the pelvic organs.
- Final activation of vitamin D, which is involved in calcium and phosphorus metabolism, occurs in the kidneys.
- Urination controlled by a voluntary sphincter in the urethra.

Reproductive system

- Mammary glands, which are modified sweat glands, nourish the infant in lactating women.
- Skin stretches during pregnancy to accommodate enlargement of the uterus.
- Changes in skin pigmentation are associated with pregnancy and puberty.

Muscular system

- Muscular activity generates heat, which is dissipated via an increase in blood flow to the skin's surface.
- Muscular activity and increased blood flow increases secretion from the skin's glands (e.g. sweating).

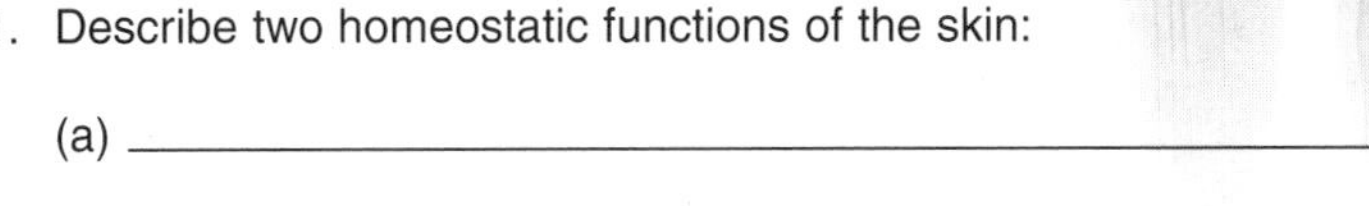

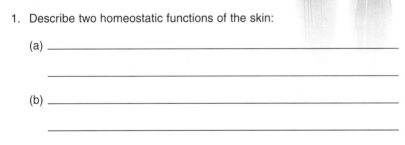

1. Describe two homeostatic functions of the skin:

   (a) ______________________________

   ____

   (b) ______________________________

   ____

   ____

2. Describe a feature of each of the skin sections (right) associated with:

   (a) Protection against wear and tear: ______________________________

   ____

   (b) Lubrication of the hair: ______________________________

   ____

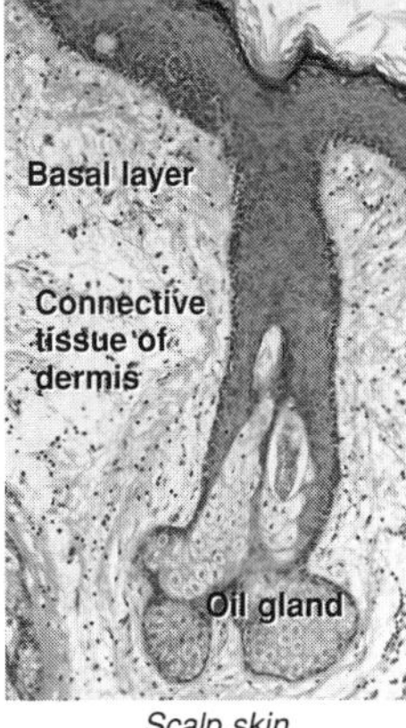

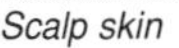

Scalp skin

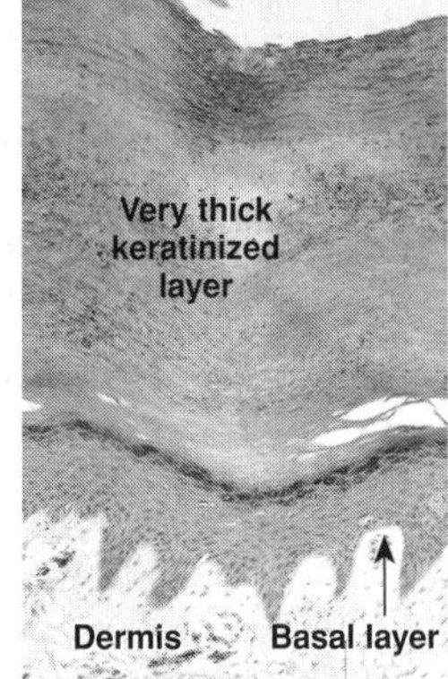

Skin: sole of foot

There are up to five layers of cells in the epidermis. The thickness of the layers, particularly the outermost heavily keratinized layer, varies depending on where the skin is. Keratin protects the deeper cell layers, and skin subjected to regular wear and tear is heavily keratinized.

Related activities: *Thermoregulation in Humans*

Thermoregulation in Humans

In humans and other placental mammals, the temperature regulation center of the body is in the **hypothalamus**. In humans, it has a '**set point**' temperature of 36.7°C. The hypothalamus responds directly to changes in core temperature and to nerve impulses from receptors in the skin. It then coordinates appropriate nervous and hormonal responses to counteract the changes and restore normal body temperature. Like a thermostat, the hypothalamus detects a return to normal temperature and the corrective mechanisms are switched off (**negative feedback**). Toxins produced by pathogens, or substances released from some white blood cells, cause the set point to be set to a higher temperature. This results in **fever** and is an important defense mechanism in the case of infection.

Counteracting Heat Loss

Heat promoting center* in the hypothalamus monitors fall in skin or core temperature below 35.8°C and coordinates responses that generate and conserve heat. These responses are mediated primarily through the **sympathetic nerves** of the autonomic nervous system.

Thyroxine (together with epinephrine) **increases metabolic rate**.

Under conditions of *extreme* cold, epinephrine and thyroxine increase the energy releasing activity of the liver. Under normal conditions, the liver is thermally neutral.

Muscular activity (including *shivering*) produces internal heat.

Erector muscles of hairs contract to raise hairs and increase insulating layer of air. Blood flow to skin decreases (**vasoconstriction**).

Factors causing heat loss

- Wind chill factor accelerates heat loss through conduction.
- Heat loss due to temperature difference between the body and the environment.
- The rate of heat loss from the body is increased by being wet, by inactivity, dehydration, inadequate clothing, or shock.

Factors causing heat gain

- Gain of heat directly from the environment through radiation and conduction.
- Excessive fat deposits make it harder to lose the heat that is generated through activity.
- Heavy exercise, especially with excessive clothing.

**NOTE: The heat promoting center is also called the cold centre and the heat losing center is also called the hot centre. We have used the terminology descriptive of the activities promoted by the center in each case.*

Counteracting Heat Gain

Heat losing center* in the hypothalamus monitors any rise in skin or core temperature above 37.5°C and coordinates responses that increase heat loss. These responses are mediated primarily through the **parasympathetic nerves** of the autonomic nervous system.

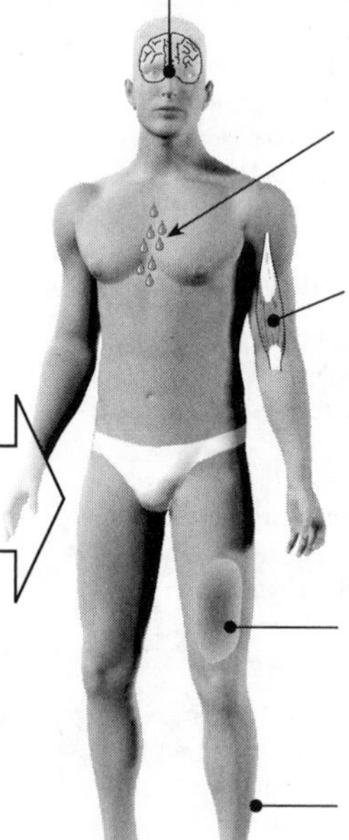

Sweating increases. Sweat cools by evaporation.

Muscle tone and **metabolic rate** are decreased. These mechanisms reduce the body's heat output.

Blood flow to skin (**vasodilation**) increases. This increases heat loss.

Erector muscles of hairs relax to flatten hairs and decrease insulating air layer.

The Skin and Thermoregulation

Thermoreceptors in the dermis (probably free nerve endings) detect changes in skin temperature outside the normal range and send nerve impulses to the hypothalamus, which mediates a response. Thermoreceptors are of two types: **hot thermoreceptors** detect a rise in skin temperature above 37.5°C while the **cold thermoreceptors** detect a fall below 35.8°C. Temperature regulation by the skin involves **negative feedback** because the output is fed back to the skin receptors and becomes part of a new stimulus-response cycle.

Note that the thermoreceptors detect the temperature change, but the hair erector muscles and blood vessels are the **effectors** for mediating a response.

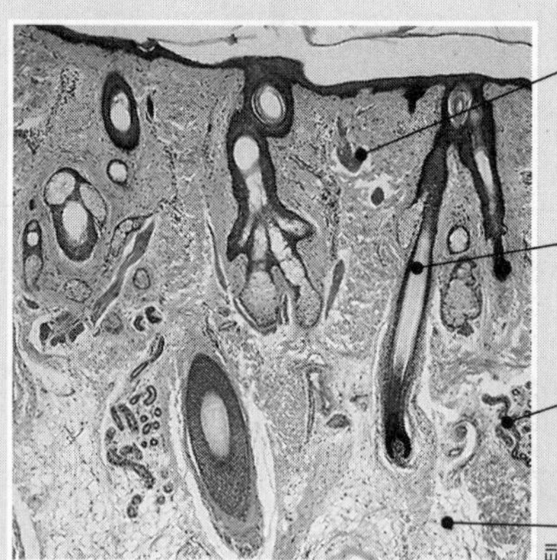

Cross section through the skin of the scalp.

Blood vessels in the dermis dilate (vasodilation) or constrict (vasoconstriction) to respectively promote or restrict heat loss.

Hairs raised or lowered to increase or decrease the thickness of the insulating air layer between the skin and the environment.

Sweat glands produce sweat in response to parasympathetic stimulation from the hypothalamus. Sweat cools through evaporation.

Fat in the subdermal layers insulates the organs against heat loss.

1. State two mechanisms by which body temperature could be reduced after intensive activity (e.g. hard exercise):

   (a) ______________________ (b) ______________________

2. Briefly state the role of the following in regulating internal body temperature:

   (a) The hypothalamus: ______________________

   (b) The skin: ______________________

   (c) Nervous input to effectors: ______________________

   (d) Hormones: ______________________

Related activities: *The Hypothalamus and Pituitary, The Skin's Role in Homeostasis*

Mechanisms of Thermoregulation

The process of controlling body temperature is called thermoregulation. For many years, animals were classified as either homeotherms (= constant body temperature) or poikilotherms (= variable body temperature). Unfortunately, these terms are not particularly accurate for many animals; for example, some mammals (typical homeotherms), may have unstable body temperatures. A more recent, thermal classification of animals is based on the source of the body heat: whether it is largely from the environment (**ectothermic**) or from metabolic activity (**endothermic**). This classification can be more accurately applied to most animals but, in reality, many animals still fall somewhere between the two extremes.

How Body Temperature Varies

Aquatic invertebrates like jellyfish are true poikilotherms: their temperature is the same as the environment.

Tuna and some of the larger sharks can maintain body temperatures up to 14°C above the water temperature.

Hibernating rodents and bats let their body temperature drop to well below what is typical for most mammals.

Most birds and mammals maintain a body temperature that varies less than 2°C: they are true homeotherms.

Increasingly homeothermic →

Poikilothermic

Body temperature varies with the environmental temperature. Traditionally includes all animals other than birds and mammals, but many reptiles, some large insects and some large fish are not true poikilotherms because they may maintain body temperatures that are different from the surrounding environment.

Homeothermic

Body temperature remains almost constant despite environmental fluctuations. Traditionally includes birds and mammals, which typically maintain body temperatures close to 37-38°C. Many reptiles are partially homeothermic and achieve often quite constant body temperatures through behavioral mechanisms.

Source of Body Heat

With a few exceptions, most fish are fully ectothermic. Unlike many reptiles they do not usually thermoregulate.

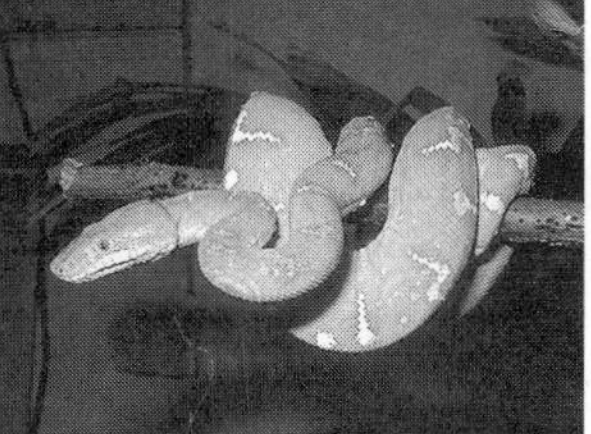

Snakes use heat energy from the environment to increase their body temperature for activity.

Some large insects like bumblebees may raise their temperature for short periods through muscular activity.

Mammals (and birds) achieve high body temperatures through metabolic activity and reduction of heat losses.

Increasingly endothermic →

Ectothermic

Ectotherms depend on the environment for their heat energy. The term ectotherm is often equated with poikilotherm, although they are not the same. Poikilotherms are also ectotherms but many ectotherms may regulate body temperature (often within narrow limits) by changing their behavior (e.g. snakes and lizards).

Endothermic

Endotherms rely largely on metabolic activity for their heat energy. Since they usually maintain a constant body temperature, most endotherms are also homeotherms. As well as birds and mammals, some fast swimming fish, like tuna, and some large insects may also use muscular activity to maintain a high body temperature.

Daily temperature variations in ectotherms and endotherms

Ectotherm: Diurnal lizard (top right)

Body temperature is regulated by behavior so that it does not rise above 40°C. Basking increases heat uptake from the sun. Activity occurs when body temperature is high. Underground burrows are used for retreat.

Endotherm: Human (bottom right)

Body temperature fluctuates within narrow limits over a 24 hour period. Exercise and eating increase body temperature for a short time. Body temperature falls during rest and is partly controlled by an internal rhythm.

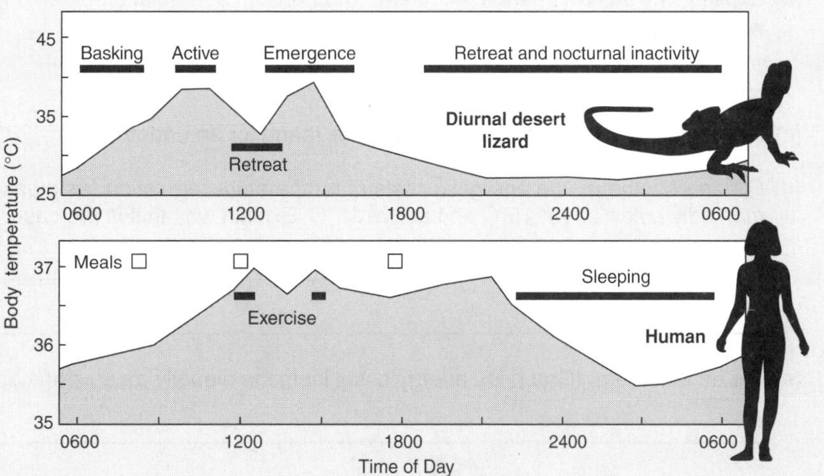

1. (a) Explain what is meant by a homeothermic endotherm:

(b) Explain why the term "poikilotherm" is not a good term for classifying many terrestrial lizards and snakes:

2. Ectotherms will often maintain high, relatively constant body temperatures for periods in spite of environmental fluctuations, yet they also tolerate marked declines in body temperature to levels lower than are tolerated by endotherms.
(a) Describe the advantages of letting body temperature fluctuate with the environment (particularly at low temperature):

(b) Suggest why ectothermy is regarded as an adaptation to low or variable food supplies:

3. Some endotherms do not always maintain a high body temperature. Some, such as small rodents, allow their body temperatures to fall during hibernation. Explain the advantage of this behavior:

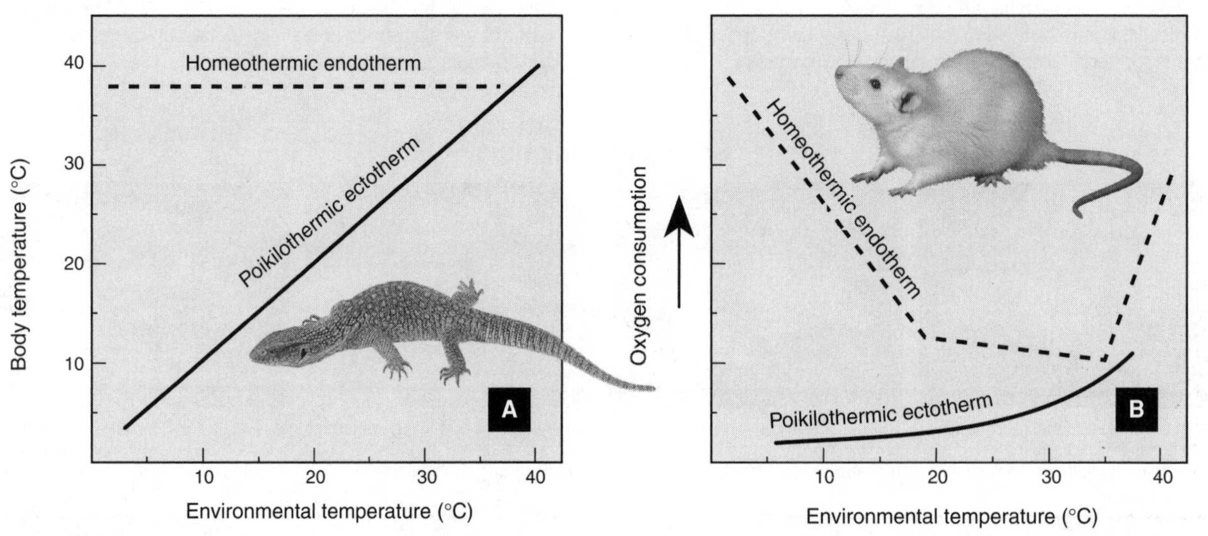

4. The two graphs above illustrate the differences in temperature regulation between a homeothermic endotherm and a poikilothermic ectotherm (such as a fish). Graph A shows change in body temperature with environmental temperature. Graph B shows change in oxygen consumption with environmental temperature. Use the graphs to answer the following:
(a) Explain how ectotherms and endotherms differ in their response to changes in environmental temperature (graph **A**):

(b) Explain why a poikilothermic ectotherm (no behavioral regulation of temperature) would be limited to environments where temperatures were below about 40°C:

(c) In graph **B**, state the optimum temperature range for an endotherm:

(d) For an endotherm, the energetic costs of temperature regulation (as measured by oxygen consumption) increase markedly below about 15°C and above 35°C. Explain why this is the case:

(e) For an ectotherm (Graph B), energy costs increase steadily as environmental temperature increases. Explain why:

Thermoregulation in Mammals

For a body to maintain a constant temperature heat losses must equal heat gains. Heat exchanges with the environment occur via three mechanisms: **conduction** (direct heat transfer), **radiation** (indirect heat transfer), and **evaporation**. The importance of each of these depends on environment. For example, there is no evaporative loss in water, but such losses in air can be very high. Coverage of temperature regulation in humans, including the role of the skin, is covered elsewhere in this chapter.

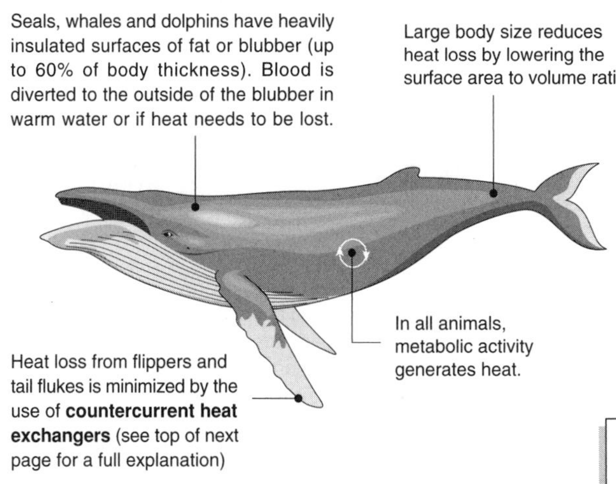

Seals, whales and dolphins have heavily insulated surfaces of fat or blubber (up to 60% of body thickness). Blood is diverted to the outside of the blubber in warm water or if heat needs to be lost.
Large body size reduces heat loss by lowering the surface area to volume ratio.
In all animals, metabolic activity generates heat.
Heat loss from flippers and tail flukes is minimized by the use of **countercurrent heat exchangers** (see top of next page for a full explanation)

Water has a great capacity to transfer heat away from organisms; its cooling power can be more than 50 times greater than that of air. For most aquatic animals (with the exception of aquatic birds and mammals and a few fish) heat retention is impossible. Instead, they carry out their metabolic activities at the ambient temperature. For most marine organisms this does not fluctuate much.

Temperature regulation mechanisms in water
- Low metabolic activity
- Heat generation from metabolic activity
- Insulation layer of blubber
- Changes in circulation patterns when swimming
- Large body size
- Heat exchange systems in limbs or high activity muscle

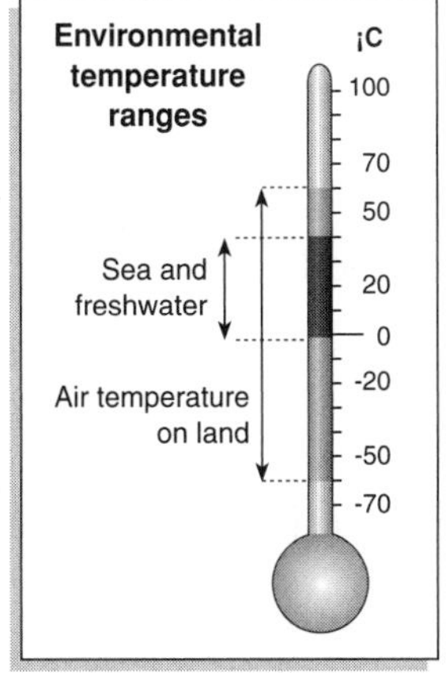

Environmental temperature ranges
¡C
100
70
50
20
0
-20
-50
-70
Sea and freshwater
Air temperature on land

Temperature regulation mechanisms in air
- Behavior or habitat choice
- Heat generation from metabolic activity
- Insulation (fat, fur, feathers)
- Changes in blood flow
- Large body size
- Sweating and panting
- Tolerance of fluctuation in body temperature

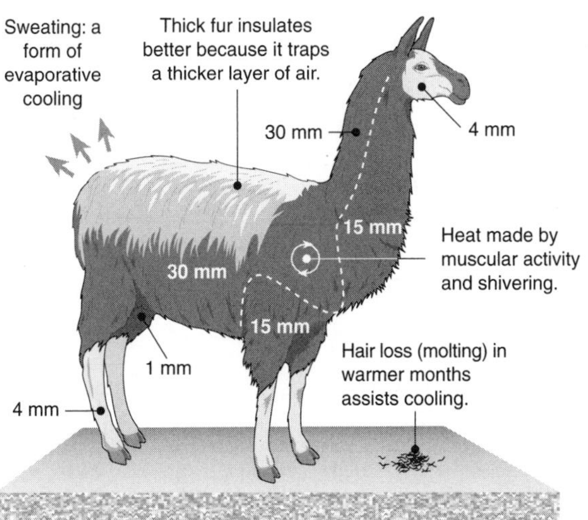

Sweating: a form of evaporative cooling
Thick fur insulates better because it traps a thicker layer of air.
30 mm
4 mm
15 mm
Heat made by muscular activity and shivering.
30 mm
15 mm
1 mm
4 mm
Hair loss (molting) in warmer months assists cooling.

For most mammals, the thickness of the fur or hair varies around the body (as indicated above). Thermoregulation is assisted by adopting body positions that expose or cover areas of thin fur (the figures above are for the llama-like guanaco).

Animals adapted to temperature extremes (hot or cold) can often tolerate large fluctuations in their body temperature before they become stressed. In camels, the body temperature may fluctuate up to 7°C (34°C to 41°C) over a 24 hour period.

Dog
Panting to lose accumulated heat is important in dogs, which have sweat glands only on the pads of their feet.

Brown bear
Thick hair, fur or wool traps air in a layer next to the skin. This insulating air layer reduces heat loss and slows heat gain.

Elephant seal
Thick blubber and large body size in seals and other marine mammals provide an effective insulation.

Musk oxen
Mammals and birds in cold climates, like the musk oxen above, cluster together to retain body heat.

1. Explain how water differs from air in the way in which it transmits heat away from the body of an organism:

____

____

2. Describe two ways in which a mammal (a endothermic homeotherm) maintains its internal body temperature in water:

(a) _______________

(b) _______________

Countercurrent Heat Exchange Systems

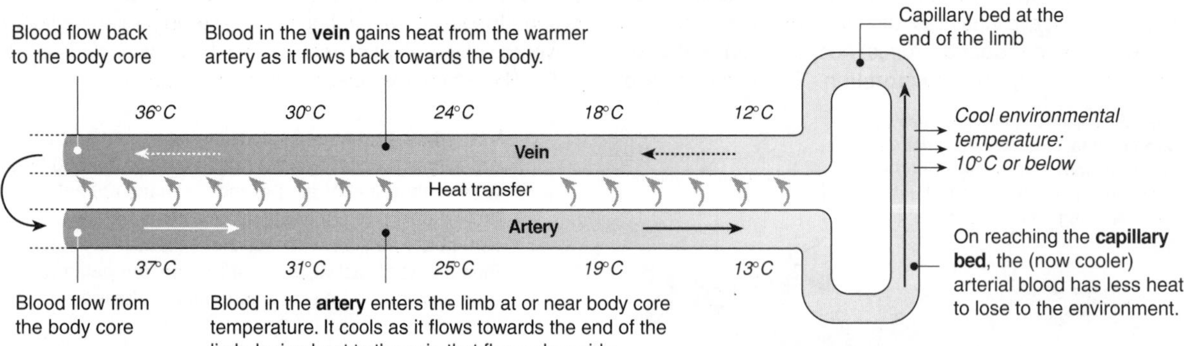
Blood flow back to the body core
Blood in the **vein** gains heat from the warmer artery as it flows back towards the body.
Capillary bed at the end of the limb
36°C
30°C
24°C
18°C
12°C
Vein
Heat transfer
Artery
Cool environmental temperature: 10°C or below
37°C
31°C
25°C
19°C
13°C
On reaching the **capillary bed**, the (now cooler) arterial blood has less heat to lose to the environment.
Blood flow from the body core
Blood in the **artery** enters the limb at or near body core temperature. It cools as it flows towards the end of the limb, losing heat to the vein that flows alongside.

Countercurrent heat exchange systems occur in both aquatic and terrestrial animals as an adaptation to maintaining a stable core temperature. The diagram illustrates the general principle of countercurrent heat exchangers: heat is exchanged between incoming and outgoing blood. In the flippers and fins of whales and dolphins, and the legs of aquatic birds, they minimize heat loss. In some terrestrial animals adapted to hot climates, the countercurrent exchange mechanism works in the opposite way to prevent the head from overheating: venous blood cools the arterial blood before it supplies the brain.

3. (a) Explain how a large body size assists in maintaining body temperature in both aquatic and terrestrial mammals:

   (b) Describe a way in which small terrestrial mammals compensate for more rapid heat loss from a high surface area:

4. (a) Explain how thick hair or fur assists in the regulation of body temperature in mammals:

   (b) Explain why fur/hair thickness varies over different regions of a mammal's body:

   (c) Explain how you would expect fur thickness to vary between related mammal species at high and low altitude:

   (d) Explain how marine mammals compensate for lack of thick hair or fur:

5. Giving an example, explain how countercurrent heat exchange systems assist in temperature regulation in mammals:

6. (a) Describe the role that group behavior plays in temperature regulation in some mammals:

   (b) Name an animal, other than musk oxen, that uses this behavior:

   (c) Describe another behavior, not reliant on a group, that is important in thermoregulation. For the behavior, suggest when and where it would occur, and comment on its adaptive value:

Hypothermia

Hypothermia is a condition experienced when the core body temperature drops below 35°C. Hypothermia is caused by exposure to low temperatures, and results from the body's inability to replace the heat being lost to the environment. The condition ranges from mild to severe depending how low the body temperature has dropped. Severe hypothermia results in severe mental confusion, including inability to speak and amnesia, organ and heart failure, and death.

Maintaining a normal body temperature of around 37°C allows the body's metabolism to function optimally. At temperatures below 35°C, metabolic reactions begin to slow, resulting in a loss of coordination, difficulty in moving, and mental fatigue. Hypothermia can result from exposure to very low temperatures for a short time or to moderately low temperatures for a long time. Exposure to cold water (even just slightly cold) will produce symptoms of hypothermia far more quickly than exposure to the same temperature of air. This is because water is much more effective than air at conducting heat away from the body.

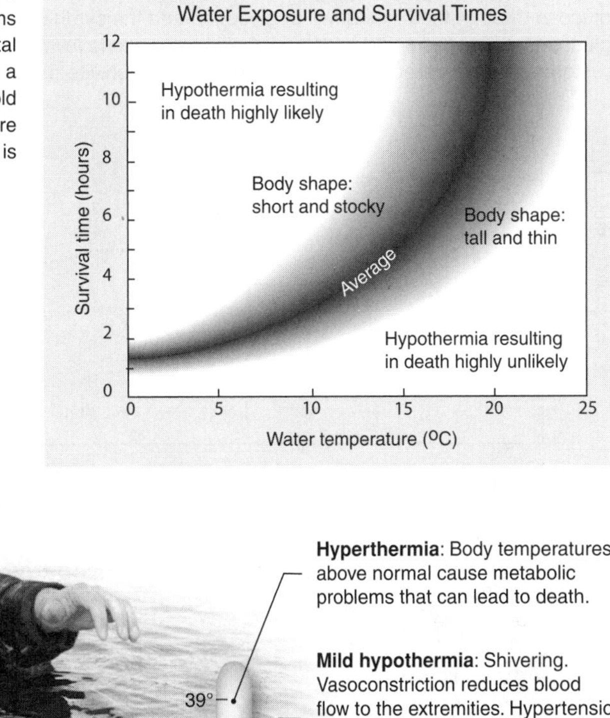

Hyperthermia: Body temperatures above normal cause metabolic problems that can lead to death.

Mild hypothermia: Shivering. Vasoconstriction reduces blood flow to the extremities. Hypertension and cold diuresis (increased urine production due to the cold).

Moderate hypothermia: Muscle coordination becomes difficult. Movements slow or laboured. Blood vessels in ears, nose, fingers, and toes constrict further resulting in these turning a blue color. Mental confusion sets in.

Severe hypothermia: Speech fails. Mental processes become irrational, victim may enter a stupor. Organs and heart eventually fail resulting in death.

Normal body temperature → 37°

39° 37° 35° 33° 31° 29° 27° °C

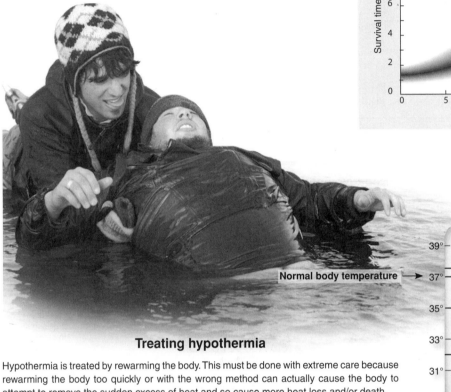

Treating hypothermia

Hypothermia is treated by rewarming the body. This must be done with extreme care because rewarming the body too quickly or with the wrong method can actually cause the body to attempt to remove the sudden excess of heat and so cause more heat loss and/or death.

Mild hypothermics can be rewarmed by **passive rewarming**, using their own body heat coupled with warm, dry, insulated clothing in a warm dry environment. Moderate hypothermia requires **active external rewarming**. This involves using warming devices such as hot water bottles or warm water baths. Severe hypothermics must be treated with **active internal** or **core warming**. Methods include delivery of warm intravenous fluids, inhaling warm moist air, or warming the blood externally by using a heart-lung machine.

1. Describe the conditions that may cause a person to become hypothermic: ____________

2. (a) With reference to the graph (above), identify which body shape has best survival at 15°C: ____________

   (b) Explain your choice: ____________

3. Describe the methods used to rewarm hypothermics and the importance of using the correct methods.

Control of Blood Glucose

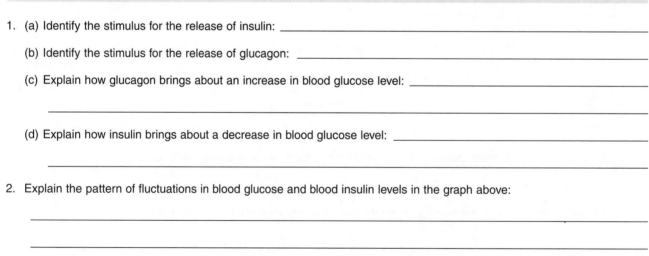

The endocrine portion of the **pancreas** (the α and β cells of the **islets of Langerhans**) produces two hormones, **insulin** and **glucagon**, which maintain blood glucose at a steady state through **negative feedback**. Insulin promotes a decrease in blood glucose by promoting cellular uptake of glucose and synthesizing glycogen. **Glucagon** promotes an increase in blood glucose through the breakdown of glycogen and the synthesis of glucose from amino acids. When normal blood glucose levels are restored, negative feedback stops hormone secretion. Regulating blood glucose to within narrow limits allows energy to be available to cells as needed. Extra energy is stored as glycogen or fat, and is mobilized to meet energy needs as required. The liver is pivotal in these carbohydrate conversions. One of the consequences of a disruption to this system is the disease **diabetes mellitus**. In type 1 diabetes, the insulin-producing β cells are destroyed as a result of autoimmune activity and insulin is not produced. In type 2 diabetes, the pancreatic cells produce insulin, but the body's cells become increasingly resistant to it.

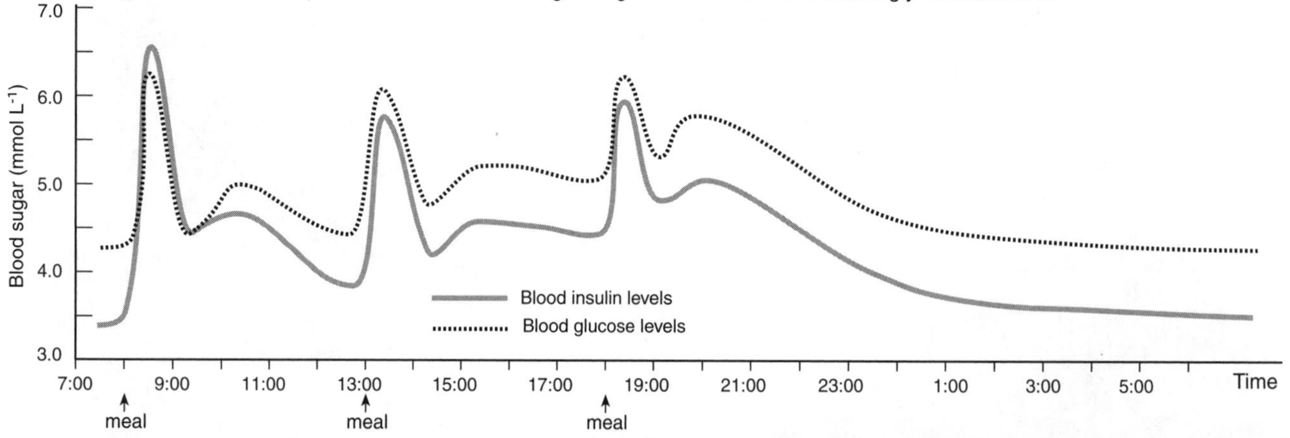

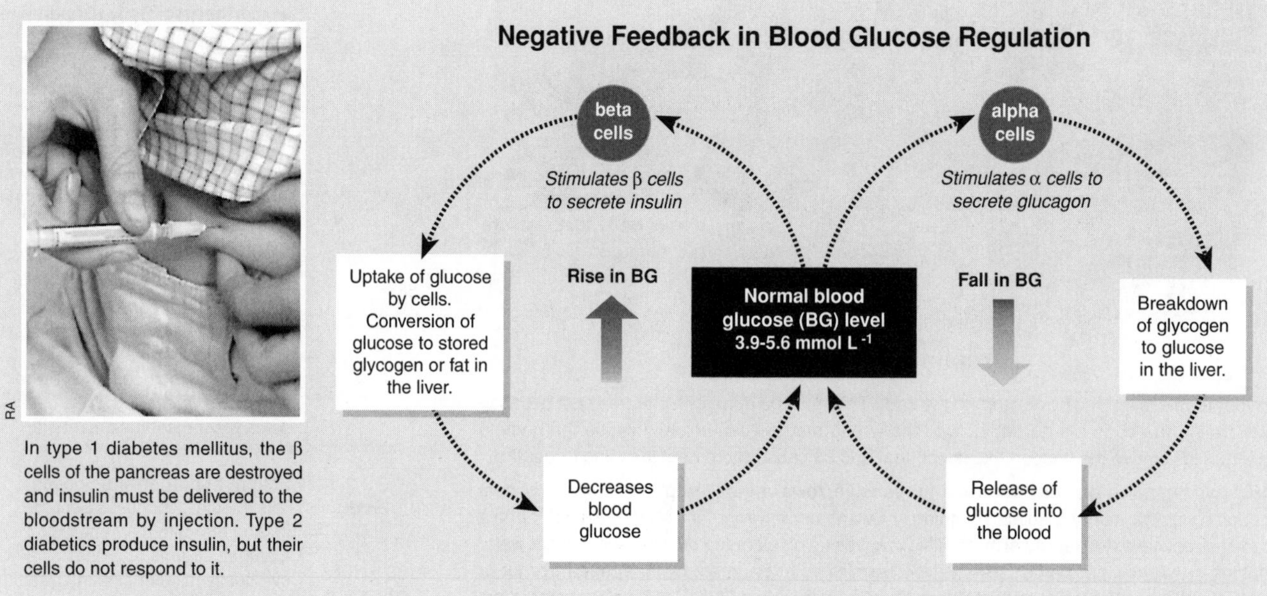

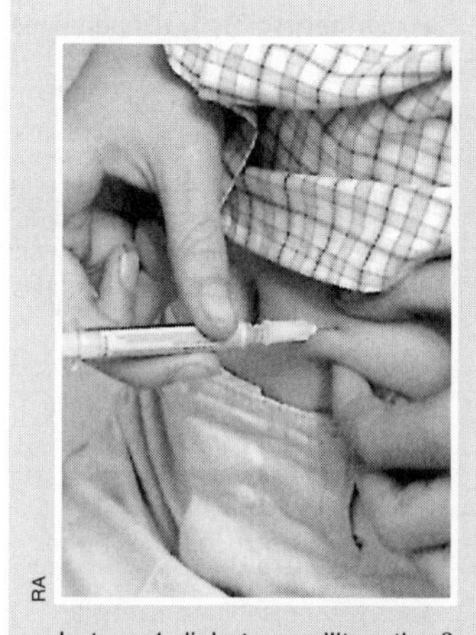

In type 1 diabetes mellitus, the β cells of the pancreas are destroyed and insulin must be delivered to the bloodstream by injection. Type 2 diabetics produce insulin, but their cells do not respond to it.

1. (a) Identify the stimulus for the release of insulin: __________

   (b) Identify the stimulus for the release of glucagon: __________

   (c) Explain how glucagon brings about an increase in blood glucose level: __________

   (d) Explain how insulin brings about a decrease in blood glucose level: __________

2. Explain the pattern of fluctuations in blood glucose and blood insulin levels in the graph above:

3. Identify the mechanism regulating insulin and glucagon secretion (humoral, hormonal, neural): __________

Related activities: *The Liver's Homeostatic Role, Diabetes Mellitus*
Web links: *Cellular Mechanisms of Diabetes*

Periodicals: *Glucose Center Stage, Food for thought*

KEY TERMS Mix and Match

INSTRUCTIONS: Test your vocab by matching each term to its correct definition, as identified by its preceding letter code.

ANTAGONISTIC HORMONES

AUTOCRINE SIGNALING

COUNTERCURRENT HEAT EXCHANGE

CYCLIC AMP

DIABETES MELLITUS

ENDOCRINE GLAND

GLUCAGON

HOMEOSTASIS

HORMONE

HUMORAL

HYPOTHALAMUS

HYPERTHERMIA

HYPOTHERMIA

INSULIN

ISLETS OF LANGERHANS

LIGAND

LIVER

NEGATIVE FEEDBACK

NERVOUS SYSTEM

NEURAL

PANCREAS

PARACRINE SIGNALING

PITUITARY GLAND

POSITIVE FEEDBACK

SECOND MESSENGER

SIGNAL TRANSDUCTION

THERMOREGULATION

A A diffuse abdominal organ with both exocrine and endocrine function.

B The condition where the body temperature is elevated above normal.

C An adjective meaning "carried in the blood".

D Cell signaling in which a cell secretes and reacts to its signal molecules.

E An organism's regulation of body temperature to within certain boundaries, against external fluctuations.

F Hormones with opposite effects on the body and which work in opposition to regulate some process (e.g. control of blood glucose).

G A large organ, with a central role in carbohydrate, protein, and fat metabolism.

H A molecule that relays signals from receptors on the cell surface to target molecules inside the cell.

I A hormone, secreted by the pancreas, that raises blood glucose levels.

J The condition where the body temperature falls below normal.

K A hormone, secreted by the pancreas, that lowers blood glucose levels.

L An organ system which includes a network of specialized cells or neurons that coordinate the actions of an animal and transmit signals between different parts of its body.

M An endocrine gland, located at the base of the brain, which secretes at least 9 hormones.

N An adjective meaning "carried by nerves".

O One of many ductless glands throughout the body that secrete and respond to hormones.

P A destabilizing mechanism in which the output of the system causes an escalation in the initial response.

Q A signal triggering molecule, which binds to a site on a target protein (receptor).

R Cell signaling in which the target cell is close to the cell producing the signal molecule.

S A mechanism by which a mechanical or chemical stimulus to a cell is converted into a specific cellular response.

T A mechanism in which the output of a system acts to oppose changes to the input of the system; the net effect is to stabilize the system against change and dampen fluctuations.

U Regulation of the internal environment to maintain a stable, constant condition.

V An important intracellular signal transduction molecule, derived from ATP.

W A thermoregulatory mechanism in which arterial and venous blood flow in opposite directions creating a heat gradient in which heat is transferred.

X A condition in which the blood glucose level is elevated above normal levels, either because the body doesn't produce enough insulin, or because the cells do not respond to the insulin that is produced.

Y A signaling molecule, produced by an endocrine gland, which is secreted into the blood and affects the metabolism specific target cells.

Z A region of the brain with an important role in linking the nervous and endocrine systems.

AA Endocrine tissue within the pancreas that secretes insulin and glucagon.

AP Topic 3B

IB SL/HL 6.1

Eating to Live

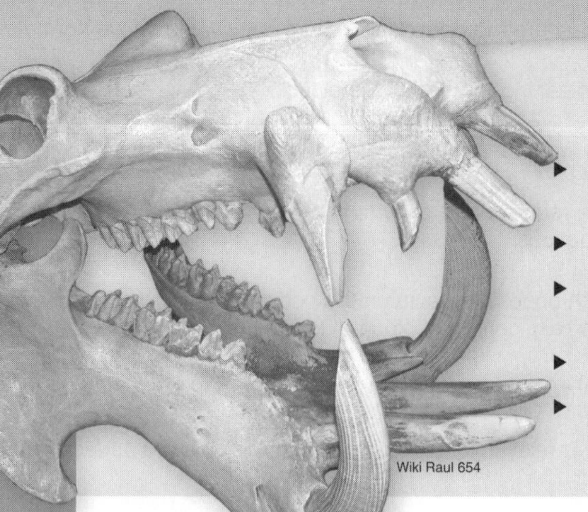

KEY CONCEPTS

- Heterotrophs rely on other organisms for their nutrition (energy and carbon).
- The principal feeding mode in animals is holozoic.
- Structural and functional diversity in animal guts is related to the volume and type of food ingested.
- Extracellular digestion relies on enzymes.
- Nutrients must be absorbed before they can be assimilated. Undigested residue is egested.

KEY TERMS

absorption
amylase
assimilation
autotrophic
bulk feeding
cholecystokinin
dentition
deposit feeding
digestion
egestion
enzyme
extracellular digestion
feeding
filter feeding
fluid feeding
gastric (*adj*)
heterotrophic
holozoic
ingestion
intracellular digestion
large intestine
lipase
lumen
mutualistic nutrition
parasitic
peptidase
protease
saprophytic
secretin
small intestine

OBJECTIVES

- ☐ 1. Use the **KEY TERMS** to help you understand and complete these objectives.

Diversity in Nutritional Modes

pages 59-63

- ☐ 2. Distinguish between **autotrophic** and **heterotrophic** nutrition.
- ☐ 3. Describe the three principal nutritional modes in heterotrophs: **parasites**, **saprophytes**, and **holozoic animals**.
- ☐ 4. Compare and contrast **intracellular digestion** and **extracellular digestion**.
- ☐ 5. Describe an example of **mutualistic nutrition**, e.g. between cellulose digesting microbes and ruminants.
- ☐ 6. Describe **parasitic nutrition** in an animal.

Diversity in Holozoic Nutrition

pages 59-60, 64-82

- ☐ 7. Describe and explain diversity in **feeding** methods in holozoic animals.
- ☐ 8. Describe and explain diversity in mouthparts or dentition in specific taxa, for example fish, birds (beak size and shape), insects, or mammals.
- ☐ 9. Describe and explain structural and functional diversity in animal digestive systems, using examples from representative taxa.

For #10-14, students may choose any one (or more) appropriate examples. <u>International Baccalaureate students must choose humans as their example</u>.

- ☐ 10. Using an annotated diagram, describe the structure and function of the digestive system in a holozoic animal. Explain the four stages involved in processing food in animals with holozoic nutrition: **ingestion, digestion**, **absorption**, and **egestion**.
- ☐ 11. Explain the role of **enzymes** in the **extracellular digestion** of ingested food in holozoic animals. Describe the source, substrate, products, and optimum pH for one **amylase**, **protease**, and **lipase** enzyme involved in digestion.
- ☐ 12. Distinguish between **absorption** and **assimilation**.
- ☐ 13. Explain the significance of increased surface area for nutrient absorption in the guts of holozoic animals. Describe specific structures to facilitate this in representative taxa (e.g. **intestinal villi** in mammals).
- ☐ 14. Describe and explain the role of **egestion** as a final stage in the processing of ingested food in holozoic animals.

Periodicals:
listings for this chapter are on page 388

Weblinks:
www.thebiozone.com/weblink/SB2-2603.html

Teacher Resource CD-ROM:
Control of Digestion

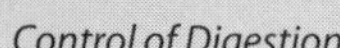

How Heterotrophs Feed

All animals are heterotrophs, they feed on other things, either dead or alive. However, they display a wide range of methods for obtaining the food they need. Animals may feed on solid or fluid food and may suck, bite, lap, or swallow it whole. The adaptations of mouthparts and other feeding appendages reflects both the diet and the way in which they obtain their food. Various behavioral adaptations also contribute to the way animals obtain food. Some animals are predators and hunt other animals. Others are herbivorous and graze or browse continually. Omnivores feed off both meat and vegetable material.

Bulk feeding and cropping

A great number of animals feed on large food masses which may be caught (actively or otherwise) and ingested whole or in pieces. Examples include snakes and most mammalian predators. Most have fangs for holding prey and/or teeth for cutting flesh. Other animals are grazing or cropping herbivores (e.g. many insects and mammalian herbivores), cutting off pieces of vegetation and using their mouthparts to chew the vegetation into pieces.

Sieve and deposit feeding

Humpback and other baleen whales use comblike plates suspended from the upper jaw to sieve shrimps and fish from large volumes of water. Food is trapped against the plates when the mouth closes and water is forced out. Earthworms are nonselective deposit feeders, moving through the soil using a powerful muscular pharynx to suck in a mix of organic and inorganic material. The undigested residue is egested as castings at the soil surface.

Fluid feeding

Fluid feeders suck or lap up fluids such as blood, plant sap, or nectar. Many insects (flies, moths, butterflies, aphids), annelids, arachnids and some mammals exploit these food sources. Fluid feeders have mouthparts and guts that enable them to obtain and process a liquid diet. Many have piercing or tubular mouthparts to obtain fluids directly. Others, like spiders, secrete enzymes into the captured prey and then suck up their liquefied remains.

Filter feeding

An extraordinary range of animals from simple sponges to large marine vertebrates, like the baleen whales, feed by filtering suspended particles out of the water. Special cells lining the body of sponges create water currents and engulf the food particles that are brought in. Many filter feeding annelids, echinoderms, and mollusks, e.g. tubeworms and feather stars, rely on mucus and cilia to trap food particles and move them to the mouth.

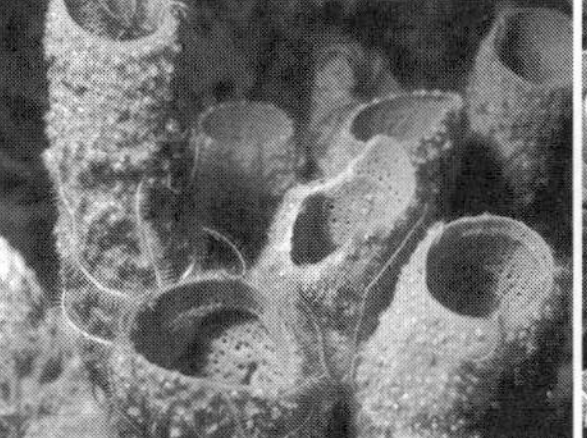

1. Explain why predators, such as lions, tend to swallow large chunks of meat instead of chewing, like cattle do:

2. Compare and contrast the feeding methods of spiders and aphids:

Related activities: Parasitic Nutrition, Insect Mouthparts, Dentition in Mammals

All animals are chemoheterotrophs, as are fungi and the majority of bacteria.

Nutritional Patterns in Organisms

Living organisms can be classified according to their source of energy and carbon. **Phototrophic** organisms use light as their main energy source, whereas **chemotrophs** use inorganic or organic compounds for energy. In terms of carbon sources, **autotrophs** are lierally 'self-feeding' and obtain carbon from carbon dioxide. In contrast, **heterotrophs** feed on others and require an organic carbon source. Most organisms are either photoautotrophs, chemoautotrophs, or chemoheterotrophs.

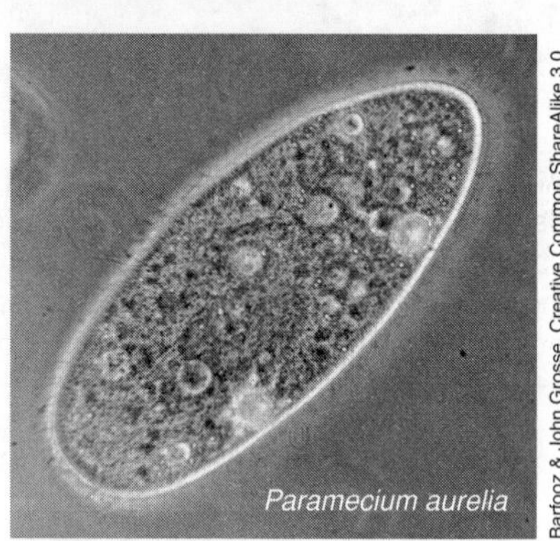

Barfooz & John Grosse, Creative Common ShareAlike 3.0

Protists are highly variable in their nutrition. many, e.g. Paramecium, are heterotrophic, but the algae are photoautotophs.

How Heterotrophs Feed

Heterotrophic organisms feed on organic material in order to obtain the energy and nutrients they require. They depend either directly on other organisms (dead or alive), or their by-products (e.g. feces, cell walls, or food stores). There are three principal modes of heterotrophic nutrition: saprotrophic, parasitic, and holozoic. Within the animal phyla, holozoic nutrition is the most common nutritional mode.

Most fungi and many bacteria are saprotrophs (also called saprophytes). They are decomposer organisms feeding off dead or decaying matter.

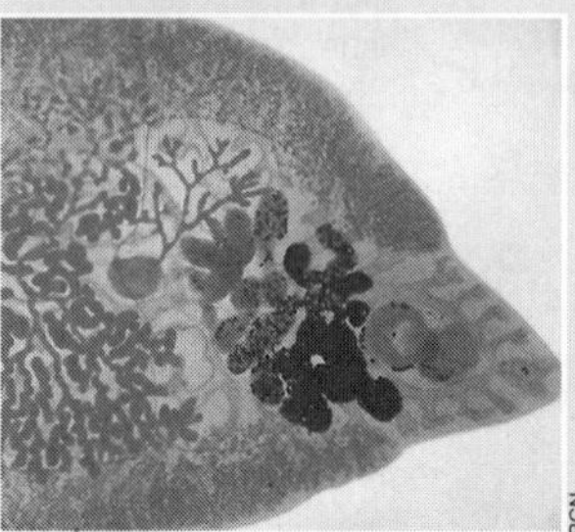

RCN

Parasites, e.g. flukes, live on or within their host for much or all of their life. Bacteria, fungi, protists, and animals all have parasitic representatives.

RCN

Holozoic means to feed on solid organic material from the bodies of other organisms. It is the main feeding mode of animals, although a few specialized plants may obtain some nutrients this way. Holozoic animals are classified according to the form of the food they take in: small or large particles, or fluid.

3. Describe one **structural** adaptation for obtaining food in each of the following:

   (a) A blood sucking mosquito: ______

   (b) A filter feeding whale: ______

   (c) A mammalian predatory carnivore: ______

   (d) A leaf chewing grasshopper: ______

   (e) An ambush predator, such as a python: ______

   (f) A filter feeding marine invertebrate: ______

   (g) *Hydra*: ______

4. Describe one **behavioural** adaptation for obtaining food in each of the following: ______

   (a) A blood sucking mosquito: ______

   (b) A filter feeding whale: ______

   (c) A mammalian predator: ______

   (d) Chimpanzees hunting for monkeys: ______

   (e) An ambush predator, such as a python: ______

   (f) A filter feeding marine invertebrate: ______

   (g) Scavenging bird (gull or vulture): ______

5. Discuss the differences between saprophytes, parasites, and holozoic animals: ______

______

______

Saprophytic Nutrition

All fungi lack chlorophyll and are **heterotrophic**, absorbing nutrients by direct absorption from the substrate. Many are **saprophytic** (also called saprotrophic or saprobiontic), feeding on dead organic matter, although some are parasitic or live in a relationship with another organism (mutualistic). Parasitic fungi are common plant pathogens, invading plant tissues through stomata, wounds, or by penetrating the epidermis. Mutualistic fungi are very important: they form lichens in association with algae or cyanobacteria, and the mutualistic mycorrhizal associations between fungi and plant roots are essential to the health of many forest plants. Saprophytic fungi, together with bacteria, are the major decomposers of the biosphere. They contribute to decay and therefore to nutrient recycling. Like all fungi, the body is composed of rapidly growing filaments called **hyphae**, which are **usually** divided by incomplete compartments called **septa**. The hyphae together form a large mass called a **mycelium** (the feeding body of the fungus). The familiar mushroom-like structures that we see are the above-ground reproductive bodies that arise from the main mycelium. The nutrition of a typical saprophyte, *Rhizopus*, is outlined below.

Bread Mold (*Rhizopus*)

Saprophytes grow best in dark, moist environments, but are found wherever organic material is available. *Rhizopus* is a common fungus, found on damp, stale bread and rotting fruit. Unlike many fungi, *Rhizopus* has hyphae that are undivided by septa.

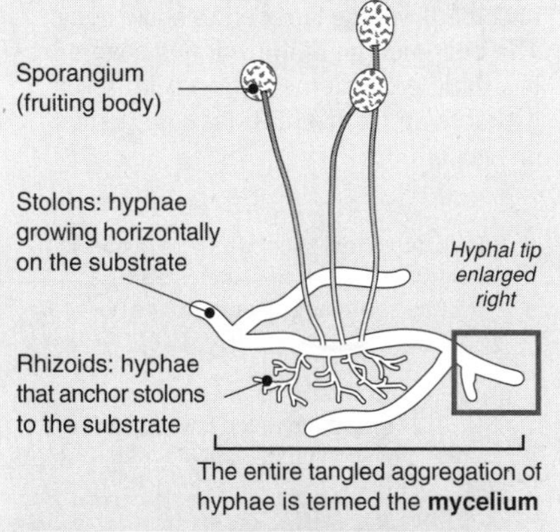
Sporangium (fruiting body)
Stolons: hyphae growing horizontally on the substrate
Hyphal tip enlarged right
Rhizoids: hyphae that anchor stolons to the substrate
The entire tangled aggregation of hyphae is termed the **mycelium**

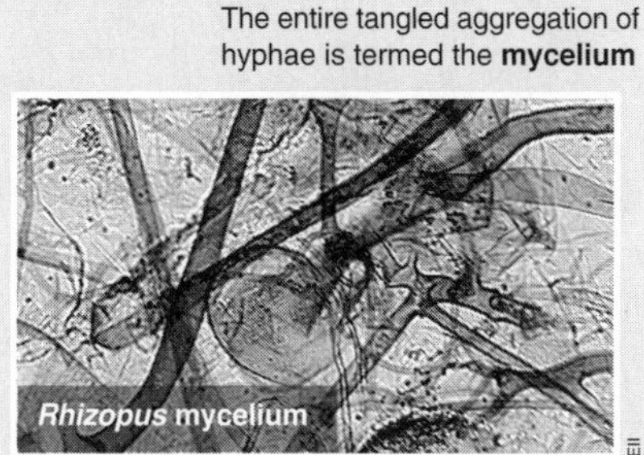
Rhizopus mycelium
EII

Saprophytic Nutrition in Bread Mold (*Rhizopus*)

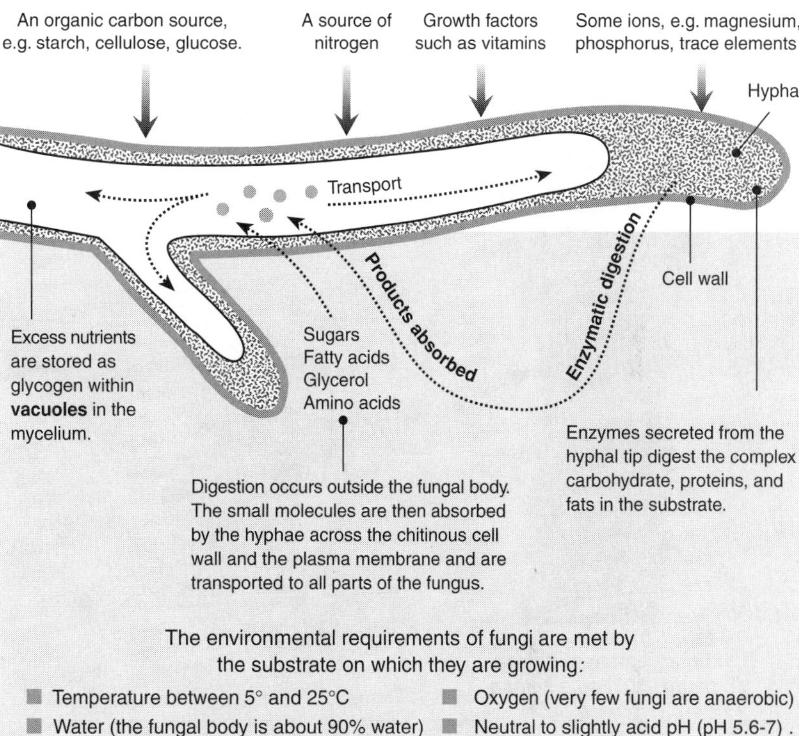
Nutrients required by most saprophytes
An organic carbon source, e.g. starch, cellulose, glucose.
A source of nitrogen
Growth factors such as vitamins
Some ions, e.g. magnesium, phosphorus, trace elements
Hypha
Transport
Cell wall
Products absorbed
Enzymatic digestion
Excess nutrients are stored as glycogen within **vacuoles** in the mycelium.
Sugars
Fatty acids
Glycerol
Amino acids
Digestion occurs outside the fungal body. The small molecules are then absorbed by the hyphae across the chitinous cell wall and the plasma membrane and are transported to all parts of the fungus.
Enzymes secreted from the hyphal tip digest the complex carbohydrate, proteins, and fats in the substrate.

The environmental requirements of fungi are met by the substrate on which they are growing:

- Temperature between 5° and 25°C
- Water (the fungal body is about 90% water)
- Oxygen (very few fungi are anaerobic)
- Neutral to slightly acid pH (pH 5.6-7) .

1. (a) Clearly describe the structure of the feeding body of a saprophytic fungus: ______________________

   (b) Explain why a moist environment is essential for fungal growth: ______________________

2. Identify four nutrients required by a saprophytic fungus:

   (a) ______________________ (c) ______________________

   (b) ______________________ (d) ______________________

3. State where these nutrients come from: ______________________

4. Describe the way in which a saprophytic fungus obtains its nutrients: ______________________

5. Contrast digestion and absorption in a saprophytic fungus and a holozoic animal: ______________________

Eating to Live

Parasitic Nutrition

Parasitism is the most common of all symbiotic relationships. Here the host is always harmed by the presence of the parasite but is usually not killed. The main benefit derived by the parasite is obtaining nutrition, but there may be secondary advantages, such as protection. Many animal groups have members that have adopted a **parasitic** lifestyle, although parasites occur more commonly in particular taxa. Insects, some annelids (e.g. leeches), and flatworms have many parasitic representatives, and two classes of flatworms are entirely parasitic. Animal parasites are highly specialized carnivores, feeding off the body fluids or skin of host species. Parasites that attach to the outside of a host are called **ectoparasites** and have mouthparts specialized for piercing and sucking blood or tissue fluids. Those that live within the body of the host are called **endoparasites**. They may obtain nutrients by sucking or absorb simple food compounds directly from the host, as in the case of the pork tapeworm shown below. All 3400 or so species of tapeworm are endoparasites and the majority are adapted for living in the guts of vertebrates. In all species, a primary host and one or more intermediate hosts are required to complete the life cycle.

The Pork Tapeworm (*Taenia solium*)

Tapeworms (cestodes) are a specialized class of flatworms adapted to a parasitic lifestyle in the intestines of vertebrates. Features of the pork tapeworm, *Taenia solium*, illustrate adaptations to obtaining nutrition at the expense of the host. Humans are the only primary host (the host for the adult stage). When the eggs are passed from humans in the feces, they may be swallowed by a pig or other mammal, whereupon they hatch and can be reingested by the primary host when the pig flesh is eaten. Tapeworms have no mouth or digestive system, and they do not produce digestive enzymes. They are protected by the host's body and bathed in the host's digested food, which is absorbed over the tapeworm's entire body. The entire length of the worm is covered in a thick, enzyme-resistant cuticle, which protects it from the host's digestive enzymes.

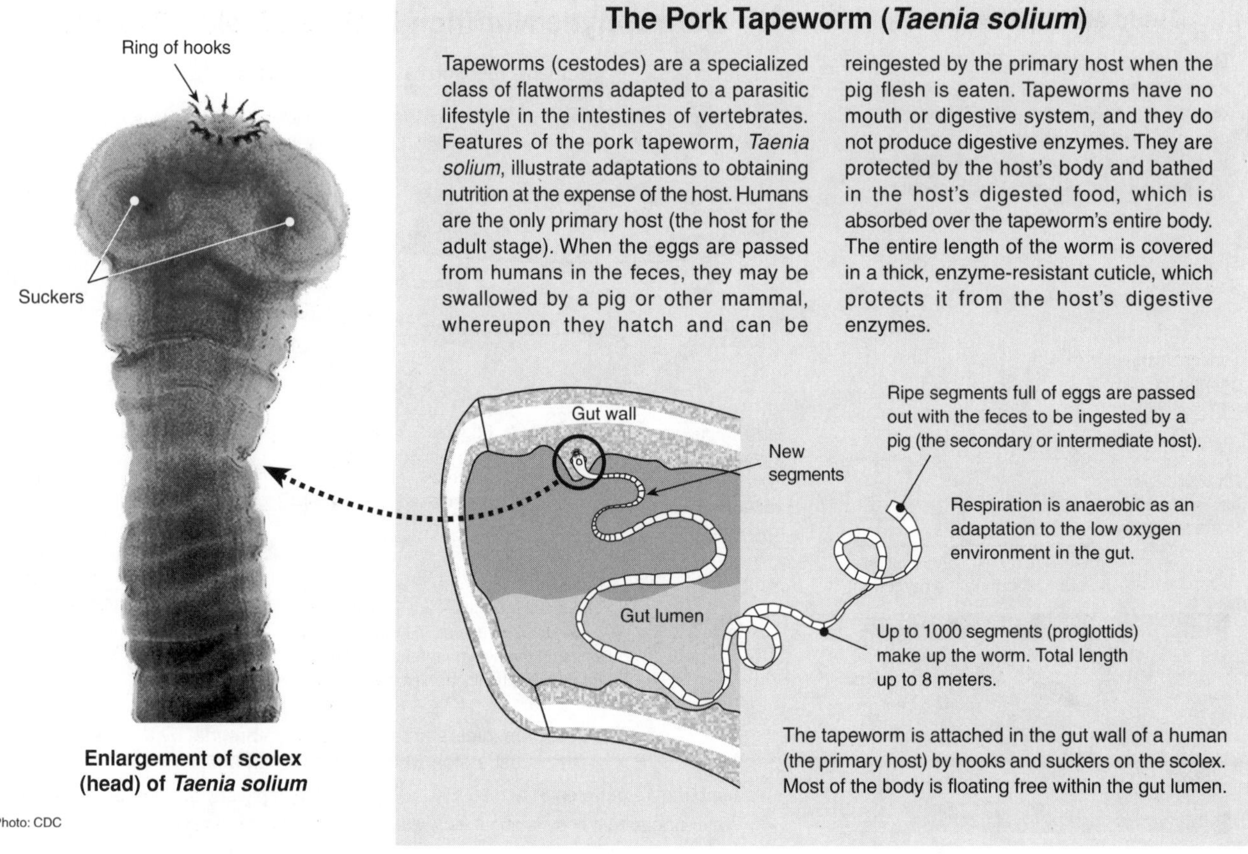
Ring of hooks
Suckers
Gut wall
New segments
Gut lumen
Ripe segments full of eggs are passed out with the feces to be ingested by a pig (the secondary or intermediate host).
Respiration is anaerobic as an adaptation to the low oxygen environment in the gut.
Up to 1000 segments (proglottids) make up the worm. Total length up to 8 meters.
The tapeworm is attached in the gut wall of a human (the primary host) by hooks and suckers on the scolex. Most of the body is floating free within the gut lumen.

Enlargement of scolex (head) of *Taenia solium*

Photo: CDC

1. Describe the way in which the pork tapeworm obtains its required nutrition: ____________________

2. Briefly describe four adaptations of the pork tapeworm for its parasitic lifestyle (include two nutritional adaptations):
   (a) ____________________
   (b) ____________________
   (c) ____________________
   (d) ____________________

3. (a) Explain what is meant by a primary host: ____________________
   (b) Explain what is meant by an intermediate host: ____________________
   (c) Name the primary host for the pork tapeworm: ____________________
   (d) Name an intermediate host for the pork tapeworm: ____________________

4. Identify a similarity between the nutrition of a tapeworm and the nutrition of a saprophytic fungus (see previous page): ____________________

5. Name another animal parasite and give its primary host: ____________________

***Related activities**: How Heterotrophs Feed*

Food Vacuoles and Simple Guts

The simplest form of digestion occurs inside cells (**intracellularly**) within food vacuoles. This process is relatively slow and digestion is exclusively intracellular only in protozoa and sponges. In animals with simple, sac-like guts, digestion begins **extracellularly** (with secretion of enzymes to the outside or into the digestive cavity) and is completed intracellularly.

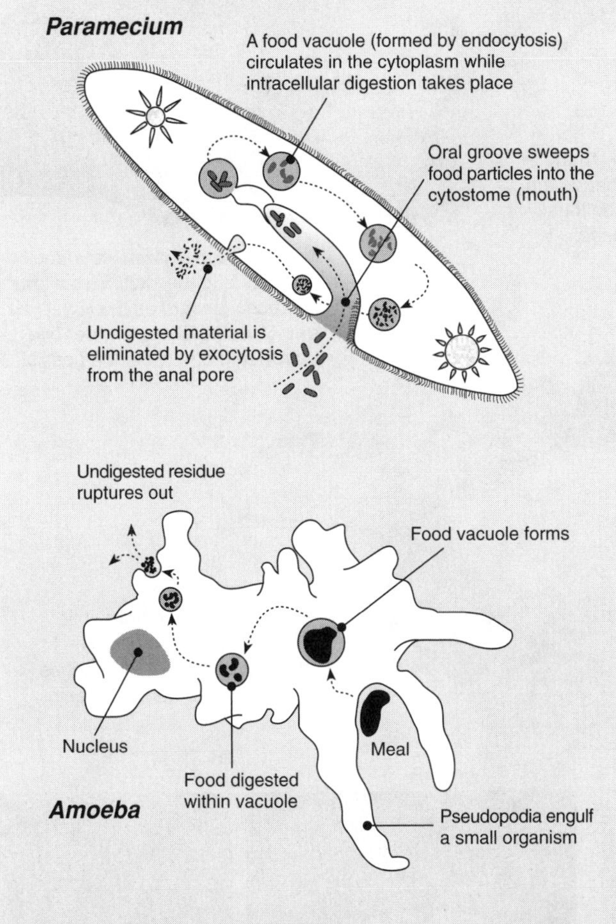

Paramecium
A food vacuole (formed by endocytosis) circulates in the cytoplasm while intracellular digestion takes place
Oral groove sweeps food particles into the cytostome (mouth)
Undigested material is eliminated by exocytosis from the anal pore
Undigested residue ruptures out
Food vacuole forms
Nucleus
Meal
Food digested within vacuole
Amoeba
Pseudopodia engulf a small organism

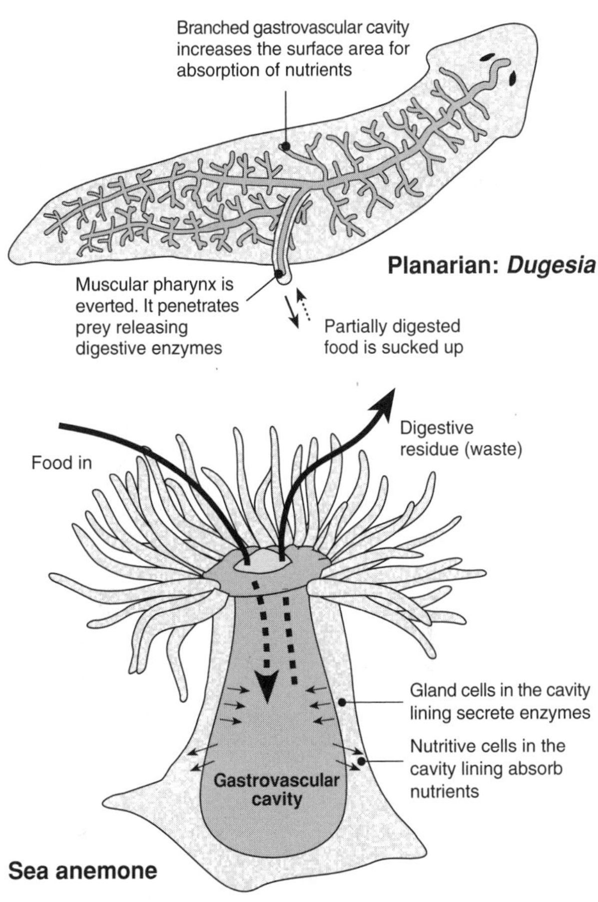

Branched gastrovascular cavity increases the surface area for absorption of nutrients
Planarian: Dugesia
Muscular pharynx is everted. It penetrates prey releasing digestive enzymes
Partially digested food is sucked up
Food in
Digestive residue (waste)
Gland cells in the cavity lining secrete enzymes
Nutritive cells in the cavity lining absorb nutrients
Gastrovascular cavity
Sea anemone

Intracellular Digestion In Food Vacuoles

EXAMPLES: *Protozoans (above), sponges*

The simplest digestive compartments are food vacuoles: organelles where a single cell can digest its food without the digestive enzymes mixing with the cell's own cytoplasm. Sponges and protozoans (e.g. *Paramecium* and *Amoeba*) digest food in this way. *Paramecium* sweeps food into a food groove, from where vacuoles form. *Amoeba* engulf food using cytoplasmic extensions called pseudopodia. Digestion is intracellular, occurring within the cell itself.

Digestion In A Gastrovascular Cavity

EXAMPLES: *Cnidarians, flatworms (above)*

Some of the simplest animals have a digestive sac or gastrovascular cavity with a single opening through which food enters and digested waste passes out. In organisms with this system, digestion is both extra- and intracellular. Digestion begins (using secreted enzymes) either in the cavity (in cnidarians) or outside it (flatworms). In both these groups, the digestion process is completed intracellularly within the vacuoles in cells.

1. Describe two ways in which simple saclike gastrovascular cavities differ from tubelike guts:

   (a) ______

   (b) ______

2. (a) Distinguish between intracellular and extracellular digestion: ______

   (b) Explain why intracellular digestion is not suitable as the only means of digestion for most animals: ______

3. State the main difference between extracellular digestion in sea anemones and *Dugesia*: ______

Related activities: *Diversity in Tube Guts, Absorbing Nutrients*

The Teeth of Fish

Fish are **homodonts**, meaning that all their teeth are identical. As a group though, fish show an extremely varied array of teeth types depending on their diet. The teeth are not always found in just the jaw. Fish may have teeth or teeth-like structures on the roof of the mouth, the tongue, and the gill arches. In most cases the teeth are curved backwards to help grip prey.

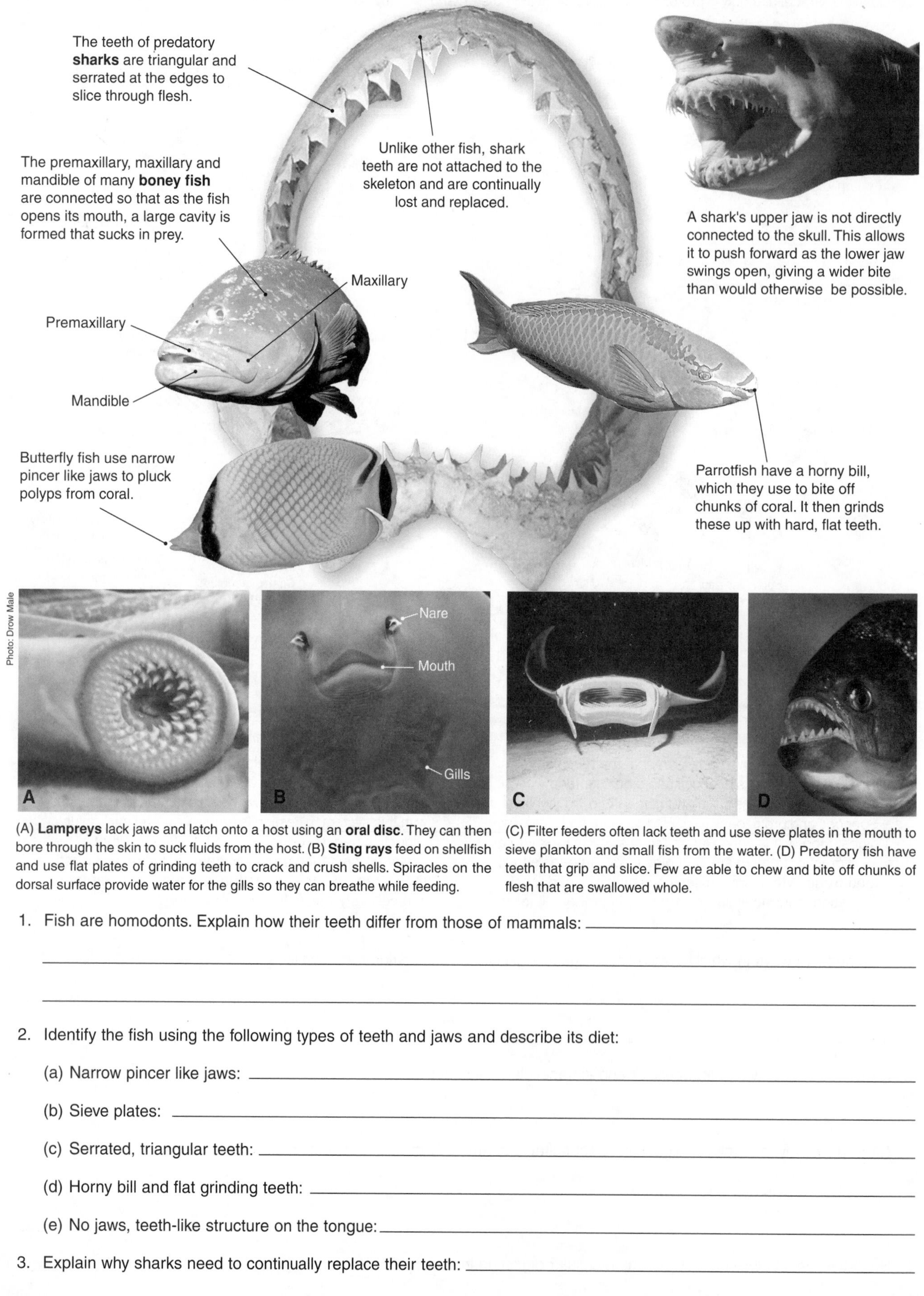

The teeth of predatory **sharks** are triangular and serrated at the edges to slice through flesh.
Unlike other fish, shark teeth are not attached to the skeleton and are continually lost and replaced.
A shark's upper jaw is not directly connected to the skull. This allows it to push forward as the lower jaw swings open, giving a wider bite than would otherwise be possible.
The premaxillary, maxillary and mandible of many **boney fish** are connected so that as the fish opens its mouth, a large cavity is formed that sucks in prey.
Maxillary
Premaxillary
Mandible
Butterfly fish use narrow pincer like jaws to pluck polyps from coral.
Parrotfish have a horny bill, which they use to bite off chunks of coral. It then grinds these up with hard, flat teeth.
Photo: Drow Male
Nare
Mouth
Gills
A
B
C
D

(A) **Lampreys** lack jaws and latch onto a host using an **oral disc**. They can then bore through the skin to suck fluids from the host. (B) **Sting rays** feed on shellfish and use flat plates of grinding teeth to crack and crush shells. Spiracles on the dorsal surface provide water for the gills so they can breathe while feeding. (C) Filter feeders often lack teeth and use sieve plates in the mouth to sieve plankton and small fish from the water. (D) Predatory fish have teeth that grip and slice. Few are able to chew and bite off chunks of flesh that are swallowed whole.

1. Fish are homodonts. Explain how their teeth differ from those of mammals: ______

2. Identify the fish using the following types of teeth and jaws and describe its diet:

   (a) Narrow pincer like jaws: ______

   (b) Sieve plates: ______

   (c) Serrated, triangular teeth: ______

   (d) Horny bill and flat grinding teeth: ______

   (e) No jaws, teeth-like structure on the tongue: ______

3. Explain why sharks need to continually replace their teeth: ______

Related activities: *How Heterotrophs Feed*
Web links: *Ichthyology Education*

Dentition in Mammals

Many of an organism's adaptations are related to the successful procurement of food. Even within the class Mammalia, there is great diversity in the adaptations for different diets. These adaptations are most obvious in the dentition and jaw musculature, as well as in the length and organization of the gut. The diversity of dentition amongst the mammals reflects the wide range of foods and feeding modes within the class; some mammals have relatively generalized dentition, while others are highly specialized, even to the extent of losing teeth entirely. This activity explores some mammalian dietary specializations by examining the dentition of different mammalian representatives.

Using the examples provided, identify the skulls and describe the diet associated with each. Describe the dental adaptations of each, looking for differences in tooth size and arrangement, and whether teeth are absent or modified.

For further help, visit *Will's skull page* web site, which provides information on the structure of mammalian skulls and their dental formulae.

Animals included in this exercise:
Lion, rabbit, mountain sheep, pig, giant anteater, black and white ruffed lemur, dolphin, gray whale, tree shrew.

Diets included in this exercise:
Rough vegetation and grasses; herbs and grasses; crustaceans; omnivorous; fish; insects and worms; leaves, fruit and flowers; meat; termites and ants.

1. Animal: ______________________

   Diet: ______________________

   Dental adaptations to diet: ______________________

   ____

   ____

   ____

1

2. Animal: ______________________

   Diet: ______________________

   Dental adaptations to diet: ______________________

   ____

   ____

   ____

2

3. Animal: ______________________

   Diet: ______________________

   Dental adaptations to diet: ______________________

   ____

   ____

   ____

3

4. Animal:

Diet:

Dental adaptations to diet:

4

5. Animal:

Diet:

Dental adaptations to diet:

5

6. Animal:

Diet:

Dental adaptations to diet:

6

7. Animal:

Diet:

Dental adaptations to diet:

7

8. Animal:

Diet:

Dental adaptations to diet:

8

9. Animal:

Diet:

Dental adaptations to diet:

9

Insect Mouthparts

Insect mouthparts consist of the labrum and three sets of modified, paired appendages known as the mandibles, maxillae, and labium. They are variously adapted to tackle different diets and, in some cases, this has involved loss or fusion of some of the paired appendages. In chewing insects, the **labrum** forms an upper lip and helps pull food into the mouth. The **mandibles** form the first pair of mouthparts and are used as jaws to chew, cut, and tear food, and may also be used to carry things, fight (see right), or to mold wax. The **maxillae** form the second pair of mouthparts and are used for food sensing and handling. The **labium** is a single structure formed from a fused pair of mouthparts. It acts as a lower lip to close the mouth. Both the maxillae and the labium may have finger-like extensions called **palps**. The particular form of the mouthparts depends on the diet, and sometimes on the life stage. In many insects, metamorphosis from the larval to adult stage involves a structural and functional change in the mouthparts associated with a change in diet.

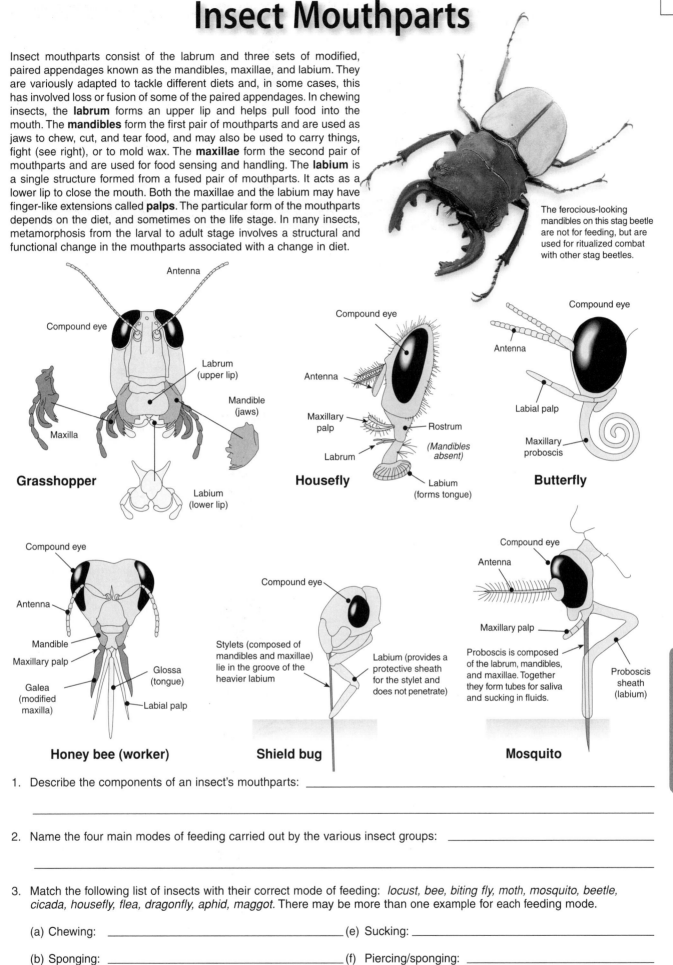

The ferocious-looking mandibles on this stag beetle are not for feeding, but are used for ritualized combat with other stag beetles.

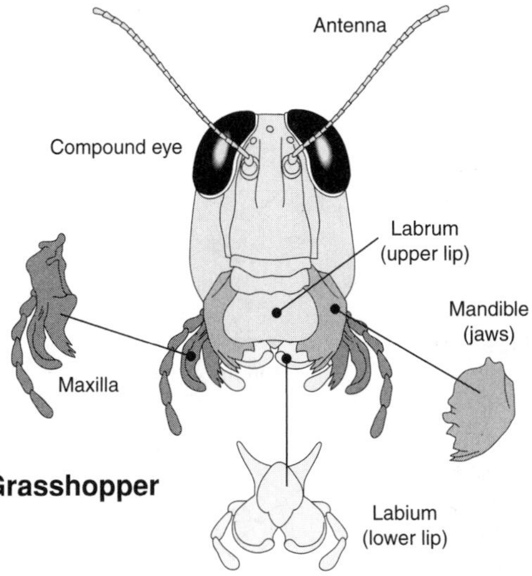

Antenna
Compound eye
Labrum (upper lip)
Mandible (jaws)
Maxilla
Labium (lower lip)

Grasshopper

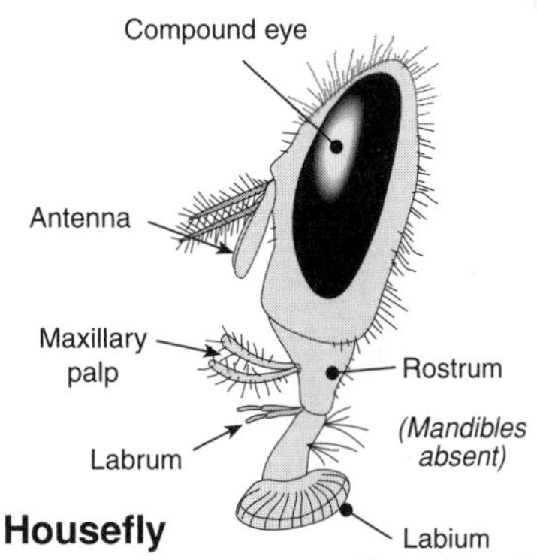

Compound eye
Antenna
Maxillary palp
Labrum
Rostrum
(Mandibles absent)
Labium (forms tongue)

Housefly

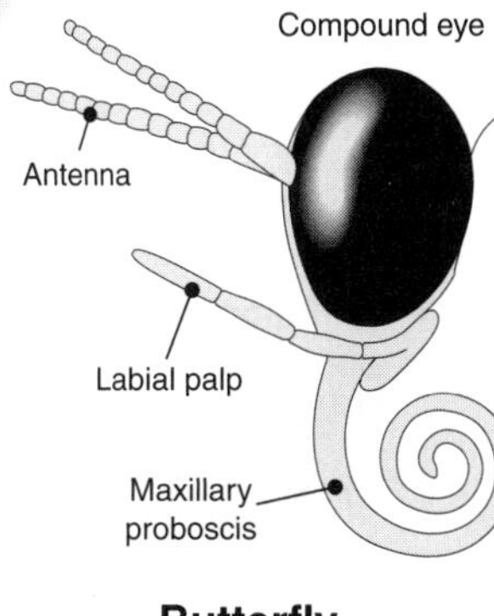

Compound eye
Antenna
Labial palp
Maxillary proboscis

Butterfly

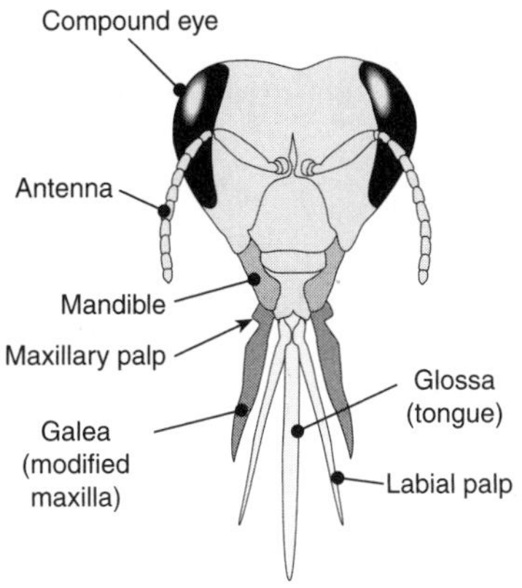

Compound eye
Antenna
Mandible
Maxillary palp
Galea (modified maxilla)
Glossa (tongue)
Labial palp

Honey bee (worker)

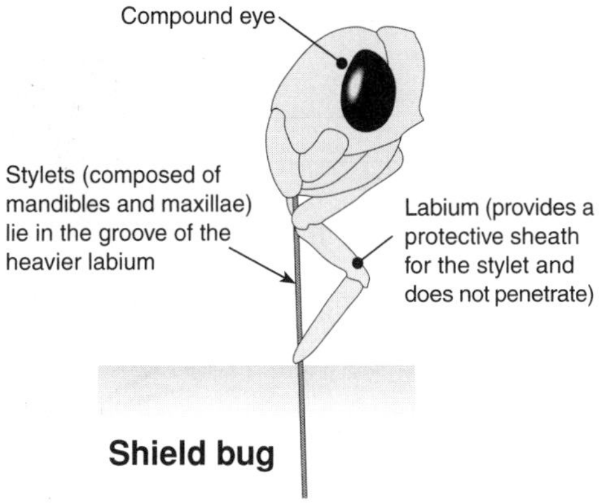

Compound eye
Stylets (composed of mandibles and maxillae) lie in the groove of the heavier labium
Labium (provides a protective sheath for the stylet and does not penetrate)

Shield bug

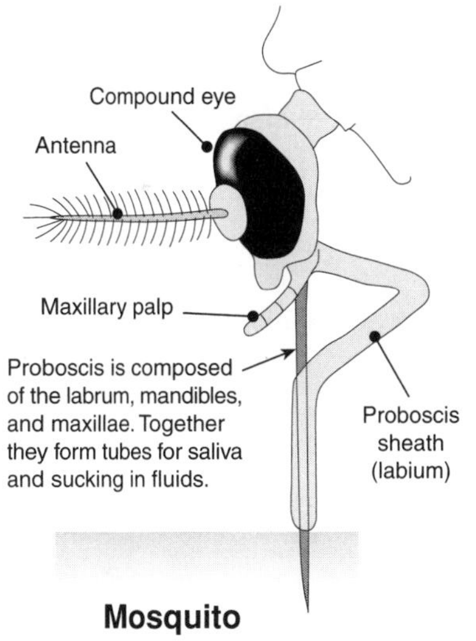

Compound eye
Antenna
Maxillary palp
Proboscis is composed of the labrum, mandibles, and maxillae. Together they form tubes for saliva and sucking in fluids.
Proboscis sheath (labium)

Mosquito

1. Describe the components of an insect's mouthparts: ______________________________

2. Name the four main modes of feeding carried out by the various insect groups: ______________________________

3. Match the following list of insects with their correct mode of feeding: *locust, bee, biting fly, moth, mosquito, beetle, cicada, housefly, flea, dragonfly, aphid, maggot.* There may be more than one example for each feeding mode.

(a) Chewing: ____________________ (e) Sucking: ____________________

(b) Sponging: ____________________ (f) Piercing/sponging: ____________________

(c) Chewing/lapping: ____________________ (g) Piercing/sucking: ____________________

(d) Seizing/chewing: ____________________ (h) Sucking with mouth hooks: ____________________

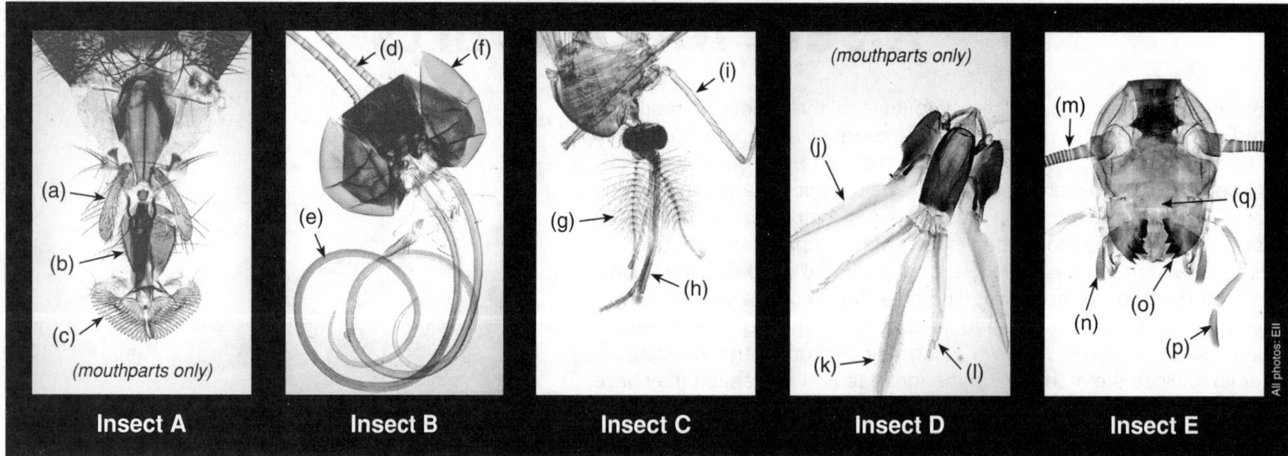

4. For each of the photographs of insects above (**A** - **E**), identify the **type of insect** and the structures labeled (a)-(q). Note that some of the labeled structures are not mouthparts:

Identity of **insect A**: ______________________

(a) ______________ (c) ______________

(b) ______________

Identity of **insect B**: ______________________

(d) ______________ (f) ______________

(e) ______________

Identity of **insect C**: ______________________

(g) ______________ (i) ______________

(h) ______________

Identity of **insect D**: ______________________

(j) ______________ (l) ______________

(k) ______________

Identity of **insect E**: ______________________

(m) ______________ (p) ______________

(n) ______________ (q) ______________

(o) ______________

5. The diagrams on the right illustrate the arrangement of the mouthparts for various insects. Use highlighter pens to create a color key and color in each type of mouthpart.

6. Many insects undergo metamorphosis at certain stages in their life cycle. Butterflies start their active life as caterpillars, after which they pass through a pupal stage, to finally emerge as butterflies. Comment on the diets and changes to the mouthparts of caterpillars and their adult forms (butterflies):

(a) Caterpillar diet: ______________________

Mouthparts: ______________________

____

(b) Butterfly diet: ______________________

Mouthparts: ______________________

____

Cross-Section Through Insect Mouthparts

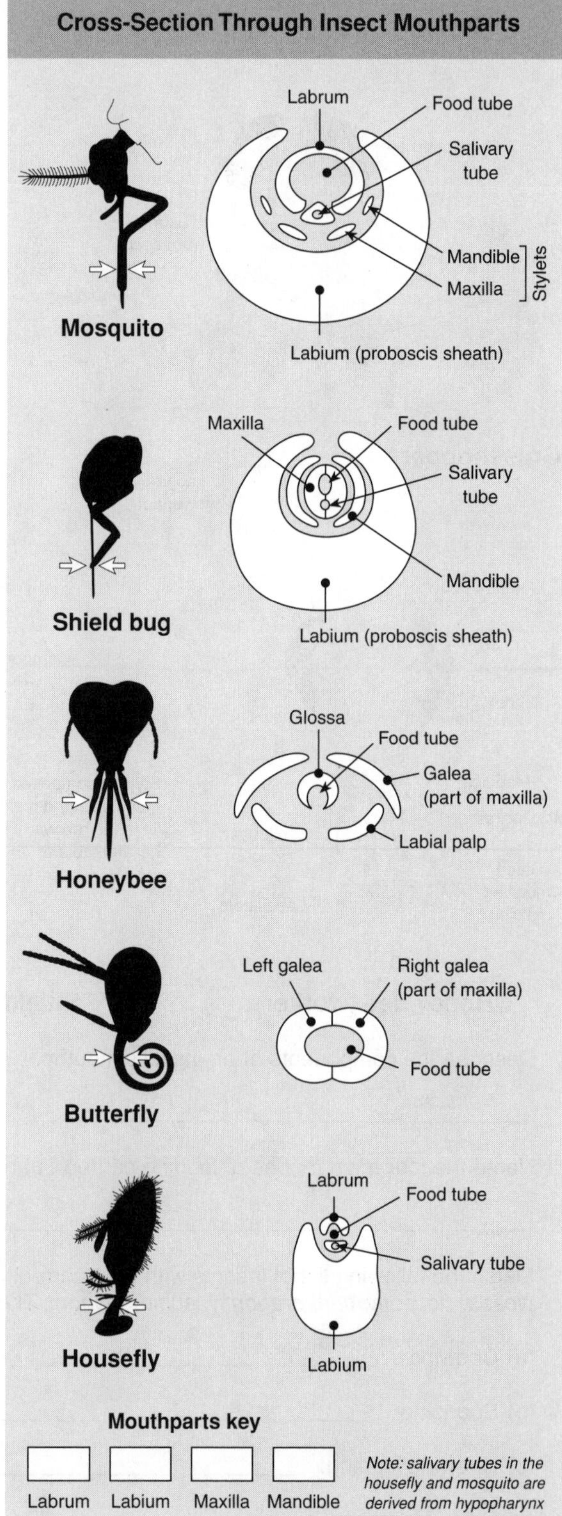

Diversity in Tube Guts

In contrast to the sac-like cavities of cnidarians and flatworms, most animals have digestive tubes running between two openings, a **mouth** and an **anus**. One-way movement of food allows the gut to become regionally specialized for processing food. Although tube guts are relatively uniform in their general structure, with regions for storing, digesting, absorbing, and eliminating the food, various specializations occur depending on the food type and how it is ingested. Usually, food ingested at the mouth and pharynx passes through an esophagus to a crop, gizzard or stomach. In the intestine, digestive enzymes break down the food molecules and nutrients are absorbed across the epithelium of the gut wall. Undigested wastes are **egested** through an anus or cloaca. Examples of gut specialization are described below and elsewhere in this topic.

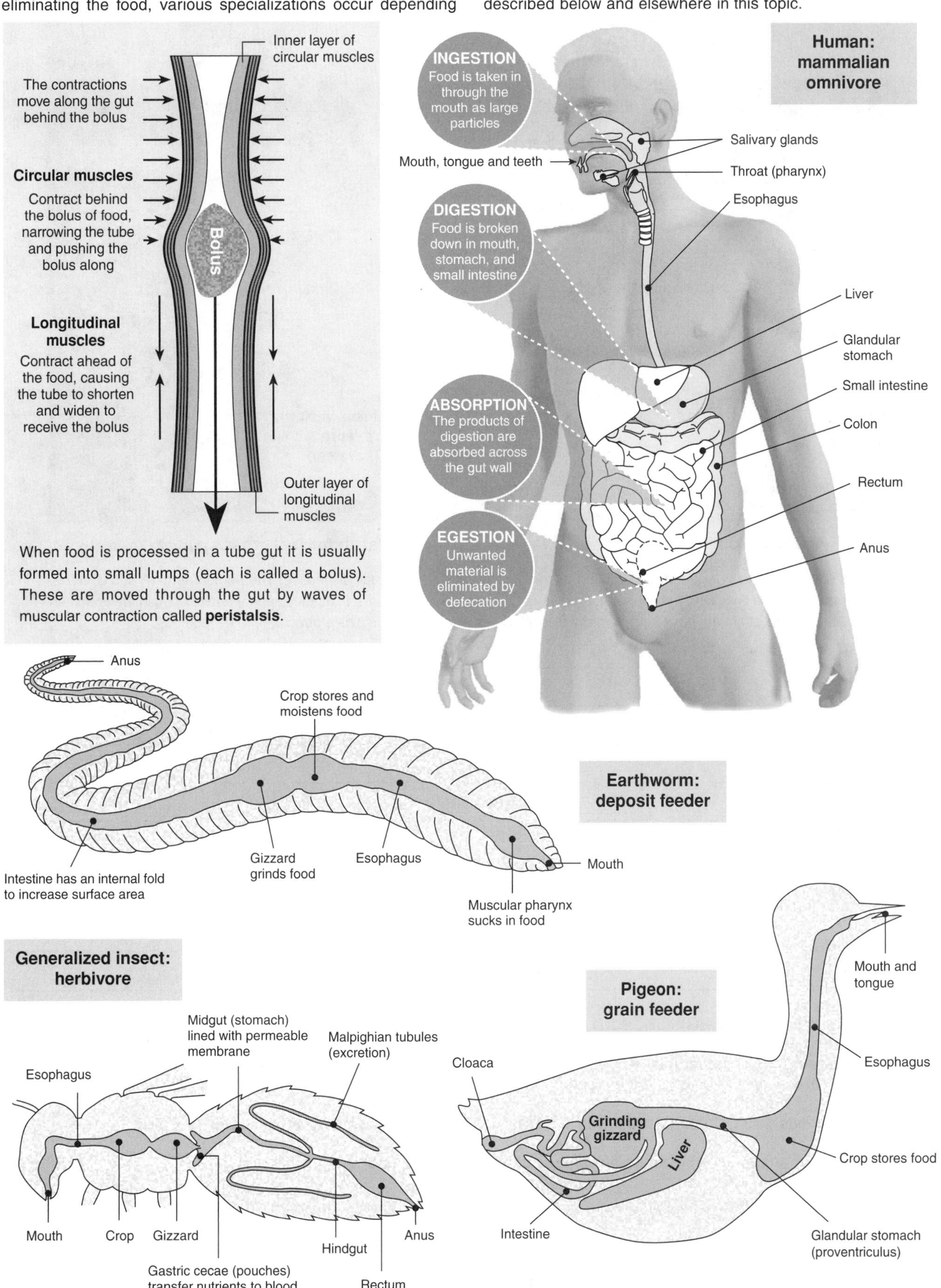

When food is processed in a tube gut it is usually formed into small lumps (each is called a bolus). These are moved through the gut by waves of muscular contraction called **peristalsis**.

***Related activities:** How Heterotrophs Feed, Mammalian Guts, The Human Digestive Tract*

Recognizing Digestive Organs in a Dissection

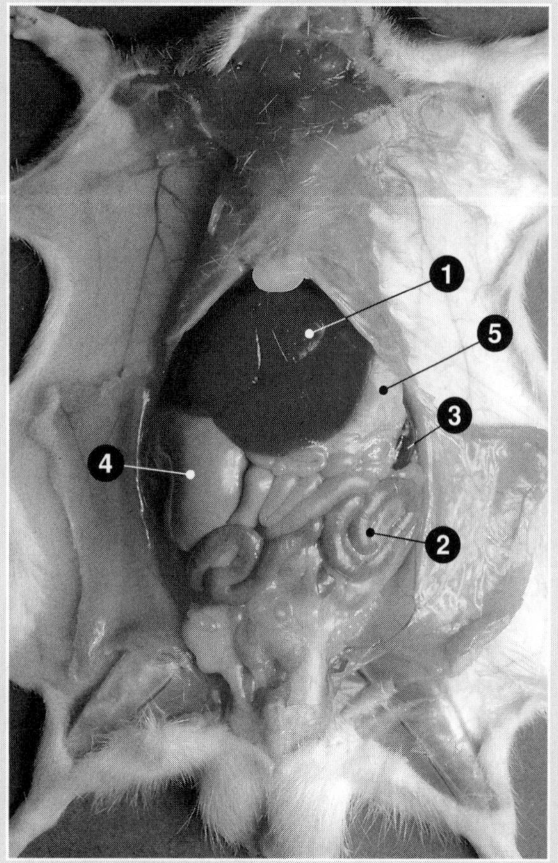

A: Rat abdominal organs *in situ*

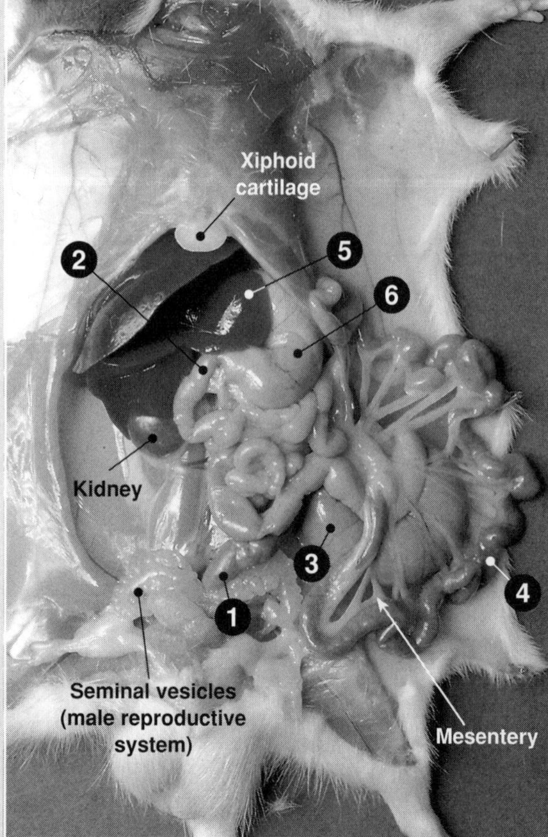

B: Rat abdominal organs *partially dissected*

In your studies of anatomy, you may be required to complete an actual, or virtual, dissection. The photographs (A and B) above show dissection of the body cavity of a laboratory white rat. **A** shows the organs *in situ*, as they appear undisturbed in the abdomen. **B** shows the organs after they have been partially dissected out. The numbers indicate structures to be labeled.

A. 1 ______________________

2 ______________________

3 ______________________

4 ______________________

5 ______________________

B. 1 ______________________

2 ______________________

3 ______________________

4 ______________________

5 ______________________

6 ______________________

1. Some structures have a similar function in different animals. State the general function of the following gut structures:

   (a) Gizzard: ______________________

   (b) Stomach or crop: ______________________

   (c) Intestine (midgut in insects): ______________________

2. (a) In the dissections of the rat (above), label each of the structures indicated in the spaces provided (photo A: 1-5, photo B: 1-6). Some structures are the same, but the same numbers do not necessarily indicate the same structure.

   (b) Of the various guts pictured on the previous page, which one does the rat gut most closely resemble: ____________

   (c) Explain your answer to (b) in terms of the structures present and absent: ______________________

   ____

   ____

   (d) State one reason why you might expect this similarity: ______________________

Mammalian Guts

Among animals, bulky, high fiber diets are harder to digest than diets containing very little plant material. Herbivores therefore tend to have longer guts with larger chambers than carnivores. Grazing mammals are dependent on symbiotic microorganisms to digest plant cellulose for them. This microbial activity may take place in the stomach (foregut fermentation) or the colon and cecum (hindgut fermentation). Some grazers are ruminants; regurgitating and rechewing partially digested food, which is then reswallowed. The diagrams below compare gut structure in representative mammals. Further detail of the adaptations of carnivores and ruminant herbivores is provided on the next page.

Omnivore

Human: *Homo sapiens*

Omnivorous diets can vary enormously and the specific structure of the gut varies accordingly. Some contain a lot of plant material, with animal flesh eaten occasionally. Pigs, bears, and some primates such as chimpanzees, are omnivores of this sort. Other omnivores forage for animal and vegetable foods about equally. The food predominating in the diet at any time will depend on seasonal availability and preference.

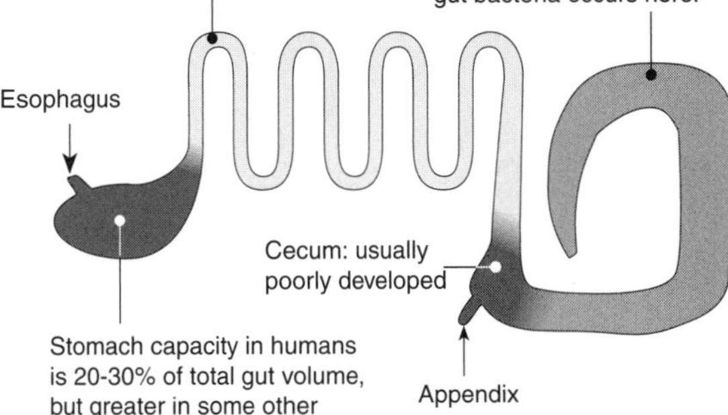

Carnivore

Dog: *Canis familiaris*

The guts of carnivores are adapted for processing animal flesh. The viscera (gut and internal organs) of killed or scavenged animals are eaten as well as the muscle, and provide valuable nutrients. Regions for microbial fermentation are poorly developed or absent. Some animals evolved as carnivores, but have since become secondarily adapted to a more omnivorous diet (bears) or a highly specialized herbivorous diet (pandas). Their guts retain the basic features of a carnivore's gut.

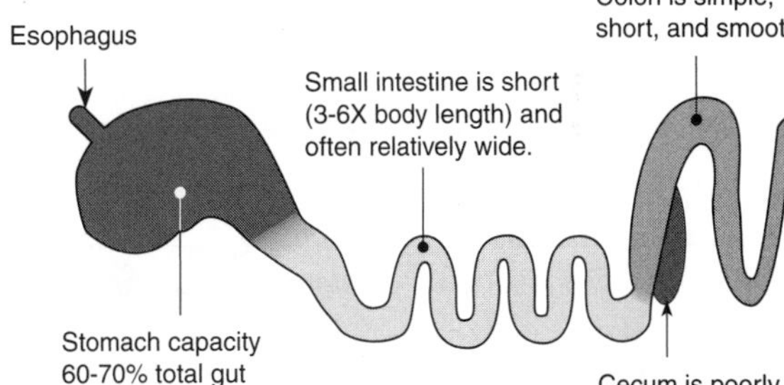

Herbivore: foregut digestion

Cattle: *Bos taurus*

Cattle, sheep, deer, and goats are ruminants. The stomach is divided into a series of large chambers, including a rumen, which contains bacteria and ciliates that digest the plant material in the diet. The division of the stomach into chambers means that the passage of food is slowed and there is time for the microorganisms to act on the plant cellulose. Volatile fatty acids released by the microbes provide energy, and digestion of the microbes themselves provides the ruminant with protein.

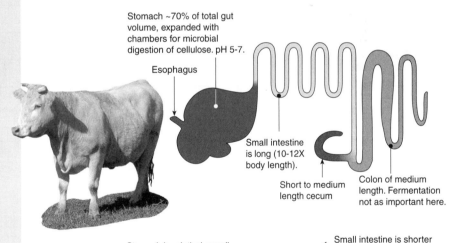

Stomach ~70% of total gut volume, expanded with chambers for microbial digestion of cellulose. pH 5-7.

Esophagus

Small intestine is long (10-12X body length).

Short to medium length cecum

Colon of medium length. Fermentation not as important here.

Herbivore: hindgut digestion

Rabbit: *Oryctolagus cuniculus*

Rabbits are specialized herbivores. The cecum is expanded into a very large chamber for digestion of cellulose. At the junction between the ileum and the colon, indigestible fiber is pushed into the colon where it forms hard feces. Digestible matter passes into the cecum where anaerobic bacteria ferment the material and more absorption takes place. Vitamins and microbial proteins from this fermentation are formed into soft fecal pellets, which pass to the anus and are reingested (coprophagy).

Stomach is relatively small with no extra chambers.

Small intestine is shorter than in foregut fermenters

Esophagus

The cecum is proportionally very large. it is greatly expanded to allow time for the microbial fermentation of plant matter.

The colon is very long and pouched. The special fecal pellets resulting from cecal fermentation are reingested directly from the anus.

Ruminant herbivore: Cattle (*Bos taurus*)

Ruminants are specialized herbivores with teeth adapted for chewing and grinding. Their nutrition is dependent on their mutualistic relationship with their microbial gut flora (bacteria and ciliates), which digest plant material and provide the ruminant with energy and protein. In return, the rumen provides the microbes with a warm, oxygen free, nutrient rich environment.

Dental adaptations

Small temporalis muscle

Horny pad

Incisors

Canine

Masseter muscle is very large to assist in chewing.

Premolars

Diastema (toothless space)

Molars are large for grinding

Flow of food in the stomach

Regurgitation, rechewing, and reswallowing

Omasum: removes water

Passage of food

Abomasum: true stomach secretes gastric juices.

Reticulum: forms the cud which is returned to the mouth.

Rumen: contains bacteria for the digestion of cellulose and ciliates to digest starch.

Carnivore: Lion (*Panthera leo*)

The teeth and guts of carnivores are superbly adapted for eating animal flesh. The canine and incisor teeth are specialized to bite down and cut, while the carnassials are enlarged, lengthened, and positioned to act as shears to slice through flesh. As meat is easier to digest than cellulose, the guts of carnivores are comparatively more uniform and shorter than those of herbivores.

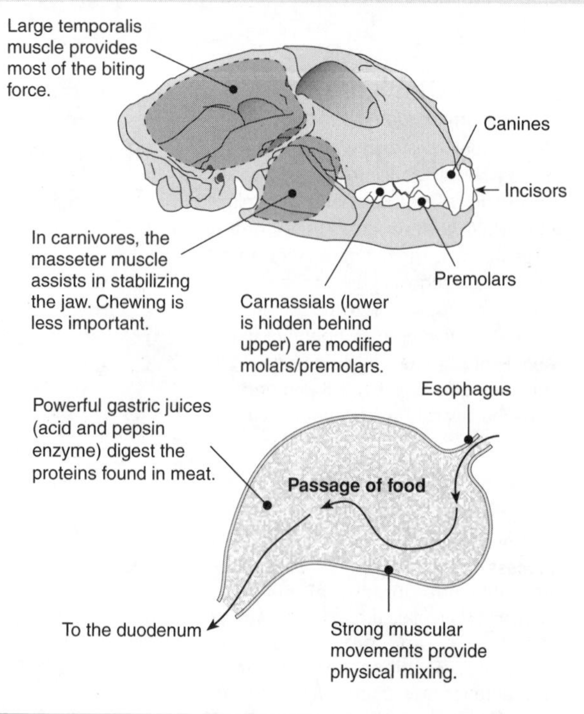

Large temporalis muscle provides most of the biting force.
Canines
Incisors
In carnivores, the masseter muscle assists in stabilizing the jaw. Chewing is less important.
Premolars
Carnassials (lower is hidden behind upper) are modified molars/premolars.
Esophagus
Powerful gastric juices (acid and pepsin enzyme) digest the proteins found in meat.
Passage of food
To the duodenum
Strong muscular movements provide physical mixing.

1. For each of the following, summarize the **structural** differences between the guts of a carnivore and a named herbivore:

   (a) Size of stomach (relative to body size): ______

   (b) Relative length of small intestine: ______

   (c) Development of hind gut (cecum and colon): ______

2. (a) Explain the role of microbial fermentation in the nutrition of foregut fermenting herbivores: ______

   (b) Describe a herbivorous diet that is less reliant on microbial fermentation: ______

3. Contrast the pH of the stomach contents in carnivores, omnivores, and herbivores, and explain the differences:

   ______

4. Identify and explain a structural difference between carnivores and ruminant herbivores with respect to:

   (a) The teeth: ______

   (b) The jaw musculature: ______

Systems for Digestion

During digestion, food is changed by physical and chemical means from its original state until its constituents are released as small, simple molecules that can be absorbed and assimilated. The content of animal diets is tremendously variable and the ways in which animals have evolved to process their food is similarly varied. Some foods, like egg and honey, are pure, concentrated nutriment. Other diets are of high nutritional value, but contain large volumes of water (e.g. blood) or are bulky and take a long time to digest (whole prey items). Other diets are not only bulky, but are also of low nutritional value (e.g. vegetation). Some adaptations for dealing with particular, specialized diets are explained below.

Diet and Problems

BLOOD

Blood is a high protein, low bulk fluid. The problems with processing it include:

- Coagulation of blood and blockage of mouthparts during ingestion.
- Storage of large quantity of fluid.
- Slowing passage of bulk food through the gut so that it can be digested.

PLANT SAP AND NECTAR

Plant sap is high volume sugary fluid. The problems with processing it include:

- Eliminating large volumes of water and obtaining sufficient protein and vitamins.
- Storage of a large quantity of fluid.
- Enzymes are diluted by the large volumes of fluid ingested.

SOIL AND DETRITUS

Soil is a nutritive material containing both organic and inorganic solids. The problems with processing it include:

- Large amount of material containing little nutrient mixed with nutritive matter.
- Large hard particles that must be broken down before digestion.
- Soil material contains large amounts of calcium ions which must be removed.

TOXIC PLANTS (EUCALYPT)

Problems with this diet are:

- The leaves contain toxins in the oils, waxes, and resins of the leaf tissue, e.g. hydrocyanic acid.
- It is low in both energy and quality (low protein).
- The diet is bulky and high volume.
- *Note: some insects, such as monarch butterflies, eat toxic plants and use the toxins for defence.*

Adaptations to Diet

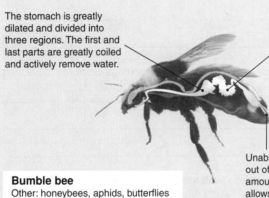

Piercing mouthparts inject powerful anticoagulants to keep the blood flowing.

The midgut ('stomach') is divided into three sequential regions that absorb water to concentrate the blood, secrete protease enzymes for digestion, and absorb nutrients.

Enlarged crop stores the blood, releasing it slowly in smaller amounts into the midgut.

Mosquito
Other: leeches, spiders, ticks

The stomach is greatly dilated and divided into three regions. The first and last parts are greatly coiled and actively remove water.

The mid region of the stomach is specialized for secretion of carbohydrase enzymes and absorption. It receives the sap only after most of the water is removed.

Unabsorbed sugars can be passed out of the hindgut in copious amounts known as 'honeydew'. This allows them to absorb enough food to meet protein requirements.

Bumble bee
Other: honeybees, aphids, butterflies

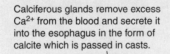

Calciferous glands remove excess Ca^{2+} from the blood and secrete it into the esophagus in the form of calcite which is passed in casts.

Crop

A muscular pharynx pushes large amount of soil into the crop for storage.

The gizzard located behind the crop grinds up larger particles. There may be several gizzards, one after the other.

Earthworm
Other: Various insect larvae, millipedes

Koalas are fastidious feeders and select only certain age leaves of certain species.

They meet their protein needs by eating large volumes.

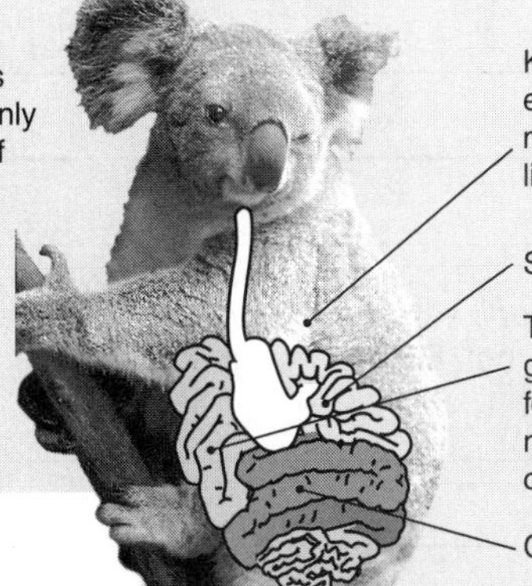

Koalas are able to detoxify eucalypt poisons by forming nontoxic compounds in the liver, which are then excreted.

Small intestine

The caecum of the hindgut is greatly expanded to form a fermentation chamber containing microorganisms which digest the cellulose of the plant material.

Colon

Koala

Related activities: *Diversity in Tube Guts, Mammalian Guts*
Web links: *Digestion Animation, Acid Secretion in the Stomach*

LARGE, IRREGULAR MEALS (MEAT)

Problems with this diet are:

- Long periods of time between meals.
- Prey may be extremely large in comparison to the snake.
- A large bulk of animal tissue, unchewed, takes a long time to break down.

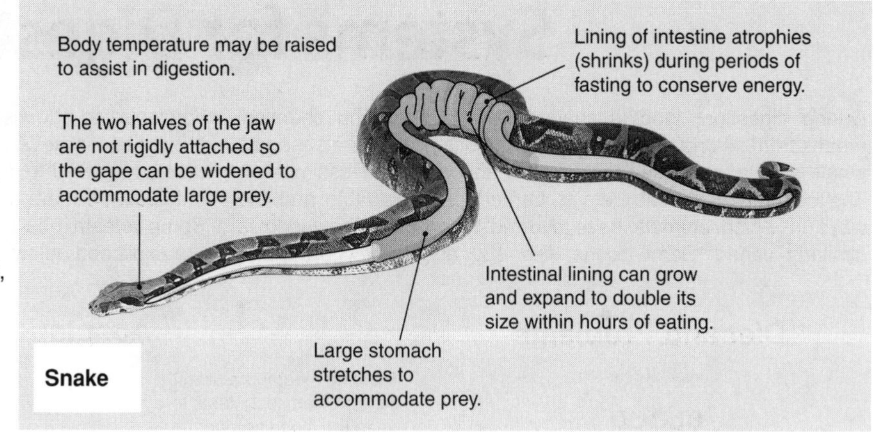

Body temperature may be raised to assist in digestion.
The two halves of the jaw are not rigidly attached so the gape can be widened to accommodate large prey.
Lining of intestine atrophies (shrinks) during periods of fasting to conserve energy.
Intestinal lining can grow and expand to double its size within hours of eating.
Large stomach stretches to accommodate prey.
Snake

1. Describe one common feature of the guts of fluid feeders: ________________

2. Explain why it is so important for fluid feeders to reduce the volume of their ingested food by absorbing water:

3. Plant sap and nectar are low in protein. Explain how sap sucking insects are able to obtain the protein they need:

4. Explain how earthworms deal with the problem of excess calcium ions absorbed from their soil diet:

5. Discuss the advantages and disadvantages of a diet consisting of a toxic plant: ________________

6. Explain how koalas (and other vertebrate herbivores) are able to digest cellulose, an extremely stable compound:

7. Discuss how snakes have developed to deal with the problem of digesting large prey items on an irregular basis:

8. Snakes are (for the most part) **ectothermic**. Although some can raise their body temperature through metabolic activity for short periods, most can not. Explain how a snake could raise its body temperature to assist in digesting a large meal:

Absorbing Nutrients

All chemical and physical digestion is aimed at the breakdown of food molecules into forms that can be absorbed across a gut lining. Absorption of the simple components of food (e.g. simple sugars, amino acids, and fatty acids) must occur before the nutrients can be **assimilated** (taken up by all the body's cells). In cnidarians, specialized cells lining the gut ingest partly digested food particles by phagocytosis. In animals with a tubular gut, the inner surface area of the gut is increased by various means (described below) to maximize the **absorption** of nutrients as the digested food passes through. After absorption, nutrients must be transported to where they are required. In vertebrates, this is facilitated by the structure of the intestinal villi. Some nutrients are absorbed directly into the bloodstream, while others are transported in the lacteals of the lymphatic system (overleaf).

Cnidarian Gastrovascular Cavity

EXAMPLE: *Hydra*

In *Hydra*, specialized cells line the gastrovascular cavity. Some of these secrete enzymes into the cavity to begin digestion. Special nutritive cells (illustrated) take in the partly digested fragments by phagocytosis, to form food vacuoles where digestion is completed. These cells have beating hair-like flagella that create currents and improve the delivery of food to the cells.

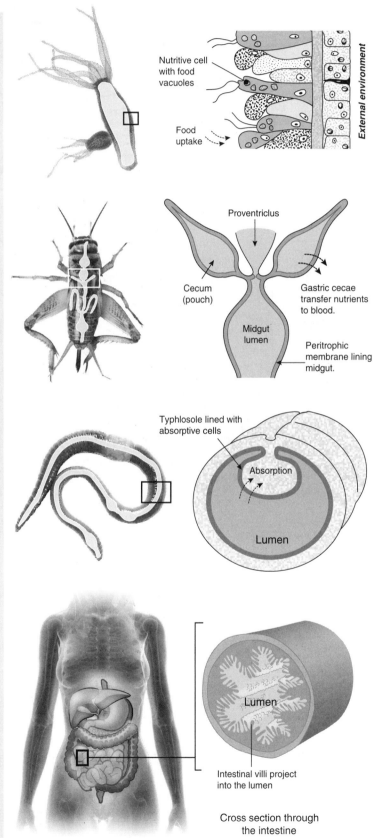

Nutritive cell with food vacuoles

Food uptake

External environment

Insect Gastric Cecae

EXAMPLE: Grasshopper or locust

In insects of the grasshopper family (and others), the gastric cecae are midgut pouches just behind the proventriculus. The cecae improve absorption by transferring nutrients into the blood. Secretion of enzymes and absorption of nutrients occurs in the midgut. Unlike the fore- and hindgut (which are lined with chitin), the midgut is lined with a permeable peritrophic membrane which allows nutrient absorption.

Proventriclus

Cecum (pouch)

Gastric cecae transfer nutrients to blood.

Midgut lumen

Peritrophic membrane lining midgut.

Annelid Typhlosole

EXAMPLE: *Lumbricus* (earthworm)

In earthworms, the entire length of the small intestine is folded into a structure called the typhlosole. Secretion of enzymes, digestion and absorption all occur in the intestine and the typhlosole increases the amount of surface area for absorption of nutrients. Not all annelids have a typhlosole, although many have similar foldings to increase surface area.

Typhlosole lined with absorptive cells

Absorption

Lumen

Mammalian intestine

EXAMPLE: Human

In mammals, the gut is broadly divided into the stomach and the intestines and associated glands. There may also be a cecum between the large and small intestines. The stomach mechanically and chemically breaks down the food before it passes into the small intestine. The wall of the intestine is folded into microscopic finger-like structures called villi (below and right), which increase the surface area for absorption of nutrients into the blood. After most of the nutrients ave been absorbed, the semi-solid waste passes to the large intestine (colon) where water and some ions and vitamins are absorbed.

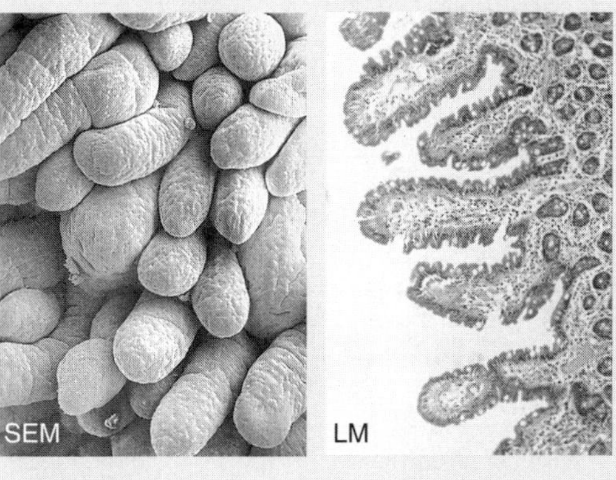

Lumen

Intestinal villi project into the lumen

Cross section through the intestine

Related activities: *Systems for Digestion, the Human Digestive Tract*
Web links: *Digestion Animation*

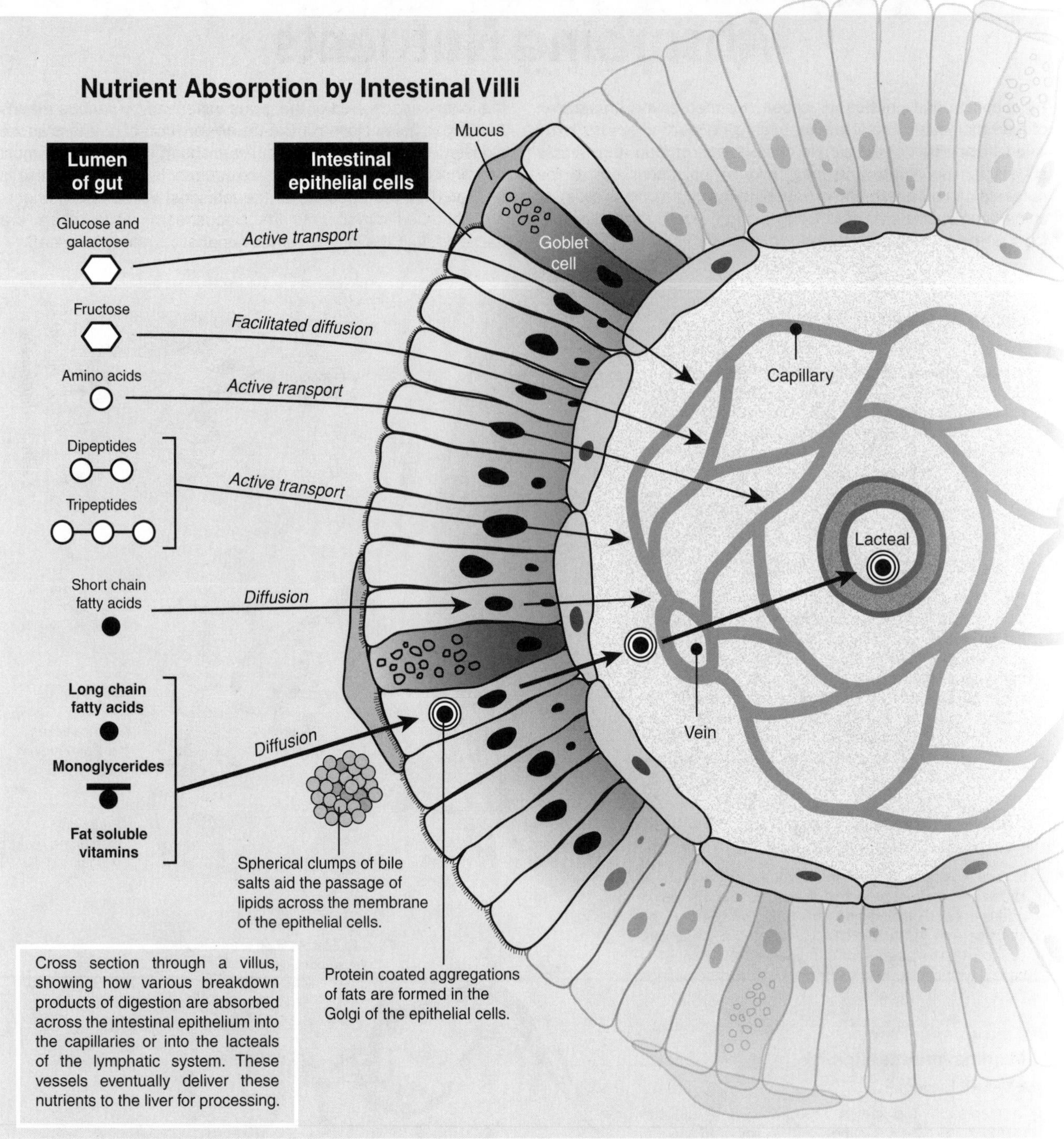

Cross section through a villus, showing how various breakdown products of digestion are absorbed across the intestinal epithelium into the capillaries or into the lacteals of the lymphatic system. These vessels eventually deliver these nutrients to the liver for processing.

1. Describe the advantage of a tubular gut over a gastrovascular cavity (as in *Hydra*): ______

2. Describe how each of the following nutrients is absorbed by the intestinal villi in mammals:

(a) Glucose and galactose: ______

(b) Fructose: ______

(c) Amino acids: ______

(d) Dipeptides: ______

(e) Tripeptides: ______

(f) Short chain fatty acids: ______

(g) Monoglycerides: ______

(h) Fat soluble vitamins: ______

3. Discuss adaptations for increasing the surface area of the absorptive surface of the gut and the advantages of this:

The Mouth and Pharynx

The mouth (**oral cavity**), is formed by the cheeks, hard and soft palate, and tongue (below right). The teeth are very hard structures, specialized for chewing food (**mastication**). The tongue, which moves food around, and the salivary glands, which produce saliva, act with the teeth to begin digestion. The structure of a tooth is illustrated by a section through a molar (below, right). The oral cavity is divided into quadrants and the number of teeth in each quadrant given by a **dental formula**. There are 32 adult (permanent) teeth, organized as shown below left. The basic structure of a tooth is described in the inset below.

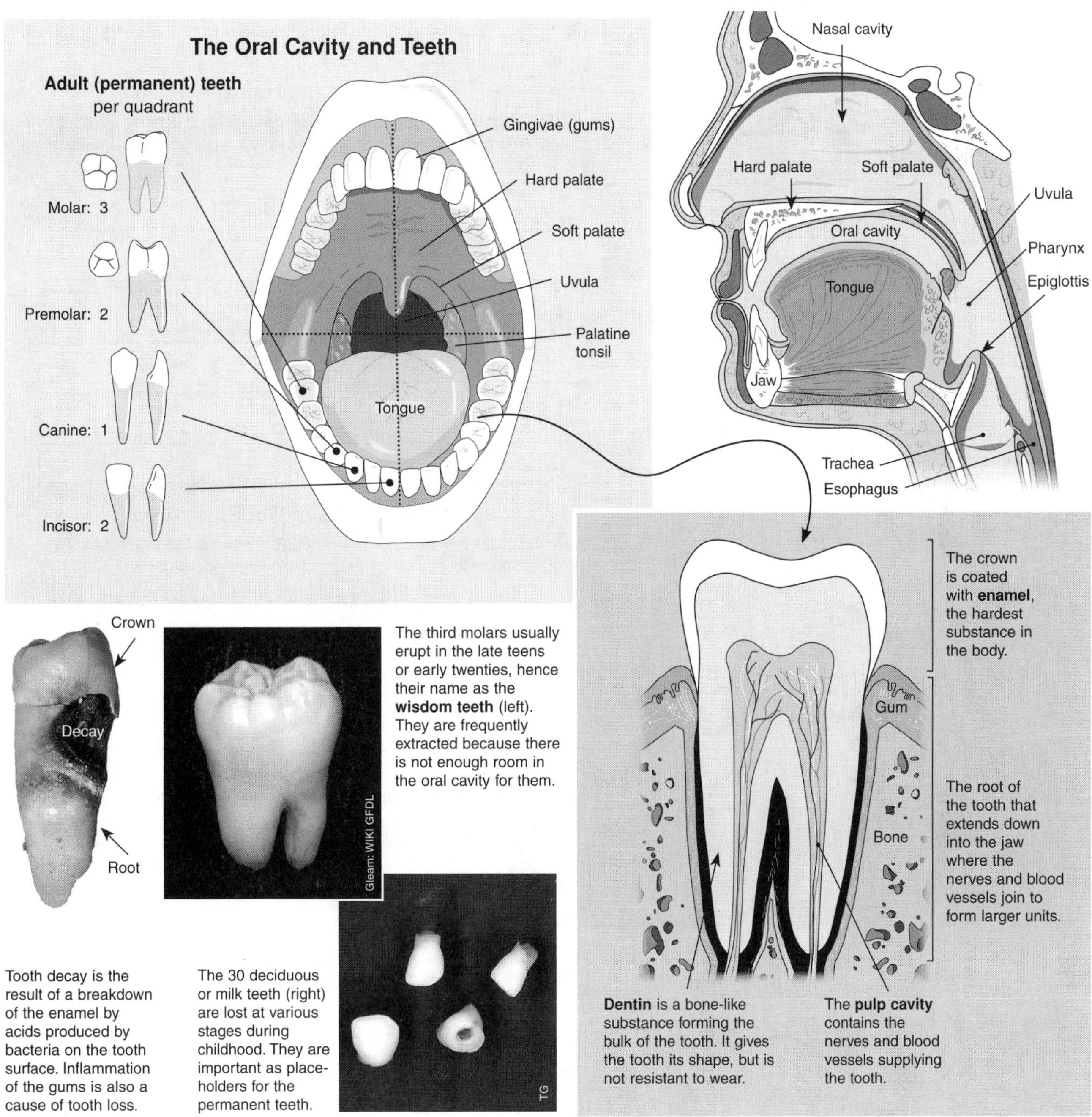

The Oral Cavity and Teeth
Adult (permanent) teeth per quadrant
Molar: 3
Premolar: 2
Canine: 1
Incisor: 2
Gingivae (gums)
Hard palate
Soft palate
Uvula
Palatine tonsil
Tongue
Nasal cavity
Hard palate
Soft palate
Uvula
Oral cavity
Pharynx
Epiglottis
Tongue
Jaw
Trachea
Esophagus
The crown is coated with **enamel**, the hardest substance in the body.
Gum
Bone
The root of the tooth that extends down into the jaw where the nerves and blood vessels join to form larger units.
Dentin is a bone-like substance forming the bulk of the tooth. It gives the tooth its shape, but is not resistant to wear.
The **pulp cavity** contains the nerves and blood vessels supplying the tooth.
Crown
Decay
Root
Gleam: WIKI GFDL
TG

The third molars usually erupt in the late teens or early twenties, hence their name as the **wisdom teeth** (left). They are frequently extracted because there is not enough room in the oral cavity for them.

Tooth decay is the result of a breakdown of the enamel by acids produced by bacteria on the tooth surface. Inflammation of the gums is also a cause of tooth loss.

The 30 deciduous or milk teeth (right) are lost at various stages during childhood. They are important as place-holders for the permanent teeth.

1. Describe two major roles of the **oral cavity** and its associated structures in digestion:

(a) ______

(b) ______

2. Based on its position projecting up behind the tongue and guarding the tracheal entrance, infer the role of the epiglottis:

______

3. Explain the protective value of having tonsils at the oral entrance to the pharynx: ______

Related activities: *Dentition in Mammals, Human Digestive Tract*

The Human Digestive Tract

An adult consumes an estimated metric tonne of food a year. Food provides the source of the energy required to maintain **metabolism**. The human digestive tract, like most tube-like guts, is regionally specialized to maximize the efficiency of physical and chemical breakdown (**digestion**), **absorption**, and **elimination**. The gut is essentially a hollow, open-ended, muscular tube, and the food within it is essentially outside the body, having contact only with the cells lining the tract. Food is physically moved through the gut tube by waves of muscular contraction, and subjected to chemical breakdown by enzymes contained within digestive secretions. The products of this breakdown are then absorbed across the gut wall. A number of organs are associated with the gut along its length and contribute, through their secretions, to the digestive process at various stages.

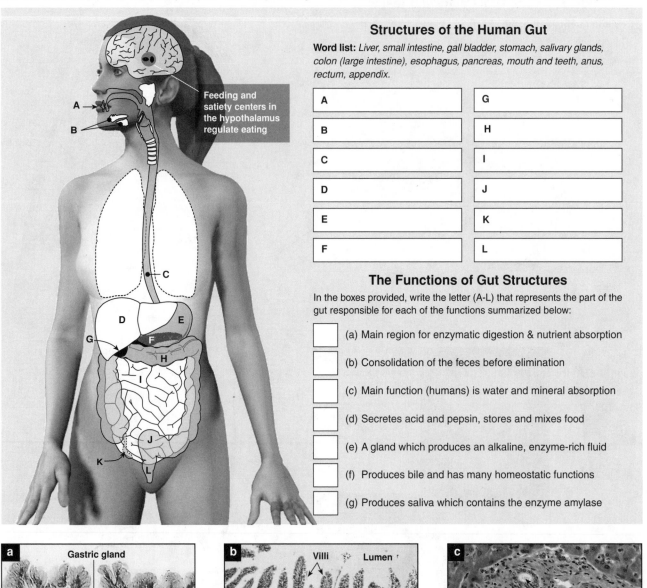

Feeding and satiety centers in the hypothalamus regulate eating

A B C D E F G H I J K L

Structures of the Human Gut

Word list: *Liver, small intestine, gall bladder, stomach, salivary glands, colon (large intestine), esophagus, pancreas, mouth and teeth, anus, rectum, appendix.*

A	G
B	H
C	I
D	J
E	K
F	L

The Functions of Gut Structures

In the boxes provided, write the letter (A-L) that represents the part of the gut responsible for each of the functions summarized below:

(a) Main region for enzymatic digestion & nutrient absorption

(b) Consolidation of the feces before elimination

(c) Main function (humans) is water and mineral absorption

(d) Secretes acid and pepsin, stores and mixes food

(e) A gland which produces an alkaline, enzyme-rich fluid

(f) Produces bile and has many homeostatic functions

(g) Produces saliva which contains the enzyme amylase

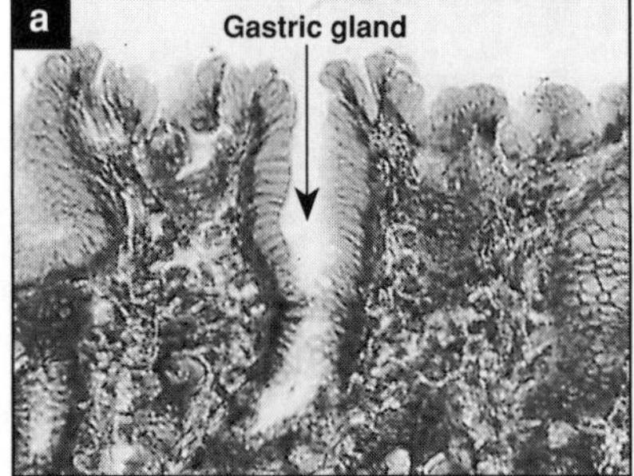

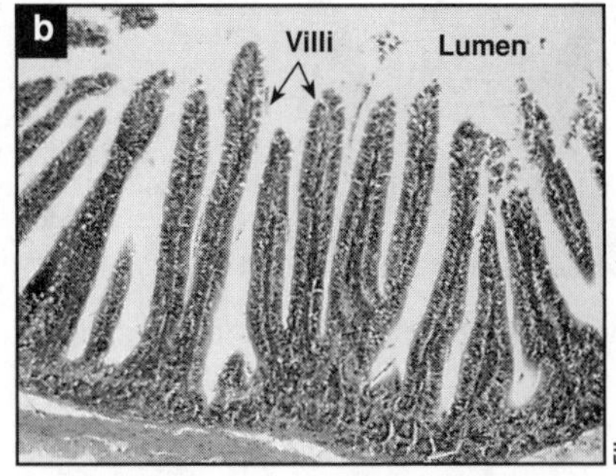

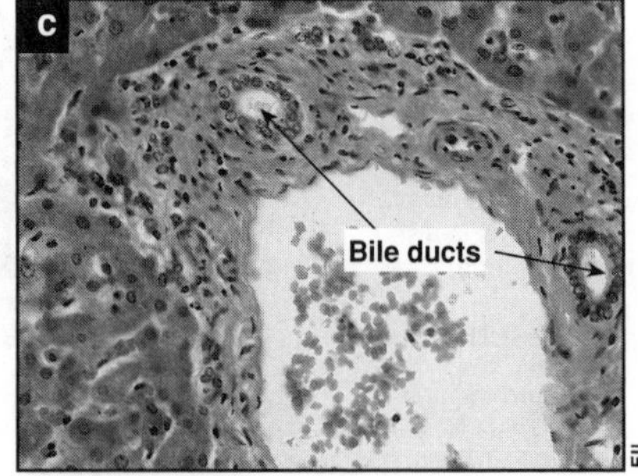

1. In the spaces provided on the diagram above, identify the parts labeled **A-L** (choose from the word list provided). Match each of the **functions** described (a)-(g) with the letter representing the corresponding structure on the diagram.

2. On the same diagram, mark with lines and labels: anal sphincter (**AS**), pyloric sphincter (**PS**), cardiac sphincter (**CS**).

3. Identify the region of the gut illustrated by the photographs (a)-(c) above. For each one, explain the identifying features:

(a) ______________________________

(b) ______________________________

(c) ______________________________

The Stomach, Duodenum, and Pancreas

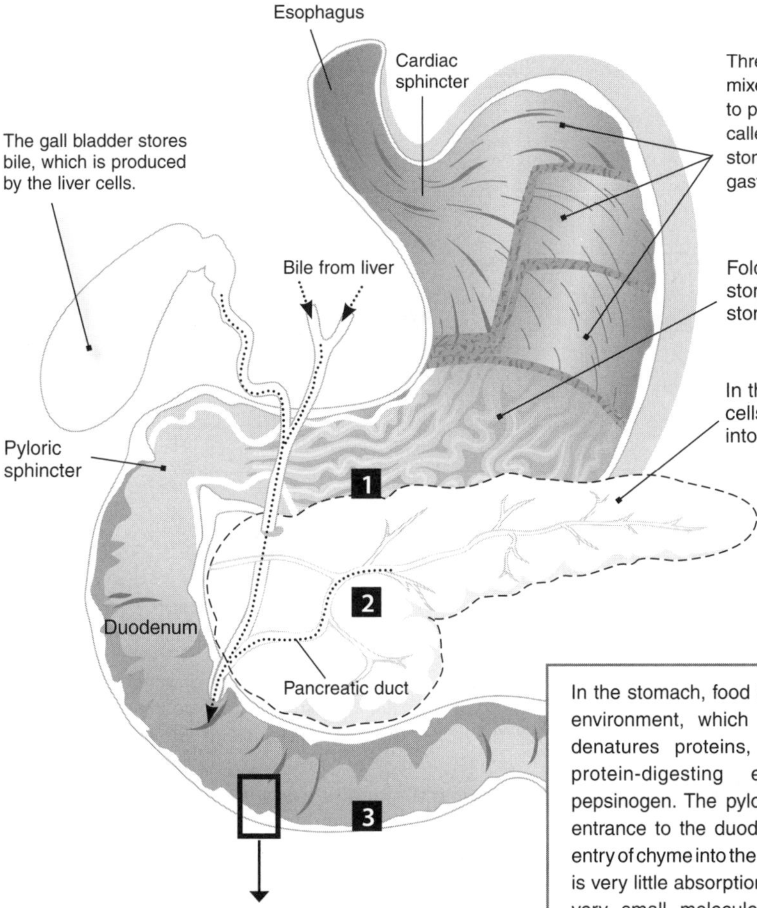

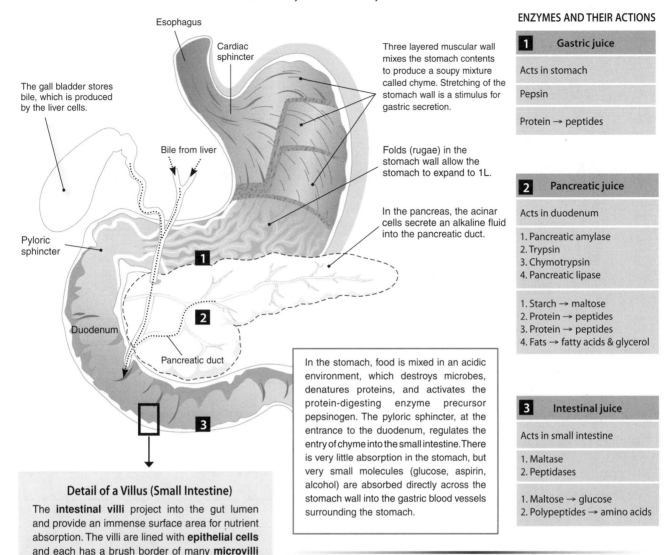

Esophagus

Cardiac sphincter

The gall bladder stores bile, which is produced by the liver cells.

Bile from liver

Three layered muscular wall mixes the stomach contents to produce a soupy mixture called chyme. Stretching of the stomach wall is a stimulus for gastric secretion.

Folds (rugae) in the stomach wall allow the stomach to expand to 1L.

In the pancreas, the acinar cells secrete an alkaline fluid into the pancreatic duct.

Pyloric sphincter

Duodenum

Pancreatic duct

1

2

3

ENZYMES AND THEIR ACTIONS

1 Gastric juice
Acts in stomach
Pepsin
Protein → peptides

2 Pancreatic juice
Acts in duodenum
1. Pancreatic amylase 2. Trypsin 3. Chymotrypsin 4. Pancreatic lipase
1. Starch → maltose 2. Protein → peptides 3. Protein → peptides 4. Fats → fatty acids & glycerol

In the stomach, food is mixed in an acidic environment, which destroys microbes, denatures proteins, and activates the protein-digesting enzyme precursor pepsinogen. The pyloric sphincter, at the entrance to the duodenum, regulates the entry of chyme into the small intestine. There is very little absorption in the stomach, but very small molecules (glucose, aspirin, alcohol) are absorbed directly across the stomach wall into the gastric blood vessels surrounding the stomach.

3 Intestinal juice
Acts in small intestine
1. Maltase 2. Peptidases
1. Maltose → glucose 2. Polypeptides → amino acids

Detail of a Villus (Small Intestine)

The **intestinal villi** project into the gut lumen and provide an immense surface area for nutrient absorption. The villi are lined with **epithelial cells** and each has a brush border of many **microvilli** which further increase the surface area.

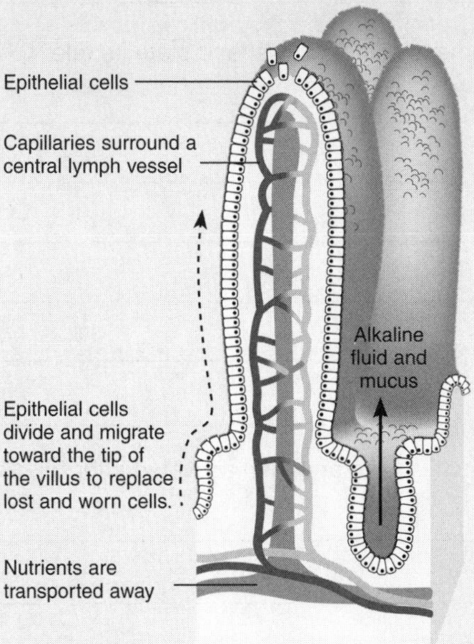

Epithelial cells

Capillaries surround a central lymph vessel

Alkaline fluid and mucus

Epithelial cells divide and migrate toward the tip of the villus to replace lost and worn cells.

Nutrients are transported away

Enzymes bound to the surfaces of the epithelial cells break down peptides and carbohydrate molecules. The breakdown products are then absorbed into the underlying blood and lymph vessels. Tubular exocrine glands and goblet cells secrete alkaline fluid and mucus into the lumen.

Detail of a Gastric Gland (Stomach Wall)

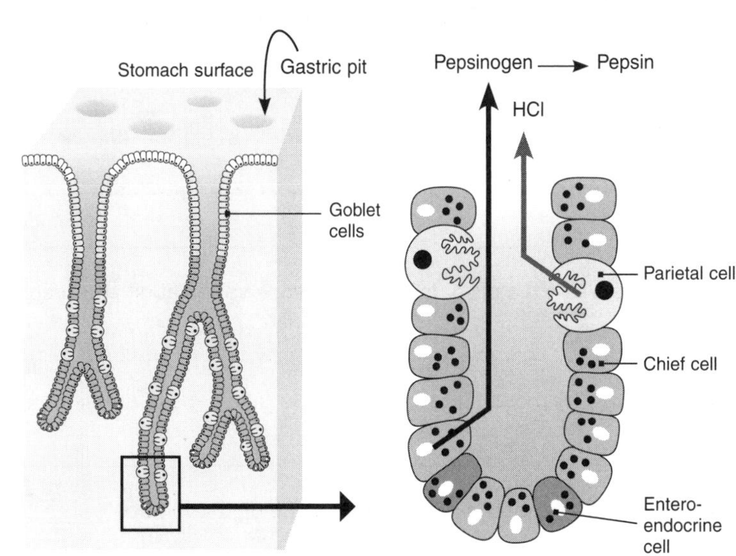

Gastric secretions are produced by **gastric glands**, which pit the lining of the stomach. Chief cells in the gland secrete pepsinogen, a precursor of the enzyme pepsin. Parietal cells produce hydrochloric acid, which activates the pepsinogen. Goblet cells at the neck of the gastric gland secrete mucus to protect the stomach mucosa from the acid. Enteroendocrine cells in the gastric gland secrete the hormone gastrin which acts on the stomach to increase gastric secretion.

The Large Intestine

After most of the nutrients have been absorbed in the small intestine, the remaining fluid contents pass into the large intestine (appendix, cecum, and colon). The fluid comprises undigested or undigestible food, bacteria, dead cells, mucus, bile, ions, and water. In humans and other omnivores, the large intestine's main role is to reabsorb water and electrolytes.

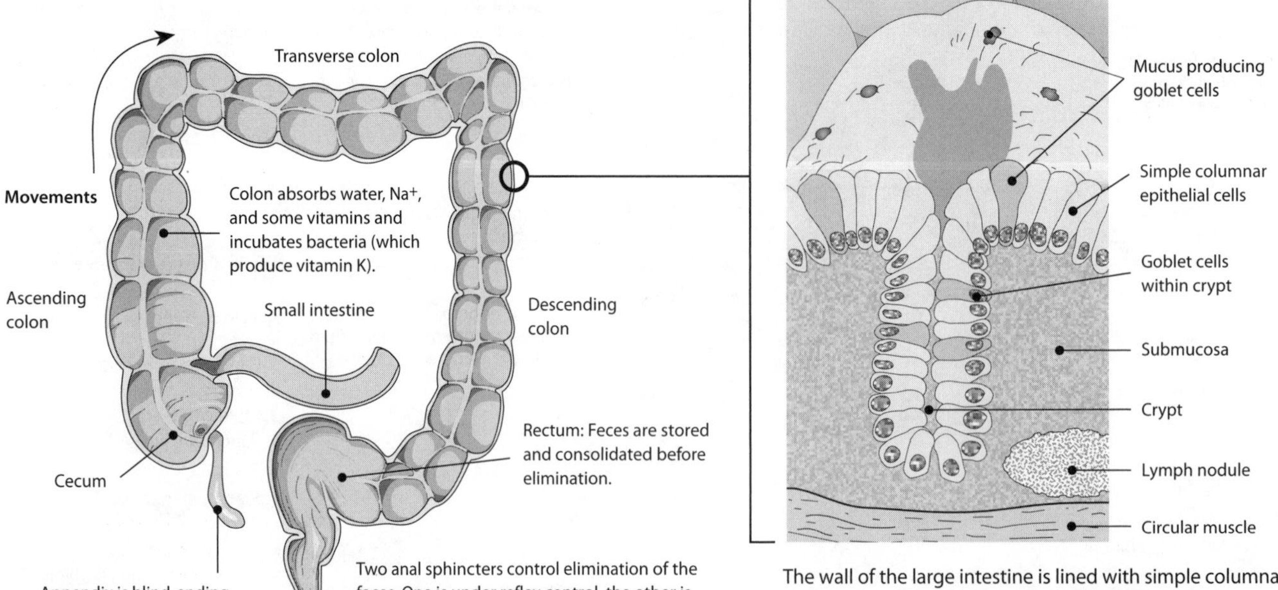

Transverse colon
Movements
Colon absorbs water, Na+, and some vitamins and incubates bacteria (which produce vitamin K).
Ascending colon
Small intestine
Descending colon
Rectum: Feces are stored and consolidated before elimination.
Cecum
Appendix is blind-ending sac that may have a minor immune system function.
Two anal sphincters control elimination of the feces. One is under reflex control, the other is under conscious control and allows the reflex activity to be modified.
Mucus producing goblet cells
Simple columnar epithelial cells
Goblet cells within crypt
Submucosa
Crypt
Lymph nodule
Circular muscle

The wall of the large intestine is lined with simple columnar epithelium. The epithelium is contains tubular glands called crypts containing mucus-secreting goblet cells. The mucus lubricates the colon wall and aids formation of the feces.

4. Summarize the structural and functional specializations in each of the following regions of the gut:

   (a) Stomach: ______________________________

   (b) Small intestine: ______________________________

   (c) Large intestine: ______________________________

5. Identify two sites for enzyme secretion in the gut, give an example of an enzyme produced there, and state its role:

   (a) Site: ____________ Enzyme: ____________

   Enzyme's role: ______________________________

   (b) Site: ____________ Enzyme: ____________

   Enzyme's role: ______________________________

6. (a) Suggest why the pH of the gut secretions varies at different regions in the gut: ______________________________

   (b) Explain why protein-digesting enzymes (e.g. pepsin) are secreted in an inactive form and then activated after release:

   ____

7. (a) Describe how food is moved through the digestive tract: ______________________________

   (b) Explain how the passage of food through the tract is regulated: ______________________________

8. (a) Predict the consequence of food moving too rapidly through the gut: ______________________________

   (b) Predict the consequence of food moving too slowly through the gut: ______________________________

The Digestive Role of the Liver

The liver is a large organ, weighing about 1.4 kg, and is well supplied with blood. It carries out several hundred different functions and has a pivotal role in the maintenance of homeostasis. Its role in the digestion of food centers around the production of the alkaline fluid, **bile**, which is secreted at a rate of 0.8-1.0 liter per day. It is also responsible for processing absorbed nutrients, which arrive at the liver via the hepatic portal system. These functions are summarized below.

Digestive Functions of the Liver

The digestive role of the liver is in the production of **bile**. Bile is a yellow, brown, or olive-green alkaline fluid (pH 7.6–8.6), consisting of water and bile salts, cholesterol, lecithin, bile pigments, and several ions. The bile salts are used in the small intestine to break up (**emulsify**) fatty molecules for easier digestion and absorption.

The high pH neutralizes the acid entering the small intestine from the stomach. Bile is also partly an excretory product; the breakdown of red blood cells in the liver produces the principal bile pigment, **bilirubin**. Bacteria act on the bile pigments, giving the brownish color to feces. The production and secretion of bile is regulated through nervous and hormonal mechanisms. Hormones (secretin and cholecystokinin) are released into the blood from the intestinal mucosa in response to the presence of food (especially fat) in the small intestine.

Liver Tissue

The liver tissue is made up of many lobules, each one comprising cords of liver cells (hepatocytes), radiating from a central vein (CV), and surrounded by branches of the hepatic artery, hepatic portal vein, and bile ductule. Bile is produced by the individual liver cells, which secrete it into canaliculi that empty into small bile ducts. The hepatocytes also process the nutrients entering the liver via the hepatic portal system.

Internal Gross Structure of the Human Liver

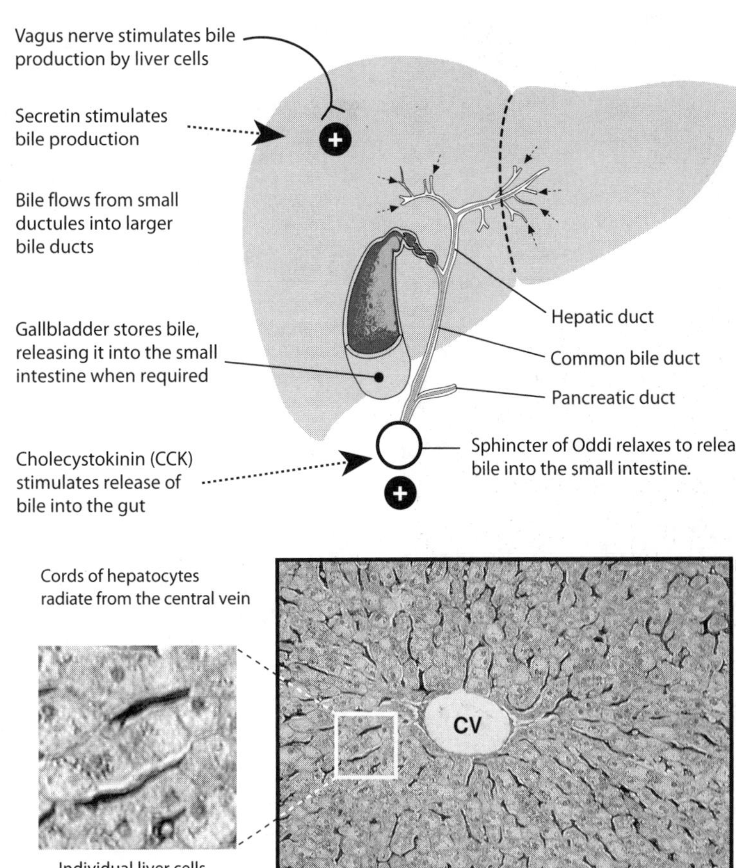
Vagus nerve stimulates bile production by liver cells
Secretin stimulates bile production
Bile flows from small ductules into larger bile ducts
Gallbladder stores bile, releasing it into the small intestine when required
Cholecystokinin (CCK) stimulates release of bile into the gut
Hepatic duct
Common bile duct
Pancreatic duct
Sphincter of Oddi relaxes to release bile into the small intestine.
Cords of hepatocytes radiate from the central vein
CV
Individual liver cells
EII

1. The liver produces bile. Describe the two main functions of bile in digestion:

   (a) ___

   (b) ___

2. Describe the two primary functions of the liver related to the processing of digestion products arriving from the gut:

   (a) ___

   (b) ___

3. Explain the role of the gall bladder in digestion: ___

4. Describe in what way bile is an excretory product as well as a digestive secretion: ___

5. Name the two principal hormones controlling the production (secretion) and release of bile, and state the effect of each:

   (a) Hormone 1: ___ Effect: ___

   (b) Hormone 2: ___ Effect: ___

6. State the stimulus for hormonal stimulation of bile secretion: ___

Diabetes Mellitus

Diabetes is a general term for a range of disorders sharing two common symptoms: production of large amounts of urine and excessive thirst. **Diabetes mellitus** is the most common form of diabetes and is characterized by **hyperglycemia** (high blood sugar). **Type 1** is characterized by a complete lack of insulin production and usually begins in childhood, while **type 2** is more typically a disease of older, overweight people whose cells develop a resistance to insulin uptake. Both types are chronic, incurable conditions and are managed differently. Type 1 is treated primarily with insulin injection, whereas type 2 sufferers manage their disease through diet and exercise in an attempt to limit the disease's long term detrimental effects.

Symptoms of Type 2 Diabetes Mellitus

a Symptoms may be mild at first. The body's cells do not respond appropriately to the insulin that is present and blood glucose levels become elevated. Normal blood glucose level is 60-110 mgL^{-1}. In diabetics, fasting blood glucose level is 126 mgL^{-1} or higher.

b Symptoms occur with varying degrees of severity:

- Cells are starved of fuel. This can lead to increased appetite and overeating and may contribute to an existing obesity problem.
- Urine production increases to rid the body of the excess glucose. Glucose is present in the urine and patients are frequently very thirsty.
- The body's inability to use glucose properly leads to muscle weakness and fatigue, irritability, frequent infections, and poor wound healing.

c Uncontrolled elevated blood glucose eventually results in damage to the blood vessels and leads to:

- coronary artery disease
- peripheral vascular disease
- retinal damage, blurred vision and blindness
- kidney damage and renal failure
- persistent ulcers and gangrene

Risk Factors

Obesity: BMI greater than 27. Distribution of weight is also important.

Age: Risk increases with age, although the incidence of type 2 diabetes is increasingly reported in obese children.

Sedentary lifestyle: Inactivity increases risk through its effects on bodyweight.

Family history: There is a strong genetic link for type 2 diabetes. Those with a family history of the disease are at greater risk.

Ethnicity: Certain ethnic groups are at higher risk of developing of type 2 diabetes.

High blood pressure: Up to 60% of people with undiagnosed diabetes have high blood pressure.

High blood lipids: More than 40% of people with diabetes have abnormally high levels of cholesterol and similar lipids in the blood.

Treating Type 2 Diabetes

Diabetes is not curable but can be managed to minimize the health effects:

- Regularly check blood glucose level
- Manage diet to reduce fluctuations in blood glucose level
- Take regular exercise
- Reduce weight
- Reduce blood pressure
- Reduce or stop smoking
- Take prescribed anti-diabetic drugs
- In time, insulin therapy may be required

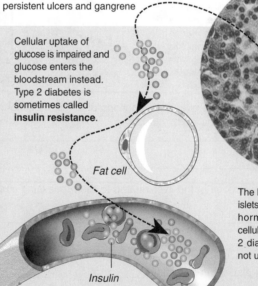

Cellular uptake of glucose is impaired and glucose enters the bloodstream instead. Type 2 diabetes is sometimes called **insulin resistance**.

The **beta cells** of the pancreatic islets (above) produce insulin, the hormone responsible for the cellular uptake of glucose. In type 2 diabetes, the body's cells do not utilize the insulin properly.

1. Distinguish between type 1 and type 2 diabetes, relating the differences to the different methods of treatment:

2. Explain what dietary advice you would give to a person diagnosed with type 2 diabetes:

3. Explain why the increase in type 2 diabetes is considered epidemic in the developed world:

Related activities: Control of Blood Glucose

Periodicals: Diabetes

KEY TERMS: What Am I?

THE OBJECT OF THIS GAME is to guess the unknown term from clues given to you by your team. Teams can be two or more, or you can play against individuals.

1) Cut out the cards below. You will need one set per team.
2) Shuffle the cards and deal them, face down, to each person in your team.
3) Affix tape to the back of the card so it can be stuck to your forehead. At no stage look at the word on the card!
4) One team starts. The members of your team give you a clue, one at a time, up to a maximum of **three** clues about what your term is. Do not use the word(s) on your card!
5) The clue should be a single point e.g. *"You feed on grass."*
6) If you guess correctly, your team receives another turn and the score is recorded. If you cannot guess, then the turn passes to the other team.
7) The game can be ended after one round or many.

Bile	Teeth	Heterotroph
Parasite	Lipase	Enzyme
Pancreas	Egestion	Villi
Carnivore	Ruminant	Gastrovascular cavity
Omnivore	Herbivore	Stomach
Stomach	Ileum	Liver

These cards have been deliberately left blank

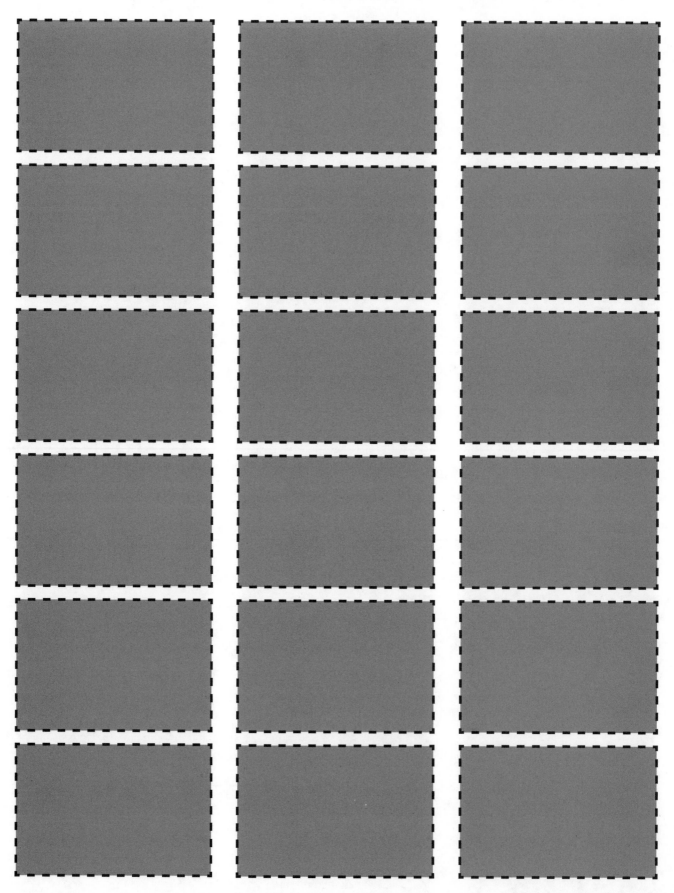

AP Topic 3B

IB SL/HL 6.4

Breath of Life

KEY CONCEPTS

- Organisms exchange respiratory gases with their environment.
- The gas exchange systems of animals reflect their adaptation to environment.
- Gas exchange surfaces have specific properties that maximize exchange rates.
- Respiratory pigments increase the oxygen carrying capacity of the circulatory fluids.

KEY TERMS

air sacs
alveolar lung
alveoli
blood
breathing
bronchi
bronchioles
carbon dioxide
cellular respiration
chlorocruorin
countercurrent flow
diaphragm
expiration (exhalation)
extraction rate
gas exchange
gills
hemerythrin
hemocyanin
hemoglobin
hemolymph
inspiration (inhalation)
lungs
oxygen
parabronchial lungs
phrenic nerve
respiration
respiratory center
respiratory gas
respiratory pigment
spiracles
spirometry
surfactant
tracheae
ventilation
ventilation rate

OBJECTIVES

☐ 1. Use the **KEY TERMS** to help you understand and complete these objectives.

Principles of Gas Exchange

pages 86, 91

☐ 2. Distinguish between **cellular respiration** and **gas exchange**. Explain why organisms need to exchange **respiratory gases** with their environment.

☐ 3. Describe the essential features of gas exchange surfaces and their significance in terms of gas exchange rates.

☐ 4. Explain, using examples, the constraints that the environment places on the gas exchange systems of animals.

Diversity in Gas Exchange Systems

pages 87-90, 92-94

☐ 5. Describe structural and functional diversity in animal gas exchange systems, relating specific features to suitability in the environment in each case.

(a) INSECTS: Describe the structure and function of **tracheae** (tracheal tubes) in insects, including the role of the **spiracles**.

(b) FISH: Describe the structure and function of **gills** in bony fish. Explain how countercurrent exchange achieves high oxygen extraction rates.

(c) AIR BREATHING VERTEBRATES: Describe the structure and function of lungs in a representative vertebrate, e.g mammal and/or bird.

☐ 6. Describe diversity in mechanisms of **ventilation** in animals. Compare breathing in animals with and without a diaphragm.

☐ 7. Explain the relationship between the gas exchange and circulatory systems.

☐ 8. Describe and explain the role of **circulatory fluids** and **respiratory pigments** in transporting gases in some animal taxa.

Gas Exchange in Humans

pages 92-100

☐ 9. Describe the structure and function of the respiratory system in humans (**trachea**, **bronchi**, **bronchioles**, **lungs**, and **alveoli**). Relate the structure of alveoli, including the role of **surfactant**, to their function in gas exchange.

☐ 10. Describe breathing in humans, including the role of the **diaphragm** and **intercostal muscles** in changing the air pressure in the lungs.

☐ 11. Describe and explain the transport of oxygen and CO_2 in the blood.

☐ 12. Explain how basic rhythm of breathing is controlled through the activity of the **respiratory center** in the medulla and its output via nerves. Explain how and why **ventilation rate** varies and explain how this is measured.

Periodicals:

listings for this chapter are on page 388

Weblinks:

www.thebiozone.com/weblink/SB2-2603.html

Teacher Resource CD-ROM:

Insects in Freshwater

Introduction to Gas Exchange

Living cells require energy for the activities of life. Energy is released in cells by the breakdown of sugars and other substances in the metabolic process called **cellular respiration**. As a consequence of this process, gases need to be exchanged between the respiring cells and the environment. In most organisms (with the exception of some bacterial groups) these gases are carbon dioxide (CO_2) and oxygen (O_2). The diagram below illustrates this process for an animal. Plant cells also respire, but their gas exchange budget is different because they also produce O_2 and consume CO_2 in photosynthesis.

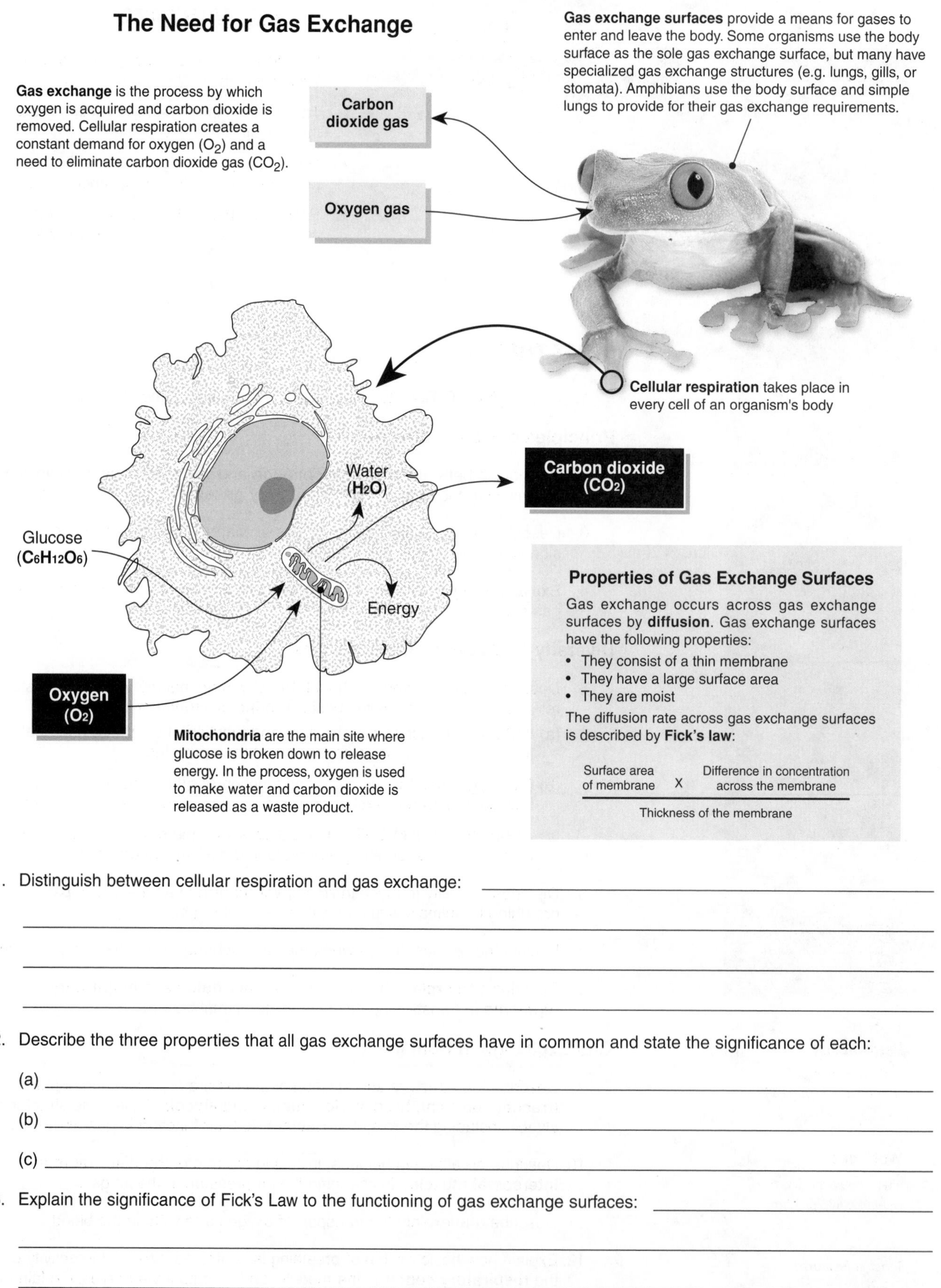

The Need for Gas Exchange
Gas exchange is the process by which oxygen is acquired and carbon dioxide is removed. Cellular respiration creates a constant demand for oxygen (O_2) and a need to eliminate carbon dioxide gas (CO_2).
Carbon dioxide gas
Oxygen gas
Gas exchange surfaces provide a means for gases to enter and leave the body. Some organisms use the body surface as the sole gas exchange surface, but many have specialized gas exchange structures (e.g. lungs, gills, or stomata). Amphibians use the body surface and simple lungs to provide for their gas exchange requirements.
Cellular respiration takes place in every cell of an organism's body
Water (H_2O)
Carbon dioxide (CO_2)
Glucose ($C_6H_{12}O_6$)
Energy
Oxygen (O_2)
Mitochondria are the main site where glucose is broken down to release energy. In the process, oxygen is used to make water and carbon dioxide is released as a waste product.

Properties of Gas Exchange Surfaces

Gas exchange occurs across gas exchange surfaces by **diffusion**. Gas exchange surfaces have the following properties:

- They consist of a thin membrane
- They have a large surface area
- They are moist

The diffusion rate across gas exchange surfaces is described by **Fick's law**:

$$\frac{\text{Surface area of membrane} \times \text{Difference in concentration across the membrane}}{\text{Thickness of the membrane}}$$

1. Distinguish between cellular respiration and gas exchange:

2. Describe the three properties that all gas exchange surfaces have in common and state the significance of each:

(a)

(b)

(c)

3. Explain the significance of Fick's Law to the functioning of gas exchange surfaces:

Gas Exchange in Fish

Fish obtain the oxygen they need from the water by means of gills: membranous structures supported by cartilaginous or bony struts. Gill surfaces are very large and as water flows over the gill surface, respiratory gases are exchanged between the blood and the water. The percentage of dissolved oxygen in a volume of water is much less than in the same volume of air; air is 21% oxygen while in water *dissolved* oxygen is about 1% by volume. High rates of oxygen extraction from the water, as achieved by gills, are therefore a necessary requirement for active organisms in an aquatic environment. In fish, ventilation of the gill surfaces to facilitate gas exchange is achieved by actively pumping water across the gill or swimming continuously with the mouth open.

Bony fish have four pairs of gills, each supported by a bony arch. The operculum (gill cover) is important in ventilation of the gills.

Cartilaginous fish have five or six pairs of gills. Water is drawn in via the mouth and spiracle and exits via the gill slits (there is no operculum).

Circulation and Gas Exchange in Fish

Fish and other vertebrates have a **closed circulatory system** where the blood is entirely contained within vessels. Fish have a **single circuit system;** the blood goes directly to the body from the gills (the gas exchange surface) and only flows once through the heart in each circulation of the body. The blood loses pressure when passing through the gills and, on leaving them, flows at low pressure around the body before returning to the heart. The gas exchange and circulatory systems are closely linked because the blood has a role in the transport of respiratory gases around the body.

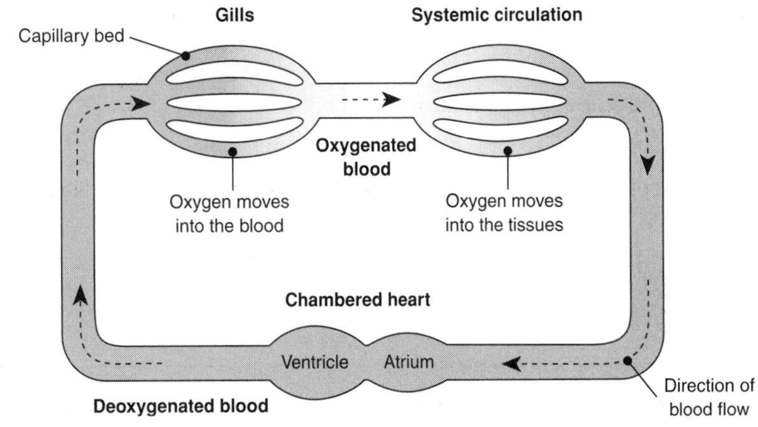

Fish Gills

The gills of fish have a great many folds, which are supported and kept apart from each other by the water. This gives them a high surface area for gas exchange. The outer surface of the gill is in contact with the water, and blood flows in vessels inside the gill. Gas exchange occurs by diffusion between the water and blood across the gill membrane and capillaries. The operculum (gill cover) permits exit of water and acts as a pump, drawing water past the gill filaments. Fish gills are very efficient and can achieve an 80% extraction rate of oxygen from water; over three times the rate of human lungs from air.

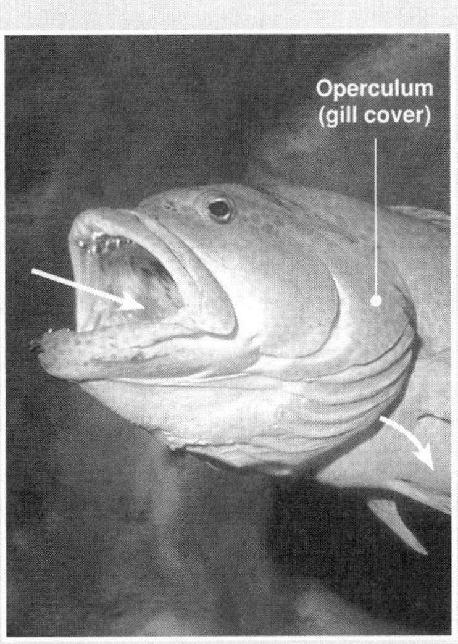

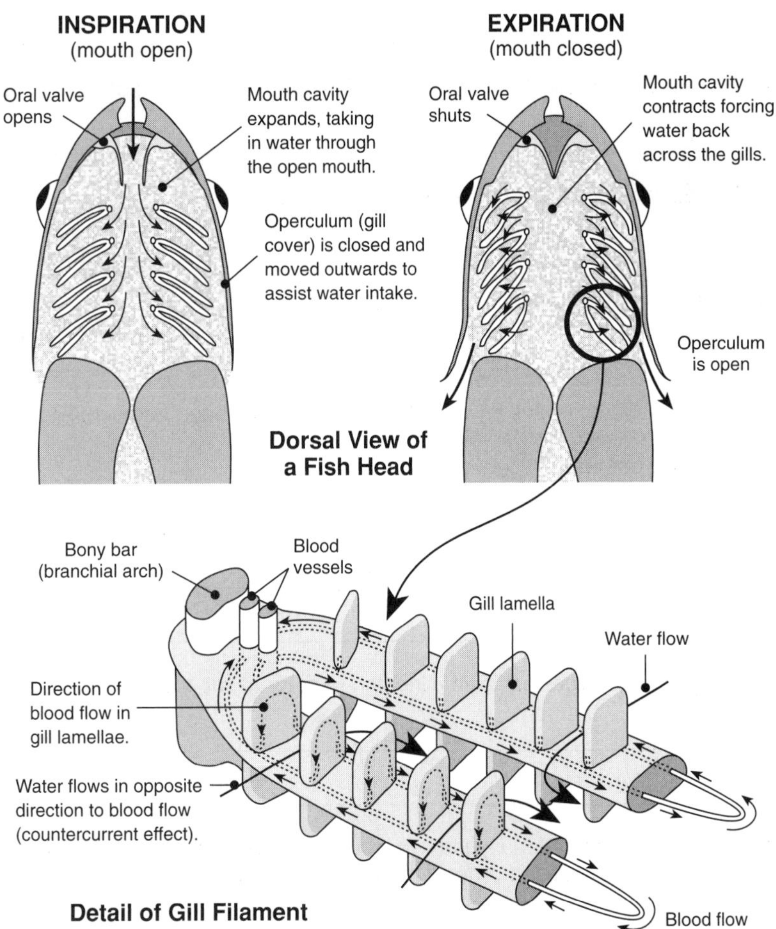

Detail of Gill Filament

Source: C.J. Clegg & D.G. McKean (1994)

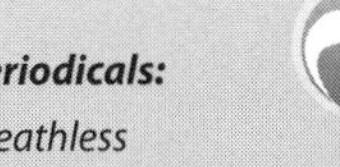

The structure of fish gills and their physical arrangement in relation to the blood flow ensure that gas exchange rates are maximized. A constant stream of oxygen-rich water flows over the gill filaments in the **opposite** direction to the direction of blood flow through the gills. This is termed **countercurrent flow** (below, left). Blood flowing through the gill capillaries therefore encounters water of increasing oxygen content. In this way, the concentration gradient (for oxygen uptake) across the gill is maintained across the entire distance of the gill lamella. A parallel current flow would not achieve the same oxygen extraction rates because the concentrations across the gill would quickly equalize (below, right).

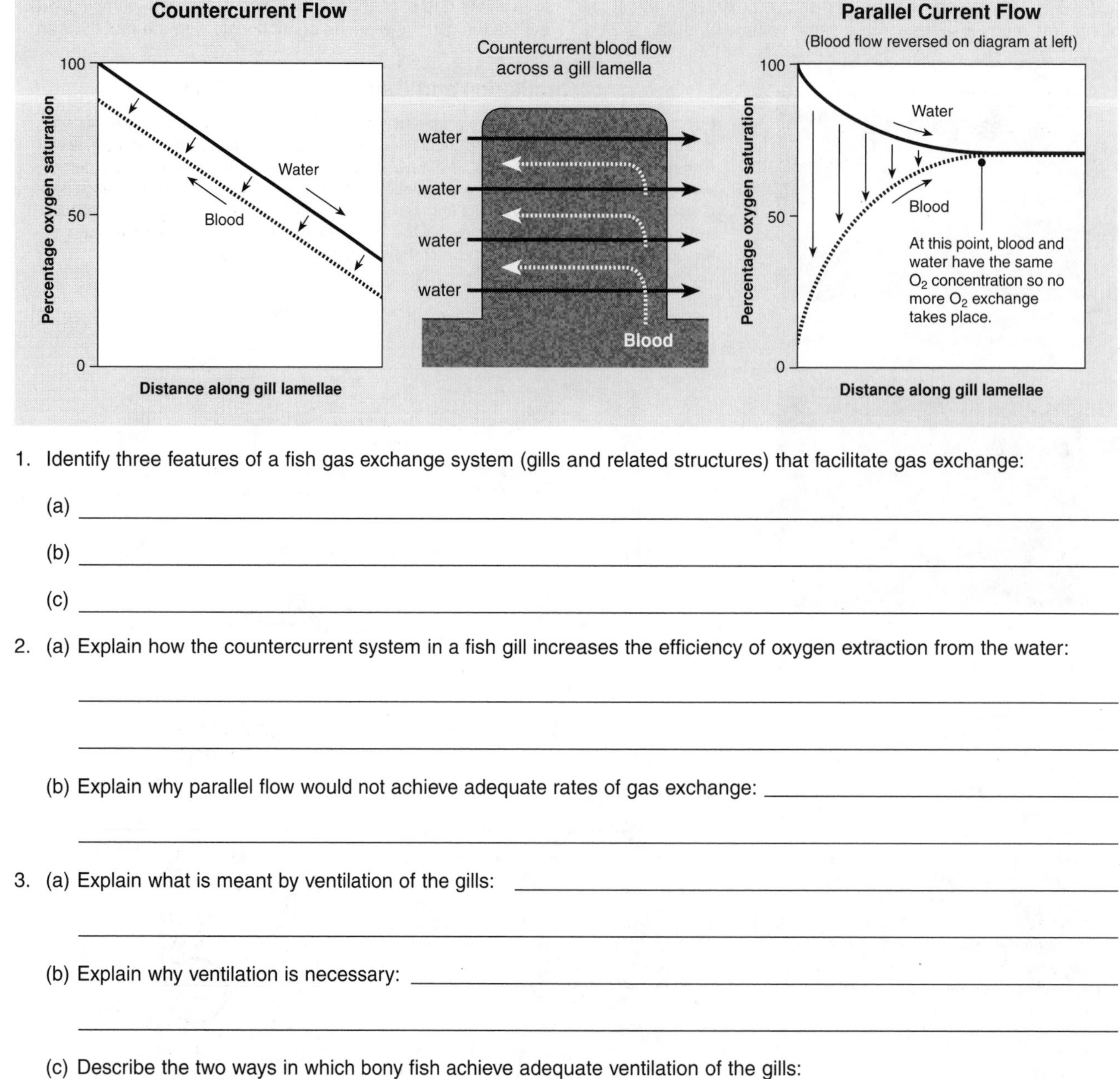

1. Identify three features of a fish gas exchange system (gills and related structures) that facilitate gas exchange:

   (a) ______

   (b) ______

   (c) ______

2. (a) Explain how the countercurrent system in a fish gill increases the efficiency of oxygen extraction from the water:

   ______

   (b) Explain why parallel flow would not achieve adequate rates of gas exchange: ______

3. (a) Explain what is meant by ventilation of the gills: ______

   (b) Explain why ventilation is necessary: ______

   (c) Describe the two ways in which bony fish achieve adequate ventilation of the gills:

   Pumping (mouth and operculum): ______

   Continuous swimming (mouth open): ______

   (d) Suggest why large, fast swimming fish (e.g. tuna) will die in aquaria that restrict continuous swimming movement:

   ______

4. In terms of the amount of oxygen available in the water, explain why fish are very sensitive to increases in water temperature or suspended organic material in the water:

   ______

Gas Exchange in Birds

In the lungs of mammals, the finest branches of the bronchi end in sac-like alveoli and air moves tidally in and out the same way. The lungs of birds are very different. Instead of alveoli, the finest branches end in open-ended tubelike structures, called **parabronchi**, which allow movement of air in only one direction. The one-way movement of air across the gas exchange surface maximizes the efficiency of oxygen uptake, especially at altitude (e.g. during flight). The lungs of birds are ventilated by a series of air sacs, which act as bellows to pump air through the lungs. Unlike mammals, birds do not have a diaphragm and their breathing movements rely on rocking motion of the sternum to create local areas of reduced pressure to supply the air sacs.

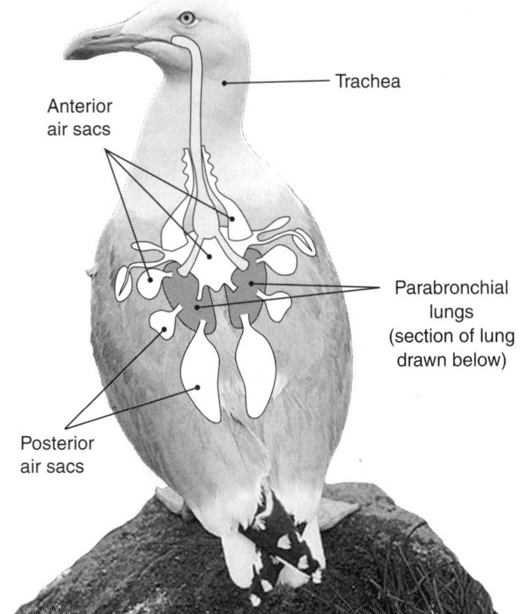

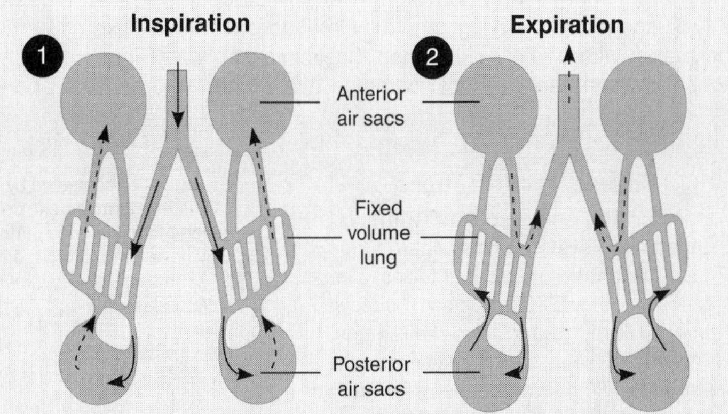

During inspiration, the sternum moves forward and down, expanding the air sacs and lowering the pressure so that air moves into them. Air from the trachea and bronchi moves into the posterior air sacs at the same time as air moves from the lungs into the anterior air sacs.

During expiration, the sternum moves back and up, reducing the volume of the air sacs and causing air to move out. Air from the posterior sacs moves into the lungs at the same time as air from the anterior sacs moves into the trachea and out of the body.

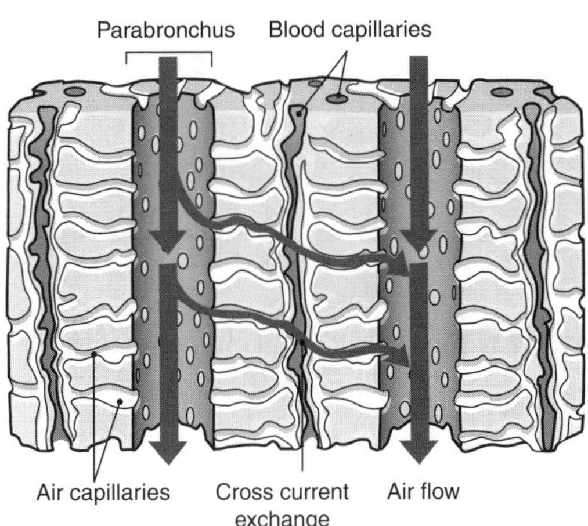

The air in a bird's lungs moves in one direction through **parabronchi**. The parabronchi are connected by tiny tubules called **air capillaries**. Blood capillaries flow past these air capillaries in a **cross current exchange**.

Fig.1: Thickness of blood-gas barrier vs body mass

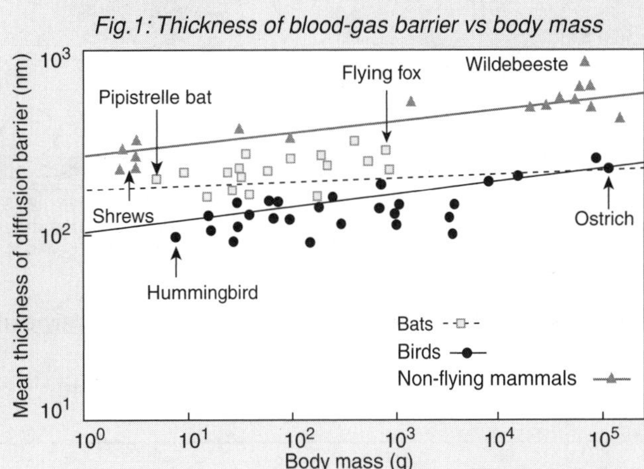

To meet the high oxygen demands of flight muscles, birds not only have efficient parabronchial lungs but also a thinner respiratory (blood-air) membrane. A thinner diffusion barrier ensures a faster rate of gas exchange relative to mammals of comparable size and metabolic rate (above).

1. Describe how the flow of air in bird lungs differs from that in mammalian lungs: ______________________

____

2. State how many full respiratory cycles are required to move air through its complete path in a bird lung: ________

3. (a) Explain how birds achieve high gas exchange efficiencies: ______________________

____

____

(b) Explain the adaptive value of this to birds in terms of their lifestyles: ______________________

____

____

Breath of Life

***Related activities**: Gas Exchange in Mammals, Flying*
***Web links**: Avian Respiration*

Gas Exchange in Insects

One advantage of gas exchange for terrestrial animals is that oxygen is proportionately more abundant in air than in water. However, terrestrial life also presents certain problems: body water can be lost easily through any exposed surface that is moist, thin, permeable, and vascular enough to serve as a respiratory membrane. Most insects are small terrestrial animals with a large surface area to volume ratio. Although they are highly susceptible to drying out, they are covered by a hard exoskeleton with a waxy outer layer that minimizes water loss. Tracheal systems are the most common gas exchange organs of terrestrial arthropods, including insects. Most body segments have paired apertures called spiracles in the lateral body wall through which air enters. Filtering devices in the spiracles prevent small particles from clogging the system, and valves control the degree to which the spiracles are open. In small insects, diffusion is the only mechanism needed to exchange gases, because it occurs so rapidly through the air-filled tubules. Larger, more active insects, such as locusts (below) have a tracheal system which includes air sacs that can be compressed and expanded to assist in moving air through the tubules.

Insect Tracheal Tubes

Insects, and some spiders, transport gases via a system of branching tubes called tracheae or tracheal tubes. The gases move by diffusion across the moist lining directly to and from the tissues. The end of each tube contains a small amount of fluid in which the respiratory gases are dissolved. The fluid is drawn into the muscle tissues during their contraction, and is released back into the tracheole when the muscle rests. Insects ventilate their tracheal system by making rhythmic body movements to help move the air in and out of the tracheae.

Spiracle openings on the abdomen

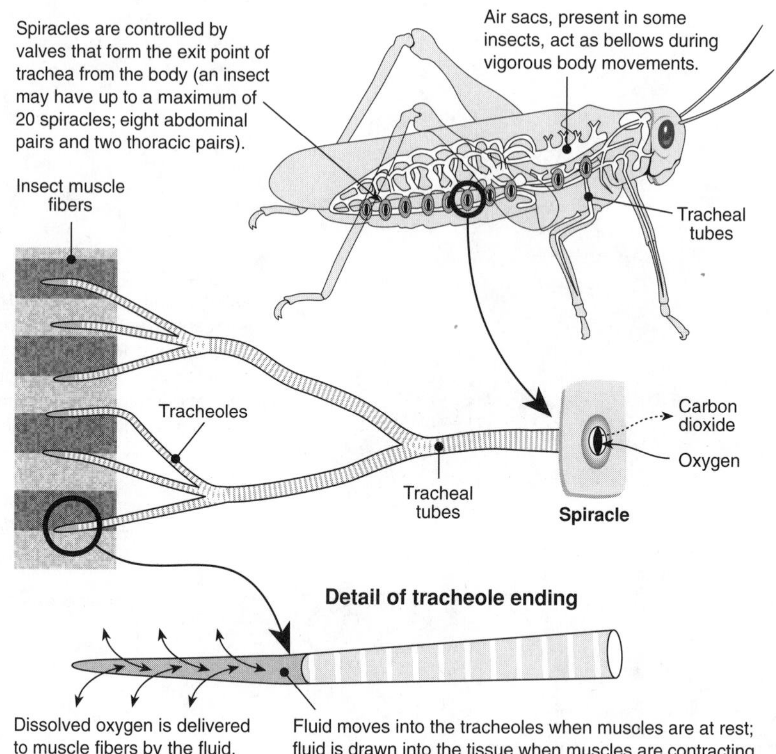
Spiracles are controlled by valves that form the exit point of trachea from the body (an insect may have up to a maximum of 20 spiracles; eight abdominal pairs and two thoracic pairs).
Air sacs, present in some insects, act as bellows during vigorous body movements.
Insect muscle fibers
Tracheal tubes
Tracheoles
Carbon dioxide
Oxygen
Tracheal tubes
Spiracle
Detail of tracheole ending
Dissolved oxygen is delivered to muscle fibers by the fluid.
Fluid moves into the tracheoles when muscles are at rest; fluid is drawn into the tissue when muscles are contracting.

1. Explain how oxygen and carbon dioxide are exchanged between the air and body tissues at the end of insect tracheoles:

2. Valves in the spiracles can regulate the amount of air entering the tracheal system. Suggest a reason for this adaptation:

3. Explain how ventilation is achieved in a terrestrial insect:

4. Even though most insects are small, they have evolved an efficient and highly developed gas exchange system that is independent of diffusion across the body surface. Suggest why this is the case:

Web links: *Bubble Breathing in Insects, Adaptations of Aquatic Insects*

Respiratory Pigments

Regardless of the gas exchange system present, the amount of oxygen that can be carried in solution in the blood is small. The efficiency of gas exchange in animals is enhanced by the presence of **respiratory pigments**. All respiratory pigments consist of proteins complexed with iron or copper. They combine reversibly with oxygen and greatly increase the capacity of blood to transport oxygen and deliver it to the tissues. For example, the amount of oxygen dissolved in the plasma in mammals is only about 2 cm^3 O_2 per liter. However the amount carried bound to hemoglobin is 100 times this. Hemoglobin is the most widely distributed respiratory pigment and is characteristic of all vertebrates and many invertebrate taxa. Other respiratory pigments include chlorocruorin, hemocyanin, and hemerythrin. Note that the precise structure and carrying capacity of any one particular pigment type varies between taxa (see the range of hemoglobins in the table below).

Respiratory Pigments

Respiratory pigments are colored proteins capable of combining reversibly with oxygen, hence increasing the amount of oxygen that can be carried by the blood. Pigments typical of representative taxa are listed below. Note that the polychaetes are very variable in terms of the pigment possessed.

Taxon	Oxygen capacity (cm^3 O_2 per 100 cm^3 blood)	Pigment
Oligochaetes	1 - 10	Hemoglobin
Polychaetes	1 - 10	Hemoglobin, chlorocruorin, or hemerythrin
Crustaceans	1 - 6	Hemocyanin
Molluscs	1 - 6	Hemocyanin
Fishes	2 - 4	Hemoglobin
Reptiles	7 - 12	Hemoglobin
Birds	20 - 25	Hemoglobin
Mammals	15 - 30	Hemoglobin

Mammalian Hemoglobin

Hemoglobin is a globular protein consisting of 574 amino acids arranged in four polypeptide sub-units: two identical ***beta chains*** and two identical ***alpha chains***. The four sub-units are held together as a functional unit by bonds. Each sub-unit has an iron-containing heme group at its center and binds one molecule of oxygen.

Chemical formula:
$C_{3032}H_{4816}O_{872}N_{780}S_8Fe_4$

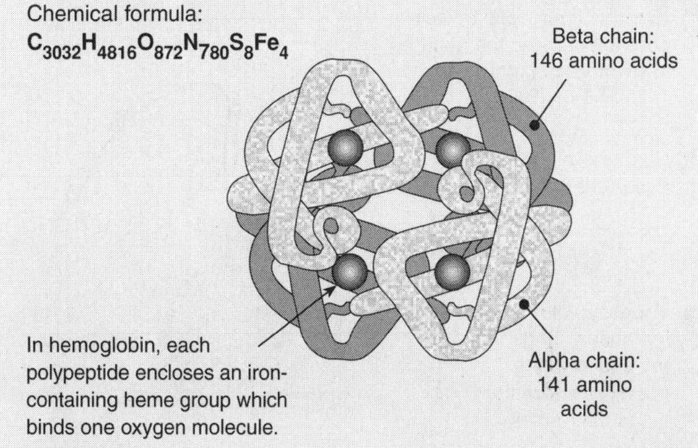

Aquatic polychaete fanworms e.g. *Sabella*, possess **chlorocruorin**.

Oligochaete annelids, such as earthworms, have **hemoglobin**.

Aquatic crustaceans e.g. crabs, possess **hemocyanin** pigment.

Vertebrates such as this fish have **hemoglobin** pigment.

Cephalopod molluscs such as *Nautilus* contain **hemocyanin**.

Birds, being vertebrates contain the pigment **hemoglobin**.

Many large active polychaetes, e.g. *Nereis,* contain **hemoglobin**.

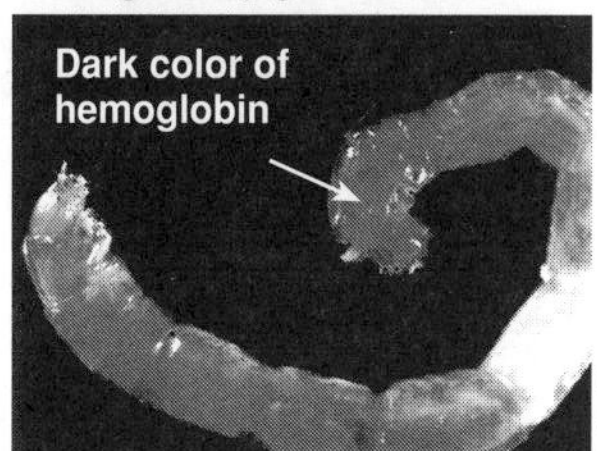

Chironomus is one of only two insect genera to contain a pigment.

1. (a) Explain how respiratory pigments increase the carrying capacity of the blood: ______

   (b) Identify which feature of a respiratory pigment determines its oxygen carrying capacity: ______

2. With reference to hemoglobin, suggest how oxygen carrying capacity is related to metabolic activity: ______

3. Suggest why larger molecular weight respiratory pigments are carried dissolved in the plasma rather than within cells: ______

Breath of Life

Related activities: *Gas Transport in Humans*

Breathing in Humans

In mammals, the mechanism of breathing (ventilation) provides a continual supply of fresh air to the lungs and helps to maintain a large diffusion gradient for respiratory gases across the gas exchange surface. Oxygen must be delivered regularly to supply the needs of respiring cells. Similarly, carbon dioxide, which is produced as a result of cellular metabolism, must be quickly eliminated from the body. Adequate lung ventilation is essential to these exchanges. The cardiovascular system participates by transporting respiratory gases to and from the cells of the body. The volume of gases exchanged during breathing varies according to the physiological demands placed on the body (e.g. by exercise). These changes can be measured using spirometry.

Inspiration (inhalation or breathing in)

During quiet breathing, inspiration is achieved by increasing the space (therefore decreasing the pressure) inside the lungs. Air then flows into the lungs in response to the decreased pressure inside the lung. Inspiration is always an active process involving muscle contraction.

1a External intercostal muscles contract causing the ribcage to expand and move up.

1b Diaphragm contracts and moves down.

2 Thoracic volume increases, lungs expand, and the pressure inside the lungs decreases.

3 Air flows into the lungs in response to the pressure gradient.

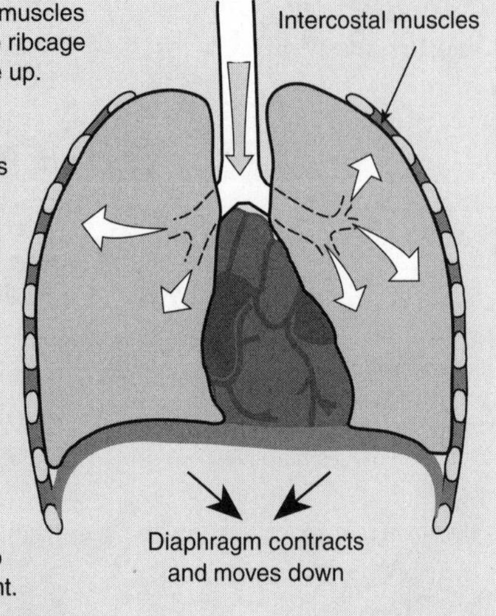

Expiration (exhalation or breathing out)

During quiet breathing, expiration is achieved passively by decreasing the space (thus increasing the pressure) inside the lungs. Air then flows passively out of the lungs to equalise with the air pressure. In active breathing, muscle contraction is involved in bringing about both inspiration and expiration.

1 In **quiet breathing**, external intercostal muscles and diaphragm relax. Elasticity of the lung tissue causes recoil.

In **forced breathing**, the internal intercostals and abdominal muscles also contract to increase the force of the expiration.

2 Thoracic volume decreases and the pressure inside the lungs increases.

3 Air flows passively out of the lungs in response to the pressure gradient.

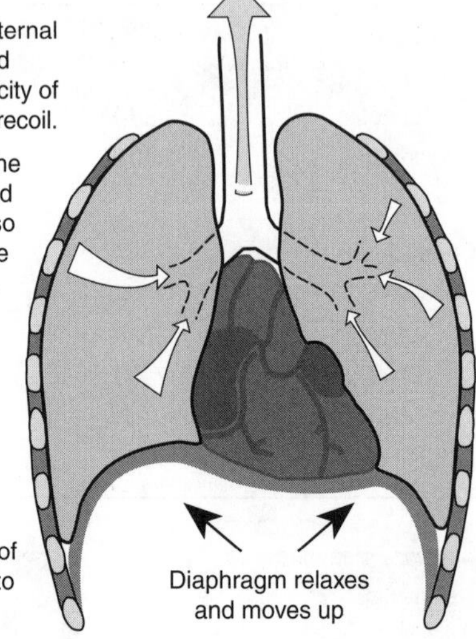

1. Explain the purpose of breathing: ______

2. (a) Describe the sequence of events involved in quiet breathing: ______

   (b) Explain the essential difference between this and the situation during heavy exercise or forced breathing: ______

3. Identify what other gas is lost from the body in addition to carbon dioxide: ______

4. Explain the role of the elasticity of the lung tissue in normal, quiet breathing: ______

5. Breathing rate is regulated through the medullary respiratory center in response to demand for oxygen. The trigger for increased breathing rate is a drop in blood pH. Suggest why this is an appropriate trigger to increase breathing rate: ______

The Human Respiratory System

The paired lungs of mammals, including humans, are located within the thorax and are connected to the outside air by way of a system of tubular passageways: the trachea, bronchi, and bronchioles. Ciliated, mucus secreting epithelium lines this system of tubules, trapping and removing dust and pathogens before they reach the gas exchange surfaces. Each lung is divided into a number of lobes, each receiving its own bronchus.

Each bronchus divides many times, terminating in the respiratory bronchioles from which arise 2-11 alveolar ducts and numerous **alveoli** (air sacs). These provide a very large surface area (around 70 m^2) for the exchange of respiratory gases by diffusion between the alveoli and the blood in the capillaries. The details of this exchange across the **respiratory membrane** are described on the next page.

Morphology of the Gas Exchange System

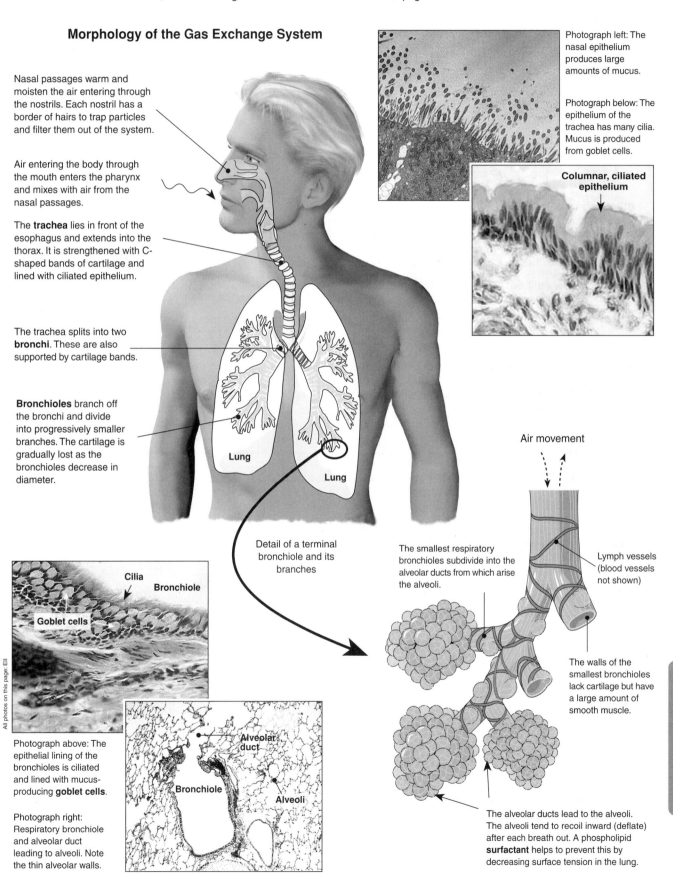

Nasal passages warm and moisten the air entering through the nostrils. Each nostril has a border of hairs to trap particles and filter them out of the system.

Air entering the body through the mouth enters the pharynx and mixes with air from the nasal passages.

The **trachea** lies in front of the esophagus and extends into the thorax. It is strengthened with C-shaped bands of cartilage and lined with ciliated epithelium.

The trachea splits into two **bronchi**. These are also supported by cartilage bands.

Bronchioles branch off the bronchi and divide into progressively smaller branches. The cartilage is gradually lost as the bronchioles decrease in diameter.

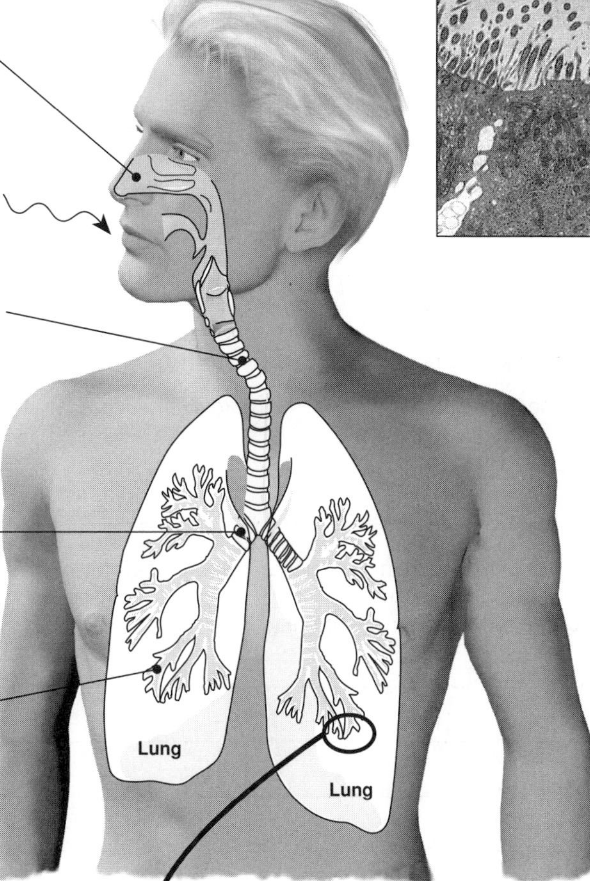

Lung

Lung

Photograph left: The nasal epithelium produces large amounts of mucus.

Photograph below: The epithelium of the trachea has many cilia. Mucus is produced from goblet cells.

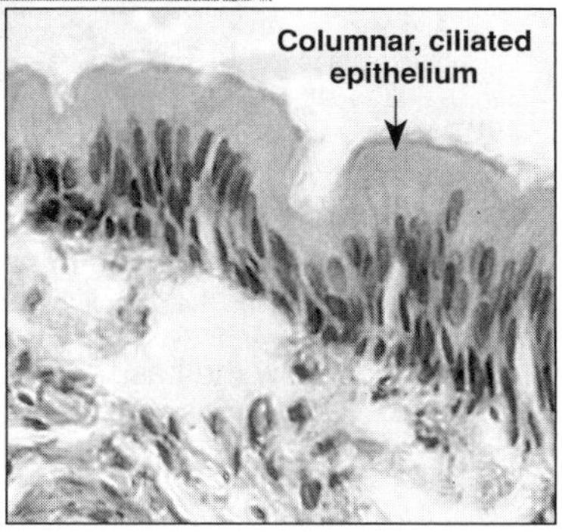

Columnar, ciliated epithelium

Detail of a terminal bronchiole and its branches

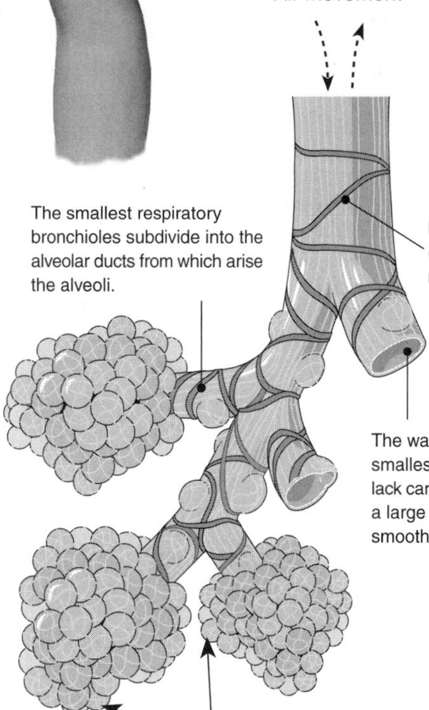

Air movement

The smallest respiratory bronchioles subdivide into the alveolar ducts from which arise the alveoli.

Lymph vessels (blood vessels not shown)

The walls of the smallest bronchioles lack cartilage but have a large amount of smooth muscle.

The alveolar ducts lead to the alveoli. The alveoli tend to recoil inward (deflate) after each breath out. A phospholipid **surfactant** helps to prevent this by decreasing surface tension in the lung.

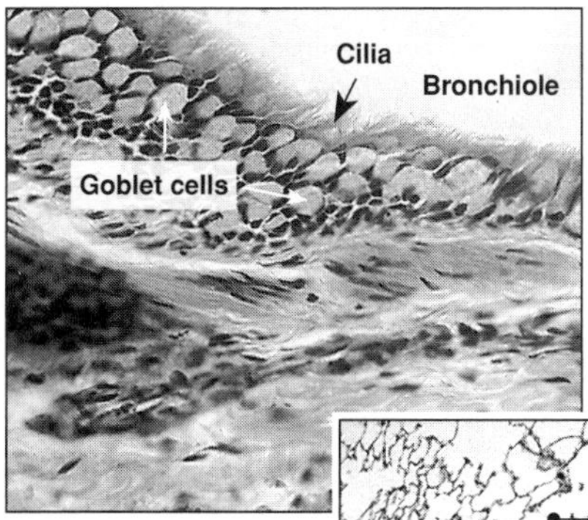

Cilia

Bronchiole

Goblet cells

Alveolar duct

Bronchiole

Alveoli

Photograph above: The epithelial lining of the bronchioles is ciliated and lined with mucus-producing **goblet cells**.

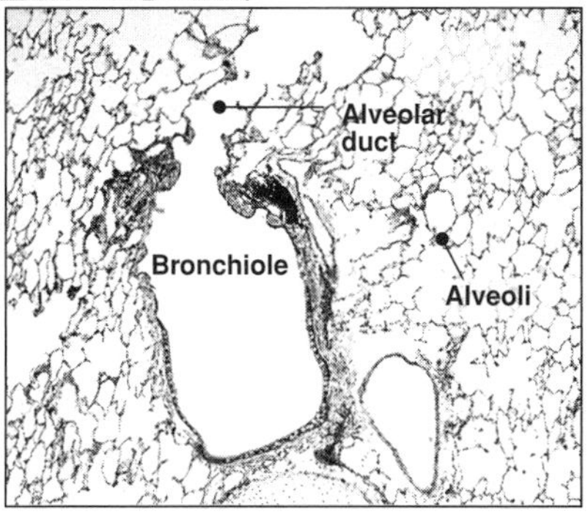

Photograph right: Respiratory bronchiole and alveolar duct leading to alveoli. Note the thin alveolar walls.

All photos on this page: EII

An Alveolus

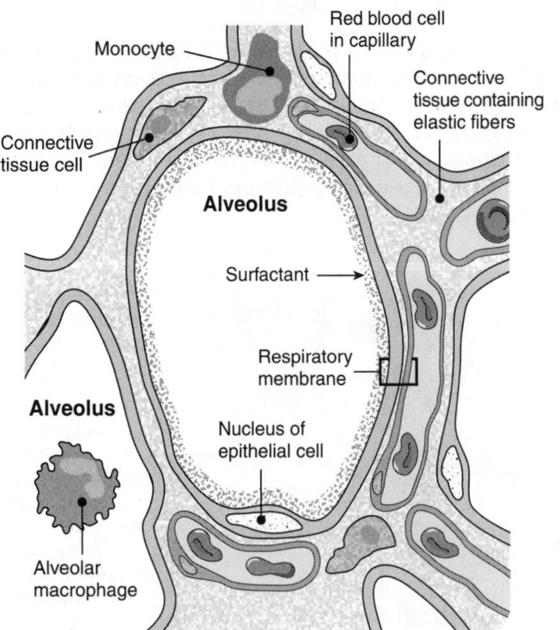

The diagram above illustrates the physical arrangement of the alveoli to the capillaries through which the blood moves. Phagocytic monocytes and macrophages are also present to protect the lung tissue. Elastic connective tissue gives the alveoli their ability to expand and recoil.

The Respiratory Membrane

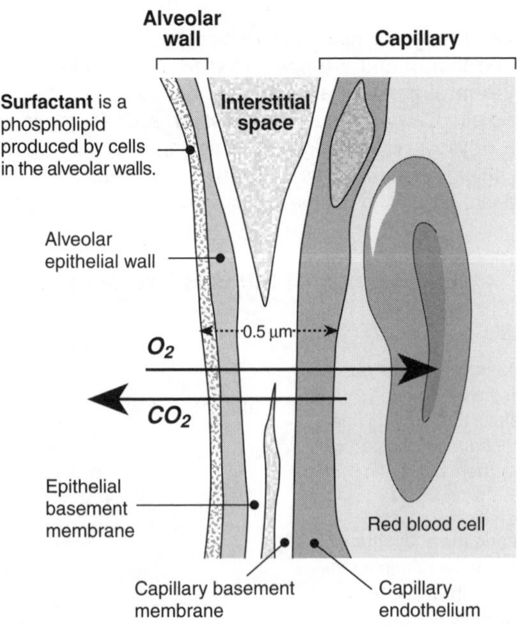

The **respiratory membrane** is the term for the layered junction between the alveolar epithelial cells, the endothelial cells of the capillary, and their associated basement membranes (thin, collagenous layers that underlie the epithelial tissues). Gases move freely across this membrane.

1. (a) Explain how the basic structure of the human respiratory system provides such a large area for gas exchange:

   (b) Identify the general region of the lung where exchange of gases takes place:

2. Describe the structure and purpose of the respiratory membrane:

3. Describe the role of the surfactant in the alveoli:

4. Using the information above and on the previous page, complete the table below summarizing the **histology of the respiratory pathway**. Name each numbered region and use a tick or cross to indicate the presence or absence of particular tissues.

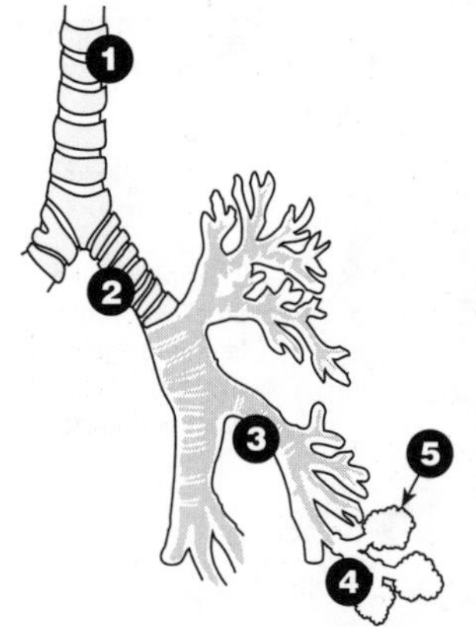

	Region	Cartilage	Ciliated epithelium	Goblet cells (mucus)	Smooth muscle	Connective tissue
1						✓
2						
3		gradually lost				
4	Alveolar duct		✗	✗		
5					very little	

5. Babies born prematurely are often deficient in surfactant. This causes respiratory distress syndrome; a condition where breathing is very difficult. From what you know about the role of surfactant, explain the symptoms of this syndrome:

Measuring Lung Function

Changes in lung volume can be measured using a technique called **spirometry**. Total adult lung capacity varies between 4 and 6 litres (L or dm3) and is greater in males. The **vital capacity**, which describes the volume exhaled after a maximum inspiration, is somewhat less than this because of the residual volume of air that remains in the lungs even after expiration. The exchange between fresh air and the residual volume is a slow process and the composition of gases in the lungs remains relatively constant. Once measured, the tidal volume can be used to calculate the **pulmonary ventilation rate** or PV, which describes the amount of air exchanged with the environment per minute. Measures of respiratory capacity provide one way in which a reduction in lung function can be assessed (for example, as might occur as result of disease or an obstructive lung disorder such asthma).

Determining changes in lung volume using spirometry

The apparatus used to measure the amount of air exchanged during breathing and the rate of breathing is a **spirometer** (also called a respirometer). A simple spirometer consists of a weighted drum, containing oxygen or air, inverted over a chamber of water. A tube connects the air-filled chamber with the subject's mouth, and soda lime in the system absorbs the carbon dioxide breathed out. Breathing results in a trace called a spirogram, from which lung volumes can be measured directly.

During inspiration
Air is removed from the chamber, the drum sinks, and an upward deflection is recorded on the paper on the rotating drum.

During expiration
Air is added to the chamber, the drum rises, and a downward deflection is recorded.

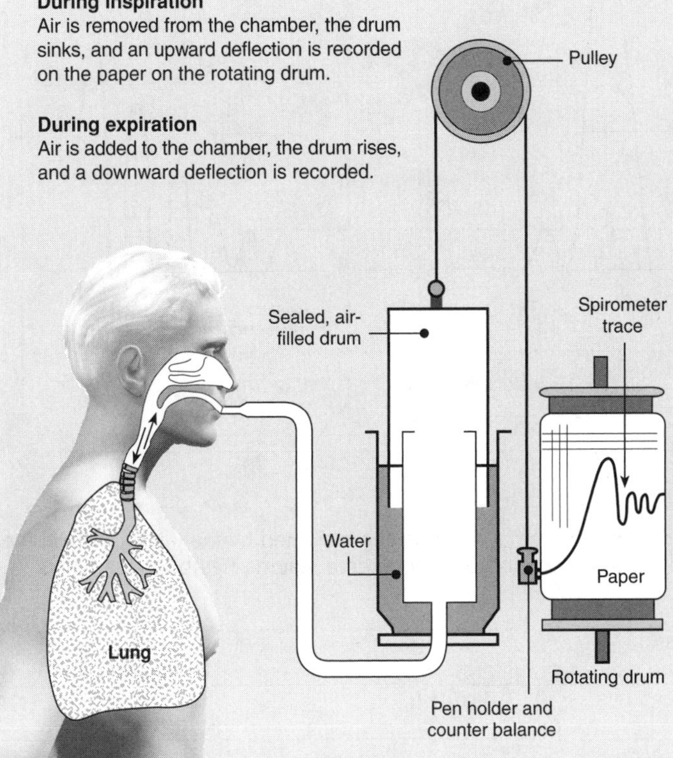

Lung Volumes and Capacities

The air in the lungs can be divided into volumes. Lung capacities are combinations of volumes.

DESCRIPTION OF VOLUME	Vol / L
Tidal volume (TV) Volume of air breathed in and out in a single breath	0.5
Inspiratory reserve volume (IRV) Volume breathed in by a maximum inspiration at the end of a normal inspiration	3.3
Expiratory reserve volume (ERV) Volume breathed out by a maximum effort at the end of a normal expiration	1.0
Residual volume (RV) Volume of air remaining in the lungs at the end of a maximum expiration	1.2
DESCRIPTION OF CAPACITY	
Inspiratory capacity (IC) = TV + IRV Volume breathed in by a maximum inspiration at the end of a normal expiration	3.8
Vital capacity (VC) = IRV + TV + ERV Volume that can be exhaled after a maximum inspiration.	4.8
Total lung capacity (TLC) = VC + RV The total volume of the lungs. Only a fraction of TLC is used in normal breathing	6.0

PRIMARY INDICATORS OF LUNG FUNCTION

Forced expiratory volume in 1 second (FEV_1)
The volume of air that is maximally exhaled in the first second of exhalation.

Forced vital capacity (FVC)
The total volume of air that can be forcibly exhaled after a maximum inspiration.

1. Describe how each of the following might be expected to influence values for lung volumes and capacities obtained using spirometry:

   (a) Height: ______

   (b) Gender: ______

   (c) Age: ______

2. A percentage decline in FEV_1 and FVC (to <80% of normal) are indicators of impaired lung function, e.g in asthma:

   (a) Explain why a forced volume is a more useful indicator of lung function than tidal volume:

   ______

   (b) Asthma is treated with drugs to relax the airways. Suggest how spirometry could be used during asthma treatment:

   ______

Respiratory gas	Approximate percentages of O_2 and CO_2		
	Inhaled air	Air in lungs	Exhaled air
O_2	21.0	13.8	16.4
CO_2	0.04	5.5	3.6

Above: The percentages of respiratory gases in air (by volume) during normal breathing. The percentage volume of oxygen in the alveolar air (in the lung) is lower than that in the exhaled air because of the influence of the **dead air volume** (the air in the spaces of the nose, throat, larynx, trachea and bronchi). This air (about 30% of the air inhaled) is unavailable for gas exchange.

Left: During exercise, the breathing rate, tidal volume, and PV increase up to a maximum (as indicated below).

Spirogram for a male during quiet and forced breathing, and during exercise

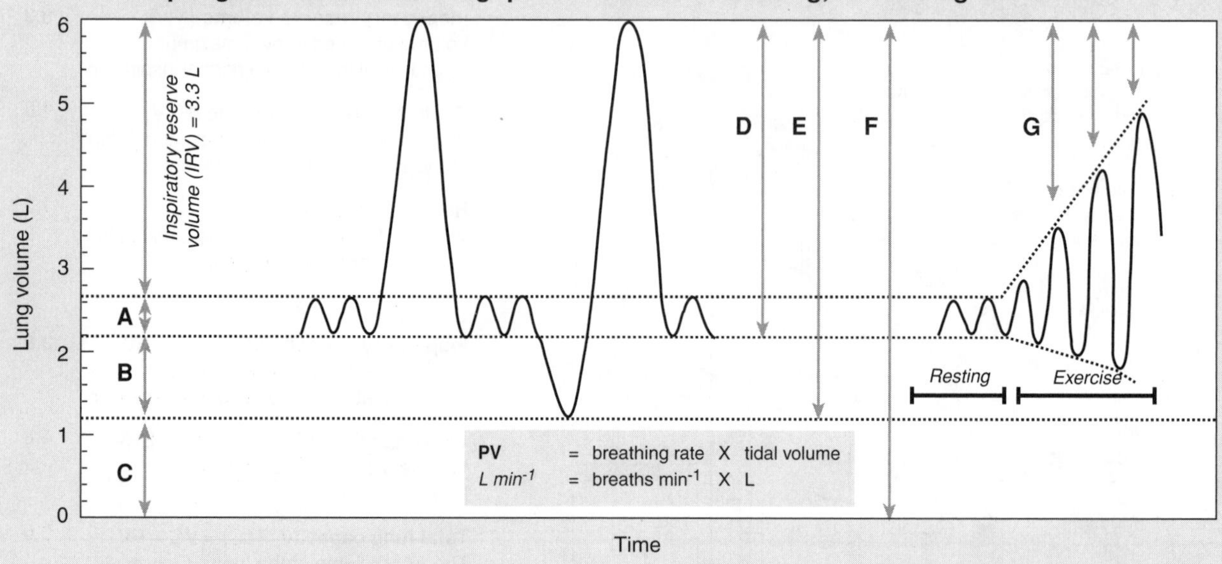

3. Using the definitions given on the previous page, identify the volumes and capacities indicated by the letters **A**-**F** on the spirogram above. For each, indicate the volume (vol) in liters (L). The inspiratory reserve volume has been identified:

(a) A: ______________________ Vol: ________ (d) D: ______________________ Vol: ________

(b) B: ______________________ Vol: ________ (e) E: ______________________ Vol: ________

(c) C: ______________________ Vol: ________ (f) F: ______________________ Vol: ________

4. Explain what is happening in the sequence indicated by the letter **G**: ______________________

5. Calculate PV when breathing rate is 15 breaths per minute and tidal volume is 0.4 L: ______________________

6. (a) Describe what would happen to PV during strenuous exercise: ______________________

(b) Explain how this is achieved: ______________________

7. The table above gives approximate percentages for respiratory gases during breathing. Study the data and then:

(a) Calculate the difference in CO_2 between inhaled and exhaled air: ______________________

(b) Explain where this 'extra' CO_2 comes from: ______________________

(c) Explain why the dead air volume raises the oxygen content of exhaled air above that in the lungs: ______________________

Gas Transport in Humans

The transport of respiratory gases around the body is the role of the blood and its respiratory pigments. Oxygen is transported throughout the body chemically bound to the respiratory pigment **hemoglobin** inside the red blood cells. In the muscles, oxygen from hemoglobin is transferred to and retained by **myoglobin**, a molecule that is chemically similar to hemoglobin except that it consists of only one heme-globin unit. Myoglobin has a greater affinity for oxygen than hemoglobin and acts as an oxygen store within muscles, releasing the oxygen during periods of prolonged or extreme muscular activity. If the myoglobin store is exhausted, the muscles are forced into oxygen debt and must respire anaerobically. The waste product of this, lactic acid, accumulates in the muscle and is transported (as lactate) to the liver where it is metabolized under aerobic conditions.

Gas Exchange and Transport

Air movement

Bronchiole

Alveoli

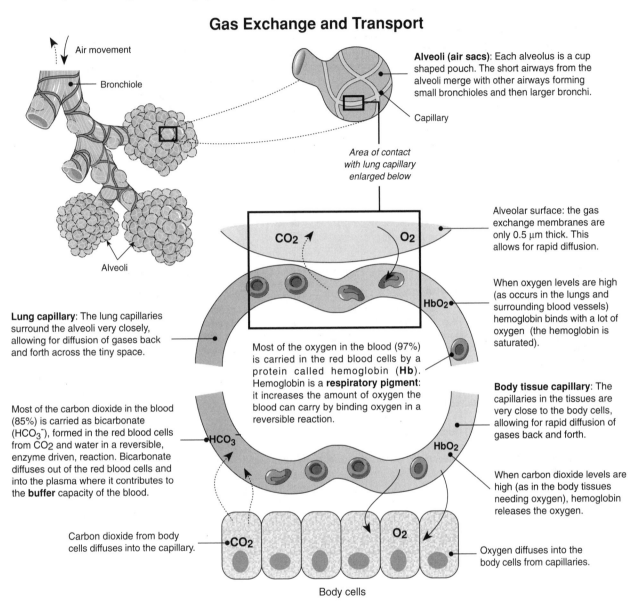

Alveoli (air sacs): Each alveolus is a cup shaped pouch. The short airways from the alveoli merge with other airways forming small bronchioles and then larger bronchi.

Capillary

Area of contact with lung capillary enlarged below

CO_2

O_2

Alveolar surface: the gas exchange membranes are only 0.5 µm thick. This allows for rapid diffusion.

When oxygen levels are high (as occurs in the lungs and surrounding blood vessels) hemoglobin binds with a lot of oxygen (the hemoglobin is saturated).

HbO_2

Lung capillary: The lung capillaries surround the alveoli very closely, allowing for diffusion of gases back and forth across the tiny space.

Most of the oxygen in the blood (97%) is carried in the red blood cells by a protein called hemoglobin (**Hb**). Hemoglobin is a **respiratory pigment**: it increases the amount of oxygen the blood can carry by binding oxygen in a reversible reaction.

Most of the carbon dioxide in the blood (85%) is carried as bicarbonate (HCO_3^-), formed in the red blood cells from CO_2 and water in a reversible, enzyme driven, reaction. Bicarbonate diffuses out of the red blood cells and into the plasma where it contributes to the **buffer** capacity of the blood.

HCO_3^-

Body tissue capillary: The capillaries in the tissues are very close to the body cells, allowing for rapid diffusion of gases back and forth.

HbO_2

When carbon dioxide levels are high (as in the body tissues needing oxygen), hemoglobin releases the oxygen.

Carbon dioxide from body cells diffuses into the capillary.

CO_2

O_2

Oxygen diffuses into the body cells from capillaries.

Body cells

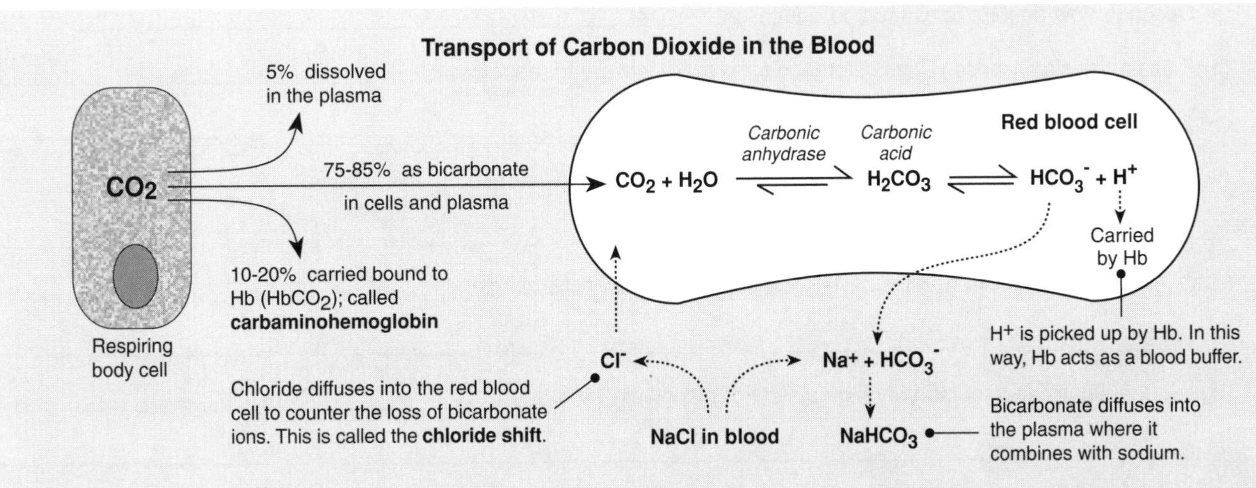

Oxygen does not dissolve easily in blood, but is carried in chemical combination with hemoglobin (Hb) in red blood cells. The most important factor determining how much oxygen is carried by Hb is the level of oxygen in the blood. The greater the oxygen tension, the more oxygen will combine with Hb. This relationship can be illustrated with an oxygen-hemoglobin dissociation curve as shown below (Fig. 1). In the lung capillaries, (high O_2), a lot of oxygen is picked up and bound by Hb. In the tissues, (low O_2), oxygen is released. In skeletal muscle, myoglobin picks up oxygen from hemoglobin and therefore serves as an oxygen store when oxygen tensions begin to fall. The release of oxygen is enhanced by the **Bohr effect** (Fig. 2).

Respiratory Pigments and the Transport of Oxygen

Fig. 1: Dissociation curves for hemoglobin and myoglobin at normal body temperature for fetal and adult human blood.

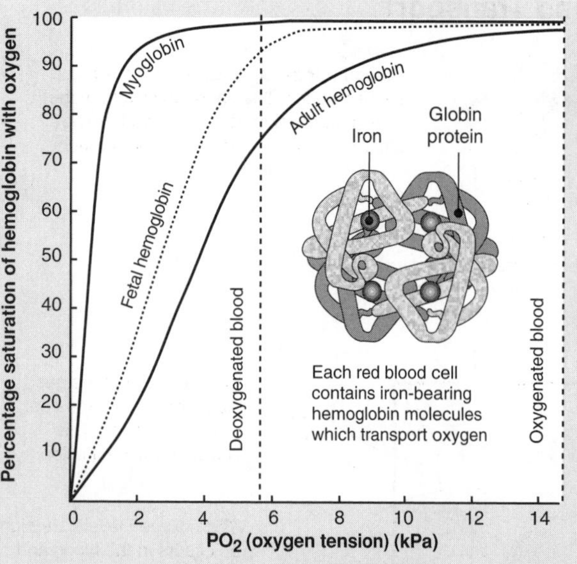

As oxygen level increases, more oxygen combines with hemoglobin (Hb). Hb saturation remains high, even at low oxygen tensions. Fetal Hb has a high affinity for oxygen and carries 20-30% more than maternal Hb. Myoglobin in skeletal muscle has a very high affinity for oxygen and will take up oxygen from hemoglobin in the blood.

Fig. 2: Oxygen-hemoglobin dissociation curves for human blood at normal body temperature at different blood pH.

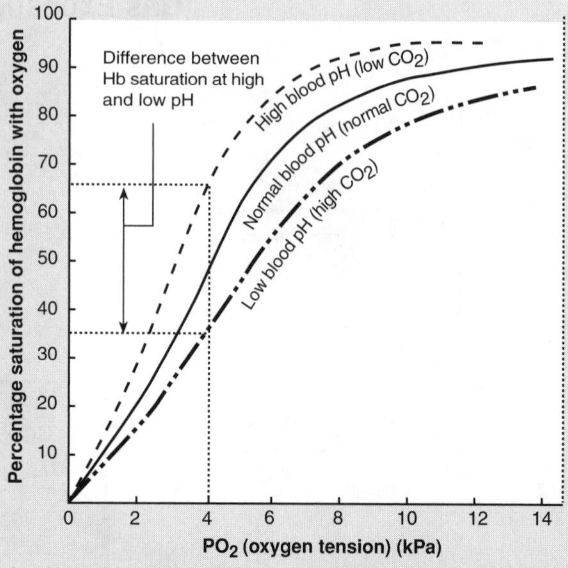

As pH increases (lower CO_2), more oxygen combines with Hb. As the blood pH decreases (higher CO_2), Hb binds less oxygen and releases more to the tissues (**the Bohr effect**). The difference between Hb saturation at high and low pH represents the amount of oxygen released to the tissues.

1. (a) Identify two regions in the body where oxygen levels are very high: ____________________

   (b) Identify two regions where carbon dioxide levels are very high: ____________________

2. Explain the significance of the **reversible binding** reaction of hemoglobin (Hb) to oxygen: ____________________

3. (a) Hemoglobin saturation is affected by the oxygen level in the blood. Describe the nature of this relationship:

   (b) Comment on the significance of this relationship to oxygen delivery to the tissues: ____________________

4. (a) Describe how fetal Hb is different to adult Hb: ____________________

   (b) Explain the significance of this difference to oxygen delivery to the fetus: ____________________

5. At low blood pH, less oxygen is bound by hemoglobin and more is released to the tissues:

   (a) Name this effect: ____________________

   (b) Comment on its significance to oxygen delivery to respiring tissue: ____________________

6. Explain the significance of the very high affinity of myoglobin for oxygen: ____________________

7. Identify the two main contributors to the buffer capacity of the blood: ____________________

Control of Breathing

The basic rhythm of breathing is controlled by the **respiratory center**, a cluster of neurons located in the medulla oblongata. This rhythm is adjusted in response to the physical and chemical changes that occur when we carry out different activities. Although the control of breathing is involuntary, we can exert some degree of conscious control over it. The diagram below illustrates these controls.

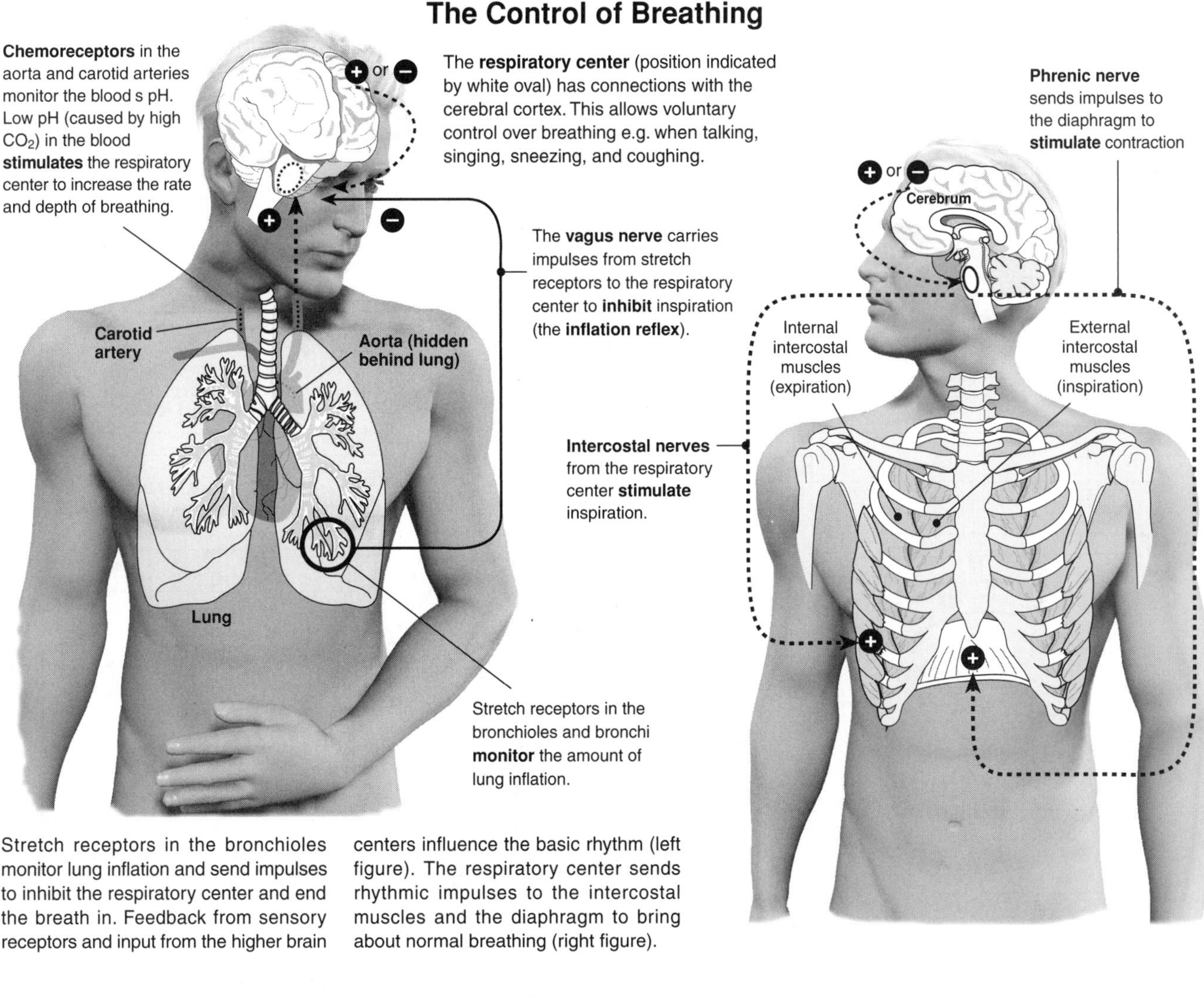
The Control of Breathing
Chemoreceptors in the aorta and carotid arteries monitor the blood s pH. Low pH (caused by high CO_2) in the blood stimulates the respiratory center to increase the rate and depth of breathing.
or
The respiratory center (position indicated by white oval) has connections with the cerebral cortex. This allows voluntary control over breathing e.g. when talking, singing, sneezing, and coughing.
The vagus nerve carries impulses from stretch receptors to the respiratory center to inhibit inspiration (the inflation reflex).
Carotid artery
Aorta (hidden behind lung)
Lung
Stretch receptors in the bronchioles and bronchi monitor the amount of lung inflation.
Phrenic nerve sends impulses to the diaphragm to stimulate contraction
or
Cerebrum
Internal intercostal muscles (expiration)
External intercostal muscles (inspiration)
Intercostal nerves from the respiratory center stimulate inspiration.

Stretch receptors in the bronchioles monitor lung inflation and send impulses to inhibit the respiratory center and end the breath in. Feedback from sensory receptors and input from the higher brain centers influence the basic rhythm (left figure). The respiratory center sends rhythmic impulses to the intercostal muscles and the diaphragm to bring about normal breathing (right figure).

1. Explain how the basic rhythm of breathing is controlled: ______________________________

____

2. Describe the role of each of the following in the regulation of breathing:

   (a) Phrenic nerve: ______________________________

   (b) Intercostal nerves: ______________________________

   (c) Vagus nerve: ______________________________

   (d) Inflation reflex: ______________________________

3. (a) Describe the effect of low blood pH on the rate and depth of breathing: ______________________________

   ____

   (b) Explain how this effect is mediated: ______________________________

   ____

   (c) Suggest why blood pH is a good mechanism by which to regulate breathing rate: ______________________________

   ____

   ____

Review of Lung Function

The respiratory system in humans and other air breathing vertebrates includes the lungs and the system of tubes through which the air reaches them. Breathing (ventilation) provides a continual supply of fresh air to the lungs and helps to maintain a large diffusion gradient for respiratory gases across the gas exchange surface. The basic rhythm of breathing is controlled by the respiratory center in the medulla of the hindbrain. The volume of gases exchanged during breathing varies according to the physiological demands placed on the body. These changes can be measured using spirometry. The following activity summarizes the key features of respiratory system structure and function. The stimulus material can be found in earlier exercises in this topic.

Components of the respiratory system

(a)

(b)

(c)

(d)

(e)

(f)

(g)

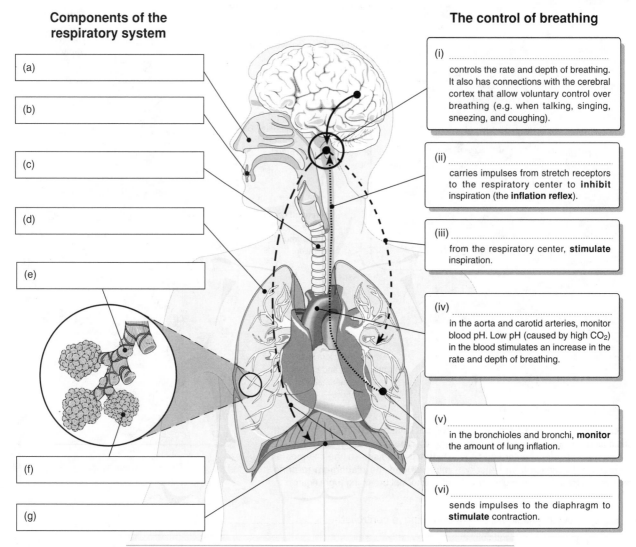

The control of breathing

(i) ....
controls the rate and depth of breathing. It also has connections with the cerebral cortex that allow voluntary control over breathing (e.g. when talking, singing, sneezing, and coughing).

(ii) ....
carries impulses from stretch receptors to the respiratory center to **inhibit** inspiration (the **inflation reflex**).

(iii) ....
from the respiratory center, **stimulate** inspiration.

(iv) ....
in the aorta and carotid arteries, monitor blood pH. Low pH (caused by high CO_2) in the blood stimulates an increase in the rate and depth of breathing.

(v) ....
in the bronchioles and bronchi, **monitor** the amount of lung inflation.

(vi) ....
sends impulses to the diaphragm to **stimulate** contraction.

1. On the diagram above, label the components of the respiratory system (a-g) and the components that control the rate of breathing (i - vi).

2. Identify the volumes and capacities indicated by the letters A - E on the diagram of a spirogram below.

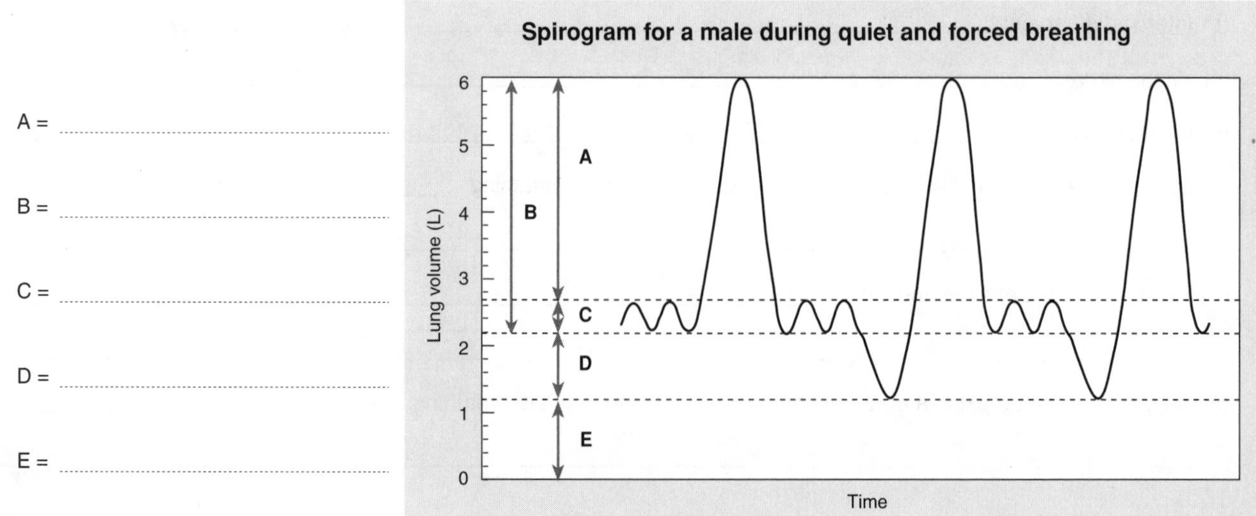

Related activities: *Control of Breathing, Measuring Lung Function*

The Effects of High Altitude

The air at high altitudes contains less oxygen than the air at sea level. Air pressure decreases with altitude so the pressure (therefore amount) of oxygen in the air also decreases. Sudden exposure to an altitude of 2000 m would make you breathless on exertion and above 7000 m most people would become unconscious. The effects of altitude on physiology are related to this lower oxygen availability. Humans and other animals can make some physiological adjustments to life at altitude; this is called acclimatization. Some of the changes to the cardiovascular and respiratory systems to high altitude are outlined below.

Mountain Sickness

Altitude sickness or mountain sickness is usually a mild illness associated with trekking to altitudes of 5000 meters or so. Common symptoms include headache, insomnia, poor appetite and nausea, vomiting, dizziness, tiredness, coughing and breathlessness. The best way to avoid mountain sickness is to ascend to altitude slowly (no more than 300 m per day above 3000 m). Continuing to ascend with mountain sickness can result in more serious illnesses: accumulation of fluid on the brain (cerebral edema) and accumulation of fluid in the lungs (pulmonary edema). These complications can be fatal if not treated with oxygen and a rapid descent to lower altitude.

Physiological Adjustment to Altitude

Effect	Minutes	Days	Weeks
Increased heart rate	←——	——→	
Increased breathing	←——	————	→
Concentration of blood		←——→	
Increased red blood cell production		←——	——→
Increased capillary density			←——→

The human body can make adjustments to life at altitude. Some of these changes take place almost immediately: breathing and heart rates increase. Other adjustments may take weeks (see above). These responses are all aimed at improving the rate of supply of oxygen to the body's tissues. When more permanent adjustments to physiology are made (increased blood cells and capillary networks) heart and breathing rates can return to normal.

People who live permanently at high altitude, e.g. Tibetans, Nepalese, and Peruvian Indians, have physiologies adapted (genetically, through evolution) to high altitude. Their blood volumes and red blood cell counts are high, and they can carry heavy loads effortlessly despite a small build. In addition, their metabolism uses oxygen very efficiently.

Llamas, vicunas, and Bactrian camels are well suited to high altitude life. Vicunas and llamas, which live in the Andes, have high blood cell counts and their red blood cells live almost twice as long as those in humans. Their hemoglobin also picks up and offloads oxygen more efficiently than the hemoglobin of most mammals.

1. (a) Describe the general effects of high altitude on the body: __________

   (b) Name the general term given to describe these effects: __________

2. (a) Name one short term physiological adaptation that humans make to high altitude: __________

   (b) Explain how this adaptation helps to increase the amount of oxygen the body receives: __________

3. (a) Describe one longer term adaptation that humans can make to living at high altitude: __________

   (b) Explain how this adaptation helps to increase the amount of oxygen the body receives: __________

Breath of Life

KEY TERMS Mix and Match

INSTRUCTIONS: Test your vocab by matching each term to its correct definition, as identified by its preceding letter code.

AIR SACS ....

ALVEOLI ....

BREATHING ....

BRONCHI ....

BRONCHIOLES ....

CARBON DIOXIDE ....

CELLULAR RESPIRATION ....

CIRCULATORY FLUIDS ....

COUNTERCURRENT FLOW ....

DIAPHRAGM ....

EXPIRATION ....

EXTRACTION RATE ....

GAS EXCHANGE ....

GILLS ....

HEMOGLOBIN ....

INSPIRATION ....

LUNGS ....

OXYGEN ....

PHRENIC NERVE ....

SURFACTANT ....

PARABRONCHI ....

RESPIRATORY GAS ....

RESPIRATORY PIGMENT ....

SPIRACLES ....

SPIROMETRY ....

TRACHEAE ....

VENTILATION ....

A An internal muscle separating the thorax and abdomen of mammals and involved in enabling thoracic volume changes during breathing.

B A substance carried in blood that is able to bind oxygen for transport to cells. Examples include hemoglobin and hemocyanin.

C The rate at which oxygen is removed by the lungs from the inspired air.

D Large air tubes that branch from the trachea to enter the lungs.

E Small air tubes that divide from the bronchi and become progressively smaller.

F The exterior opening of the tracheae in arthropods.

G Series of tube like structures that allow gas to be conducted into the body in insects.

H A method of measuring volume changes in the lung.

I A term describing the flow of fluids and/or air in opposite directions so that diffusion gradients are maintained between the two media and exchanges between them are maximized.

J Fluids that move through the body, either freely or within vessels and may transport any of nutrients, wastes, and respiratory gases.

K A large iron-containing protein, which transports oxygen in the blood of vertebrates.

L A general term for movement of air or water across an organism's gas exchange surfaces.

M The act of breathing in or filling the lungs with air.

N A gas produced as a waste product of metabolism.

O Membranous structures in birds that do not take part in gas exchanges but are important in ventilation of the lungs. Also a common name for alveoli.

P The catabolic process in which the chemical energy in complex organic molecules is coupled to ATP production.

Q The exchange of oxygen and carbon dioxide across the respiratory membrane.

R Gas required for aerobic respiration. Levels average 21% in the atmosphere.

S Internal gas exchange structures found in vertebrates.

T The act of breathing out or removing air from the lungs.

U Air flows in one direction through these fine airways in the lungs of birds.

V Any gas that takes part in the respiratory process; usually oxygen or carbon dioxide.

W This nerve sends impulses to the diaphragm to stimulate contraction.

X The respiratory organs of most aquatic animals (although not aquatic mammals).

Y The act of inhaling air into and exhaling air from the lungs.

Z Phospholipid substance responsible for reducing surface tension in the lung tissue.

AA Microscopic structures in the lungs of most vertebrates that form the terminus of the bronchioles. The site of gas exchange.

AP Topic 3B

IB SL/HL 6.2

Life Blood

KEY CONCEPTS

- An internal transport system is a requirement in most multicellular animals.
- Animal circulatory systems commonly include a heart, vessels, and a circulatory fluid.
- Circulatory systems may be open or closed.
- Heart structure reflects function in relation to circulatory and metabolic requirements.
- Human heart rate is precisely controlled to meet the changing demands of metabolism.

KEY TERMS

artery (pl. arteries)
atrio-ventricular node
atrium (pl. atria)
blood
blood vessels
bulk flow (=mass flow)
capillary (pl. capillaries)
cardiac cycle
cardiovascular control center
closed circulatory system
coronary arteries
diastole
double circulatory system
hemolymph
heart
heart valves
lymph
microcirculation
myogenic
open circulatory system
portal circulation
pulmonary circulation
respiratory pigment
single circulatory system
sino-atrial node
sinus
sinus venosus
surface area: volume
systemic circulation
systole
tissue fluid
vein
ventricle

Periodicals:
listings for this chapter are on page 388

Weblinks:
www.thebiozone.com/weblink/SB2-2603.html

OBJECTIVES

- ☐ 1. Use the **KEY TERMS** to help you understand and complete these objectives.

Principles of Internal Transport — pages 104, 113-115

- ☐ 2. Describe the **surface area: volume** relationship in different animal taxa and relate this to the presence or absence of an internal transport system. Describe the components and functions of a **transport system** in animals.

Diversity in Circulatory Systems — pages 105-112

- ☐ 3. Describe and explain diversity in the structure and function of **blood vessels** in vertebrates, including **arteries**, **capillaries**, and **veins**.
- ☐ 4. Explain the production and functional role of **tissue fluid**.
- ☐ 5. Describe diversity in the structure and function of circulatory systems in animals, including:
  (a) **Open circulatory systems** (arthropods and most mollusks).
  (b) **Closed circulatory systems** (vertebrates and some invertebrates).
- ☐ 6. Describe diversity in the structure and function of closed circulatory systems in representative taxa:
  (a) Closed circulatory system in invertebrates (annelids, cephalopods)
  (b) **Single circulatory system** in fish
  (c) **Double circulatory systems** in vertebrates other than fish
- ☐ 7. Compare and contrast the structure and function of vertebrate hearts:
  (a) Three chambers in series (fish)
  (b) Three chambers with two atria and an undivided ventricle (amphibian)
  (c) Four chambers separating systemic and pulmonary circulation (mammal)
- ☐ 8. Explain how the type of circulatory system and the functional efficiency of transport is related to the metabolic needs and environment.

The Human Circulatory System — pages 116-126

- ☐ 9. Describe the human circulatory system, including reference to the structure and function of the **blood**, major blood vessels, and the **heart**.
- ☐ 10. Annotate a diagram of the heart, and describe the function of the four chambers, associated blood vessels, **coronary arteries**, and **heart valves**.
- ☐ 11. Describe the **cardiac cycle**, relating stages in the cycle (atrial **systole**, ventricular systole, and **diastole**) to the blood flow through the heart.
- ☐ 12. Describe the control of heartbeat, including the intrinsic regulation through the **sinoatrial node**, and extrinsic regulation via the medulla through autonomic nerves. Describe the response of heart rate to exercise and explain how this is mediated.

Transport and Exchange Systems

Living cells require a constant supply of nutrients and oxygen, and continuous removal of wastes. Simple, small organisms can achieve this through **diffusion** across moist body surfaces without requiring specialized transport or exchange systems. Larger, more complex organisms require systems to facilitate exchanges as their surface area to volume ratio decreases. **Mass transport** (also known as mass flow or **bulk flow**) describes the movement of materials at equal rates or as a single mass. Mass transport accounts for the long distance transport of fluids in living organisms. It includes the movement of blood in the circulatory systems of animals and the transport of water and solutes in the xylem and phloem of plants. In the diagram below, exchanges by diffusion are compared with mass transport to specific exchange sites.

Exchanges Across a Body Surface

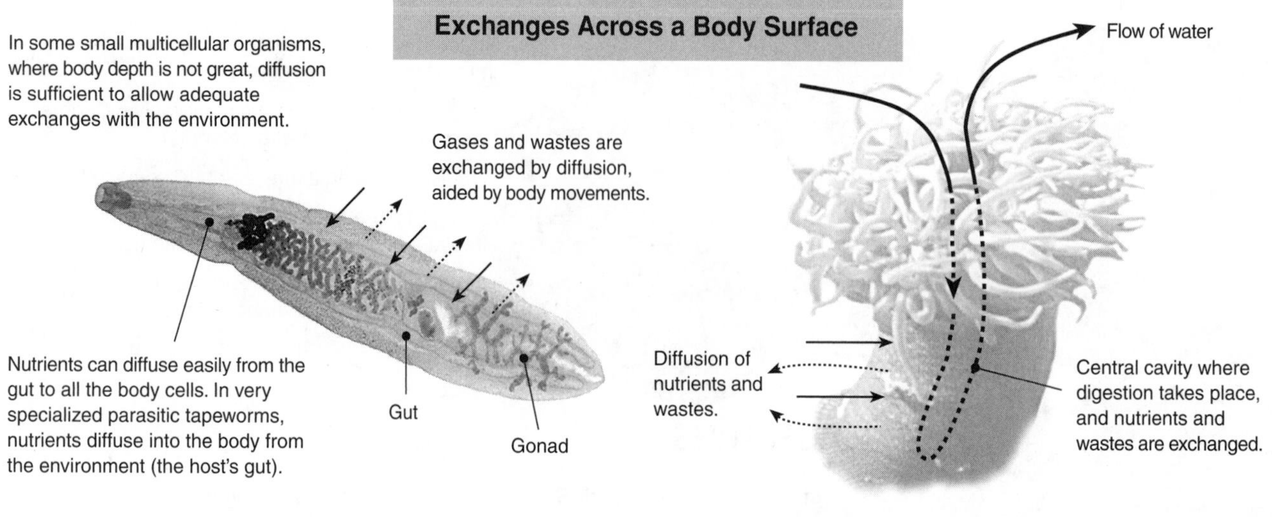

In some small multicellular organisms, where body depth is not great, diffusion is sufficient to allow adequate exchanges with the environment.
Gases and wastes are exchanged by diffusion, aided by body movements.
Nutrients can diffuse easily from the gut to all the body cells. In very specialized parasitic tapeworms, nutrients diffuse into the body from the environment (the host's gut).
Gut
Gonad
Flow of water
Diffusion of nutrients and wastes.
Central cavity where digestion takes place, and nutrients and wastes are exchanged.

Platyhelminthes (liver fluke) **Cnidarians (sea anemone)**

Systems for Exchange and Transport

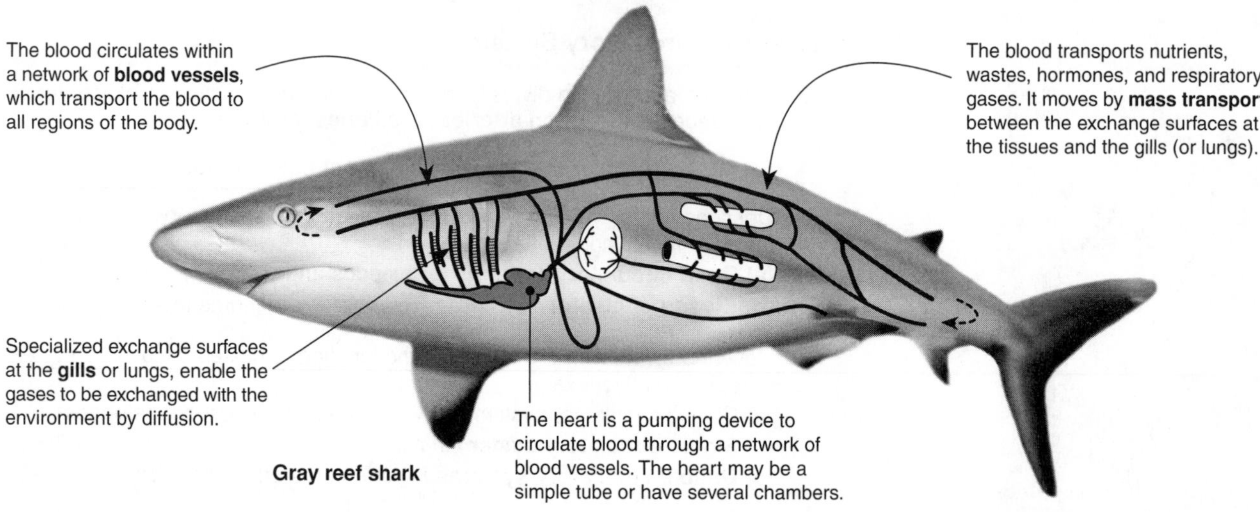

The blood circulates within a network of **blood vessels**, which transport the blood to all regions of the body.
The blood transports nutrients, wastes, hormones, and respiratory gases. It moves by **mass transport** between the exchange surfaces at the tissues and the gills (or lungs).
Specialized exchange surfaces at the **gills** or lungs, enable the gases to be exchanged with the environment by diffusion.
The heart is a pumping device to circulate blood through a network of blood vessels. The heart may be a simple tube or have several chambers.

Gray reef shark

1. Explain why animals above a certain size or level of complexity require specialized systems for transport and exchange:

2. (a) Describe how materials move within the circulatory system of a vertebrate:

   (b) Contrast this with how materials are transported in a flatworm or single celled eukaryote:

   (c) Identify two exchange sites in a vertebrate:

***Related activities**: Open Circulatory Systems, Closed Circulatory Systems*

Blood Vessels

The blood vessels of the circulatory system connect the fluid environment of the body's cells to the organs that exchange gases, absorb nutrients, and dispose of wastes. In vertebrates, arteries carry blood away from the heart to the capillaries within the tissues. The large arteries leaving the heart branch repeatedly to form distributing arteries, which themselves divide to form small **arterioles** within the tissues and organs. Arterioles deliver blood to the capillaries connecting the arterial and venous systems. Capillaries enable the exchange of nutrients and wastes between the blood and tissues, and they form large networks, especially in tissues and organs with high metabolic rates. The structural differences between blood vessels are related to their functional roles. While vessels close to the heart exhibit all the layers typical of the vessel's type, one or more layers may be absent in vessels more distant from the heart. Capillaries, whose functional role is exchange, consist only of a thin endothelium.

Artery structure

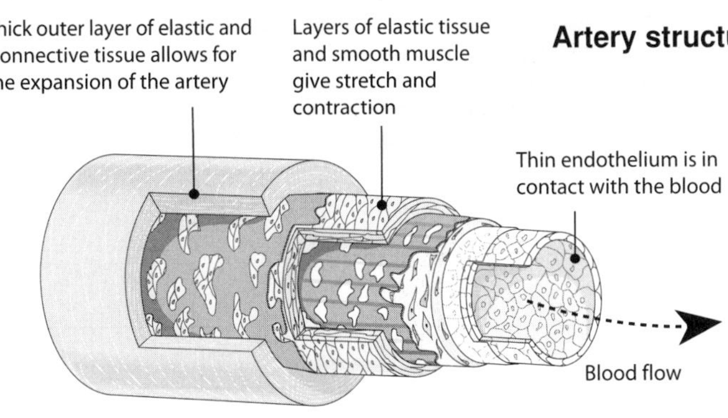
Thick outer layer of elastic and connective tissue allows for the expansion of the artery
Layers of elastic tissue and smooth muscle give stretch and contraction
Thin endothelium is in contact with the blood
Blood flow

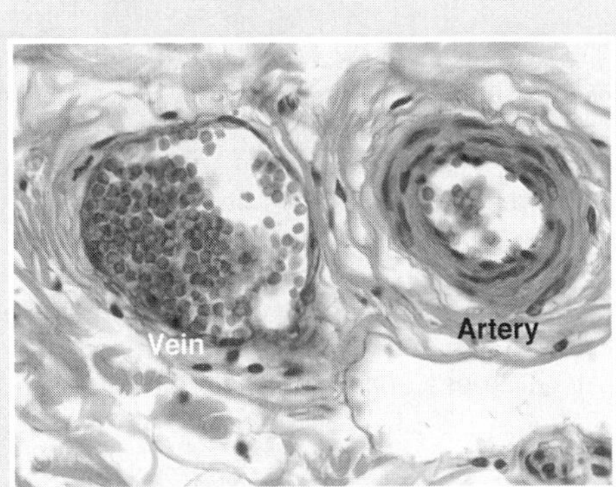
Vein
Artery

Arteries are made up of three layers; an inner layer of thin epthelium called the endothelium, a stretchy middle layer, and a thick outer layer. This structure enables arteries to withstand and maintain high blood pressure. **Veins** are similar in structure to arteries, but have less elastic and muscle tissue. Although veins are less elastic than arteries, they can still expand enough to adapt to changes in the pressure and volume of the blood passing through them.

Vein structure

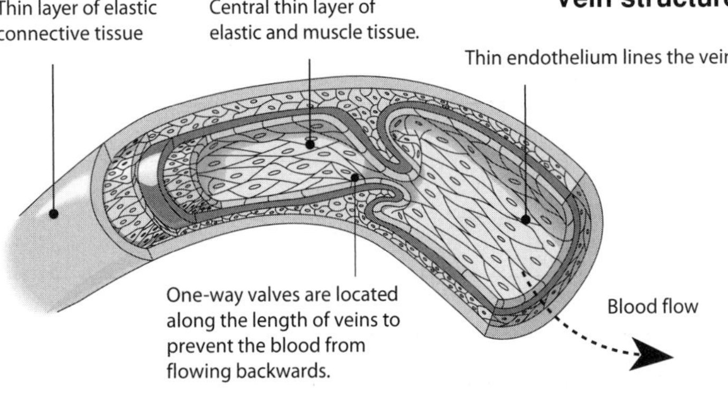
Thin layer of elastic connective tissue
Central thin layer of elastic and muscle tissue.
Thin endothelium lines the vein.
One-way valves are located along the length of veins to prevent the blood from flowing backwards.
Blood flow

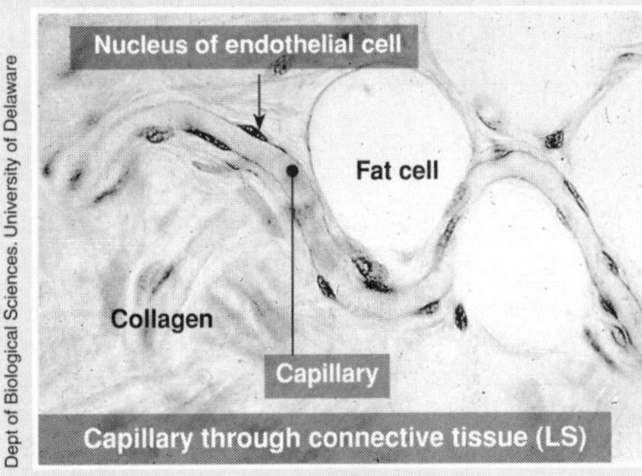
Nucleus of endothelial cell
Fat cell
Collagen
Capillary
Capillary through connective tissue (LS)

Dept of Biological Sciences, University of Delaware

Capillaries are very small blood vessels (4-10 µm diameter) made up of only a single layer of flattened (squamous) epithelial cells. Capillaries form a vast network of vessels that penetrate all parts of the body and are so numerous that no cell is more than 25 µm from any capillary. It is in the capillaries that the exchange of materials between the body cells and the blood takes place.

Capillary structure

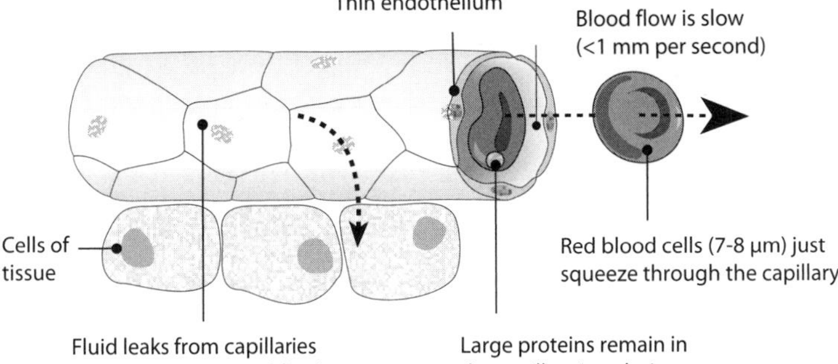
Thin endothelium
Blood flow is slow (<1 mm per second)
Cells of tissue
Red blood cells (7-8 µm) just squeeze through the capillary
Fluid leaks from capillaries to form the interstitial fluid that bathes the tissues.
Large proteins remain in the capillary in solution.

Formation of Tissue Fluid

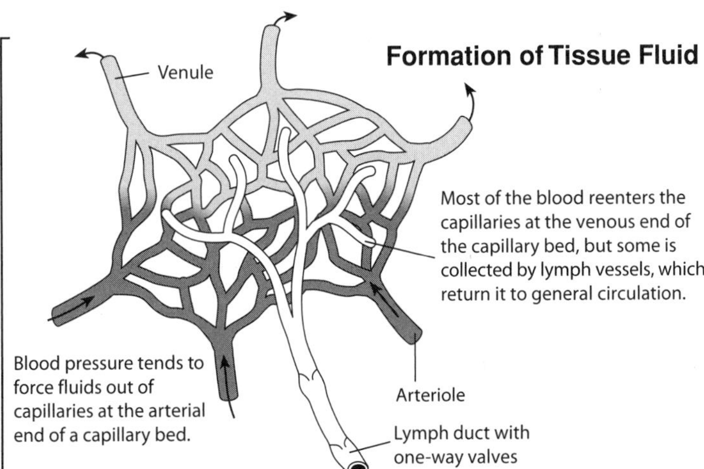
Capillary bed and associated lymphatic vessels
Venule
Most of the blood reenters the capillaries at the venous end of the capillary bed, but some is collected by lymph vessels, which return it to general circulation.
Blood pressure tends to force fluids out of capillaries at the arterial end of a capillary bed.
Arteriole
Lymph duct with one-way valves

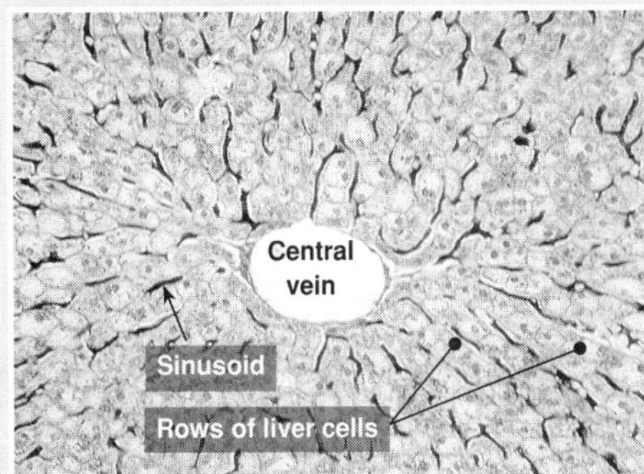
Central vein
Sinusoid
Rows of liver cells

EII

Tiny blood vessels in dense organs, such as the liver (above), are called **sinusoids**. They are wider than capillaries and follow a more convoluted path through the tissue. Instead of the usual endothelial lining, they are lined with phagocytic cells. Sinusoids, like capillaries, transport blood from arterioles to venules.

Periodicals: *Cunning plumbing, Blood pressure*

Related activities: *Capillary Networks*
Web links: *Arteries, Veins*

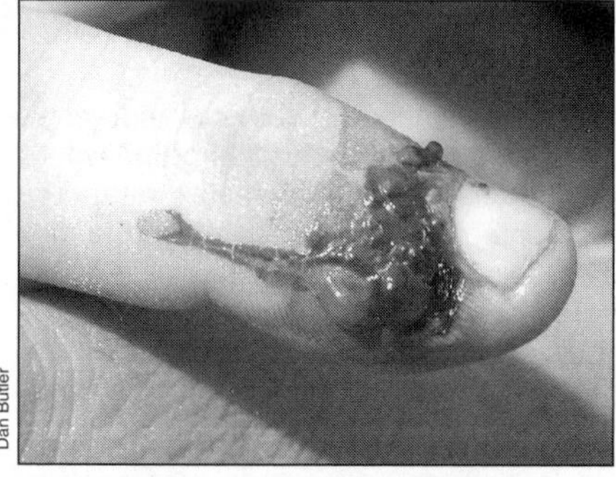

If a vein is cut, as in the severe finger wound shown above, the blood oozes out slowly in an even flow, and usually clots quickly as it leaves. In contrast, if a cut is made into an artery, the arterial blood spurts rapidly and requires pressure to staunch the flow.

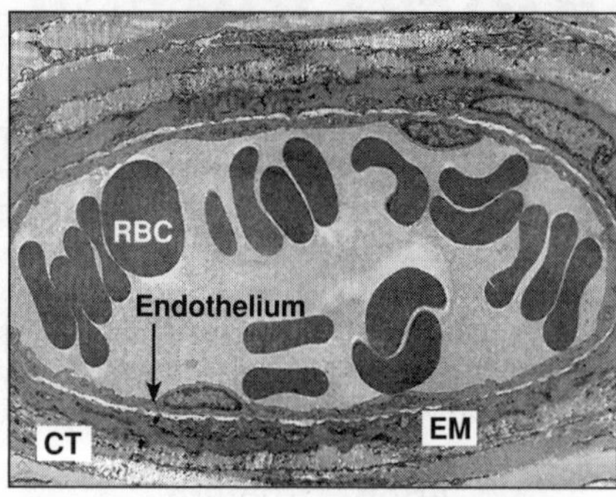

This TEM shows the structure of a typical vein. Note the red blood cells (RBC) in the lumen of the vessel, the inner layer of of epithelial cells (the endothelium), the central layer of elastic and muscle tissue (EM), and the outer connective tissue (CT) layer.

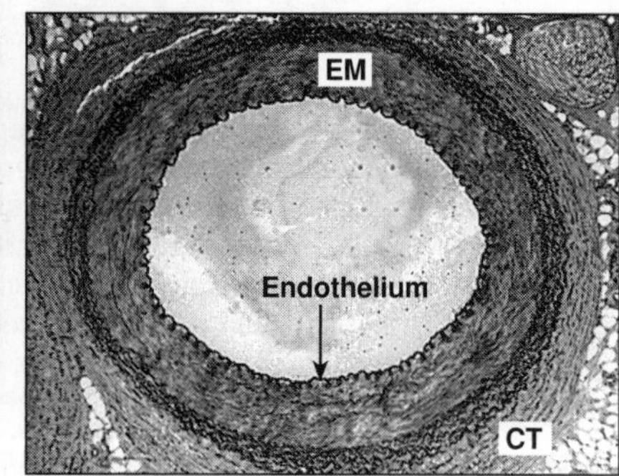

Arteries have a thick central layer of elastic and smooth muscle tissue (EM). Near the heart, arteries have more elastic tissue. This enables them to withstand high blood pressure. Arteries further from the heart have more smooth muscle; this helps them to maintain blood pressure.

1. Describe the contrasting structure of veins and arteries for each of the following properties:

   (a) Thickness of muscle and elastic tissue: ______

   (b) Size of the lumen (inside of the vessel): ______

2. Explain the reasons for the differences you have described above: ______

3. (a) Describe the structure of capillaries, explaining how it differs from that of veins and arteries: ______

   (b) Explain the reasons for these differences: ______

4. Compare the rate and force of blood flow in arteries, veins, and capillaries, explaining reasons for the differences:

5. Describe the role of the valves in assisting the veins to return blood back to the heart: ______

6. Explain why blood oozes from a venous wound, rather than spurting as it does from an arterial wound:

7. Explain why capillaries form dense networks in tissues with a high metabolic rate: ______

Capillary Networks

Capillaries form branching networks where exchanges between the blood and tissues take place. The flow of blood through a capillary bed is called **microcirculation**. In most parts of the body, there are two types of vessels in a capillary bed: the **true capillaries**, where exchanges take place, and a vessel called a **vascular shunt**, which connects the arteriole and venule at either end of the bed. The shunt diverts blood past the true capillaries when the metabolic demands of the tissue are low (e.g. vasoconstriction in the skin when conserving body heat). When tissue activity increases, the entire network fills with blood.

1. Describe the structure of a capillary network:

2. Explain the role of the smooth muscle sphincters and the vascular shunt in a capillary network:

3. (a) Describe a situation where the capillary bed would be in the condition labeled A:

   (b) Describe a situation where the capillary bed would be in the condition labeled B:

4. Explain how a portal venous system differs from other capillary systems:

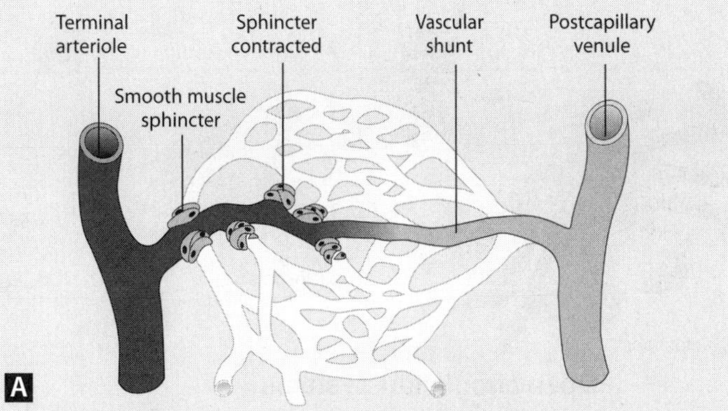

A When the sphincters contract (close), blood is diverted via the vascular shunt to the postcapillary venule, bypassing the exchange capillaries.

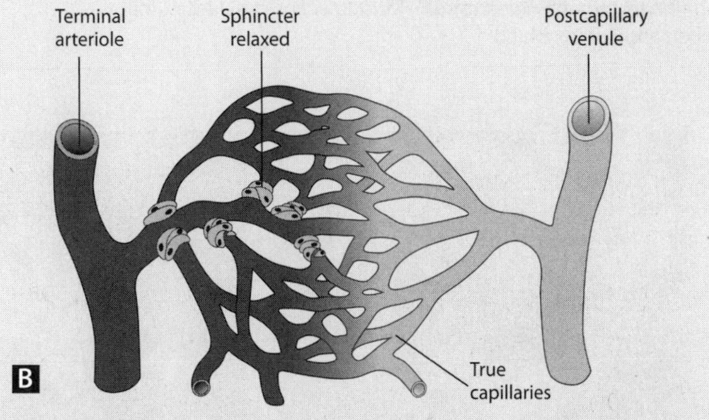

B When the sphincters are relaxed (open), blood flows through the entire capillary bed allowing exchanges with the cells of the surrounding tissue.

Connecting Capillary Beds

the role of portal venous systems

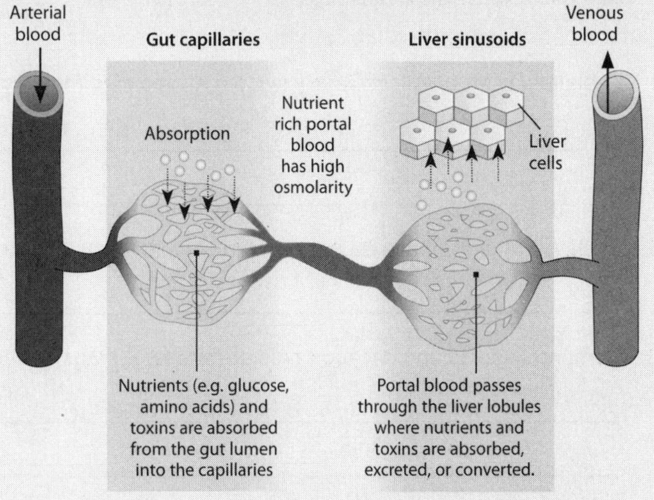

A portal venous system occurs when a capillary bed drains into another capillary bed through veins, without first going through the heart. Portal systems are relatively uncommon; most capillary beds drain into veins which then drain into the heart, not into another capillary bed. The diagram above depicts the hepatic portal system, which includes both capillary beds and the blood vessels connecting them.

Open Circulatory Systems

Two basic types of circulatory systems have evolved in animals. Many invertebrates have an **open circulatory system**, while vertebrates have a **closed circulatory system**, consisting of a heart and a network of tube-like vessels. The circulatory systems of arthropods are open but varied in complexity. Insects, unlike most other arthropods, do not use a circulatory system to transport oxygen, which is delivered directly to the tissues via the system of tracheal tubes. In addition to its usual transport functions, the circulatory system may also be important in hydraulic movements of the whole body (as in many molluscs) or its component parts (e.g. newly emerged butterflies expand their wings through hydraulic pressure).

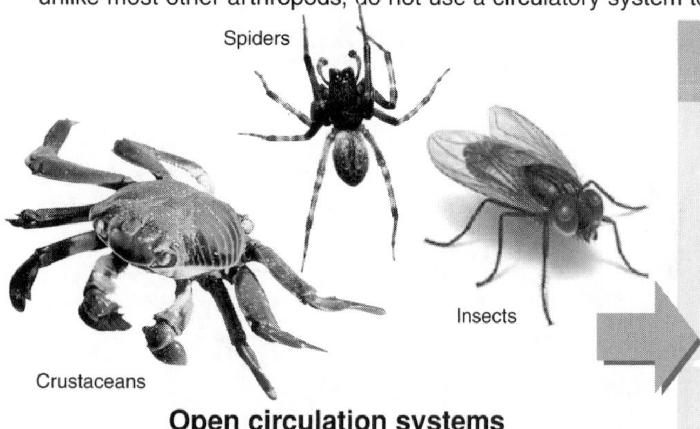

Open circulation systems

Arthropods and mollusks (except cephalopods) have open circulatory systems in which the blood is pumped by a tubular, or sac-like, heart through short vessels into large spaces in the body cavity. The blood bathes the cells before reentering the heart through holes (**ostia**). Muscle action may assist the circulation of the blood.

Open Circulatory System

Tubular heart on the dorsal (top) surface of the animal. Circulating fluids are pumped towards the head.

Ostium (hole) for the uptake of blood

One way valves ensure the blood flows in the forward direction.

Head

TUBULAR HEART

Abdomen

Body fluids flow freely within the body cavity

The circulatory system of crabs is best described as incompletely closed. The thoracic heart has three pairs of ostia a number of arteries, which leave the heart and branch extensively to supply various organs before draining into discrete channel-like sinuses.

In spiders, arteries from the dorsal heart empty the hemolymph into tissue spaces and then into a large ventral sinus that bathes the book lungs where gas exchange takes place. Venous channels conduct the hemolymph back to the heart.

The hemolymph occupies up to 40% of the body mass of an insect and is usually under low pressure due its lack of confinement in vessels. The circulation of the hemolymph is aided by body movements such as the ventilating movements of the abdomen.

1. Explain how an open circulatory system moves fluid (hemolymph) about the body: ________________

2. Explain why arthropods do not bleed in a similar way to vertebrates: ________________

3. Compare insects and decapod crustaceans (e.g. crabs) in the degree to which the circulatory system is closed:

4. (a) Explain why the crab's circulatory system is usually described as an open system: ________________

   (b) Explain in what way this description is not entirely accurate: ________________

Closed Circulatory Systems

Closed circulatory systems are used by vertebrates, annelids (earthworms) and cephalopods (octopus and squid). The blood is pumped by a heart through a series of arteries and veins. Oxygen is transported around the body by the blood and diffuses through capillary walls into the body cells. Closed circulatory systems are useful for large, active animals where oxygen can not easily be transported to the interior of the body. They also allow the animal more control over the distribution of blood flow by contracting or dilating blood vessels. Closed systems are the most developed in vertebrates where a chambered heart pumps the blood into blood vessels at high pressure. The system can also be divided into two separate regions, the pulmonary region taking up oxygen and the systematic region pumping oxygenated blood to the rest of the body.

INVERTEBRATE CLOSED SYSTEMS

Wiki: Hans Hillewaert

The closed systems of many annelids (e.g. earthworms) circulate blood through a series of vessels before returning it to the heart. In annelids, the dorsal and ventral blood vessels are connected by lateral vessels in every segment (right). The dorsal vessel receives blood from the lateral vessels and carries it towards the head. The ventral vessel carries blood posteriorly and distributes it to the segmental vessels. The dorsal vessel is contractile and is the main method of propelling the blood, but there are also several contractile aortic arches ('hearts') which act as accessory organs for blood propulsion.

VERTEBRATE CLOSED SYSTEMS

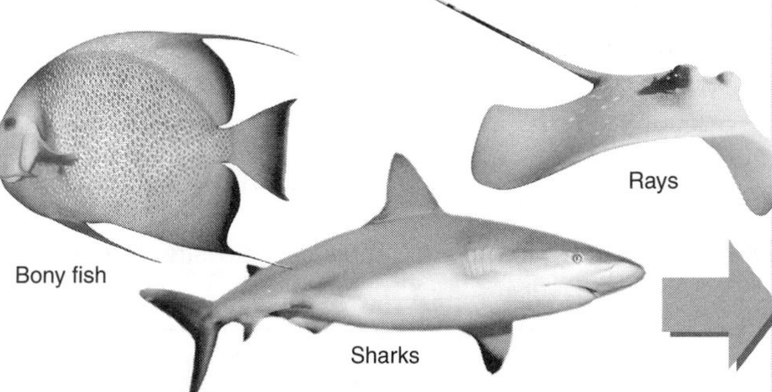

Closed, single circuit systems

In closed circulation systems, the blood is contained within vessels and is returned to the heart after every circulation of the body. Exchanges between the blood and the fluids bathing the cells occurs by diffusion across capillaries. In single circuit systems, typical of fish, the blood goes directly from the gills to the body. The blood loses pressure at the gills and flows at low pressure around the body.

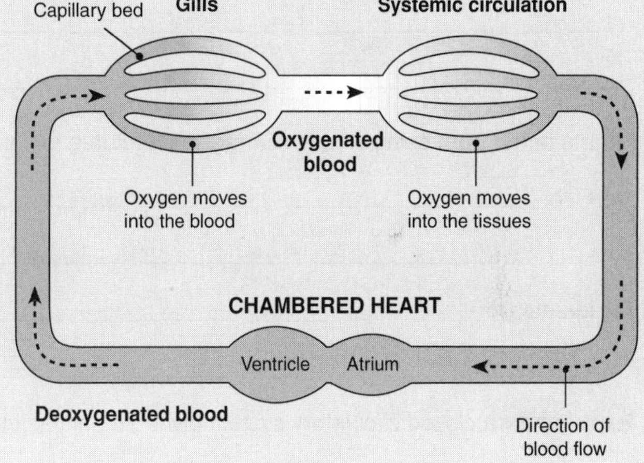

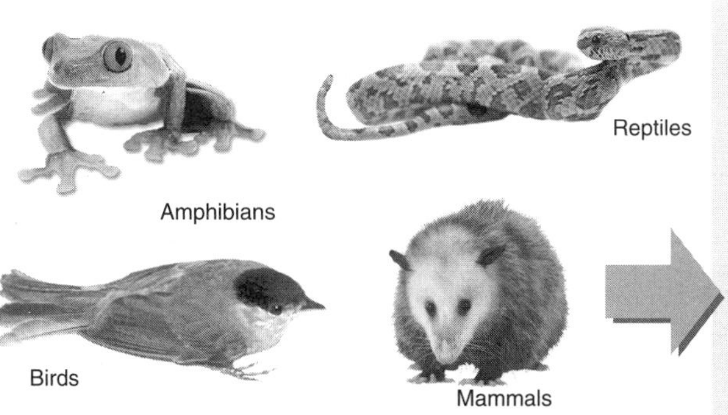

Closed, double circuit systems

Double circulation systems occur in all vertebrates other than fish. The blood is pumped through a pulmonary circuit to the lungs, where it is oxygenated. The blood returns to the heart, which pumps the oxygenated blood, through a systemic circuit, to the body. In amphibians and most reptiles, the heart is not completely divided and there is some mixing of oxygenated and deoxygenated blood. In birds and mammals, the heart is fully divided and there is no mixing.

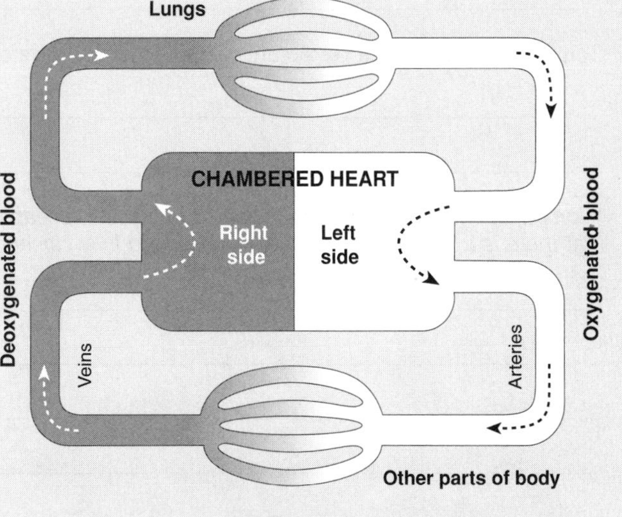

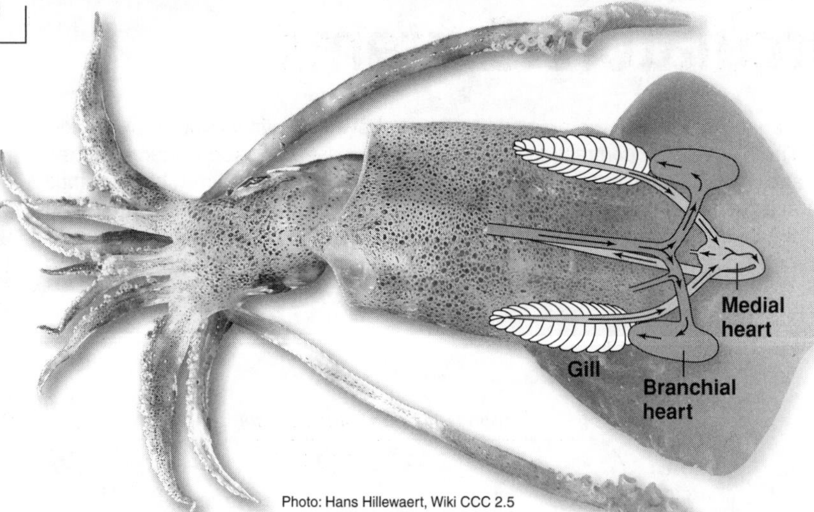

Photo: Hans Hillewaert, Wiki CCC 2.5

CEPHALOPOD MOLLUSCS: HIGH PERFORMING INVERTEBRATES

The circulatory system is largely closed in all cephalopod molluscs (nautilus, cuttlefish, squid, and octopus). It has an extensive system of vessels making it the most complex and efficient system of all the molluscs, and enables cephalopods to be active, intelligent predators.

The circulatory system consists of one systemic heart, two branchial hearts, and blood vessels. The branchial hearts, which sit at the gill base, collect deoxygenated blood from all the body parts and direct it through the gills. The blood returns to the medial systemic heart and is pumped to the body via an anterior and posterior aorta, through smaller vessels and into tissue capillaries.

1. Describe the main difference between closed and open systems of circulation: __________

2. Describe where the blood flows to immediately after it has passed through the gills in a fish: __________

3. Describe where the blood flows immediately after it has passed through the lungs in a mammal: __________

4. Explain the higher functional efficiency of a double circuit system, relative to a single circuit system: __________

5. Hearts range from being simple contractile structures to complex chambered organs. Describe basic heart structure in:

   (a) Fish: __________

   (b) Mammals: __________

6. Explain how a closed circulatory system gives an animal finer control over the distribution of blood to tissues and organs:

7. Compare and contrast a vertebrate closed circulatory system with the circulatory system of an annelid: __________

8. "*Comparisons of the circulatory systems of insects, decapods (e.g. crabs), annelids, and cephalopod molluscs indicates that there is a gradient between fully open and fully closed circulatory systems*". Discuss this statement:

The Heart as a Pump

In vertebrates, the heart shows a sequential increase in complexity from fish through to mammals. In fish, the heart is linear and contains two major chambers in series, and on the venous side there is an enlarged chamber or **sinus** on the vein (the sinus venosus). In mammals, the heart comprises four chambers - two pumps side by side - and there are large pressure differences between the pulmonary (lung) and systemic (body) circulations. The three chambered heart of amphibians reflects, in part, their incomplete shift to terrestrial life. Although the ventricle is undivided, a baffle-like spiral valve at the exit point of the ventricle helps to separate the arterial and venous flows and there is limited mixing of oxygenated and deoxygenated blood. The pulmonary circuit in amphibians also sends branches to the skin, reflecting its importance in oxygen uptake.

Fish Heart

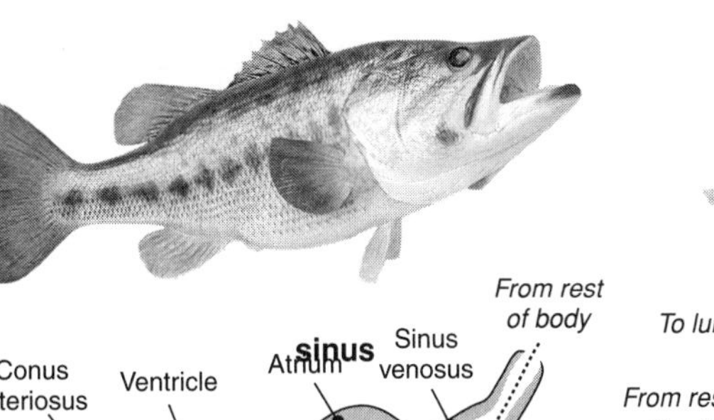

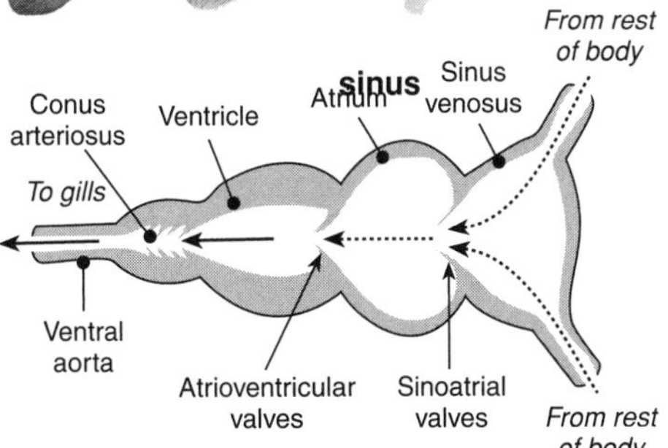

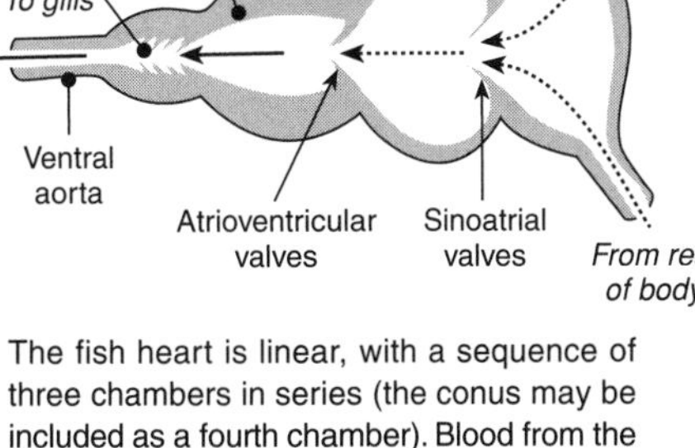

The fish heart is linear, with a sequence of three chambers in series (the conus may be included as a fourth chamber). Blood from the body first enters the heart through the sinus venosus, then passes into the atrium and the ventricle. A series of one-way valves between the chambers prevents reverse blood flow. Blood leaving the heart travels to the gills.

Amphibian Heart

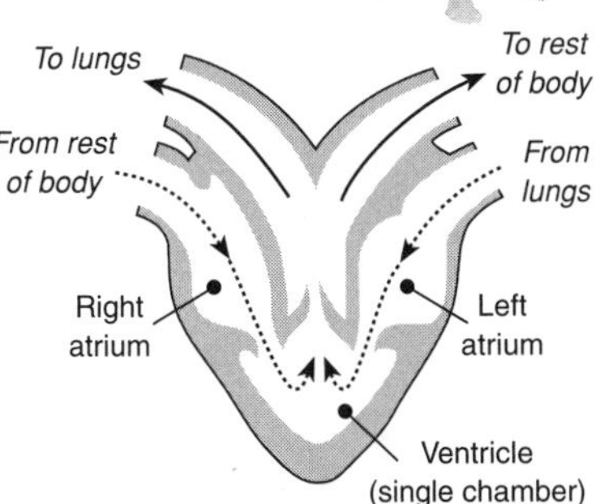

Amphibian hearts are three chambered. The atrium is divided into left and right chambers, but the ventricle lacks an internal dividing wall. Although this allows mixing of oxygenated and deoxygenated blood, the spongy nature of the ventricle reduces mixing. Amphibians are able to tolerate this because much of their oxygen uptake occurs across their moist skin, and not their lungs.

Mammalian Heart

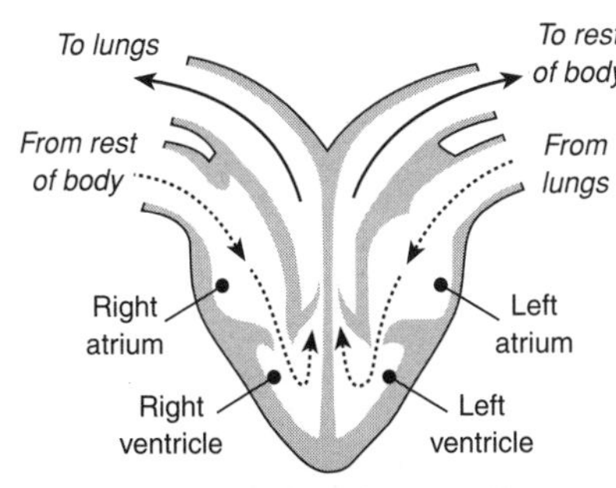

In birds and mammals, the heart is fully partitioned into two halves, resulting in four chambers. Blood circulates through two circuits, with no mixing of the two. Oxygenated blood from the lungs is kept separated from the deoxygenated blood returning from the rest of the body.

Heart Size and Rate in Mammals

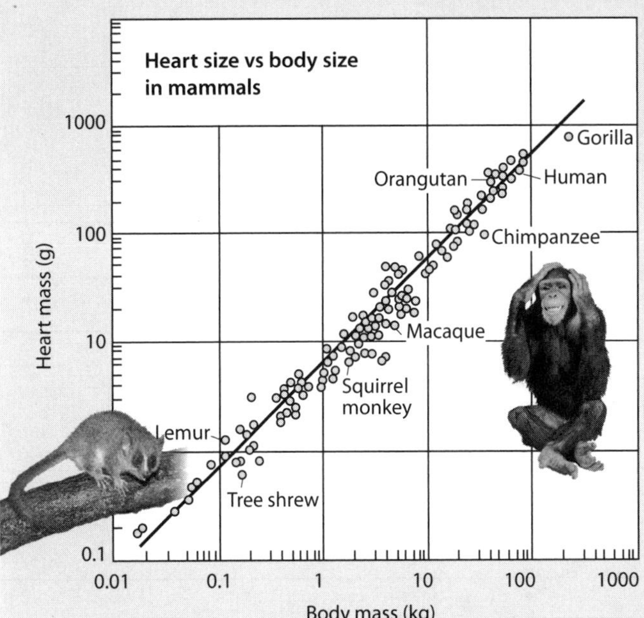

Heart size in mammals (above) increases with body size, but relative to body size, small and large mammals have the same heart size. Irrespective of body size, the size of the heart in mammals is 0.59% of body size. Humans and other primates fall within the range of other mammals.

Adapted from Schmidt-Nielsen, 1979: Animal Physiology, Adaptation and Environment

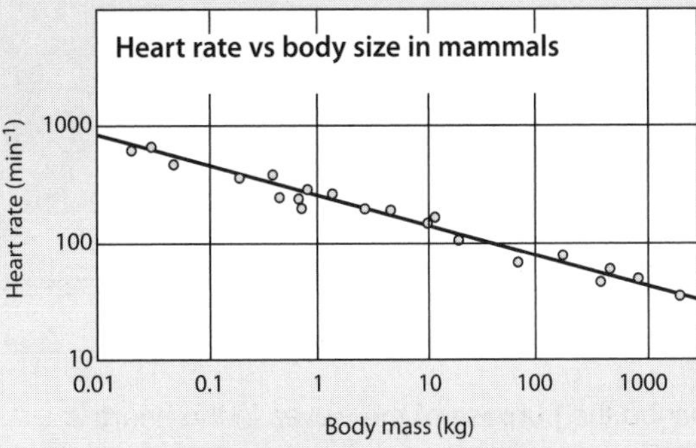

In contrast to the heart size-body size relationship, heart rate (above) is inversely related to body size. An elephant (3000 kg) has a resting pulse of around 25 beats per minute, compared with a 3 g shrew (the smallest mammal), which has a resting heart rate of over 600 beats per minute. The relationship is identical to that between body mass and oxygen consumption per unit body weight. The information from these two figures tell us that, in mammals:

1. The size of the heart (the pump) remains a constant percentage of body size, and...
2. The increase in heart rate in smaller mammals is in exact proportion to the need for oxygen.

1. In the schematic diagram of the human heart (below), label the four chambers and the main vessels entering and leaving them. Arrows indicate the direction of blood flow. Use large circles to mark the position of each of the four valves.

2. Use the diagram on the previous page to draw a similar labeled schematic of an amphibian heart.

Schematic Diagrams of Heart Structure

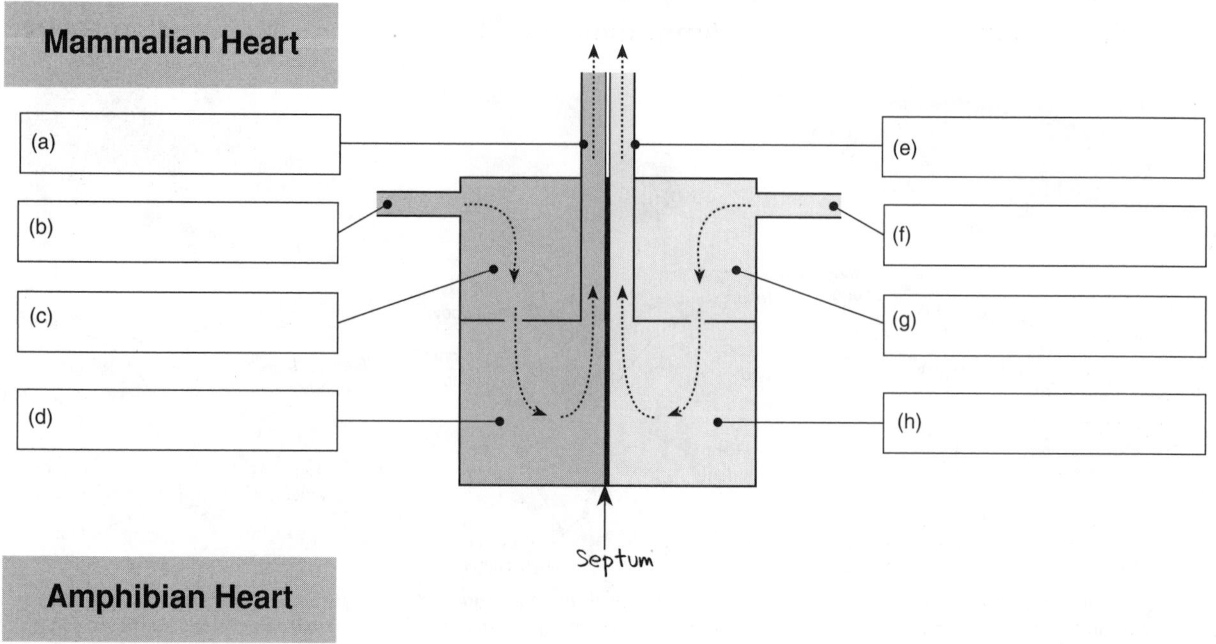

Amphibian Heart

3. Describe the function of the aorta: ______

4. Describe the function of the vena cava: ______

5. Describe the purpose of the valves in the heart: ______

6. Discuss the structure of the heart in at least two vertebrate classes, relating features of the heart's structure and function to the animal's size, metabolic rate, or environment in each case:

Blood

Blood makes up about 8% of body weight in humans. Blood is a complex liquid tissue comprising cellular components suspended in plasma. If a blood sample is taken, the cells can be separated from the plasma by centrifugation. The cells (formed elements) settle as a dense red pellet below the transparent, straw-colored plasma. Blood performs many functions: it transports nutrients, respiratory gases, hormones, and wastes; it has a role in thermoregulation through the distribution of heat; it defends against infection; and its ability to clot protects against blood loss. The examination of blood is also useful in diagnosing disease. The cellular components of blood are normally present in particular specified ratios. A change in the morphology, type, or proportion of different blood cells can therefore be used to indicate a specific disorder or infection (see the next page).

Non-Cellular Blood Components

The non-cellular blood components form the plasma. Plasma is a watery matrix of ions and proteins and makes up 50-60% of the total blood volume.

Water

The main constituent of blood and lymph.
Role: Transports dissolved substances. Provides body cells with water. Distributes heat and has a central role in thermoregulation. Regulation of water content helps to regulate blood pressure and volume.

Mineral ions

Sodium, bicarbonate, magnesium, potassium, calcium, chloride.
Role: Osmotic balance, pH buffering, and regulation of membrane permeability. They also have a variety of other functions, e.g. Ca^{2+} is involved in blood clotting.

Plasma proteins

7-9% of the plasma volume.
Serum albumin
Role: Osmotic balance and pH buffering, Ca^{2+} transport.
Fibrinogen and prothrombin
Role: Take part in blood clotting.
Immunoglobulins
Role: Antibodies involved in the immune response.
α-globulins
Role: Bind/transport hormones, lipids, fat soluble vitamins.
β-globulins
Role: Bind/transport iron, cholesterol, fat soluble vitamins.
Enzymes
Role: Take part in and regulate metabolic activities.

Substances transported by non-cellular components

Products of digestion
Examples: sugars, fatty acids, glycerol, and amino acids.
Excretory products
Example: urea
Hormones and vitamins
Examples: insulin, sex hormones, vitamins A and B_{12}.
Importance: These substances occur at varying levels in the blood. They are transported to and from the cells dissolved in the plasma or bound to plasma proteins.

Cellular Blood Components

The cellular components of the blood (also called the formed elements) float in the plasma and make up 40-50% of the total blood volume.

Erythrocytes (red blood cells or RBCs)

5-6 million per mm^3 blood; 38-48% of total blood volume.
Role: RBCs transport oxygen (O_2) and a small amount of carbon dioxide (CO_2). The oxygen is carried bound to hemoglobin (Hb) in the cells. Each Hb molecule can bind four molecules of oxygen.

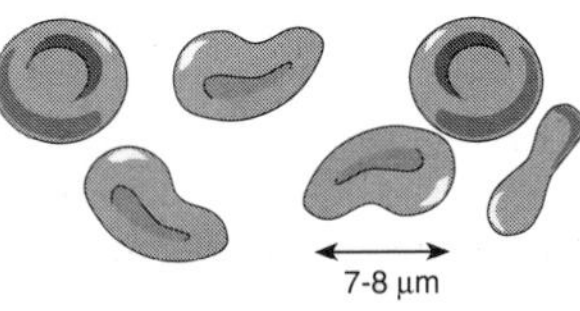

Platelets

2 μm

Small, membrane bound cell fragments derived from bone marrow cells; about 1/4 the size of RBCs.
0.25 million per mm^3 blood.
Role: To start the blood clotting process.

Leukocytes (white blood cells)

5-10 000 per mm^3 blood
2-3% of total blood volume.
Role: Involved in internal defense. There are several types of white blood cells (see below).

Lymphocytes
T and B cells.
24% of the white cell count.
Role: Antibody production and cell mediated immunity.

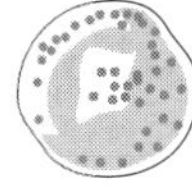

Neutrophils
Phagocytes.
70% of the white cell count.
Role: Engulf foreign material.

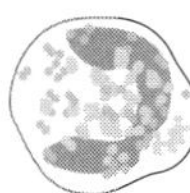

Eosinophils
Rare leukocytes; normally 1.5% of the white cell count.
Role: Mediate allergic responses such as hayfever and asthma.

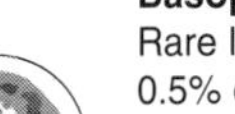

Basophils
Rare leukocytes; normally 0.5% of the white cell count.
Role: Produce heparin (an anti-clotting protein), and histamine. Involved in inflammation.

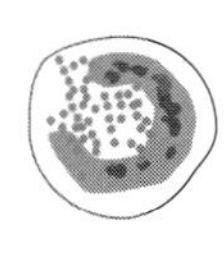

Periodicals:
Red blood cells

Related activities: Gas Transport in Humans, The Body's Defenses, Hemostasis, Thermoregulation in Humans

The Examination of Blood

Different types of microscopy give different information about blood. A SEM (right) shows the detailed external morphology of the blood cells. A fixed smear of a blood sample viewed with a light microscope (far right) can be used to identify the different blood cell types present, and their ratio to each other. Determining the types and proportions of different white blood cells in blood is called a **differential white blood cell count**. Elevated counts of particular cell types indicate allergy or infection.

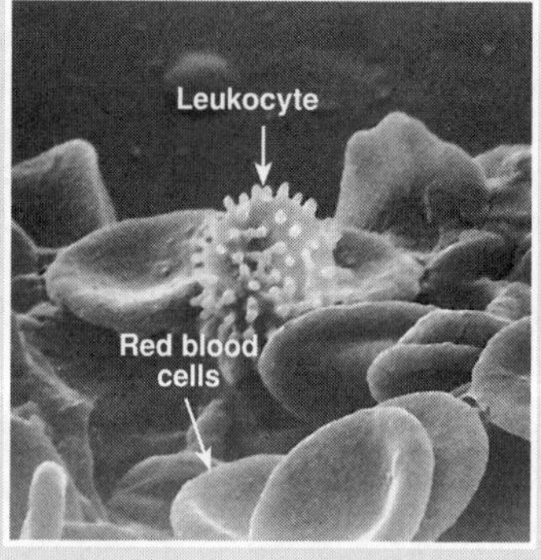

SEM of red blood cells and a leukocytes.

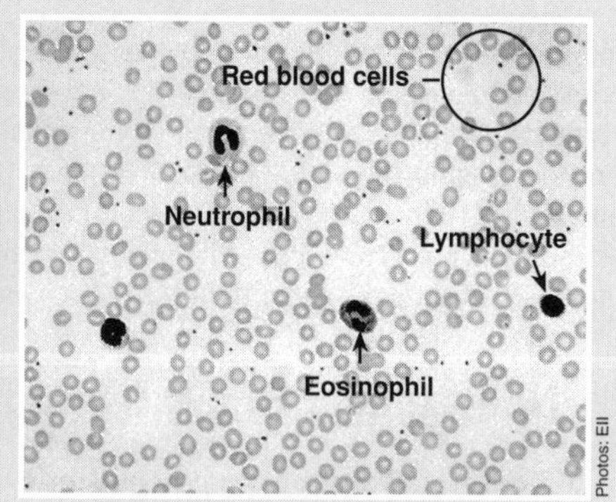

Light microscope view of a fixed blood smear.

1. For each of the following blood functions, identify the component (or components) of the blood responsible and state how the function is carried out (the mode of action). The first one is done for you:

   (a) **Temperature regulation**. *Blood component*: Water component of the plasma

   Mode of action: Water absorbs heat and dissipates it from sites of production (e.g. organs)

   (b) **Protection against disease**. *Blood component*: ______

   Mode of action: ______

   (c) **Communication between cells, tissues, and organs**. *Blood component*: ______

   Mode of action: ______

   (d) **Oxygen transport**. *Blood component*: ______

   Mode of action: ______

   (e) **CO_2 transport**. *Blood components*: ______

   Mode of action: ______

   (f) **Buffer against pH changes**. *Blood components*: ______

   Mode of action: ______

   (g) **Nutrient supply**. *Blood component*: ______

   Mode of action: ______

   (h) **Tissue repair**. *Blood components*: ______

   Mode of action: ______

   (i) **Transport of hormones, lipids, and fat soluble vitamins**. *Blood component*: ______

   Mode of action: ______

2. Identify a feature that distinguishes red and white blood cells: ______

3. Explain two physiological advantages of red blood cell structure (lacking nucleus and mitochondria):

   (a) ______

   (b) ______

4. Suggest what each of the following results from a differential white blood cell count would suggest:

   (a) Elevated levels of eosinophils (above the normal range): ______

   (b) Elevated levels of neutrophils (above the normal range): ______

   (c) Elevated levels of basophils (above the normal range): ______

   (d) Elevated levels of lymphocytes (above the normal range): ______

Circulatory Fluids in Invertebrates

The internal transport system of most animals includes a circulating fluid. In animals with closed systems, the fluid in the blood vessels is distinct from the tissue fluid outside the vessels and is called **blood**. Blood can have many different appearances, depending on the animal group, but it usually consists of cells and cell fragments suspended in a watery fluid. In animals with open systems, there is no difference between the fluid in the vessels and that in the sinuses **(hemocoel)** so the circulating fluid is called **hemolymph.** In insects, the hemolymph carries nutrients but not respiratory gases.

Hemolymph is a blood-like substance found in all invertebrates with open circulatory systems. The hemolymph fills the hemocoel and surrounds all cells.

About 90% of insect hemolymph is plasma, a watery fluid, which is usually clear. Compared to vertebrate blood, it contains relatively high concentrations of amino acids, proteins, sugars, and inorganic ions.

The remaining 10% of hemolymph volume is made up of various cell types (hemocytes). These are involved in clotting and internal defence. Unlike vertebrate blood, insect hemolymph lacks red blood cells and (with a few exceptions) lacks respiratory pigment, because oxygen is delivered directly to tissues by the tracheal system.

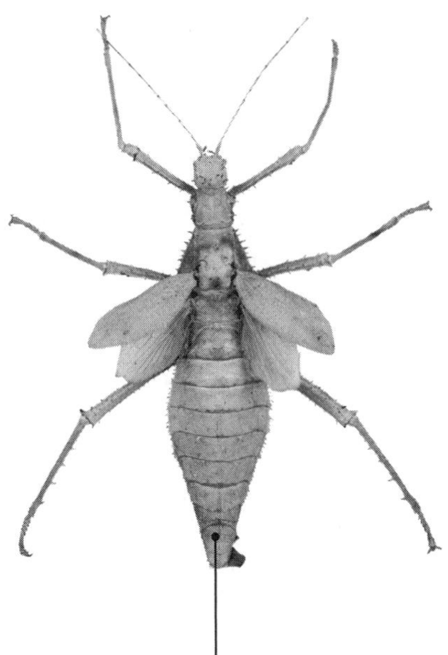

Hemolymph may make up between 11% and 40% of the total body mass of an insect

Fluid pressure is used to facilitate molting in insects (above and right). Some insects, such as New Zealand's alpine weta (below) can tolerate freezing during winter, when the osmotic pressure of the hemolymph almost doubles.

New Zealand's alpine weta can freeze solid. Their hemolymph contains the disaccharide trehalose, which acts as a cryoprotectant.

Pressure of the hemolymph enables arthropods to expand the soft cuticle of the body segments before they harden (sclerotize). This enables them to grow.

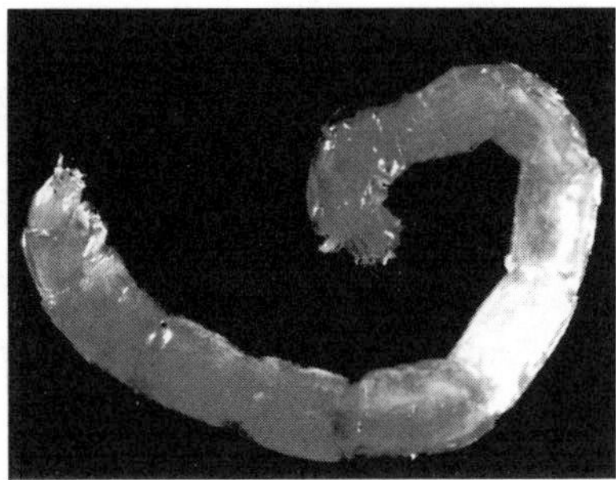

A few insects (but not many), like this midge larva, possess hemoglobin as an adaptation to living in low-oxygen substrates.

1. Describe two common functions of mammalian blood and insect hemolymph:

   (a) ______________________________

   (b) ______________________________

2. Describe one function of mammalian blood not commonly performed by insect hemolymph: ______________________________

3. Describe one function of insect hemolymph not performed by mammalian blood: ______________________________

4. Contrast the proportions of cellular and non-cellular components in blood and hemolymph: ______________________________

Related activities: *Blood, Respiratory Pigments*

Hemostasis

Apart from its transport role, **blood** has a role in the body's defense against infection and **hemostasis** (the prevention of bleeding and maintenance of blood volume). The tearing or puncturing of a blood vessel initiates **clotting**. Clotting is normally a rapid process that seals off the tear, preventing blood loss and the invasion of bacteria into the site. Clot formation is triggered by the release of clotting factors from the damaged cells at the site of the tear or puncture. A hardened clot forms a scab, which acts to prevent further blood loss and acts as a mechanical barrier to the entry of pathogens.

Blood Clotting

When tissue is wounded, the blood quickly coagulates to prevent further blood loss and maintain the integrity of the circulatory system. For external wounds, clotting also prevents the entry of pathogens. Blood clotting involves a cascade of reactions involving at least twelve clotting factors in the blood. The end result is the formation of an insoluble network of fibers, which traps red blood cells and seals the wound.

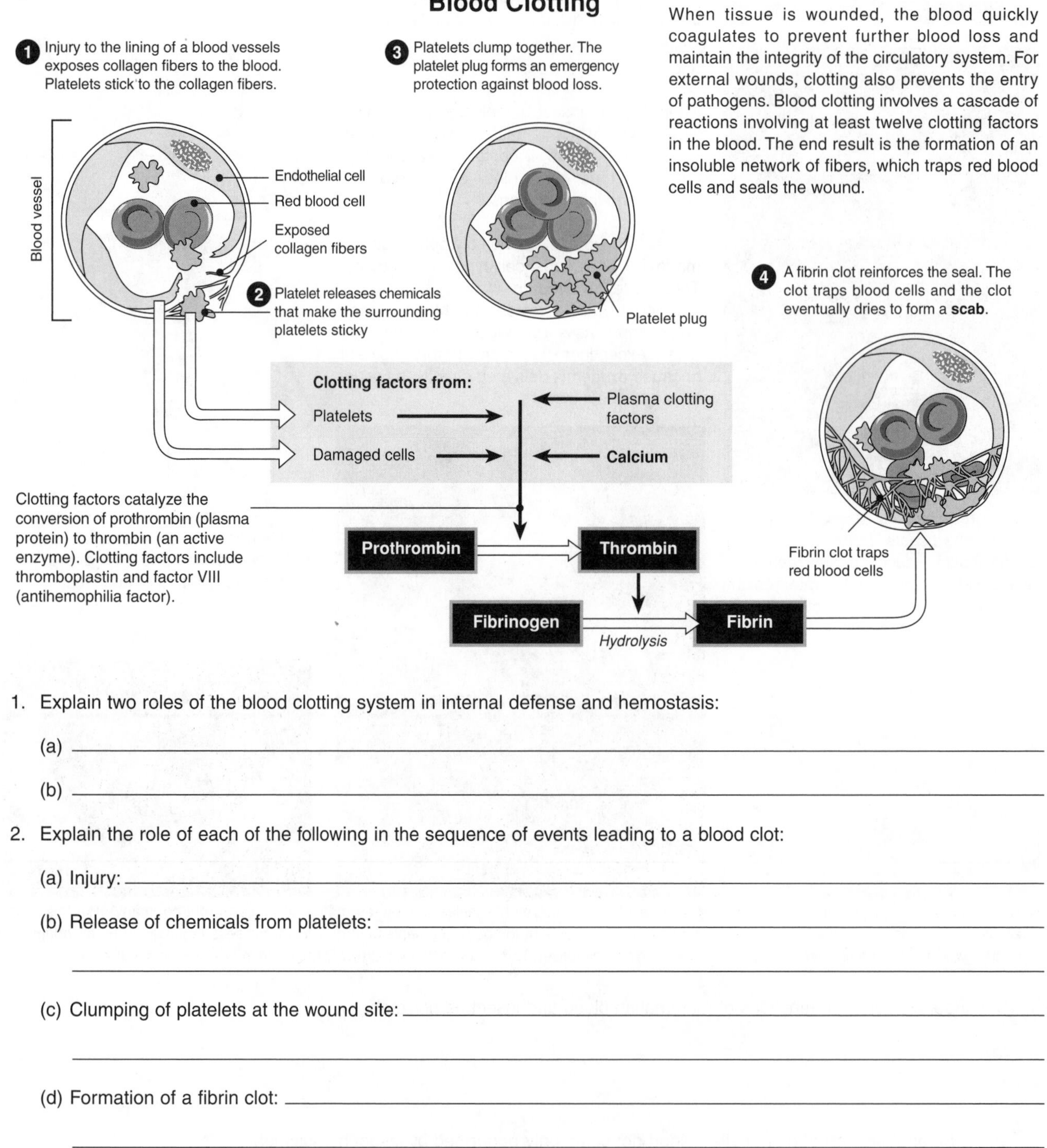
1 Injury to the lining of a blood vessels exposes collagen fibers to the blood. Platelets stick to the collagen fibers.
Blood vessel
Endothelial cell
Red blood cell
Exposed collagen fibers
2 Platelet releases chemicals that make the surrounding platelets sticky
3 Platelets clump together. The platelet plug forms an emergency protection against blood loss.
Platelet plug
4 A fibrin clot reinforces the seal. The clot traps blood cells and the clot eventually dries to form a scab.
Clotting factors from:
Platelets
Damaged cells
Plasma clotting factors
Calcium
Clotting factors catalyze the conversion of prothrombin (plasma protein) to thrombin (an active enzyme). Clotting factors include thromboplastin and factor VIII (antihemophilia factor).
Prothrombin
Thrombin
Fibrinogen
Fibrin
Hydrolysis
Fibrin clot traps red blood cells

1. Explain two roles of the blood clotting system in internal defense and hemostasis:

   (a) ____________________

   (b) ____________________

2. Explain the role of each of the following in the sequence of events leading to a blood clot:

   (a) Injury: ____________________

   (b) Release of chemicals from platelets: ____________________

   (c) Clumping of platelets at the wound site: ____________________

   (d) Formation of a fibrin clot: ____________________

3. (a) Explain the role of clotting factors in the blood in formation of the clot: ____________________

   (b) Explain why these clotting factors are not normally present in the plasma: ____________________

4. (a) Name one inherited disease caused by the absence of a clotting factor: ____________________

   (b) Name the clotting factor involved: ____________________

The Search for Blood Substitutes

Blood's essential homeostatic role is evident when considering the problems encountered when large volumes of blood are lost. Transfusion of whole blood (see photograph below) or plasma is an essential part of many medical procedures, e.g. after trauma or surgery, or as a regular part of the treatment for some disorders (e.g. thalassemia). This makes blood a valuable commodity. A blood supply relies on blood donations, but as the demand for blood increases, the availability of donors continues to decline. This decline is partly due to more stringent screening of donors for diseases such as HIV/AIDS, hepatitis, and variant CJD. The inadequacy of blood supplies has made the search for a safe, effective blood substitute the focus of much research. Despite some possibilities, no currently available substitute reproduces all of blood's many homeostatic functions.

Essential criteria for a successful blood substitute

- The substitute should be non-toxic and free from diseases.
- It should work for all blood types.
- It should not cause an immune response.
- It should remain in circulation until the blood volume is restored and then it should be safely excreted.
- It must be easily transported and suitable for storage under normal refrigeration.
- It should have a long shelf life.
- It should perform some or all of blood tasks.

A shortfall in blood supplies, greater demand, and public fear of contaminated blood, have increased the need for a safe, effective blood substitute. Such a substitute must fulfil strict criteria (above).

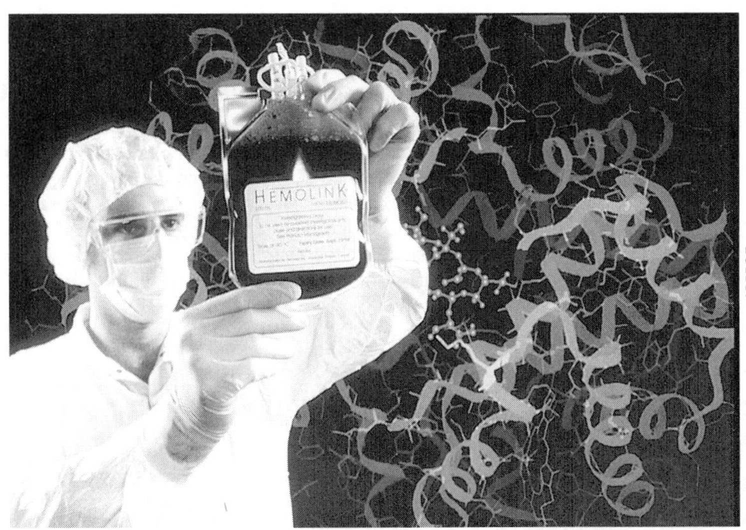
HEMOLINK

A researcher displays a hemoglobin based artificial blood product, developed by Defense R&D Canada, Toronto and now produced under license by Hemosol Inc. Human testing and marketing has now progressed successfully into advanced trials. A human hemoglobin molecule is pictured in the background. Photo with permission from Hemosol inc.

Photo Hemosol Inc. Information courtesy of DRDC Toronto

Chemical based

These rely on synthetic oxygen-carrying compounds called **perfluorocarbons** (PFCs). PFCs are able to dissolve large quantities of gases. They do not dissolve freely in the plasma, so they must be emulsified with an agent that enables them to be dispersed in the blood.

Advantages: PFCs can transport a lot of oxygen, and transfer gases quickly.

Disadvantages: May result in oxygen accumulation in the tissues, which can lead to damage.

Examples: Oxygent™: Produced in commercial quantities using PFC emulsion technology; Perflubon (a PFC), water, a surfactant, and salts, homogenized into a stable, biologically compatible emulsion.

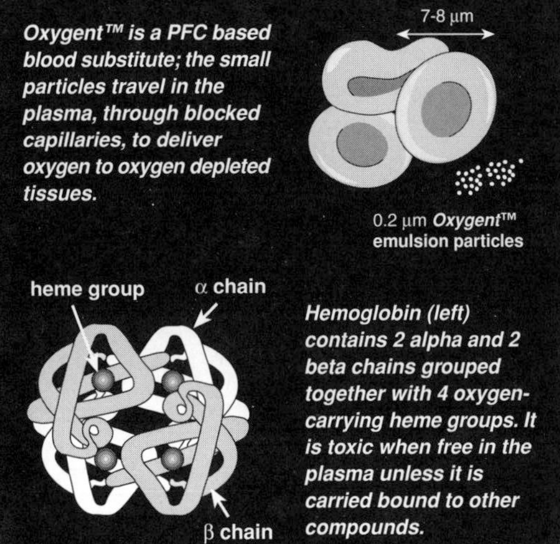
Oxygent™ is a PFC based blood substitute; the small particles travel in the plasma, through blocked capillaries, to deliver oxygen to oxygen depleted tissues.
7-8 μm
0.2 μm *Oxygent™* emulsion particles
heme group
α chain
β chain
Hemoglobin (left) contains 2 alpha and 2 beta chains grouped together with 4 oxygen-carrying heme groups. It is toxic when free in the plasma unless it is carried bound to other compounds.

Hemoglobin based

These rely on hemoglobin (Hb), modified by joining it to a polymer (polyethylene glycol) to make it larger.

Advantages: Modified hemoglobin should better be able to approximate the various properties of blood.

Disadvantages: Hb is toxic unless carried within RBCs; it requires modification before it can be safely transported free in the plasma. Substitutes made from human Hb use outdated blood as the Hb source. Bovine Hb may transmit diseases (e.g. BSE).

Examples: Hemolink™, a modified human Hb produced by Hemosol Inc. in California. Research is focused on developing cell culture lines with the ability to produce Hb.

1. Describe two essential features of a successful blood substitute, identifying briefly why the feature is important:

   (a) ____________________

   (b) ____________________

2. Identify the two classes of artificial blood substitutes: ____________________

3. Discuss the advantages and risks associated with the use of blood substitutes: ____________________

The Human Circulatory System

The blood vessels of the circulatory system form a vast network of tubes that carry blood away from the heart, transport it to the tissues of the body, and then return it to the heart. The arteries, arterioles, capillaries, venules, and veins are organized into specific routes to circulate the blood throughout the body. The figure below shows a number of the basic **circulatory routes** through which the blood travels. Humans, like all mammals have a double circulatory system: a **pulmonary circulation**, which carries blood between the heart and lungs, and a **systemic circulation**, which carries blood between the heart and the rest of the body. The systemic circulation has many subdivisions. Two important subdivisions are the coronary (cardiac) circulation, which supplies the heart muscle, and the **hepatic portal circulation**, which runs from the gut to the liver.

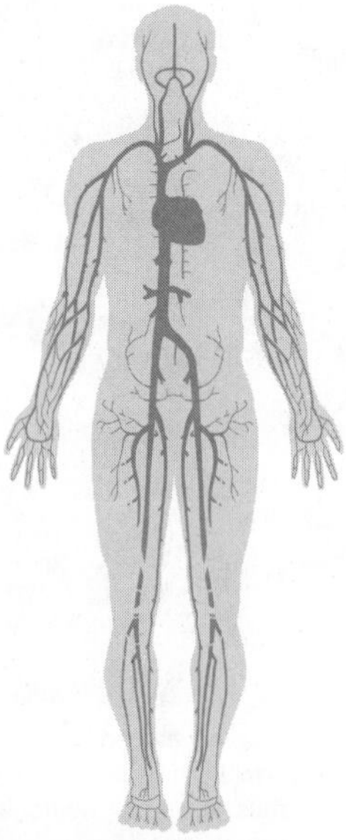

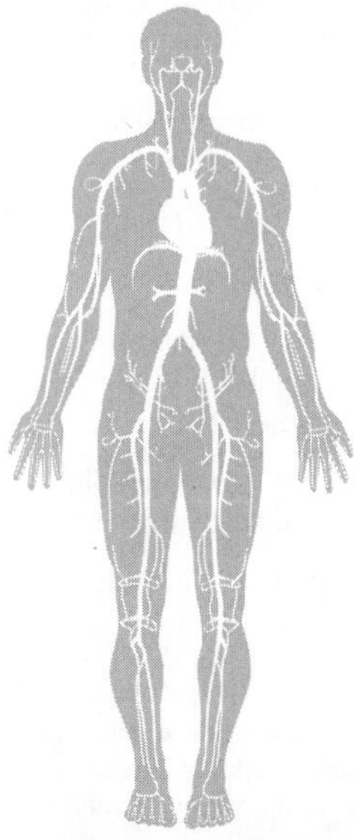

Schematic Overview of the Human Circulatory System

Deoxygenated blood (colored gray below) travels to the right side of the heart via the vena cavae. The heart pumps the deoxygenated blood to the lungs where it releases carbon dioxide and receives oxygen. The oxygenated blood (colored white below) travels via the pulmonary vein back to the heart from where it is pumped to all parts of the body. The **venous system** (figure, left) returns blood from the capillaries to the heart. The **arterial system** (figure right) carries blood from the heart to the capillaries. **Portal systems** carry blood between two capillary beds.

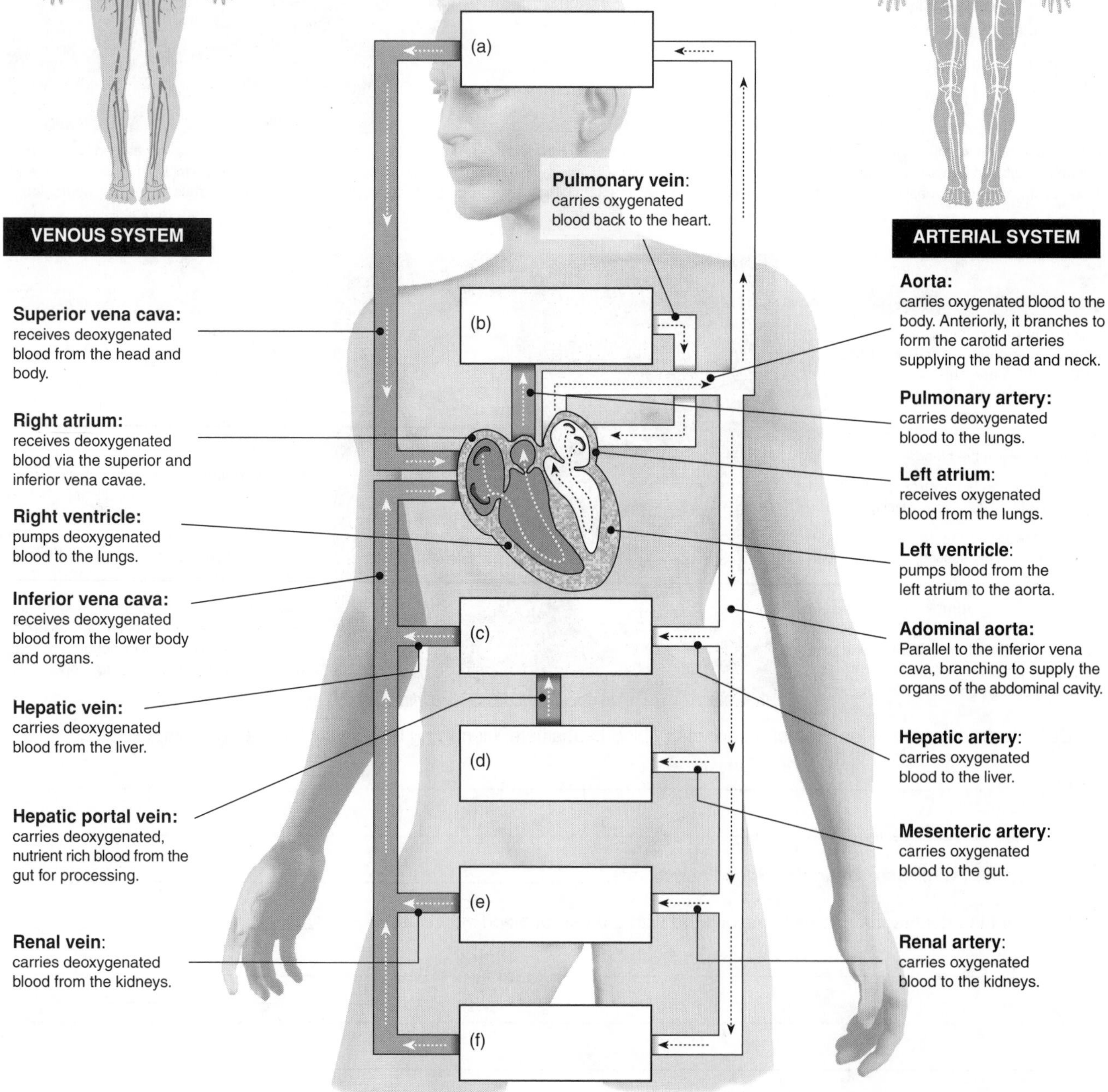

VENOUS SYSTEM

ARTERIAL SYSTEM

(a)

Pulmonary vein: carries oxygenated blood back to the heart.

(b)

Superior vena cava: receives deoxygenated blood from the head and body.

Aorta: carries oxygenated blood to the body. Anteriorly, it branches to form the carotid arteries supplying the head and neck.

Right atrium: receives deoxygenated blood via the superior and inferior vena cavae.

Pulmonary artery: carries deoxygenated blood to the lungs.

Left atrium: receives oxygenated blood from the lungs.

Right ventricle: pumps deoxygenated blood to the lungs.

Left ventricle: pumps blood from the left atrium to the aorta.

Inferior vena cava: receives deoxygenated blood from the lower body and organs.

Adominal aorta: Parallel to the inferior vena cava, branching to supply the organs of the abdominal cavity.

(c)

Hepatic vein: carries deoxygenated blood from the liver.

Hepatic artery: carries oxygenated blood to the liver.

(d)

Hepatic portal vein: carries deoxygenated, nutrient rich blood from the gut for processing.

Mesenteric artery: carries oxygenated blood to the gut.

(e)

Renal vein: carries deoxygenated blood from the kidneys.

Renal artery: carries oxygenated blood to the kidneys.

(f)

1. Complete the diagram above by labeling the boxes with the organs or structures they represent.

The Human Heart

The heart is at the center of the human cardiovascular system. It is a hollow, muscular organ, weighing on average 342 grams. Each day it beats over 100 000 times to pump 3780 liters of blood through 100 000 kilometers of blood vessels. It comprises a system of four muscular chambers (two **atria** and two **ventricles**) that alternately fill and empty of blood, acting as a double pump. The left side pumps blood to the body tissues and the right side pumps blood to the lungs. The heart lies between the lungs, to the left of the body's midline, and it is surrounded by a double layered **pericardium** of tough fibrous connective tissue. The pericardium prevents over-distension of the heart and anchors the heart within the **mediastinum**.

Human Heart Structure

(sectioned, anterior view)

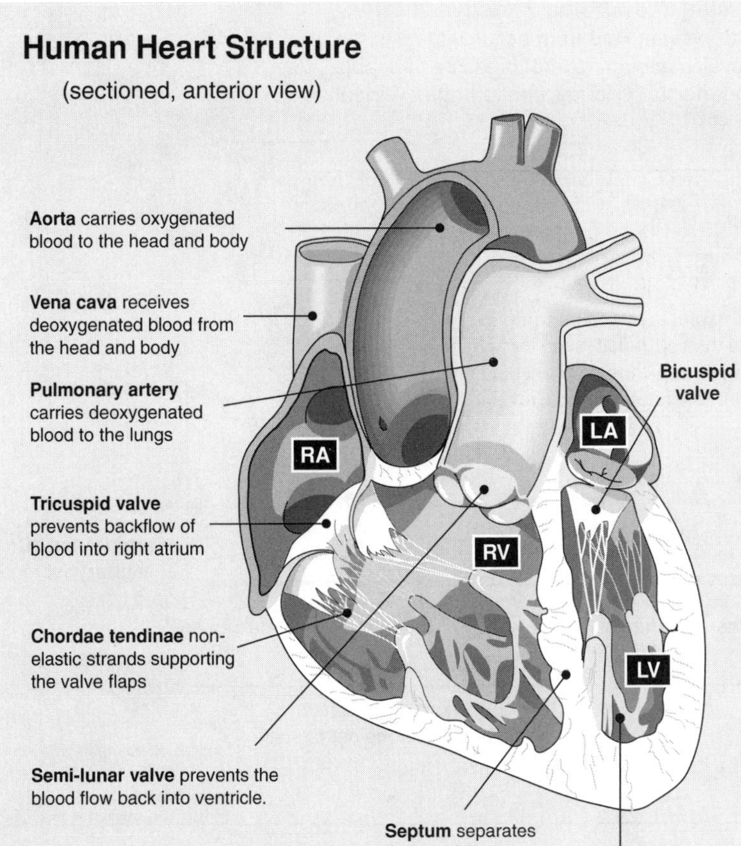

The heart is not a symmetrical organ. Although the quantity of blood pumped by each side is the same, the walls of the left ventricle are thicker and more muscular than those of the right ventricle. The difference affects the shape of the ventricular cavities, so the right ventricle is twisted over the left.

Key to abbreviations

- **RA** Right atrium; receives deoxygenated blood via anterior and posterior vena cavae
- **RV** Right ventricle; pumps deoxygenated blood to the lungs via the pulmonary artery
- **LA** Left atrium; receives blood returning to the heart from the lungs via the pulmonary veins
- **LV** Left ventricle; pumps oxygenated blood to the head and body via the aorta

Top view of a heart in section, showing valves

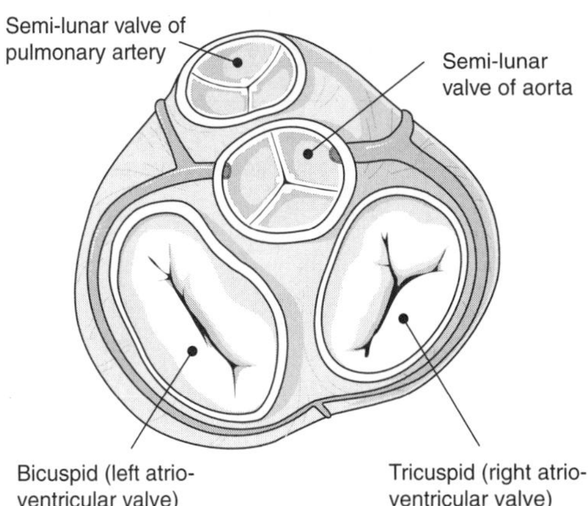

Posterior view of heart

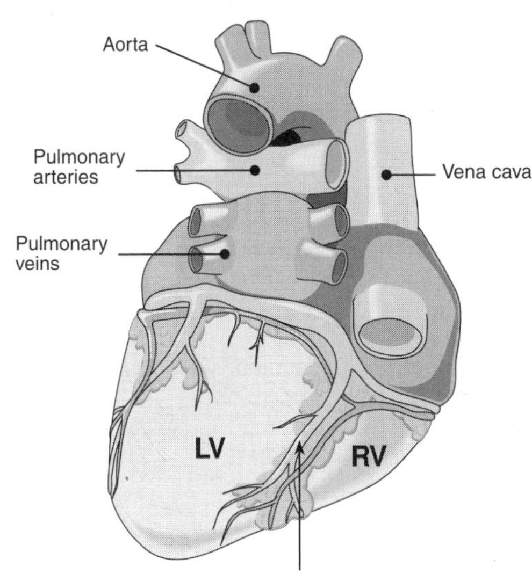

Coronary arteries: The high oxygen demands of the heart muscle are met by a dense capillary network. Coronary arteries arise from the aorta and spread over the surface of the heart supplying the cardiac muscle with oxygenated blood. Deoxygenated blood is collected by cardiac veins and returned to the right atrium via a large coronary sinus.

Coronary Artery Disease and Atherosclerosis

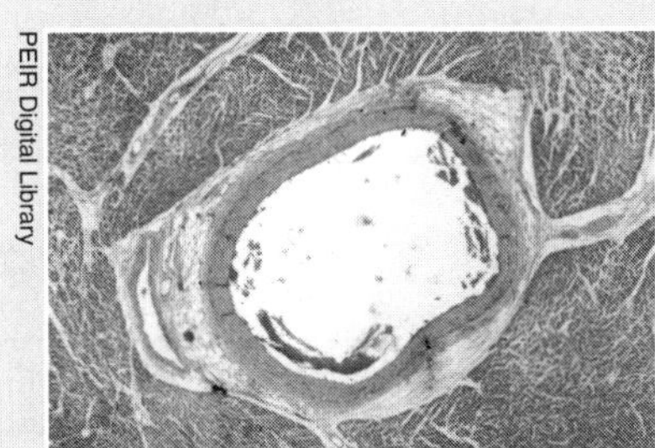

Normal unobstructed coronary artery. Blood flow through the lumen (inside space) of the vessels is unobstructed.

Atherosclerosis is a disease of the arteries caused by fatty deposits (**atheromas**) on the inner walls of the arteries. An atheroma is made up of cells (mostly macrophages) or cell debris, with associated fatty acids, cholesterol, calcium, and varying amounts of fibrous connective tissue. The lining of the arteries degenerates due to the accumulation of fat and plaques. Atheromas weaken the arterial walls and eventually restrict blood flow through the arteries, increasing the risk of **aneurysm,** thrombosis, heart attack, and stroke.

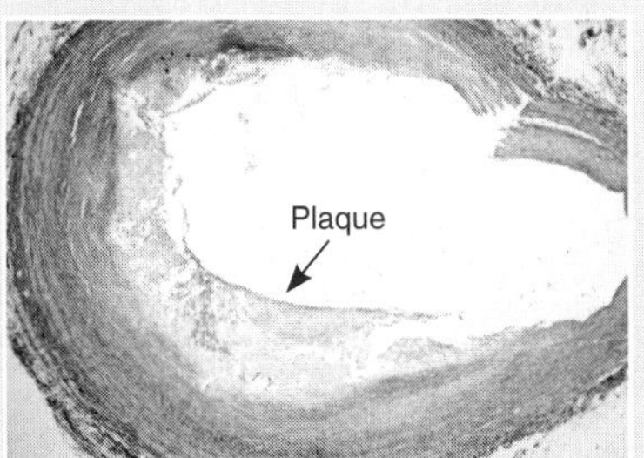

A coronary artery with moderately severe atheroma. Note the formation of the plaque on the inside surface of the artery.

Pressure Changes and the Asymmetry of the Heart

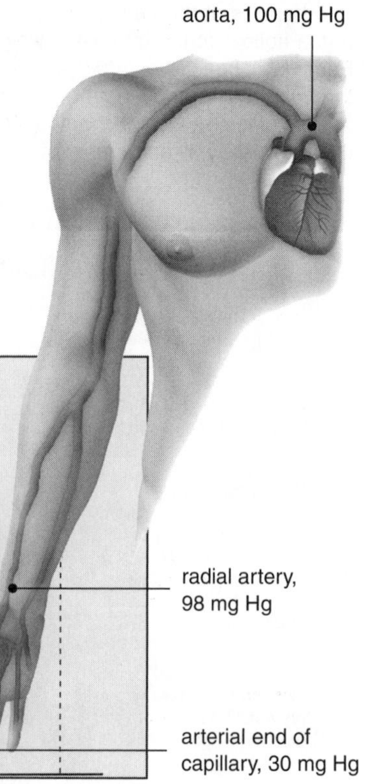

The heart is not a symmetrical organ.The left ventricle and its associated arteries are thicker and more muscular than the corresponding structures on the right side. This asymmetry is related to the necessary pressure differences between the pulmonary (lung) and systemic (body) circulations (not to the distance over which the blood is pumped per se). The graph below shows changes blood pressure in each of the major blood vessel types in the systemic and pulmonary circuits (the horizontal distance not to scale). The pulmonary circuit must operate at a much lower pressure than the systemic circuit to prevent fluid from accumulating in the alveoli of the lungs. The left side of the heart must develop enough "spare" pressure to enable increased blood flow to the muscles of the body and maintain kidney filtration rates without decreasing the blood supply to the brain.

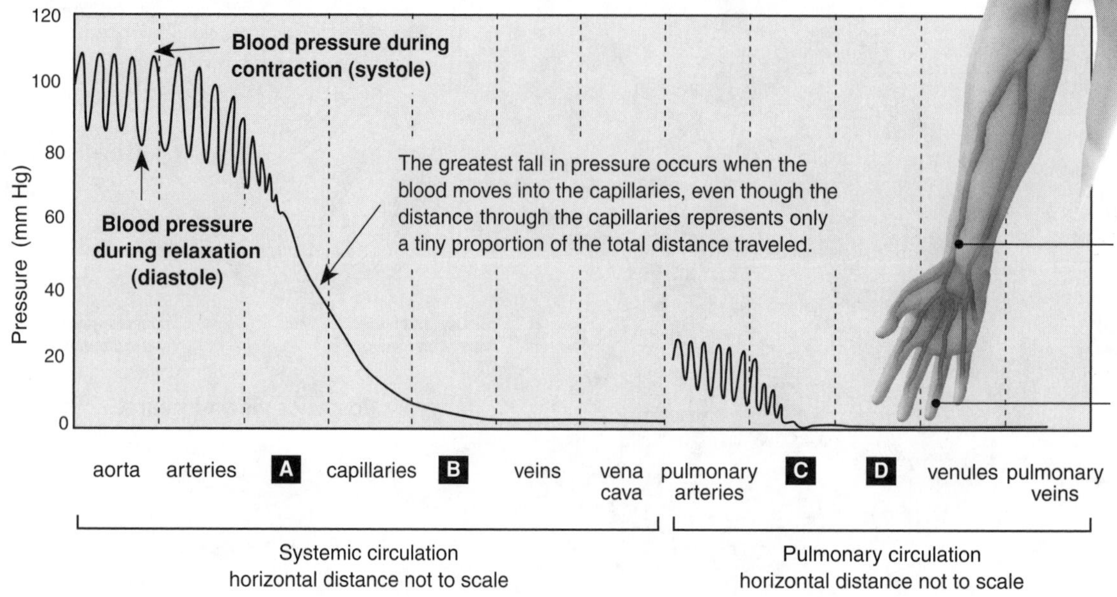

1. Explain the purpose of the valves in the heart: ______

2. The heart is full of blood. Suggest two reasons why, despite this, it needs its own blood supply:

   (a) ______

   (b) ______

3. Predict the effect on the heart if blood flow through a coronary artery is restricted or blocked: ______

4. Identify the vessels corresponding to the letters **A-D** on the graph above:

   A: ______ B: ______ C: ______ D: ______

5. (a) Explain why the pulmonary circuit must operate at a lower pressure than the systemic system: ______

   (b) Relate this to differences in the thickness of the wall of the left and right ventricles of the heart: ______

6. Explain what you are recording when you take a pulse: ______

Control of Heart Activity

When removed from the body, cardiac muscle continues to beat. This indicates that the origin of the heartbeat is **myogenic**; the contractions arise as an intrinsic property of the cardiac muscle itself. The heartbeat is regulated by a special conduction system consisting of the pacemaker (**sinoatrial node**) and specialized conduction fibers called **Purkinje fibers**. The pacemaker sets a basic rhythm for the heart, but this rate is influenced by the cardiovascular control center in the medulla in response to sensory information from pressure receptors in the walls of the heart and blood vessels, and by higher brain functions. Changing the rate and force of heart contraction is the main mechanism for controlling cardiac output in order to meet changing demands.

Generation of the Heartbeat

The basic rhythmic heartbeat is **myogenic**. The nodal cells (SAN and atrioventricular node) spontaneously generate rhythmic action potentials without neural stimulation. The normal resting rate of self-excitation of the SAN is about 50 beats per minute.

The amount of blood ejected from the left ventricle per minute is called the **cardiac output**. It is determined by the **stroke volume** (the volume of blood ejected with each contraction) and the **heart rate** (number of heart beats per minute).

Cardiac output
= **stroke volume** x **heart rate**

Cardiac muscle responds to stretching by contracting more strongly. The greater the blood volume entering the ventricle, the greater the force of contraction. This relationship is known as **Starling's Law.**

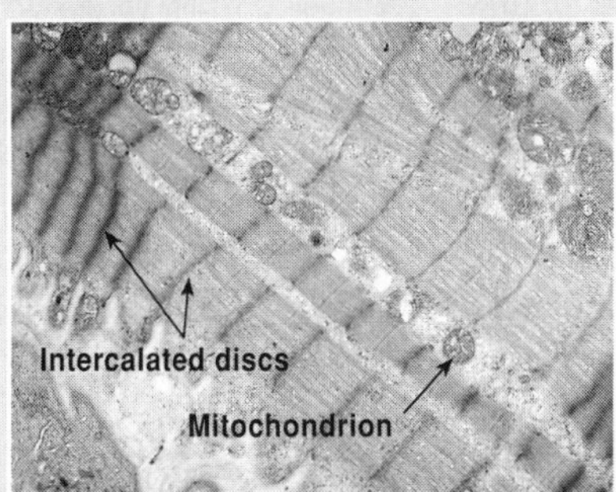

TEM of cardiac muscle showing branched fibers (muscle cells). Each fiber has one or two nuclei and many large mitochondria. **Intercalated discs** are specialized regions between neighboring cells that support synchronized contraction of the muscle. They contain **gap junctions**, specialized electrical synapses that allow very rapid spread of nerve impulses through the heart muscle.

Sinoatrial node (SAN) is also called the **pacemaker**. It is a mass of specialized muscle cells near the opening of the superior vena cava. The pacemaker initiates the cardiac cycle, spontaneously generating action potentials that cause the atria to contract. The SAN sets the basic pace of the heart rate, although this rate is influenced by hormones and impulses from the autonomic nervous system.

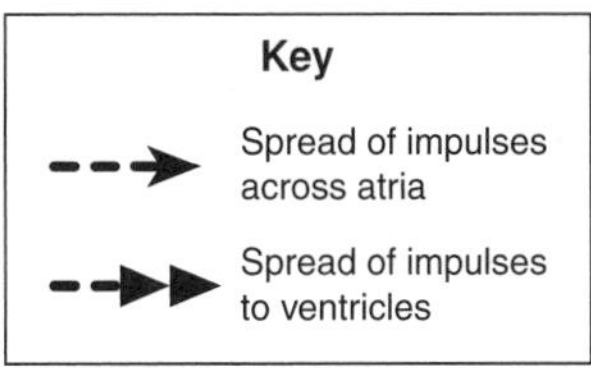

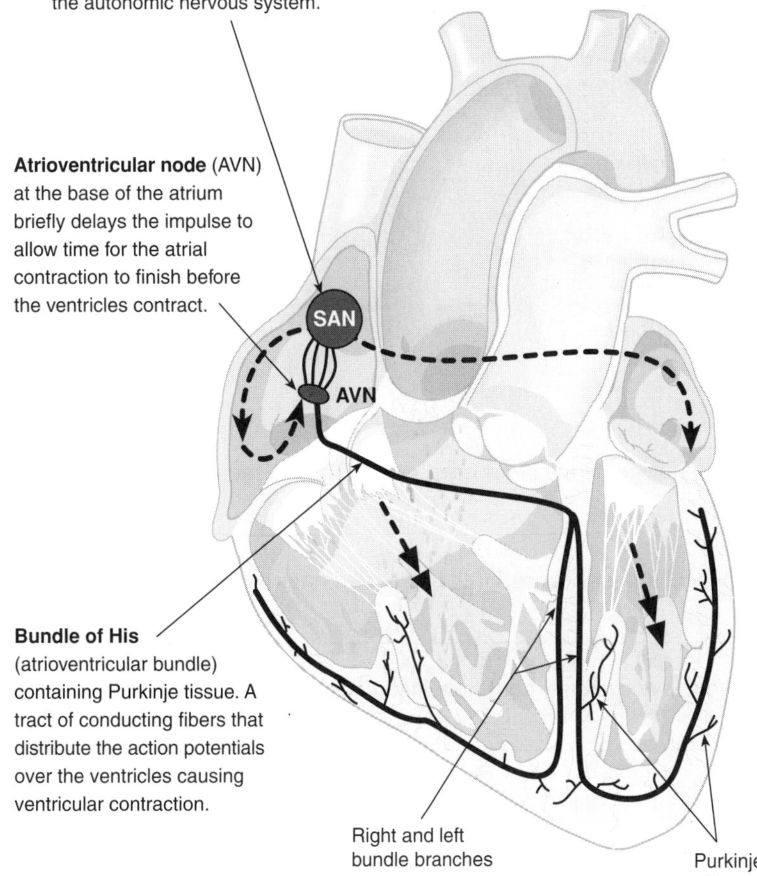

Atrioventricular node (AVN) at the base of the atrium briefly delays the impulse to allow time for the atrial contraction to finish before the ventricles contract.

Bundle of His (atrioventricular bundle) containing Purkinje tissue. A tract of conducting fibers that distribute the action potentials over the ventricles causing ventricular contraction.

1. Identify the role of each of the following in heart activity:

   (a) The sinoatrial node: ______________________________

   (b) The atrioventricular node: ______________________________

   (c) The bundle of His: ______________________________

2. Explain the significance of the delay in impulse conduction at the AVN: ______________________________

   ____

3. (a) Calculate the **cardiac output** when stroke volume is 70 cm^3 and the heart rate is 70 beats per minute:

   ____

   (b) Trained endurance athletes have a very high cardiac output. Suggest how this is achieved: ______________________________

   ____

Autonomic Nervous System Control of Heartbeat

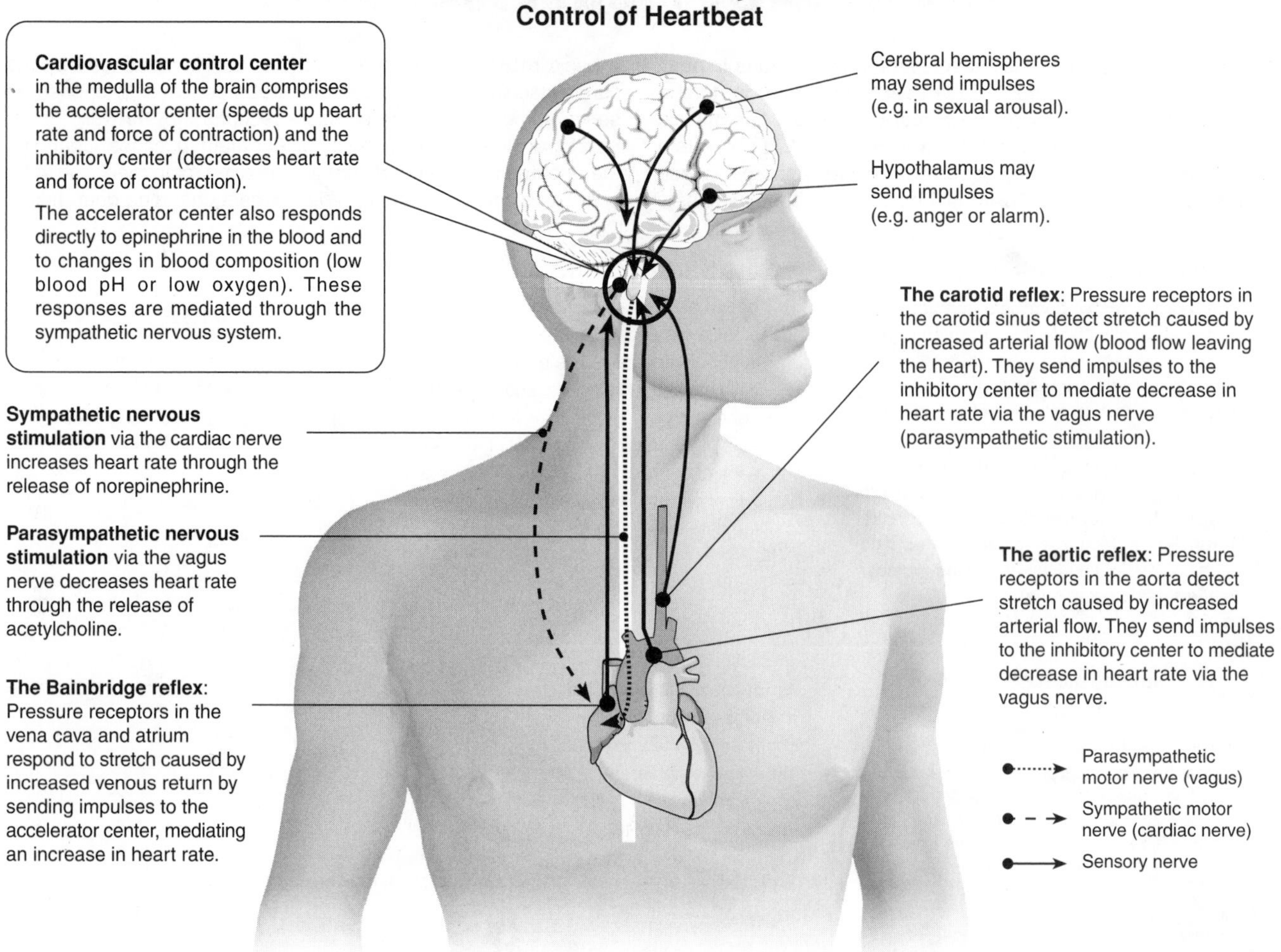

Cardiovascular control center in the medulla of the brain comprises the accelerator center (speeds up heart rate and force of contraction) and the inhibitory center (decreases heart rate and force of contraction).

The accelerator center also responds directly to epinephrine in the blood and to changes in blood composition (low blood pH or low oxygen). These responses are mediated through the sympathetic nervous system.

Cerebral hemispheres may send impulses (e.g. in sexual arousal).

Hypothalamus may send impulses (e.g. anger or alarm).

The carotid reflex: Pressure receptors in the carotid sinus detect stretch caused by increased arterial flow (blood flow leaving the heart). They send impulses to the inhibitory center to mediate decrease in heart rate via the vagus nerve (parasympathetic stimulation).

Sympathetic nervous stimulation via the cardiac nerve increases heart rate through the release of norepinephrine.

Parasympathetic nervous stimulation via the vagus nerve decreases heart rate through the release of acetylcholine.

The aortic reflex: Pressure receptors in the aorta detect stretch caused by increased arterial flow. They send impulses to the inhibitory center to mediate decrease in heart rate via the vagus nerve.

The Bainbridge reflex: Pressure receptors in the vena cava and atrium respond to stretch caused by increased venous return by sending impulses to the accelerator center, mediating an increase in heart rate.

Parasympathetic motor nerve (vagus)

Sympathetic motor nerve (cardiac nerve)

Sensory nerve

4. (a) With respect to the heart beat, explain what is meant by **myogenic**: ________________

   (b) Describe the evidence for the myogenic nature of the heart beat: ________________

5. During heavy exercise, heart rate increases. Describe the mechanisms that are involved in bringing about this increase:

6. (a) Identify a stimulus for a decrease in heart rate: ________________

   (b) Explain how this change in heart rate is brought about: ________________

7. Identify two pressure receptors involved in control of heart rate and state what they respond to:

   (a) ________________

   (b) ________________

8. Guarana is a chemical found in many energy drinks. A group of students designed an experiment to test whether guarana stimulates a cardiovascular response. The test subjects had their pulses recorded before and after drinking an energy drink containing a known amount of guarana.

   (a) Suggest two reasons why the test subjects may respond in different ways: ________________

   (b) Describe a suitable control for this experiment: ________________

The Cardiac Cycle

The heart pumps with alternate contractions (**systole**) and relaxations (**diastole**). The **cardiac cycle** refers to the sequence of events of a heartbeat and involves three main stages: atrial systole, ventricular systole, and complete cardiac diastole. Pressure changes within the heart's chambers generated by the cycle of contraction and relaxation are responsible for blood movement and cause the heart valves to open and close, preventing the backflow of blood. The noise of the blood when the valves open and close produces the heartbeat sound (**lubb-dupp**). The heart beat occurs in response to electrical impulses, which can be recorded as a trace, called an **electrocardiogram** or **ECG**. The ECG pattern is the result of the different impulses produced at each phase of the cardiac cycle, and each part is identified with a letter code. An ECG provides a useful method of monitoring changes in heart rate and activity and detection of heart disorders. The electrical trace is accompanied by volume and pressure changes (below).

The Cardiac Cycle

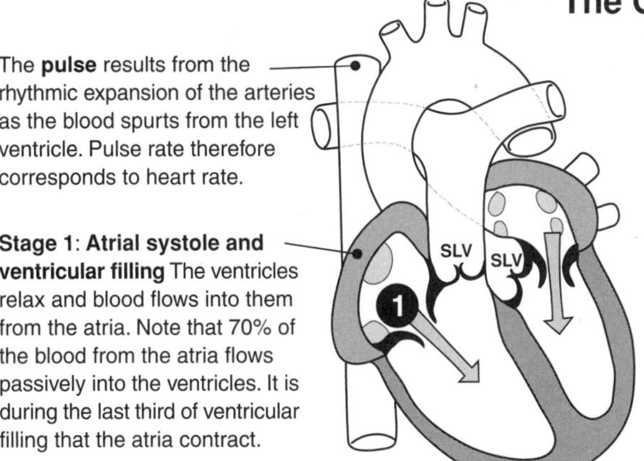

The **pulse** results from the rhythmic expansion of the arteries as the blood spurts from the left ventricle. Pulse rate therefore corresponds to heart rate.

Stage 1: **Atrial systole and ventricular filling** The ventricles relax and blood flows into them from the atria. Note that 70% of the blood from the atria flows passively into the ventricles. It is during the last third of ventricular filling that the atria contract.

Heart during ventricular filling

Stage 2: **Ventricular systole** The atria relax, the ventricles contract, and blood is pumped from the ventricles into the aorta and the pulmonary artery. The start of ventricular contraction coincides with the first heart sound.

Stage 3: (not shown) There is a short period of atrial and ventricular relaxation (diastole). Semilunar valves (**SLV**) close to prevent backflow into the ventricles (see diagram, left). The cycle begins again. For a heart beating at 75 beats per minute, one cardiac cycle lasts about 0.8 seconds.

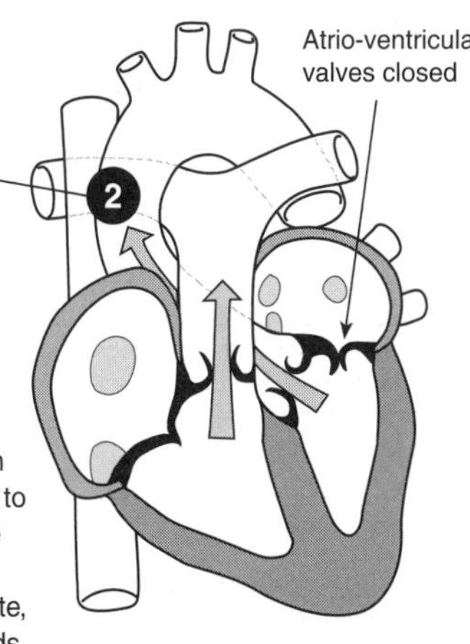

Heart during ventricular contraction

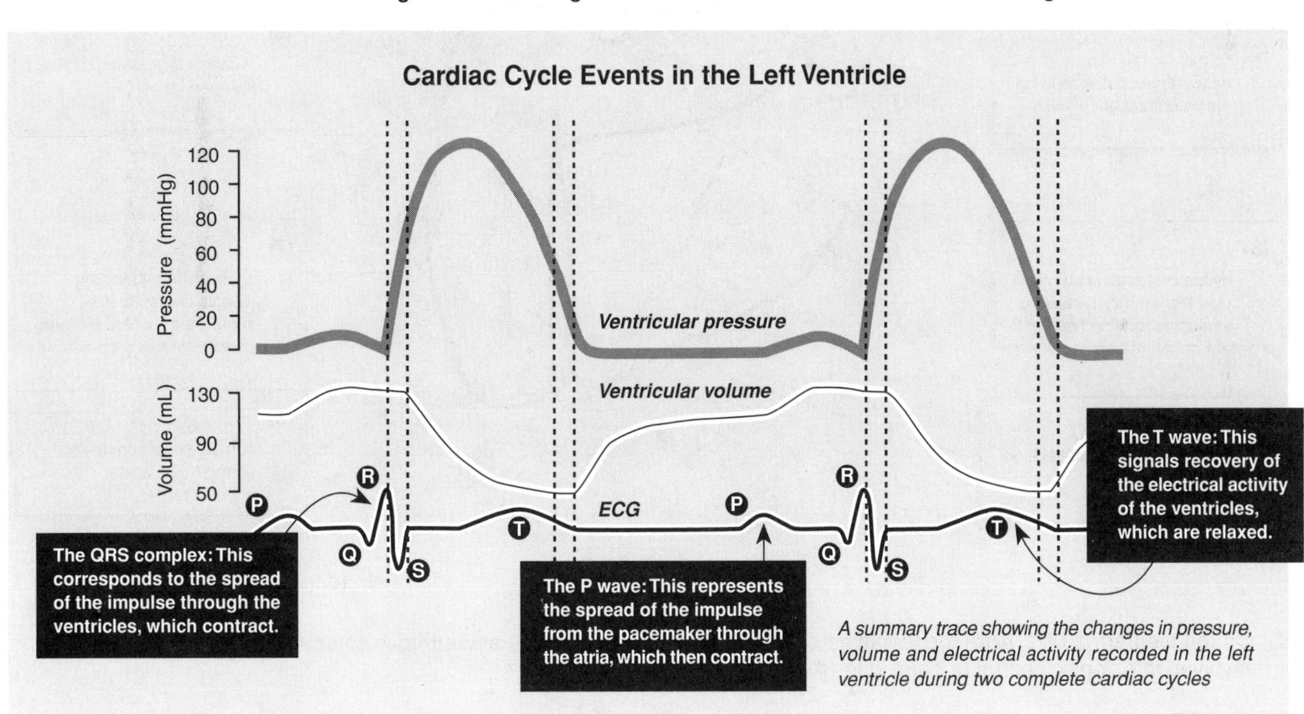

A summary trace showing the changes in pressure, volume and electrical activity recorded in the left ventricle during two complete cardiac cycles

1. Identify each of the following phases of an ECG by its international code:

   (a) Excitation of the ventricles and ventricular systole: ____________________

   (b) Electrical recovery of the ventricles and ventricular diastole: ____________________

   (c) Excitation of the atria and atrial systole: ____________________

2. Suggest the physiological reason for the period of electrical recovery experienced each cycle (the T wave):

   ____

3. Using the letters indicated, mark the points on the trace above corresponding to each of the following:

   (a) E: Ejection of blood from the ventricle

   (b) AVC: Closing of the atrioventricular valve

   (c) FV: Filling of the ventricle

   (d) AVO: Opening of the atrioventricular valve

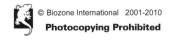

Review of the Human Heart

A circulatory system is required to transport materials because diffusion is too inefficient and slow to supply all the cells of the body adequately. The circulatory system in humans transports nutrients, respiratory gases, wastes, and hormones, aids in regulating body temperature and maintaining fluid balance, and has a role in internal defence. The circulatory system comprises a network of vessels, a circulatory fluid (blood), and a heart. This activity summarizes key features of the structure and function of the human heart. The necessary information can be found in earlier activities in this topic.

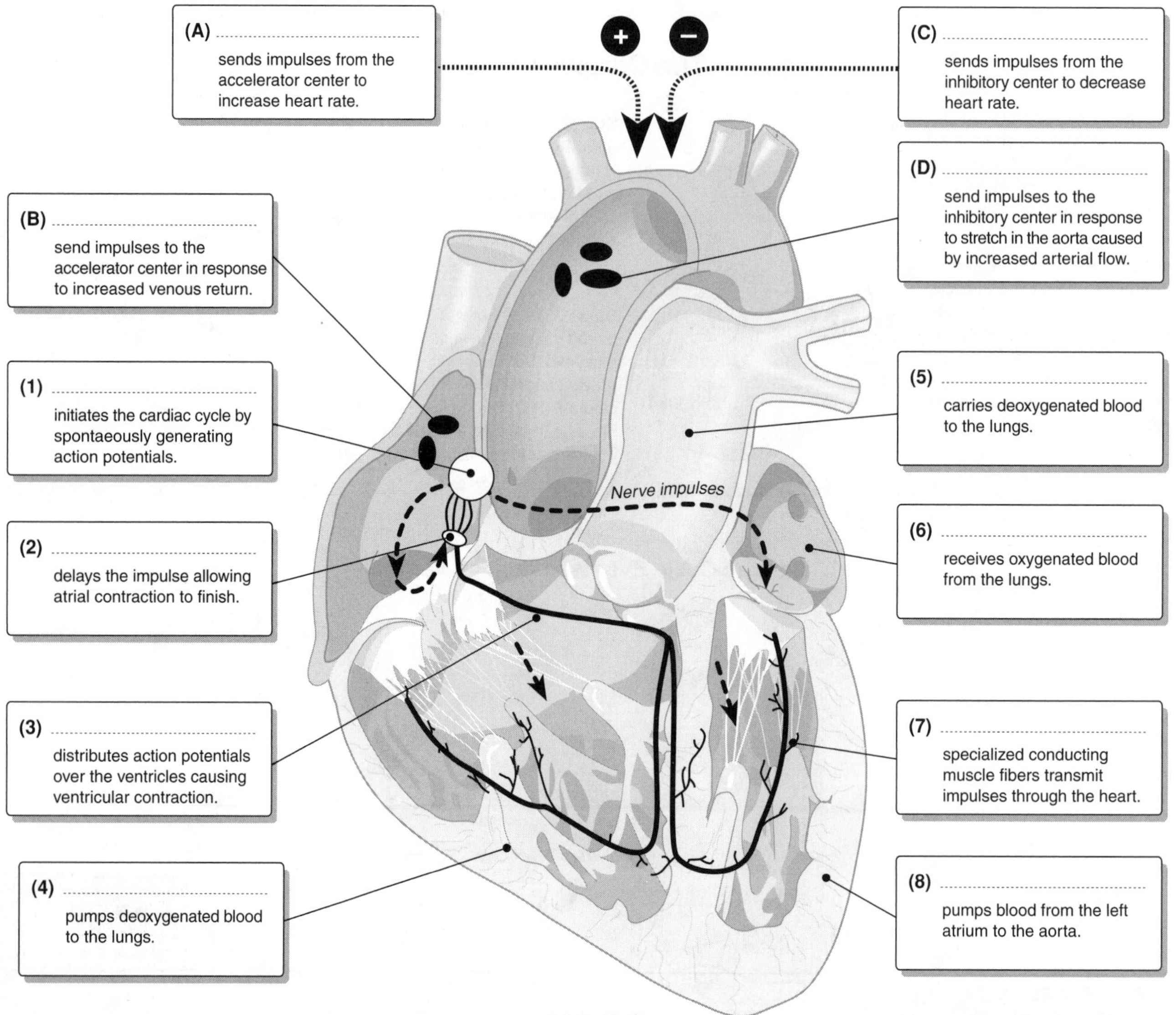

(A) ....
sends impulses from the accelerator center to increase heart rate.
+
−
(C) ....
sends impulses from the inhibitory center to decrease heart rate.
(B) ....
send impulses to the accelerator center in response to increased venous return.
(D) ....
send impulses to the inhibitory center in response to stretch in the aorta caused by increased arterial flow.
(1) ....
initiates the cardiac cycle by spontaeously generating action potentials.
(5) ....
carries deoxygenated blood to the lungs.
Nerve impulses
(2) ....
delays the impulse allowing atrial contraction to finish.
(6) ....
receives oxygenated blood from the lungs.
(3) ....
distributes action potentials over the ventricles causing ventricular contraction.
(7) ....
specialized conducting muscle fibers transmit impulses through the heart.
(4) ....
pumps deoxygenated blood to the lungs.
(8) ....
pumps blood from the left atrium to the aorta.

1. On the diagram above, label the identified components of heart structure and intrinsic control (**1**-**8**), and the components involved in extrinsic control of heart rate (**A**-**D**).

2. An **ECG** is the result of different impulses produced at each phase of the **cardiac cycle** (the sequence of events in a heartbeat). For each electrical event indicated in the ECG below, describe the corresponding event in the cardiac cycle:

A ....
The spread of the impulse from the pacemaker (sinoatrial node) through the atria.

B ....
The spread of the impulse through the ventricles.

C ....
Recovery of the electrical activity of the ventricles.

A
B
C

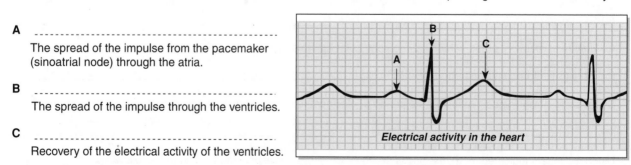

Electrical activity in the heart

3. Describe one treatment that may be indicated when heart rhythm is erratic or too slow: ____________________

____

Related activities: *The Human Heart, Control of Heart Activity, The Cardiac Cycle*

Exercise and Blood Flow

Exercise promotes health by improving the rate of blood flow back to the heart (venous return). This is achieved by strengthening all types of muscle and by increasing the efficiency of the heart. During exercise blood flow to different parts of the body changes in order to cope with the extra demands of the muscles, the heart and the lungs.

1. The following table gives data for the **rate** of blood flow to various parts of the body at rest and during strenuous exercise. **Calculate** the **percentage** of the total blood flow that each organ or tissue receives under each regime of activity.

Organ or tissue	At rest		Strenuous exercise	
	$cm^3\ min^{-1}$	% of total	$cm^3\ min^{-1}$	% of total
Brain	700	14	750	4.2
Heart	200		750	
Lung tissue	100		200	
Kidneys	1100		600	
Liver	1350		600	
Skeletal muscles	750		12 500	
Bone	250		250	
Skin	300		1900	
Thyroid gland	50		50	
Adrenal glands	25		25	
Other tissue	175		175	
TOTAL	5000	**100**	17 800	**100**

2. Explain how the body increases the rate of blood flow during exercise: ______________________

3. (a) State approximately how many times the total rate of blood flow increases between rest and exercise: ____________

   (b) Explain why the increase is necessary: ______________________

4. (a) Identify which organs or tissues show no change in the rate of blood flow with exercise: ____________

   (b) Explain why this is the case: ______________________

5. (a) Identify the organs or tissues that show the most change in the rate of blood flow with exercise: ____________

   (b) Explain why this is the case: ______________________

Related activities: *Homeostasis During Exercise*

Endurance refers to the ability of the muscles and the cardiovascular and respiratory systems to carry out sustained activity. Muscular strength and short term muscular endurance allows sprinters to run fast for a short time or body builders and weight lifters to lift an immense weight and hold it. Cardiovascular and respiratory endurance refer to the body as a whole: the ability to endure a high level of activity over a prolonged period. This type of endurance is seen in marathon runners, and long distance swimmers and cyclists. Different sports ("short burst sports" compared with endurance type sports) require different training methods and the physiologies (muscle bulk and cardiovascular fitness) of the athletes can be quite different.

The human heart and circulatory system make a number of adjustments in response to aerobic or endurance training. These include:

- **Heart size**: Increases. The left ventricle wall becomes thicker and its chamber bigger.
- **Heart rate**: Heart rate (at rest and during exercise) decreases markedly from non-trained people.
- **Recovery**: Recovery after exercise (of breathing and heart rate) is faster in trained athletes.
- **Stroke volume**: The volume of blood pumped with each heart beat increases with endurance training.
- **Blood volume**: Endurance training increases blood volume (the amount of blood in the body).

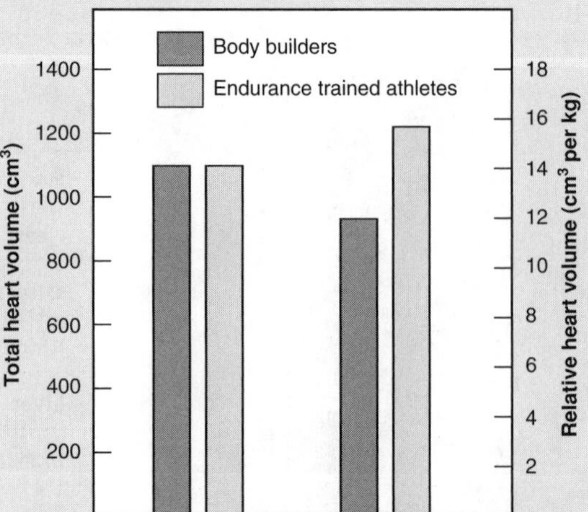

Difference in heart size of highly trained body builders and endurance athletes. Total heart volume is compared to heart volume as related to body weight. Average weights as follows: Body builders = 90.1 kg. Endurance athletes = 68.7 kg.

Weightlifters have high muscular strength and short term muscular endurance; they can lift extremely heavy weights and hold them for a short time. Typical sports requiring these attributes are sprinting, weight lifting, body building, boxing and wrestling.

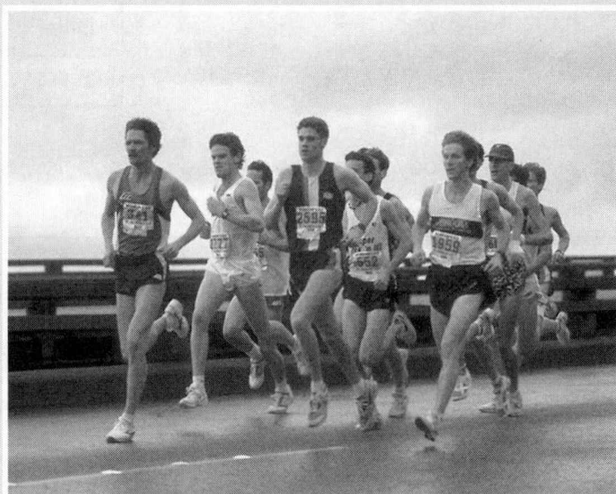

Distance runners have very good cardiovascular, respiratory, and muscular endurance; they sustain high intensity exercise for a long time. Typical sports needing overall endurance are distance running, cycling, and swimming (triathletes combine all three).

6. Suggest a reason why heart size increases with endurance activity:

7. In the graph above right, explain why the relative heart volume of endurance athletes is greater than that of body builders, even though their total heart volumes are the same:

8. Heart stroke volume increases with endurance training. Explain how this increases the efficiency of the heart as a pump:

9. Resting heart rates are much lower in trained athletes compared with non-active people. Explain the health benefits of a lower resting heart rate:

KEY TERMS Word Find

Use the clues below to find the relevant key terms in the WORD FIND grid

```
D V A W Y P U L M O N A R Y C I R C U L A T I O N
I Q R S U R F A C E A R E A V O L U M E S K M R F
O Z T C P C W G U C P X F X N H E Q Y I H E A R T
H F E O O L Y M P H M N S V G P W G O C C I B N P
N R R D X R E S P I R A T O R Y P I G M E N T R L
F C Y Z B E J F F V P R A F C T Z M T Z C B M Z Q
D C S H L H U M G E W I Q M T L W I B Z L U N T S
I D Y B Y X K J B L O O D P A Z E M V M I H K N X
V R U T I S S U E F L U I D L G P X W R M G V V N
B O O R W R H E C L O S E D A H X A T C Q G I E V
D D Y Y I T L E E B U L K F L O W A D G K B P N E
H P I S R G E H A E M O L Y M P H P I O F O M W I
W W P H N B L O O D V E S S E L M B Y M R Y Q U N
R Z I I C S Y S T E M I C C I R C U L A T I O N N
D I S P I I C A P I L L A R Y C N O F L G F A G W
S V E N T R I C L E J Z R H N U L S E D P S H L E
S Q D M C Q Z Q I P S C S O P R A W V X X K G B T
```

A large blood vessel with a thick, muscled wall which carries blood away from the heart.

A chamber of the heart that receives blood from the body or lungs.

Circulatory fluid comprising numerous cell types, which moves respiratory gases and nutrients around the body.

Vessel that carries the blood. Includes arteries, veins, and capillaries.

The process by which fluids move as a single mass around the body. The movement may be induced by pressure and/or gravity.

The smallest blood vessel. May have a wall only one or two cells thick.

The combination of the interstitial fluid and blood in insects and other invertebrates.

Muscular organ used to pump blood about the body. In vertebrates this may have up to four chambers while in invertebrates it may consist only of specialized contractile blood vessels.

A clear fluid contained within the lymphatic system. Similar in composition to the interstitial fluid.

Circulatory system in which blood travels from the heart to lungs and back before being pumped to the rest of the body.

A circulatory system in which the blood and interstitial fluid mix freely in the hemocoel.

That part of a double circulatory system in air breathing vertebrates that transports blood from the heart to the lungs and back.

A ratio that expresses the surface area of a structure relative to its volume.

That part of a double circulatory system that transports blood from the heart to the body and back.

A fluid derived from the blood plasma by leakage through capillaries. it bathes the tissues and is also called interstitial fluid.

Large blood vessel that returns blood to the heart.

A chamber of the heart that pumps blood into arteries.

Circulatory system in which blood travels from the heart to the gills and then to the body without first returning to the heart.

A substance present in blood that is able to bind oxygen for transport to cells. Includes hemoglobin and hemocyanin.

A circulatory system in which the blood is fully contained within vessels.

AP Topic 3B
IB SL/HL 6.3
AHL 11.1

Defending Against Disease

KEY CONCEPTS

- The body can distinguish self from non-self.
- The body can defend itself against pathogens.
- Non-specific defenses target any foreign material.
- The immune response targets specific antigens and has a memory for antigens previously encountered.
- Some pathogens, through their specific mode of action, cause immune system failure.

KEY TERMS

active immunity
AIDS
antibody (=immunoglobulin)
antigen
autoimmune disease
B cell (=B lymphocyte)
cell-mediated immunity
clonal selection
disease
fever
hemostasis
HIV
humoral immunity
immune response
immunity
immunological memory
infection
inflammation
interferon
leukocyte
lymphocyte
macrophage
MHC
monoclonal antibody
non-specific defense
passive immunity
pathogen
phagocyte
primary response
secondary response
specific defense
T cell (=T lymphocyte)
thymus
vaccination (=immunization)

OBJECTIVES

☐ 1. Use the **KEY TERMS** to help you understand and complete these objectives.

Pathogens and Disease

pages 129-130, 150

☐ 2. Define **pathogen** and identify common pathogens. Describe how pathogens are transmitted, identifying portals of entry and modes of transmission.

☐ 3. Describe how **antibiotics** work against bacterial infections. Explain why antibiotics are ineffective against viral infections.

The Body's Defenses

pages 116, 131-146

☐ 4. Explain how the body distinguishes self from non-self and why this is important. Describe situations when this recognition system fails.

☐ 5. Describe the process of **blood clotting**, and explain its role in **hemostasis** and in restricting entry of pathogens after injury.

☐ 6. Describe **non-specific defenses** in humans, describing the nature and role of each of the following in protecting against pathogens:

(a) Skin (including sweat and sebum production) and mucous membranes.
(b) Body secretions (tears, urine, saliva, gastric juice).
(c) Natural anti-bacterial and anti-viral proteins, e.g. interferon.
(d) The inflammatory response, fever, and cell death.
(e) Phagocytosis by phagocytes.

☐ 7. Describe the **immune response**, including the importance of both specificity and memory. Distinguish between naturally acquired and artificially acquired immunity and between **active** and **passive immunity**.

☐ 8. Describe **cell-mediated immunity** and **humoral** (antibody-mediated) immunity, identifying the specific white blood cells involved in each case.

☐ 9. Describe **clonal selection** and the basis of **immunological memory**. Explain how the immune system is able to respond to the large and unpredictable range of potential antigens.

☐ 10. Explain **antibody** production, including how B cells bring about humoral (antibody-mediated) immunity to specific **antigens**.

☐ 11. Explain the principles of **vaccination**, including reference to the **primary** and **secondary response** to infection and the role of these.

☐ 12. Describe the production and applications of **monoclonal antibodies**.

Immune Dysfunction and Disease

pages 147-149

☐ 13. Describe the effects of **HIV** on the immune system, including the reduction in the number of active lymphocytes and the loss of immune function.

☐ 14. Describe the cause, transmission, and social impact of AIDS.

Periodicals:
listings for this chapter are on page 389

Weblinks:
www.thebiozone.com/weblink/SB2-2603.html

Teacher Resource CD-ROM:
Asthma & Hypersensitivity

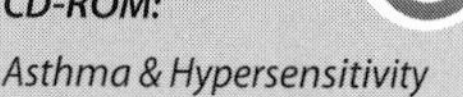

Infection and Disease

Infectious disease refers to disease caused by a **pathogen** (an infectious agent). In 1861, **Louis Pasteur** demonstrated experimentally that microorganisms can be present in non-living matter and can contaminate seemingly sterile solutions. He also showed conclusively that microbes can be destroyed by heat; a discovery that formed the basis of modern-day **aseptic technique**. The development of the germ theory of disease followed Pasteur's discoveries and, in 1876-1877, **Robert Koch** established a sequence of experimental steps (known as **Koch's postulates**) for directly relating a specific microbe to a specific disease. During the past 100 years, the postulates have been invaluable in determining the specific agents of many diseases.

Koch's Postulates

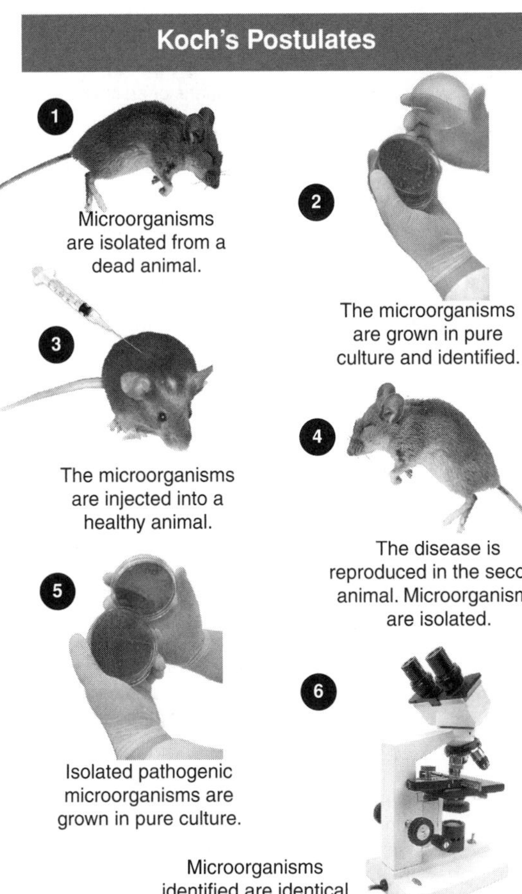
1
Microorganisms are isolated from a dead animal.
2
The microorganisms are grown in pure culture and identified.
3
The microorganisms are injected into a healthy animal.
4
The disease is reproduced in the second animal. Microorganisms are isolated.
5
Isolated pathogenic microorganisms are grown in pure culture.
6
Microorganisms identified are identical to the first cultures.

Koch isolated bacteria from a diseased animal, then injected them into a healthy animal, causing it to exhibit identical symptoms to the first. This demonstrated that a specific infectious disease (e.g. anthrax) was caused by a specific microorganism (*Bacillus anthracis*). Koch used the procedure to identify the bacteria that caused anthrax and tuberculosis.

Koch's findings are summarised as **Koch's postulates**:

1. The same pathogen must be present in every case of the disease.
2. The pathogen must be isolated from the diseased host and grown in pure culture.
3. The pathogen from the pure culture must then cause the disease when it is inoculated into a healthy, susceptible animal.
4. The pathogen must be isolated from the inoculated animal and shown to be the original organism.

Types of Pathogens

CDC

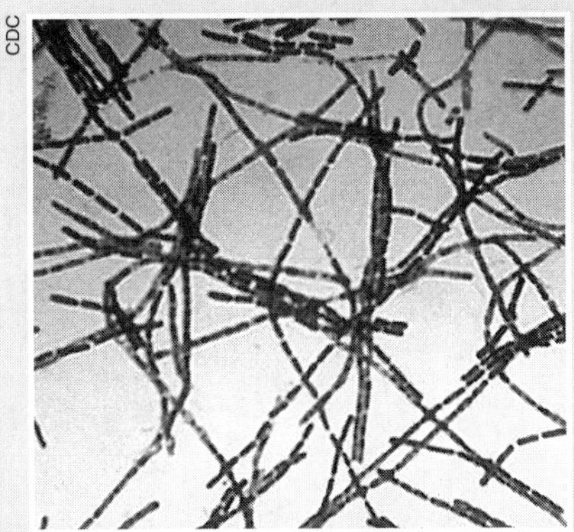

Bacillus anthracis bacterium causes anthrax. The anthrax bacillum can form long-lived spores.

Bacteria: All bacteria are prokaryotes, and are categorised according to the properties of their cell walls and features such as cell shape and arrangement, oxygen requirement, and motility. Many bacteria are useful, but the relatively few species that are pathogenic are responsible for enormous social and economic cost. This is especially so since the rise in incidence of antibiotic resistance.

CDC

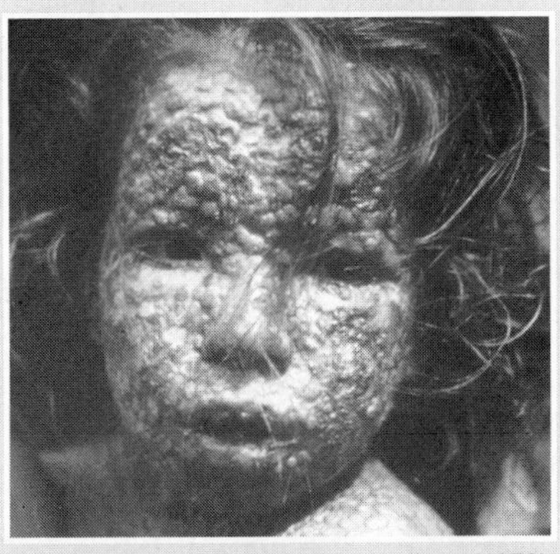

Photo: Bangladeshi girl with smallpox (1973). Smallpox was eradicated from the country in 1977.

Viral pathogens: Viruses are responsible for many everyday diseases (e.g. the common cold), as well as more dangerous diseases, such as Ebola, and some diseases that have since been eradicated as a result of vaccination programmes (e.g. the *Variola* virus, which causes smallpox, above). Viruses are obligate intracellular parasites and need living host cells in order to multiply.

Photo: Ute Frevert

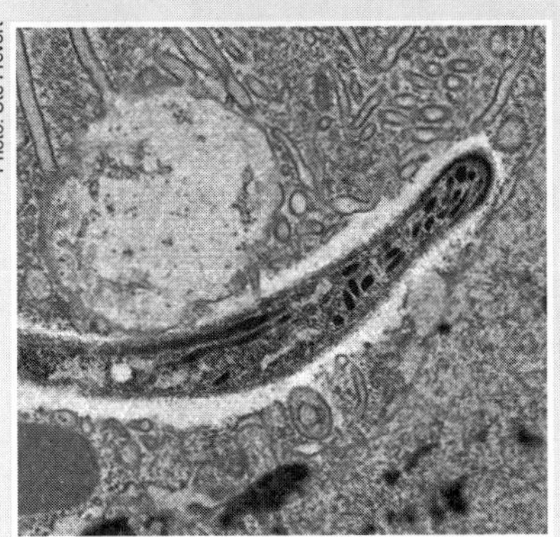

Malaria sporozoite moving through gut epithelia. The parasite is carried by a mosquito vector.

Eukaryotic pathogens: Eukaryotic pathogens (fungi, algae, protozoa, and parasitic worms) include those responsible for malaria and schistosomiasis. Many are highly specialised parasites with a number of hosts. The malaria parasite for example has a mosquito and a human host. Like many other pathogens, this parasite has become resistant to the drugs used to treat it.

1. Explain the contribution of Robert Koch to the **etiology** of disease: ____________________

2. From your knowledge of pathogens, describe a possible exception to Koch's postulates: ____________________

3. Explain why diseases caused by **intracellular parasites** are difficult to control and treat: ____________________

Transmission of Disease

The human body, like that of other large animals, is under constant attack from a wide range of potential parasites and pathogens. Once inside us, these organisms seek to reproduce and exploit us for food. Pathogens may be transferred from one individual to another by a number of methods (below). The transmission of infectious diseases can be virtually eliminated by observing appropriate personal hygiene procedures, by chlorinating drinking water, and by providing adequate sanitation.

The Body Under Assault

Portals of Entry

Respiratory tract
The mouth and nose are major entry points for pathogens, particularly airborne viruses, which are inhaled from other people's expelled mucus.

Examples: diphtheria, meningococcal meningitis, tuberculosis, whooping cough, influenza, measles, rubella, chickenpox.

Gastrointestinal tract
Food is often contaminated with microorganisms, but most of these are destroyed in the stomach.

Examples: cholera, typhoid fever, mumps, hepatitis A, poliomyelitis, bacillary dysentery, salmonellosis.

Breaking the skin surface
The skin provides an effective barrier to most pathogens, but cuts and abrasions allow pathogens to penetrate.

Examples: tetanus, gas gangrene, hepatitis B, rabies, malaria, and HIV.

Urinogenital openings
Urinogenital openings provide entry points for the pathogens responsible for sexually transmitted infections (STIs) and other opportunistic infections (i.e. thrush).

Examples: gonorrhoea, HIV.

Modes of Transmission

Contact transmission
The agent of disease may occur by contact with other infected humans or animals:

Droplet transmission: Mucus droplets are discharged into the air by coughing, sneezing, laughing, or talking within a radius of 1 m.

Direct contact: Direct transmission of an agent by physical contact between its source and a potential host. Includes touching, kissing, and sexual intercourse. May be person to person, or between humans and other animals.

Indirect contact: Includes touching objects that have been in contact with the source of infection. Examples include: eating utensils, drinking cups, bedding, toys, money, and used syringes.

Vehicle transmission
Agents of disease may be transmitted by a medium such as food, blood, water, intravenous fluids (e.g. drugs), and air. Airborne transmission refers to the spread of fungal spores, some viruses, and bacteria that are transported on dust particles.

Animal Vectors
Some pathogens are transmitted between hosts by other animals. Bites from arthropods (e.g. mosquitoes, fleas) and mammals may introduce pathogens. In 1897, **Ronald Ross** identified the *Anopheles* mosquito as the vector for malaria. He was the first to implicate insects in the transmission of disease.

1. State how pathogens benefit from invading a host: ______________________

____

2. Describe two personal hygiene practices that would minimize the risk of transmitting an infectious disease:

(a) ______________________

(b) ______________________

3. Identify the common **mode of transmission** and the **portal of entry** for the following pathogens:

(a) Protozoan causing malaria: ______________________

(b) Tetanus bacteria: ______________________

(c) Cholera bacteria: ______________________

(d) Common cold virus: ______________________

(e) Tuberculosis bacteria: ______________________

(f) HIV (AIDS) virus: ______________________

(g) Gonorrhoea bacteria: ______________________

The Body's Defenses

If microorganisms never encountered resistance from our body defenses, we would be constantly ill and would eventually die of various diseases. Fortunately, in most cases our defenses prevent this from happening. Some of these defenses are designed to keep microorganisms from entering the body. Other defenses remove the microorganisms if they manage to get inside. Further defenses attack the microorganisms if they remain inside the body. The ability to ward off disease through the various defense mechanisms is called **resistance**. The lack of resistance, or vulnerability to disease, is known as **susceptibility**. One form of defense is referred to as **non-specific resistance**, and includes defenses that protect us from any pathogen. This includes a first line of defense such as the physical barriers to infection (skin and mucous membranes) and a second line of defense (phagocytes, inflammation, fever, and antimicrobial substances). **Specific resistance** is a third line of defense that forms the **immune response** and targets specific pathogens. Specialized cells of the immune system, called lymphocytes, produce specific proteins called antibodies which are produced against specific antigens.

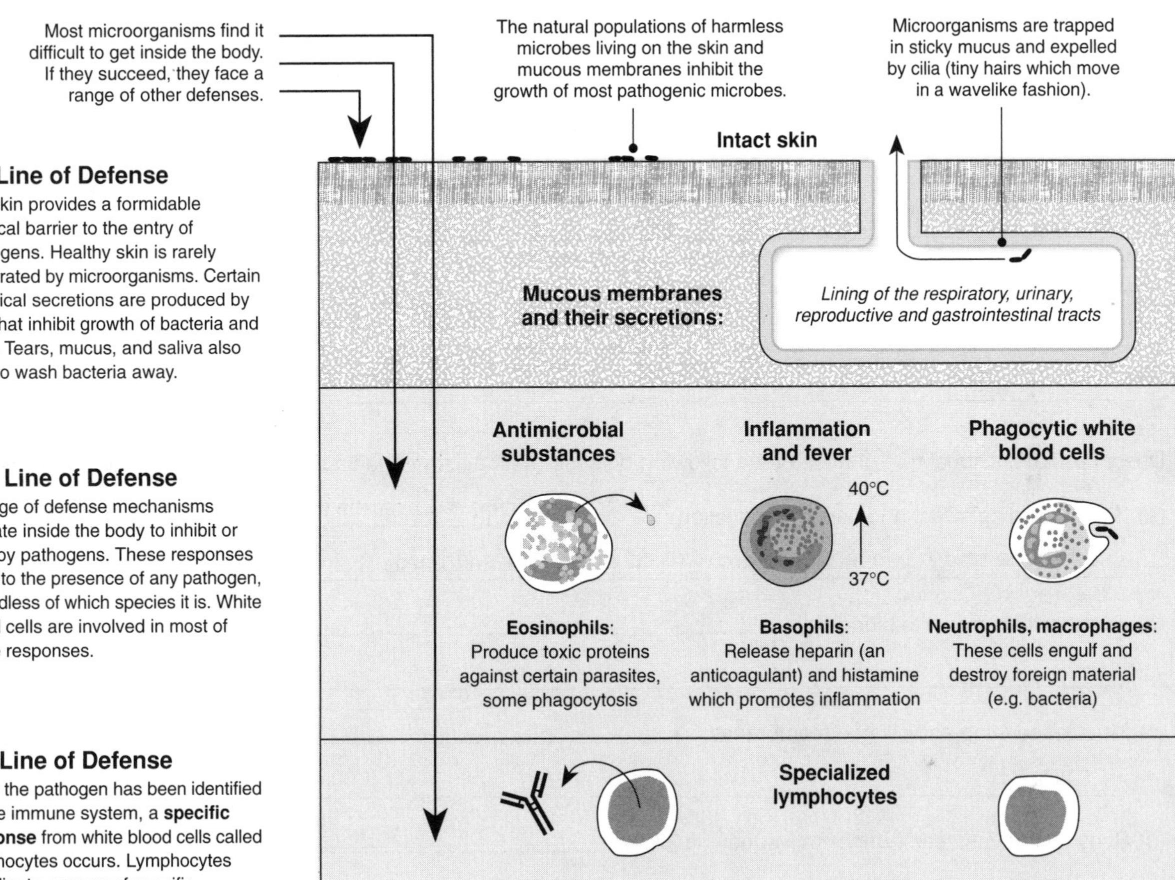

Most microorganisms find it difficult to get inside the body. If they succeed, they face a range of other defenses.
The natural populations of harmless microbes living on the skin and mucous membranes inhibit the growth of most pathogenic microbes.
Microorganisms are trapped in sticky mucus and expelled by cilia (tiny hairs which move in a wavelike fashion).
Intact skin
1st Line of Defense
The skin provides a formidable physical barrier to the entry of pathogens. Healthy skin is rarely penetrated by microorganisms. Certain chemical secretions are produced by skin that inhibit growth of bacteria and fungi. Tears, mucus, and saliva also help to wash bacteria away.
Mucous membranes and their secretions:
Lining of the respiratory, urinary, reproductive and gastrointestinal tracts
2nd Line of Defense
A range of defense mechanisms operate inside the body to inhibit or destroy pathogens. These responses react to the presence of any pathogen, regardless of which species it is. White blood cells are involved in most of these responses.
Antimicrobial substances
Inflammation and fever
Phagocytic white blood cells
40°C
37°C
Eosinophils: Produce toxic proteins against certain parasites, some phagocytosis
Basophils: Release heparin (an anticoagulant) and histamine which promotes inflammation
Neutrophils, macrophages: These cells engulf and destroy foreign material (e.g. bacteria)
3rd Line of Defense
Once the pathogen has been identified by the immune system, a specific response from white blood cells called lymphocytes occurs. Lymphocytes coordinate a range of specific responses to the pathogen.
Specialized lymphocytes
B cell: Antibody production
T cell: Cell-mediated immunity

1. Compare and contrast the type of response against pathogens carried out by each of the three levels of defense:

Periodicals: *Skin, scabs, and scars, Fight for your life!*

Related activities: *The Action of Phagocytes, Inflammation, Fever, The Immune System* ***Web links:*** *Immunoanimations*

2. Distinguish between specific and non-specific resistance:

3. Describe features of the different types of white blood cells and explain how these relate to their role in the second line of defense:

4. Describe the functional role of each of the following defense mechanisms (the first one has been completed for you):

(a) Skin (including sweat and sebum production): Skin helps to prevent direct entry of pathogens into the body. Sebum slows growth of bacteria and fungi.

(b) Phagocytosis by white blood cells:

(c) Mucus-secreting and ciliated membranes:

(d) Body secretions: tears, urine, saliva, gastric juice:

(e) Natural antimicrobial proteins (e.g. interferon):

(f) Antibody production:

(g) Fever:

(h) Cell-mediated immunity:

(i) The inflammatory response:

5. Infection with HIV results in the progressive destruction of T lymphocytes. Suggest why this leads to an increasing number of opportunistic infections in AIDS sufferers:

Targets for Defense

In order for the body to present an effective defense against pathogens, it must first be able to recognize its own tissues (self) and ignore the body's normal microflora (e.g. the bacteria of the skin and gastrointestinal tract). In addition, the body needs to be able to deal with abnormal cells which, if not eliminated, may become cancerous. Failure of self/non-self recognition can lead to autoimmune disorders, in which the immune system mistakenly attacks its own tissues. The body's ability to recognize its own molecules has implications for procedures such as tissue grafts, organ transplants, and blood transfusions. Incompatible tissues (identified as foreign) are attacked by the body's immune system (**rejected**). Even a healthy pregnancy involves suppression of specific features of the self recognition system, allowing the mother to tolerate a nine month gestation with the fetus.

The Body's Natural Microbiota

After birth, normal and characteristic microbial populations begin to establish themselves on and in the body. A typical human body contains 1 X 10^{13} body cells, yet harbors 1 X 10^{14} bacterial cells. These microorganisms establish more or less permanent residence but, under normal conditions, do not cause disease. In fact, this normal microflora can benefit the host by preventing the overgrowth of harmful pathogens. They are not found throughout the entire body, but are located in certain regions.

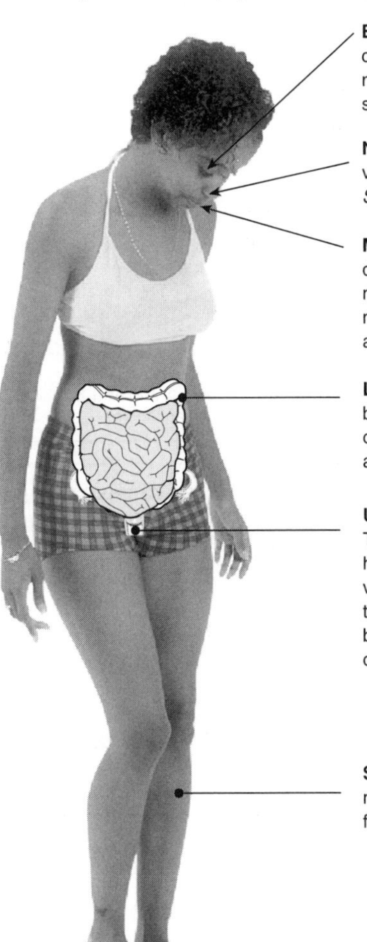

Eyes: The conjuctiva, a continuation of the skin or mucous membrane, contains a similar microbiota to the skin.

Nose and throat: Harbors a variety of microorganisms, e.g. *Staphylococcus spp.*

Mouth: Supports a large and diverse microbiota. It is an ideal microbial environment; high in moisture, warmth, and nutrient availability.

Large intestine: Contains the body's largest resident population of microbes because of its available moisture and nutrients.

Urinary and genital systems: The lower urethra in both sexes has a resident population; the vagina has a particular acid-tolerant population of microbes because of the low pH nature of its secretions.

Skin: Skin secretions prevent most of the microbes on the skin from becoming residents.

Distinguishing Self from Non-Self

The human immune system achieves self-recognition through the **major histocompatibility complex** (MHC). This is a cluster of tightly linked genes on chromosome 6 in humans. These genes code for protein molecules (MHC antigens) that are attached to the surface of body cells. They are used by the immune system to recognize its own or foreign material. **Class I MHC** antigens are located on the surface of virtually all human cells, but **Class II MHC** antigens are restricted to macrophages and the antibody-producing B-lymphocytes.

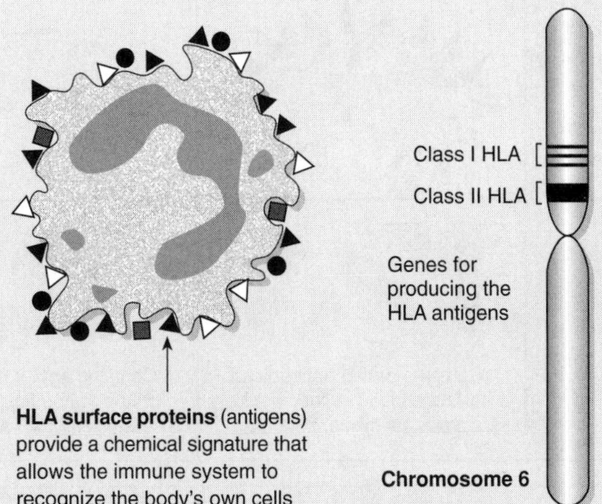
Class I HLA
Class II HLA
Genes for producing the HLA antigens
HLA surface proteins (antigens) provide a chemical signature that allows the immune system to recognize the body's own cells
Chromosome 6

Tissue Transplants

The MHC is responsible for the rejection of tissue grafts and organ transplants. Foreign MHC molecules are antigenic, causing the immune system to respond in the following way:

- T cells directly lyse the foreign cells
- Macrophages are activated by T cells and engulf foreign cells
- Antibodies are released that attack the foreign cell
- The complement system injures blood vessels supplying the graft or transplanted organ

To minimize this rejection, attempts are made to match the MHC of the organ donor to that of the recipient as closely as possible.

1. Explain why it is healthy to have a natural population of microbes on and inside the body: ______

2. (a) Explain the nature and purpose of the **major histocompatibility complex** (MHC): ______

   (b) Explain the importance of such a self-recognition system: ______

3. Identify two situations when the body's recognition of 'self' is undesirable: ______

Periodicals: *What is the human microbiome?*

Related activities: *The Body's Defenses, The Immune System*

Blood Group Antigens

Blood groups classify blood according to the different marker proteins on the surface of red blood cells (RBCs). These marker proteins act as **antigens** and affect the ability of RBCs to provoke an immune response. The **ABO blood group** is the most important blood typing system in medical practice, because of the presence of anti-A and anti-B antibodies in nearly all people who lack the corresponding red cell antigens (these antibodies are carried in the plasma and are present at birth). If a patient is to receive blood from a blood donor, that blood must be compatible otherwise the red blood cells of the donated blood will clump together (agglutinate), break apart, and block capillaries. There is a small margin of safety in certain blood group combinations, because the volume of donated blood is usually relatively small and the donor's antibodies are quickly diluted in the plasma. In practice, blood is carefully matched, not only for ABO types, but for other types as well. Although human RBCs have more than 500 known antigens, fewer than 30 (in 9 blood groups) are regularly tested for when blood is donated for transfusion. The blood groups involved are: *ABO, Rh, MNS, P, Lewis, Lutheran, Kell, Duffy,* and *Kidd.* The ABO and rhesus (Rh) are the best known. Although blood typing has important applications in medicine, it can also be used to rule out individuals in cases of crime (or paternity) and establish a list of potential suspects (or fathers).

	Blood Type A	Blood Type B	Blood Type AB	Blood Type O
Antigens* present on the *red blood cells	antigen ***A***	antigen ***B***	antigens ***A*** and ***B***	Neither antigen ***A*** nor ***B***
Anti-bodies* present in the *plasma	Contains **anti-B** antibodies; but no antibodies that would attack its own antigen ***A***	Contains **anti-A** antibodies; but no antibodies that would attack its own antigen ***B***	Contains neither **anti-A** nor **anti-B** antibodies	Contains both **anti-A** and **anti-B** antibodies

Blood type	Frequency in US Rh^+	Rh^-	Antigen	Antibody	Can donate blood to:	Can receive blood from:
A	34%	6%	A	anti-B	A, AB	A, O
B	9%	2%				
AB	3%	1%				
O	38%	7%				

1. Complete the table above to show the antibodies and antigens in each blood group, and donor/recipient blood types:

2. In a hypothetical murder case, blood from both the victim and the murderer was left at the scene. There were five suspects under investigation:

   (a) Describe what blood typing could establish about the guilt or innocence of the suspects: ________________

   (b) Identify what a blood typing could not establish: ________________

   (c) Suggest how the murderer's identity could be firmly established (assuming that s/he was one of the five suspects):

   (d) Explain why blood typing is not used forensically to any great extent: ________________

3. Explain why the discovery of the ABO system was such a significant medical breakthrough: ________________

Related activities: *Antibodies, Blood*
Web links: *Blood Typing Game*

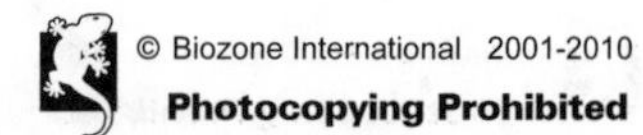

The Action of Phagocytes

Human cells that ingest microbes and digest them by the process of **phagocytosis** are called **phagocytes**. All are types of white blood cells. During many kinds of infections, especially bacterial infections, the total number of white blood cells increases by two to four times the normal number. The ratio of various white blood cell types changes during the course of an infection.

How a Phagocyte Destroys Microbes

1 Detection
Phagocyte detects microbes by the chemicals they give off (chemotaxis) and sticks the microbes to its surface.

2 Ingestion
The microbe is engulfed by the phagocyte wrapping pseudopodia around it to form a vesicle.

3 Phagosome forms
A phagosome (phagocytic vesicle) is formed, which encloses the microbes in a membrane.

4 Fusion with lysosome
Phagosome fuses with a lysosome (which contains powerful enzymes that can digest the microbe).

5 Digestion
The microbes are broken down by enzymes into their chemical constituents.

6 Discharge
Indigestible material is discharged from the phagocyte cell.

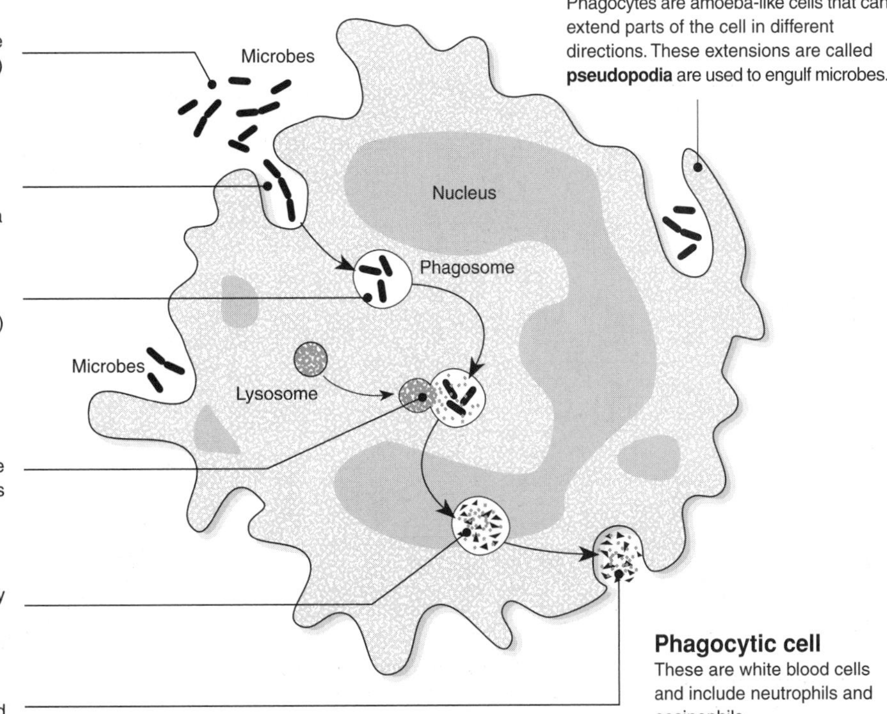

Phagocytes are amoeba-like cells that can extend parts of the cell in different directions. These extensions are called **pseudopodia** are used to engulf microbes.

Phagocytic cell
These are white blood cells and include neutrophils and eosinophils.

The Interaction of Microbes and Phagocytes

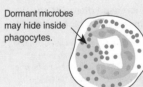

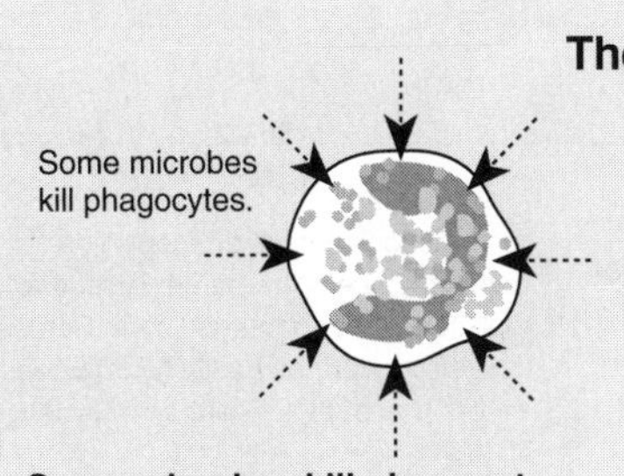

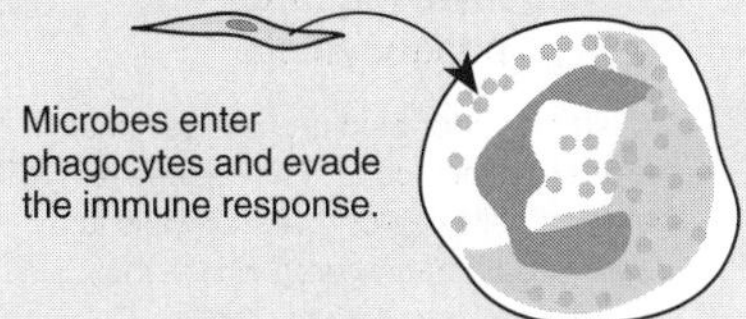

Microbes enter phagocytes and evade the immune response.

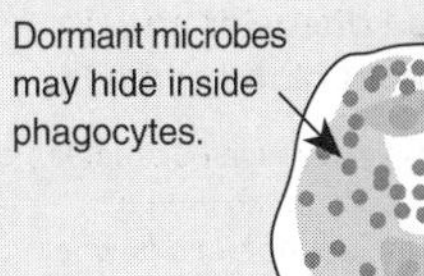

Some microbes kill phagocytes
Some microbes produce toxins that can actually kill phagocytes, e.g. toxin-producing staphylococci and the dental plaque-forming bacteria *Actinobacillus*.

Microbes evade immune system
Some microbes can evade the immune system by entering phagocytes. The microbes prevent fusion of the lysosome with the phagosome and multiply inside the phagocyte, almost filling it. Examples include *Chlamydia, Mycobacterium tuberculosis, Shigella*, and malarial parasites.

Dormant microbes hide inside
Some microbes can remain dormant inside the phagocyte for months or years at a time. Examples include the microbes that cause brucellosis and tularemia.

1. Identify the white blood cells capable of phagocytosis: ______________________

____

2. Describe how a blood sample from a patient may be used to determine whether they have a microbial infection (without looking for the microbes themselves):

____

____

3. Explain how some microbes are able to overcome phagocytic cells and use them to their advantage:

____

____

Periodicals:
Looking out for danger

***Related activities:** The Body's Defenses, Blood*
***Web links:** Phagocytosis and Bacterial Pathogens*

Inflammation

Damage to the body's tissues can be caused by physical agents (e.g. sharp objects, heat, radiant energy, or electricity), microbial infection, or chemical agents (e.g. gases, acids and bases). The damage triggers a defensive response called **inflammation**. It is usually characterized by four symptoms: pain, redness, heat and swelling. The inflammatory response is beneficial and has the following functions: (1) to destroy the cause of the infection and remove it and its products from the body; (2) if this fails, to limit the effects on the body by confining the infection to a small area; (3) replacing or repairing tissue damaged by the infection. The process of inflammation can be divided into three distinct stages. These are described below.

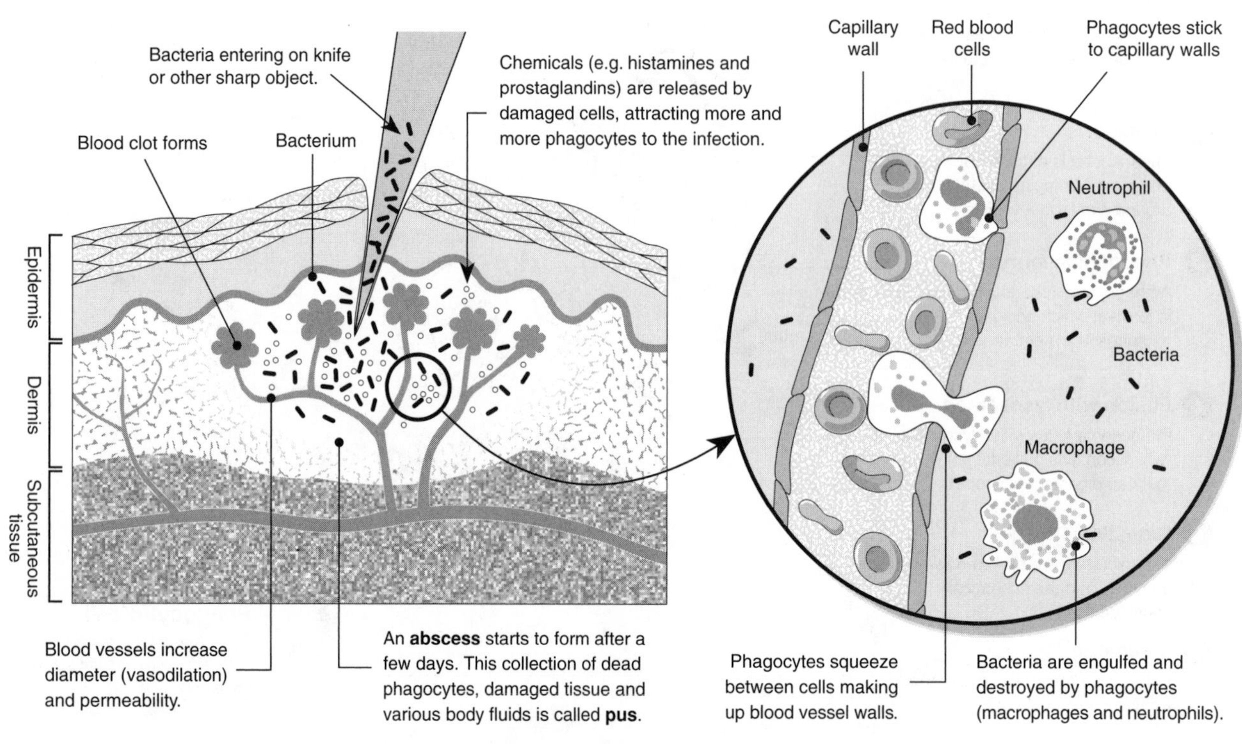
Bacteria entering on knife or other sharp object.
Blood clot forms
Bacterium
Chemicals (e.g. histamines and prostaglandins) are released by damaged cells, attracting more and more phagocytes to the infection.
Epidermis
Dermis
Subcutaneous tissue
Capillary wall
Red blood cells
Phagocytes stick to capillary walls
Neutrophil
Bacteria
Macrophage
Blood vessels increase diameter (vasodilation) and permeability.
An **abscess** starts to form after a few days. This collection of dead phagocytes, damaged tissue and various body fluids is called **pus**.
Phagocytes squeeze between cells making up blood vessel walls.
Bacteria are engulfed and destroyed by phagocytes (macrophages and neutrophils).

Stages in inflammation

Increased diameter and permeability of blood vessels

Blood vessels increase their diameter and permeability in the area of damage. This increases blood flow to the area and allows defensive substances to leak into tissue spaces.

→

Phagocyte migration and phagocytosis

Within one hour of injury, phagocytes appear on the scene. They squeeze between cells of blood vessel walls to reach the damaged area where they destroy invading microbes.

→

Tissue repair

Functioning cells or supporting connective cells create new tissue to replace dead or damaged cells. Some tissue regenerates easily (skin) while others do not at all (cardiac muscle).

1. Outline the three stages of inflammation and identify the beneficial role of each stage:

(a) ____________________

____

(b) ____________________

____

(c) ____________________

____

2. Identify two features of phagocytes important in the response to microbial invasion: ____________________

____

3. State the role of histamines and prostaglandins in inflammation: ____________________

4. Explain why pus forms at the site of infection: ____________________

____

Fever

To a point, fever is beneficial, because it assists a number of the defense processes. The release of the protein **interleukin-1** helps to reset the thermostat of the body to a higher level and increases production of **T cells** (lymphocytes). High body temperature also intensifies the effect of **interferon** (an antiviral protein) and may inhibit the growth of some bacteria and viruses. Because high temperatures speed up the body's **metabolic reactions**, it may promote more rapid tissue repair. Fever also increases heart rate so that white blood cells are delivered to sites of infection more rapidly. The normal body temperature range for most people is 36.2 to 37.2°C. Fevers of less than 40°C do not need treatment for **hyperthermia**, but excessive fever requires prompt attention (particularly in children). Death usually results if body temperature rises above 44.4 to 45.5°C.

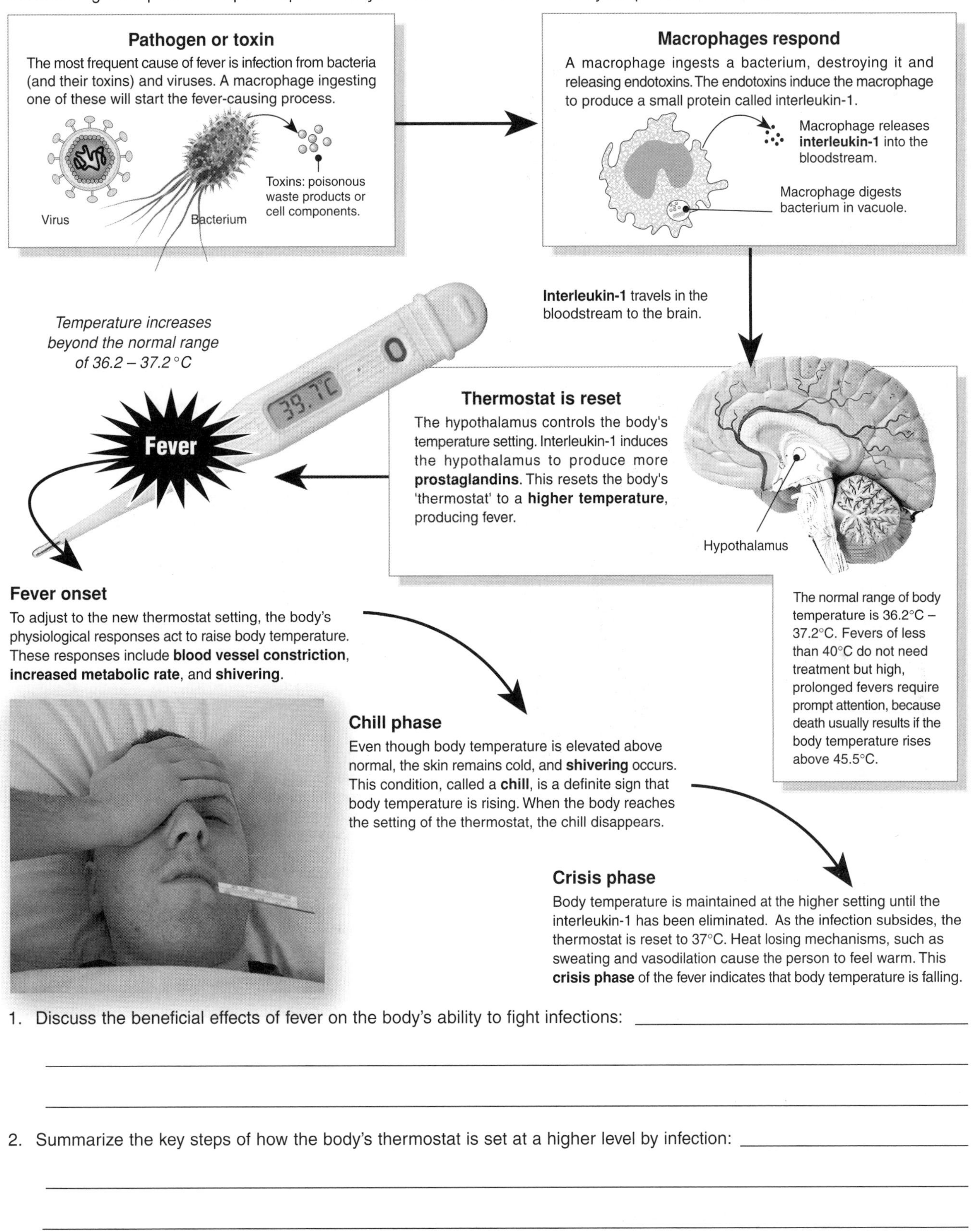

Pathogen or toxin
The most frequent cause of fever is infection from bacteria (and their toxins) and viruses. A macrophage ingesting one of these will start the fever-causing process.
Virus
Bacterium
Toxins: poisonous waste products or cell components.
Macrophages respond
A macrophage ingests a bacterium, destroying it and releasing endotoxins. The endotoxins induce the macrophage to produce a small protein called interleukin-1.
Macrophage releases interleukin-1 into the bloodstream.
Macrophage digests bacterium in vacuole.
Interleukin-1 travels in the bloodstream to the brain.
Temperature increases beyond the normal range of 36.2 – 37.2 °C
39.7°C
Fever
Thermostat is reset
The hypothalamus controls the body's temperature setting. Interleukin-1 induces the hypothalamus to produce more prostaglandins. This resets the body's 'thermostat' to a higher temperature, producing fever.
Hypothalamus
The normal range of body temperature is 36.2°C – 37.2°C. Fevers of less than 40°C do not need treatment but high, prolonged fevers require prompt attention, because death usually results if the body temperature rises above 45.5°C.
Fever onset
To adjust to the new thermostat setting, the body's physiological responses act to raise body temperature. These responses include blood vessel constriction, increased metabolic rate, and shivering.
Chill phase
Even though body temperature is elevated above normal, the skin remains cold, and shivering occurs. This condition, called a chill, is a definite sign that body temperature is rising. When the body reaches the setting of the thermostat, the chill disappears.
Crisis phase
Body temperature is maintained at the higher setting until the interleukin-1 has been eliminated. As the infection subsides, the thermostat is reset to 37°C. Heat losing mechanisms, such as sweating and vasodilation cause the person to feel warm. This crisis phase of the fever indicates that body temperature is falling.

1. Discuss the beneficial effects of fever on the body's ability to fight infections: ______________________

2. Summarize the key steps of how the body's thermostat is set at a higher level by infection: ______________________

Related activities: *Principles of Homeostasis, The Body's Defenses*

The Lymphatic System

Fluid leaks out from capillaries and forms the tissue fluid, which is similar in composition to plasma but lacks large proteins. This fluid bathes the tissues, supplying them with nutrients and oxygen, and removing wastes. Some of the tissue fluid returns directly into the capillaries, but some drains back into the blood circulation through a network of lymph vessels. This fluid, called **lymph**, is similar to tissue fluid, but contains more leukocytes. Apart from its circulatory role, the lymphatic system also has an important function in the immune response. Lymph nodes are the primary sites where the destruction of pathogens and other foreign substances occurs. A lymph node that is fighting an infection becomes swollen and hard as the lymph cells reproduce rapidly to increase their numbers. The thymus, spleen, and bone marrow also contribute leukocytes to the lymphatic and circulatory systems.

Tonsils: Tonsils (and adenoids) comprise a collection of large lymphatic nodules at the back of the throat. They produce lymphocytes and antibodies and are well-placed to protect against invasion of pathogens.

Thymus gland: The thymus is a two-lobed organ located close to the heart. It is prominent in infants and diminishes after puberty to a fraction of its original size. Its role in immunity is to help produce **T cells** that destroy invading microbes directly or indirectly by producing various substances.

Spleen: The oval spleen is the largest mass of lymphatic tissue in the body, measuring about 12 cm in length. It stores and releases blood in case of demand (e.g. in cases of bleeding), produces mature **B cells**, and destroys bacteria by phagocytosis.

Bone marrow: Bone marrow produces red blood cells and many kinds of leukocytes: monocytes (and macrophages), neutrophils, eosinophils, basophils, and lymphocytes (B cells and T cells).

Lymphatic vessels: When tissue fluid is picked up by lymph capillaries, it is called **lymph**. The lymph is passed along lymphatic vessels to a series of lymph nodes. These vessels contain one-way valves that move the lymph in the direction of the heart until it is reintroduced to the blood at the subclavian veins.

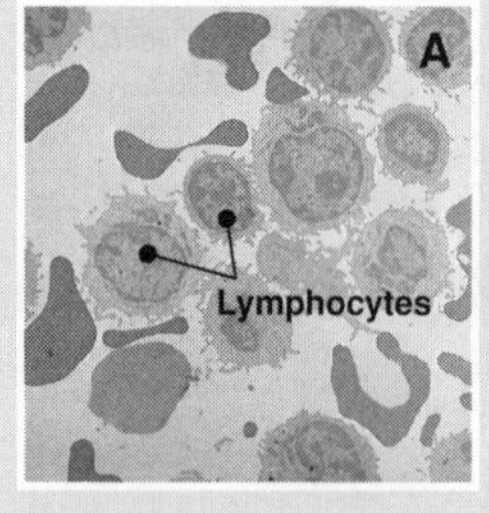

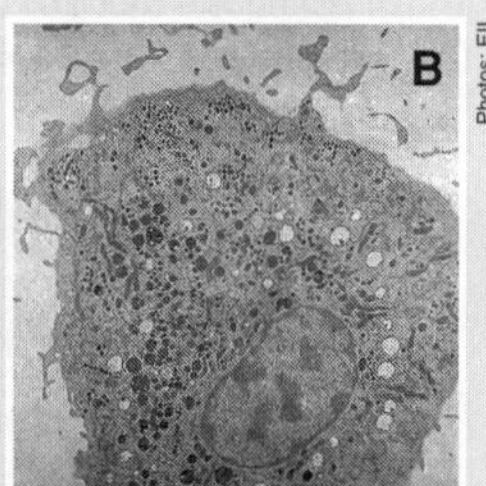

Many types of leukocytes are involved in internal defense. The photos above illustrate examples of leukocytes. **A** shows a cluster of **lymphocytes**. **B** shows a single **macrophage**: large, phagocytic cells that develop from monocytes and move from the blood to reside in many organs and tissues, including the spleen and lymph nodes.

Lymph node

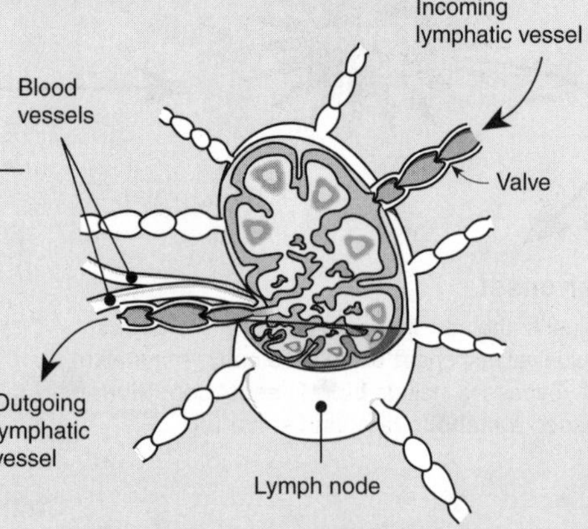

Lymph nodes are oval or bean-shaped structures, scattered throughout the body, usually in groups, along the length of lymphatic vessels. As lymph passes through the nodes, it filters foreign particles (including pathogens) by trapping them in fibers. Lymph nodes are also a "store" of **lymphocytes**, which may circulate to other parts of the body. Once trapped, macrophages destroy the foreign substances by phagocytosis. T cells may destroy them by releasing various products, and/or B cells may release antibodies that destroy them.

1. Briefly describe the composition of lymph: ______________________________

____

2. Discuss the various roles of lymph: ______________________________

____

____

3. Describe one role of each of the following in the lymphatic system:

   (a) Lymph nodes: ______________________________

   (b) Bone marrow: ______________________________

Related activities: *Capillary Networks, The Immune System*

Acquired Immunity

We have natural or **innate resistance** to certain illnesses, including most diseases of other animal species. **Acquired immunity** refers to the protection we develop during our lifetime against microbes and foreign substances. Immunity can be acquired either passively or actively. **Active immunity** develops when a person is exposed to foreign substances or to microorganisms (e.g. through infection) and the immune system responds. **Passive immunity** is acquired when antibodies are transferred from one person to another. Recipients do not make the antibodies themselves and the effect lasts only as long as the antibodies are present (usually several weeks or months). Either type of immunity may also be **naturally acquired** through natural exposure to microbes, or **artificially acquired** as a result of medical treatment.

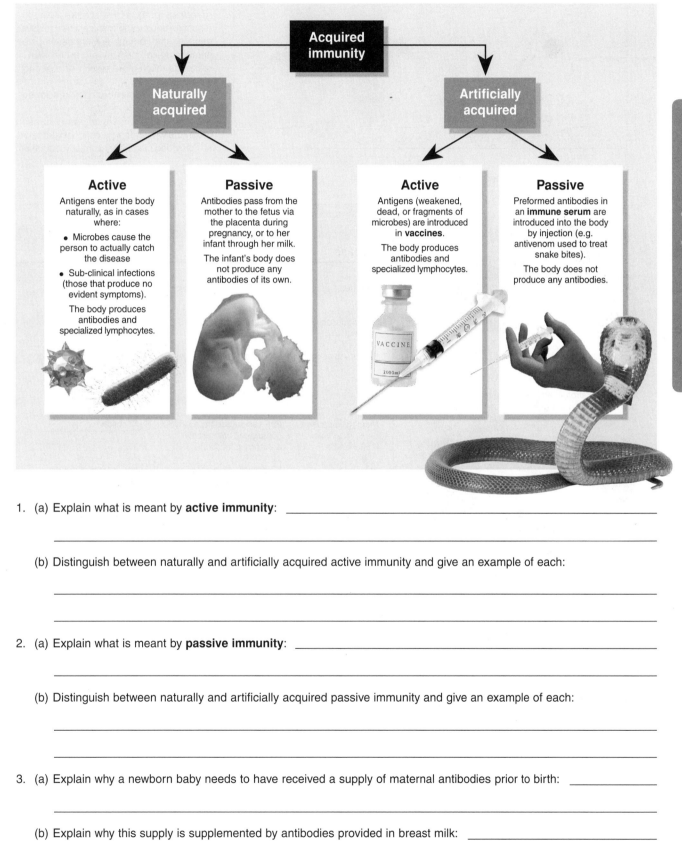

Acquired immunity

Naturally acquired

Active

Antigens enter the body naturally, as in cases where:

- Microbes cause the person to actually catch the disease
- Sub-clinical infections (those that produce no evident symptoms).

The body produces antibodies and specialized lymphocytes.

Passive

Antibodies pass from the mother to the fetus via the placenta during pregnancy, or to her infant through her milk.

The infant's body does not produce any antibodies of its own.

Artificially acquired

Active

Antigens (weakened, dead, or fragments of microbes) are introduced in **vaccines**.

The body produces antibodies and specialized lymphocytes.

Passive

Preformed antibodies in an **immune serum** are introduced into the body by injection (e.g. antivenom used to treat snake bites).

The body does not produce any antibodies.

1. (a) Explain what is meant by **active immunity**:

   (b) Distinguish between naturally and artificially acquired active immunity and give an example of each:

2. (a) Explain what is meant by **passive immunity**:

   (b) Distinguish between naturally and artificially acquired passive immunity and give an example of each:

3. (a) Explain why a newborn baby needs to have received a supply of maternal antibodies prior to birth:

   (b) Explain why this supply is supplemented by antibodies provided in breast milk:

Primary and Secondary Responses to Antigens

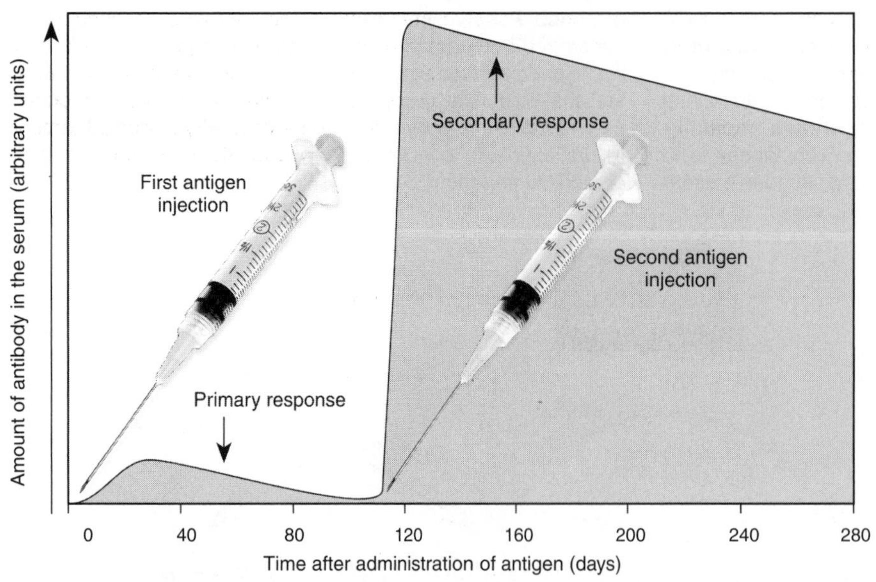

When the B cells encounter antigens and produce antibodies, the body develops **active immunity** against that antigen.

The initial response to antigenic stimulation, caused by the sudden increase in B cell clones, is called the **primary response**. Antibody levels as a result of the primary response peak a few weeks after the response begins and then decline. However, because the immune system develops an immunological memory of that antigen, it responds much more quickly and strongly when presented with the same antigen subsequently (the **secondary response**).

This forms the basis of immunization programs where one or more booster shots are provided following the inital vaccination.

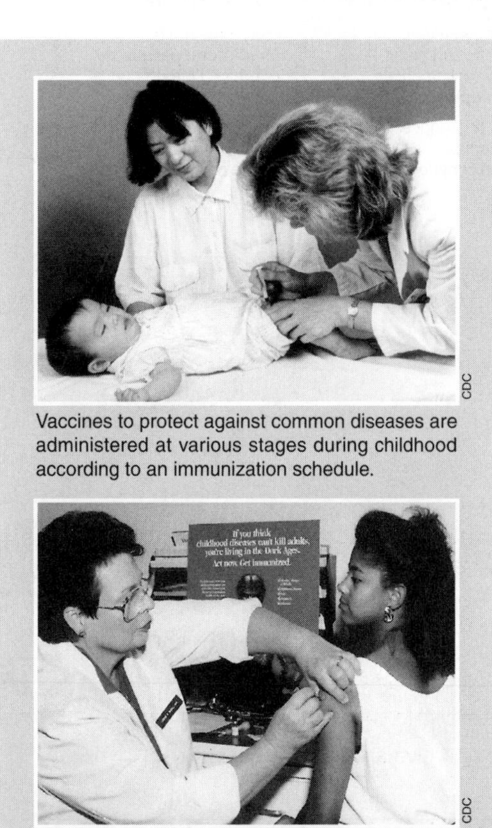

Vaccines to protect against common diseases are administered at various stages during childhood according to an immunization schedule.

While most vaccinations are given in childhood, adults may be vaccinated against specific diseases (e.g. tuberculosis) if they are in a high risk group or if they are traveling to a region in the world where a disease is prevalent.

Selected Vaccines Used To Prevent Diseases In Humans

Disease	Type of vaccine	Recommendation
Diphtheria	Purified diphtheria toxoid	From early childhood and every 10 years for adults
Meningococcal meningitis	Purified polysaccharide of *Neisseria menigitidis*	For people with substantial risk of infection
Whooping cough	Killed cells or fragments of *Bordetella pertussis*	Children prior to school age
Tetanus	Purified tetanus toxoid	14-16 year olds with booster every 10 years
Meningitis caused by *Hemophilus influenzae* b	Polysaccharide from virus conjugated with protein to enhance effectiveness	Early childhood
Influenza	Killed virus (vaccines using genetically engineered antigenic fragments are also being developed)	For chronically ill people, especially with respiratory diseases, or for healthy people over 65 years of age
Measles	Attenuated virus	Early childhood
Mumps	Attenuated virus	Early childhood
Rubella	Attenuated virus	Early childhood; for females of child-bearing age who are not pregnant

Vaccine development is an important part of public health. Immunization programs have been behind the eradication of some debilitating human diseases, such as smallpox. Many childhood diseases for which immunization programs exist are kept at a low level because of the phenomenon of ***herd immunity****. If a large proportion of the population is immune, those that are not immunized may be protected because the disease becomes uncommon.*

4. (a) Describe two differences between the primary and secondary responses to antigen presentation:

(b) Explain why the secondary response is so different from the primary response:

The Immune System

The efficient internal defense provided by the immune system is based on its ability to respond specifically against a foreign substance and its ability to hold a memory of this response. There are two main components of the immune system: the humoral and the cell-mediated responses. They work separately and together to protect us from disease. The **humoral immune response** is associated with the serum (non-cellular part of the blood) and involves the action of **antibodies** secreted by B cell lymphocytes. Antibodies are found in extracellular fluids including lymph, plasma, and mucus secretions. The humoral response protects the body against circulating viruses, and bacteria and their toxins. The **cell-mediated immune response** is associated with the production of specialized lymphocytes called **T cells**. It is most effective against bacteria and viruses located within host cells, as well as against parasitic protozoa, fungi, and worms. This system is also an important defense against cancer, and is responsible for the rejection of transplanted tissue. Both B and T cells develop from stem cells located in the liver of fetuses and the bone marrow of adults. T cells complete their development in the thymus, whilst the B cells mature in the bone marrow.

Lymphocytes and their Functions

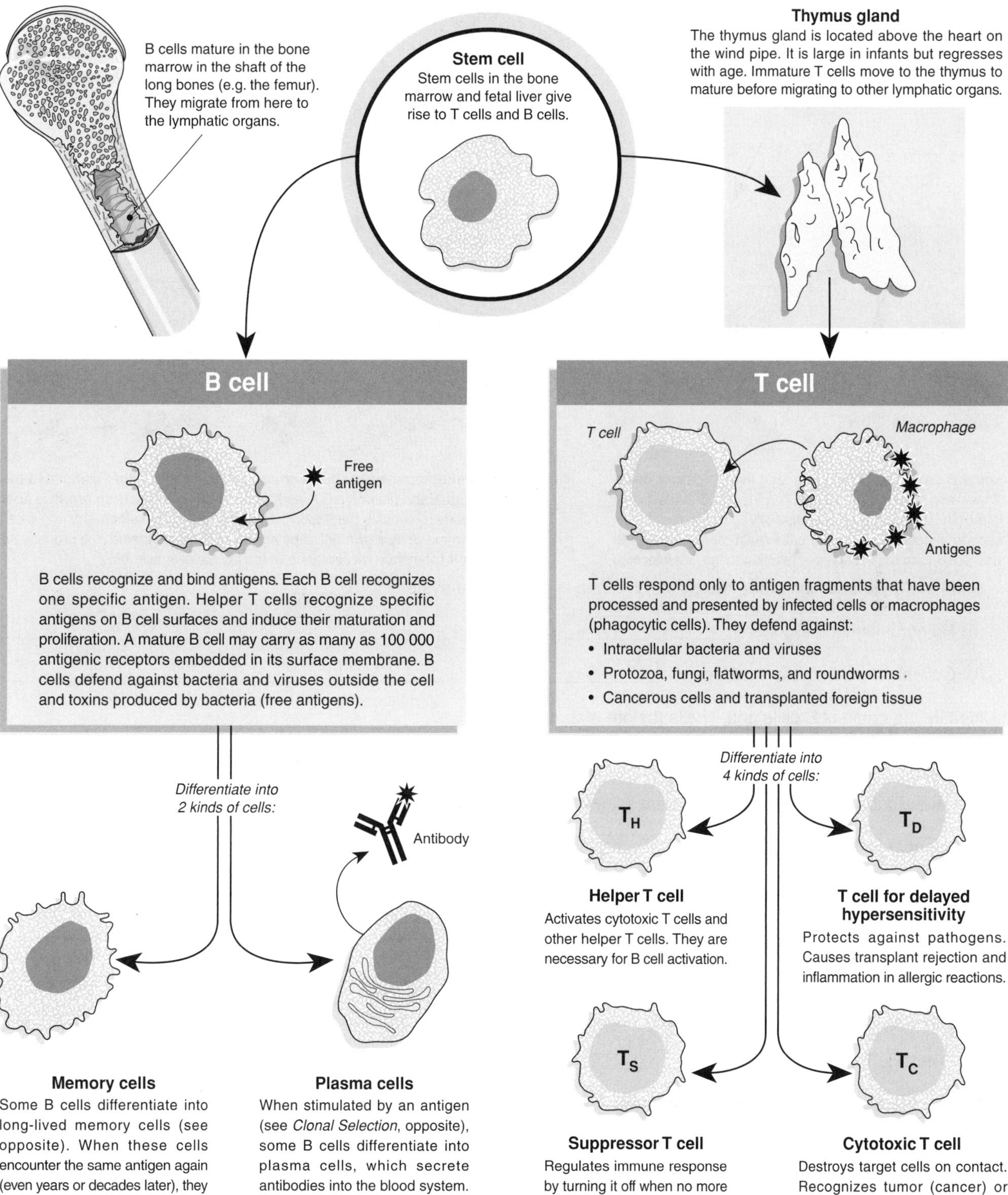

B cells mature in the bone marrow in the shaft of the long bones (e.g. the femur). They migrate from here to the lymphatic organs.
Stem cell
Stem cells in the bone marrow and fetal liver give rise to T cells and B cells.
Thymus gland
The thymus gland is located above the heart on the wind pipe. It is large in infants but regresses with age. Immature T cells move to the thymus to mature before migrating to other lymphatic organs.
B cell
Free antigen
B cells recognize and bind antigens. Each B cell recognizes one specific antigen. Helper T cells recognize specific antigens on B cell surfaces and induce their maturation and proliferation. A mature B cell may carry as many as 100 000 antigenic receptors embedded in its surface membrane. B cells defend against bacteria and viruses outside the cell and toxins produced by bacteria (free antigens).
T cell
T cell
Macrophage
Antigens
T cells respond only to antigen fragments that have been processed and presented by infected cells or macrophages (phagocytic cells). They defend against:
• Intracellular bacteria and viruses
• Protozoa, fungi, flatworms, and roundworms
• Cancerous cells and transplanted foreign tissue
Differentiate into 2 kinds of cells:
Antibody
Differentiate into 4 kinds of cells:
T_H
T_D
T_S
T_C
Helper T cell
Activates cytotoxic T cells and other helper T cells. They are necessary for B cell activation.
T cell for delayed hypersensitivity
Protects against pathogens. Causes transplant rejection and inflammation in allergic reactions.
Memory cells
Some B cells differentiate into long-lived memory cells (see opposite). When these cells encounter the same antigen again (even years or decades later), they rapidly differentiate into antibody-producing plasma cells.
Plasma cells
When stimulated by an antigen (see Clonal Selection, opposite), some B cells differentiate into plasma cells, which secrete antibodies into the blood system. The antibodies then inactivate the circulating antigens.
Suppressor T cell
Regulates immune response by turning it off when no more antigen is present.
Cytotoxic T cell
Destroys target cells on contact. Recognizes tumor (cancer) or virus infected cells by their surface (antigens and MHC markers).

Periodicals: *Lymphocytes - the heart of the immune system*

Related activities: *The Lymphatic System,* ***Web links:*** *The Immune System Overview, The Humoral Response, Introducing... Specific Immunity,*

The immune system has the ability to respond to the large and unpredictable range of potential antigens encountered in the environment. The diagram below explains how this ability is based on **clonal selection** after antigen exposure. The example illustrated is for B cell lymphocytes. In the same way, a T cell stimulated by a specific antigen will multiply and develop into different types of T cells. Clonal selection and differentiation of lymphocytes provide the basis for **immunological memory**.

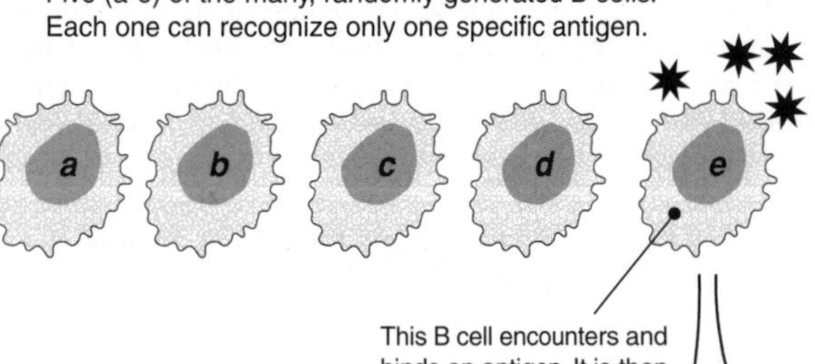

Five (a-e) of the many, randomly generated B cells. Each one can recognize only one specific antigen.
a
b
c
d
e
This B cell encounters and binds an antigen. It is then stimulated to proliferate.

Clonal Selection Theory

Millions of randomly generated B cells form during development. Collectively, they can recognize many antigens, including those that have never been encountered. Each B cell makes an antibody specific to the type of antigenic receptor on its surface. The receptor reacts only to that specific antigen. When a B cell encounters its antigen, it responds by proliferating and producing many clones all with the same kind of antibody. This is called **clonal selection** because the antigen selects the B cells that will proliferate.

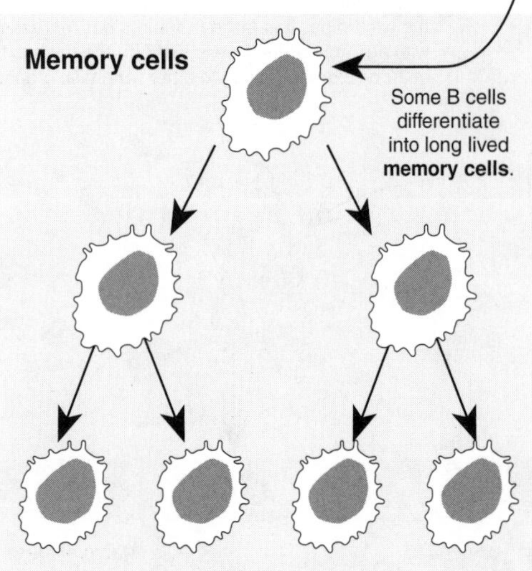

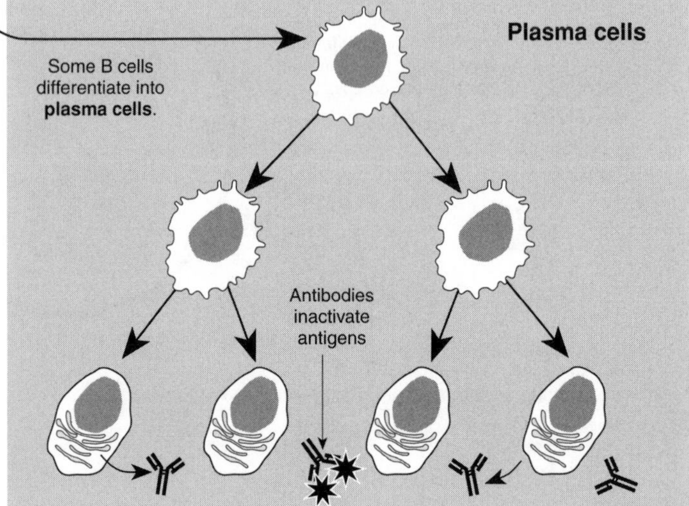

Some B cells differentiate into long lived **memory cells**. These are retained in the lymph nodes to provide future immunity (**immunological memory**). In the event of a second infection, B-memory cells react more quickly and vigorously than the initial B-cell reaction to the first infection.

Plasma cells secrete antibodies specific to the antigen that stimulated their development. Each plasma cell lives for only a few days, but can produce about 2000 antibody molecules per second. Note that during development, any B cells that react to the body's own antigens are selectively destroyed in a process that leads to **self tolerance** (acceptance of the body's own tissues).

1. State the general action of the two major divisions in the immune system:

   (a) Humoral immune system: ______________________________

   (b) Cell-mediated immune system: ______________________________

2. Identify the origin of B cells and T cells (before maturing): ______________________________

3. (a) State where B cells mature: ____________________ (b) State where T cells mature: ____________________

4. Briefly describe the function of each of the following cells in the immune system response:

   (a) Memory cells: ______________________________

   (b) Plasma cells: ______________________________

   (c) Helper T cells: ______________________________

   (d) Suppressor T cells: ______________________________

   (e) Delayed hypersensitivity T cells: ______________________________

   (f) Cytotoxic T cells: ______________________________

5. Explain the basis of **immunological memory**: ______________________________

____

____

____

Antibodies

Antibodies and antigens play key roles in the response of the immune system. Antigens are foreign molecules that are able to bind to antibodies (or T cell receptors) and provoke a specific immune response. Antigens include potentially damaging microbes and their toxins (see below) as well as substances such as pollen grains, blood cell surface molecules, and the surface proteins on transplanted tissues. **Antibodies** (also called immunoglobulins) are proteins that are made in response to antigens. They are secreted into the plasma where they circulate and can recognize, bind to, and help to destroy antigens. There are five classes of **immunoglobulins**. Each plays a different role in the immune response (including destroying protozoan parasites, enhancing phagocytosis, protecting mucous surfaces, and neutralizing toxins and viruses). The human body can produce an estimated 100 million antibodies, recognizing many different antigens, including those it has never encountered. Each type of antibody is highly specific to only one particular antigen. The ability of the immune system to recognize and ignore the antigenic properties of its own tissues occurs early in development and is called **self-tolerance**. Exceptions occur when the immune system malfunctions and the body attacks its own tissues, causing an **autoimmune disorder**.

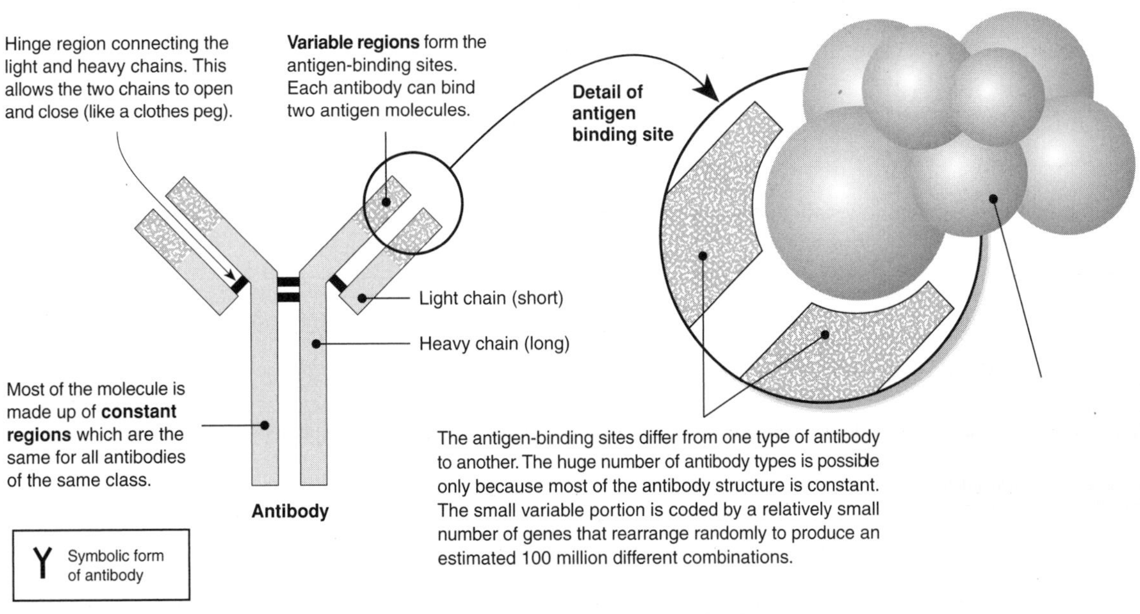

Hinge region connecting the light and heavy chains. This allows the two chains to open and close (like a clothes peg).
Variable regions form the antigen-binding sites. Each antibody can bind two antigen molecules.
Detail of antigen binding site
Light chain (short)
Heavy chain (long)
Most of the molecule is made up of constant regions which are the same for all antibodies of the same class.
Antibody
Symbolic form of antibody
The antigen-binding sites differ from one type of antibody to another. The huge number of antibody types is possible only because most of the antibody structure is constant. The small variable portion is coded by a relatively small number of genes that rearrange randomly to produce an estimated 100 million different combinations.

How Antibodies Inactivate Antigens

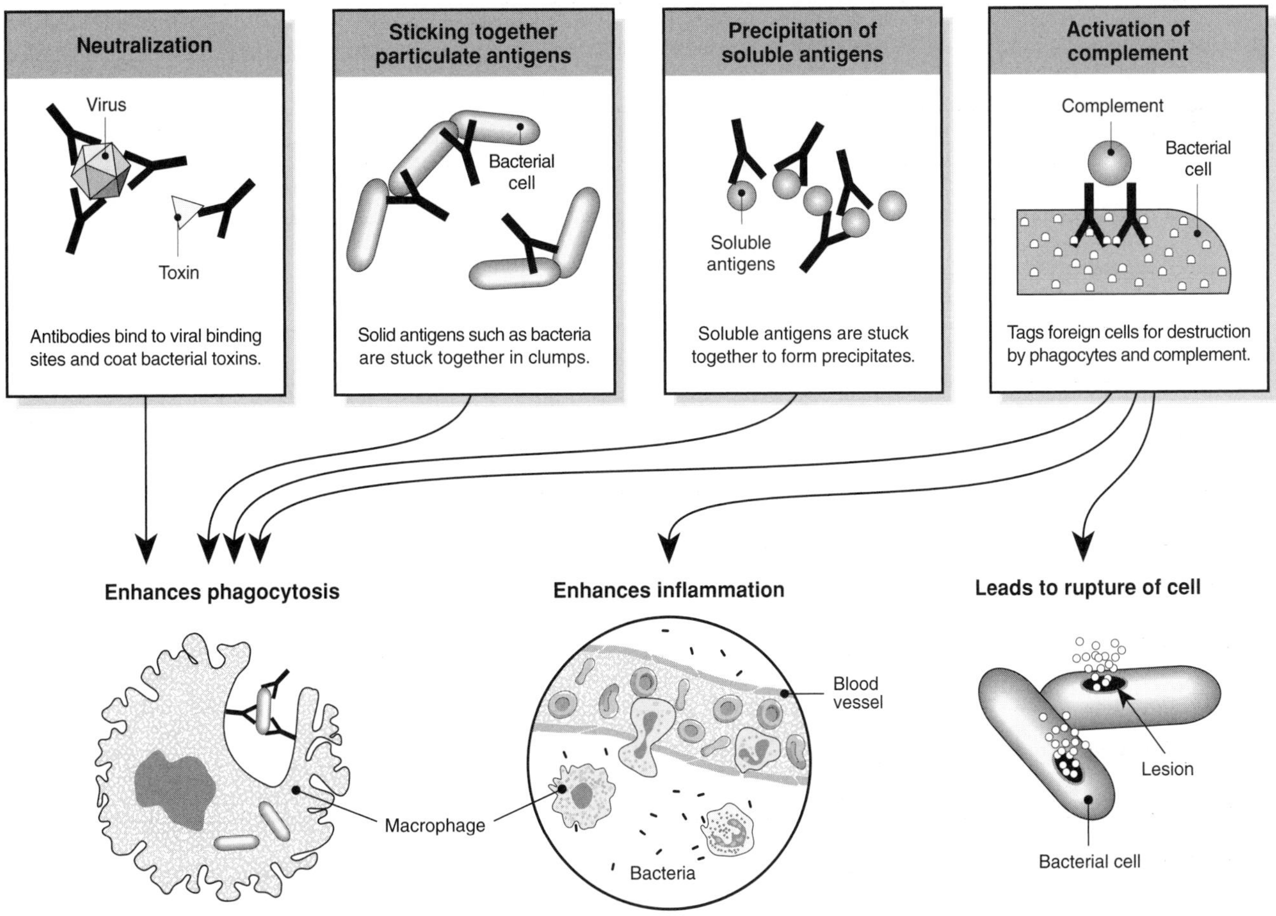

Neutralization

Virus

Toxin

Antibodies bind to viral binding sites and coat bacterial toxins.

Sticking together particulate antigens

Bacterial cell

Solid antigens such as bacteria are stuck together in clumps.

Precipitation of soluble antigens

Soluble antigens

Soluble antigens are stuck together to form precipitates.

Activation of complement

Complement

Bacterial cell

Tags foreign cells for destruction by phagocytes and complement.

Enhances phagocytosis

Macrophage

Enhances inflammation

Blood vessel

Bacteria

Leads to rupture of cell

Lesion

Bacterial cell

Periodicals:
Antibodies

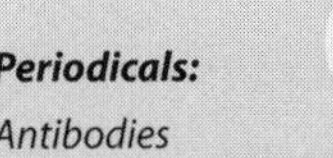

Related activities: *Targets for Defense, The Immune System*
Web links: *How Lymphocytes Produce Antibodies*

1. Distinguish between an antibody and an antigen:

2. It is necessary for the immune system to clearly distinguish cells and proteins made by the body, from foreign ones.

   (a) Explain why this is the case:

   (b) In simple terms, explain how **self tolerance** develops (see the activity *The Immune System* if you need help):

   (c) Name the type of disorder that results when this recognition system fails:

   (d) Describe two examples of disorders that are caused in this way, identifying what happens in each case:

3. Discuss the ways in which antibodies work to inactivate antigens:

4. Explain how antibody activity enhances or leads to:

   (a) Phagocytosis:

   (b) Inflammation:

   (c) Bacterial cell lysis:

Monoclonal Antibodies

A **monoclonal antibody** is an artificially produced antibody that neutralizes only one specific protein (antigen). Monoclonal antibodies are produced in the laboratory by stimulating the production of B-cell in mice injected with the antigen. These B-cells produce an antibody against the antigen. When isolated and made to fuse with immortal tumor cells, they can be cultured indefinitely in a suitable growing medium (as illustrated below). Monoclonal antibodies are useful for three reasons: they are totally uniform (i.e. clones), they can be produced in large quantities, and they are highly specific. The uses of antibodies produced by this method have ranged from use as diagnostic tools to treatments for infections and cancer. The therapeutic use of monoclonal antibodies has been somewhat limited because they are currently produced by non-human cells and the immune systems of some people have reacted against these foreign proteins (remember that antibodies are proteins). It is hoped in the future to produce monoclonal antibodies derived from human cells, which will probably cause fewer reactions.

Making Monoclonal Antibodies

A mouse is injected with a foreign protein (antigen) that will stimulate the mouse to produce antibodies against it.

The mouse's B- cells (lymphocytes) have developed an antibody to recognize the foreign protein (antigen).

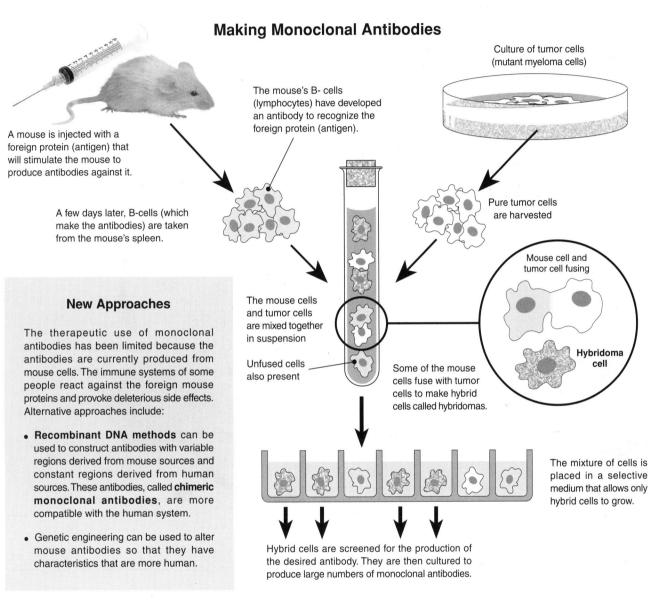

Culture of tumor cells (mutant myeloma cells)

A few days later, B-cells (which make the antibodies) are taken from the mouse's spleen.

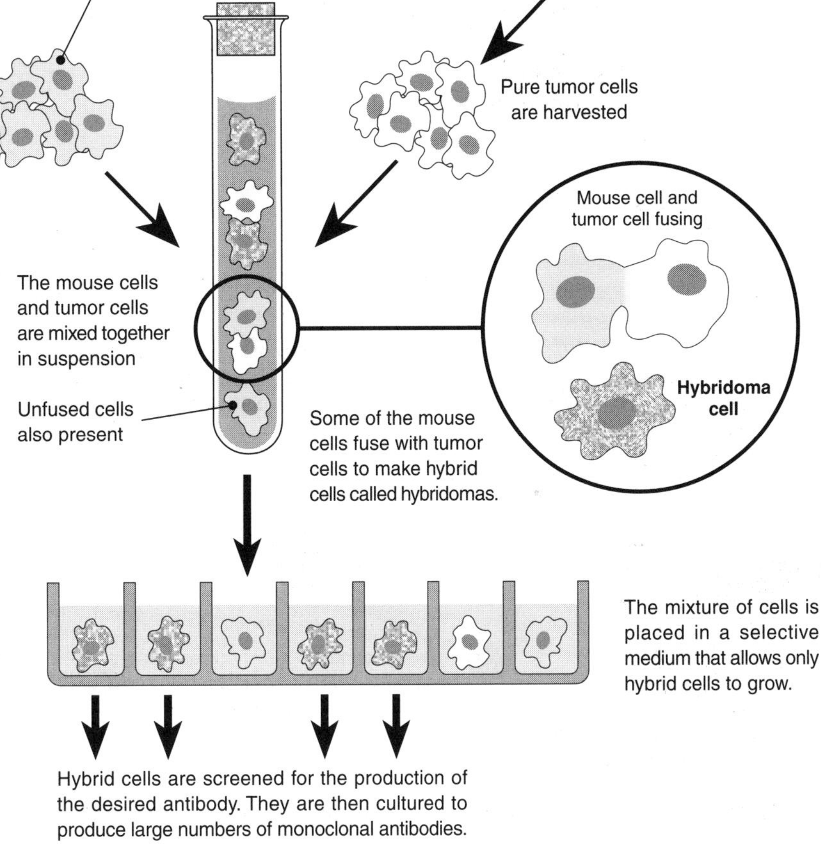

Pure tumor cells are harvested

The mouse cells and tumor cells are mixed together in suspension

Unfused cells also present

Mouse cell and tumor cell fusing

Hybridoma cell

Some of the mouse cells fuse with tumor cells to make hybrid cells called hybridomas.

The mixture of cells is placed in a selective medium that allows only hybrid cells to grow.

Hybrid cells are screened for the production of the desired antibody. They are then cultured to produce large numbers of monoclonal antibodies.

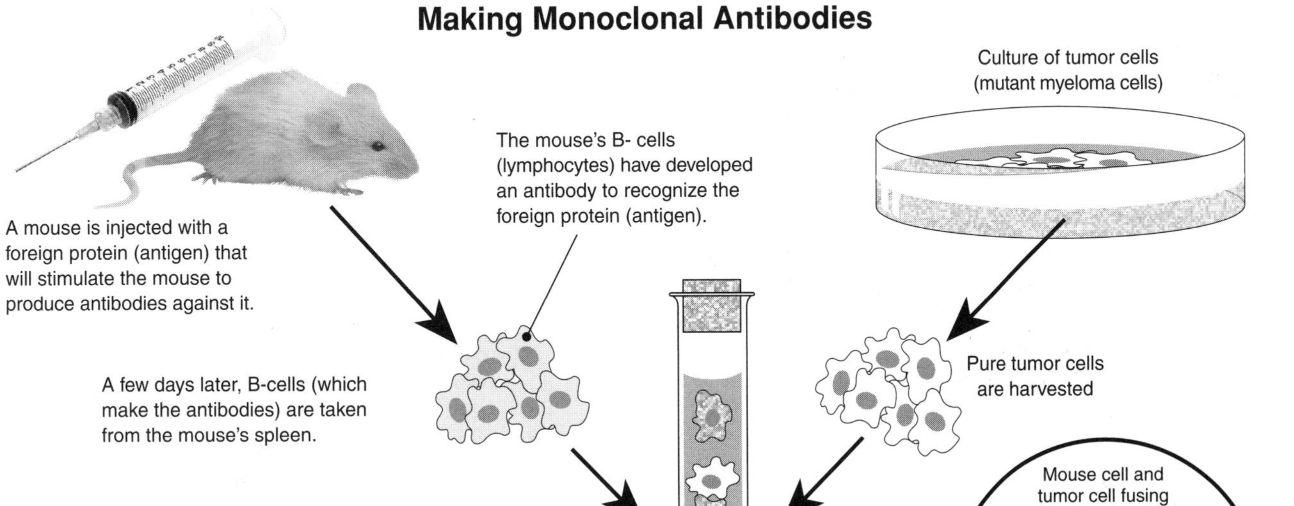

New Approaches

The therapeutic use of monoclonal antibodies has been limited because the antibodies are currently produced from mouse cells. The immune systems of some people react against the foreign mouse proteins and provoke deleterious side effects. Alternative approaches include:

- **Recombinant DNA methods** can be used to construct antibodies with variable regions derived from mouse sources and constant regions derived from human sources. These antibodies, called **chimeric monoclonal antibodies**, are more compatible with the human system.
- Genetic engineering can be used to alter mouse antibodies so that they have characteristics that are more human.

1. Identify the mouse cells used to produce the monoclonal antibodies: ______________________

2. Describe the characteristic of tumor cells that allows an ongoing culture of antibody-producing lymphocytes to be made:

____

3. Discuss the benefits of using monoclonal antibodies for diagnostic and therapeutic purposes:

____

____

____

____

____

Herceptin is the patented name of a **monoclonal antibody** for the targeted treatment of breast cancer. This drug (chemical name Trastuzumab) recognizes and is specific to the receptor proteins on the outside of cancer cells that are produced by the **HER2 gene**. The HER2 (**H**uman **E**pidermal growth factor **R**eceptor **2**) gene codes for proteins on the cell surface that signal to the cell when it should divide. Cancerous cells contain 20-30% more of the HER2 gene than non-cancerous cells and this causes **over-expression** of the HER2 gene, producing large amounts of HER2 protein. The overexpression causes the cell to divide more often than normal, producing a tumor. Cancerous cells are designated **HER2+** indicating receptor protein over-expression. The immune system fails to destroy these cells because they are not recognized as being abnormal. Herceptin's role is to recognize and bind to the HER2 protein on the surface of the cancerous cell. The immune system can then identify the antibodies as foreign and destroy the cell. The antibody also has the effect of blocking the cell's signaling pathway and thus stops the cell from dividing.

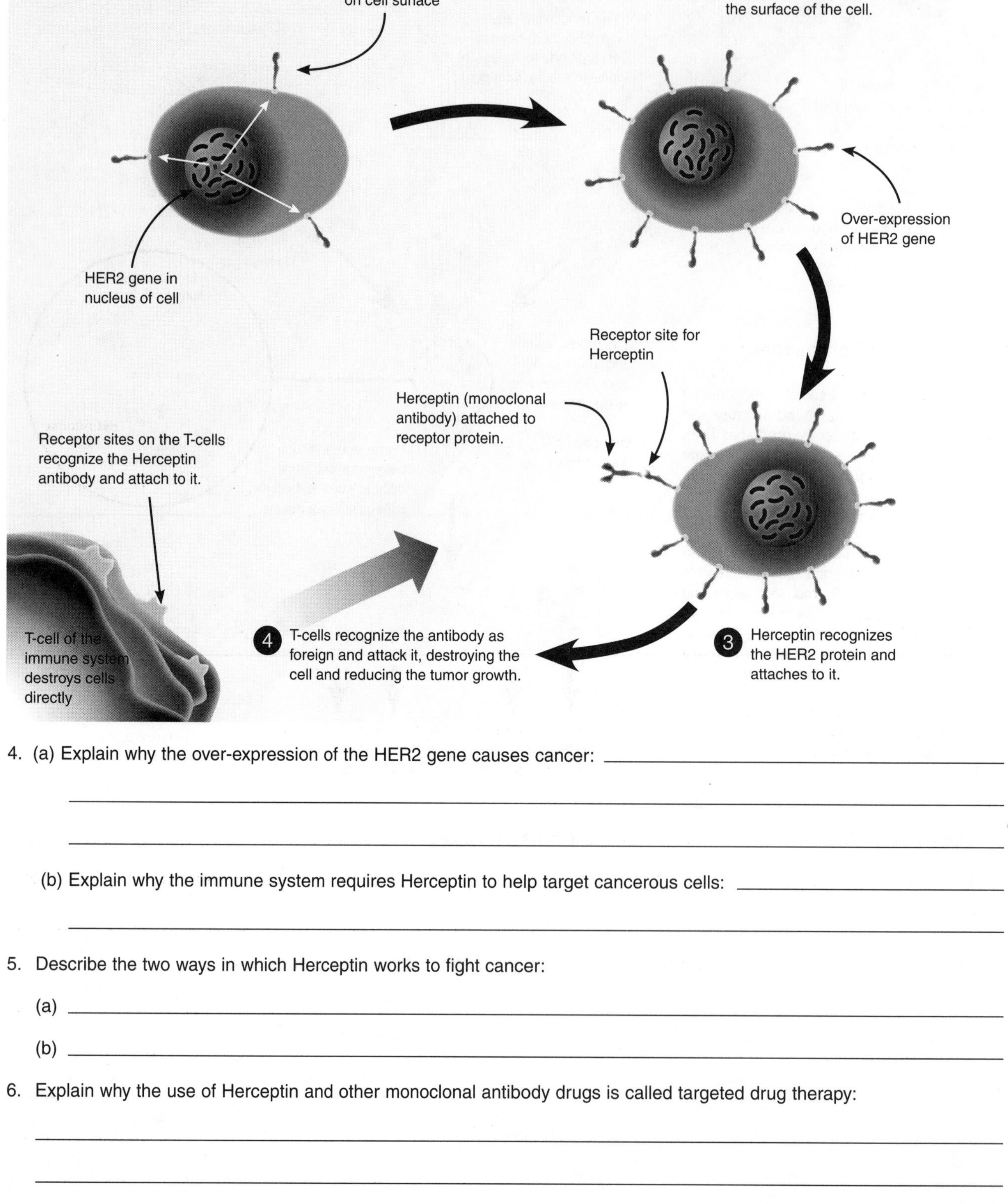

Herceptin Targeted Destruction of Cancer Cells
1 HER2 gene produces proteins that migrate to the surface of the affected cell.
HER2 protein on cell surface
2 The cancerous cell over-expresses the HER2 gene, producing large amounts of the HER2 protein on the surface of the cell.
Over-expression of HER2 gene
HER2 gene in nucleus of cell
Receptor site for Herceptin
Herceptin (monoclonal antibody) attached to receptor protein.
Receptor sites on the T-cells recognize the Herceptin antibody and attach to it.
T-cell of the immune system destroys cells directly
4 T-cells recognize the antibody as foreign and attack it, destroying the cell and reducing the tumor growth.
3 Herceptin recognizes the HER2 protein and attaches to it.

4. (a) Explain why the over-expression of the HER2 gene causes cancer: ______

(b) Explain why the immune system requires Herceptin to help target cancerous cells: ______

5. Describe the two ways in which Herceptin works to fight cancer:

(a) ______

(b) ______

6. Explain why the use of Herceptin and other monoclonal antibody drugs is called targeted drug therapy:

______

HIV and AIDS

AIDS (acquired immune deficiency syndrome) was first reported in the US in 1981. By 1983, the pathogen had been identified as a retrovirus that selectively infects **helper T cells**. It has since been established that HIV arose by the recombination of two simian viruses. It has probably been endemic in some central African regions for decades, as HIV has been found in blood samples from several African nations from as early as 1959. The disease causes a massive deficiency in the immune system due to infection with **HIV** (human immunodeficiency virus). HIV is a **retrovirus** (RNA, not DNA) and is able to splice its genes into the host cell's chromosome. As yet, there is no cure or vaccine, and the disease has taken the form of a **pandemic**, spreading to all parts of the globe and killing more than a million people each year. In southern Africa, AIDS is widespread through the heterosexual community, partly as a result of social resistance to condom use and the high incidence of risky, polygamous behavior.

Capsid
Protein coat that protects the nucleic acids (RNA) within.

Viral envelope
A piece of the cell membrane budded off from the last human host cell.

Nucleic acid
Two identical strands of RNA contain the genetic blueprint for making more HIV viruses.

Reverse transcriptase
Two copies of this important enzyme convert the RNA into DNA once inside a host cell.

Surface proteins
These spikes allow HIV to attach to receptors on the host cells (T cells and macrophages).

The structure of HIV

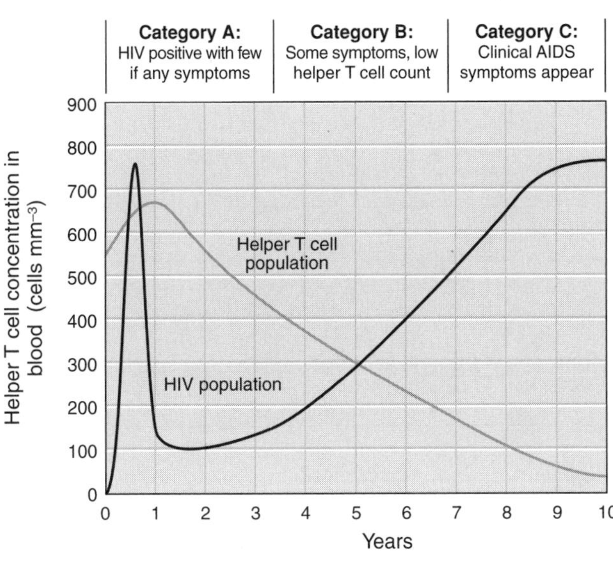

The stages of an HIV infection

AIDS is actually only the end stage of an HIV infection. Shortly after the initial infection, HIV antibodies appear within the blood. The progress of infection has three clinical categories shown on the graph above.

HIV/AIDS

Individuals affected by the human immunodeficiency virus (HIV) may have no symptoms, while medical examination may detect swollen lymph glands. Others may experience a short-lived illness when they first become infected (resembling infectious mononucleosis). The range of symptoms resulting from HIV infection is huge, and is not the result of the HIV infection directly. The symptoms arise from an onslaught of secondary infections that gain a foothold in the body due to the suppressed immune system (due to the few helper T cells). These infections are from normally rare fungal, viral, and bacterial sources. Full blown AIDS can also feature some rare forms of cancer. Some symptoms are listed below:

Fever, lymphoma (cancer) and toxoplasmosis of the brain, dementia.

Eye infections (*Cytomegalovirus).*

Skin inflammation (dermatitis) particularly affecting the face.

Oral thrush (*Candida albicans*) of the esophagus, bronchi, and lungs.

A variety of opportunistic infections, including: chronic or persistent *Herpes simplex*, tuberculosis (TB), pneumocystis pneumonia, shingles, shigellosis and salmonellosis.

Diarrhea caused by *Isospora* or *Cryptosporidium.*

Marked weight loss.

A number of autoimmune diseases, especially destruction of platelets.

Kaposi's sarcoma: a highly aggressive malignant skin tumor consisting of blue-red nodules, usually start at the feet and ankles, spreading to the rest of the body later, including respiratory and gastrointestinal tracts.

1. Explain why the HIV virus has such a devastating effect on the human body's ability to fight disease:

2. Consult the graph above showing the stages of HIV infection (remember, HIV infects and destroys helper T cells).

   (a) Describe how the virus population changes with the progression of the disease:

Transmission, Diagnosis, Treatment, and Prevention of HIV

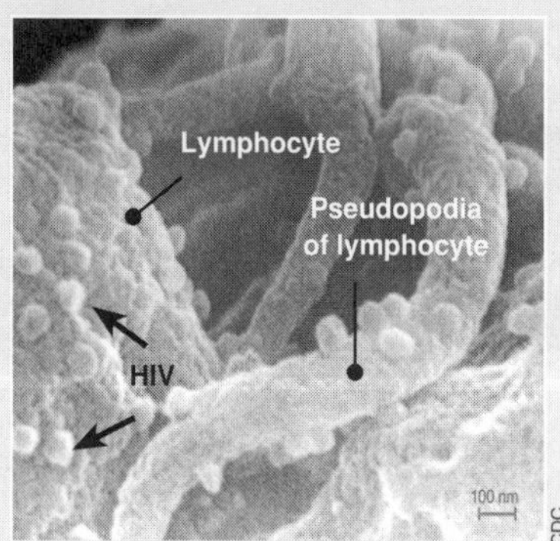

A SEM shows spherical HIV-1 virions on the surface of a human lymphocyte.

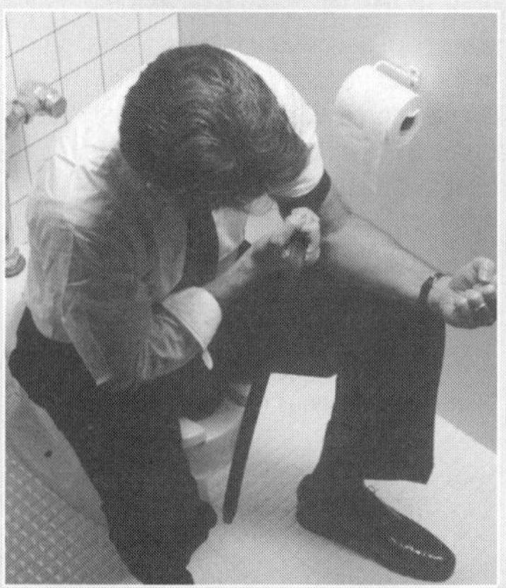

HIV is easily transmitted between intravenous drug users who share needles.

Modes of Transmission

1. HIV is transmitted in blood, vaginal secretions, semen, breast milk, and across the placenta.
2. In developed countries, blood transfusions are no longer a likely source of infection because blood is tested for HIV antibodies.
3. Historically, transmission of HIV in developed countries has been primarily through intravenous drug use and homosexual activity, but heterosexual transmission is increasing.
4. Transmission via heterosexual activity has been particularly important to the spread of HIV in Asia and southern Africa, partly because of the high prevalence of risky sexual behavior in these regions.

Treatment and Prevention

Improving the acceptance and use of safe sex practices and condoms are crucial to reducing HIV infection rates. Condoms are protective irrespective of age, the scope of sexual networks, or the presence of other sexually transmitted infections. HIV's ability to destroy, evade, and hide inside the cells of the human immune system make it difficult to treat. Research into vaccination and chemotherapy is ongoing. The first chemotherapy drug to show promise was the nucleotide analogue AZT, which inhibits reverse transcriptase. Protease inhibitors are also used. These work by blocking the HIV protease so that HIV makes copies of itself that cannot infect other cells. An effective vaccine is still some time away, although recent vaccines based on monoclonal antibody technology appear to be promising.

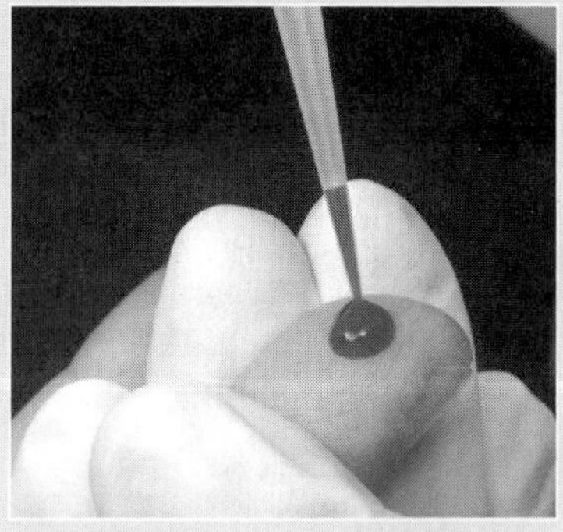

Diagnosis of HIV is possible using a simple antibody-based test on a blood sample.

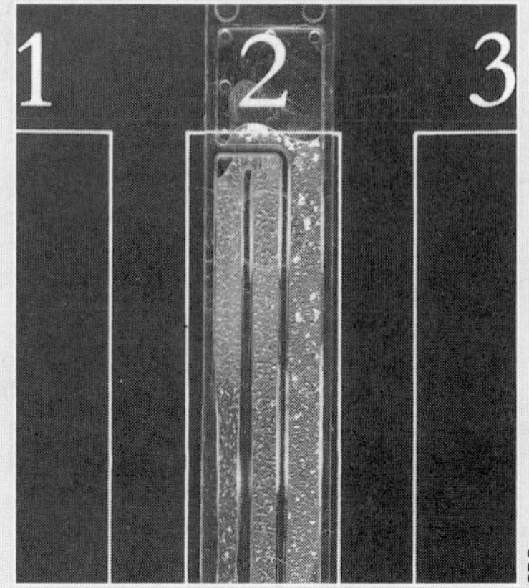

A positive HIV rapid test result shows clumping (aggregation) where HIV antibodies have reacted with HIV protein-coated latex beads.

(b) Describe how the helper T cells respond to the infection: ______

______

______

3. Describe three common ways in which HIV can be transmitted from one person to another: ______

______

______

4. Discuss the social factors that have contributed to a high prevalence of HIV among heterosexuals in southern Africa:

______

______

______

______

______

5. Explain why it has been so difficult to develop a **vaccine** for HIV: ______

______

______

6. In a rare number of cases, people who have been HIV positive for many years still have no apparent symptoms. comment on the potential importance of this observation and its likely potential in the search for a cure for AIDS:

______

______

Autoimmune Diseases

Any of numerous disorders, including **rheumatoid arthritis**, insulin dependent **diabetes mellitus**, and **multiple sclerosis**, are caused by an individual's immune system reaction to their own cells or tissues. The immune system normally distinguishes self from non-self. Some lymphocytes are capable of reacting against self, but these are generally suppressed. **Autoimmune diseases** occur when there is some interruption of the normal control process, allowing lymphocytes to escape from suppression, or when there is an alteration in some body tissue so that it is no longer recognized as self. The exact mechanisms behind autoimmune malfunctions are not fully understood but pathogens or drugs may play a role in triggering an autoimmune response in someone who already has a genetic predisposition. The reactions are similar to those that occur in allergies, except that in autoimmune disorders, the hypersensitivity response is to the body itself, rather than to an outside substance.

Multiple Sclerosis

MS is a progressive inflammatory disease of the central nervous system in which scattered patches of **myelin** (white matter) in the brain and spinal cord are destroyed. Myelin is the fatty connective tissue sheath surrounding conducting axons and its destruction results in the symptoms of MS: numbness, tingling, muscle weakness and **paralysis**.

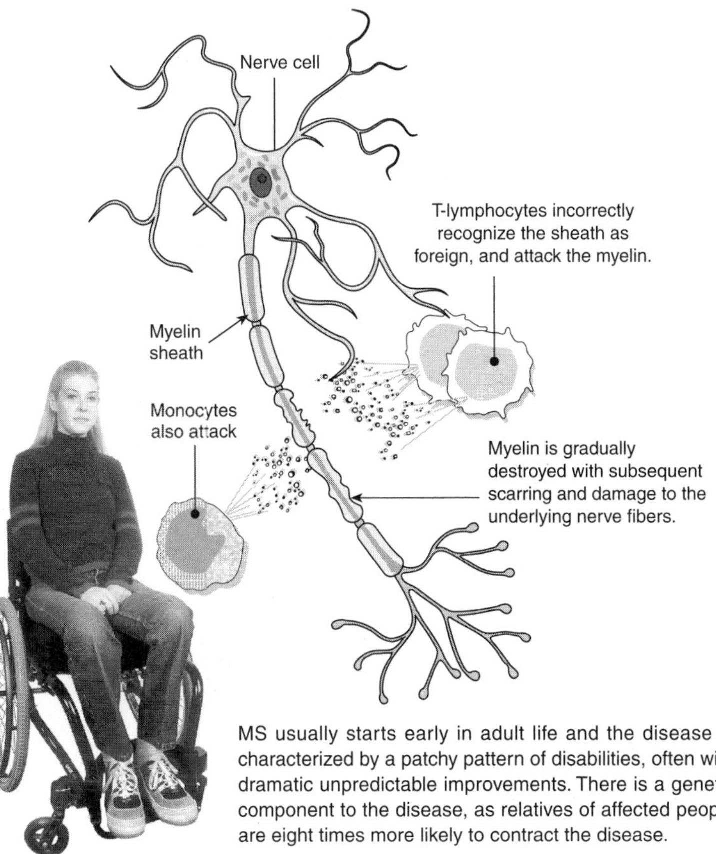

Nerve cell
T-lymphocytes incorrectly recognize the sheath as foreign, and attack the myelin.
Myelin sheath
Monocytes also attack
Myelin is gradually destroyed with subsequent scarring and damage to the underlying nerve fibers.

MS usually starts early in adult life and the disease is characterized by a patchy pattern of disabilities, often with dramatic unpredictable improvements. There is a genetic component to the disease, as relatives of affected people are eight times more likely to contract the disease.

Other Immune System Disorders

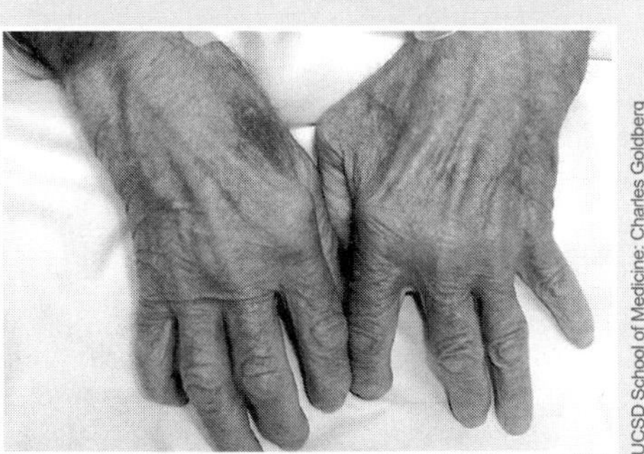

UCSD School of Medicine: Charles Goldberg

Rheumatoid arthritis is a type of joint inflammation, usually in the hands and feet, which results in destruction of cartilage and painful, swollen joints. The disease often begins in adulthood, but can also occur in children or the elderly. Rheumatoid arthritis affects more women than men and is treated with anti-inflammatory and immunosuppressant drugs, and physiotherapy.

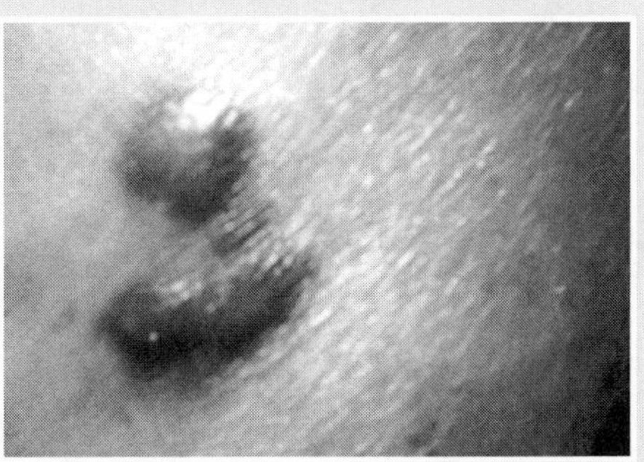

CDC

Lacking a sufficient immune response is called **immune deficiency**, and may be either **congenital** (present at birth) or **acquired** as a result of drugs, cancer, or infectious agents (e.g. HIV infection). HIV causes AIDS, which results in a steady destruction of the immune system. Sufferers then succumb to opportunistic infections and rare cancers such as Kaposi s sarcoma (above).

1. Explain the basis of the following autoimmune diseases:

   (a) Multiple sclerosis: ______________________

   (b) Rheumatoid arthritis: ______________________

2. Suggest why autoimmune diseases are difficult to treat effectively: ______________________

3. Explain why sufferers of immune deficiencies, such as AIDS, develop a range of debilitating infections:

   ____

Antibiotics

An **antibiotic** is a chemotherapeutic agent that inhibits or prevents microbial growth. Antibiotics are produced naturally by bacteria and fungi, but some synthetic (manufactured) **antimicrobial drugs** are also effective against microbial infections. Antimicrobial drugs interfere with the growth of microorganisms (see diagram below) by either killing microbes directly (**bactericidal**) or preventing them from growing (**bacteriostatic**). To be effective, they must often act inside the host, so their effect on the host's cells and tissues is important. The ideal antimicrobial drug has **selective toxicity**, killing the pathogen without damaging the host. Some antimicrobial drugs have a narrow **spectrum of activity**, and affect only a limited number of microbial types. Others are **broad-spectrum drugs** and affect a large number of microbial species. When the identity of a pathogen is not known, a broad-spectrum drug may be prescribed in order to save valuable time. There is a disadvantage with this, because broad spectrum drugs target not just the pathogen, but much of the host's normal microflora also. The normal microbial community usually controls the growth of pathogens and other microbes by competing with them. By selectively removing them with drugs, certain microbes in the community that do not normally cause problems, may flourish and become **opportunistic pathogens**.

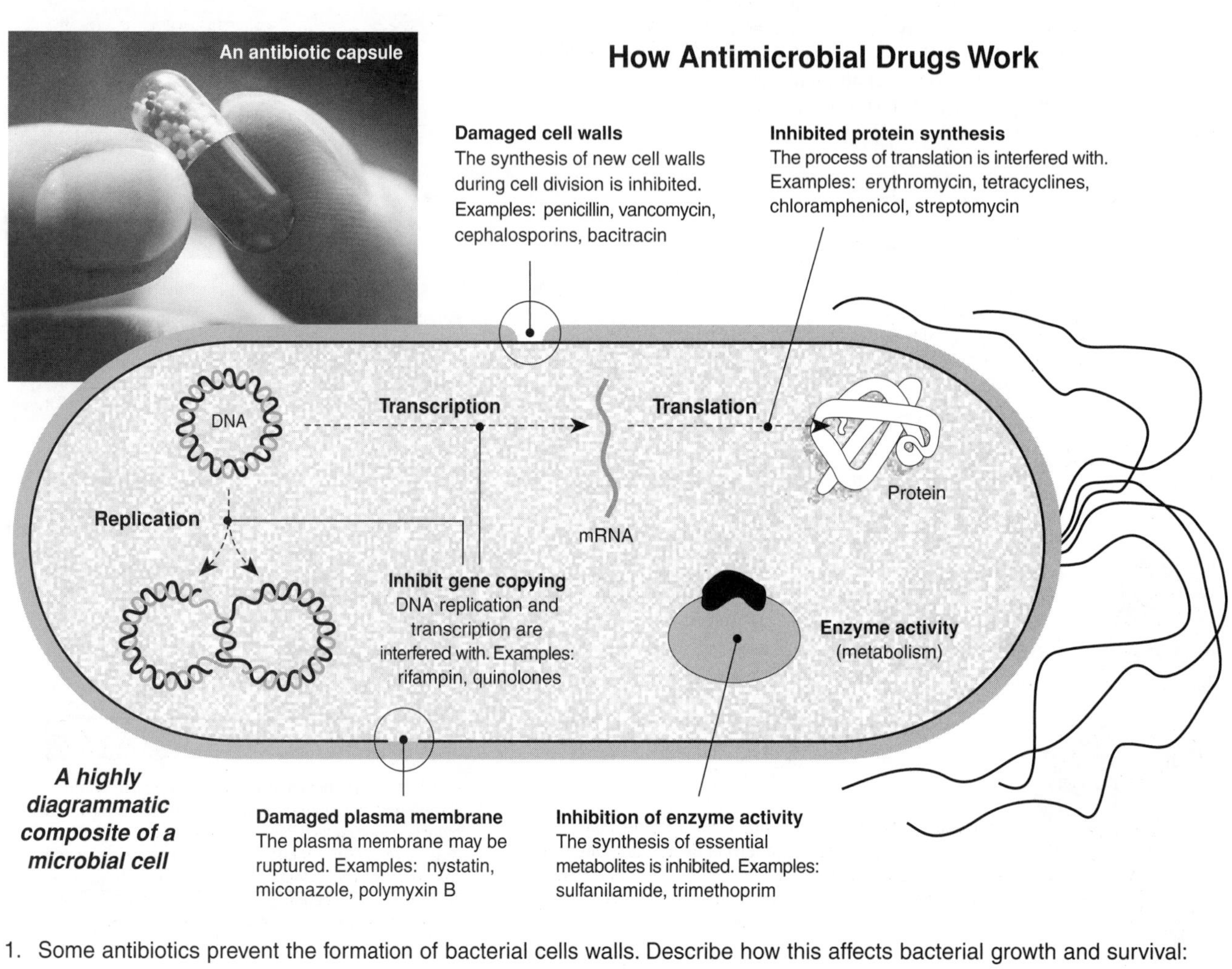

1. Some antibiotics prevent the formation of bacterial cells walls. Describe how this affects bacterial growth and survival:

2. Explain the advantages and disadvantages of using a **broad-spectrum drug** on an unidentified bacterial infection:

3. Discuss the requirements of an "ideal" anti-microbial drug, and explain in what way antibiotics satisfy these requirements:

***Web links**: Microbe Library: How Antibiotics Work*

KEY TERMS Mix and Match

INSTRUCTIONS: Test your vocab by matching each term to its correct definition, as identified by its preceding letter code.

ACTIVE IMMUNITY

ANTIBODIES

ANTIGEN

AUTOIMMUNE DISEASE

B CELL

CELL MEDIATED IMMUNITY

CLONAL SELECTION

DISEASE

FEVER

HEMOSTASIS

HIV

HUMORAL IMMUNITY

IMMUNITY

IMMUNOLOGICAL MEMORY

INFECTION

INFLAMMATION

INTERFERON

LEUKOCYTES

LYMPHOCYTE

MACROPHAGE

MONOCLONAL ANTIBODIES

NON-SPECIFIC DEFENSES

PASSIVE IMMUNITY

PATHOGEN

PHAGOCYTE

PRIMARY RESPONSE

SECONDARY RESPONSE

VACCINATION

A A rise in body temperature above the normal range of as a result of an increase in the body temperature regulatory set-point.

B The initial response of the immune system to exposure to an antigen.

C Resistance of an organism to infection or disease.

D Immune response involving the activation of macrophages, specific T cells, and cytokines against antigens.

E A retrovirus which infects immune system cells and causes the immune system to fail.

F The more rapid and stronger response of the immune system to an antigen that it has encountered before.

G The delivery by of antigenic material (the vaccine) to produce immunity to a disease.

H White blood cells, including lymphocytes, and macrophages and other phagocytic cells.

I Lymphocytes that make antibodies against specific antigens.

J A complex process which causes wounds to be sealed off and bleeding to stop.

K Specific white blood cells involved in the adaptive immune response.

L Gamma globulin proteins in the blood or other bodily fluids, which identify and neutralize foreign material, such as bacteria and viruses.

M Large white blood cells within tissues, produced by the differentiation of monocytes.

N A disease resulting from an overactive immune response against substances and tissues normally present in the body.

O Immunity that is induced in the host itself by the antigen, and is long-lasting.

P An abnormal condition of the body when bodily functions are impaired.

Q The ability of the immune system to respond rapidly in the future to antigens encountered in the past

R Generalized defense mechanisms against pathogens, e.g. physical barriers, secretions, and phagocytosis.

S Antibodies made by one type of immune cell that are all clones of a unique parent cell.

T Immune response that is mediated by secreted antibodies.

U A disease-causing organism.

V White blood cells that destroy foreign material, e.g. bacteria, by ingesting them.

W Anti-microbial protein made and released by lymphocytes in response to the presence of pathogens.

X The protective response of vascular tissues to harmful stimuli, such as irritants, pathogens, or damaged cells.

Y A molecule that is recognized by the immune system as foreign.

Z Immunity gained by the receipt of ready-made antibodies.

AA The invasion of a host organism by a pathogen to the detriment of the host.

BB A model for how B and T cells are selected to target specific antigens invading the body.

AP Topic 3B

IB AHL 11.3

Regulating Fluid and Removing Wastes

KEY CONCEPTS

- Maintenance of fluid and ion balance ensures optimum conditions for metabolism.
- The problems of fluid and ion balance are different in different environments.
- Diversity in excretory systems is reflective of the range of excretory products and the environments in which animals live.
- Urine production in the kidney is the result of ultrafiltration, secretion, and reabsorption.

KEY TERMS

absorption
ammonia
antennal gland
bladder
collecting duct
convoluted tubule
excretion
flame cells
gills
glomerulus
ion balance
kidney
loop of Henle
malpighian tubules
nephridia
nephron
nitrogenous waste
osmoregulation
protonephridia
renal dialysis
secretion
ultrafiltration
urea
ureter
urethra
uric acid
urine

OBJECTIVES

☐ 1. Use the **KEY TERMS** to help you understand and complete these objectives.

Balancing Fluid and Ions
pages 153-156, 161, 163

☐ 2. Distinguish between **excretion** and **osmoregulation** and explain the link between these processes.

☐ 3. Describe the problems associated with maintaining water and salt (ion) balance in marine and fresh water.

☐ 4. Describe diversity in mechanisms and strategies for salt and water balance in different environments and in representative taxa.

☐ 5. Evaluate water budgets for a desert adapted mammal and a human and relate these differences to adaptation to environment.

Diversity in Excretory Systems
pages 157-160

☐ 6. Identify the form of the **nitrogenous waste** excreted by different animal taxa and relate this to life history and environment.

☐ 7. Describe diversity in the structure and function of systems for excretion and osmoregulation in representative animal taxa. Examples include:

(a) **Protonephridia** (flatworms) or **nephridia** (annelids)

(b) **Malpighian tubules** (insects)

(c) **Antennal glands** and **gills** (crustaceans) and **gills** in fish.

(d) **Kidneys** (all vertebrates) and associated structures

Excretion in Humans
pages 162, 164-169

☐ 8. Describe the origin of waste products in humans. Describe the structure of the urinary system including kidneys, **ureters**, **bladder**, and **urethra**.

☐ 9. Describe the ultrastructure of the human kidney including the role of the **nephron** and the significance of nephron orientation in the kidney.

☐ 10. Explain the processes involved in filtration of the blood and production of urine in the kidney nephron, including:

- **ultrafiltration** in the **glomerulus**
- **secretion** of unwanted ions and toxins
- **reabsorption** in the **tubules** and **collecting duct**

☐ 11. Describe the role of **ADH** in regulating urine output. Explain how ADH secretion is controlled via negative feedback in response to blood osmolarity.

☐ 12. Describe the symptoms of **renal failure** and discuss options for treatment, including kidney transplant and **renal dialysis**.

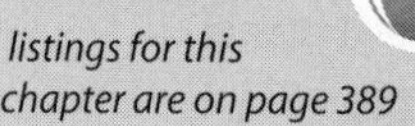

Periodicals:
listings for this chapter are on page 389

Weblinks:
www.thebiozone.com/weblink/SB2-2603.html

Teacher Resource CD-ROM:
Urine Analysis

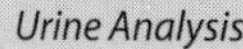

Birds With Runny Noses

"*Water, water, everywhere and not a drop to drink*".... The line from the 'Rime of the Ancient Mariner' clearly states the need to drink, but in an ocean, none can be found. However it may have been a lot easier for the albatross that accompanied the ship on its voyage. It, and other marine adapted animals can, literally, drink seawater.

There are around 35 g of dissolved salts (mostly sodium chloride with some potassium salts) in every liter of seawater. Human kidneys have an excretory efficiency such that it would take 1.5 L of water from the body to rid the body of the salt in 1 L of seawater. Consequently, drinking seawater causes humans to become more dehydrated than not drinking at all. In addition, in the period of time it takes for the kidneys to excrete that volume of water, the excess salt accumulates in the blood and creates potentially fatal ion imbalances.

The kidneys of birds and reptiles can produce urine with salt concentrations twice that of their blood, but they are less efficient than mammalian kidneys. Seabirds (and marine animals in general) have a special problem with salt loading because they eat food in an environment where the intake of seawater is inevitable. Seabirds and marine reptiles solve this problem by excreting the salt though a different mechanism; salt glands in the head. In birds, the salt glands excrete salt via the nares (nostrils); in turtles, the salt is excreted by glands near the eyes, so that turtles often appear to be weeping. The salt glands in birds likely evolved from the nasal glands of reptiles in the late Palaeozoic era. They work by active transport via Na-K pump that moves salt from the blood into the collecting duct of the gland, where it can be excreted as a concentrated solution. The glands regulate salt balance, and allow marine vertebrates to drink seawater. The salt glands have an extraordinarily high capacity for excreting salt. One experiment found that a gull given 10% of its body weight in seawater (equivalent to a 70 kg human drinking 7 L of seawater) was able to excrete the salt in around 3 hours. Concentrations of up to 64 gL^{-1} of salt have been recorded in the salt gland secretions of petrels; nearly twice the concentration of seawater. Marine iguanas can produce nearly 70 gL^{-1}.

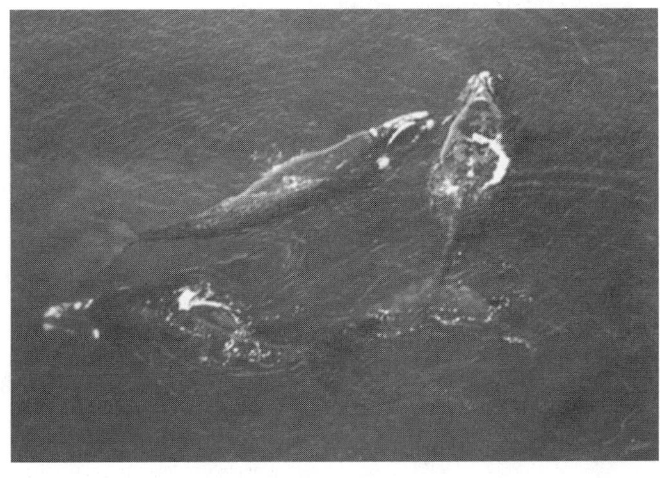

In contrast to marine birds and reptiles, marine mammals lack specialised glands to excrete salt, despite the fact that their incidental salt intake (with food) is high. All salt excretion is through the kidney and their kidneys are highly specialised to handle large volumes of electrolytes and protein. Evaporative losses are low for aquatic species so there is little need to drink, and the efficient kidneys, with their long loops of Henle, can concentrate the urine to twice the concentration of seawater and therefore easily deal with the incidental consumption of salt through the diet. The urine of a humpback whale may contain around 44 gL^{-1} salt; well above the concentration of seawater.

1. Explain why humans can not drink seawater to quench their thirst: ______________________

2. Describe and explain the role of the salt glands in marine birds and reptiles: ______________________

3. Discuss the efficiency of the salt glands in birds and reptiles and the kidneys in marine mammals:

Regulating Fluid and Removing Wastes

Related activities: *Nitrogenous Wastes in Animals*

Osmoregulation in Water

Animals living in aquatic environments have quite different osmoregulatory problems to those on land. Animals in freshwater tend to gain water osmotically and must prevent dilution of body fluids, so all are **osmoregulators**. Marine species have fewer problems and are either **osmoconformers**, or counter osmotic losses by drinking seawater and excreting the excess ions.

Osmoregulators vs Osmoconformers

The body fluids of most marine invertebrates, e.g. sea anemones (left), fluctuate with the environment. These animals are **osmoconformers** and cannot regulate salt and water balance. Some intertidal species can tolerate frequent dilutions of normal seawater and some actively take up ions across the gills to compensate for ion loss via the urine.

Bony fish

Animals that regulate their salt and water fluxes, such as fish and marine mammals, are termed **osmoregulators**. Bony fish lose water osmotically and counter the loss by drinking salt water and then excreting the excess salt across the gill surfaces.

Ray

Marine sharks and rays generate osmotic concentrations in their body fluids similar to seawater by tolerating high urea. Excess salt from the diet is excreted via a salt gland in the rectum. Marine mammals produce a urine that is high in both salt and urea.

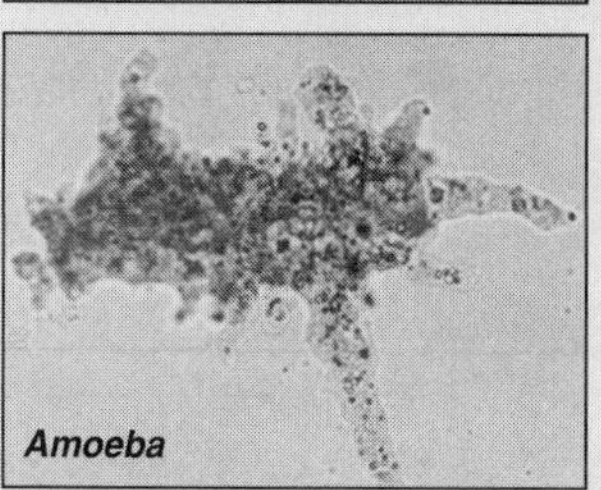

Amoeba

Freshwater animals have body fluids that are osmotically more concentrated than the water they live in and are all osmoregulators. Water tends to enter their tissues by osmosis and must be expelled to avoid flooding the body. Simple protozoans (*Amoeba*, left) use contractile vacuoles to collect the excess water and expel it.

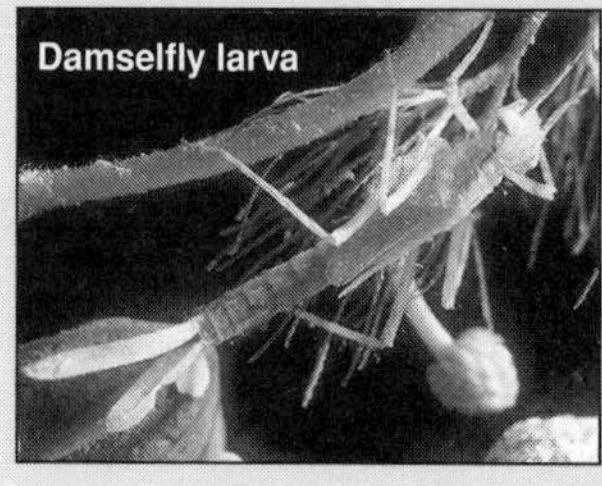

Damselfly larva

Other invertebrates expel water and nitrogenous wastes using simple nephridial organs. Bony fish and aquatic arthropods produce dilute urine (containing ammonia) and actively take up salts across their gill surfaces.

Response to Seawater Dilution in Rock Crabs

Species of intertidal crabs vary widely in their ability to regulate their salt and water levels in the face of environmental fluctuations. As shown in the graph below, intertidal crabs face an osmotic influx of water when placed into dilute salinities. The excess water can be excreted in the urine (via the antennal glands), but this is accompanied by a loss of valuable ions. In many species, after a period of adjustment, this loss is met by active ion uptake across the gills.

In an experiment, a student investigated the effect of increasing seawater dilution on the cumulative weight gain of a common rock crab. Six crabs in total were used in the experiment. Three were placed in seawater dilution of 75:25 (75% seawater) and three were placed in a seawater dilution of 50:50 (50% seawater). Cumulative weight gain in each of the six crabs was measured at regular intervals over a period of 30 minutes. The results are plotted below and a line of best fit has been drawn for each set of data.

Weight gain in crabs at two seawater dilutions

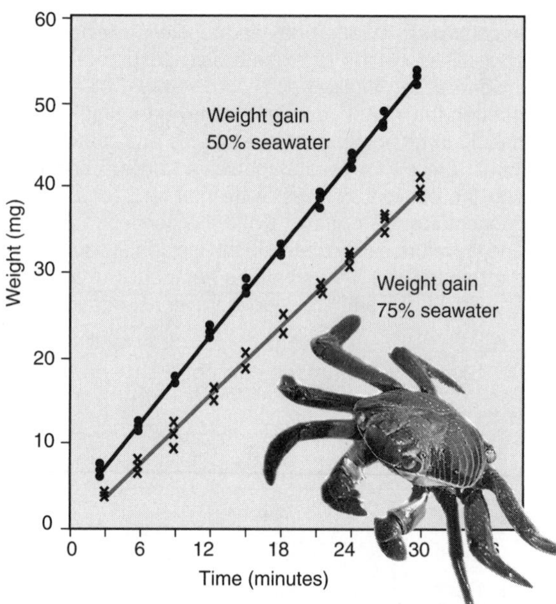

1. Describe one way in which freshwater animals can compensate for ions lost during excretion of excess water:

2. Explain the difference between an **osmoregulator** and an **osmoconformer**:

3. (a) Explain the difference in the two lines plotted on the graph of crab weight gain (above right):

   (b) Explain what this experiment suggests about the osmoregulatory ability of this crab species:

Managing Fluid Balance on Land

Many aspects of metabolism, including enzyme activity, membrane transport, and nerve conduction, are dependent on particular concentrations of ions and metabolites. To achieve this ion balance, the salt and fluid content of the internal body fluids must be regulated; a process called **osmoregulation**. **Terrestrial** animals show specific adaptations for obtaining and conserving water in an environment where (to varying degrees) water loss is a constant problem. These adaptations are most well developed in desert animals. Despite the lack of available water in deserts, a large a number of animal taxa have desert-dwelling representatives. The most arid tolerant are even capable of surviving without drinking.

Obtaining Water

Most animals obtain the majority of their water by **drinking.** Some, such as camels, can retain relatively large volumes of water in the gut, but most will need to regularly visit a water supply.

Obtaining water from **food** is important in dry environments where free standing water is limited. Many large predators obtain a large amount of their water in this way.

US Fish and Wildlife

Some desert animals, such as the kangaroo rat, do not need to drink. 90% of their water comes from metabolism (oxidation of food). The rest comes from the small amount of water present in the food.

Ian Smith

Amphibians can take up water directly through their skin. The skin is water permeable, so they can osmotically acquire water when they need it when submerged or resting on a damp surface.

Conserving and Losing Water

Water losses through airways

INVERTEBRATES
- Waxy cuticle reduces water loss
- Excrete nitrogenous wastes as uric acid paste to conserve water
- Seek out shade to reduce evaporative losses
- May take up water from the atmosphere

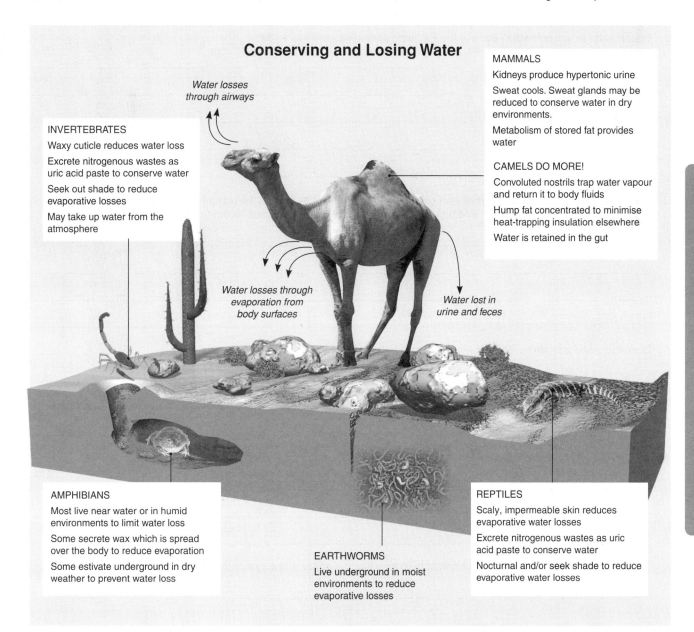

MAMMALS
- Kidneys produce hypertonic urine
- Sweat cools. Sweat glands may be reduced to conserve water in dry environments.
- Metabolism of stored fat provides water

CAMELS DO MORE!
- Convoluted nostrils trap water vapour and return it to body fluids
- Hump fat concentrated to minimise heat-trapping insulation elsewhere
- Water is retained in the gut

Water losses through evaporation from body surfaces

Water lost in urine and feces

AMPHIBIANS
- Most live near water or in humid environments to limit water loss
- Some secrete wax which is spread over the body to reduce evaporation
- Some estivate underground in dry weather to prevent water loss

EARTHWORMS
- Live underground in moist environments to reduce evaporative losses

REPTILES
- Scaly, impermeable skin reduces evaporative water losses
- Excrete nitrogenous wastes as uric acid paste to conserve water
- Nocturnal and/or seek shade to reduce evaporative water losses

Periodicals: *Countercurrent exchange mechanisms*

Related activities: *Vertebrate Excretory Systems, Osmoregulation in Water*

The Loop of Henle and Water Conservation

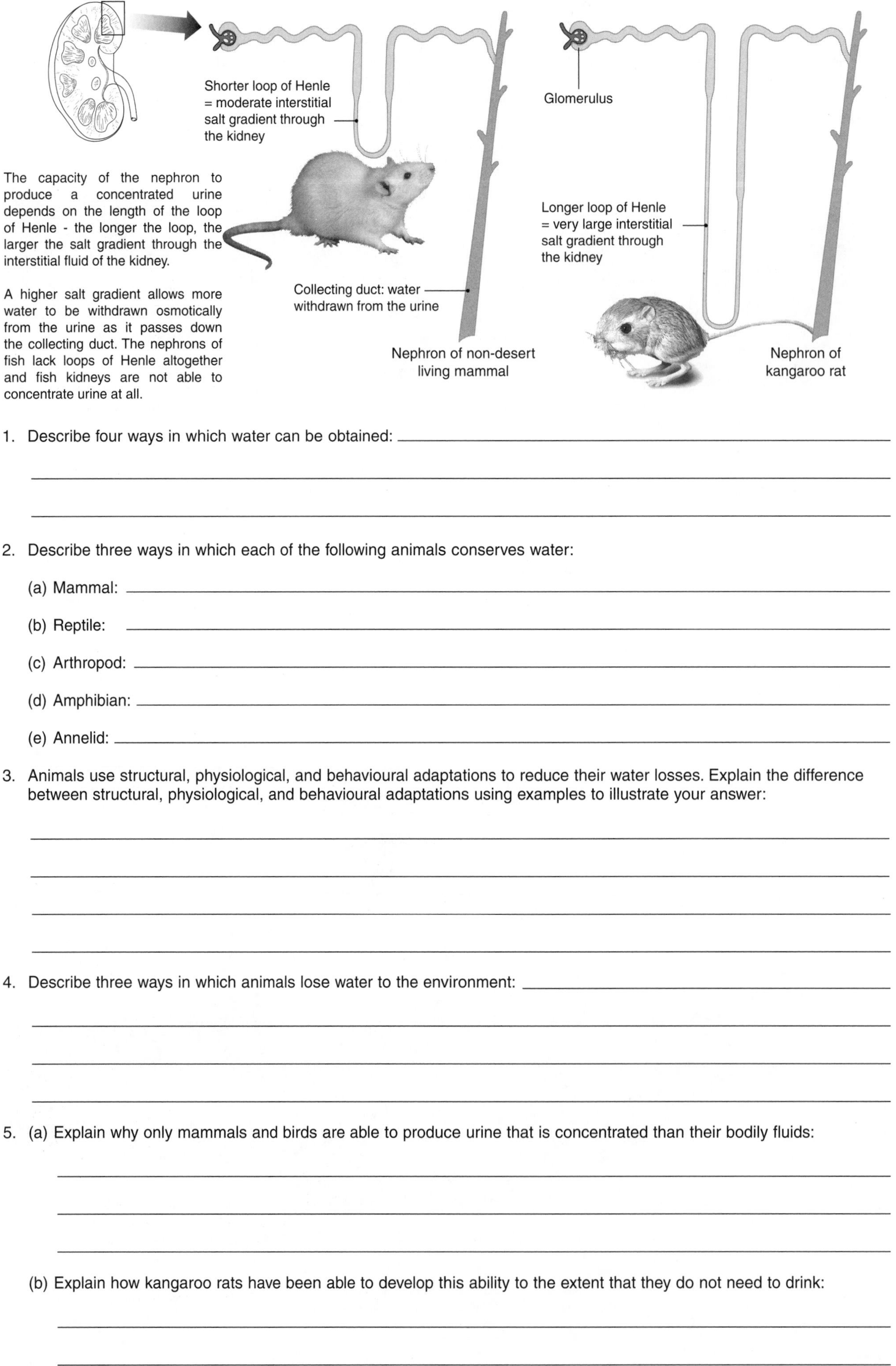

The capacity of the nephron to produce a concentrated urine depends on the length of the loop of Henle - the longer the loop, the larger the salt gradient through the interstitial fluid of the kidney.

A higher salt gradient allows more water to be withdrawn osmotically from the urine as it passes down the collecting duct. The nephrons of fish lack loops of Henle altogether and fish kidneys are not able to concentrate urine at all.

1. Describe four ways in which water can be obtained: ______

2. Describe three ways in which each of the following animals conserves water:

   (a) Mammal: ______

   (b) Reptile: ______

   (c) Arthropod: ______

   (d) Amphibian: ______

   (e) Annelid: ______

3. Animals use structural, physiological, and behavioural adaptations to reduce their water losses. Explain the difference between structural, physiological, and behavioural adaptations using examples to illustrate your answer:

4. Describe three ways in which animals lose water to the environment: ______

5. (a) Explain why only mammals and birds are able to produce urine that is concentrated than their bodily fluids:

   (b) Explain how kangaroo rats have been able to develop this ability to the extent that they do not need to drink:

Nitrogenous Wastes in Animals

Waste materials are generated by the metabolic activity of cells. If allowed to accumulate, they would reach toxic concentrations and so must be continually removed. Excretion is the process of removing waste products and other toxins from the body. Waste products include carbon dioxide and water, and the nitrogenous (nitrogen containing) wastes that result from the breakdown of amino acids and nucleic acids. The simplest breakdown product of nitrogen containing compounds is ammonia, a small molecule that cannot be retained for long in the body because of its high toxicity. Most aquatic animals excrete ammonia immediately into the water where it is washed away. Other animals convert the ammonia to a less toxic form that can remain in the body for a short time before being excreted via special excretory organs. The form of the excretory product in terrestrial animals (urea or uric acid) depends on the type of organism and its life history. Terrestrial animals that lay eggs produce uric acid rather than urea, because it is non-toxic and very insoluble. It remains as an inert solid mass in the egg until hatching.

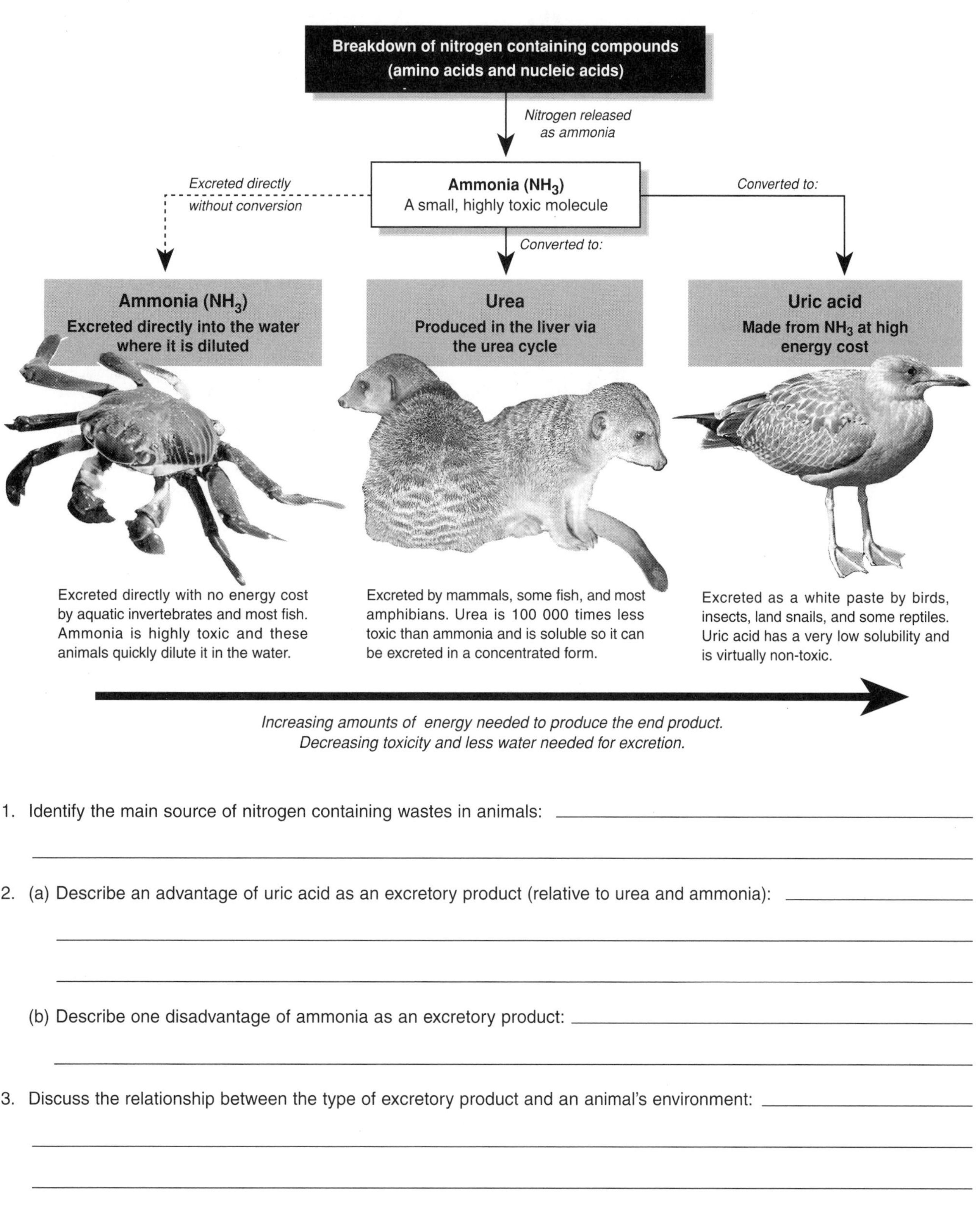

Breakdown of nitrogen containing compounds (amino acids and nucleic acids)
Nitrogen released as ammonia
Ammonia (NH_3)
A small, highly toxic molecule
Excreted directly without conversion
Converted to:
Converted to:
Ammonia (NH_3)
Excreted directly into the water where it is diluted
Urea
Produced in the liver via the urea cycle
Uric acid
Made from NH_3 at high energy cost
Excreted directly with no energy cost by aquatic invertebrates and most fish. Ammonia is highly toxic and these animals quickly dilute it in the water.
Excreted by mammals, some fish, and most amphibians. Urea is 100 000 times less toxic than ammonia and is soluble so it can be excreted in a concentrated form.
Excreted as a white paste by birds, insects, land snails, and some reptiles. Uric acid has a very low solubility and is virtually non-toxic.
Increasing amounts of energy needed to produce the end product.
Decreasing toxicity and less water needed for excretion.

1. Identify the main source of nitrogen containing wastes in animals: ______________________

2. (a) Describe an advantage of uric acid as an excretory product (relative to urea and ammonia): ______________________

   (b) Describe one disadvantage of ammonia as an excretory product: ______________________

3. Discuss the relationship between the type of excretory product and an animal's environment: ______________________

Regulating Fluid and Removing Wastes

Invertebrate Excretory Systems

Metabolism produces toxic by-products. The most troublesome of these to eliminate from the body is nitrogenous waste from the metabolism of proteins and nucleic acids. The simplest and most common type of excretory organs, widely distributed in invertebrates, are simple tubes (**protonephridia** and **nephridia**) opening to the outside through a pore. The **malpighian tubules** of insects are highly efficient, removing nitrogenous wastes from the blood, and also functioning in **osmoregulation**. Note that all three forms of nitrogenous waste are represented here: ammonia (flatworms, annelids), urea (annelids), and uric acid (insects).

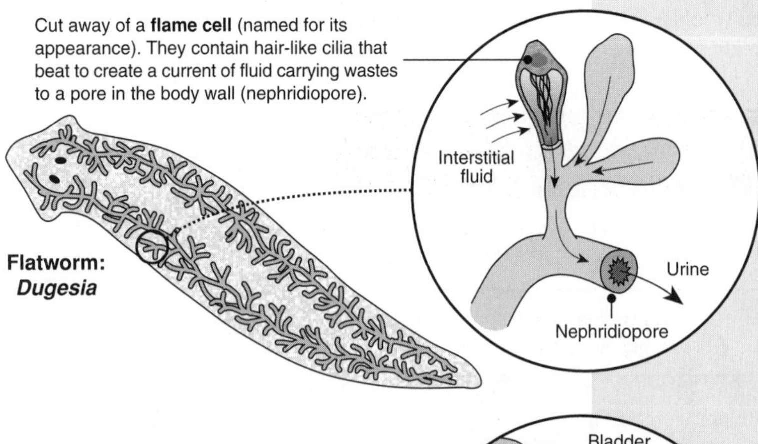

Platyhelminthes (flatworms)

Excretory system: **protonephridia**

Protonephridia are very simple excretory structures. Each protonephridium comprises a branched tubule ending in a number of blind capillaries called **flame cells**. **Ammonia** is excreted directly into the moist environment. Flatworms do not have a circulatory system or fluid-filled inner body spaces. They use their branching network of flame cells to regulate the composition of the fluid bathing the cells (interstitial fluid). Interstitial fluid enters the flame cell and is propelled along the tubule, away from the blind end, by beating cilia. Tubules merge into ducts that expel the urine through **nephridiopores**.

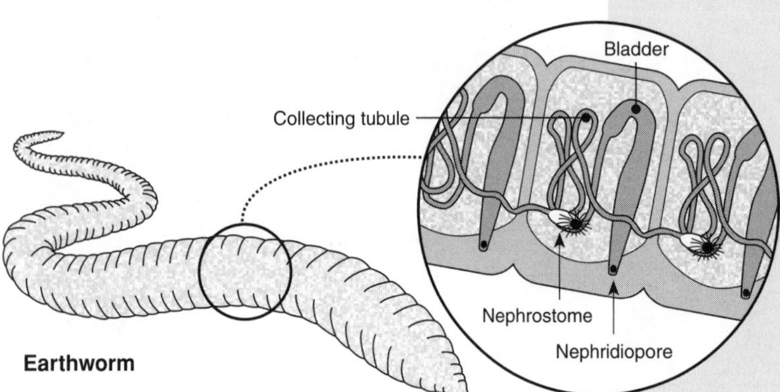

Annelids (segmented worms)

Excretory system: **nephridia**

In earthworms, each segment has a pair of excretory organs called **nephridia**, which drain the next segment in front. Fluid enters the nephrostome and passes through the collecting tubule. These tubules are surrounded by a capillary network of blood vessels (not shown here) which recover valuable salts from the developing urine. The collecting tubule empties into a storage bladder which expels the dilute urine (a mix of **ammonia** and **urea**) to the outside through the nephridiopore.

Insects

Excretory system: **malpighian tubules**

Insects have two to several hundred **malpighian tubules** projecting from the junction of the midgut and hindgut. They bathe in the clear fluid (hemolymph) of the insect's body cavity where they actively pump K^+ and Na^+ into the tubule. Water, uric acid salts, and several other substances follow by passive transport. Water and some ions are reabsorbed in the hindgut, while **uric acid** precipitates out as a paste and is passed out of the anus along with the fecal material. The ability to conserve water by excreting solid uric acid has enabled insects to colonize very arid environments.

1. For each of the following, name the organs for excreting nitrogenous waste and state the form of the waste product:

   (a) Flatworm: ______________________ Waste: ____________

   (b) Insect: ______________________ Waste: ____________

   (c) Earthworm: ______________________ Waste: ____________

2. Explain briefly how insects concentrate their nitrogenous waste into a paste: ______________________

   ____

   ____

3. For one of the above animals, relate the form of the excretory product to the environment in which the animal lives:

   ____

   ____

Vertebrate Excretory Systems

In vertebrates, the excretory units (**nephrons**) are collected into discrete organs called **kidneys**. The kidneys of most vertebrates are similar in that they produce urine (excretory fluid or paste) by **filtering** the body fluids and then modifying the filtrate by **reabsorption** and **secretion** of ions. The kidneys of a few bony (teleost) fish differ from this general pattern by lacking a filtration mechanism and producing urine solely through the secretion of ions. Whilst all vertebrates have kidneys and can produce urine, only birds and mammals can produce a urine that is more concentrated than the body fluids. The kidneys of bony fish have few nephrons and can only create a urine that is the same concentration, or less concentrated, than the blood. Nearly all their nitrogenous waste is excreted by the gills, which also have an important role in salt balance. Mammals and birds have highly efficient kidneys that can produce a concentrated urine, excreting nitrogenous wastes whilst conserving water and valuable ions. The mechanism of urine formation in mammals is provided in more detail in the activity *The Physiology of the Kidney*.

Freshwater Bony Fish

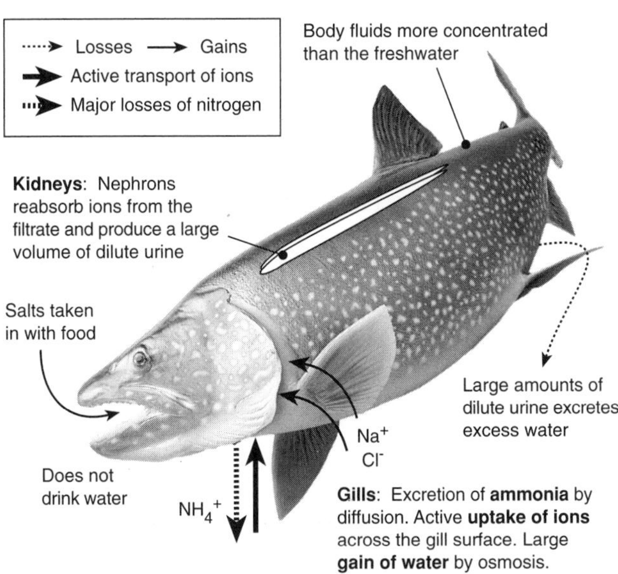

Fish in freshwater must excrete **excess water** gained through osmosis and they must excrete **nitrogenous waste**. Their kidneys excrete large amounts of dilute urine; valuable ions are lost because of the large urine volumes produced. The kidneys reabsorb salts from the filtrate through active transport mechanisms, and the gills take up ions from the water.

Marine Bony Fish

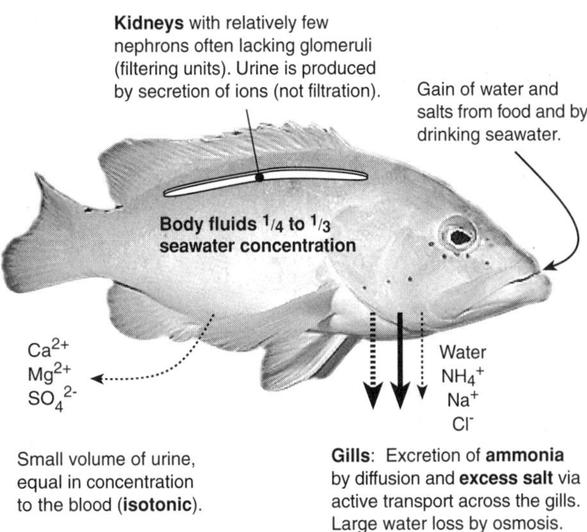

Fish in seawater must excrete **excess salt** gained through diet as well as **nitrogenous waste**. The urine is isotonic, and excess salts are actively excreted across the gill surface into the water (against a concentration gradient). Note that, unlike bony fish, sharks and rays tolerate high urea levels in their tissues and excrete excess salts via a salt gland in the rectum.

Terrestrial Mammal

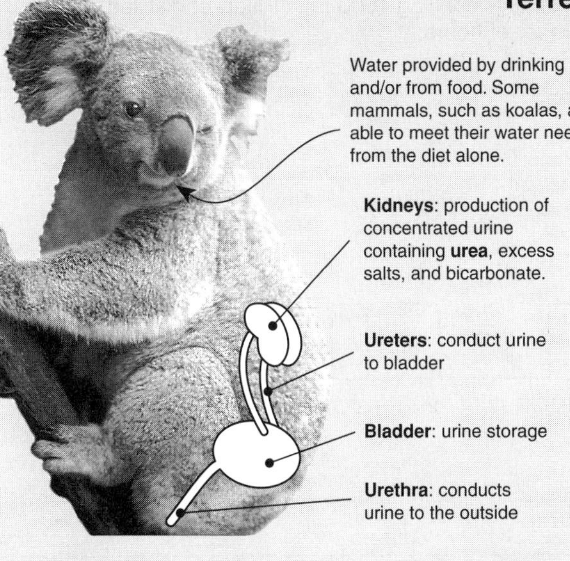

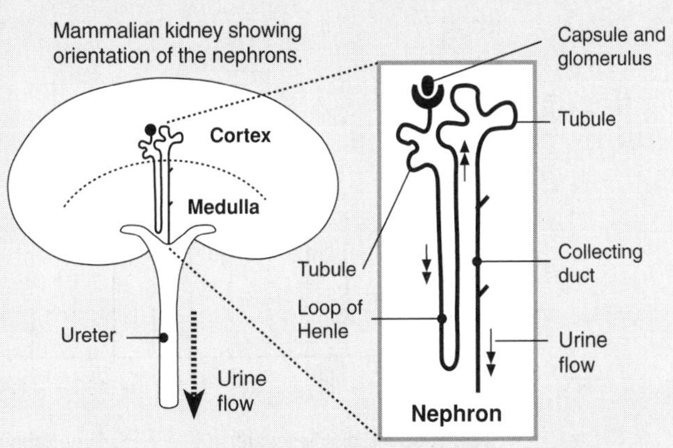

Water loss is a major problem for most mammals. The degree to which urine can be concentrated (and water conserved) depends on the number of nephrons present and the length of the loop of Henle. The highest urine concentrations are found in desert-adapted mammals.

Mammalian **kidneys** each contain more than one million **nephrons**: the selective filter elements which regulate the composition of the blood and excrete wastes. In the nephron, the initial urine is formed by **filtration** in the glomerulus and Bowman's capsule. The filtrate is modified by **secretion** and **reabsorption** of ions and water. These processes create a salt gradient in the fluid around the nephron, which allows water to be withdrawn from the urine in the collecting duct.

Periodicals: The kidney

Related activities: Fluid Regulation on Land, Osmoregulation in Water

1. Name the primary organ for nitrogen excretion in fish: ______

2. (a) Discuss the problems of excretion and osmoregulation for bony fish in fresh and salt water environments:

______

(b) Explain how fish in these two environments overcome these difficulties: ______

3. (a) Name the functional excretory unit of the mammalian kidney: ______

(b) Explain how mammals are able to produce a concentrated urine: ______

4. Describe the two factors that determine the degree to which mammalian urine can be concentrated and explain why they are important:

(a) ______

(b) ______

5. Discuss the functional and structural differences between mammalian kidneys and the kidneys of fish:

______

6. The graph below shows the volume of urine collected from a subject after drinking 1000 mL (1 liter) of distilled water. The subject's urine was collected at 25 minute intervals over a number of hours.

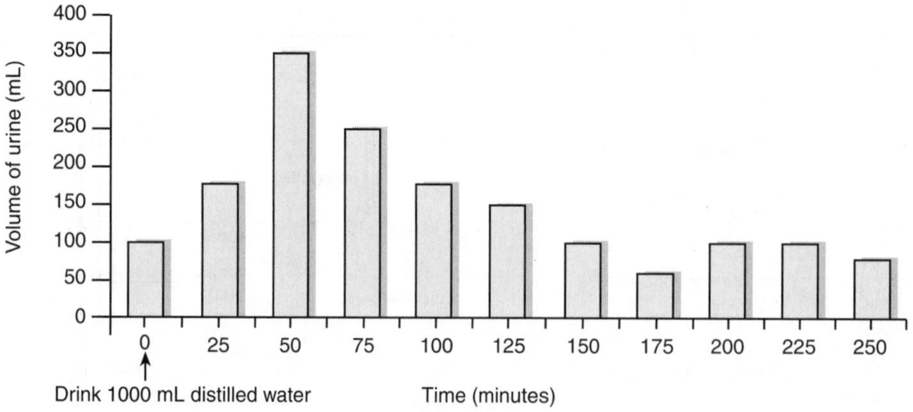

(a) Describe the changes in urine output during the experiment: ______

(b) Explain the difference in the volume of urine collected at 25 minutes and 50 minutes: ______

Water Balances in Desert Mammals

Water loss is a major problem for most mammals. The behavioral, physiological, and structural adaptations of mammals adapted to desert or arid regions enables them to minimize their water losses and reduce the amount of water they need to drink. Arid-adapted species typically produce very concentrated urine (their kidney nephrons have long loops of Henle), water losses through this route are minimal. In addition, the metabolic breakdown of food contributes a large proportion of daily water needs. Some, like kangaroo rats (below, left) do not need to drink at all, and obtain most of their water this way.

Adaptations of Arid Adapted Rodents

Most desert-dwelling mammals are adapted to tolerate a low water intake. Arid adapted rodents, such as jerboas and kangaroo rats, conserve water by reducing losses to the environment and obtain the balance of their water needs from the oxidation of dry foods (respiratory metabolism). The table below shows the water balance in a kangaroo rat after eating 100 g of dry pearl barley. Note the high urine to plasma concentration ratio (17) which is more than four times that of a human (4).

Water balance in a kangaroo rat (*Dipodomys spectabilis*)

Water gains		Water losses	
Absorbed from food	6.0 ml	Breathing	43.9 ml
From metabolism	54.0 ml	Urination	13.5 ml
		Defecation	2.6 ml

Urine/plasma concentration ratio = 17

Adaptations of kangaroo rats

Kangaroo rats, and other arid-adapted rodents, tolerate long periods without drinking, meeting their water requirements from the metabolism of dry foods. They dispose of nitrogenous wastes with very little output of water and they neither sweat nor pant to keep cool.

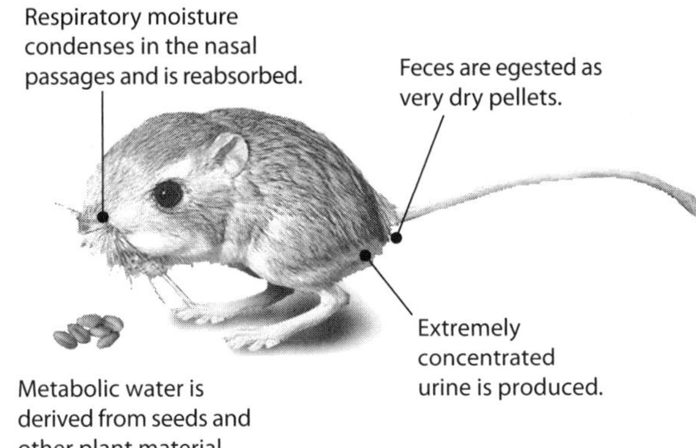

Respiratory moisture condenses in the nasal passages and is reabsorbed.
Feces are egested as very dry pellets.
Extremely concentrated urine is produced.
Metabolic water is derived from seeds and other plant material.

Large Desert Mammals

Camels are well adapted to desert life. Their nasal passages are highly convoluted to trap moisture when exhaling, and sweating only begins when the body temperature exceeds 41°C. The urine is highly concentrated and syrupy and the feces are dry enough to be used as fuel. Metabolism of the fat in the hump also produces water. Camels can also withstand large losses of water without circulatory disturbance because their oval-shaped blood cells can flow even when they are dehydrated.

A suite of adaptations enable kangaroos to exploit the vast arid environment of Australia's semi-deserts. Water is taken in with food but is also derived from metabolism. An elongated large intestine maximizes the removal of water from undigested material, producing very dry feces. The urine is also concentrated, reducing water loss. The hopping mode of locomotion is also very energy efficient, and enables them to cover large expanses of country in search of food and water.

1. Explain why most mammals need to drink regularly: ______________________________

____

2. Use the tabulated data for the kangaroo rat (above) to graph the water gains and losses in the spaces provided below.

3. Describe three physiological adaptations of desert adapted rodents to low water availability:

   (a) ______________________________

   (b) ______________________________

   (c) ______________________________

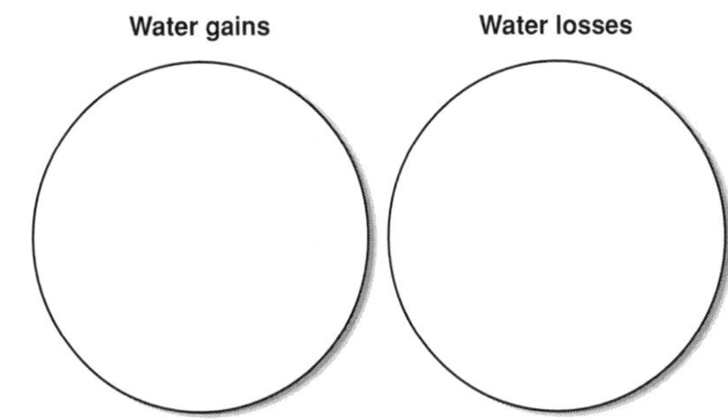

4. If kangaroo rats neither pant nor sweat, suggest how they might keep cool during the heat of the day

____

Waste Products in Humans

In humans and other mammals, a number of organs are involved in the excretion of the waste products of metabolism: mainly the kidneys, lungs, skin, and gut. The liver is a particularly important organ in the initial treatment of waste products, particularly the breakdown of hemoglobin and the formation of urea from ammonia. Excretion should not be confused with the elimination or egestion of undigested and unabsorbed food material from the gut. Note that the breakdown products of hemoglobin (blood pigment) are excreted in bile and pass out with the feces, but they are not the result of digestion.

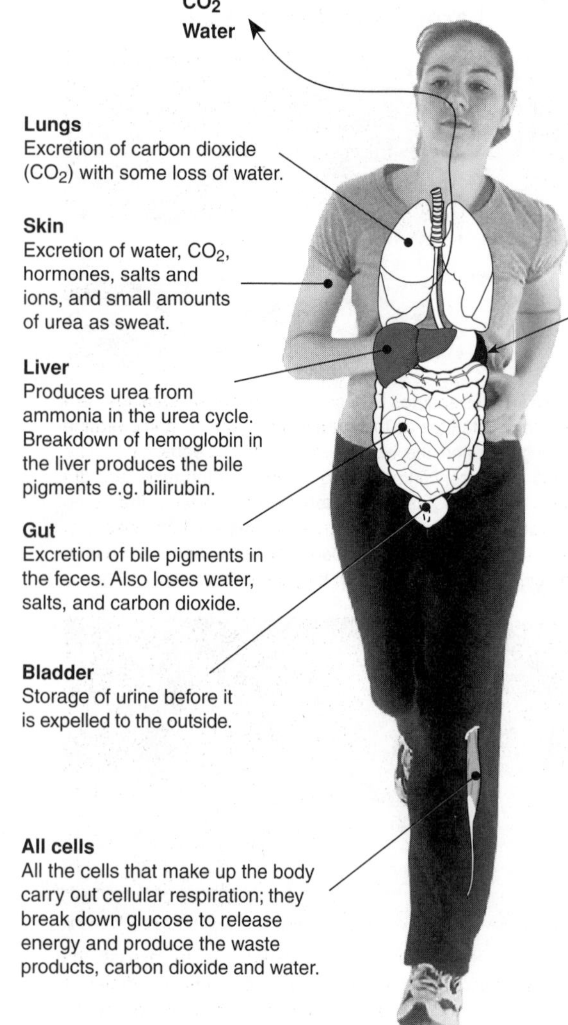

Excretion In Humans

In mammals, the kidney is the main organ of excretion, although the skin, gut, and lungs also play important roles. As well as ridding the body of nitrogenous wastes, the kidneys are also able to excrete many unwanted poisons and drugs that are taken in from the environment. Usually these are ingested with food or drink, or inhaled. As long as these are not present in toxic amounts, they can usually be slowly eliminated from the body.

Kidney
Filtration of the blood to remove urea. Unwanted ions, particularly hydrogen (H^+) and potassium (K^+), and some hormones are also excreted by the kidneys. Some poisons and drugs (e.g. penicillin) are also excreted by active secretion into the urine. Water is lost in excreting these substances and extra water may be excreted if necessary.

Substance	Origin*	Organ(s) of excretion
Carbon dioxide		
Water		
Bile pigments		
Urea		
Ions (K^+, HCO_3^-, H^+)		
Hormones		
Poisons		
Drugs		

* Origin refers to from where in the body each substance originates

1. Complete the table above summarizing the origin of excretory products and the main organ(s) of excretion for each.
2. Explain the role of the liver in excretion, even though it is not primarily an organ of excretion: ______________________
3. Tests for pregnancy are sensitive to an excreted substance in the urine. Suggest what type of substance this might be:
4. In people suffering renal failure, the kidneys cease to produce filtrate. Based on your knowledge of the central role of the kidneys in fluid and electrolyte balance, as well as nitrogen excretion, describe the typical symptoms of kidney failure:

Water Budget in Humans

We cannot live without water for more than about 100 hours and adequate water is a requirement for physiological function and health. Body water content varies between individuals and through life, from above about 90% of total weight as a fetus to 74% as an infant, 60% as a child, and around 50-59% in adults, depending on gender and age. Gender differences (males usually have a higher water content than females) are the result of differing fat levels. Water intake and output are highly variable but closely matched to less than 0.1% over an extended period. Typical values for water gains and losses, as well as daily water transfers are given below. Men need more water than women due to their higher (on average) fat-free mass and energy expenditure. Infants and young children need more water in proportion to their body weight as they cannot concentrate their urine as efficiently as adults. They also have a greater surface area relative to weight, so water losses from the skin are greater.

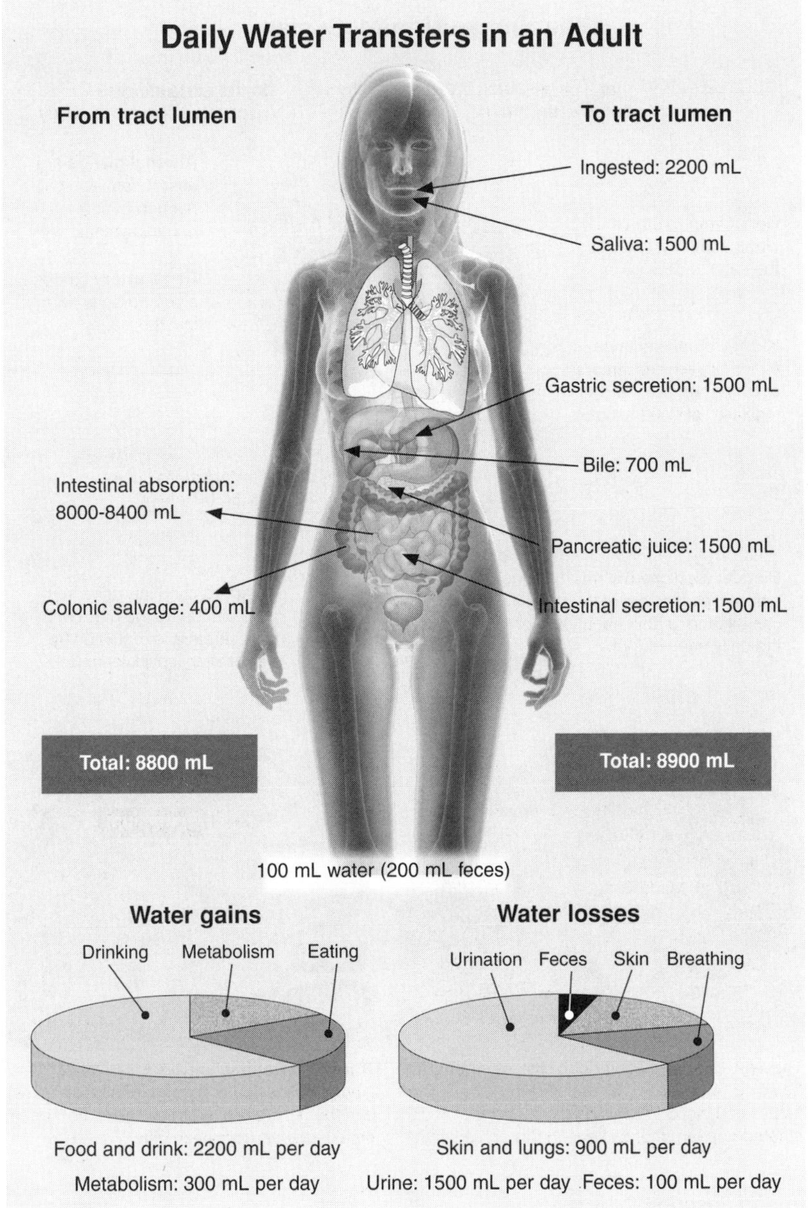

About 63% of our daily requirement for water is met through drinking fluids, 25% is obtained from food, and the remaining 12% comes from metabolism (the oxidation of glucose to ATP, CO_2, and water) .

Typically, we lose 60% of body water through urination, 36% through the skin and lungs, and 4% in feces. Losses through the skin and from the lungs (breathing) average about 900 mL per day or more during heavy exercise. These are called **insensible losses**.

1. Explain how metabolism provides water for the body's activities: ______________________

____

2. Describe four common causes of physiological dehydration:

(a) ______________________ (c) ______________________

(b) ______________________ (d) ______________________

3. Some recent sports events have received media coverage because athletes have collapsed after excessive water intakes. This condition, called **hyponatremia** or water intoxication, causes nausea, confusion, diminished reflex activity, stupor, and eventually coma. From what you know of fluid and electrolyte balances in the body, explain these symptoms:

____

____

The Urinary System

The mammalian urinary system consists of the kidneys and bladder, and their associated blood vessels and ducts. The kidneys have a plentiful blood supply from the renal artery. The blood plasma is filtered by the kidneys to form urine. Urine is produced continuously, passing along the ureters to the bladder, a hollow muscular organ lined with smooth muscle and stretchable epithelium. Each day the kidneys filter about 180 liters of plasma. Most of this is reabsorbed, leaving a daily urine output of about 1 liter. By adjusting the composition of the fluid excreted, the kidneys help to maintain the body's internal chemical balance. All vertebrates have kidneys, but their efficiency in producing a concentrated urine varies considerably. Mammalian kidneys are very efficient, producing a urine that is concentrated to varying degrees depending on requirements.

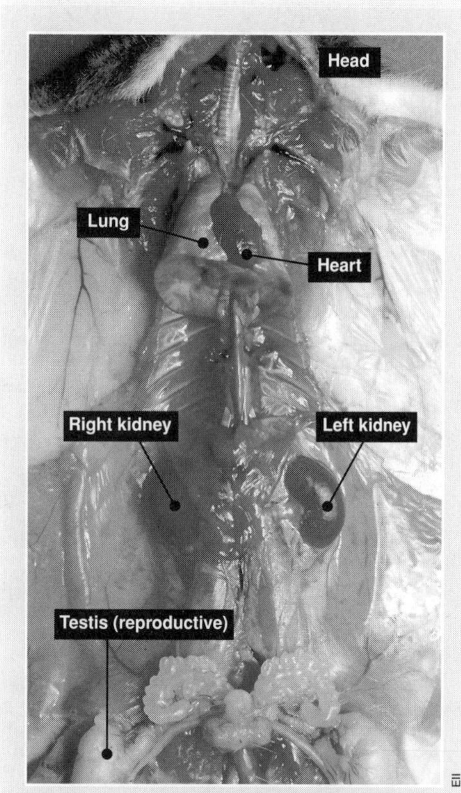

The kidneys of **rats** (above), humans, and other mammals are distinctive, bean shaped organs that lie at the back of the abdominal cavity to either side of the spine. The kidneys lie outside the peritoneum of the abdominal cavity and are partly protected by the lower ribs. Each kidney is surrounded by three layers of tissue. The innermost renal capsule is a smooth fibrous membrane that acts as a barrier against trauma and infection. The two outer layers comprise fatty tissue and fibrous connective tissue. These act to protect the kidney and anchor it firmly in place.

The Human Urinary System

Vena cava returns blood to the heart.

Dorsal aorta supplies oxygenated blood to the body.

Adrenal glands are associated with, but not part of, the urinary system.

Renal vein returns the blood from the kidney to the venous circulation.

Renal artery carries blood from the aorta into the kidney.

Kidney produces urine (blood filtration, the removal of waste products, and the regulation of blood volume).

Ureter carries urine to the bladder.

Bladder (sectioned) stores the urine before it passes out of the body. It can expand to hold about 80% of the daily urine output.

Urethra conducts urine from the bladder to the outside. The urethra is regulated by a voluntary sphincter muscle.

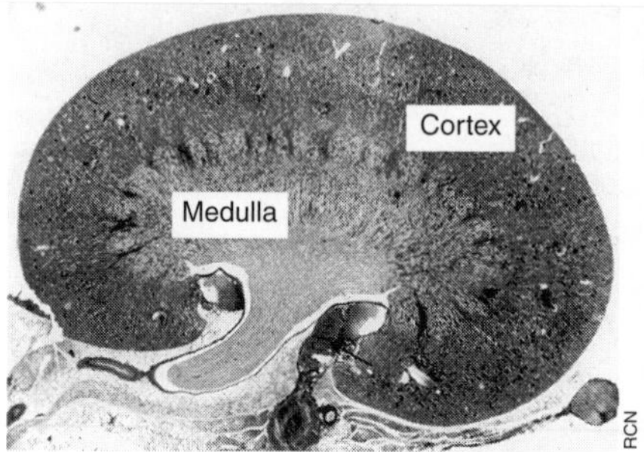

The very precise alignment of the nephrons (the filtering elements of the kidney) and their associated blood vessels gives the kidney tissue a striated appearance, as seen in this cross section.

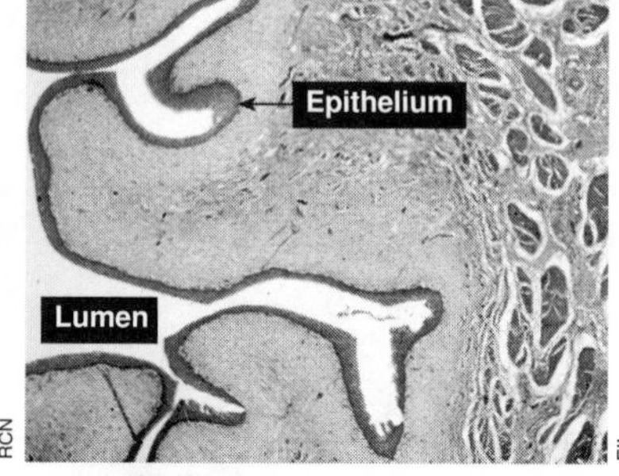

Transitional epithelium lines the bladder. This type of epithelium is layered, or stratified, and can be stretched without the outer cells breaking apart from each other.

1. Identify the components of the urinary system and describe their functions: ______________________________

2. Calculate the percentage of the plasma reabsorbed by the kidneys: ______________________________

3. The kidney receives blood at a higher pressure than other organs. Explain why this is the case: ______________________________

4. Suggest why the kidneys are surrounded by fatty connective tissue: ______________________________

The Physiology of the Kidney

The functional unit of the kidney, the **nephron**, is a selective filter element, comprising a renal tubule and its associated blood vessels. Filtration, i.e. forcing fluid and dissolved substances through a membrane by pressure, occurs in the first part of the nephron, across the membranes of the capillaries and the glomerular capsule. The passage of water and solutes into the nephron and the formation of the glomerular filtrate depends on the pressure of the blood entering the afferent arteriole (below). If it increases, filtration rate increases; when it falls, glomerular filtration rate also falls. This process is so precisely regulated that, in spite of fluctuations in arteriolar pressure, glomerular filtration rate per day stays constant. After formation of the initial filtrate, the **urine** is modified through secretion and tubular reabsorption according to physiological needs at the time.

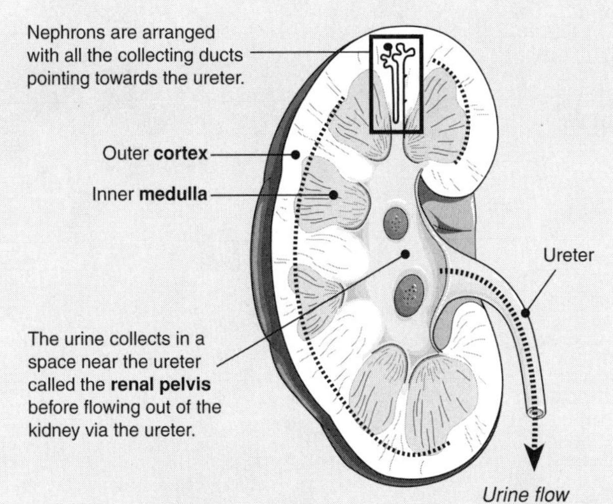
Nephrons are arranged with all the collecting ducts pointing towards the ureter.
Outer **cortex**
Inner **medulla**
Ureter
The urine collects in a space near the ureter called the **renal pelvis** before flowing out of the kidney via the ureter.
Urine flow

Internal Structure of the Human Kidney

Human kidneys are about 100-120 mm long and 25 mm thick. The functional unit of the kidney is the **nephron**. The other parts of the urinary system are primarily passageways and storage areas. The inner tissue of the kidney appears striated (striped), because of the orientation and alignment of the nephrons and blood vessels in the kidney tissue. It is this precise arrangement that makes it possible to fit in all the filtering units required.

Each kidney contains more than 1 million nephrons. By selectively filtering the blood plasma, the nephrons regulate blood composition and pH, and excrete wastes (e.g. urea) and toxins. The initial fluid is formed by **filtration** through the specialized epithelium of the glomerulus. This filtrate is then modified as it passes through the tubules of the nephron. The resulting **urine**, passes out the ureter.

Nephron Structure and Function

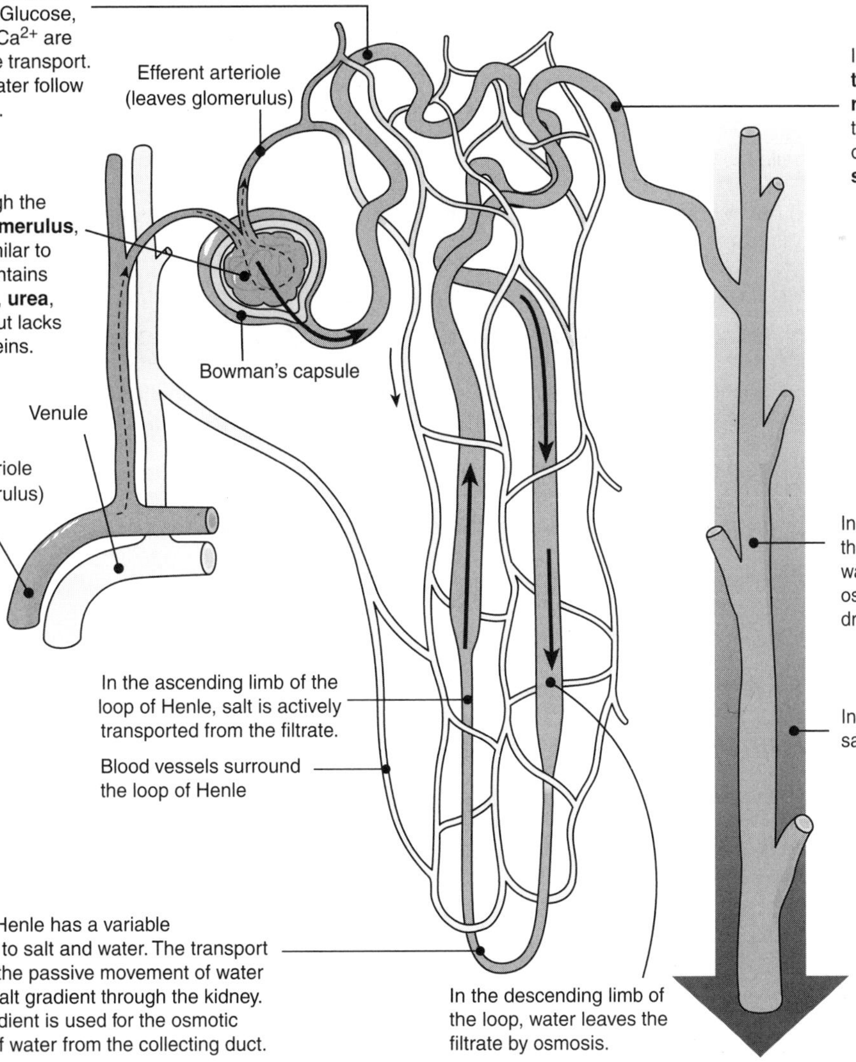

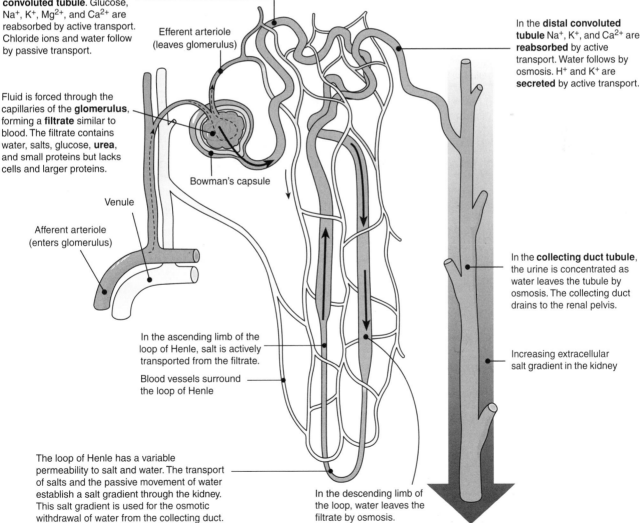

About 90% of the filtrate is **reabsorbed** in the **proximal convoluted tubule**. Glucose, Na^+, K^+, Mg^{2+}, and Ca^{2+} are reabsorbed by active transport. Chloride ions and water follow by passive transport.

Efferent arteriole (leaves glomerulus)

Fluid is forced through the capillaries of the **glomerulus**, forming a **filtrate** similar to blood. The filtrate contains water, salts, glucose, **urea**, and small proteins but lacks cells and larger proteins.

Bowman's capsule

Venule

Afferent arteriole (enters glomerulus)

In the **distal convoluted tubule** Na^+, K^+, and Ca^{2+} are **reabsorbed** by active transport. Water follows by osmosis. H^+ and K^+ are **secreted** by active transport.

In the **collecting duct tubule**, the urine is concentrated as water leaves the tubule by osmosis. The collecting duct drains to the renal pelvis.

Increasing extracellular salt gradient in the kidney

In the ascending limb of the loop of Henle, salt is actively transported from the filtrate.

Blood vessels surround the loop of Henle

The loop of Henle has a variable permeability to salt and water. The transport of salts and the passive movement of water establish a salt gradient through the kidney. This salt gradient is used for the osmotic withdrawal of water from the collecting duct.

In the descending limb of the loop, water leaves the filtrate by osmosis.

Related activities: *The Urinary System* ***Web links:*** *Interactive Kidney Quiz, Kidney Vascular System, The Juxtaglomerular Apparatus*

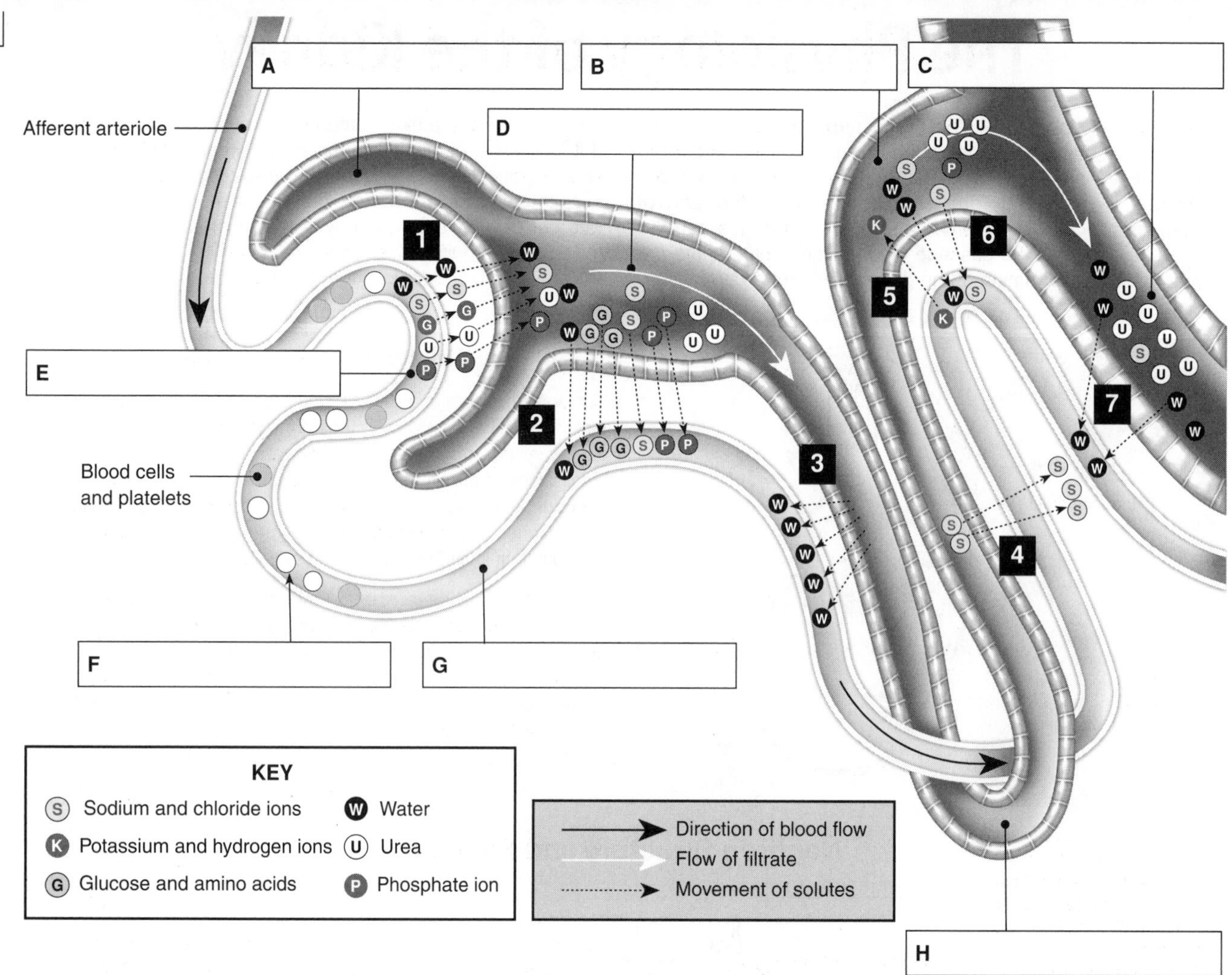

The diagram above presents an overview of the structures and processes involved in the formation of urine in the kidney. The structures involved are labelled with letters (**A-H**), while the major processes are identified with numbers (**1-7**).

1. Using the word list provided below, identify each of the structures marked with a letter. Write the name of the structure in the space provided on the diagram.

   distal convoluted tubule, efferent arteriole, glomerulus, Bowman's capsule,
   proximal convoluted tubule, loop of Henle, large blood proteins, collecting duct

2. Match each of the processes (identified on the diagram with numbers 1-7) to the correct summary of the process provided below. Write the process number next to the appropriate sentence.

   .... Active transport of salt (Na+ and Cl-) from the ascending limb of the loop of Henle.

   .... Filtration through the membranes of the glomerulus. Glucose, water, and ions pass through.

   .... Reabsorption of glucose and ions by active transport. Water follows by diffusion.

   .... Reabsorption of water by osmosis from the descending limb of the loop of Henle.

   .... Active secretion (into the filtrate) of H+ and K+ (NH3 also diffuses into the filtrate).

   .... Concentration of the urine by osmotic withdrawal of water from the flitrate.

   .... Reabsorption of Na+ and Cl- by active transport and water by osmosis.

3. There is marked gradient in salt concentration in the extracellular fluid of the medulla, produced by the transport of salt out of the filtrate. Explain the purpose of this salt gradient:

Control of Urine Output

Variations in salt and water intake, and in the environmental conditions to which we are exposed, contribute to fluctuations in blood volume and composition. The primary role of the kidneys is to regulate blood volume and composition (including the removal of nitrogenous wastes), so that homeostasis is maintained. This is achieved through varying the volume and composition of the urine. Two hormones, **antidiuretic hormone** (ADH) and **aldosterone**, are involved in the process.

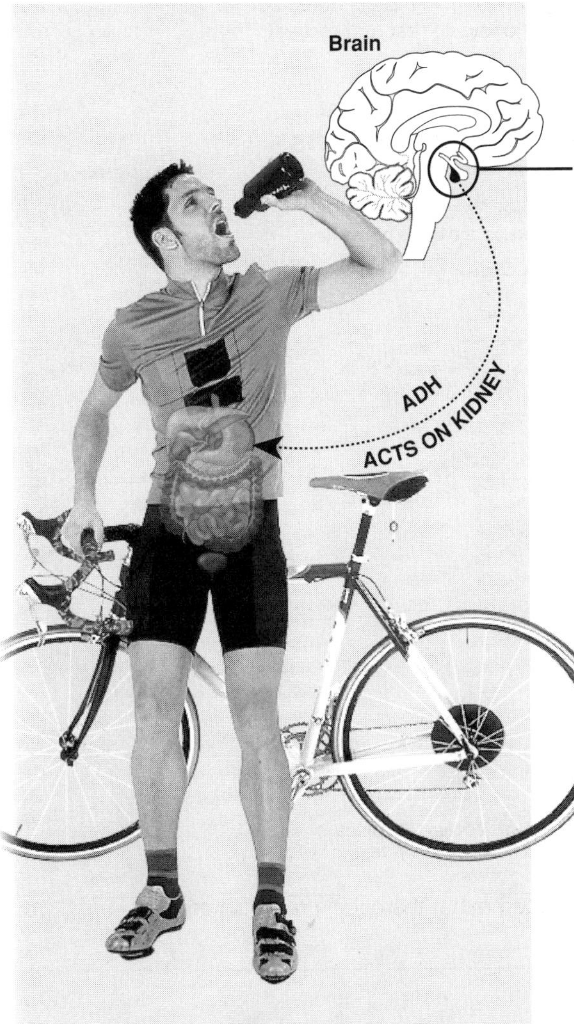

Control of Urine Output

Brain

ADH ACTS ON KIDNEY

Osmoreceptors in the **hypothalamus** detect a fall in the concentration of water in the blood. They stimulate **neurosecretory cells** in the hypothalamus to synthesize and secrete the hormone ADH (antidiuretic hormone).

ADH passes from the hypothalamus to the posterior pituitary where it is released into the blood. ADH increases the permeability of the kidney collecting duct to water so that more water is reabsorbed and urine volume decreases.

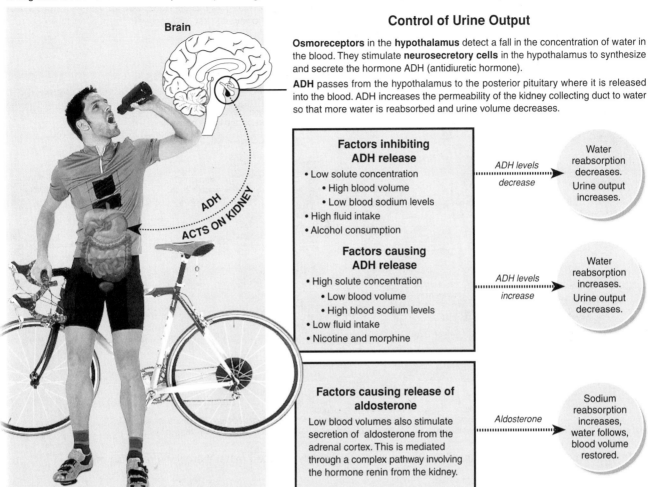

Factors inhibiting ADH release

- Low solute concentration
  - High blood volume
  - Low blood sodium levels
- High fluid intake
- Alcohol consumption

ADH levels decrease → Water reabsorption decreases. Urine output increases.

Factors causing ADH release

- High solute concentration
  - Low blood volume
  - High blood sodium levels
- Low fluid intake
- Nicotine and morphine

ADH levels increase → Water reabsorption increases. Urine output decreases.

Factors causing release of aldosterone

Low blood volumes also stimulate secretion of aldosterone from the adrenal cortex. This is mediated through a complex pathway involving the hormone renin from the kidney.

Aldosterone → Sodium reabsorption increases, water follows, blood volume restored.

1. (a) **Diabetes insipidus** is a disease caused by a lack of ADH. Based on what you know of the role of ADH in regulating urine volumes, describe the symptoms of this disease:

   (b) Suggest how this disorder might be treated:

2. Explain why alcohol consumption (especially to excess) causes dehydration and thirst:

3. Explain how negative feedback mechanisms operate to regulate blood volume and urine output:

4. **Diuretics** are drugs that increase urine volume. Many work by inhibiting the active transport of sodium and chloride in the nephron. Explain how this would lead to an increase in urine volume:

Regulating Fluid and Removing Wastes

Related activities: *Water Budget in a Human, The Physiology of the Kidney, Hormones of the Pituitary* ***Web links:*** *Adrenaline and ADH*

Renal Dialysis

A dialysis machine is a machine designed to remove wastes from the blood. It is used when the kidneys fail, or when blood acidity, urea, or potassium levels increase much above normal. In kidney dialysis, blood flows through a system of tubes composed of partially permeable membranes. Dialysis fluid (dialyzate) has a composition similar to blood except that the concentration of wastes is low. It flows in the opposite direction to the blood on the outside of the dialysis tubes. Consequently, waste products like urea diffuse from the blood into the dialysis fluid, which is constantly replaced. The dialysis fluid flows at a rate of several 100 cm^3 per minute over a large surface area. For some people dialysis is an ongoing procedure, but for others dialysis just allows the kidneys to rest and recover.

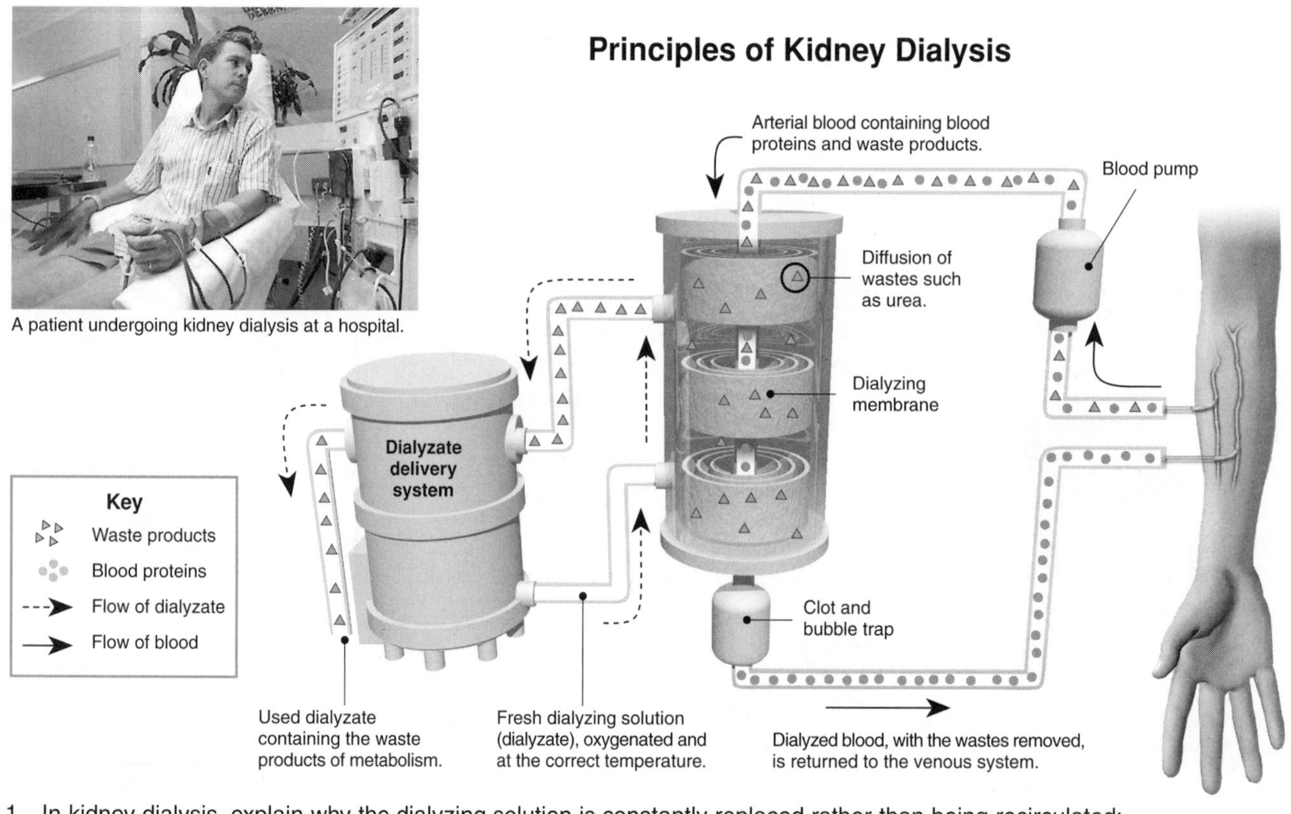

A patient undergoing kidney dialysis at a hospital.

1. In kidney dialysis, explain why the dialyzing solution is constantly replaced rather than being recirculated:

2. Explain why ions such as potassium and sodium, and small molecules like glucose do not diffuse rapidly from the blood into the dialyzing solution along with the urea:

3. Explain why the urea passes from the blood into the dialyzing solution:

4. Describe the general transport process involved in dialysis:

5. Give a reason why the dialyzing solution flows in the opposite direction to the blood:

6. Explain why a clot and bubble trap is needed after the blood has been dialyzed but before it re-enters the body:

Kidney Transplants

Kidney failure (also called renal failure) arises when the kidneys fail to function adequately and filtrate formation decreases or stops. In cases of renal failure, normal blood volume levels and electrolyte balances are not maintained, and waste products build up in the body. Kidney failure is classified as **acute** (rapid onset) or **chronic** (developing over a period of months or years). There are many causes of kidney failure including decreased blood supply, drug overdose, chemotherapy, infection, and poorly controlled diabetic or hypertensive conditions. Recovery from acute renal failure is possible, but chronic renal damage can not be reversed. If kidney deterioration is ignored, the kidneys will fail completely. In some cases diet and medication can be used to treat kidney failure, but when the damage is extensive, **kidney dialysis** or a **kidney transplant** are required to keep the patient alive.

Renal Failure

Kidney (renal) failure is indicated by levels of **serum creatinine**, as well as by kidney size on ultrasound and the presence of anemia (chronic kidney disease generally leads to anemia and small kidney size). Creatinine is a break-down product of creatine phosphate in muscle, and is usually produced at a fairly constant rate by the body (depending on muscle mass). It is chiefly filtered out of the blood by the kidneys, although a small amount is actively secreted by the kidneys into the urine. A rise in blood creatinine levels is observed only with marked damage to functioning nephrons.

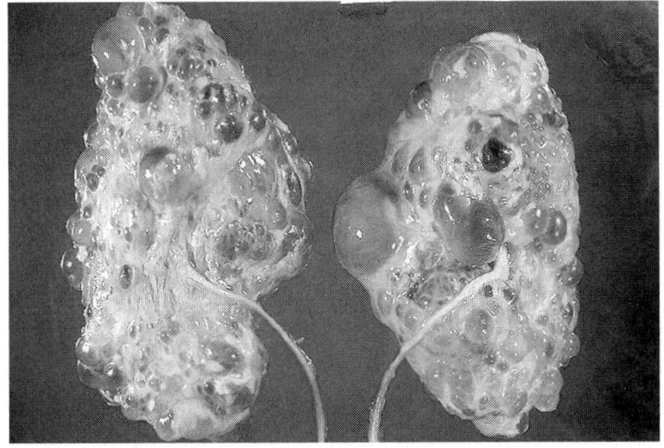

Acute renal failure (ARF) is characterized by decreased urine production (<400mL per day), and commonly arises because of low blood volume (blood loss), dehydration, or widespread infection. In contrast, chronic renal failure, which develops over months or years, is commonly the result of poorly controlled diabetes, poorly controlled high blood pressure, or **polycystic kidney disease**, a genetic disorder characterized by the growth of numerous cysts in the kidneys (above).

Kidney Transplants

Transplantation of a healthy kidney from an organ donor is the preferred treatment for end-stage kidney failure. The organ is usually taken from a person who has just died, although kidneys can also be taken from living donors. The failed organs are left in place and the new kidney transplanted into the lower abdomen. Provided recipients comply with medical requirements (e.g. correct diet and medication) over 85% of kidney transplants are successful.

There are two major problems associated with kidney transplants: lack of donors and tissue rejection. Cells from donor tissue have different antigens to that of the recipient, and are not immunocompatible. Tissue-typing and the use of immunosuppressant drugs helps to decrease organ rejection rates. In the future, xenotransplants of genetically modified organs from other species may help to solve both the problems of supply and immune rejection.

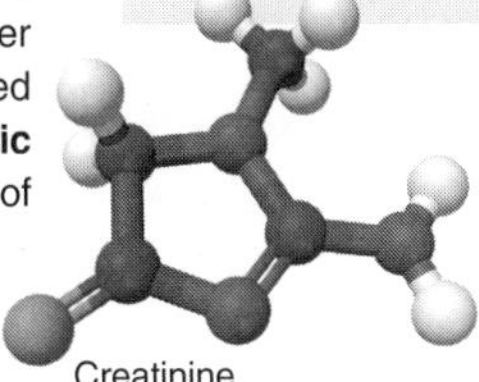

Creatinine

Creatinine levels in both blood and urine is used to calculate the creatinine clearance (CrCl), which reflects the glomerular filtration rate (GFR). The GFR is a clinically important measurement of renal function and more accurate than serum creatinine alone, since serum creatinine only rises when nephron function is very impaired.

1. Distinguish between acute and chronic renal failure and contrast their causes:

2. (a) Explain why a rise in blood (serum) levels of creatinine would indicate a failure of nephron function:

   (b) Explain why a creatinine clearance is a more accurate indicator of renal function than a serum creatinine test alone:

3. Describe some of the advantages and disadvantages of kidney transplantation over a life-time of kidney dialysis:

AP Topic 3B

IB SL/HL-OPT E

Responding to the Environment

KEY CONCEPTS

- Organisms maintain homeostasis through negative feedback mechanisms.
- Animals respond to the environment using nervous and hormonal mechanisms.
- Reflexes are central to simple behaviors.
- Sensory receptors act as biological transducers converting the stimulus energy into an electrochemical signal.

KEY TERMS

action potential
autonomic nervous system
biological transducer
brain
central nervous system
cephalization
cholinergic
courtship
depolarization
ganglion (pl. ganglia)
homing
imprinting
innate behavior
insight
learning
migration
myelin / myelinated
navigation
neuron (motor, relay, sensory)
neurotransmitter
operant learning
peripheral nervous system
pheromone
photoreceptor
reflex
releaser
resting state (of nerve)
saltatory conduction
sensory receptor
sign stimulus
social behavior
stereotyped
stimulus (pl. stimuli)
synapse
threshold potential

Periodicals:
listings for this chapter are on page 389

Weblinks:
www.thebiozone.com/weblink/SB2-2603.html

Teacher Resource CD-ROM:
Migratory Navigation in Birds

OBJECTIVES

☐ 1. Use the **KEY TERMS** to help you understand and complete these objectives.

Nervous Systems
pages 171-180

☐ 2. Compare the structure of invertebrate and vertebrate nervous systems.

☐ 3. Compare brain structure in different vertebrates and relate any differences to the animal's evolutionary history, environment, and sensory requirements.

☐ 4. Describe the structure and organization of the mammalian nervous system, including the functional roles of the two major divisions (CNS and PNS).

☐ 5. Describe the gross structure of the human **brain**, including the functional role of each of its major regions. Describe an example of brain malfunction.

☐ 6. Describe the structure of the autonomic nervous system: the sympathetic and parasympathetic neurons (nervous systems). Describe their roles and use examples to show their generally antagonistic effects.

Neuron Structure and Function
pages 181-186

☐ 7. Describe the structure and function of a **neuron** and distinguish between **sensory neurons**, **relay neurons**, and **motor neurons**. Describe **reflexes** and explain their role in nervous systems and adaptive behavior.

☐ 8. Describe and explain the **resting state** of a neuron. Explain how an **action potential** is propagated along a non-myelinated neuron and (if required) in a **myelinated nerve** by **saltatory conduction**.

☐ 9. Describe transmission across a **cholinergic synapse**.

Sensory Perception
page 187-192

☐ 10. Describe the **stimuli** to which animals respond and describe diversity in the sense organs responsible for receiving stimuli in different taxa.

☐ 11. Using examples, explain how sensory receptors receive and respond to stimuli. Explain the role of sense organs as **biological transducers**.

Animal Behavior
page 193-210

☐ 12. Describe the signals and **displays** used by animals for **communication**.

☐ 13. Describe and explain **innate behavior** and **learned behavior**. Explain the role of a **sign stimulus** (releaser) in producing a **fixed action pattern** and describe the **stereotyped** nature of such behaviors.

☐ 14. Describe and explain examples of behavior, including **social behavior**, **courtship**, **migration** and **navigation**, **homing**, **imprinting**, habituation, **operant learning**, and **insight**. Describe the adaptive value of the behavior in each case. Explain the role of behavior in species recognition.

Invertebrate Nervous Systems

Animals respond to stimuli in order to survive. In one-celled organisms the entire cell surface is sensitive to stimuli and the cell responds by moving towards or away from the stimulus. More sophisticated responses require more complex nervous control. Some invertebrates (e.g. *Hydra*) have only a loosely organised nervous system. Others have more complex systems with a nerve cord, a brain, and usually a head with sensory structures. The nerve cord is on the ventral (lower) surface in arthropods and annelids whereas it is on the dorsal (upper surface) in vertebrates. In insects and vertebrates, the nervous and endocrine systems work together to regulate many activities. This allows both rapid and slower, longer lasting responses.

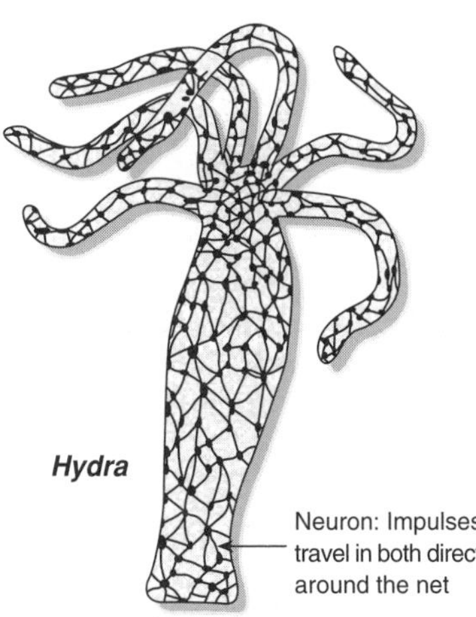

Nerve Net (left)

EXAMPLES: *Cnidarians, e.g. Hydra*

A loosely organized system of nerves with no central control (no central nervous system). The neurons are located in the outer ectoderm and mesoglea (middle jelly-like layer). The strength of the nervous system response to a stimulus depends directly on the stimulus strength. The sensory receptors are cells within the outer body layer, the effectors that bring about body movement are the contractile elements of individual cells.

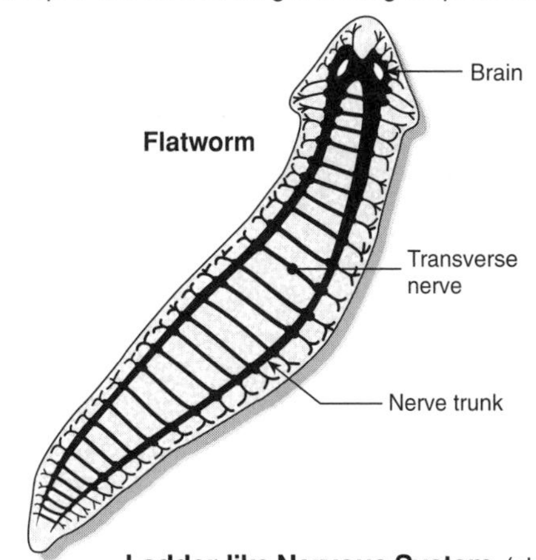

Ladder-like Nervous System (above)

EXAMPLE: *Platyhelminthes (flatworms)*

Flatworms are the simplest animals to show the development of a central nervous system, with a simple brain and nerve trunks that act as thoroughfares for information. Transverse nerves connect the main nerve trunks.

Radial Nervous System (right)

EXAMPLES: *Echinoderms, e.g. starfish*

Radial nerves run through each arm from the central nerve ring around the mouth. Branches of the radial nerves form an interconnected network similar to the nerve net of *Hydra*. It is believed that a nerve center exists at the junction of each radial nerve with the nerve ring, and that this nerve center allows one or other arm to dominate during directional movement.

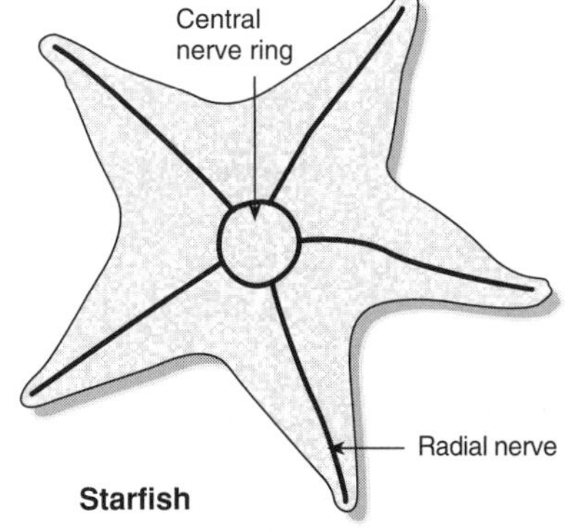

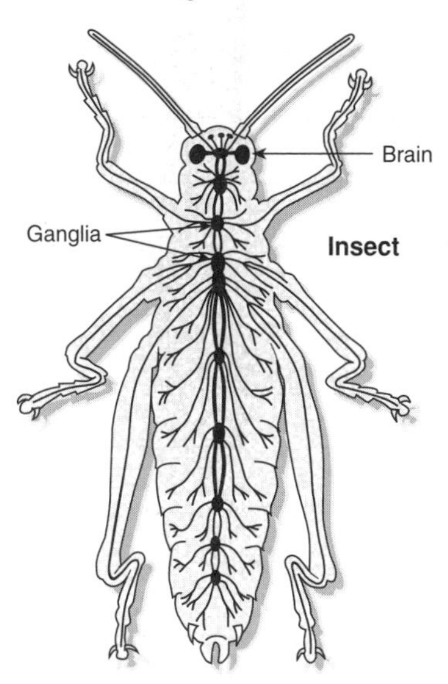

Ventral Nerve Cord and Brain (left and right)

EXAMPLES: *Annelids (earthworms, leeches), arthropods (crustaceans, insects, spiders, scorpions)*

The nerve cords of these segmented animals often contain ganglia (clumps of nerve cells) in each segment. These ganglia coordinate the actions of each segment. The brains are larger and more complex than in flatworms. In insects, there has been an even greater trend towards concentration of ganglia in the head, thorax and abdomen, associated with the fusion of segments and different activities in these regions. In insects, there is also prominent **cephalization**; formation of a distinctive head and brain.

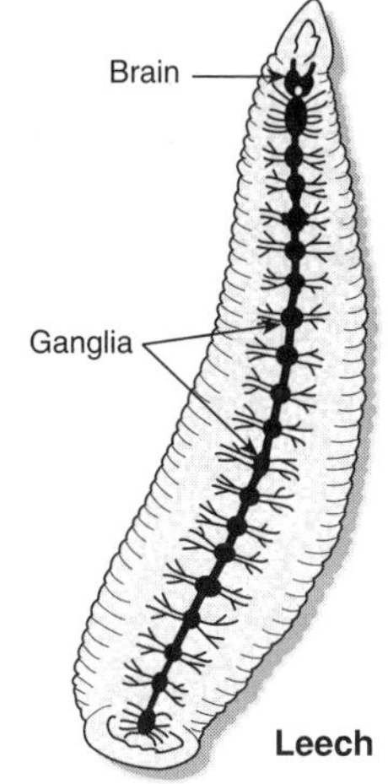

Annelid: earthworm

Arthropod: grasshopper

Cnidarian: sea anemone

Echinoderm: sea star

The more highly organized nervous systems of annelids and arthropods are associated with a number of sensory organs of varying complexity. Those in annelids are generally very simple structures in the epidermis, consisting of individual cells or small groups of cells that are sensitive to touch, pressure, chemical stimuli, and temperature. Insect sense organs include simple eyes (ocelli), complex compound eyes, and a number of epidermal sense organs.

The sensory receptors of cnidarians are restricted largely to the outer (ectoderm) layers. They are simple sensory cells, concentrated around the mouth and on the tentacles, and are used for prey detection and capture.

The primary sensory receptors of sea stars and other echinoderms are contained within the epidermis. They are simple sensory cells and function in the reception of light, contact, and chemical stimuli.

1. Briefly describe the arrangement of the nervous system in the following invertebrates:

   (a) Cnidarians e.g. *Hydra*, sea anemone:

   (b) Platyhelminthes:

   (c) Echinoderms:

   (d) Annelids:

   (e) Arthropods:

2. (a) Contrast the speed and efficiency of nerve impulse transmission in cnidarians and arthropods:

   (b) Explain a reason for these differences:

3. (a) Describe what is meant by **cephalization**:

   (b) Explain the significance of cephalization to the type of responses and behavior possible in invertebrates:

   (c) Describe the general trend seen in nervous system organisation, from cnidarians through to insects:

   (d) Describe what you would regard as the main evolutionary (selection) pressure for the development of a brain:

4. Describe the advantages of combining segmental nervous control (via ganglia) with control via a distinctive brain:

Detecting Changing States

A **stimulus** is any physical or chemical change in the environment capable of provoking a response in an organism. Animals respond to stimuli in order to survive. This response is adaptive; it acts to maintain the organism's state of homeostasis. Stimuli may be either external (outside the organism) or internal (within its body). Some of the stimuli to which humans and other mammals respond are described below, together with the sense organs that detect and respond to these stimuli. Note that sensory receptors respond only to specific stimuli. The sense organs an animal possesses therefore determine how it perceives the world.

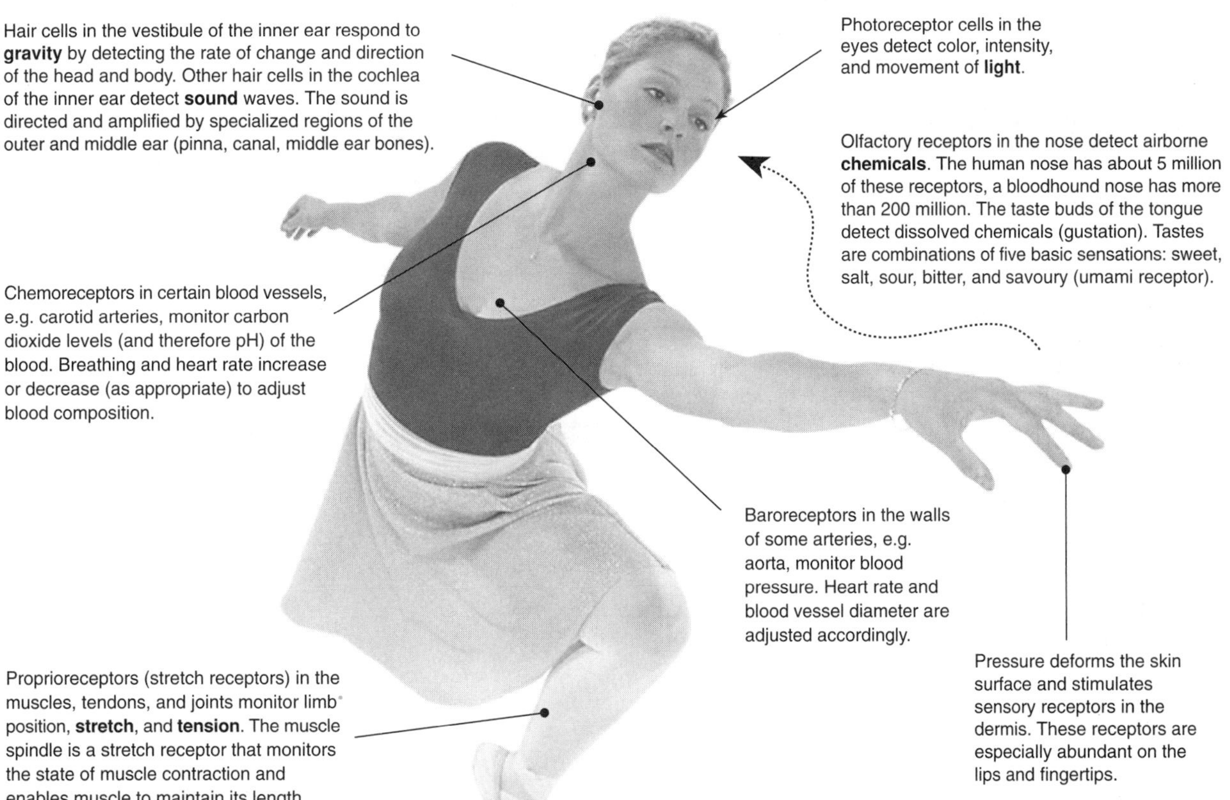

Hair cells in the vestibule of the inner ear respond to **gravity** by detecting the rate of change and direction of the head and body. Other hair cells in the cochlea of the inner ear detect **sound** waves. The sound is directed and amplified by specialized regions of the outer and middle ear (pinna, canal, middle ear bones).
Photoreceptor cells in the eyes detect color, intensity, and movement of **light**.
Olfactory receptors in the nose detect airborne **chemicals**. The human nose has about 5 million of these receptors, a bloodhound nose has more than 200 million. The taste buds of the tongue detect dissolved chemicals (gustation). Tastes are combinations of five basic sensations: sweet, salt, sour, bitter, and savoury (umami receptor).
Chemoreceptors in certain blood vessels, e.g. carotid arteries, monitor carbon dioxide levels (and therefore pH) of the blood. Breathing and heart rate increase or decrease (as appropriate) to adjust blood composition.
Baroreceptors in the walls of some arteries, e.g. aorta, monitor blood pressure. Heart rate and blood vessel diameter are adjusted accordingly.
Pressure deforms the skin surface and stimulates sensory receptors in the dermis. These receptors are especially abundant on the lips and fingertips.
Proprioreceptors (stretch receptors) in the muscles, tendons, and joints monitor limb position, **stretch**, and **tension**. The muscle spindle is a stretch receptor that monitors the state of muscle contraction and enables muscle to maintain its length.

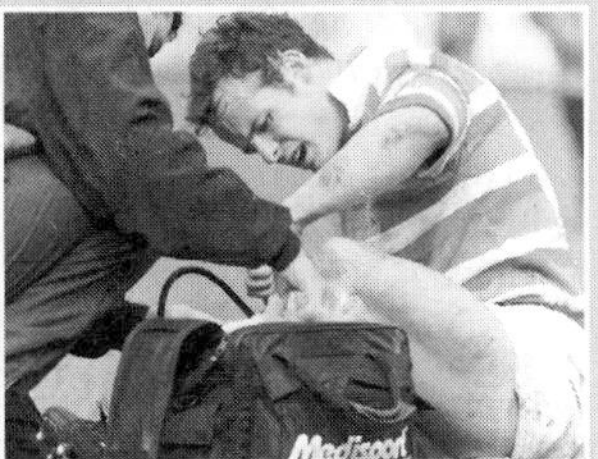
Pain and temperature are detected by simple nerve endings in the skin. Deep tissue injury is sometimes felt on the skin as referred pain.

Humans rely heavily on their hearing when learning to communicate; without it, speech and language development are more difficult.

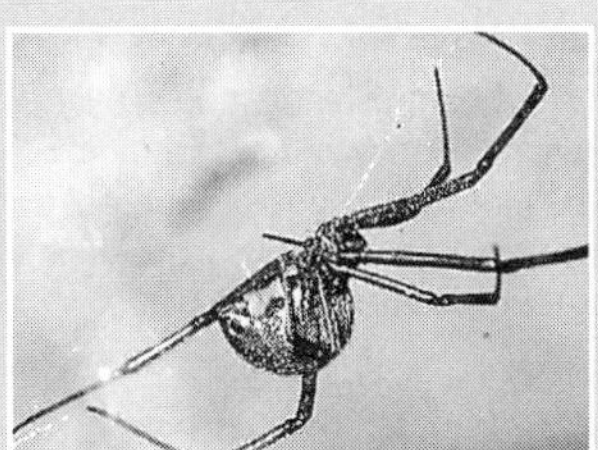
The vibration receptors in the limbs of arthropods are sensitive to movement: either sound or vibration (as caused by struggling prey).

The chemosensory Jacobson's organ in the roof of the mouth of reptiles (e.g. snakes) enables them to detect chemical stimuli.

Breathing and heart rates are regulated in response to sensory input from chemoreceptors.

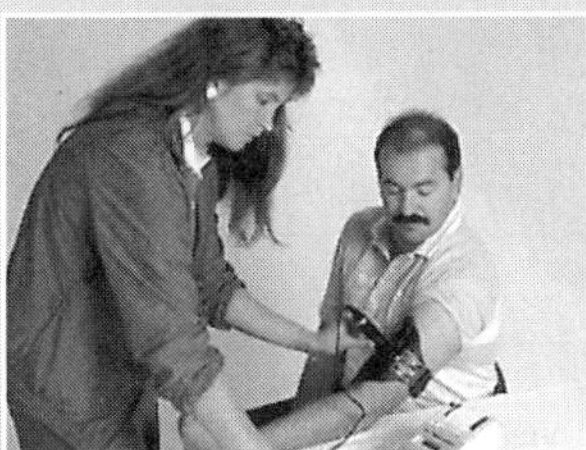
Baroreceptors and osmoreceptors act together to keep blood pressure and volume within narrow limits.

Many insects, such as these ants, rely on chemical sense for location of food and communication.

Jacobson's organ is also present in mammals and is used to detect sexual receptivity in potential mates.

1. Provide a concise definition of a stimulus: ______________________________

____

2. (a) Name one external stimulus and its sensory receptor: ______________________________

(b) Name one internal stimulus and its sensory receptor: ______________________________

Related activities: *The Basis of Sensory Perception*

The Vertebrate Brain

The vertebrate brain develops as an expansion of the anterior end of the neural tube in embryos. The forebrain, midbrain, and hindbrain can be seen very early in development, with further differentiation as development continues. The brains of fish and amphibians are relatively unspecialized, with a rudimentary cerebrum and cerebellum. The reptiles show the first real expansion of the cerebrum, with the grey matter external in the cortex. In the birds and mammals, the brain is relatively large, with well developed cerebral and cerebellar regions. In primitive vertebrates, the cerebral regions act primarily as olfactory centres. In higher vertebrates, the cerebrum takes over the many of the functions of other regions of the brain (e.g. the optic lobes), becoming the primary integration centre of the brain. The cerebellum also becomes more important as locomotor and other muscular activities increase in complexity. The relative sizes of different regions of vertebrate brains are shown below.

Vertebrate Brains

All vertebrates, from fish and amphibians, to humans and other mammals, have brains with the same basic structure. The brain develops from a hollow tube and comprises the forebrain, midbrain, and hindbrain (which runs into the spinal cord). During the course of vertebrate evolution some parts of the brain (e.g. the medulla) have remained largely unchanged, retaining their primitive functions. Other parts (e.g. the cerebrum of the forebrain) have expanded and taken on new functions.

Key to Brain Regions and their Functions

- **Olfactory bulb**: Receives and processes olfactory signals.
- **Cerebrum**: Behavior, complex thought and reasoning.
- **Cerebellum**: Center for controlling movement and balance.
- **Medulla**: Reflex functions and relay for sensory information.
- **Optic lobe**: Receives and processes visual information.

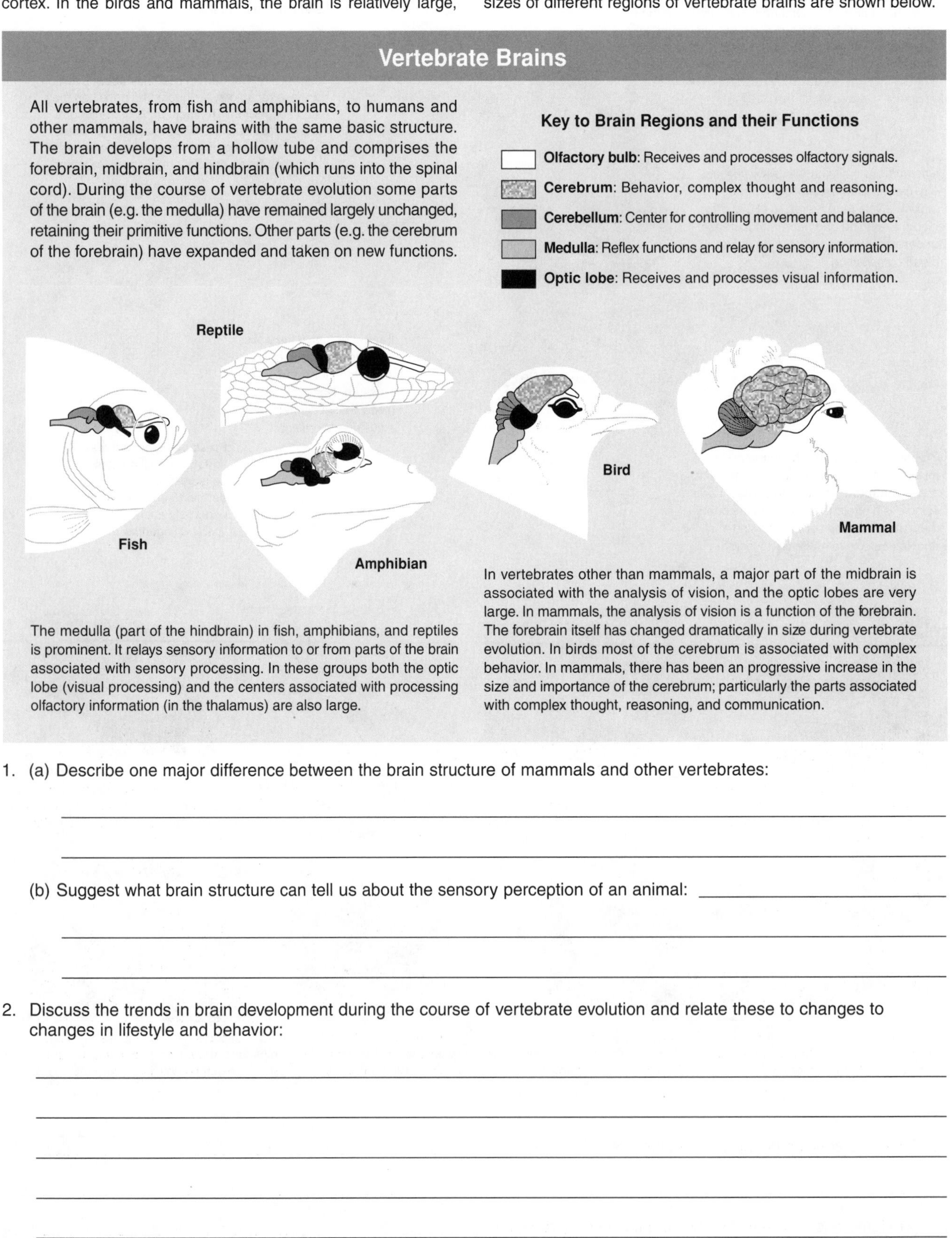

The medulla (part of the hindbrain) in fish, amphibians, and reptiles is prominent. It relays sensory information to or from parts of the brain associated with sensory processing. In these groups both the optic lobe (visual processing) and the centers associated with processing olfactory information (in the thalamus) are also large.

In vertebrates other than mammals, a major part of the midbrain is associated with the analysis of vision, and the optic lobes are very large. In mammals, the analysis of vision is a function of the forebrain. The forebrain itself has changed dramatically in size during vertebrate evolution. In birds most of the cerebrum is associated with complex behavior. In mammals, there has been an progressive increase in the size and importance of the cerebrum; particularly the parts associated with complex thought, reasoning, and communication.

1. (a) Describe one major difference between the brain structure of mammals and other vertebrates:

   (b) Suggest what brain structure can tell us about the sensory perception of an animal:

2. Discuss the trends in brain development during the course of vertebrate evolution and relate these to changes to changes in lifestyle and behavior:

The Mammalian Nervous System

The **nervous system** is the body's control and communication center. It has three broad functions: detecting stimuli, interpreting them, and initiating appropriate responses. Its basic structure is outlined below. Further detail is provided in the following pages.

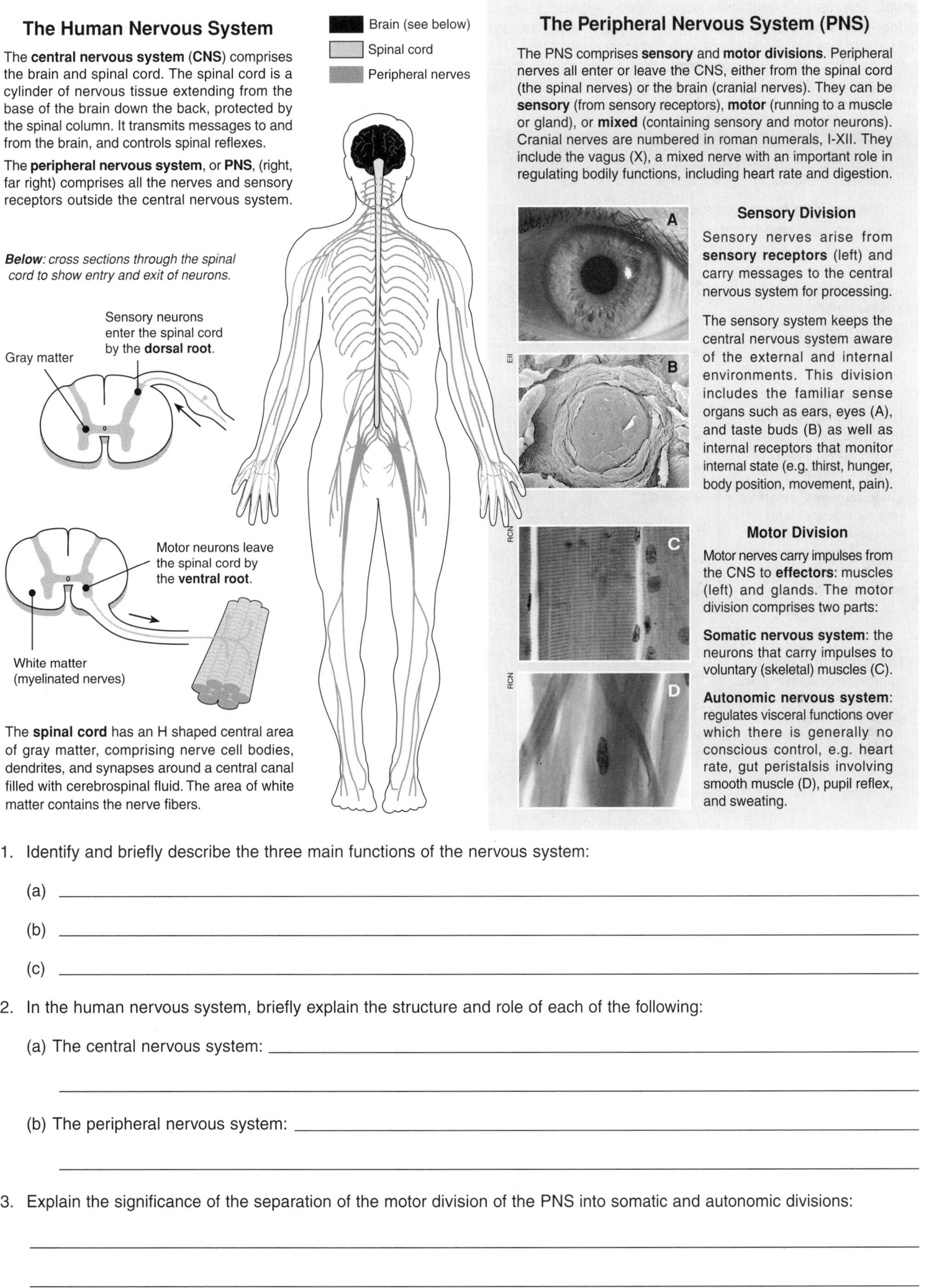

The Human Nervous System

The **central nervous system** (**CNS**) comprises the brain and spinal cord. The spinal cord is a cylinder of nervous tissue extending from the base of the brain down the back, protected by the spinal column. It transmits messages to and from the brain, and controls spinal reflexes.

The **peripheral nervous system**, or **PNS**, (right, far right) comprises all the nerves and sensory receptors outside the central nervous system.

***Below**: cross sections through the spinal cord to show entry and exit of neurons.*

The **spinal cord** has an H shaped central area of gray matter, comprising nerve cell bodies, dendrites, and synapses around a central canal filled with cerebrospinal fluid. The area of white matter contains the nerve fibers.

The Peripheral Nervous System (PNS)

The PNS comprises **sensory** and **motor divisions**. Peripheral nerves all enter or leave the CNS, either from the spinal cord (the spinal nerves) or the brain (cranial nerves). They can be **sensory** (from sensory receptors), **motor** (running to a muscle or gland), or **mixed** (containing sensory and motor neurons). Cranial nerves are numbered in roman numerals, I-XII. They include the vagus (X), a mixed nerve with an important role in regulating bodily functions, including heart rate and digestion.

Sensory Division

Sensory nerves arise from **sensory receptors** (left) and carry messages to the central nervous system for processing.

The sensory system keeps the central nervous system aware of the external and internal environments. This division includes the familiar sense organs such as ears, eyes (A), and taste buds (B) as well as internal receptors that monitor internal state (e.g. thirst, hunger, body position, movement, pain).

Motor Division

Motor nerves carry impulses from the CNS to **effectors**: muscles (left) and glands. The motor division comprises two parts:

Somatic nervous system: the neurons that carry impulses to voluntary (skeletal) muscles (C).

Autonomic nervous system: regulates visceral functions over which there is generally no conscious control, e.g. heart rate, gut peristalsis involving smooth muscle (D), pupil reflex, and sweating.

1. Identify and briefly describe the three main functions of the nervous system:

   (a) ______________________________

   (b) ______________________________

   (c) ______________________________

2. In the human nervous system, briefly explain the structure and role of each of the following:

   (a) The central nervous system: ______________________________

   (b) The peripheral nervous system: ______________________________

3. Explain the significance of the separation of the motor division of the PNS into somatic and autonomic divisions:

____

***Related activities:** Nervous Regulatory Systems, The Autonomic Nervous System*

The Autonomic Nervous System

The **autonomic nervous system** (ANS) regulates involuntary visceral functions by means of **reflexes**. Although most autonomic nervous system activity is beyond our conscious control, voluntary control over some basic reflexes (such as bladder emptying) can be learned. Most visceral effectors have dual innervation, receiving fibers from both branches of the ANS. These two branches, the **parasympathetic** and **sympathetic** divisions, have broadly opposing actions on the organs they control (excitatory or inhibitory). Nerves in the parasympathetic division release acetylcholine. This neurotransmitter is rapidly deactivated at the synapse and its effects are short lived and localized. Most sympathetic postganglionic nerves release noradrenaline, which enters the bloodstream and is deactivated slowly. Hence, sympathetic stimulation tends to have more widespread and long lasting effects than parasympathetic stimulation. Aspects of autonomic nervous system structure and function are illustrated below. The arrows indicate nerves to organs or ganglia (concentrations of nerve cell bodies).

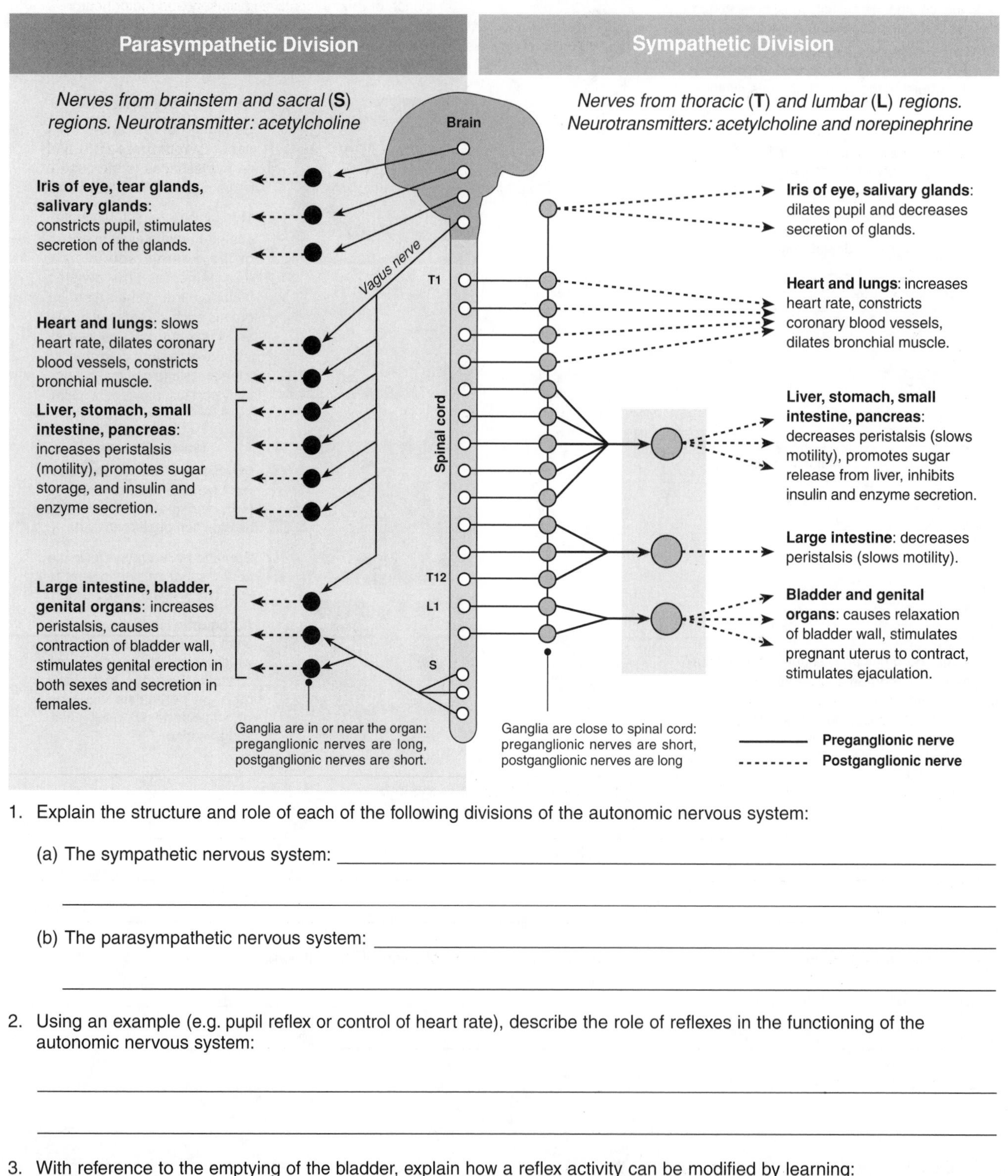

1. Explain the structure and role of each of the following divisions of the autonomic nervous system:

   (a) The sympathetic nervous system: ______________________________

   ____

   (b) The parasympathetic nervous system: ______________________________

   ____

2. Using an example (e.g. pupil reflex or control of heart rate), describe the role of reflexes in the functioning of the autonomic nervous system:

   ____

   ____

3. With reference to the emptying of the bladder, explain how a reflex activity can be modified by learning:

   ____

   ____

   ____

The Human Brain

The brain is one the largest organs in the body. It is protected by the skull, the **meninges**, and the **cerebrospinal fluid** (CSF). The brain is the body's control center. It receives a constant flow of sensory information, but responds only to what is important at the time. Some responses are very simple (e.g. cranial reflexes), whilst others require many levels of processing. The human brain is noted for its large, well developed cerebral region, with its prominent folds (**gyri**) and grooves (**sulci**). Each cerebral hemisphere has an outer region of gray matter and an inner region of white matter, and is divided into four lobes by deep sulci or fissures. These lobes: temporal, frontal, occipital, and parietal, correspond to the bones of the skull under which they lie.

Primary Structural Regions of the Brain

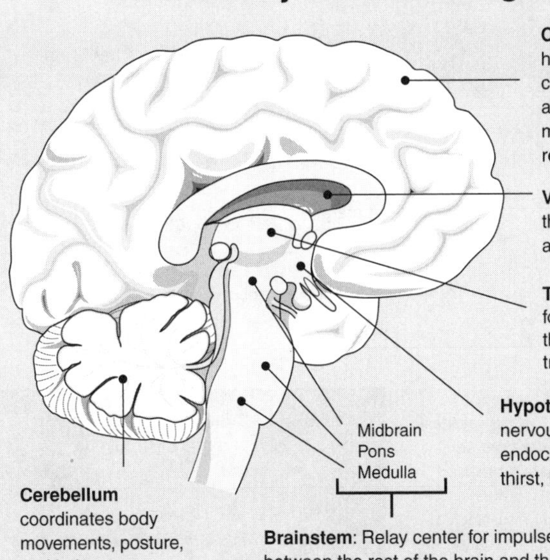

Cerebrum: Divided into two cerebral hemispheres. Many, complex roles. It contains sensory, motor, and association areas, and is involved in memory, emotion, language, reasoning, and sensory processing.

Ventricles: Cavities containing the CSF, which absorbs shocks and delivers nutritive substances.

Thalamus is the main relay center for all sensory messages that enter the brain, before they are transmitted to the cerebrum.

Hypothalamus controls the autonomic nervous system and links nervous and endocrine systems. Regulates appetite, thirst, body temperature, and sleep.

Cerebellum coordinates body movements, posture, and balance.

Brainstem: Relay center for impulses between the rest of the brain and the spinal cord. Controls breathing, heartbeat, and the coughing and vomiting reflexes.

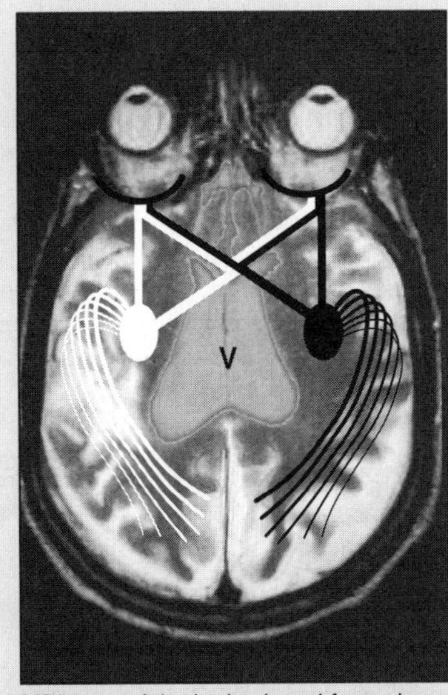

MRI scan of the brain viewed from above. The visual pathway has been superimposed on the image. Note the crossing of some sensory neurons to the opposite hemisphere and the fluid filled ventricles (V) in the center.

Sensory and Motor Regions in the Cerebrum

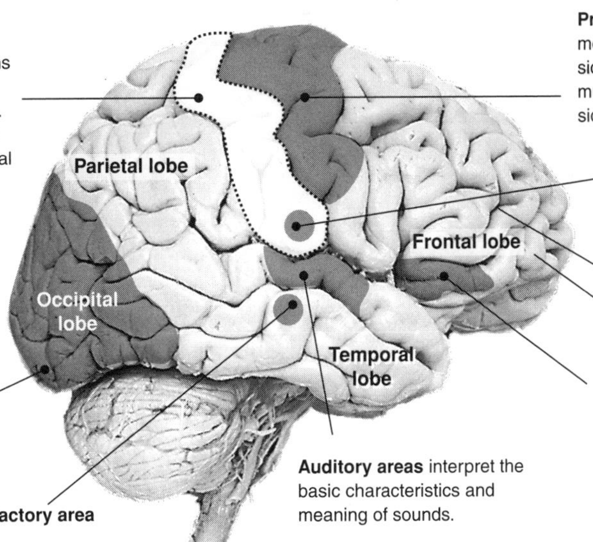

Primary somatic sensory area receives sensations from receptors in the skin, muscles and viscera, allowing recognition of pain, temperature, or touch. Sensory information from receptors on one side of the body crosses to the opposite side of the cerebral cortex where conscious sensations are produced. The size of the sensory region for different body parts depends on the number of receptors in that particular body part.

Visual areas within the occipital lobe receive, interpret, and evaluate visual stimuli. In vision, each eye views both sides of the visual field but the brain receives impulses from left and right visual fields separately (see photo caption above). The visual cortex combines the images into a single impression or **perception** of the image.

Olfactory area

Auditory areas interpret the basic characteristics and meaning of sounds.

Primary motor area controls muscle movement. Stimulation of a point one side of the motor area results in muscular contraction on the opposite side of the body.

Primary gustatory area interprets sensations related to taste.

Sulci (grooves)

Gyri (elevated folds)

Language areas: The motor speech area (Broca's area) is concerned with speech production. The sensory speech area (Wernicke's area) is concerned with speech recognition and coherence.

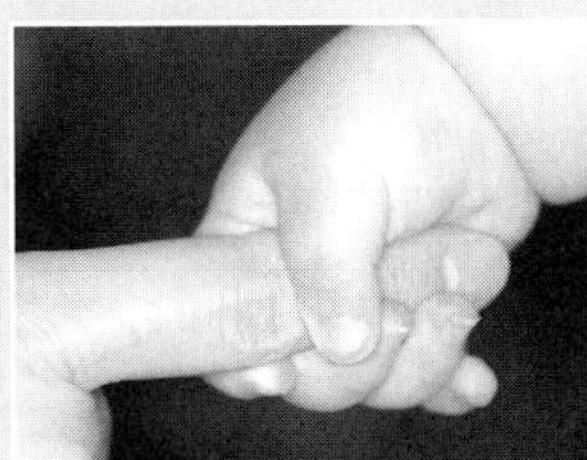

Touch is interpreted in the primary somatic sensory area. The fingertips and the lips have a relatively large amount of area devoted to them.

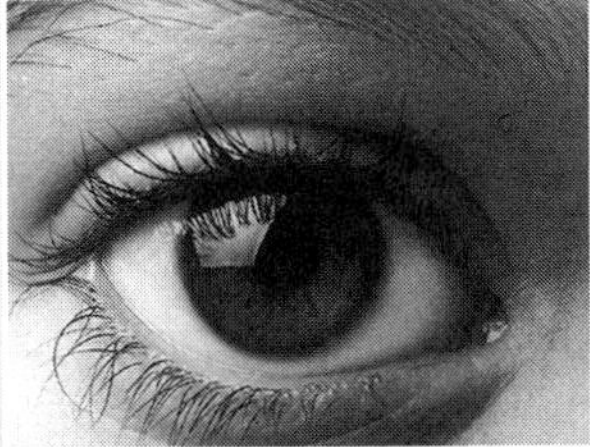

Humans rely heavily on vision. The importance of this **special sense** in humans is indicated by the large occipital region of the brain.

The olfactory tract connects the olfactory bulb with the cerebral hemispheres where olfactory information is interpreted.

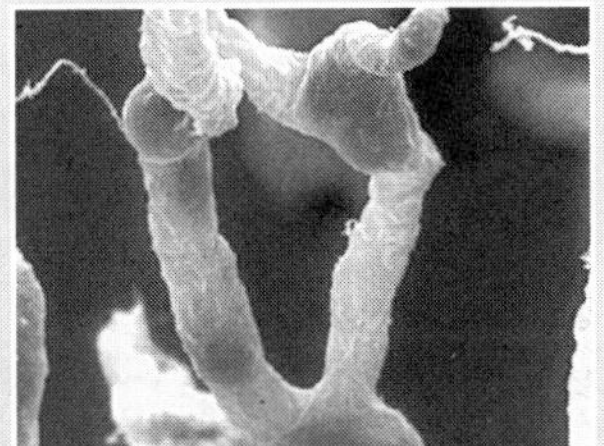

The endothelial tight junctions of the capillaries supplying the brain form a protective **blood-brain barrier** against toxins and infection.

The Ventricles and CSF

The delicate nervous tissue of the brain and spinal cord is protected against damage by the **bone** of the skull and vertebral column, the membranes overlying the brain (the **meninges**), and the watery but nutritive **cerebrospinal fluid** (CSF), which lies between the inner two of the meningeal layers.

The meninges are collectively three membranes: a tough double-layered outer **dura mater**, a web-like middle **arachnoid mater**, and an inner delicate **pia mater** that adheres to the surface of the brain. The CSF is formed from the blood by clusters of capillaries on the roof of each of the brain's ventricles (choroid plexuses). The CSF is constantly circulated through the ventricles of the brain (and into the spinal cord), returning to the blood via specialized projections of the middle meningeal layer (the arachnoid).

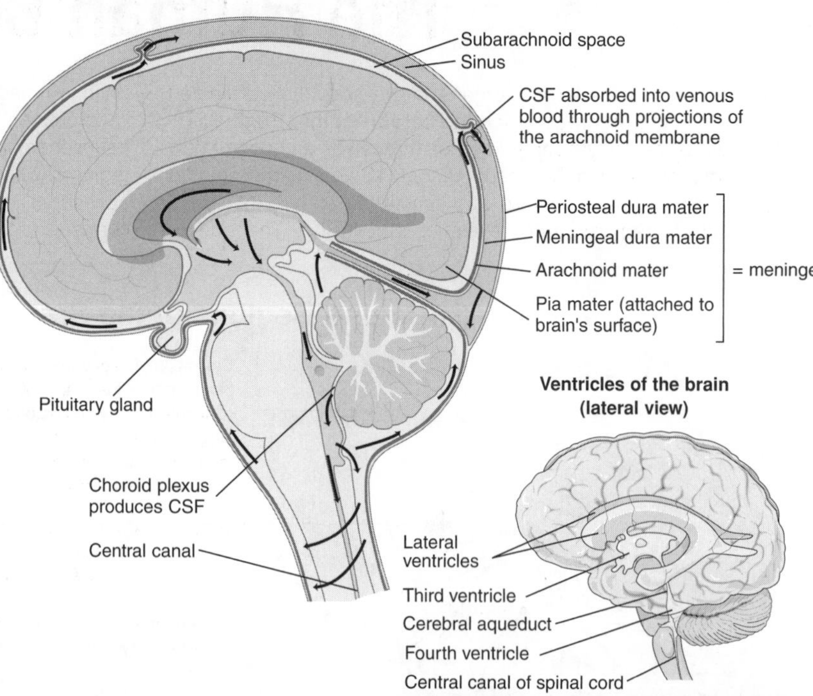

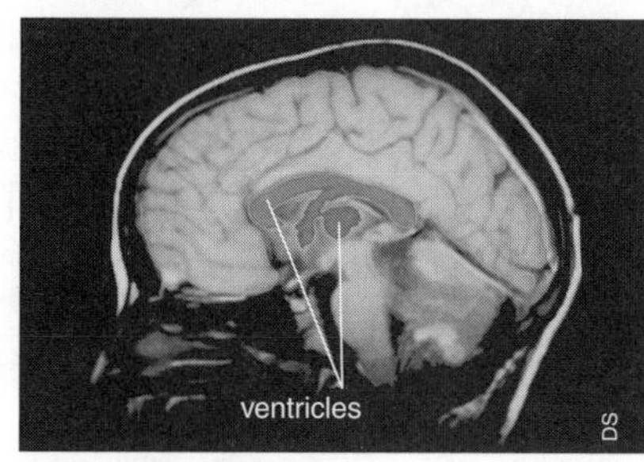

If the passages that normally allow the CSF to exit the brain become blocked, the CSF accumulates within the brain's ventricles causing a condition called hydrocephalus

The accumulated fluid can be seen in this MRI scan.

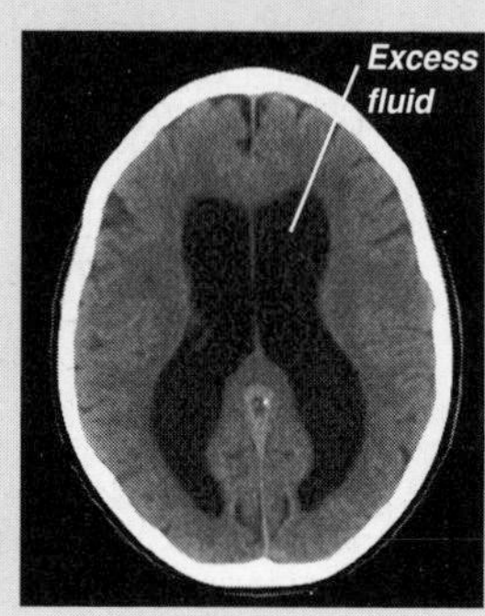

MRI scanning is a powerful technique to visualize the structure and function of the body. It provides much greater contrast between the different soft tissues than computerized tomography (CT) does, making it especially useful in neurological (brain) imaging, especially for indicating the presence of tumors or fluid, and showing up abnormalities in blood supply. In the scan pictured right, the fluid within the lateral and third ventricles is clearly visible.

ventricles

DS

1. For each of the following bodily functions, identify the region(s) of the brain involved in its control:

   (a) Breathing and heartbeat: ______

   (b) Memory and emotion: ______

   (c) Posture and balance: ______

   (d) Autonomic functions: ______

   (e) Visual processing: ______

   (f) Body temperature: ______

   (g) Language: ______

   (h) Muscular movement: ______

2. Explain how the brain is protected against physical damage and infection: ______

3. (a) Describe where CSF is produced and how the CSF returns to the blood: ______

   (b) Explain the consequences of blocking this return flow of CSF: ______

The Malfunctioning Brain: Alzheimer's

Alzheimer's disease (AD) is a disabling neurological disorder affecting about 5% of the population over 65. Although its causes are largely unknown, people with a family history of Alzheimer's have a greater risk, implying that a genetic factor is involved. Some of the cases of Alzheimer's with a familial (inherited) pattern involve a mutation of the gene for amyloid precursor protein (APP), found on chromosome 21 and nearly all people with Down syndrome (trisomy 21) who live into their 40s develop the disease. The gene for the protein apoE, which has an important role in lipid transport, degeneration and regulation in nervous tissue, is also a risk factor that may be involved in modifying the age of onset. Sufferers of Alzheimer's have trouble remembering recent events and they become confused and forgetful. In the later stages of the disease, people with Alzheimer's become very disorientated, lose past memories, and may become paranoid and moody. Dementia and loss of reason occur at the end stages of the disease. The effects of the disease are irreversible and it has no cure.

The Effects of Alzheimer's Disease

Alzheimer's is associated with accelerated loss of neurons, particularly in regions of the brain that are important for memory and intellectual processing, such as the cerebral cortex and hippocampus. The disease has been linked to abnormal accumulations of protein-rich **amyloid** plaques and tangles, which invade the brain tissue and interfere with synaptic transmission.

Cerebral cortex: Conscious thought, reasoning, and language. Alzheimer's sufferers show considerable loss of function from this region.

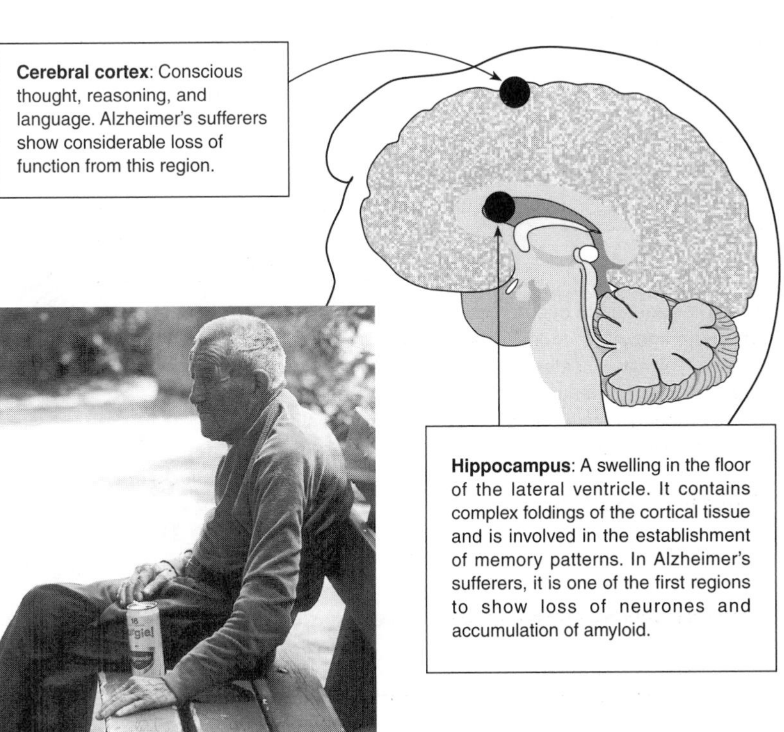

Hippocampus: A swelling in the floor of the lateral ventricle. It contains complex foldings of the cortical tissue and is involved in the establishment of memory patterns. In Alzheimer's sufferers, it is one of the first regions to show loss of neurones and accumulation of amyloid.

It is not uncommon for Alzheimer's sufferers to wander and become lost and disorientated.

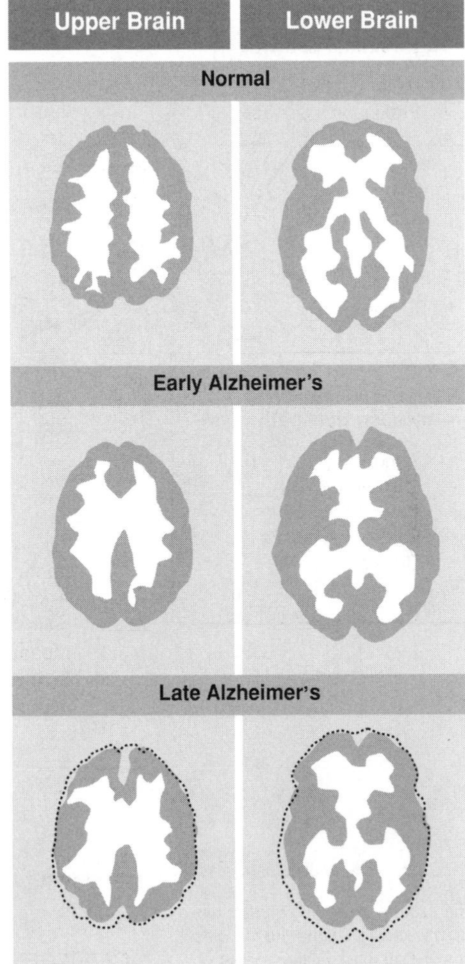

The brain scans above show diminishing brain function in certain areas of the brain in Alzheimer's sufferers. Note, particularly in the two lower scans, how much the brain has shrunk (original size indicated by the dotted line). Light areas indicate brain activity.

1. Describe the biological basis behind the degenerative changes associated with Alzheimer's disease:

2. Describe the evidence for the Alzheimer's disease having a genetic component in some cases:

3. Some loss of neuronal function occurs normally as a result of ageing. Identify the features distinguishing Alzheimer's disease from normal age related loss of neuronal function:

Dopamine, the Brain, and Behavior

The role of genes in some aspects of human behavior, although controversial, is becoming more apparent as research is carried out on more genes and their functions. The dopamine receptor D4 (DRD4) has received a lot of research attention, because of its apparent link to substance dependence and sensation seeking. Dopamine pathways are neural pathways in the brain which transmit the neurotransmitter dopamine from one region of the brain to another. Dopamine has many functions in the brain, including important roles in behavior and cognition, motivation and reward, mood, attention, and learning. In cases of deficiency, dopamine cannot be administered as a drug itself, because it cannot cross the blood-brain barrier.

Dopamine Pathways in the Brain

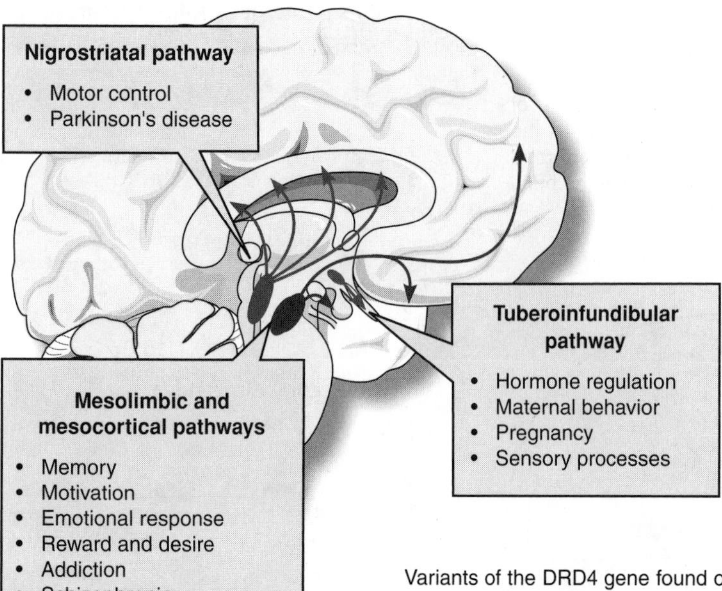

Effects of dopamine in the mesolimbic and mesocortical pathways

Level	Effect
Very low	Inability to focus. Associated with Attention Deficient Hyperactivity Disorder (ADHD).
Low	Addiction: the euphoric feeling caused by certain drugs is similar to the effects of dopamine and can leave the body wanting more.
Normal	Focus constant.
High	Hyperstimulation. Focus becomes narrowed to highly specific objects. Increase in perception of senses.
Very high	Paranoia and hallucination. Extreme situations can lead to schizophrenia.

Variants of the DRD4 gene found of chromosome 11 have been much studied. A section of the gene known as exon III has a repeating section of 48 base pairs. This section is typically repeated between 2 and 10 times. The number of repeats is divided into short repeats (from 2-5 repeats) and long repeats (from 6-10 repeats). Long repeats cause the receptor to be inefficient in binding with dopamine, leaving the body dopamine deficient.

Structure of DRD4 gene

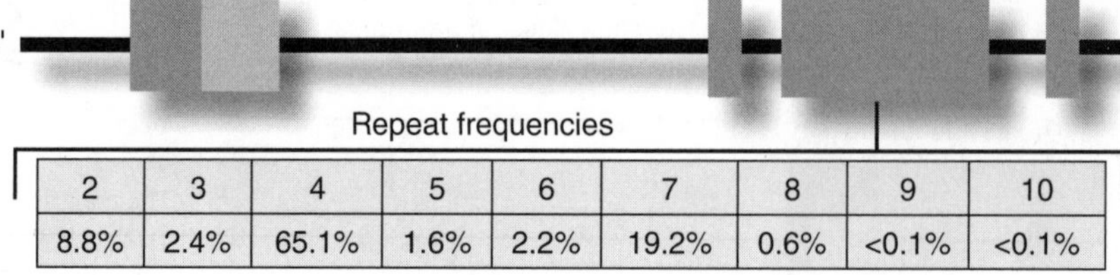

2	3	4	5	6	7	8	9	10
8.8%	2.4%	65.1%	1.6%	2.2%	19.2%	0.6%	<0.1%	<0.1%

The occurrence of long repeats has been linked with higher frequencies of ADHD and risk taking. "DRD4 knock-out" mice (mice without the DRD4 gene) show a deficiency in exploration of novel situations.

1. Explain why it is difficult to attribute behavior to a single gene: ______

2. Contrast the effects of various levels of dopamine in the brain on behavior. ______

3. Explain the link between the DRD4 gene and behavior: ______

4. Discuss how this link helps us understand other aspects of human behavior: ______

Neuron Structure and Function

As described earlier, homeostasis depends on the nervous system detecting, interpreting, and responding appropriately to both internal and external stimuli. Many of these responses are involuntary and are achieved through **reflexes**. Information, in the form of electrochemical impulses, is transmitted along nerve cells (**neurons**) from receptors to effectors. The speed of impulse conduction depends primarily on the axon diameter and whether or not the axon is **myelinated**. Within the tolerable physiological range, an increase in temperature also increases the speed of impulse conduction. In cool environments, impulses travel faster in endothermic than in ectothermic vertebrates. Neurons typically consist of a cell body, dendrites, and an axon (below). The principle behind increasing impulse speed through **saltatory conduction** is described overleaf.

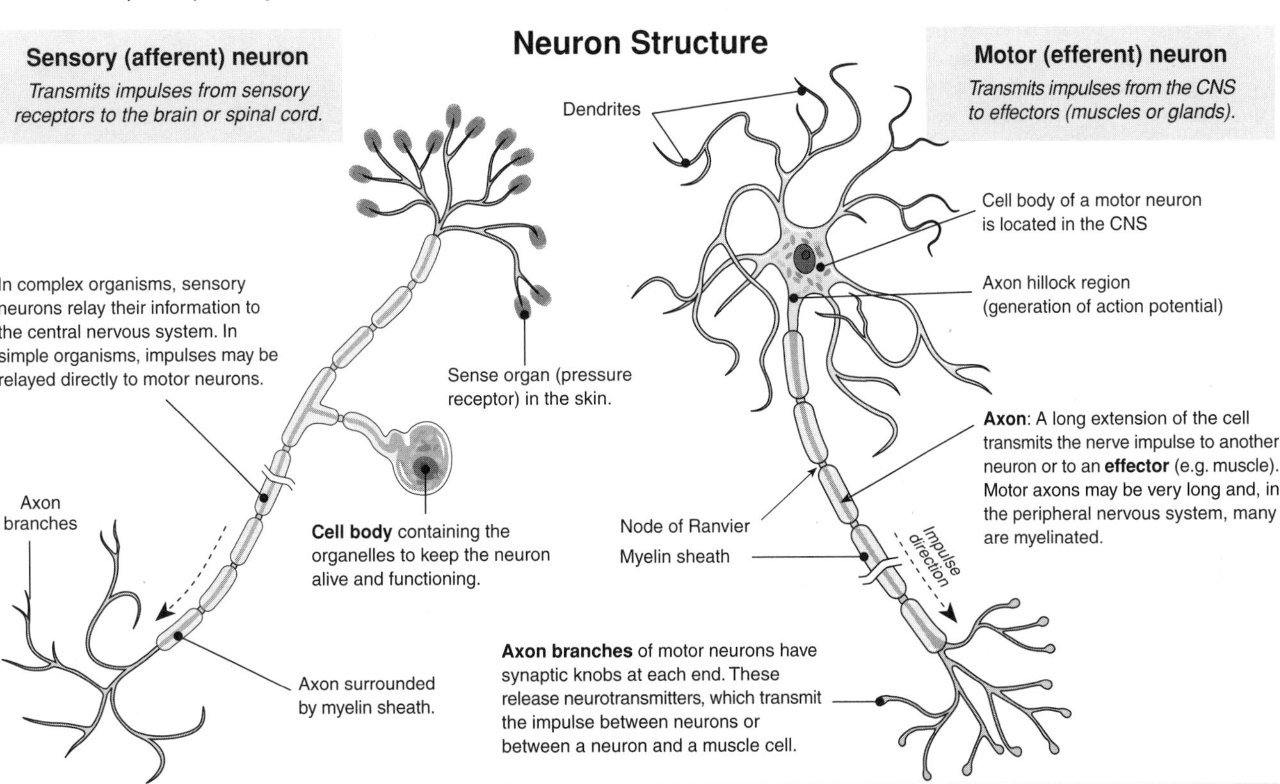
Neuron Structure
Sensory (afferent) neuron
Transmits impulses from sensory receptors to the brain or spinal cord.
Motor (efferent) neuron
Transmits impulses from the CNS to effectors (muscles or glands).
Dendrites
Cell body of a motor neuron is located in the CNS
Axon hillock region (generation of action potential)
In complex organisms, sensory neurons relay their information to the central nervous system. In simple organisms, impulses may be relayed directly to motor neurons.
Sense organ (pressure receptor) in the skin.
Axon: A long extension of the cell transmits the nerve impulse to another neuron or to an effector (e.g. muscle). Motor axons may be very long and, in the peripheral nervous system, many are myelinated.
Axon branches
Cell body containing the organelles to keep the neuron alive and functioning.
Node of Ranvier
Myelin sheath
Impulse direction
Axon surrounded by myelin sheath.
Axon branches of motor neurons have synaptic knobs at each end. These release neurotransmitters, which transmit the impulse between neurons or between a neuron and a muscle cell.

Where conduction speed is important, the axons of neurons are sheathed within a lipid and protein rich substance called **myelin**. Myelin is produced by **oligodendrocytes** in the central nervous system (CNS) and by **Schwann cells** in the peripheral nervous system (PNS). At intervals along the axons of myelinated neurons, there are gaps between neighbouring Schwann cells and their sheaths. These are called **nodes of Ranvier**. Myelin acts as an insulator, increasing the speed at which nerve impulses travel because it prevents ion flow across the neuron membrane and forces the current to "jump" along the axon from node to node.

Myelinated Neurons

Diameter: 1-25 μm

Conduction speed: 6-120 ms^{-1}

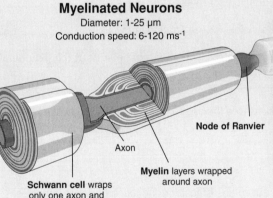

Node of Ranvier

Axon

Myelin layers wrapped around axon

Schwann cell wraps only one axon and produces myelin

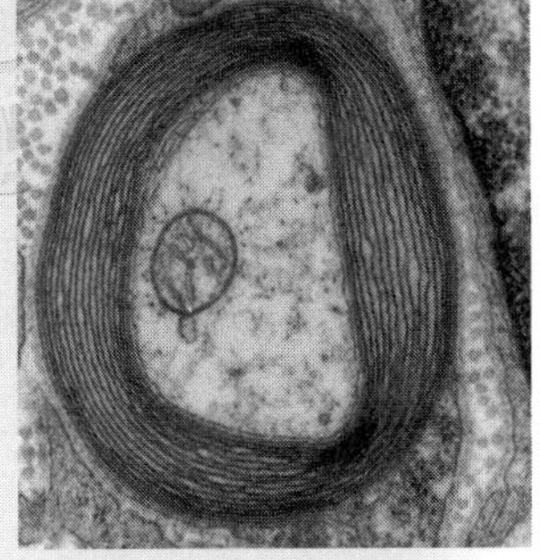
WIKI

TEM cross section through a myelinated axon

Non-myelinated axons are relatively more common in the CNS where the distances travelled are less than in the PNS. Here, the axons are encased within the cytoplasmic extensions of oligodendrocytes or Schwann cells, rather than within a myelin sheath. The speed of impulse conduction is slower than in myelinated neurons because the nerve impulse is propagated along the entire axon membrane, rather than jumping from node to node as occurs in myelinated neurons. Conduction speeds are slower than in myelinated neurons, although they are faster in larger neurons (there is less ion leakage from a larger diameter axon).

Non-myelinated Neurons

Diameter: <1 μm in vertebrates

Conduction speed: 0.2-0.5 ms^{-1}

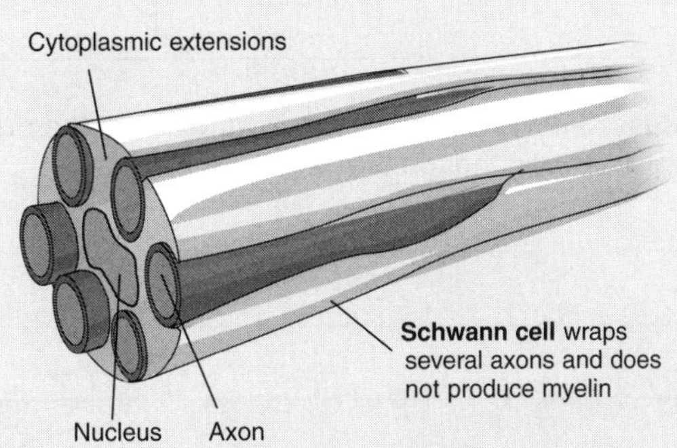
Cytoplasmic extensions
Schwann cell wraps several axons and does not produce myelin
Nucleus
Axon

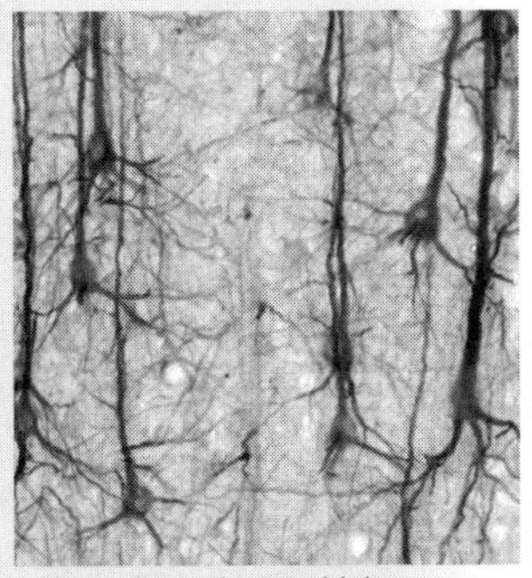
UC Regents David campus

Unmyelinated pyramidal neurons of the cerebral cortex

Axon myelination is a characteristic feature of vertebrate nervous systems and it enables them to achieve very rapid speeds of nerve conduction. Myelinated neurons conduct impulses by **saltatory conduction**, a term that describes how the impulse jumps along the fibre. In saltatory conduction, only the nodes of Ranvier are involved in action potential generation. In myelinated (insulated) regions, there is no leakage of ions across the neuron membrane and the action potential at one node is sufficient to trigger an action potential in the next node.

Apart from increasing the speed of the nerve impulse, the myelin sheath helps in reducing energy expenditure because the area of depolarization is decreased (and therefore also the number of sodium and potassium ions that need to be pumped to restore the resting potential).

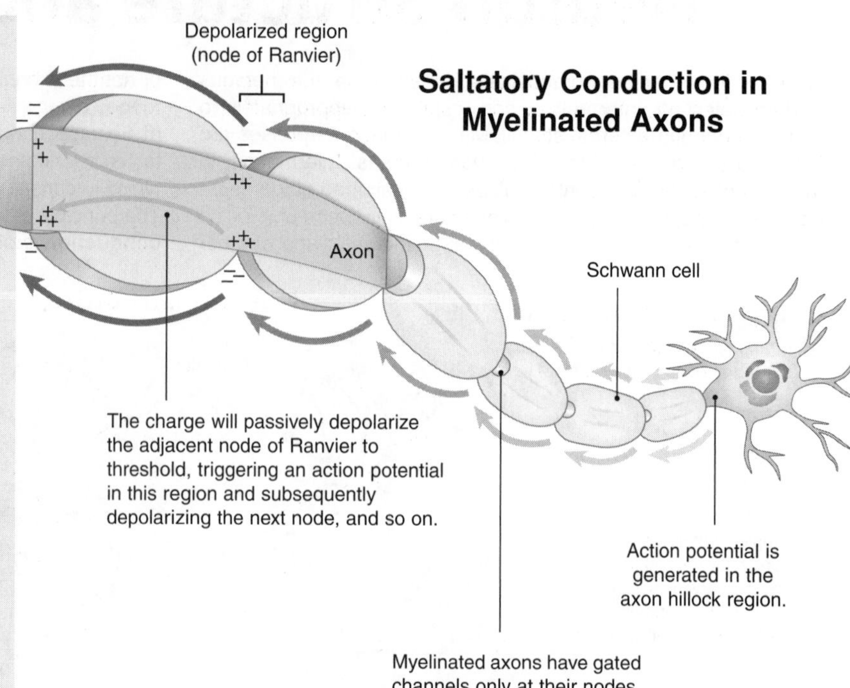

Saltatory Conduction in Myelinated Axons
Depolarized region (node of Ranvier)
Axon
Schwann cell
The charge will passively depolarize the adjacent node of Ranvier to threshold, triggering an action potential in this region and subsequently depolarizing the next node, and so on.
Action potential is generated in the axon hillock region.
Myelinated axons have gated channels only at their nodes.

1. Complete the missing panels of the following table summarising structural and functional differences between neurons:

	Sensory neuron	Interneuron	Motor neuron
Length of fibers		Short dendrites, long or short axon	
Location	Dendrites outside the spinal cord, cell body in spinal ganglion	Entirely within the CNS	Dendrites and cell body in the spinal cord; axon outside the spinal cord.
Function		Connect sensory & motor neurons	

2. (a) Describe the function of myelination in neurons: ____________

   (b) Name the cell type responsible for myelination in the CNS: ____________

   (c) Name the cell type responsible for myelination in the PNS: ____________

   (d) Explain why myelination is a typically a feature of neurons in the peripheral nervous system: ____________

3. Explain how myelination increases the speed of nerve impulse conduction: ____________

4. (a) Describe the adaptive advantage of faster conduction of nerve impulses: ____________

   (b) Explain why increasing the axon diameter also increases the speed of impulse conduction: ____________

5. Multiple sclerosis (MS) is a disease involving progressive destruction of the myelin sheaths around axons. Explain why MS impairs nervous system function even though the axons are still intact:

Reflexes

A reflex is an automatic response to a stimulus involving a small number of neurons and a central nervous system (CNS) processing point (usually the spinal cord, but sometimes the brain stem). This type of circuit is called a **reflex arc**. Reflexes permit rapid responses to stimuli. They are classified according to the number of CNS synapses involved; **monosynaptic reflexes** involve only one CNS synapse (e.g. knee jerk reflex), **polysynaptic reflexes** involve two or more (e.g. pain withdrawal reflex). Both are spinal reflexes. The pupil reflex (opening and closure of the pupil) is an example of a cranial reflex.

Pain Withdrawal: A Polysynaptic Reflex Arc

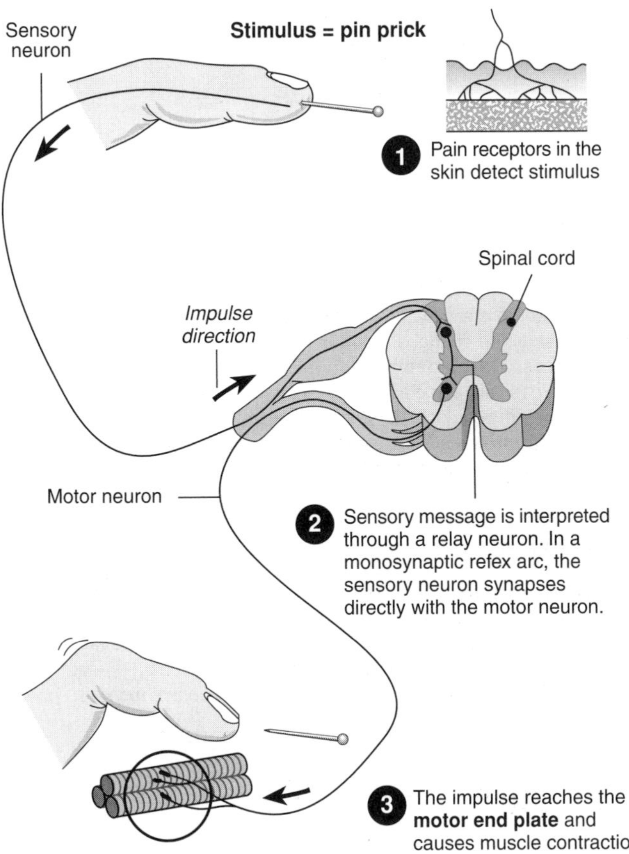
Sensory neuron
Stimulus = pin prick
1 Pain receptors in the skin detect stimulus
Spinal cord
Impulse direction
Motor neuron
2 Sensory message is interpreted through a relay neuron. In a monosynaptic refex arc, the sensory neuron synapses directly with the motor neuron.
3 The impulse reaches the **motor end plate** and causes muscle contraction.
Response = withdraw finger

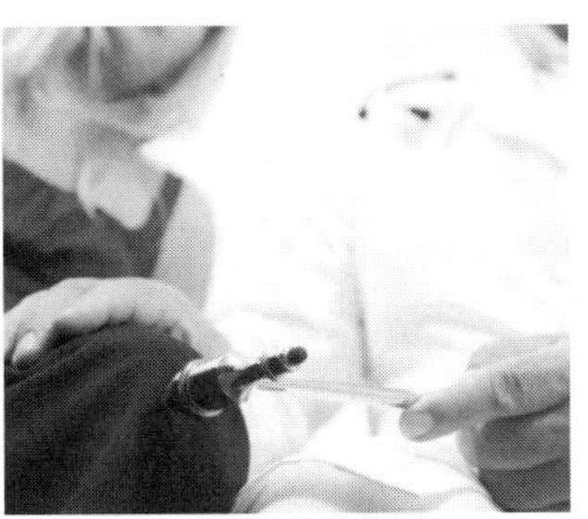

The patella (knee jerk) reflex is a simple deep tendon reflex that is used to test the function of the femoral nerve and spinal cord segments L2-L4. It helps to maintain posture and balance when walking.

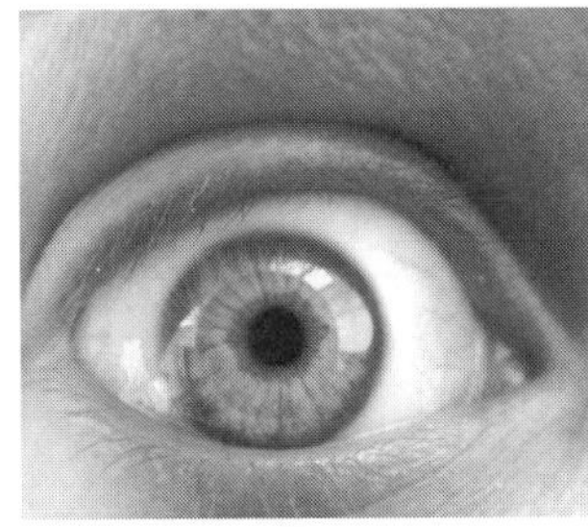

The pupillary light reflex refers to the rapid expansion or contraction of the pupils in response to the intensity of light falling on the retina. It is a polysynaptic cranial reflex and can be used to test for brain death.

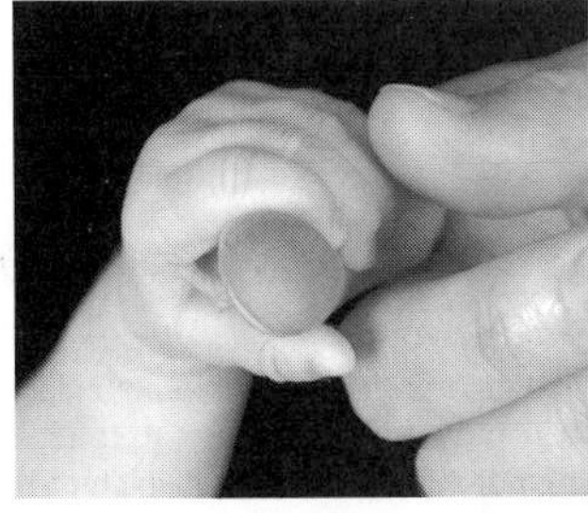

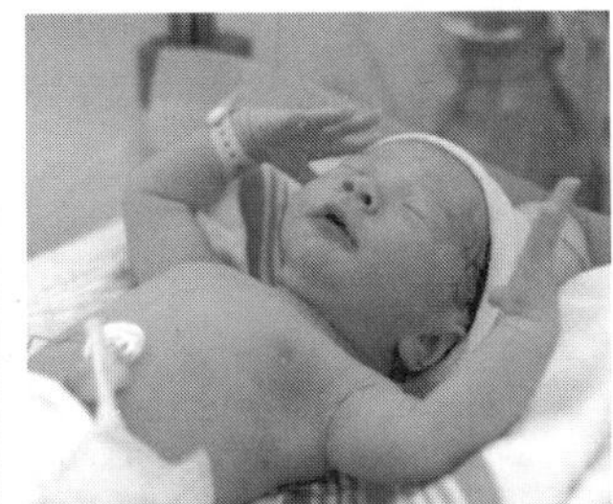

Normal newborns exhibit a number of primitive reflexes in response to particular stimuli. These reflexes disappear within a few months of birth as the child develops. Primitive reflexes include the grasp reflex (above left) and the startle or Moro reflex (right) in which a sudden noise will cause the infant to throw out its arms, extend the legs and head, and cry. The rooting and sucking reflexes are other examples of primitive reflexes.

1. Explain why higher reasoning or conscious thought are not necessary or desirable features of reflex behaviors: ________

2. Distinguish between a spinal reflex and a cranial reflex and give an example of each: ________

3. (a) Distinguish between a monosynaptic and a polysynaptic reflex arc and give an example of each: ________

   (b) Given similar length sensory and motor pathways, identify which would produce the most rapid response and why: ________

4. (a) With reference to specific examples, describe the adaptive value of primitive reflexes in newborns: ________

   (b) Explain why newborns are tested for the presence of these reflexes: ________

Related activities: *The Mammalian Nervous System*
Web links: *Parasympathetic Eye Responses, Knee Jerk Reflex*

Transmission of Nerve Impulses

Neurons, like all cells, contain ions or charged atoms. Those of special importance include sodium (Na^+), potassium (K^+), and negatively charged proteins. Neurons are **electrically excitable** cells: a property that results from the separation of ion charge either side of the neuron membrane. They may exist in either a resting or stimulated state. When stimulated, neurons produce electrical impulses that are transmitted along the axon. These impulses are transmitted between neurons across junctions called **synapses**. Synapses enable the transmission of impulses rapidly all around the body.

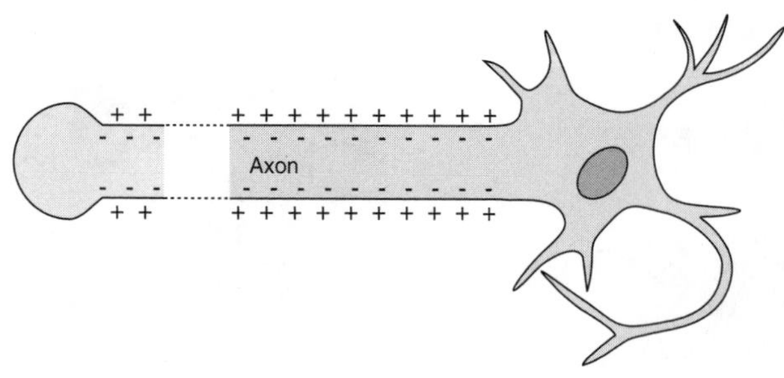

Impulse travels in this direction

Na^+ Na^+

Area of impulse

Next area to be stimulated

Area returning to resting state

The Resting Neuron

When a neuron is not transmitting an impulse, the inside of the cell is negatively charged compared with the outside of the cell. The cell is said to be electrically polarized, because the inside and the outside of the cell are oppositely charged. The potential difference (voltage) across the membrane is called the resting potential and for most nerve cells is about -70 mV. Nerve transmission is possible because this membrane potential exists.

The Nerve Impulse

When a neuron is stimulated, the distribution of charges on each side of the membrane changes. For a millisecond, the charges reverse. This process, called **depolarization**, causes a burst of electrical activity to pass along the axon of the neuron. As the charge reversal reaches one region, local currents depolarize the next region. In this way the impulse spreads along the axon. An impulse that spreads this way is called an **action potential**.

The Action Potential

The depolarization described above can be illustrated as a change in membrane potential (in millivolts). In order for an action potential to be generated, the stimulation must be strong enough to reach the **threshold** potential; this is the potential (voltage) at which the depolarization of the membrane becomes "unstoppable" and the action potential is generated. The action potential is **all or none** in its generation. Either the **threshold** is reached and the action potential is generated or the nerve does not fire. The resting potential is restored by the movement of potassium ions (K^+) out of the cell. During this **refractory period**, the nerve cannot respond.

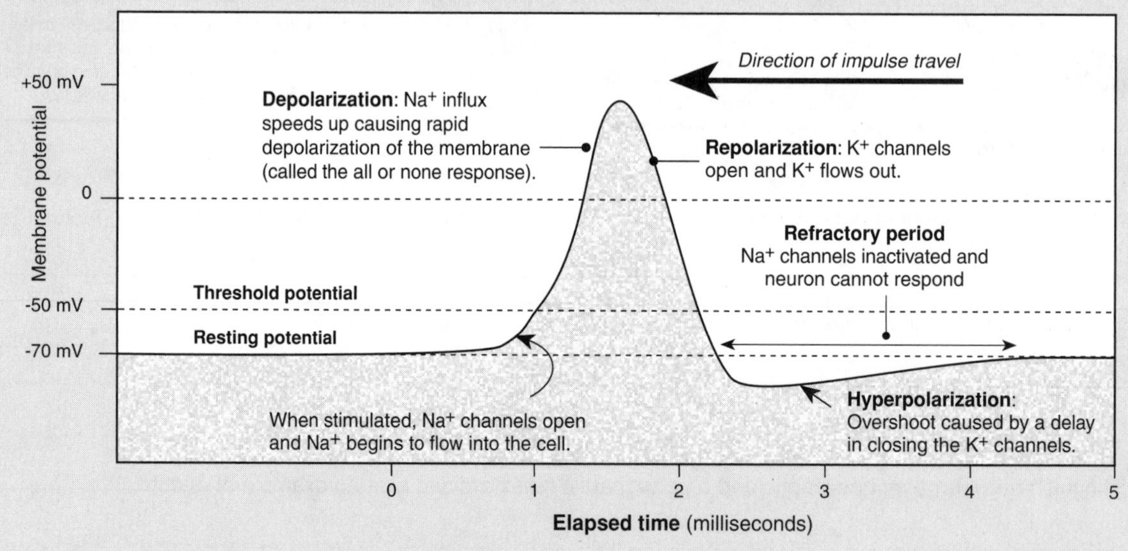

1. Explain how an action potential is able to pass along a nerve: ______________________

____

2. Explain how the refractory period influences the direction in which an impulse will travel: ______________________

____

3. Action potentials themselves are indistinguishable from each other. Explain how the nervous system is able to interpret the impulses correctly and bring about an appropriate response:

____

____

Chemical Synapses

Action potentials are transmitted between neurons across synapses: junctions between the end of one axon and the dendrite or cell body of a receiving neuron. **Chemical synapses** are the most widespread type of synapse in nervous systems. The axon terminal is a swollen knob, and a small gap separates it from the receiving neuron. The synaptic knobs are filled with tiny packets of chemicals called **neurotransmitters**. Transmission involves the diffusion of the neurotransmitter across the gap, where it interacts with the receiving membrane and causes an electrical response. The response of a receiving cell to the arrival of a neurotransmitter depends on the nature of the cell itself, on its location in the nervous system, and on the neurotransmitter involved. Synapses that release acetylcholine (ACh) are termed **cholinergic**. In the example below, ACh causes membrane depolarization and the generation of an action potential (termed excitation or an excitatory response).

A Cholinergic Synapse

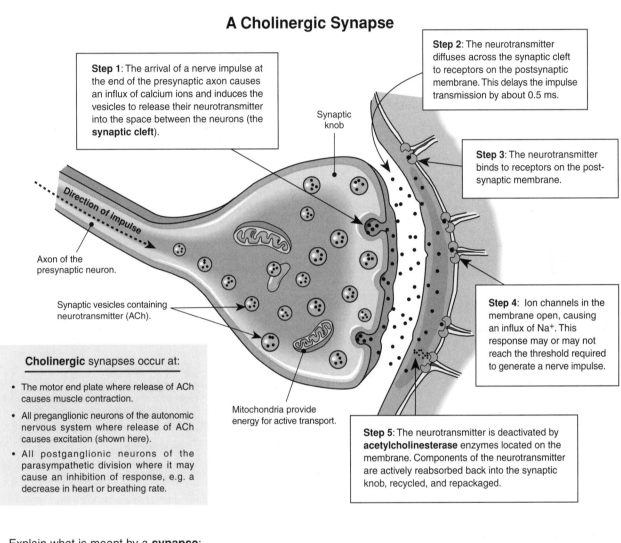

Step 1: The arrival of a nerve impulse at the end of the presynaptic axon causes an influx of calcium ions and induces the vesicles to release their neurotransmitter into the space between the neurons (the **synaptic cleft**).

Step 2: The neurotransmitter diffuses across the synaptic cleft to receptors on the postsynaptic membrane. This delays the impulse transmission by about 0.5 ms.

Step 3: The neurotransmitter binds to receptors on the post-synaptic membrane.

Step 4: Ion channels in the membrane open, causing an influx of Na^+. This response may or may not reach the threshold required to generate a nerve impulse.

Step 5: The neurotransmitter is deactivated by **acetylcholinesterase** enzymes located on the membrane. Components of the neurotransmitter are actively reabsorbed back into the synaptic knob, recycled, and repackaged.

Synaptic knob

Direction of impulse

Axon of the presynaptic neuron.

Synaptic vesicles containing neurotransmitter (ACh).

Mitochondria provide energy for active transport.

Cholinergic synapses occur at:

- The motor end plate where release of ACh causes muscle contraction.
- All preganglionic neurons of the autonomic nervous system where release of ACh causes excitation (shown here).
- All postganglionic neurons of the parasympathetic division where it may cause an inhibition of response, e.g. a decrease in heart or breathing rate.

1. Explain what is meant by a **synapse**: ______________________

2. Explain what causes the release of neurotransmitter into the synaptic cleft: ______________________

____

3. State why there is a brief delay in transmission of an impulse across the synapse: ______________________

____

4. (a) State how the neurotransmitter is deactivated: ______________________

   (b) Explain why it is important for the neurotransmitter substance to be deactivated soon after its release:

____

5. Consult a reference source to identify one function of acetylcholine in the nervous system: ______________________

____

6. Suggest one factor that might influence the strength of the response in the receiving cell: ______________________

____

Integration at Synapses

Synapses play a pivotal role in the ability of the nervous system to respond appropriately to stimulation and to adapt to change. The nature of synaptic transmission allows the **integration** (interpretation and coordination) of inputs from many sources. These inputs need not be just excitatory (causing depolarization). Inhibition results when the neurotransmitter released causes negative chloride ions (rather than sodium ions) to enter the postsynaptic neuron. The postsynaptic neuron then becomes more negative inside (hyperpolarized) and an action potential is less likely to be generated. At synapses, it is the sum of **all** inputs (excitatory and inhibitory) that leads to the final response in a postsynaptic cell. Integration at synapses makes possible the various responses we have to stimuli. It is also the most probable mechanism by which learning and memory are achieved.

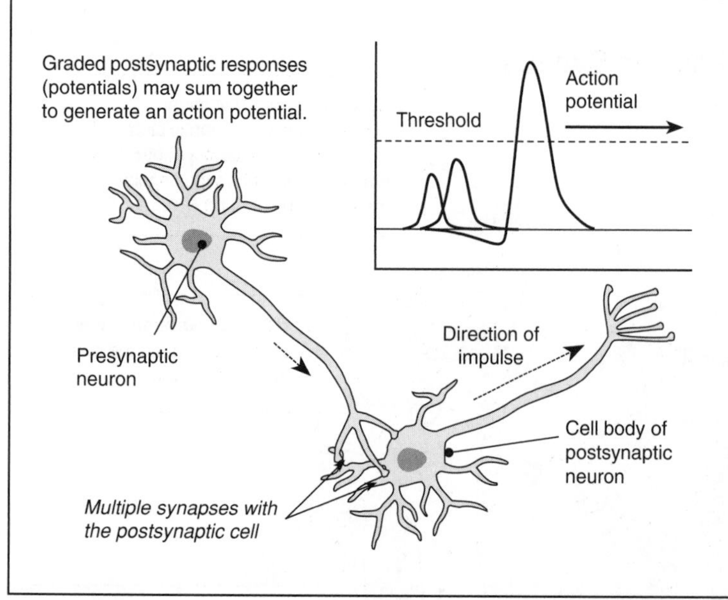

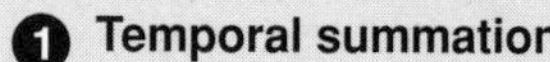

Synapses and Summation

Nerve transmission across chemical synapses has several advantages, despite the delay caused by neurotransmitter diffusion. Chemical synapses transmit impulses in one direction to a precise location and, because they rely on a limited supply of neurotransmitter, they are subject to fatigue (inability to respond to repeated stimulation). This protects the system against overstimulation.

Synapses also act as centers for the **integration** of inputs from many sources. The response of a postsynaptic cell is often graded; it is not strong enough on its own to generate an action potential. However, because the strength of the response is related to the amount of neurotransmitter released, subthreshold responses can sum to produce a response in the post-synaptic cell. This additive effect is termed **summation**. Summation can be **temporal** or **spatial** (below). A neuromuscular junction (photo below) is a specialized form of synapse between a motor neuron and a skeletal muscle fiber. Functionally, it is similar to any excitatory cholinergic synapse.

1 Temporal summation | **2 Spatial summation**

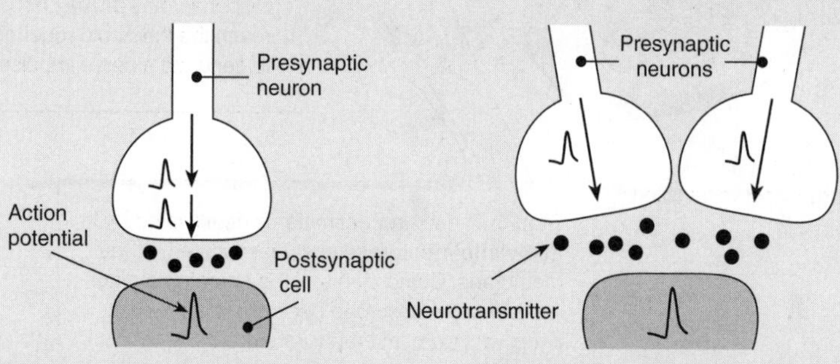

Several impulses may arrive at the synapse in quick succession from a single axon. The individual responses are so close together in time that they sum to reach threshold and produce an action potential in the postsynaptic neuron.

Individual impulses from spatially separated axon terminals may arrive **simultaneously** at different regions of the same postsynaptic neuron. The responses from the different places sum to reach threshold and produce an action potential.

3 Neuromuscular junction

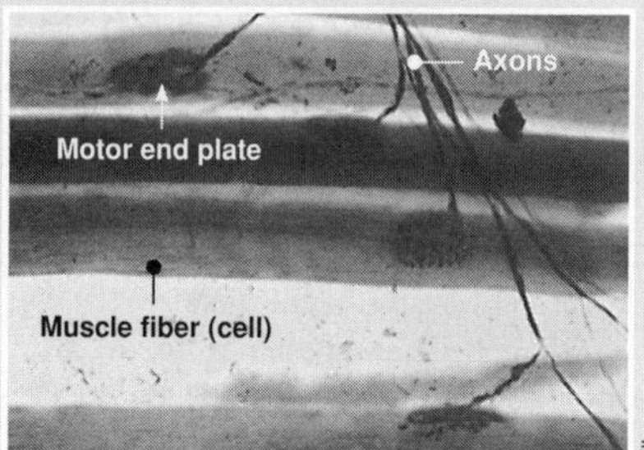

The arrival of an impulse at the neuromuscular junction causes the release of acetylcholine from the synaptic knobs. This causes the muscle cell membrane (sarcolemma) to depolarize, and an action potential is generated in the muscle cell.

1. Explain the purpose of nervous system integration: __________

2. (a) Explain what is meant by **summation**: __________

(b) In simple terms, distinguish between temporal and spatial summation: __________

3. Describe two ways in which a neuromuscular junction is similar to any excitatory cholinergic synapse:

(a) __________

(b) __________

The Basis of Sensory Perception

Sensory receptors are specialized to detect stimuli and respond by producing an electrical discharge. In this way they act as **biological transducers**, converting the energy from a stimulus into an electrochemical signal. Stimulation of a sensory receptor cell results in an electrical impulse with specific properties. The frequency of impulses produced by the receptor cell encodes information about the strength of the stimulus; a stronger stimulus produces more frequent impulses. Sensory receptors also show **sensory adaptation** and will cease responding to a stimulus of the same intensity. The simplest sensory receptors consist of a single sensory neuron (e.g. free nerve endings). More complex sense cells form synapses with their sensory neurons (e.g. taste buds). Sensory receptors are classified according to the stimuli to which they respond (for example, photoreceptors respond to light). The response of a simple **mechanoreceptor**, the Pacinian corpuscle, to a stimulus (pressure) is described below.

The Pacinian Corpuscle

Pacinian corpuscles are pressure receptors occurring in deep subcutaneous tissues all over the body. They are relatively large and simple in structure, consisting of a sensory nerve ending (dendrite) surrounded by a capsule of layered connective tissue. Pressure deforms the capsule, stretching the nerve ending and leading to a localized depolarization. Once a **threshold** value is reached, an **action potential** propagates along the axon.

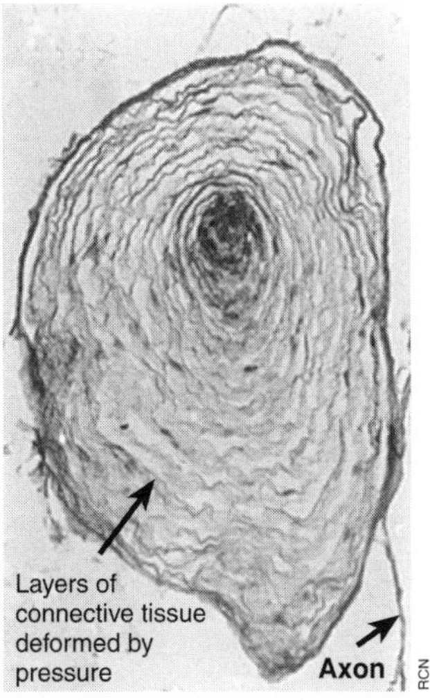

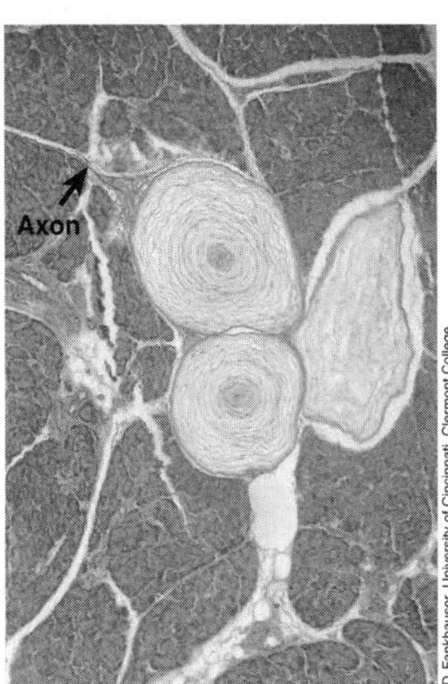

Pacinian corpuscle (above, left), illustrating the distinctive layers of connective tissue. The photograph on the right shows corpuscles grouped together in the pancreas. Pacinian corpuscles are rapidly adapting receptors; they fire at the beginning and end of a stimulus, but do not respond to unchanging pressure.

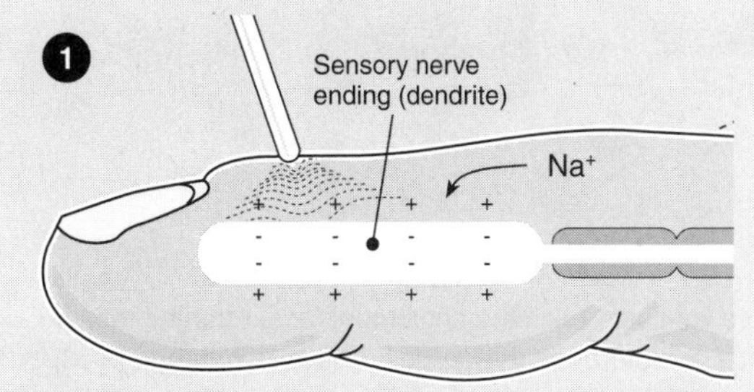

Deforming the corpuscle leads to an increase in the permeability of the nerve to sodium. Na^+ diffuses into the nerve ending creating a localized depolarization. This depolarization is called a **generator potential**.

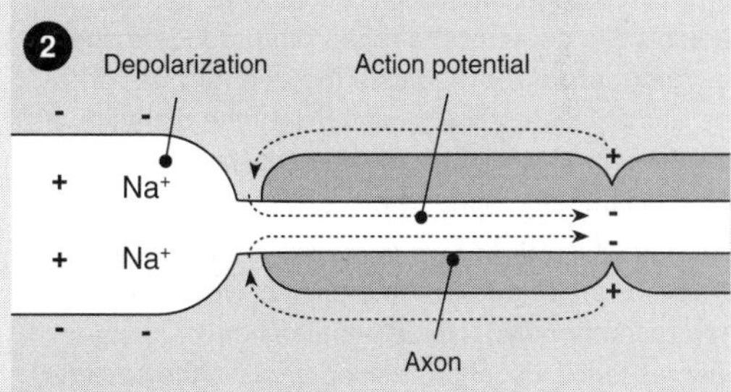

A volley of **action potentials** is triggered once the generator potential reaches or exceeds a **threshold value**. These action potentials are conducted along the sensory axon. A strong stimulus results in a high frequency of impulses.

1. Explain why sensory receptors are termed 'biological transducers': __________

2. Explain the significance of linking the magnitude of a sensory response to stimulus intensity: __________

3. Explain the physiological importance of sensory adaptation: __________

4. (a) Describe the properties of a generator potential: __________

   (b) Suggest why a simple mechanoreceptor, such as the Pacinian corpuscle, does not fire action potentials unless a stimulus of threshold value is reached:

Periodicals: *Infinite sensation, What is a Pacinian corpuscle?*

Related activities: *The Physiology of Vision* ***Web links:*** *Hair Cell Transduction Neuron Information Coding & Transfer*

The Structure of the Human Eye

The eye is a complex and highly sophisticated sense organ specialized to detect light. The adult eyeball is about 25 mm in diameter. Only the anterior one-sixth of its total surface area is exposed; the rest lies recessed and protected by the **orbit** into which it fits. The eyeball is protected and given shape by a fibrous tunic. The posterior part of this structure is the **sclera** (the white of the eye), while the anterior transparent portion is the **cornea**, which covers the colored iris.

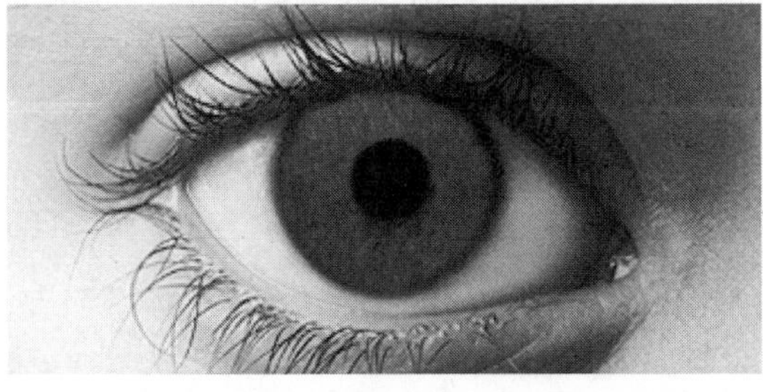

The Structure and Function of the Mammalian Eye

The human eye is essentially a three layered structure comprising an outer fibrous layer (the sclera and cornea), a middle vascular layer (the choroid, ciliary body, and iris), and inner **retina** (neurones and **photoreceptor cells**). The shape of the eye is maintained by the fluid filled cavities (aqueous and vitreous humours), which also assist in light refraction. Eye colour is provided by the pigmented iris. The iris also regulates the entry of light into the eye through the contraction of circular and radial muscles.

Forming a Visual Image

Before light can reach the photoreceptor cells of the retina, it must pass through the cornea, aqueous humour, pupil, lens, and vitreous humour. For vision to occur, light reaching the photoreceptor cells must form an image on the retina. This requires **refraction** of the incoming light, **accommodation** of the lens, and **constriction** of the pupil.

The anterior of the eye is concerned mainly with **refracting** (bending) the incoming light rays so that they focus on the retina (below left). Most refraction occurs at the cornea. The lens adjusts the degree of refraction to produce a sharp image. **Accommodation** adjusts the eye for near or far objects. Constriction of the pupil narrows the diameter of the hole through which light enters the eye, preventing light rays entering from the periphery.

The point at which the nerve fibres leave the eye as the optic nerve, is the **blind spot** (the point at which there are no photoreceptor cells). Nerve impulses travel along the optic nerves to the visual processing areas in the cerebral cortex. Images on the retina are inverted and reversed by the lens but the brain interprets the information it receives to correct for this image reversal.

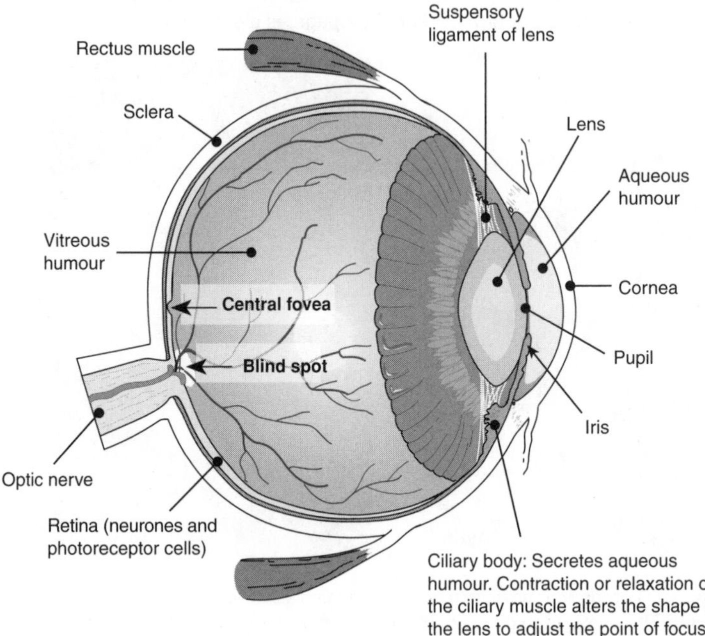

1. Identify the function of each of the structures of the eye listed below:

   (a) Cornea: ______

   (b) Ciliary body: ______

   (c) Iris: ______

2. (a) The first stage of vision involves forming an image on the retina. Explain what this involves: ______

   (b) Explain how accommodation is achieved: ______

***Related activities:** The Physiology of Vision*
***Web links:** Anatomy of the Eye, Label the Eye*

Periodicals:
Remodelling the eye

The Physiology of Vision

Vision involves essentially two stages: formation of the image on the retina (see previous activity), and generation and conduction of nerve impulses. When light reaches the retina, it is absorbed by the photosensitive pigments associated with the membranes of the photoreceptor cells (the rods and cones). The pigment molecules are altered by the absorption of light in such a way as to lead to the generation of nerve impulses. It is these impulses that are conducted via nerve fibres to the visual processing centre of the cerebral cortex (see the activity *The Human Brain* for the location of this region).

The Structure and Arrangement of Photoreceptor Cells in the Retina

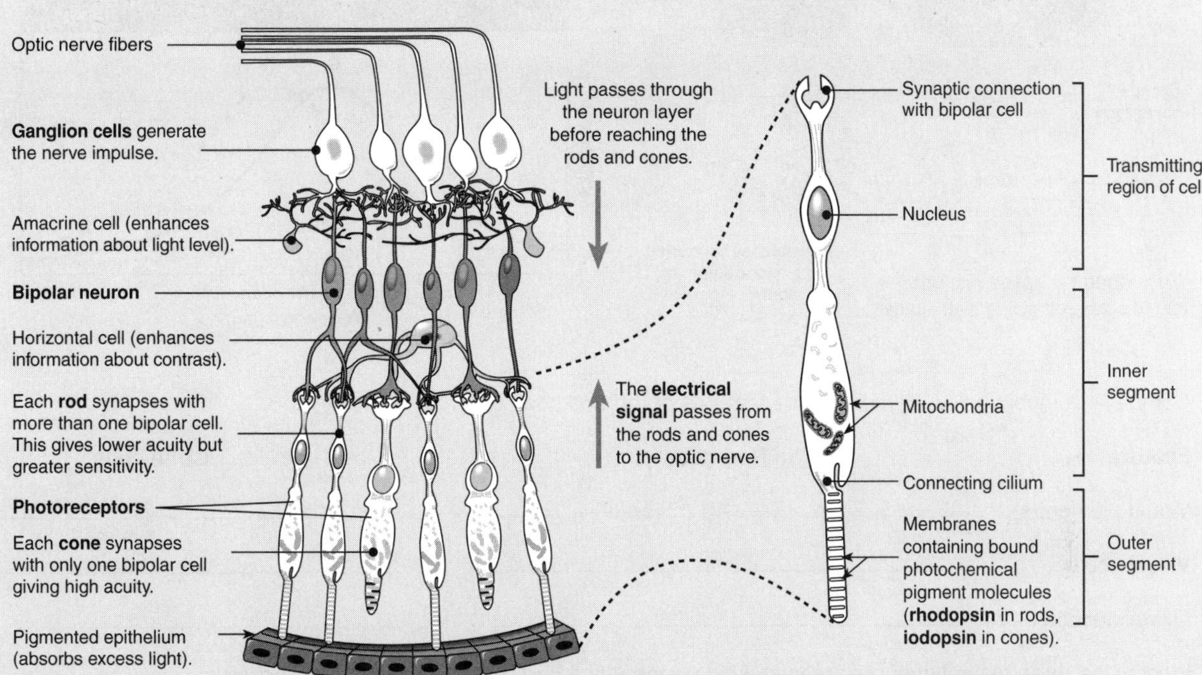

Arrangement of photoreceptors and neurons in the retina

Structure of a rod photoreceptor cell

The photoreceptor cells of the mammalian retina are the **rods** and **cones**. Rods are specialized for vision in dim light, whereas cones are specialized for color vision and high visual acuity. Cone density and visual acuity are greatest in the **central fovea** (rods are absent here). After an image is formed on the retina, light impulses must be converted into nerve impulses. The first step is the development of **generator potentials** by the rods and cones. Light induces structural changes in the **photochemical pigments** (or photopigments) of the rod and cone membranes. The generator potential that develops from the pigment breakdown in the rods and cones is different from the generator potentials that occur in other types of sensory receptors because stimulation results in a **hyperpolarization** rather than a depolarization (in other words, there is a net loss of Na^+ from the photoreceptor cell). Once generator potentials have developed, the graded changes in membrane conductance spread through the photoreceptor cell. Each photoreceptor makes synaptic connection with a bipolar neuron, which transmits the potentials to the **ganglion cells**. The ganglion cells become **depolarized** and initiate nerve impulses which pass through the optic chiasma and eventually to the visual areas of the cerebral cortex. The frequency and pattern of impulses in the optic nerve conveys information about the changing visual field.

1. Describe the structure and the function of each of the structures listed below:

   (a) Retina: ______

   (b) Optic nerve: ______

2. Contrast the structure of the blind spot and the central fovea: ______

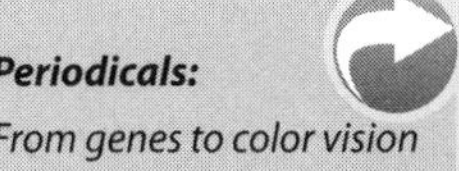

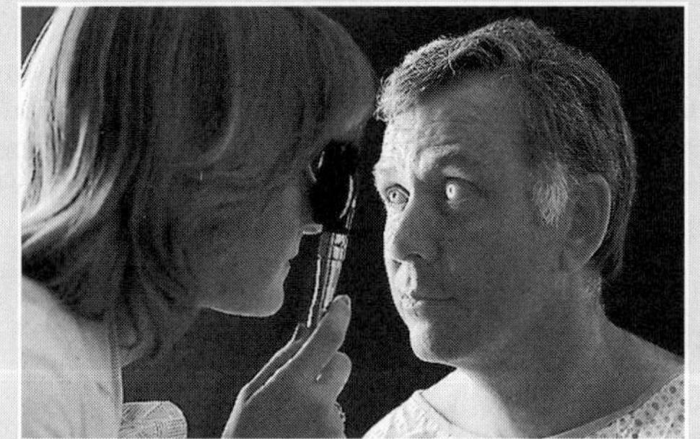

The Basis of Trichromatic Vision

There are three classes of **cones**, each with a maximal response in either short (blue), intermediate (green) or long (yellow-green) wavelength light (coded B, G, and R below). The yellow-green cone is also sensitive to the red part of the spectrum and is often called the red cone (R). The differential responses of the cones to light of different wavelengths provides the basis of trichromatic color vision.

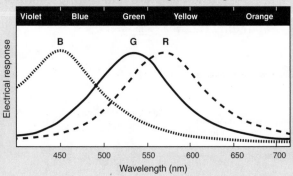

Synaptic connection
Nucleus
Mitochondrion
Membranes containing bound **iodopsin** pigment molecules.

Each **cone** synapses with only one bipolar cell giving high acuity.

Cone response to light wavelengths

Violet
Blue
Green
Yellow
Orange
B
G
R
Electrical response
450
500
550
600
650
700
Wavelength (nm)

3. Complete the table below, comparing the features of rod and cone cells:

Feature	Rod cells	Cone cells
Visual pigment(s):		
Visual acuity:		
Overall function:		

4. Identify the three major types of neuron making up the retina and describe their basic function:

   (a)

   (b)

   (c)

5. Identify two types of accessory neurons in the retina and describe their basic function:

   (a)

   (b)

6. Account for the differences in acuity and sensitivity between rod and cone cells:

7. (a) Explain clearly what is meant by the term photochemical pigment (photopigment):

   (b) Identify two photopigments and their location:

8. In your own words, explain how light is able to produce a nerve impulse in the ganglion cells:

Adaptations for Vision

Humans rely on light more than any other stimulus when gathering information about their surroundings. Our eyes detect only a small portion (400-700 nm) of the total electromagnetic spectrum. Within this band, we can detect and color code the information that we receive. Our complex visual perception of the world is a function of the amount of brain processing power devoted to the interpretation of visual information and it reflects the importance of sight in our lives. Other animals depend much less on vision, and the visual processing regions of their brains are less well developed. Although mammalian eyes are relatively uniform in their basic structure, there is considerable adaptation to niche. This is particularly evident in the eyes of nocturnal species. Some of the general patterns in visual adaptation evident in mammals are described below.

Adaptations for Vision in Mammals

Quite apart from the physiology of perceiving light, an animal's view of the world in influenced by its **field of view** and its **visual overlap**. The field of view refers to the area within which an object may be observed, whereas the visual overlap between what each eye sees provides stereoscopic vision. Some extent of color vision is present in most mammals, although it is poorly developed in nocturnal species. The position of an animal's eyes provides a good indication of whether it is prey or predator. Some examples of the differences between animals are illustrated below. The lighter gray indicates the full extent of the field of view, whereas the darker gray indicates the amount of visual overlap.

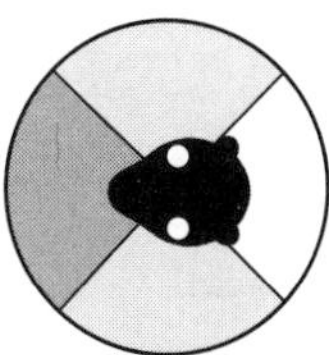

Prey animals

Location of eyes: Side of the head.
Field of view: Wide, some have a full 360°.
Amount of visual overlap: Small.
Adaptiveness: Habitat surveillance and predator detection.

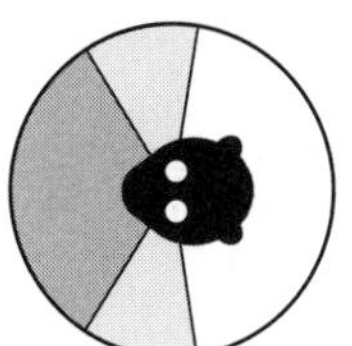

Predators

Location of eyes: At the front of the head.
Field of view: Limited.
Amount of visual overlap: Large.
Adaptiveness: Focus on prey and good distance judgment for prey capture.

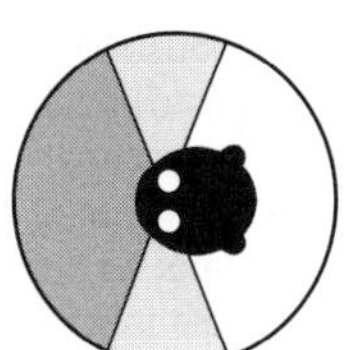

Humans and other primates

Location of eyes: At the front of the head.
Field of view: Limited.
Amount of visual overlap: Large.
Adaptiveness: Good depth perception and judgment of distance for arboreal lifestyle.

Stereoscopic vision

The excellent stereoscopic vision of **primates** is typical of animals that hunt or forage for specific foods. It provides good depth perception and judgment of distance. Many primates are diurnal and have developed excellent color vision, probably as an adaptation for the visual detection of ripe fruit.

Nocturnal vision

The eyes of nocturnal animals (e.g. the **bushbaby**) are highly specialized to capture low intensity light. The eyes are very large and may have a **tapetum** behind the retina, which reflects light back through the receptor cells. The eyes of nocturnal animals have few cones and they see mostly in monochrome.

Wide field of view

In contrast to predators, prey species, such as **chipmunks**, have their eyes positioned to the side of their heads. They need to scan as much of the habitat around them as possible in order to detect danger. Squirrels also need to manipulate their food so they have much more visual overlap than a grazing animal.

1. Describe two adaptations for nocturnal vision. For each, give an example and explain how the adaptation assists vision:

   (a) ______________________________

   (b) ______________________________

2. Explain why the density of rods is very high in nocturnal mammals (refer to the previous activity for help):

   ____

3. (a) Explain what is meant by the field of view: ______________________________

   (b) Contrast the fields of view for predator and prey mammals and suggest a reason for the difference:

   ____

4. (a) Suggest one advantage of color vision to a mammal: ______________________________

   (b) State whether a mammal with good color vision is likely to be nocturnal or diurnal: ______________________________

Periodicals: *Under color of darkness*

Related activities: *The Structure of the Eye, The Physiology of Vision*
Web links: *Night Creatures*

Invertebrate Vision

Amongst the invertebrates, there is a great diversity in the complexity of eyes found in different groups, from the very simple eyes (called ocelli) of bivalve mollusks, to the compound eyes of arthropods and the highly developed single-lens eyes of cephalopod mollusks. The structure and basic function of compound eyes in insects and single-lens eyes in cephalopods are described below. Note the similarity of structure and function between the cephalopod eye and the vertebrate eye in the previous activity. The compound eye of insects is very different, both in its structure and in the type of image it produces.

Cephalopod Single Lens Eye

Insect Compound Eye

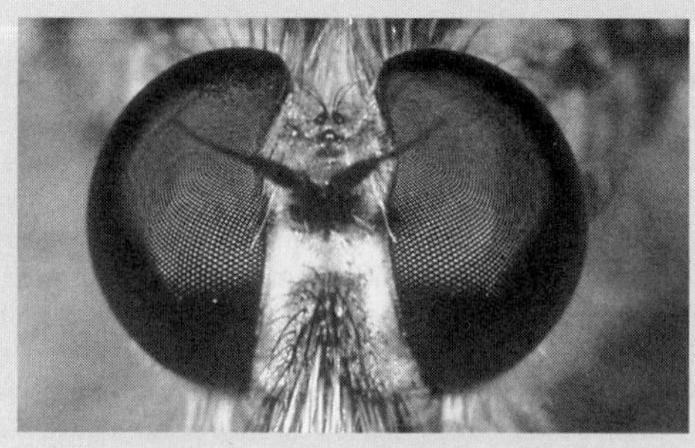

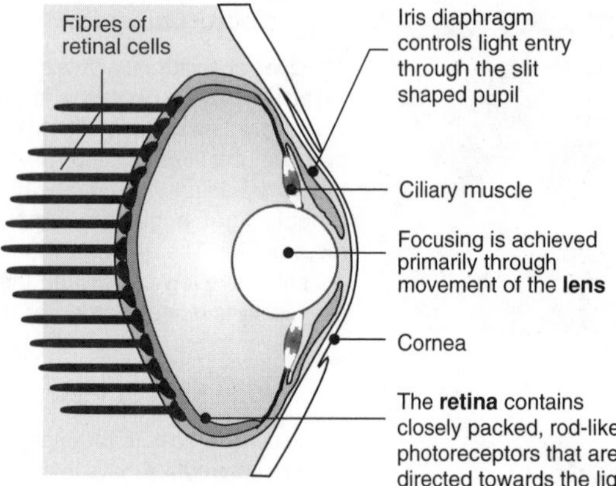
Fibres of retinal cells
Iris diaphragm controls light entry through the slit shaped pupil
Ciliary muscle
Focusing is achieved primarily through movement of the **lens**
Cornea
The **retina** contains closely packed, rod-like photoreceptors that are directed towards the light

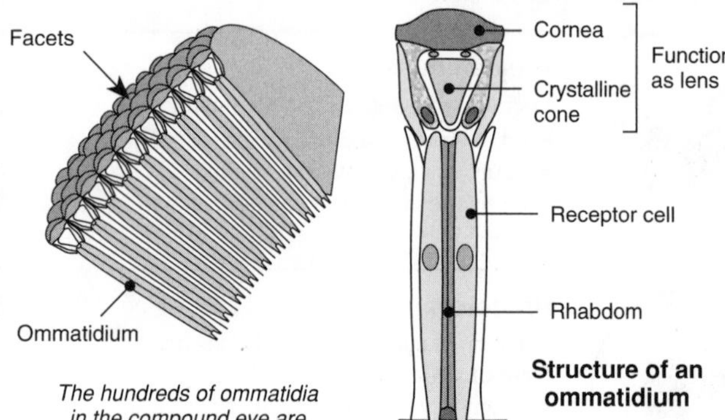
Facets
Ommatidium
Cornea
Crystalline cone
Function as lens
Receptor cell
Rhabdom

The hundreds of ommatidia in the compound eye are seen as an array of facets

Structure of an ommatidium

The eyes of cephalopod mollusks (squid, cuttlefish, and octopods) are well developed and strikingly similar in appearance to those of vertebrates, although their origins are quite different. The single lens is suspended by a ciliary muscle. The cornea contributes very little to focusing because the eye is surrounded by water and there is little refraction at its surface. The photoreceptors are connected to retinal cells that send fibres to an optic ganglion where processing occurs.

The compound eyes of many invertebrates consist of thousands of identical visual units called **ommatidia**. Each unit has its own lens structure, light sensitive retinal cell and nerve fiber. The cornea and crystalline cone function together as a lens that focuses light through the rhabdom, a stack of pigmented plates on the inside of a circle of receptor cells. The image produced is a mosaic of dots that has low acuity but is very sensitive to movement and pattern changes.

1. The eye of mammals and those of the octopus have evolved a similar structure quite independently of each other. Briefly describe three structural similarities:

   (a) ______

   (b) ______

   (c) ______

2. Light rays are bent (refracted) when they enter the eye. State which structure is primarily responsible for refraction in:

   (a) An octopus eye: ______ (b) A mammalian eye: ______

3. Explain why the image produced by a compound eye is highly sensitive to movement and changes in pattern:

4. Cephalopods are active hunters. Explain how the eye structure of cephalopods is well suited to predatory behavior:

The Components of Behavior

Behavior in animals can be attributed to two components: **innate behavior** that has a genetic basis, and **learned behavior**, which results from the experiences of the animal. Together they combine to produce the total behavior exhibited by the animal. Experience may also modify some innate behaviors. Animals behave in fixed, predictable ways in many situations. The innate behavior follows a classical pathway called a **fixed-action pattern** (FAP) where an innate behavioral program is activated by a stimulus or **releaser** to direct a certain behavioral response. Innate behaviors are generally adaptive and are performed for a variety of reasons. Learning, which involves the modification of behavior by experience, occurs in various ways.

Innate Behaviors

Reflex behavior
Simplest type of animal behavior. A sudden stimulus induces an automatic, involuntary and stereotyped response. Many reflexes are protective.

Kinesis
Random movement of an animal in which the rate of movement is related to the intensity of the stimulus, but not to its direction.

Taxis
A movement in response to the direction of a stimulus. Movement towards a stimulus are positive while those away from a stimulus are negative.

Stereotyped behavior
Occurs when the same response is given to the same stimulus on different occasions. This behavior shows fixed patterns of coordinated movements called fixed action patterns.

Learned Behaviors

Classical conditioning
Also called associative learning. Animals come to associate one stimulus with another.

Habituation
Response to a stimulus wanes when it is repeated with no apparent effect.

Insight behavior
Correct behavior on the first attempt where the animal has no prior experience.

Imprinting behavior
During a critical period, an animal can adopt a behavior by latching on to its first stimulus.

Operant conditioning
Also called trial and error learning, an animal is rewarded or punished after chance behavior.

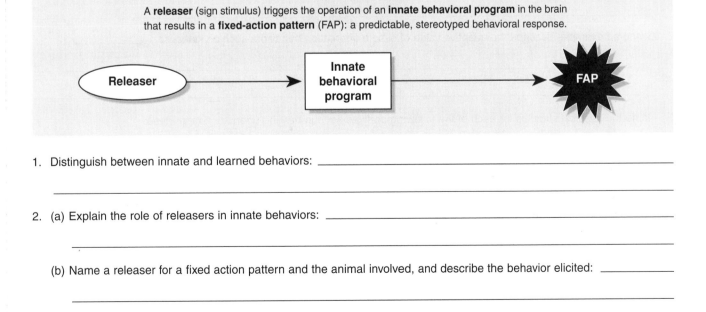

The complex behavior patterns exhibited by an animal

Fixed Action Pattern

A **releaser** (sign stimulus) triggers the operation of an **innate behavioral program** in the brain that results in a **fixed-action pattern** (FAP): a predictable, stereotyped behavioral response.

Releaser → Innate behavioral program → FAP

1. Distinguish between innate and learned behaviors: ______________________________

____

2. (a) Explain the role of releasers in innate behaviors: ______________________________

____

(b) Name a releaser for a fixed action pattern and the animal involved, and describe the behavior elicited: ______________

____

____

Related activities: *Simple Behaviors, Learned Behavior*

Simple Behaviors

Taxes and kineses are adaptive locomotory behaviors and involve movements in response to external stimuli such as gravity, light, chemicals, and temperature. A **kinesis** (pl. kineses) is a non-directional response to a stimulus in which the speed of movement or the rate of turning is proportional to the stimulus intensity. Kineses do not involve orientation directly to the stimulus and are typical of many invertebrates and protozoa. In contrast, **taxes** involve orientation and movement in response to a directional stimulus or a gradient in stimulus intensity. Taxes often involve moving the head (bearing sensory receptors) from side to side until the sensory input from both sides is equal (a **klinotaxic** response). Many taxic responses are complicated by a simultaneous response to more than one stimulus. For example, fish orientate dorsal side up in response to both light and gravity. Male moths orientate positively to **pheromones** but use the wind to judge the direction of the odor source (the female moth).

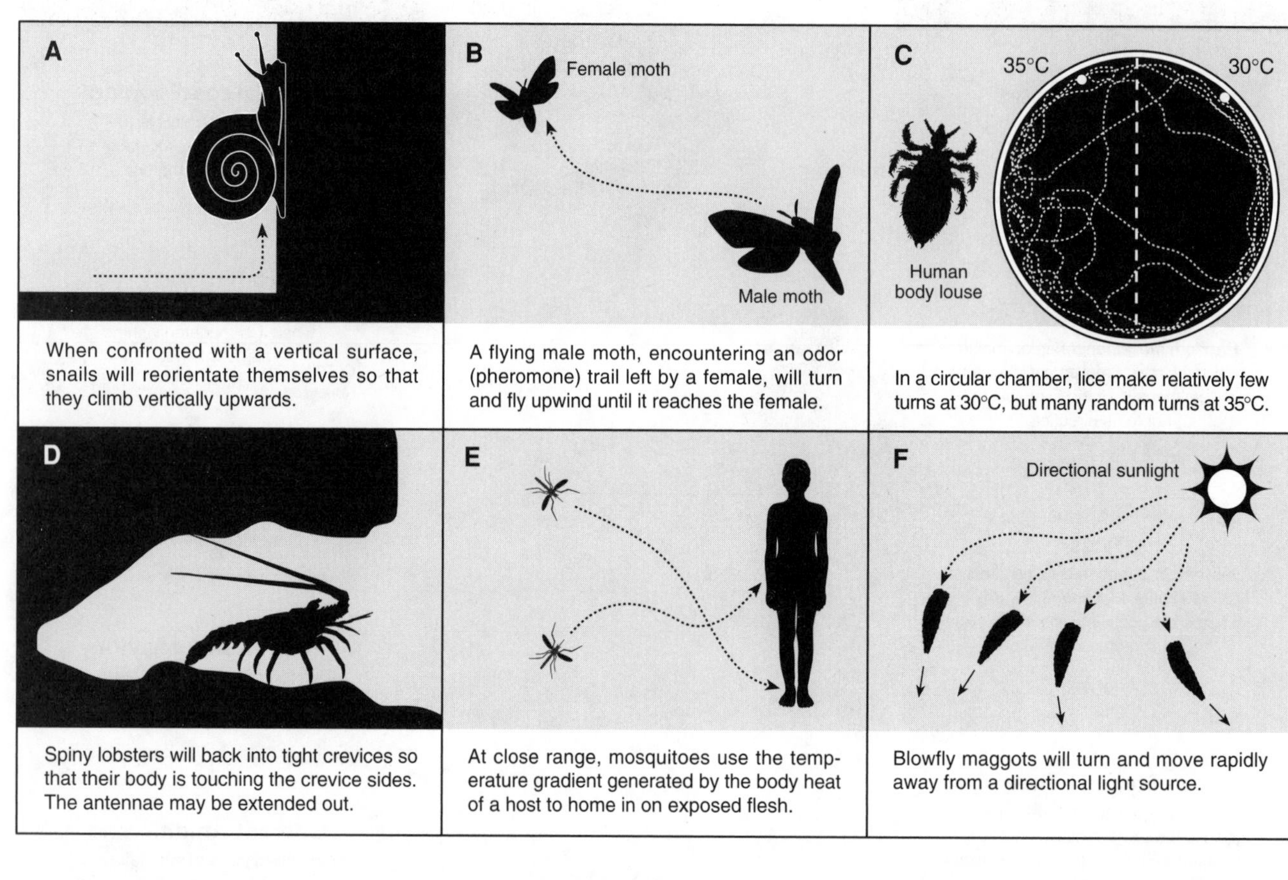

A: When confronted with a vertical surface, snails will reorientate themselves so that they climb vertically upwards.

B: A flying male moth, encountering an odor (pheromone) trail left by a female, will turn and fly upwind until it reaches the female.

C: In a circular chamber, lice make relatively few turns at 30°C, but many random turns at 35°C.

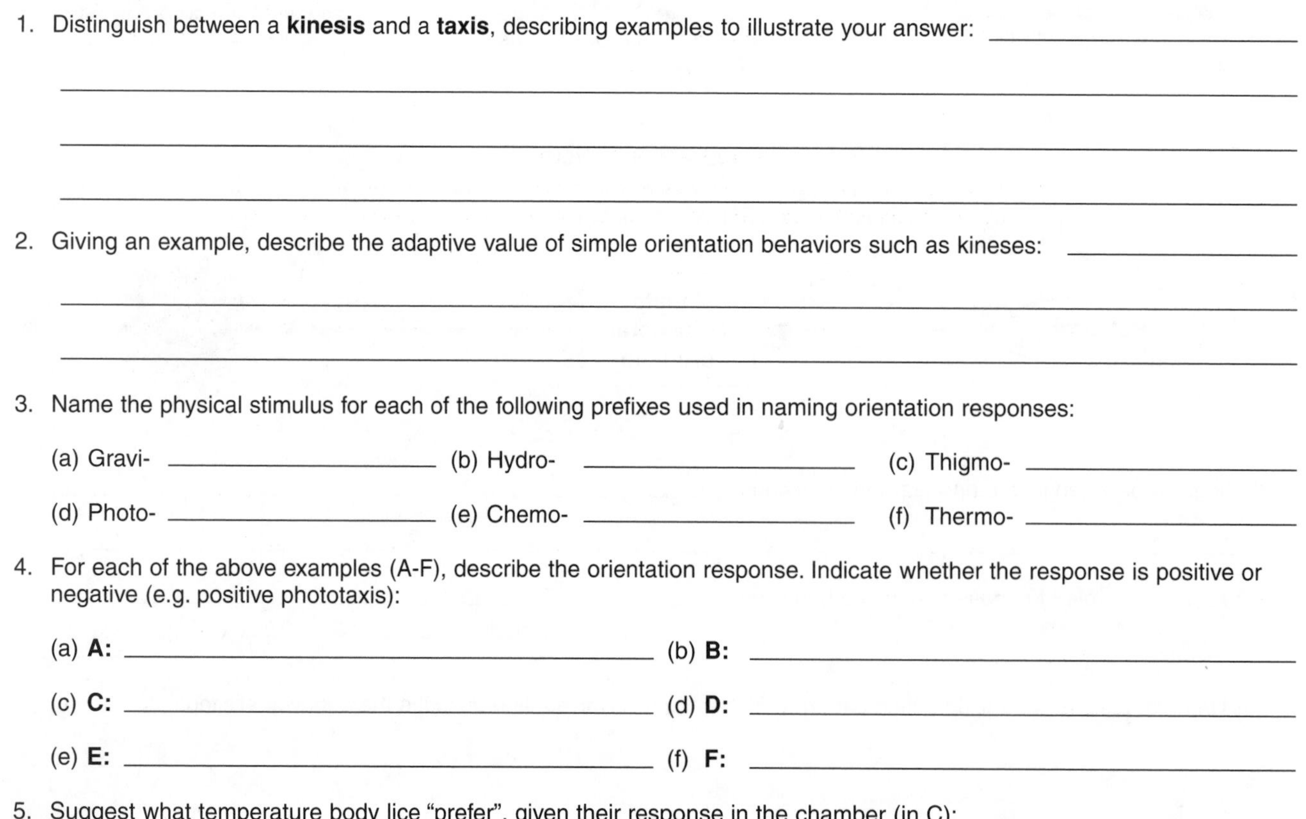

D: Spiny lobsters will back into tight crevices so that their body is touching the crevice sides. The antennae may be extended out.

E: At close range, mosquitoes use the temperature gradient generated by the body heat of a host to home in on exposed flesh.

F: Blowfly maggots will turn and move rapidly away from a directional light source.

1. Distinguish between a **kinesis** and a **taxis**, describing examples to illustrate your answer: ____________________

____

____

____

2. Giving an example, describe the adaptive value of simple orientation behaviors such as kineses: ____________________

____

____

3. Name the physical stimulus for each of the following prefixes used in naming orientation responses:

(a) Gravi- ____________ (b) Hydro- ____________ (c) Thigmo- ____________

(d) Photo- ____________ (e) Chemo- ____________ (f) Thermo- ____________

4. For each of the above examples (A-F), describe the orientation response. Indicate whether the response is positive or negative (e.g. positive phototaxis):

(a) **A:** ____________ (b) **B:** ____________

(c) **C:** ____________ (d) **D:** ____________

(e) **E:** ____________ (f) **F:** ____________

5. Suggest what temperature body lice "prefer", given their response in the chamber (in C): ____________

Related activities: *The Components of Behavior, Pheromones*

Animal Communication

Communication (the transmission of (understood) information) between animals of the same species is essential to their survival and reproductive success. Effective communication relies on being correctly understood so the messages conveyed between animals are often highly **ritualized** and therefore not easily misinterpreted. Communication involves a range of signals, which commonly include visual, chemical, auditory, and tactile perception. Which of these signals is adopted will depend on the activity pattern and habitat of the animal. Visual displays, for example, are ineffective at night or in heavy undergrowth.

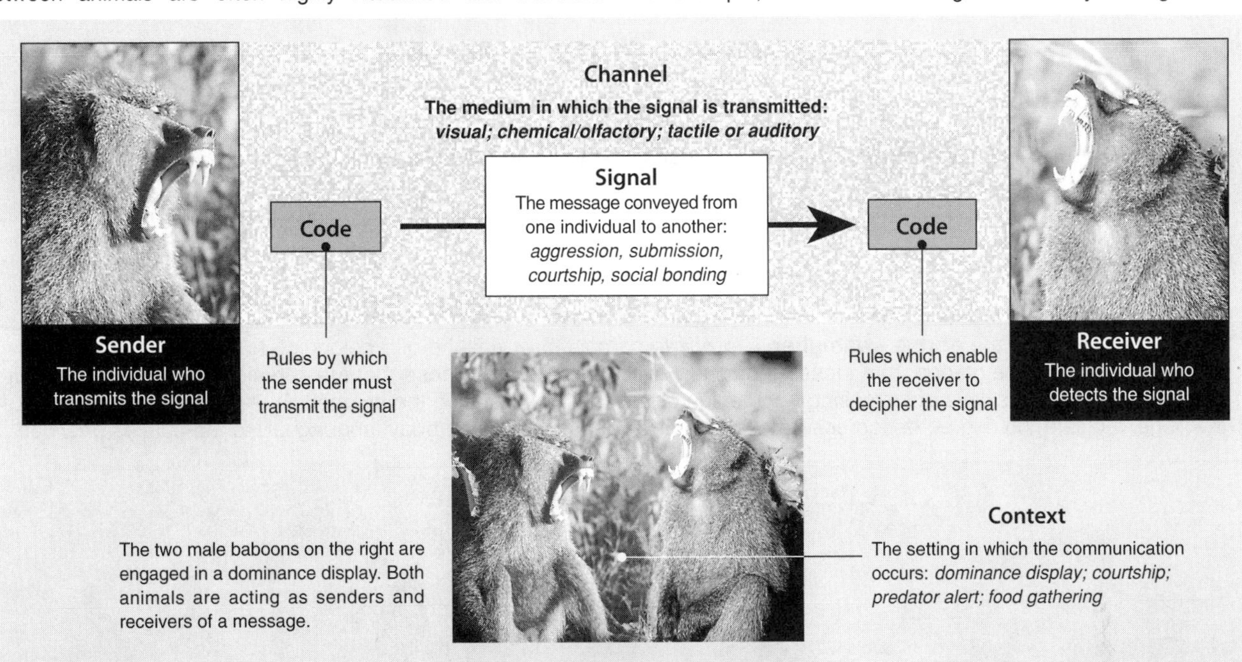

Channel
The medium in which the signal is transmitted: visual; chemical/olfactory; tactile or auditory
Signal
The message conveyed from one individual to another: aggression, submission, courtship, social bonding
Code
Code
Sender
The individual who transmits the signal
Rules by which the sender must transmit the signal
Rules which enable the receiver to decipher the signal
Receiver
The individual who detects the signal
The two male baboons on the right are engaged in a dominance display. Both animals are acting as senders and receivers of a message.
Context
The setting in which the communication occurs: dominance display; courtship; predator alert; food gathering

RA

Olfactory Messages

Some animals produce scents that are carried long distances by the wind. This may advertise for a potential mate, or warn neighboring competitors to avoid a territory. In some cases, mammals use their urine and feces to mark territorial boundaries. Sniffing genitals is common among mammals.

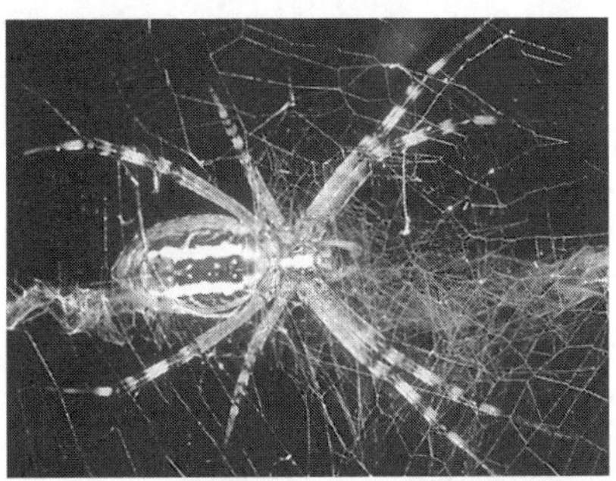

Tactile Messages

The touching of one animal by another may be a cooperative interaction or an aggressive one. Grooming behavior between members of a primate group communicates social bonding. Vibrations sent along a web by a male spider communicates to a potential female mate not to eat him.

Auditory Messages

Sound can be used to communicate over great distances. Birds keep rivals away and advertise for mates with song. Fin whales can send messages over thousands of kilometers of ocean. Calls made by mammals can be used to attract mates, contact other members of a group, or threaten competitors.

Visual Messages

Many animals convey information to other members of the species through their colors and adornments, and through gestures and body language. Visual displays can deliver threats, show submission, attract a mate, and even exert control over a social group.

Warning

Animals may communicate a warning to other animals through visual displays. Many wasp species (like those above) have brightly colored black and yellow markings to tell potential predators that they risk being stung if attacked.

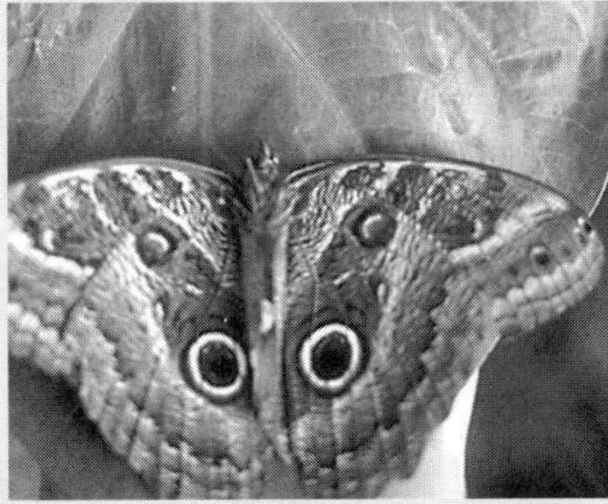

Deception

Animals may seek to deceive other animals about their identity. As an alternative to camouflage, animals may use visual markings that startle or deter potential predators. The eye spots on this moth may confuse a predator.

Attraction

Some animals produce a stunning visual display in order to attract a mate. The plumage of some bird species can be extremely colorful and elaborate, such as the peacock (above), the birds of paradise, and the lyrebird.

Related activities: *Pheromones, Cooperative Behavior*
Web links: *Norway Rat Behavior Repertoire*

1. (a) Explain why much of animal communication is ritualized: ______

(b) Discuss the benefits to animals of effective communication over both long and short distances: ______

2. Describe and explain the communication methods best suited to nocturnal animals in a forest habitat: ______

3. Postures provide a very important form of communication between animals. The **ethogram** below illustrates the various postures exhibited by the **purple swamphen** (*Porphyrio porphyrio*), a wetland bird belonging to the rail family. Social behaviors can often be graded from those that display overt aggression to those that are submissive. Swamphens have a **graded range** of displays of increasing aggression or submission. By using the symbols (+, – and 0) at the bottom of the page, indicate the degree of aggressiveness, submissiveness, or neutral body language. Use the spaces provided:

Ethogram for Swamphen Behavior

1B
1A
Fighting. One bird jumping with feet ready for clawing and beak open for pecking.

2A 2B
Fighting. One bird in aggressive upright posture with wings and tail raised and feet raised. The other bird is in the aggressive upright but not attacking.

3
Full bow. Submissive wings and tail fully up.

4A 4B
Fighting. Both birds in aggressive uprights and using feet to attack.

5A 5B
Facing away. Submissive display to an aggressive upright bird.

6A 6B
Fighting. Both birds jumping with feet ready for clawing and beak open for pecking.

7
Aggressive upright. Wings down. Tail horizontal.

8
Move away. Submissive display. Wings exaggerated. Tail fully up to uncover white feathers.

9A 9B
Crouch. Submissive display to an aggressive upright bird.

10
Horizontal forward. Aggressive display but not as aggressive as an upright.

11A 11B
Head flagging. Submissive display. Head held low and moved from side to side.

12
Head flick. Submissive display. Usually at end of encounter. Wings exaggerated, tail fully up. Beak held too high to peck at other bird.

Range of aggressive/submissive behaviors

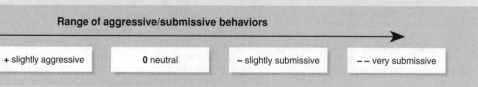

++ very aggressive | + slightly aggressive | **0** neutral | – slightly submissive | – – very submissive

Social Organization

All behavior appears to have its roots in the underlying genetic program of the individual. These innate behaviors may be modified by interactions of the individual with its environment, such as the experiences it is exposed to and its opportunities for learning. The behavioral adaptations of organisms affect their fitness (their ability to survive and successfully reproduce) and so are the products of natural selection. A behavior that leads to greater reproductive success should become more common in a species over time. Few animal species lead totally solitary lives. Many live in cooperative groups for all or part of their lives. Social animals comprise groups of individuals of the same species, living together in an organized fashion. They divide resources and activities between them and are mutually dependent (i.e. they do not survive or successfully reproduce outside the group).

Tigers are solitary and territorial animals, living and hunting alone. A male will remain with a female for 3-5 days during the mating season. A female may have 3 or 4 cubs which will stay with their mother for more than 2 years.

Many invertebrates (e.g. hermit crabs) are solitary animals, with occasional, random encounters. Some animals may be drawn together at feeding sites. Wind or currents may also cause aggregations.

Schooling fish and herds of mammals are examples of animals that form groups of a loose association. There is no set structure or hierarchy to the group. The grouping is often to provide mutual protection.

Family groups may consist of one or more parents with offspring of various ages. The relationship between parents may be a temporary, seasonal one or may be life-long.

Some insects (e.g. ants, termites, some wasp and many bee species) form colonies. The social structure of these colonies ranges from simple to complex, and may involve castes that provide for division of labor.

Primates such as chimpanzees and baboons have evolved complex social structures. Organized in terms of dominance hierarchies, higher ranked animals within the group have priority access to food and other resources.

Advantages of large social groupings

1. Protection from physical factors
2. Protection from predators
3. Assembly for mate selection
4. Locating and obtaining food
5. Defense of resources against other groups
6. Division of labor amongst specialists
7. Richer learning environment
8. Population regulation

Possible disadvantages of large social groupings

1. Increased competition between group members for resources as group size increases.
2. Increased chance of the spread of diseases and parasites.
3. Interference with reproduction (e.g. cheating in parental care; infanticide by non-parents).

1. Briefly describe two ways in which behavior may be passed on between generations:

   (a) ____________________

   (b) ____________________

2. Explain how large social groupings confer an advantage by providing:

   (a) Richer learning environment: ____________________

The effect of the number of adults in the family on pup survival for black-backed jackals

Number of pups surviving: 0, 1, 2, 3, 4, 5, 6

Number of adults: 1, 2, 3, 4, 5

SOURCE: Drickamer & Vessey, Animal behavior (3rd Ed) PWS, 1992

Black-backed jackal *(Canis mesomelas)*

Black-backed jackals live in the brushland of Africa. Monogamous pairs (single male and female parents) hunt cooperatively, share food and defend territories. Offspring from the previous year's litter frequently help rear their siblings by regurgitating food for the lactating mother and for the pups themselves. The pup survival results of 15 separate jackal groups are shown in the graph on the left.

(b) Division of labor among specialists: ______________________________

(c) Assembly for mate selection: ______________________________

3. The graph at the top of this page shows how the survival of black-backed jackal pups is influenced by the number of adult helpers in the group.

   (a) Draw an approximate 'line of best fit' on the graph (by eye) and describe the general trend: ______________________________

   (b) Describe two ways in which additional adult helpers may increase the survival prospects of pups:

4. Explain how a social behavior that is beneficial to individuals in a species may become more common over time:

Biological Rhythms

Environmental cues, such as daylength, timing and the height of tides, temperature, and phase of the moon are often used by plants and animals to establish and maintain a pattern of activity. Regular environmental cues assist in survival by synchronizing important events in the life cycle of an organism; events such as pollination, mating, birth, germination, rearing of offspring, collection and storage of food reserves and body fat, and periods of torpor. **Biological rhythms** (biorhythms) in direct response to environmental stimuli are said to be **exogenous**, because the rhythm is controlled by an environmental stimulus that is external to the organism. Those rhythms that continue in the absence of external cues are said to be **endogenous**.

Rhythm: tidal.
Period: ~ 12.4 h (coincident with tidal flows).
Examples: In mud crabs, locomotion and feeding occurs when covered by tidal waters.

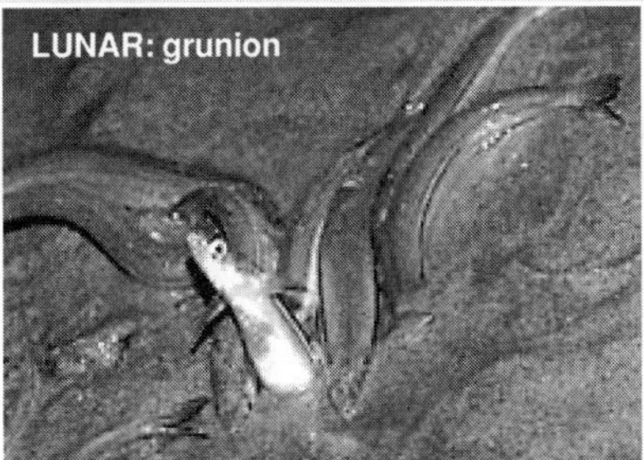

Rhythm: lunar.
Period: ~ 29.5 days (a month).
Example: In grunions, egg laying above the high water mark coincides with a new moon.

Rhythm: daily
Period: ~ 24 h.
Examples: In kokako, general activity and feeding occurs during daylight hours.

Rhythm: annual.
Period: ~ a year.
Example: In many domestic livestock and antelope species, young are born in spring.

Rhythm: annual.
Period: ~ a year.
Example: NZ long-tailed bats hibernate for 4-5 months during the autumn and winter.

Rhythm: intermittent.
Period: does not apply.
Example: In some shiners (a minnow-like fish), reproduction is triggered by flooding.

1. Use the examples provided above to determine a definition of each of the following terms describing biological rhythms. For each type of rhythm describe one other example to illustrate how the behavior follows the astronomical cycle:

   (a) **Daily** rhythm: ______

   Example: ______

   (b) **Lunar** rhythm: ______

   Example: ______

   (c) **Annual** rhythm: ______

   Example: ______

   (d) **Tidal** rhythm: ______

   Example: ______

2. For each of the examples below, describe an **environmental cue** that might be used to induce or maintain the activity:

   (a) Hedgehog's hibernation in winter: ______

   (b) Blackbird's foraging and social behavior during daylight: ______

   (c) Kiwi's activity of hunting for soil organisms at night: ______

   (d) Coordinated flowering of plants in spring: ______

3. Explain what is meant by an exogenous rhythm: ______

Migration Patterns

In many animals, migration is an important response to environmental change. True migrations are those where animals travel from one well-defined region to another, for a specific purpose such as overwintering, breeding, or seeking food. Migrations often involve very large distances and usually involve a return journey. They are initiated by the activity of internal clocks or timekeepers in response to environmental cues such as change in temperature or daylength. Some mass movements of animals are not truly migrations in that they do not involve a return journey and they are governed by something other than an internal biological clock (e.g. depletion of a food resource). Such movements are best described as dispersals and are typical of species such as 'migratory' locusts and some large African mammals which relocate in response to changing food supplies.

Dispersal: one-way migration

Some migrations of animals involve a one-way movement. In such cases, the animal does not return to its original home range. This is typical of population dispersal. This often occurs to escape deteriorating habitats and to colonise new ones.

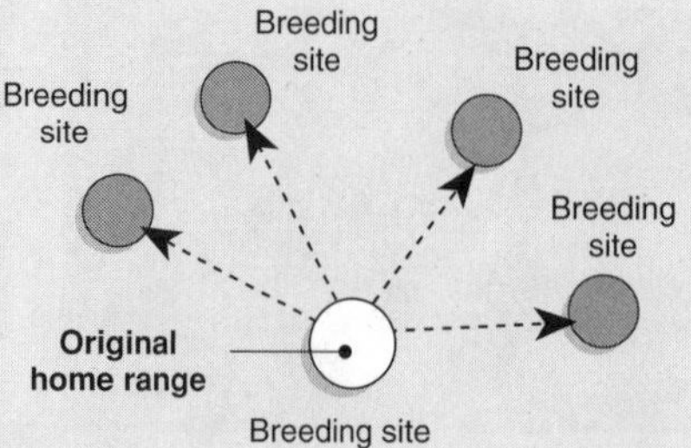

Return migration

Animals that move to a winter feeding ground are making one leg of a return migration. The same animals return to their home range in the spring which is where they have their breeding sites. Sometimes they follow different routes on the return journey.

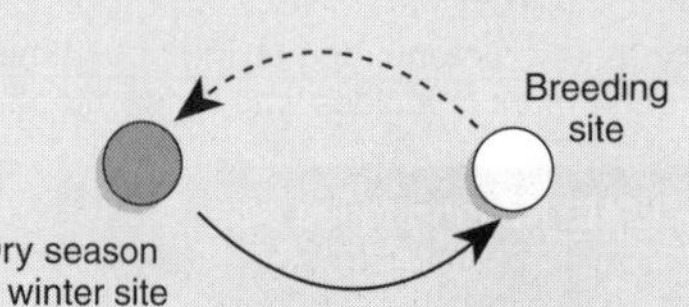

Nomadic migration

Similar to one-way migration but individuals may breed at several locations during their lifetimes. These migrations are apparently directionless, with no set pattern. Each stopover point is a potential breeding site. There may also be temporary non-breeding stopovers for the winter or dry season.

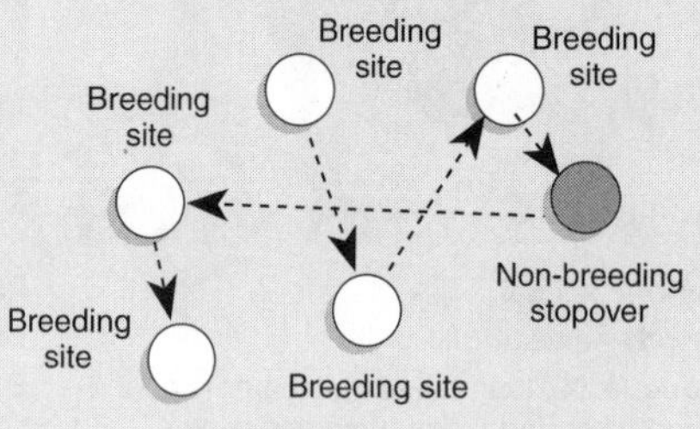

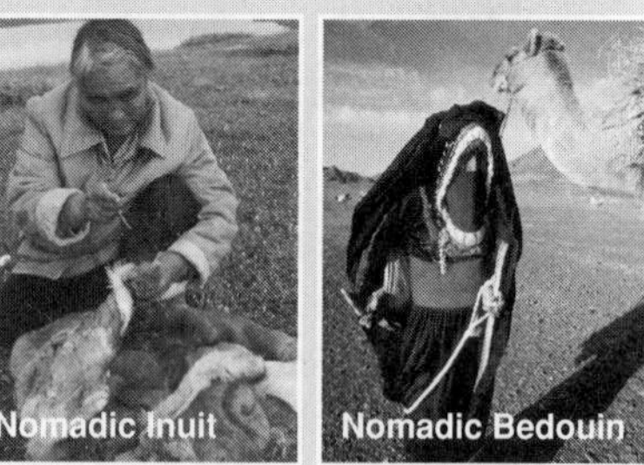

Remigration circuits

In some populations, the return leg of a migration may have stopovers and may be completed by one or more subsequent generations. In addition to winter or dry season areas, there may be stops at feeding areas by juveniles or adults. Also included are closed circuits where animals die after breeding.

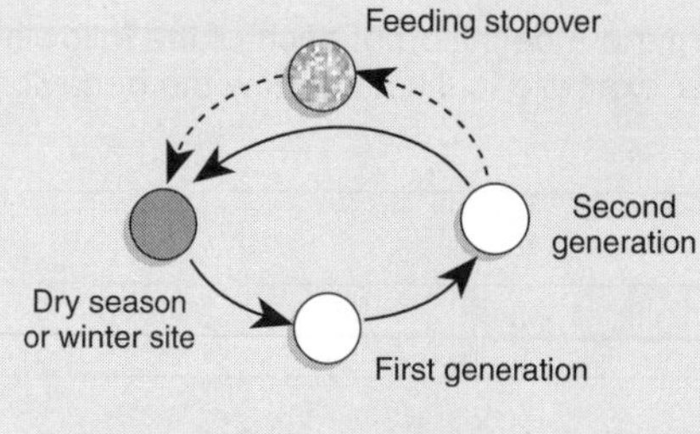

1. Giving an example, describe the conditions under which nomadic migration behavior might be necessary:

2. Identify which of the above forms of migration would lead to further dispersal of a population. Explain your answer:

3. Describe an environmental cue important in the regular migratory behavior of a named species:

4. Discuss the adaptive value of migratory behavior:

Sun Compass Navigation

Honeybees are able to navigate by using a sun compass, and they are able to communicate the location of a food source to other bees in the hive. Up to 12 000 bees from a hive of 45 000 will leave each sunny day in search of pollen and nectar.

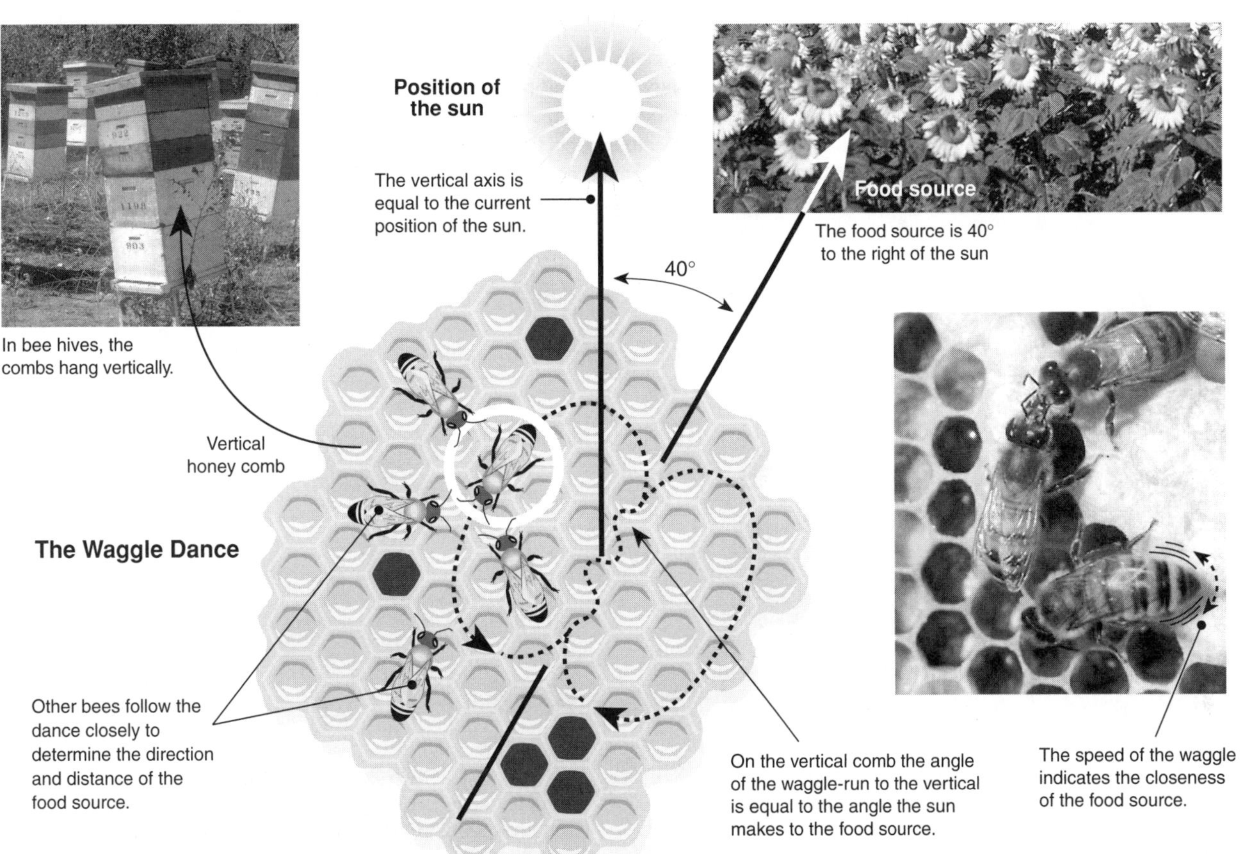

The Waggle Dance

In the 'waggle dance' (**above**), the bee on the vertical comb (white circle) repeats the figure-eight dance with the waggle equal to the angle the sun makes with the food source. Other bees will be in close attendance to monitor the dance and learn the location of the new food source. The dance is occasionally performed on the horizontal entrance board of the hive, where the waggle run 'points' directly towards the food source.

The Round Dance

If the food source is very close by, say within 50 metres, the honey bee will carry out another kind of dance. The honey bee stimulates other workers to leave the hive and search within 50 metres for a food source, by performing a 'round dance' (**see right box**).

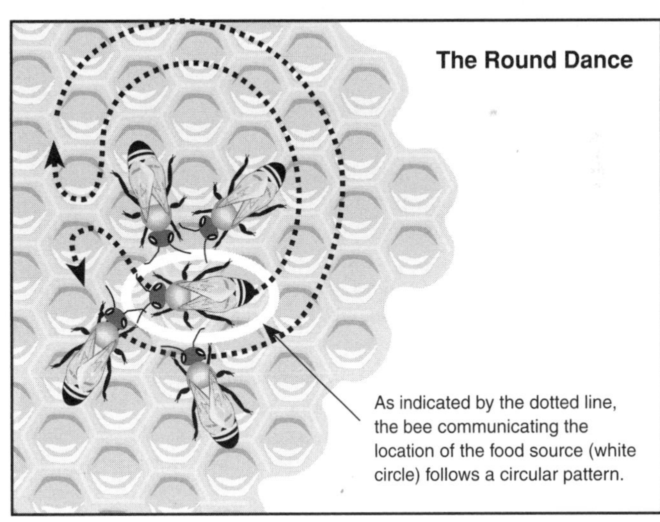

The Round Dance
As indicated by the dotted line, the bee communicating the location of the food source (white circle) follows a circular pattern.

1. Name the environmental reference used by honey bees to orientate for navigation: ______________________

2. Describe how a bee communicates the proximity of a food source in the waggle dance: ______________________

3. Explain how the bee compensates for the time it takes between finding the food and delivering its message to the hive:

4. Describe the circumstances under which the round dance is used: ______________________

Homing in Insects

Unlike migration, which is the periodic movement from one location and climate to another, **homing** refers to the ability of an animal to return to its home site or locality after being displaced. **Navigation** is involved in both migration and homing, but homing behavior often relies on the recognition of familiar landmarks, especially where the distances involved are relatively short.

The **digger wasp**, *Philanthus triangulum*, builds a nest burrow in sand. It captures and paralyses an insect grub as a food supply for the wasp's larvae to feed on during development. The paralysed grub is then taken back to the wasp's underground nest, where it lays its eggs in the still living body. In a well-known experiment to test the homing behaviour of this wasp, a research scientist by the name of Tinbergen, carried out the following 2-step procedure. (After Tinbergen, 1951. The Study of Instinct. Oxford University Press, London)

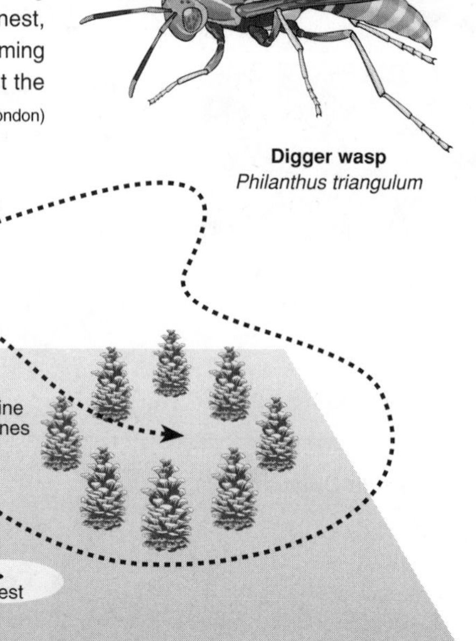
Digger wasp
Philanthus triangulum
Wasp
Prey
Pine cones
Nest

Wasp

Nest

Pine cones

Step 1: Orientation Flight

While the female wasp was in the burrow, Tinbergen placed a circle of pine cones around the entrance. When she emerged, the wasp reacted to the situation by carrying out a wavering orientation flight before flying off.

Step 2: Return Flight

During her absence, the pine cones were moved away from the burrow leading to the nest. Returning to the nest with prey, the wasp orientated to the circle of pine cones, and not the entrance to the nest.

Foraging Ant Trails

A foraging ant is able to leave a chemical trail so that it may return to its nest after having searched widely for food sources. The ant secretes a pheromone from its abdomen, that other ants will follow. Soon many ants will follow a well-established trail to and from the food. Later, when the food is gone, the ants stop secreting the pheromone, and the chemical soon diffuses from the area. The trail finally vanishes and the movement of the ants becomes random again as they search for new food sources.

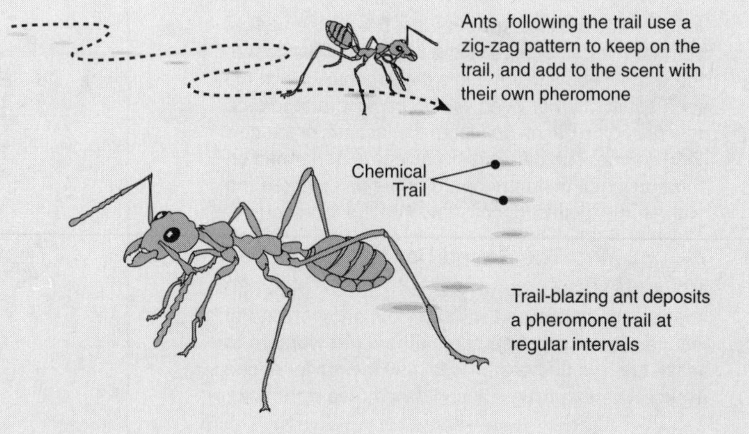
Ants following the trail use a zig-zag pattern to keep on the trail, and add to the scent with their own pheromone
Chemical Trail
Trail-blazing ant deposits a pheromone trail at regular intervals

1. Describe a method of navigation used by each insect listed below in the following contexts:

   (a) Honey bees relocating a rich food source (flowers): ______

   (b) Moths finding a mate at night: ______

   (c) Digger wasps locating their nest: ______

   (d) Ants following other foraging ants: ______

   (e) Mosquito locating a meal (blood from a mammal): ______

2. Explain how **homing** is different from migration: ______

Related activities: Pheromones

Learned Behavior

Learning describes a relatively permanent modification of behavior that occurs as a result of practice or experience. Learning is a critical process that affects the behavior of animals of all ages, across many taxa. Learning behaviors vary widely. The simplest are **habituation** and **imprinting**, when an animal learns to make a particular response only to one type of animal or object. Like most behaviors, they are adaptive in that they enhance survival by ensuring appropriate responses in the given environment. More complex behaviors arise through **conditioning** and observational learning, such as imitation. Latent learning and insight are not readily demonstrated but have been shown experimentally in a range of species.

Digger wasp, *Philanthus triangulum*, with prey

© Karen Nichols

Filial (Parent) Imprinting

Filial imprinting is the process by which animals develop a social attachment. It differs from most other kinds of learning (including other types of imprinting), in that it normally can occur only at a specific time during an animal's life. This **critical period** is usually shortly after hatching (about 12 hours) and may last for several days. Ducks and geese have no innate ability to recognize *mother* or even their own species. They simply respond to, and identify as mother, the first object they encounter that has certain characteristics.

Sexual Identity Imprinting

Individuals learn to direct their sexual behavior at some stimulus objects, but not at others. Termed **sexual imprinting**, it may serve as a species identifying and species isolating mechanism. The mate preferences of birds have been shown to be imprinted according to the stimulus they were exposed to (other birds) during early rearing. Sexual imprinting generally involves longer periods of exposure to the stimulus than filial imprinting (*see left*).

Latent Learning

Latent learning describes an association made without reinforcement and expressed later. Tinbergen and Kruyt tested latent learning in predatory digger wasps by providing landmarks around the wasps' nest burrows. On emerging, the wasps survey the area and fly off to forage. If the landmarks were removed or rearranged, the returning wasps became disorientated, supporting the conclusion that the returning wasps use the entire configuration of landmarks as a guide to relocating the burrow.

Habituation

Habituation is a very simple type of learning involving a loss of a response to a repeated stimulus when it fails to provide any form of reinforcement (reward or punishment). Habituation is different to fatigue, which involves loss of efficiency in a repeated activity, and arises as a result of the nature of sensory reception itself. An example of habituation is the waning response of a snail attempting to cross a platform that is being tapped at regular time intervals. At first, the snail retreats into its shell for a considerable period after each tap. As the tapping continues, the snail stays in its shell for a shorter duration, before resuming its travel.

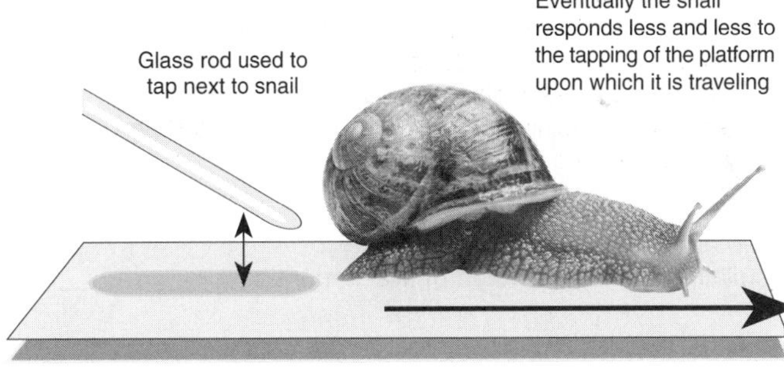

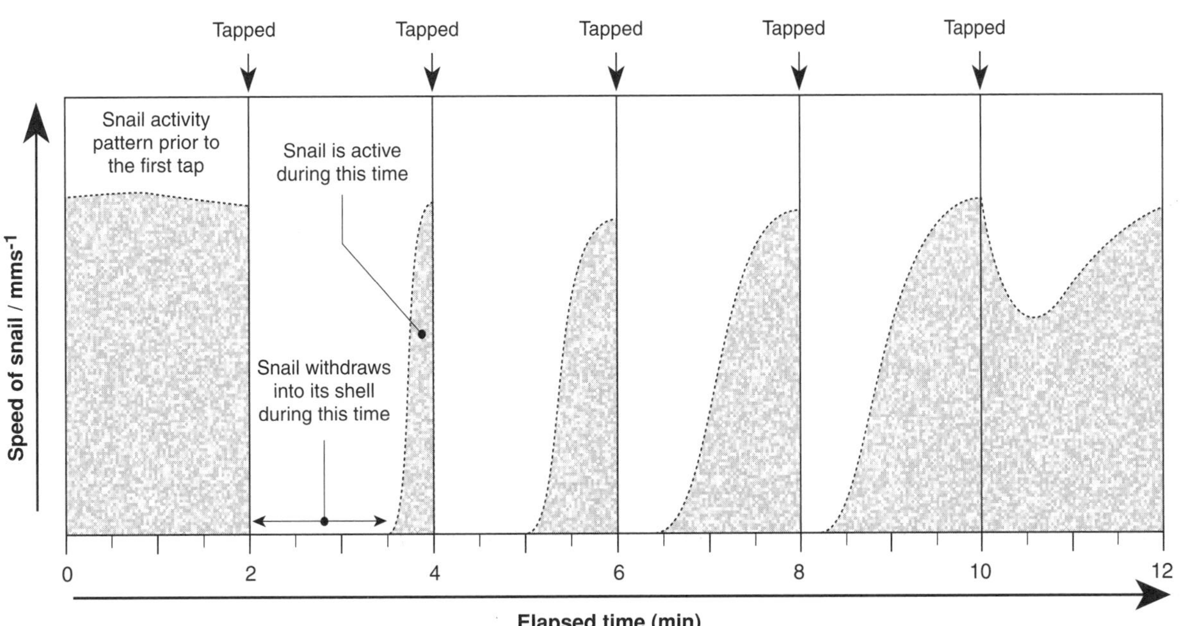

Related activities: *The Components of Behavior, Homing in Insects, Learning to Sing*
Web links: *Foraging Behavior, Optimal Foraging*

Operant Conditioning

Operant conditioning describes situations where an animal learns to associate a particular behavior with a reward. The appearance of the reward is dependent on the appearance of the behavior. Burrhus Skinner studied operant conditioning using an apparatus he invented called a Skinner box (below). When an animal (usually a pigeon or rat) pushed a particular button it was rewarded with food. The animals learned to associate the pushing of the button with obtaining food (the reward). The behavioral act that leads the animal to push the button in the first place is thought to be generated spontaneously (by accident or curiosity). This type of learning is also called instrumental learning and it is the predominant learning process found in animals.

A Skinner box typically contains one or more buttons, which can be pressed to obtain a reward.
Food is delivered when the correct button is pushed.

Operant Learning in Sparrows

Common house sparrows provide a good example of operant conditioning. They have learned to gain access to restaurants and cafés through automatic doors by triggering the motion sensors that control their opening. The birds will flutter in front of the sensor until the door opens, or perch on top of the sensor and lean over until it is triggered. Presumably, after accidentally triggering the sensor and gaining access, the birds learned which behaviors will bring them a reward.

Insight Behavior

Insight behavior involves using reason to form conclusions or solve a new problem. It is not based on past experiences of a similar problem, but does involve linking together isolated experiences from different problems to reach an appropriate response. Insight learning is common in higher mammals (e.g. apes) and there is good evidence for its occurrence in many other animals including dogs, pigeons and ravens.

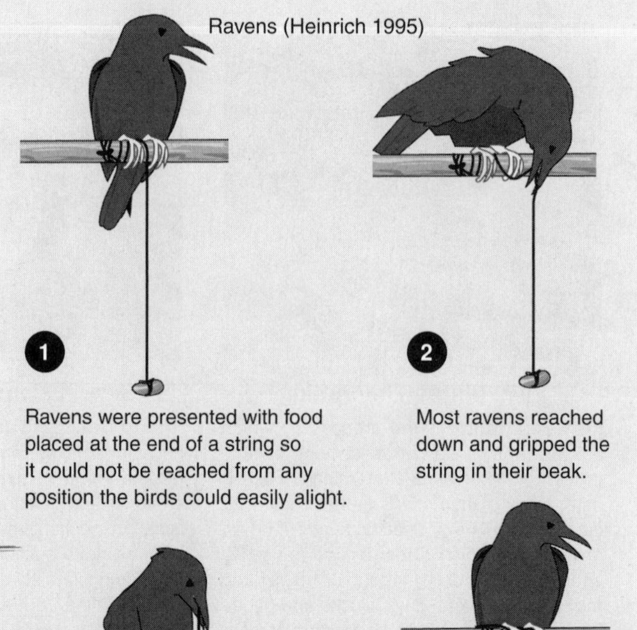

Ravens (Heinrich 1995)
1
Ravens were presented with food placed at the end of a string so it could not be reached from any position the birds could easily alight.
2
Most ravens reached down and gripped the string in their beak.

3 They then pulled the string up, moving the food to a higher position.

4 The ravens then held the string in place with their foot while reaching down to the string again. After repeating this several times, the ravens were able to reach the food.

1. Explain the adaptive value of filial imprinting: ____________________

2. Explain in what way habituation is adaptive: ____________________

3. Suggest when latent learning might be important to an animal's survival: ____________________

4. (a) Describe the basic features of operant conditioning: ____________________

   (b) Explain why operant conditioning is likely to be the predominant learning process in animals: ____________________

5. Explain why it is difficult to prove conclusively that an animal is using insight learning when solving a given problem:

Learning to Sing

Song birds use vocalizations (songs and calls) as a way to communicate, establish territories, and attract mates. The characteristics of a song may also be an indicator of fitness, as it has been shown that parasites and disease affect the song produced. While singing is instinctive, learning to sing the correct song is a learned behavior, and without it, a song bird is unlikely to gain a territory or a mate. Analyses of many bird species show that there are at least two major strategies for song development: (1) imitation of other birds, particularly adults of the same species, and (2) invention or improvisation. These strategies overlie the genetic template for the song learning process. The window during which a song can be learned varies between species. In some species, the inherited song pattern can be modified by learning only during early life. In others, the song is modified according to experience for at least another year, and some (e.g. blackbirds) modify their songs throughout life.

Learning to Sing

The songs of different bird species vary but are generally characteristic of the species. The structure of bird song is studied using a technique called **acoustic spectroscopy**, a technique which produces a graphical representation of the sounds being made. This enables song patterns to be compared between individuals and has led to experimental work to establish how birds learn song and how much of the song is genetically determined.

Blackbirds modify their songs throughout their life

The sound spectrographs of **chaffinch** song (below) illustrate how the final song that is produced can be altered by exposure to the songs of same-species individuals during the first three months when the song is learned. The upper trace shows the song of a normal male, while the lower trace is the song of a male reared in isolation from the nest. The isolated male's song is the right pitch and relatively normal in length, but it is simpler and lacks the acoustic 'flourish' at the end (arrow), which is typical of the chaffinch's song.

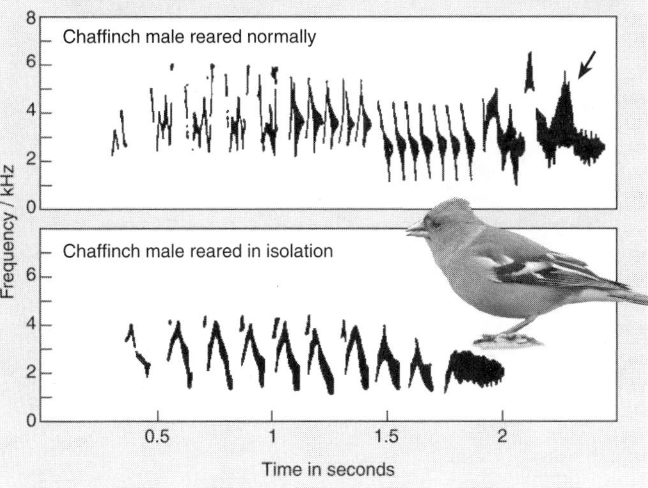

All data modified after A. Manning: An Introduction to Animal Behaviour (1979)

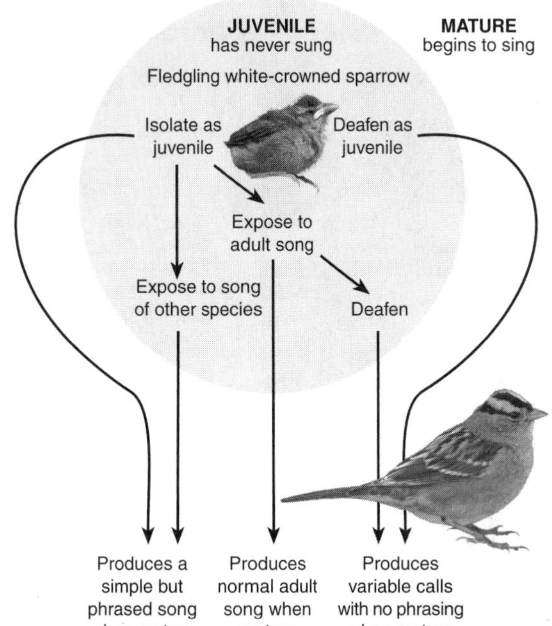

The diagram above summarizes experiments investigating song development in **white-crowned sparrows**. This small finch has a wide range on the Pacific Coast of America and birds from different regions have different **song dialects**. The experiments found:

- Isolated males will eventually sing similar and simplified versions of the normal song, regardless of which region they come from.
- Isolated males can be trained to sing their own dialect by playing them tape recorded songs of birds from their region.
- After 4 months of age, birds are unreceptive to further learning.
- A bird needs to be able to hear itself in order to produce the normal song. It requires auditory feedback to adjust the notes.
- Once birds have learned their normal adult song, they can continue to sing normally even if deafened.

1. From the studies of song development in white-crowned sparrows, describe the evidence for the following statements:

   (a) The inherited song pattern can be modified by learning only in the early stages of life: ____________________

   (b) Young birds need to hear themselves sing to produce a normal song as adults: ____________________

   (c) The song produced by an adult bird is determined by a genetic predisposition and what it hears when it is a juvenile:

   ____

2. Discuss the possible evolutionary significance of modifying a basic (genetically determined) song pattern by learning:

   ____

Related activities: *Animal Communication, Behavior and Species Recognition*
Web links: *Starling Talk*

Pheromones

A **pheromone** is a chemical produced by an animal and released into the external environment where it has an effect on the physiology or behavior of members of the same species. Hundreds of pheromones, some of which are sex attractants, are known. They are especially common amongst insects and mammals, and commonly relate to reproductive behavior. Many mammals, including canids and all members of the cat family, commonly use scent marking to mark territorial boundaries and to advertise their sexual receptivity to potential mates. Other mammals, including rabbits, release a mammary pheromone that triggers nursing behavior in the young. Pheromones are also used as signalling molecules in social insects such as bees, wasps, and ants. They may be used to mark a scent trail to a food source or to signal alarm. Species specific odor cues are now widely used as bait in traps when controlling insect or mammalian pests or to capture animals for study.

Pheromones and Animal Communication

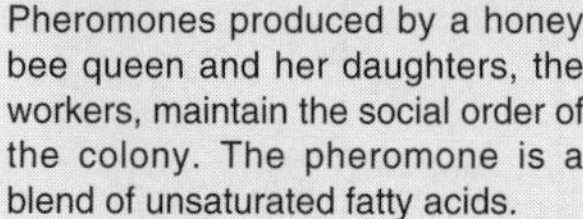

Pheromones produced by a honey bee queen and her daughters, the workers, maintain the social order of the colony. The pheromone is a blend of unsaturated fatty acids.

Like mammals, reptiles detect chemicals with the vomeronasal (Jacobson's) organ. The flicking of a snake's tongue allows the snake to chemically sample its environment.

In mammals, pheromones are used to signal sexual receptivity and territorial presence, or to synchronize group behavior. Pheromone detection relies on the vomeronasal organ (VNO), an area of receptor tissue in the nasal cavity. Mammals use a flehmen response, in which the upper lip is curled up, to better expose the VNO to the chemicals of interest.

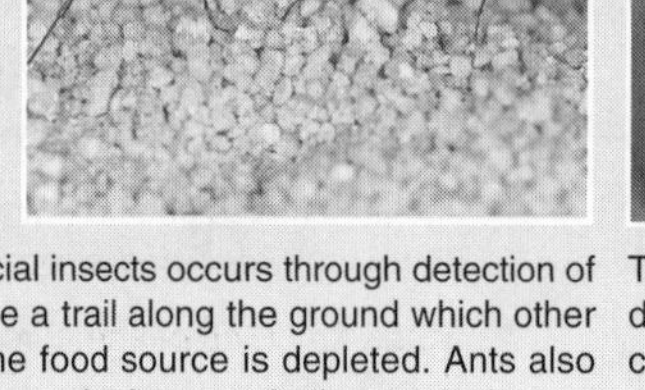

Communication in ants and other social insects occurs through detection of pheromones. Foraging ants will leave a trail along the ground which other ants will follow and reinforce until the food source is depleted. Ants also release alarm substances, which will send other ants in the vicinity into an attack frenzy. These signals dissipate rapidly if not reinforced.

Photo courtesy of Cereal Research Centre, AAFC

The feathery antennae of male moths are stereochemically specialised to detect the pheromone released by a female moth. Male moths can detect concentrations as low as 2ppm and will fly upwind toward the source to mate with the female. This sex attractant property of pheromones is exploited in pheromone traps, which are widely used to trap insect pests in orchards.

1. (a) Distinguish between hormones and **pheromones**: ______________________________

   (b) Explain the significance of pheromones being species specific: ______________________________

2. Giving examples, briefly describe the role of pheromones in three aspects of animals behavior:

   (a) ______________________________

   (b) ______________________________

   (c) ______________________________

3. From what you know of pheromone activity, suggest how a pheromone trap would operate to control an insect pest:

____

Cooperative Behavior

Individuals both within and between species may cooperate with each other for many reasons: for mutual defense and protection, to enhance food acquisition, or to rear young. To explain the evolution of cooperative behavior, it has been suggested that individuals benefit their own survival or the survival of their genes (offspring) by cooperating. **Kin selection** is a form of selection that favors altruistic (self-sacrificing) behavior towards relatives. In this type of behavior an individual will sacrifice its own opportunity to reproduce for the benefit of its close relatives. Individuals may also cooperate and behave altruistically if there is a chance that the "favor" may be returned at a later time. **Altruistic behavior** towards non-relatives is usually explained in terms of trade-offs, where individuals weigh up the costs and benefits of helpful behavior. Cooperation will evolve in systems where, in the long term, individuals all derive some benefit.

Many mammalian predators live in well organized social groups. These are formed for the purposes of cooperative hunting and defense and they facilitate offspring survival. In the gray wolves above, territories are marked by scent. Howling promotes group bonding and helps to keep neighboring packs away.

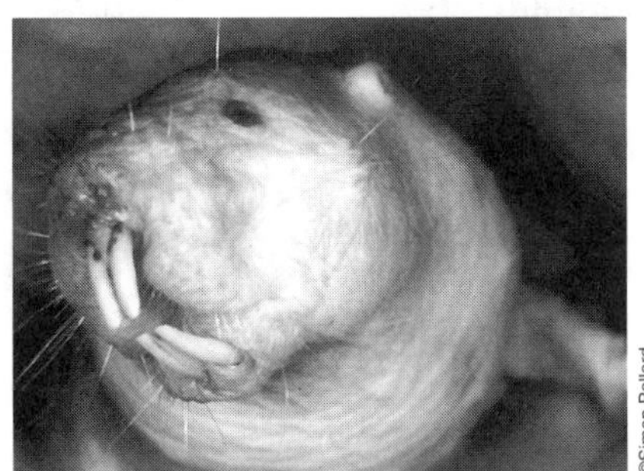

©Simon Pollard

Naked mole rats, from the arid regions of Kenya, are unique among mammals in having a social organization similar to that of social insects. Up to 300 of the rodents spend their lives underground in a **colony** with a **caste system**, with workers, soldiers, infertile females, and one breeding queen.

©Dr. M. Soper

The males of many species help their mates collect enough food to meet reproductive needs. In some species, especially amongst birds, non-breeding individuals, e.g. older siblings, may assist in rearing the offspring by protecting or feeding them. This type of altruism may arise through **kin selection**.

JW

Herding is an effective defensive behavior, providing a great number of eyes to detect approaching predators. Although they have horns for defense, when wildebeest (above) detect danger, their first reaction is to run as a group (at up to 80 kmh^{-1}).

South African meerkats live in communities in earth burrows. They are vulnerable to attack from land and aerial predators (especially vultures). The group maintains a constant surveillance by posting sentinels to warn the rest of the group of danger.

JW

Cooperative (mutualistic) associations can occur between different species. Cape buffalos are warned of approaching predators by cattle egrets and maribou storks which in turn feed on insects disturbed by the buffalo as it grazes.

1. Using examples, discuss the difference between **altruistic behavior** and **kin selection**: ______

2. Explain (in evolutionary terms) why an animal would raise the offspring of a close relative rather than their own:

3. Describe two ways in which members of a herd or a shoal reduce their likelihood of being attacked by a predator:

   (a) ______

   (b) ______

Periodicals: *Relative distance*

Related activities: *Animal Communication, Social Organization*

Behavior and Species Recognition

Many of the barriers observed in animals are associated with reproduction, reflecting the importance of this event in an individual's life cycle. Many types of behavior are aimed at facilitating successful reproduction. These include **courtship behaviors**, which may involve attracting a mate to a particular breeding site. Courtship behaviors are aimed at reducing conflict between the sexes and are often stereotyped or **ritualized**. They rely on sign stimuli to elicit specific responses in potential mates. In addition, there are other reproductive behaviors which are associated with assessing the receptivity of a mate, defending mates against competitors, and rearing the young. Behavioral (ethological) differences between species are a type of **prezygotic** isolating mechanism to help preserve the uniqueness of a species gene pool.

Courtship and Species Recognition

Accurate species recognition when choosing a mate is vital for successful reproduction and species survival. Failure to choose a mate of the same species would result in reproductive failure or hybrid offspring which are infertile or unable to survive. Birds exhibit a wide range of species-specific courtship displays to identify potential mates of the same species who are physiologically ready to reproduce. They may use simple visual or auditory stimuli, or complex stimuli involving several modes of communication specific to the species.

Peacock courtship (left) involves a visually elaborate tail display to attract female attention. The male raises and fans his tail to display the bright colors and eye-spot patterns. Peahens tend to mate with peacocks displaying the best quality tail display which includes the quantity, size and distribution of eye-spots.

Bird song is an important behavioral isolation method for many species including eastern and western meadowlarks. Despite the fact that they look very similar and share the same habitat, they have remained as two separate species. Differences between the songs of the two species enables them to recognize individuals of their own species and mate only with them. This maintains the species isolation.

Eastern meadowlark

Some species use chemical cues as mating signals and to determine mate choice. The crested auklet (left) secretes aldehydes which smell like tangerines. Birds rub their bills in the scented nape of a partner during courtship. This "ruff-sniff" behavior allows mate evaluation based on chemical potency. A potential partner might be seen as fitter and more attractive if it produces more aldehydes, because the chemical repels ectoparasites.

Courtship Behavior is a Necessary Precursor to Successful Mating

Courtship behavior occurs as a prelude to mating. One of its functions is to synchronize the behaviors of the male and female so that mating can occur, and to override attack or escape behavior. Here, a male greater frigatebird calls, spreads its wings, and inflates its throat pouch to court a female.

In many bird and arthropod species, the male will provide an offering, such as food, to the female. These **rituals** reduce aggression in the male and promote appeasement behavior by the female. For some **monogamous** species, e.g. the blue-footed boobies (left), the pairing begins a long term breeding partnership.

Courtship: Galapagos albatrosses

Although courtship rituals may be complex, they are very stereotyped and not easily misinterpreted. Males display, usually through exaggerated physical posturing, and the females then select their mates. Courtship displays are species specific and may include ritualized behavior such as dancing, feeding, and nest-building.

1. (a) Suggest why courtship behavior may be necessary prior to mating:

   (b) Explain why courtship behavior is often ritualized and involves stereotyped displays:

2. In terms of species continuity, explain the significance of courtship behavior in species recognition:

Breeding Behavior

In the animal world, humans are unusual in being sexually receptive most of the time. Most animals breed on an annual basis and show no reproductive behavior outside the breeding season. Photoperiod is an important cue for triggering the onset of breeding behavior in many species. Species with a relatively short gestation (e.g. most birds) respond to increasing daylength, whereas those with a long gestation (e.g. large mammals) respond to decreasing daylength. The short time period that most sexually reproducing animals have in which to breed creates strong selective pressure for clearly understood patterns of behavior that improve the chances of reproductive success. A breeding scenario typical for many insects is described below.

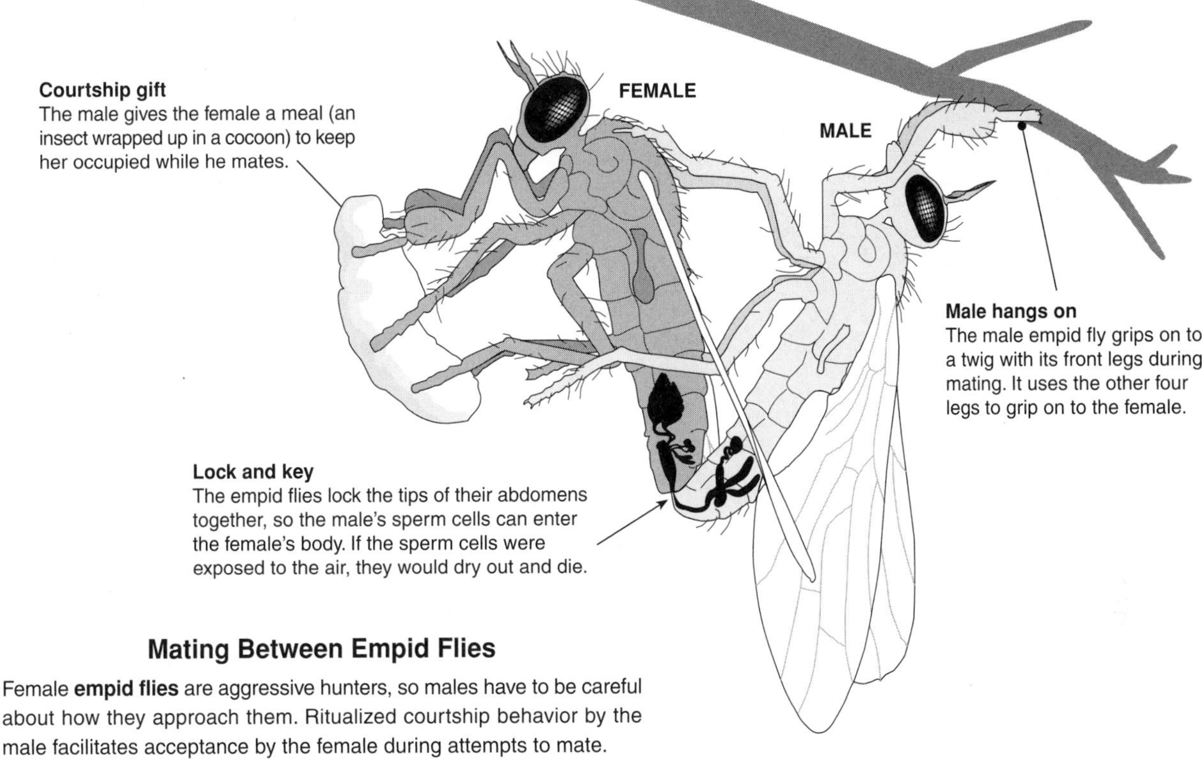

Courtship gift
The male gives the female a meal (an insect wrapped up in a cocoon) to keep her occupied while he mates.
FEMALE
MALE
Male hangs on
The male empid fly grips on to a twig with its front legs during mating. It uses the other four legs to grip on to the female.
Lock and key
The empid flies lock the tips of their abdomens together, so the male's sperm cells can enter the female's body. If the sperm cells were exposed to the air, they would dry out and die.

Mating Between Empid Flies

Female **empid flies** are aggressive hunters, so males have to be careful about how they approach them. Ritualized courtship behavior by the male facilitates acceptance by the female during attempts to mate.

Territories and Reproductive Behavior

A territory is any area that is defended against members of the same species. The territory, which usually contains access to valuable resources, is usually defended by acts of aggression or ritualized signals (e.g. vocal, visual or chemical signals).

Resource availability often determines territory size and the population becomes spread out accordingly. Gannet territories are relatively small, with hens defending only the area they can reach while sitting on their nest.

Topi, a lek species

Lek systems involve gatherings of males for competitive mating displays. In a lek system, the only area defended by a male is the space where mating occurs. Females compete for the highest ranked males, as judged by their position on the lek.

1. Describe two aspects of mating behavior in empid flies that help to ensure successful mating:

   (a) ___

   (b) ___

2. Explain the role of photoperiod in determining the onset of reproductive behavior in many species: ___

3. Describe the role a territory could play in an individual's reproductive success: ___

Aggression, Hierarchies, and Resources

Aggression is a complex phenomenon often associated with competition for resources but also including predatory behavior. **Agonistic behavior** is a related but more precise term for behavior associated with **conflict** among members of the same species. Agonistic behavior includes all aspects of conflict, including threats, submissions, chases, and physical combat, but it specifically excludes predatory aggression. Conflict behavior rarely results in serious injury and, often, ritualized behavior (e.g. submissive behavior or ritualized threat) within a **hierarchy** will resolve conflict without physical combat being necessary. Olive baboons (*Papio anubis*) live in the African savannah and have a highly organized hierarchical structure. The hierarchy promotes division of labor and maximizes the efficiency with which the troop can search for food and defend itself. Each of the different baboon troops occupies an area called a **home range**, which provides all the resources the troop needs for its survival. Home ranges differ from territories in that they may overlap in places and are not necessarily defended exclusively.

Baboon Home Ranges in Nairobi Park

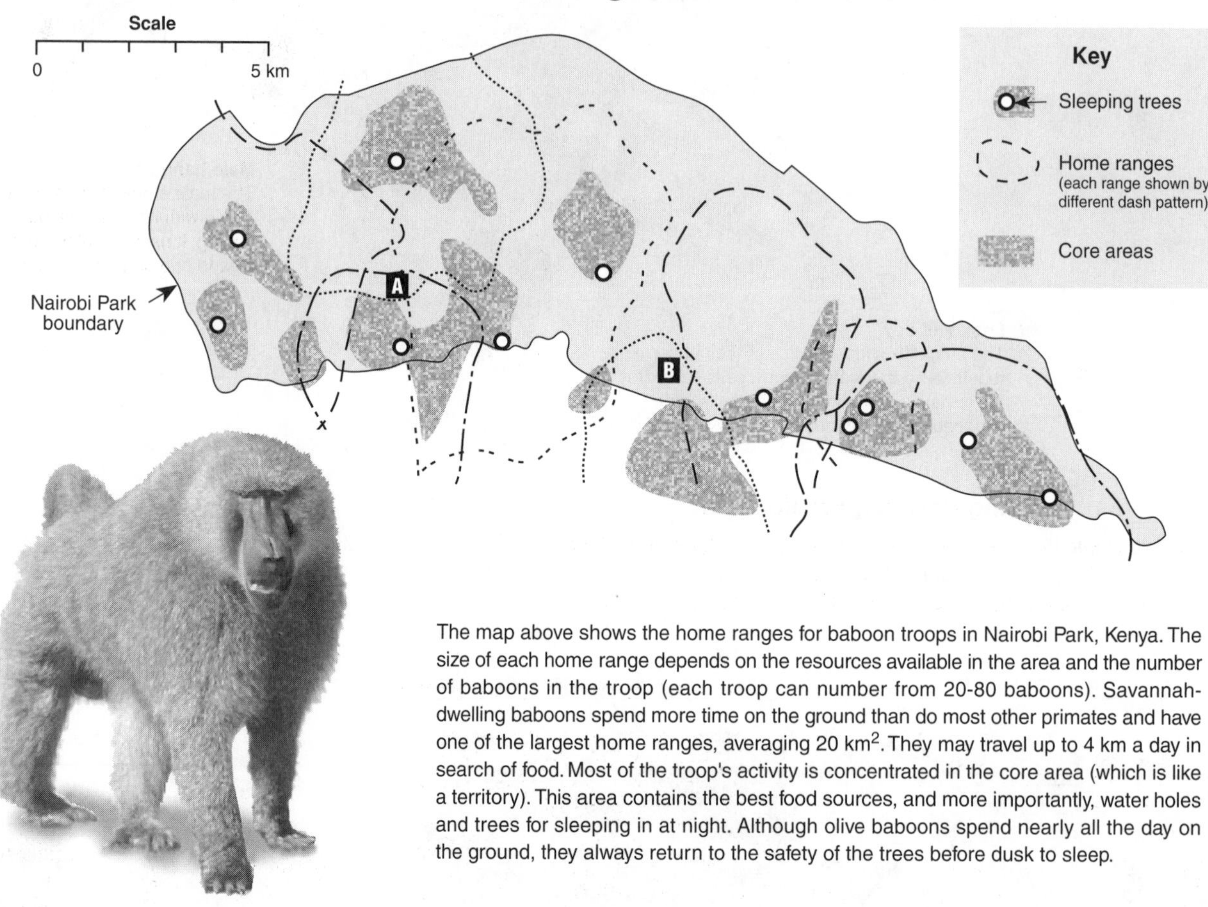

The map above shows the home ranges for baboon troops in Nairobi Park, Kenya. The size of each home range depends on the resources available in the area and the number of baboons in the troop (each troop can number from 20-80 baboons). Savannah-dwelling baboons spend more time on the ground than do most other primates and have one of the largest home ranges, averaging 20 km^2. They may travel up to 4 km a day in search of food. Most of the troop's activity is concentrated in the core area (which is like a territory). This area contains the best food sources, and more importantly, water holes and trees for sleeping in at night. Although olive baboons spend nearly all the day on the ground, they always return to the safety of the trees before dusk to sleep.

1. Describe one advantage to a baboon troop in having a troop hierarchy: ______________________

2. Describe the factors that might determine the size of the home range in any given area: ______________________

3. Baboons defend the core areas aggressively. Suggest why they would do this: ______________________

4. (a) State how many home ranges are represented on the map: ______________________

   (b) State how many home ranges overlap at the following points on the map: Point A: __________ Point B: __________

   (c) Contrast the distribution of **home ranges** and **core areas** of neighboring troops: ______________________

5. Provide an example of a territorial animal and suggest a purpose for the territoriality: ______________________

KEY TERMS Crossword

Complete the crossword below, which will test your understanding of key terms in this chapter and their meanings

Clues Across

1. This region of the brain connects the brain to the spinal cord. (2 words: 5, 4).
6. A temporary change in membrane potential caused by influx of sodium ions.
8. A type of learning, occurring at a particular life stage, that is rapid and independent of consequences.
10. Chemicals released into the external environment which act as a signal to another animal.
11. The largest and most anterior part of the human brain.
15. A branch of the peripheral nervous system that controls involuntary visceral functions (3 words: 9, 7, 6).
17. The region of the human brain which coordinates body movement, posture and balance.
19. Along with the spinal cord, this organ comprises the central nervous system.
21. A cell specialized to transmit electrical impulses.
24. The tendency for nervous tissue to become concentrated at the anterior end of an animal.
25. The membrane potential to which a membrane must be depolarized to initiate an action potential (2 words: 9, 9).

Clues Down

2. The region of the eye containing the photoreceptors.
3. The gap between neighboring neurons or between a neuron and an effector.
4. This lipid-rich substance surrounds and insulates the axons of nerves in the peripheral nervous system.
5. A physical or chemical change in the environment which causes a response in an organism.
7. A predictable behavioral response activated by a releaser (3 words: 5, 7, 7).
9. These chemicals relay signals between a neuron and another cell.
12. A term describing a mass of nerve cell bodies.
13. Behavior associated with selecting or attracting a mate.
14. A self propagating nerve impulse (2 words: 6, 9).
16. Annual movement of animals from one biome to another.
18. The process by which an animal uses various cues to determine its position.
20. An involuntary reaction to a stimulus.
22. The sense organs that respond to light.
23. Behavior that has a genetic basis and does not require learning is said to be this.

AP Topic 3B
IB AHL 11.2

Muscles and Movement

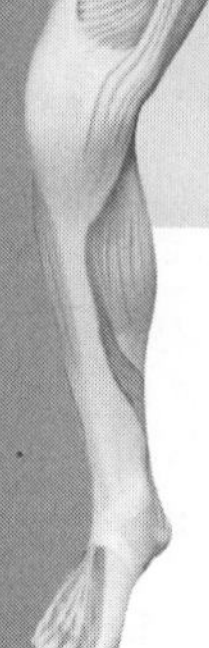

KEY CONCEPTS

- The diversity of adaptations for locomotion reflects adaptation to life in water, on land, and in the air.
- In vertebrates and arthropods, muscles generate movement by contracting against a rigid skeleton.
- The muscular system is organized into discrete muscles, which work as antagonistic pairs.
- In muscle, the movement of actin filaments against myosin filaments creates contraction. ATP is required.

KEY TERMS

actin
agonist
antagonist
appendicular skeleton
axial skeleton
bone
buoyancy
contraction
creatine phosphate
drag
endoskeleton
exoskeleton
extension
fatigue
flexion
hydrostatic skeleton
insertion (of muscle)
joint
lift
ligament
muscle
muscle fiber
myofibril
myoglobin
myosin
neuromuscular junction
origin (of muscle)
sarcomere
sliding filament theory
streamlining
synovial joint
tension
thrust

OBJECTIVES

☐ 1. Use the **KEY TERMS** to help you understand and complete these objectives.

Support and Movement in Animals
pages 213-216, 224

☐ 2. Describe diversity in animal skeletons and explain the various roles of the skeleton in different animal phyla.

☐ 3. Describe aspects of the aquatic and terrestrial environment important to how animals move. Include reference to density of the medium and drag.

☐ 4. Describe diversity in the way the animals move through their environments:
(a) Discuss adaptations for reducing **drag** (**streamlining**), maintaining **buoyancy**, and generating **thrust** in an aquatic environment.
(b) Describe diversity in the way animals move on land, including different modes of locomotion.
(c) Describe adaptations for flight, including adaptations to reduce skeletal weight, and for generating lift and overcoming drag.

☐ 5. Describe and explain the structure and function of muscles in vertebrates and invertebrates. Include reference to the **innervation** of the muscle and how muscular control and gradations in strength of contraction are achieved.

Support and Movement in Humans
pages 217-226

☐ 6. Describe the components of the musculoskeletal system and the main regions of the human **skeleton**. Describe the role of the **nerves**, **ligaments**, skeletal muscles, **tendons**, and **bones** in producing movement.

☐ 7. Describe the structure and function of the elbow joint, including reference to: cartilage, **synovial fluid**, tendons, ligaments, and named bones and **antagonistic muscles**.

☐ 8. Describe how movement is achieved by antagonistic muscle action. Identify the role of reflex inhibition in the movement of **antagonistic muscle pairs**.

☐ 9. Compare cardiac muscle, skeletal (striated) muscle, and smooth muscle, with reference to gross structure, physiology, and functional role.

☐ 10. Describe the ultrastructure of skeletal muscle **fibers**, identifying the **sarcomere** and **myofibrils**, and the composition and arrangement of the **(myo)filaments**.

☐ 11. Explain the contraction of skeletal muscle, including the role of **actin** and **myosin filaments**, ATP, the **sarcoplasmic reticulum**, and **calcium ions**.

☐ 12. Explain **muscle fatigue** and relate it to the increase in blood **lactate**, depletion of carbohydrate supplies, and decreased pH. Explain how these changes provide the stimulus for increased breathing (and heart) rates.

Periodicals:
listings for this chapter are on page 390

Weblinks:
www.thebiozone.com/weblink/SB2-2603.html

Teacher Resource CD-ROM:
Exercise Physiology

Animal Skeletons

There are three distinctly different types of skeleton. The jointed, chitinous **exoskeleton** of arthropods lies outside the body wall and covers most of the body. The bony **endoskeleton** of vertebrates is composed mostly of calcium phosphate and lies inside the body wall. In both cases, muscles generate movement by contracting and pulling on the skeleton. **Hydrostatic skeletons** (hydroskeletons) rely on fluid pressure to support body parts and generate movement. Contraction of muscles pushes the incompressible fluid and causes body movement. These examples, as well as some variations, are shown below.

Exoskeletons

Found in: *arthropods (illustrated), corals, molluscs (in conjunction with hydrostatic skeleton)*

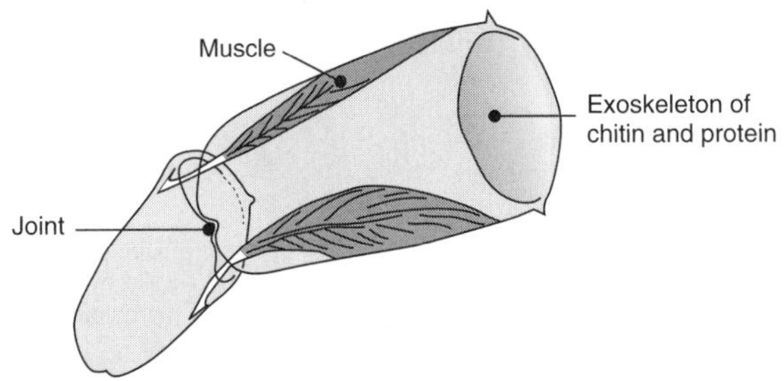

Hydrostatic Skeletons

Found in: *annelids, molluscs (sometimes with exoskeleton or shell)*

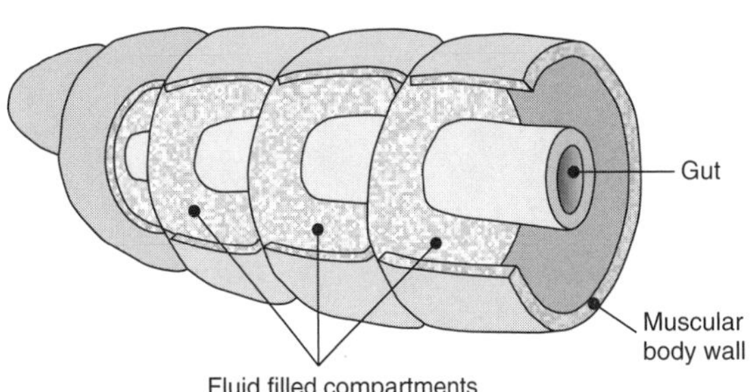

Jointed Exoskeleton

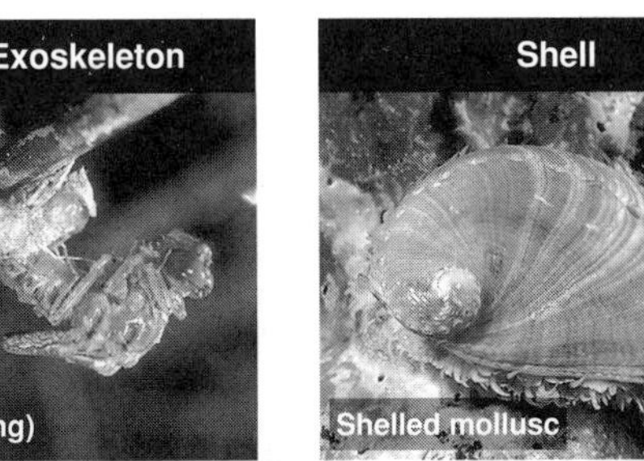

Insect (moulting)

In arthropods, the hard material is on the outside. This type of exoskeleton must be shed periodically (moulted) to allow for growth. Exoskeletons are heavy for the amount of muscle they can support and they therefore limit the size to which the animal can grow.

Shell

Shelled mollusc

Many molluscs have a particular type of exoskeleton called a shell. Unlike the exoskeleton of arthropods, the shell grows with the animal and does not need to be shed. The shell only provides protection; a hydrostatic skeleton produces movement.

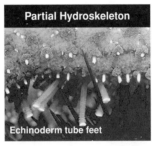

Partial Hydroskeleton

Echinoderm tube feet

As well as an endoskeleton of calcium containing ossicles, echinoderms use hydrostatic pressure to generate much of their movement. Small, fluid operated tube feet extend from the endoskeleton, gripping with suckers on the substrate to move the animal.

Annelid Hydroskeleton

Polychaete worm

RCN

The segmented body of annelids is perfectly adapted for movement using fluid pressure. As muscles in the body wall contract, they force the fluid inside the segments to move. Bristles anchor the segments that are not moving while other segments push forward.

Endoskeletons

Found in: *All vertebrates (illustrated), sponges, echinoderms*

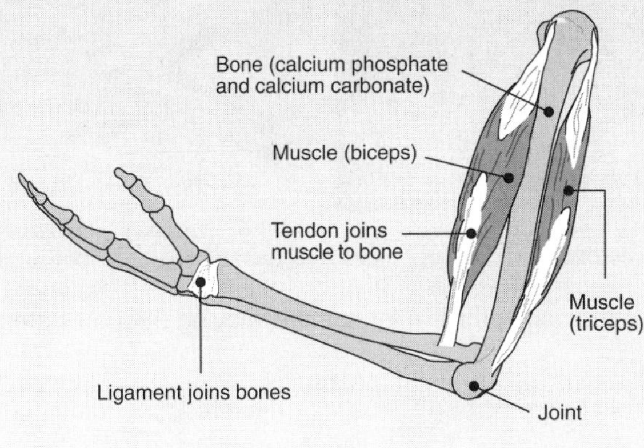

Bony Endoskeleton

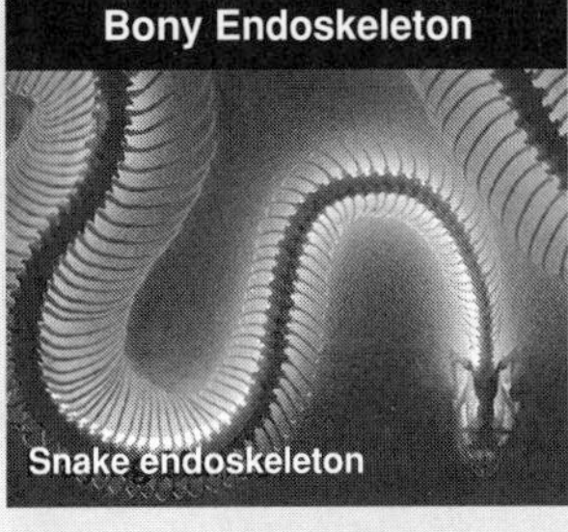

Snake endoskeleton

The bony skeleton of vertebrates provides an internal framework that protects some of the most fragile soft parts and provides for the attachment of the muscles. Animals with endoskeletons grow by adding to the hard material inside at specific places.

Endoskeleton of Ossicles

Sea egg (kina)

RA

Sea eggs at first glance appear to have an exoskeleton. They actually have a type of endoskeleton made up of small, calcareous plates (ossicles) that lie just beneath a thin layer of skin and muscle. The spines are simply projections of the endoskeleton.

1. Describe two main features of each of the following types of skeleton:

   (a) Bony endoskeleton: ______

   (b) Jointed exoskeleton: ______

   (c) Hydrostatic skeleton: ______

2. (a) Explain why exoskeletons do not enable growth to a large size in terrestrial arthropods: ______

   ______

   (b) Contrast this with the shell of molluscs: ______

Related activities: *Running, Flying, Swimming*

Swimming

Animals show a wide range of adaptations for moving through water, a medium that is much denser than air. In vertebrates, the limbs are variously modified as fins or flippers for swimming. Aquatic birds and amphibians have webbed feet to provide propulsion in water while also allowing good mobility on land. The fastest vertebrate swimmers use a powerful tail to thrust against the water and streamlining to reduce resistance. Invertebrate methods of propulsion are more diverse, and include jointed limbs and jet propulsion. This is partly a reflection of their more diverse range of support systems.

Manoeuvrability

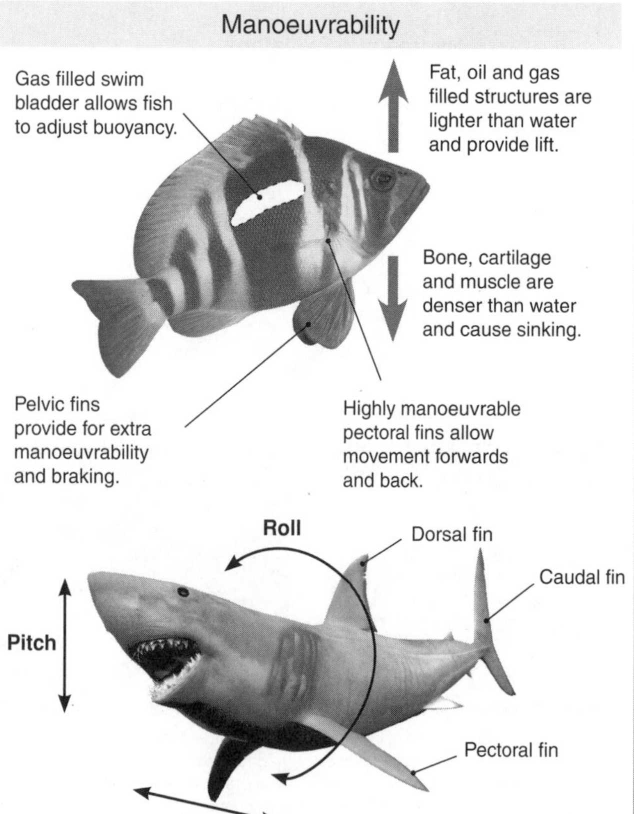

Gas filled swim bladder allows fish to adjust buoyancy.
Fat, oil and gas filled structures are lighter than water and provide lift.
Bone, cartilage and muscle are denser than water and cause sinking.
Pelvic fins provide for extra manoeuvrability and braking.
Highly manoeuvrable pectoral fins allow movement forwards and back.
Roll
Dorsal fin
Caudal fin
Pitch
Pectoral fin
Yaw

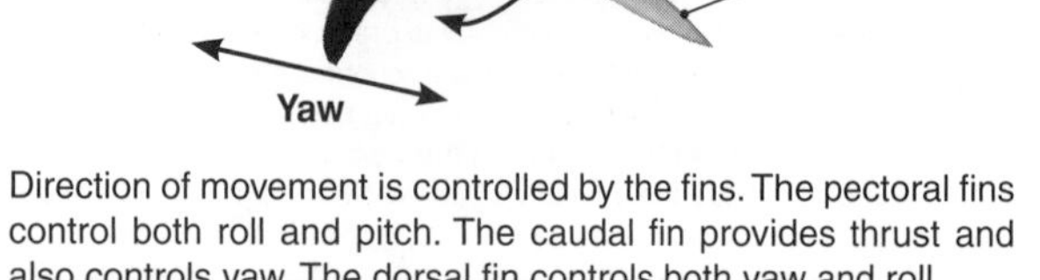

Direction of movement is controlled by the fins. The pectoral fins control both roll and pitch. The caudal fin provides thrust and also controls yaw. The dorsal fin controls both yaw and roll.

High Speed Swimming

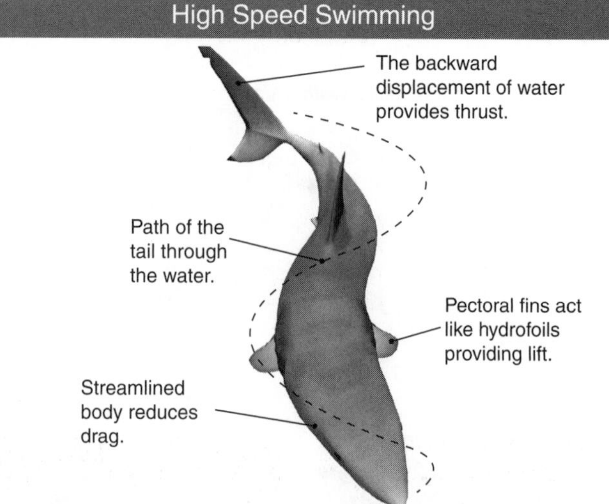

All fish swim by twisting their body from side to side to produce S-shaped movements. The tail is shaped in an upright position to make best use of this motion.

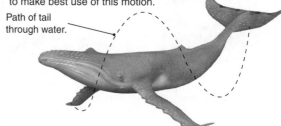

Path of tail through water.

Marine mammals swim by moving the body up and down. This manoeuvre can be traced back to land living mammal ancestors that ran with the same up-down body motion.

Broad flippers provide propulsion for sea turtles (left).

Aquatic birds and amphibians swim using webbed feet.

Cephalopods use water jet propulsion for rapid movements.

Wings modified into paddles provide propulsion for penguins.

1. Water provides most of the support for aquatic animals. Describe the main problem for animals moving through water:

2. (a) Explain what is meant by **streamlining** with respect to body shape:

   (b) Explain why streamlining is a characteristic of the fastest swimmers:

3. Describe some of the methods by which animals remain buoyant in the water:

Running

Movement on land requires an animal to physically support the body while at the same time moving it forward. This requires some limbs to provide a stable platform while others are lifted and moved. Animals have adopted various solutions to this problem. With the exception of birds, and a few mammals and reptiles, land based vertebrates use a tetrapod design. Limbs are modified depending on the degree of manoeuvrability, speed and support required by each animal. Invertebrates, which as a group are much more structurally diverse, use a wider range of methods for moving themselves through the air or water.

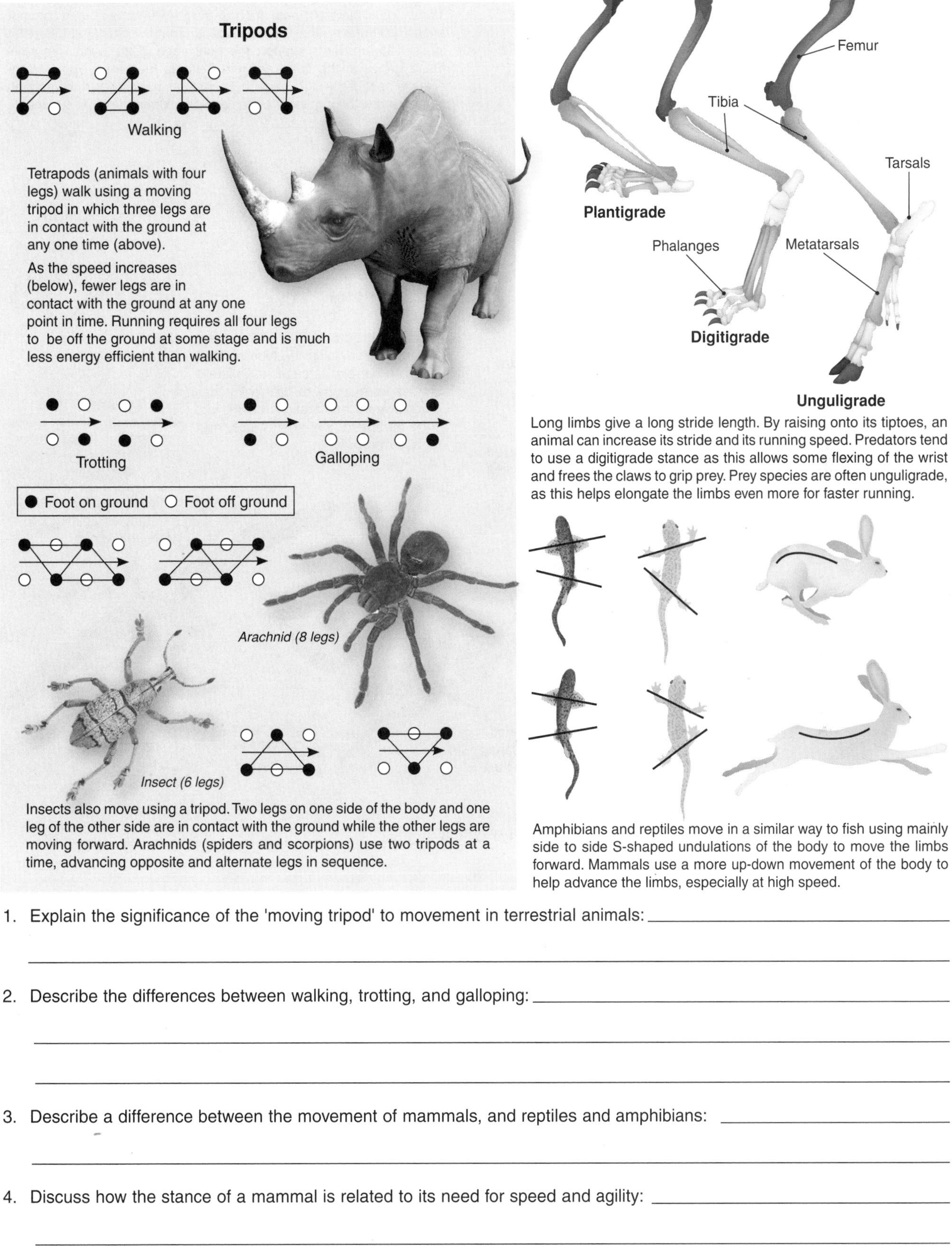

Tripods
Walking
Tetrapods (animals with four legs) walk using a moving tripod in which three legs are in contact with the ground at any one time (above).
As the speed increases (below), fewer legs are in contact with the ground at any one point in time. Running requires all four legs to be off the ground at some stage and is much less energy efficient than walking.
Trotting
Galloping
Foot on ground
Foot off ground
Arachnid (8 legs)
Insect (6 legs)
Insects also move using a tripod. Two legs on one side of the body and one leg of the other side are in contact with the ground while the other legs are moving forward. Arachnids (spiders and scorpions) use two tripods at a time, advancing opposite and alternate legs in sequence.
Femur
Tibia
Tarsals
Plantigrade
Phalanges
Metatarsals
Digitigrade
Unguligrade

Long limbs give a long stride length. By raising onto its tiptoes, an animal can increase its stride and its running speed. Predators tend to use a digitigrade stance as this allows some flexing of the wrist and frees the claws to grip prey. Prey species are often unguligrade, as this helps elongate the limbs even more for faster running.

Amphibians and reptiles move in a similar way to fish using mainly side to side S-shaped undulations of the body to move the limbs forward. Mammals use a more up-down movement of the body to help advance the limbs, especially at high speed.

1. Explain the significance of the 'moving tripod' to movement in terrestrial animals:

2. Describe the differences between walking, trotting, and galloping:

3. Describe a difference between the movement of mammals, and reptiles and amphibians:

4. Discuss how the stance of a mammal is related to its need for speed and agility:

Periodicals: Energy saving in animal movement

Related activities: Animal Skeletons

A 2

Flying

Birds, insects, and bats are the only animals to have developed true flight. Some fish, amphibians, reptiles, and other mammals have the ability to glide, but true flight requires the powered movement of the wings in order to provide both **thrust** and **lift**. Both birds and bats use modified forelimbs to provide support for the wing surface. In birds, the wing surface comprises many feathers of different types fitted together. In bats, the flight surface is a membrane of skin. Insect wings are derived from chitinous growths on the thorax and are moved in a very different way from the wing movements of either bats or birds.

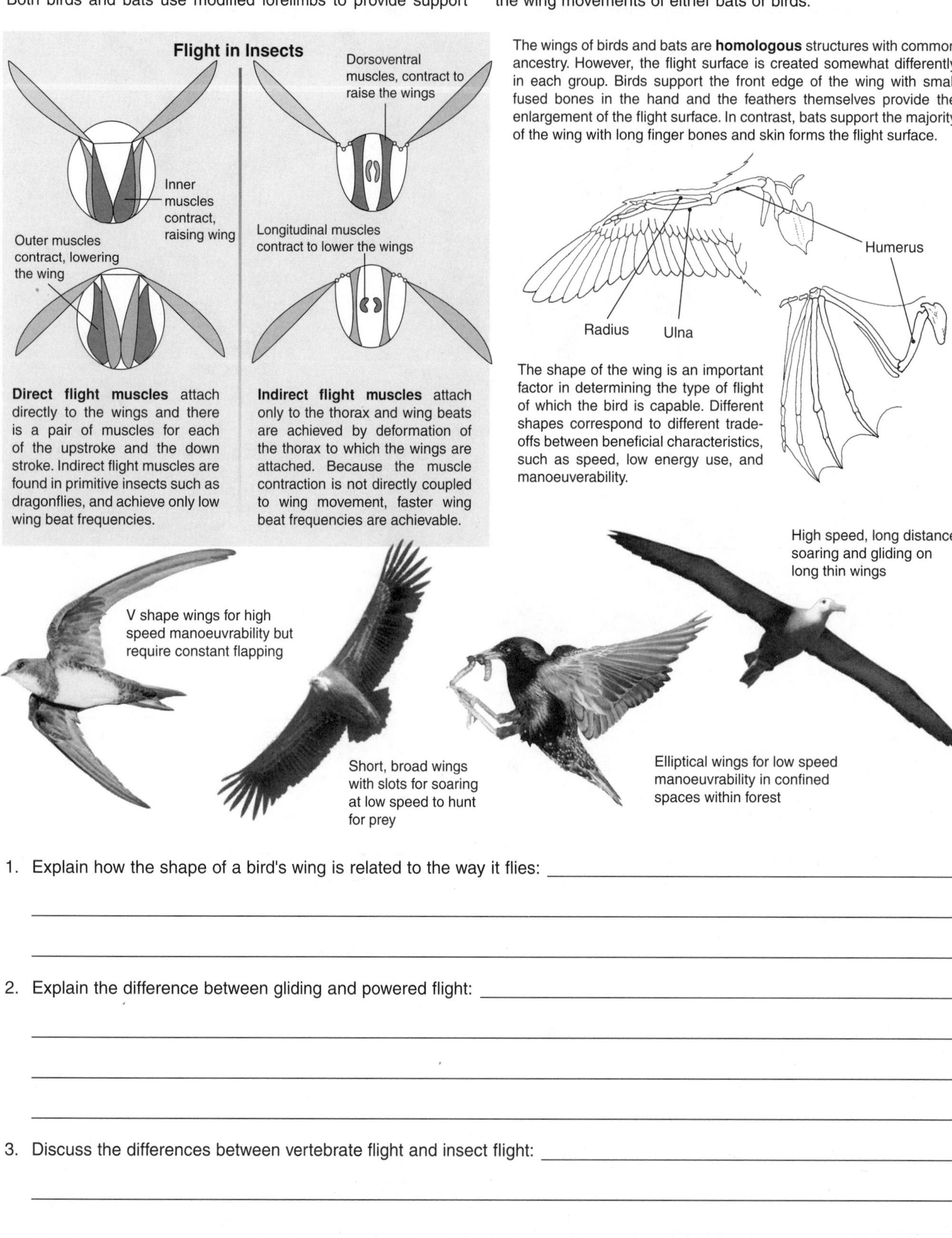

Flight in Insects
Inner muscles contract, raising wing
Outer muscles contract, lowering the wing
Dorsoventral muscles, contract to raise the wings
Longitudinal muscles contract to lower the wings
Humerus
Radius
Ulna
V shape wings for high speed manoeuvrability but require constant flapping
Short, broad wings with slots for soaring at low speed to hunt for prey
Elliptical wings for low speed manoeuvrability in confined spaces within forest
High speed, long distance soaring and gliding on long thin wings

Direct flight muscles attach directly to the wings and there is a pair of muscles for each of the upstroke and the down stroke. Indirect flight muscles are found in primitive insects such as dragonflies, and achieve only low wing beat frequencies.

Indirect flight muscles attach only to the thorax and wing beats are achieved by deformation of the thorax to which the wings are attached. Because the muscle contraction is not directly coupled to wing movement, faster wing beat frequencies are achievable.

The wings of birds and bats are **homologous** structures with common ancestry. However, the flight surface is created somewhat differently in each group. Birds support the front edge of the wing with small fused bones in the hand and the feathers themselves provide the enlargement of the flight surface. In contrast, bats support the majority of the wing with long finger bones and skin forms the flight surface.

The shape of the wing is an important factor in determining the type of flight of which the bird is capable. Different shapes correspond to different trade-offs between beneficial characteristics, such as speed, low energy use, and manoeuverability.

1. Explain how the shape of a bird's wing is related to the way it flies: ______________________________

____

____

2. Explain the difference between gliding and powered flight: ______________________________

____

____

____

3. Discuss the differences between vertebrate flight and insect flight: ______________________________

____

____

____

____

____

Related activities: *Animal Skeletons*

The Human Skeleton

The human skeleton consists of two main divisions: the **axial skeleton**, comprising the **skull**, **rib cage**, and **spine**, and the **appendicular skeleton**, made up of the limbs and the pectoral and pelvic girdles. As well as being identified by their location, bones are also described by their shape (e.g. irregular, flat, long, or short), which is related to their functional position in the skeleton. Most of the bones of the upper and lower limbs, for example, are long bones. Bones also have features such as processes, holes (foramina, *sing.* **foramen**), and depressions (**fossae**), associated with nerves, blood vessels, ligaments, and muscles. Understanding the basic organization of the skeleton, the particular features associated with its component bones, and the nature of skeletal articulations (**joints**) is essential to understanding how the movement of body parts is achieved.

WORD LIST:
phalanges, humerus, patella, scapula, tibia, clavicle, sternum, lumbar vertebra, femur, phalanges, cranium, sacrum, metacarpals, rib, ilium, fibula, carpals, tarsals, metatarsals, facial bones

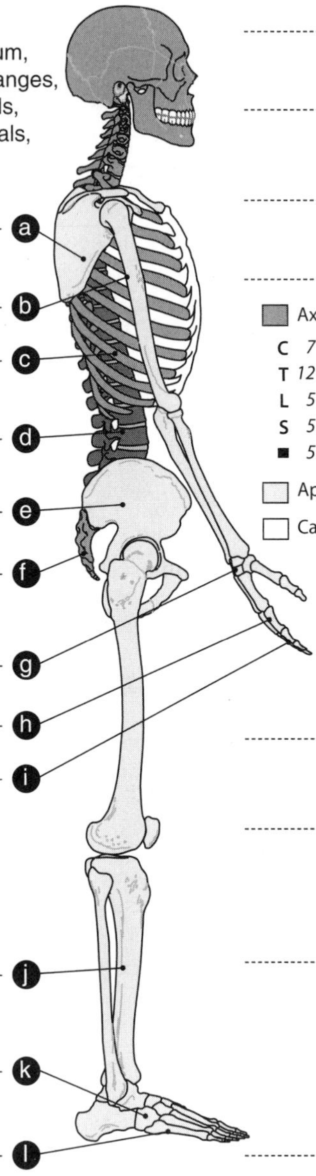

a ------------------------------------

b ------------------------------------

c ------------------------------------

d ------------------------------------

e ------------------------------------

f ------------------------------------

g ------------------------------------

h ------------------------------------

i ------------------------------------

j ------------------------------------

k ------------------------------------

l ------------------------------------

m ------------------------------------

n ------------------------------------

o ------------------------------------

p ------------------------------------

q ------------------------------------

r ------------------------------------

s ------------------------------------

t ------------------------------------

Axial skeleton
- C 7 *cervical vertebrae*
- T 12 *thoracic vertebrae*
- L 5 *lumbar vertebrae*
- S 5 *fused vertebrae = sacrum*
- ■ 5 *fused vertebrae = coccyx*

Appendicular skeleton

Cartilage

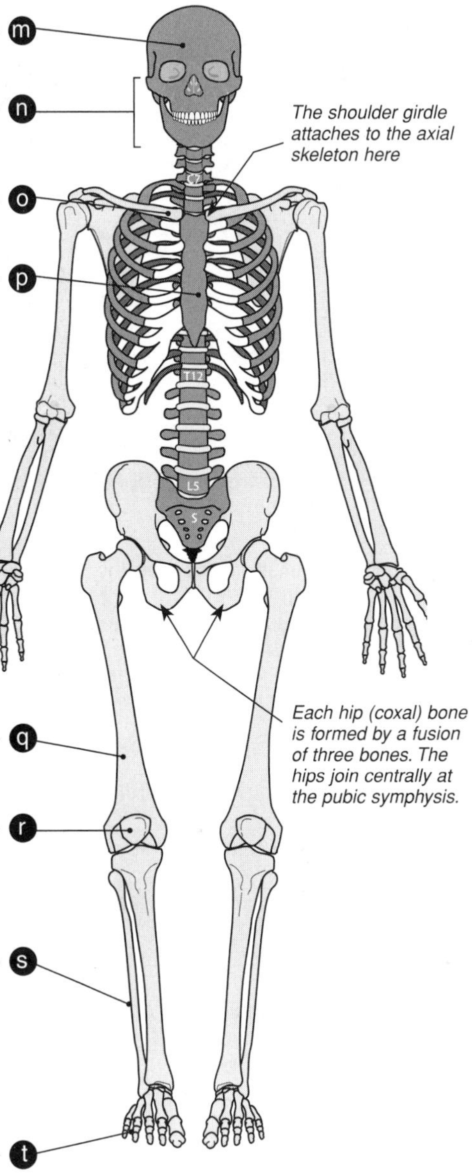

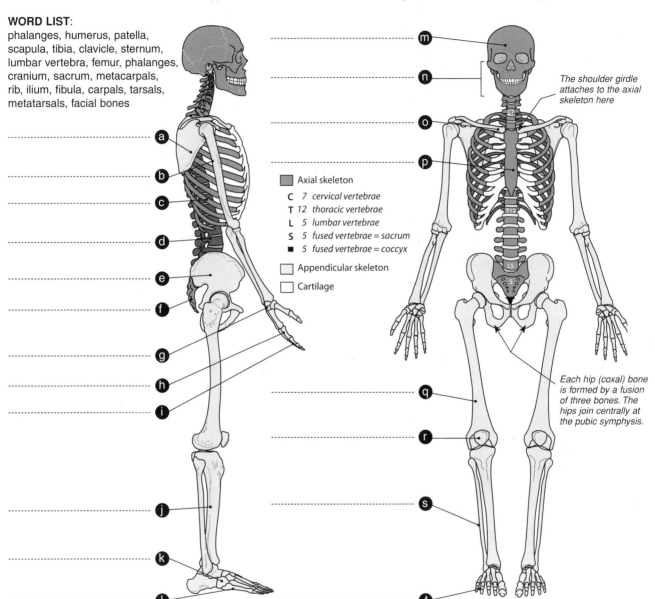

The shoulder girdle attaches to the axial skeleton here

Each hip (coxal) bone is formed by a fusion of three bones. The hips join centrally at the pubic symphysis.

Bone Shapes

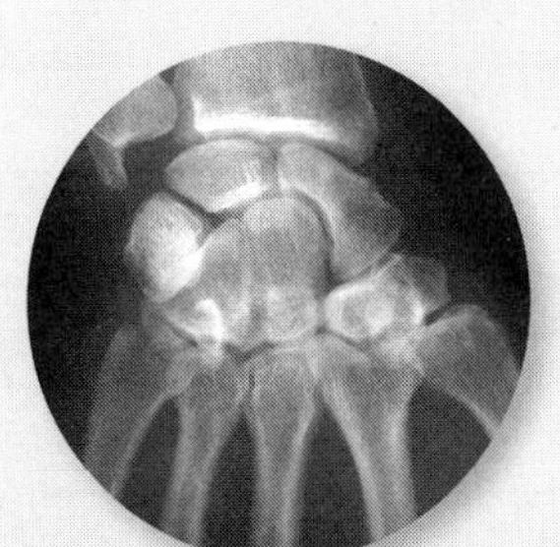

Short bones are roughly cube shaped and contain mostly spongy bone:
- carpals (above)
- tarsals
- patella

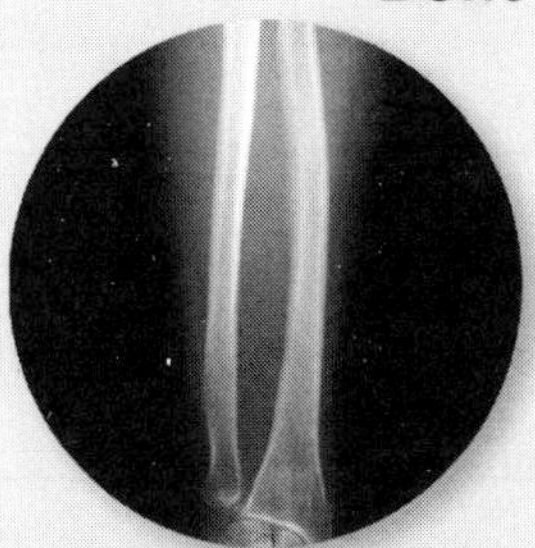

Long bones are longer than they are wide:
- most bones of the upper limbs. e.g. ulna, radius
- most bones of the lower limbs, e.g. femur, tibia

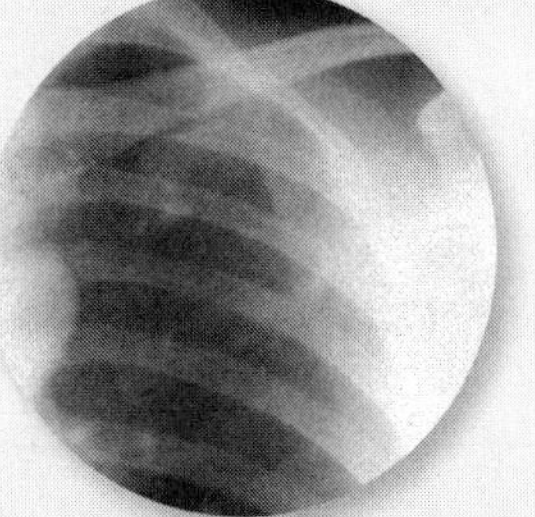

Flat bones have a thin flattened shape:
- ribs (above)
- sternum
- scapulae
- some skull bones

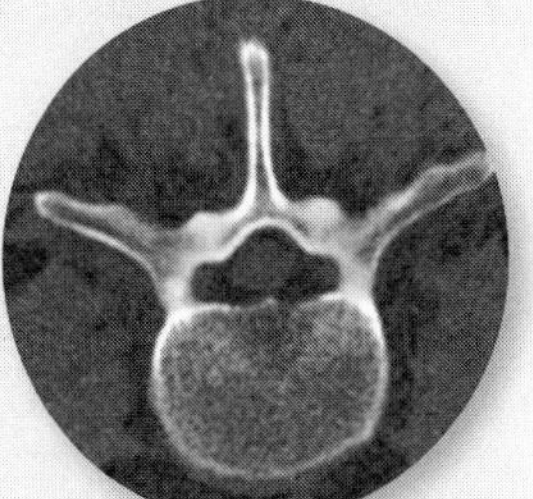

Irregular bones have an irregular shape and do not fit into the other groups:
- vertebrae (above)
- hip bones
- facial bones

Muscles and Movement

Related activities: *The Mechanics of Movement* ***Web links:*** *The Axial Skeleton, The Appendicular Skeleton, The Vertebral Column, Skeleton and Joints*

Structure of a Synovial Joint

Synovial joints (right and below) allow free movement of body parts in varying directions (one, two or three planes). The knee joint is a typical synovial joint. Like most synovial joints, it is reinforced by ligaments (most are not shown here).

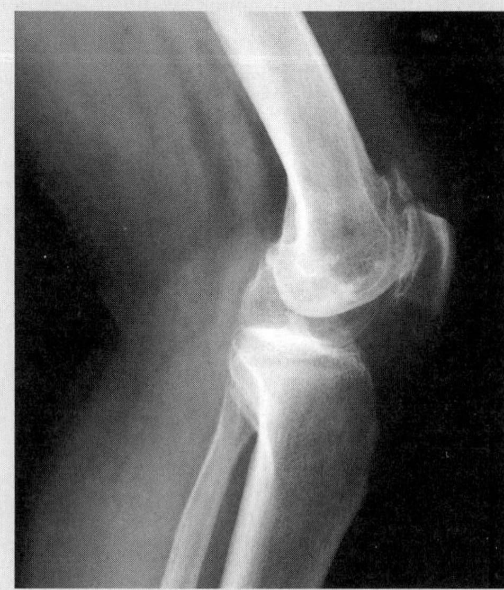

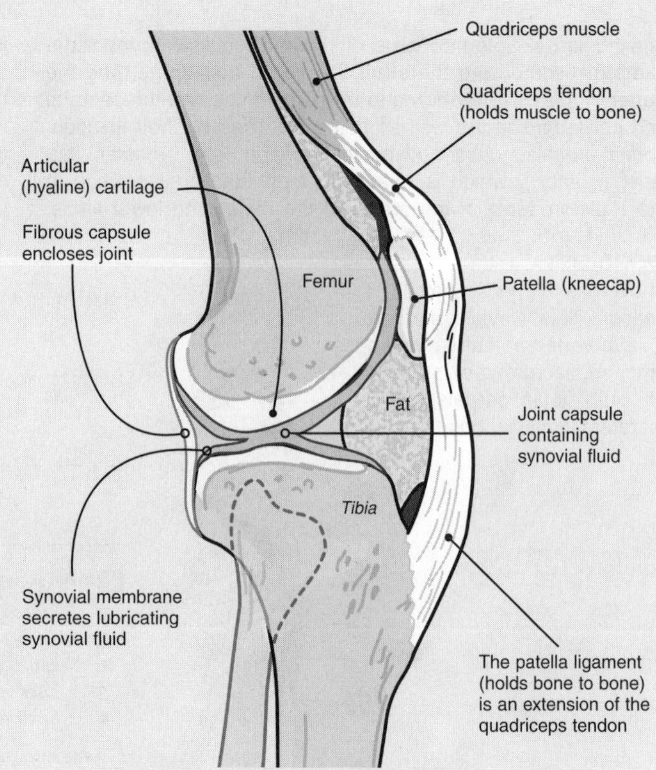

1. Use the word list provided on the previous page to label the bones (a)-(t) of the skeleton in the diagram.

2. Distinguish between the axial and the appendicular skeleton and describe the components of each: ____________________

3. (a) Identify the vertebrae with a functional role in bearing much of the spinal load: ____________________

   (b) Identify the vertebrae that form the posterior wall of the bony pelvis: ____________________

4. (a) Describe the function of the shoulder (pectoral) girdle: ____________________

   (b) Identify the single point of attachment of shoulder girdle to the axial skeleton: ____________________

5. The skull bones of babies at birth and early in infancy are not fused and some areas (the **fontanelles**) have still to be converted to bone. Describe two reasons why the skull bones are not fused into sutures until around 2 years of age:

   (a) ____________________

   (b) ____________________

6. (a) Describe the basic structure of a synovial joint: ____________________

   (b) Explain the role that synovial fluid and cartilage play in the functioning of a synovial joint: ____________________

7. Not all joints move freely. Identify two relatively immovable joints in the skeleton and state their purpose:

The Mechanics of Movement

We are familiar with the many different bodily movements achievable through the action of muscles. Contractions in which the length of the muscle shortens in the usual way are called **isotonic contractions**: the muscle shortens and movement occurs. When a muscle contracts against something immovable and does not shorten the contraction is called **isometric**. Skeletal muscles are attached to bones by tough connective tissue structures called **tendons**. They always have at least two attachments: the **origin** and the **insertion**. They create movement of body parts when they contract across **joints**. The type and degree of movement achieved depends on how much movement the joint allows and where the muscle is located in relation to the joint. Some common types of body movements are described below (left panel). Because muscles can only pull and not push, most body movements are achieved through the action of opposing sets of muscles (below, right panel).

The Action of Antagonistic Muscles

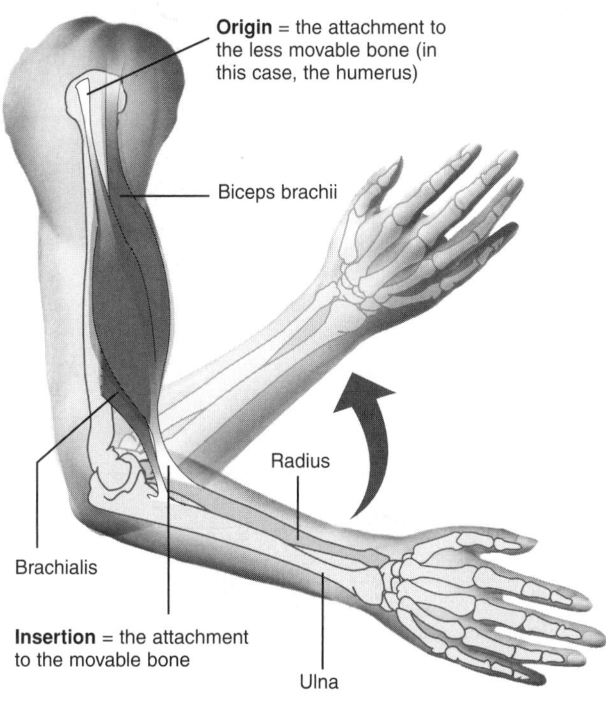

Two muscles are involved in flexing the forearm. The **brachialis**, which underlies the biceps brachii and has an origin half way up the humerus, is the **prime mover**. The more obvious **biceps brachii**, which is a two headed muscle with two origins and a common insertion near the elbow joint, acts as the synergist. During contraction, the insertion moves towards the origin.

The skeleton works as a system of levers. The joint acts as a **fulcrum** (or pivot), the muscles exert the **force**, and the weight of the bone being moved represents the **load**. The flexion (bending) and extension (unbending) of limbs is caused by the action of **antagonistic muscles**. Antagonistic muscles work in pairs and their actions oppose each other. During movement of a limb, muscles other than those primarily responsible for the movement may be involved to fine tune the movement.

Every coordinated movement in the body requires the application of muscle force. This is accomplished by the action of agonists, antagonists, and synergists. The opposing action of agonists and antagonists (working constantly at a low level) also produces muscle tone. Note that either muscle in an antagonistic pair can act as the agonist or **prime mover**, depending on the particular movement (for example, flexion or extension).

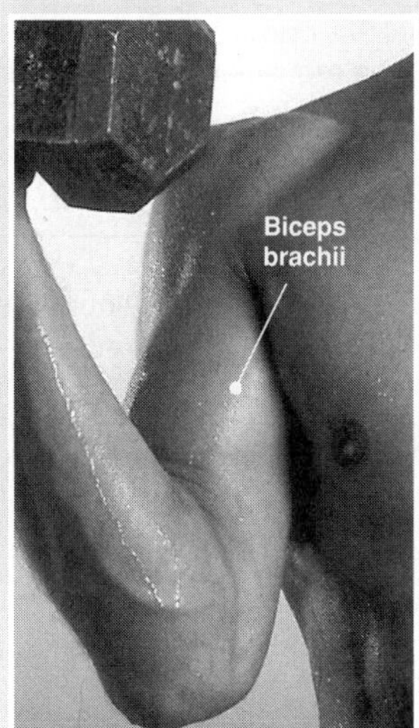

Agonists or prime movers: muscles that are primarily responsible for the movement and produce most of the force required.

Antagonists: muscles that oppose the prime mover. They may also play a protective role by preventing over-stretching of the prime mover.

Synergists: muscles that assist the prime movers and may be involved in fine-tuning the direction of the movement.

During flexion of the forearm (left) the **brachialis** muscle acts as the prime mover and the **biceps brachii** is the synergist. The antagonist, the **triceps brachii** at the back of the arm, is relaxed. During extension, their roles are reversed.

Movement at Joints

The synovial joints of the skeleton allow free movement in one or more planes. The articulating bone ends are separated by a joint cavity containing lubricating synovial fluid. Two types of synovial joint, the shoulder ball and socket joint and the hinge joint of the elbow, are illustrated below.

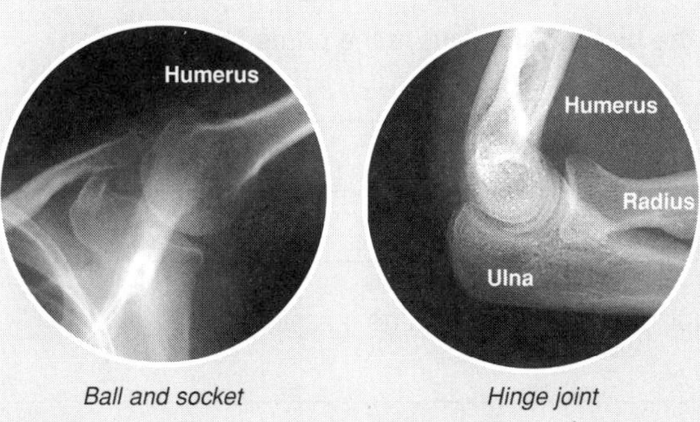

Ball and socket *Hinge joint*

Movement of the upper leg is achieved through the action of several large groups of muscles, collectively called the **quadriceps** and the **hamstrings**.

The hamstrings are actually a collection of three muscles, which act together to flex the leg.

The quadriceps at the front of the thigh (a collection of four large muscles) opposes the motion of the hamstrings and extends the leg.

When the prime mover contracts forcefully, the antagonist also contracts very slightly. This stops overstretching and allows greater control over thigh movement.

Related activities: *Muscle Structure and Function*
Web links: *Muscles in Action*

Types of Body Movement

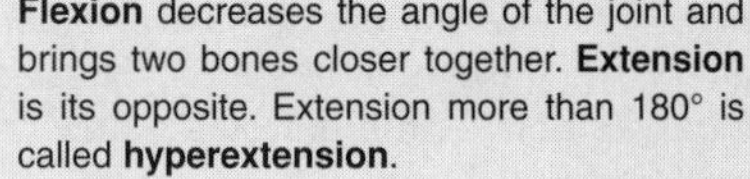

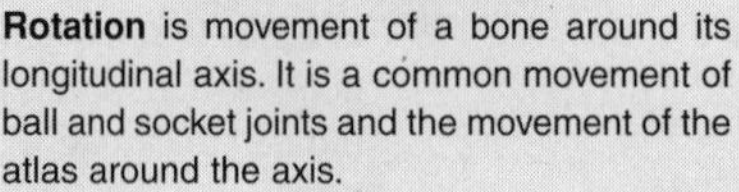

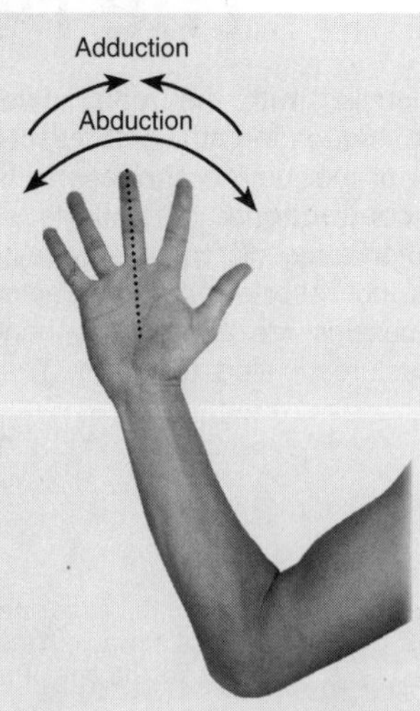

Flexion decreases the angle of the joint and brings two bones closer together. **Extension** is its opposite. Extension more than 180° is called **hyperextension**.

Rotation is movement of a bone around its longitudinal axis. It is a common movement of ball and socket joints and the movement of the atlas around the axis.

Abduction is a movement away from the midline, whereas **adduction** describes movement towards the midline. The terms also apply to opening and closing the fingers.

1. Describe the role of each of the following muscles in moving a limb:

   (a) Prime mover: ___

   (b) Antagonist: ___

   (c) Synergist: ___

2. Explain why the muscles that cause movement of body parts tend to operate as antagonist pairs: ___

3. Describe the relationship between muscles and joints Using appropriate terminology, explain how antagonistic muscles act together to raise and lower a limb:

4. Explain the role of joints in the movement of body parts: ___

5. (a) Identify the insertion for the biceps brachii during flexion of the forearm: ___

   (b) Identify the insertion of the brachialis muscle during flexion of the forearm: ___

   (c) Identify the antagonist during flexion of the forearm: ___

   (d) Given its insertion, describe the forearm movement during which the biceps brachialis is the prime mover:

6. (a) Describe a forearm movement in which the brachialis is the antagonist: ___

   (b) Identify the prime mover in this movement: ___

7. (a) Describe the actions that take place in the neck when you nod your head up and down as if saying "yes":

   (b) Describe the action being performed when a person sticks out their thumb to hitch a ride: ___

Muscle Structure and Function

There are three kinds of muscle tissue: **skeletal, cardiac**, and **smooth** muscle, each with a distinct structure. The muscles used for posture and locomotion are skeletal (voluntary) muscles and are largely under conscious control. Their distinct appearance is the result of the regular arrangement of contractile elements within the muscle cells. Muscle fibers are innervated by the branches of motor neurons, each of which terminates in a specialized cholinergic synapse called the **neuromuscular junction** (or motor end plate). A motor neuron and all the fibers it innervates (which may be a few or several hundred) are called a **motor unit**.

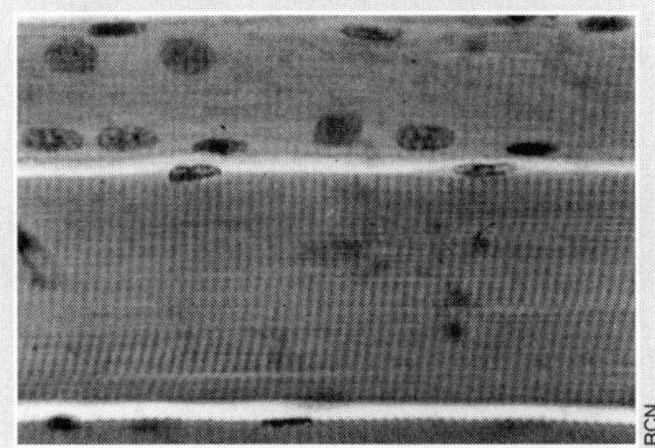
RCN

Skeletal muscle
Also called striated or striped muscle. It has a banded appearance under high power microscopy. Sometimes called voluntary muscle because it is under conscious control. The cells are large with many nuclei at the edge of each cell.

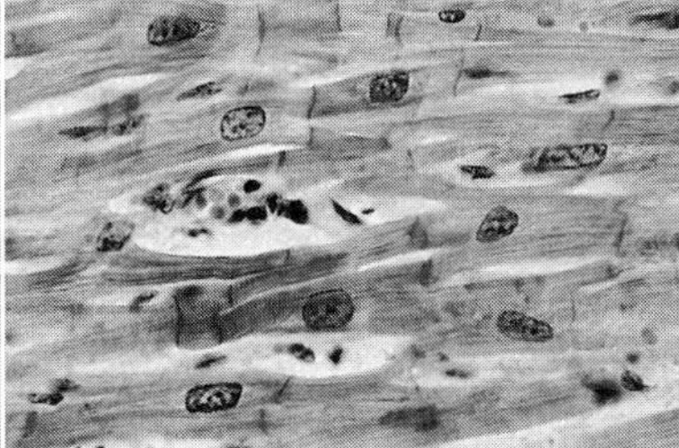

Cardiac muscle
Specialized striated muscle that does not fatigue. Cells branch and connect with each other to assist the passage of nerve impulses through the muscle. Cardiac muscle is not under conscious control (it is involuntary).

RCN

Smooth muscle
Also called involuntary muscle because it is not under conscious control. Contractile filaments are irregularly arranged so the contraction is not in one direction as in skeletal muscle. Cells are spindle shaped with one central nucleus.

Structure of Skeletal Muscle

Skeletal muscle is organized into bundles of muscle cells or **fibers**. Each fiber is a single cell with many nuclei and each fiber is itself a bundle of smaller **myofibrils** arranged lengthwise. Each myofibril is in turn composed of two kinds of **myofilaments** (thick and thin), which overlap to form light and dark bands. It is the alternation of these light and dark bands which gives skeletal muscle its striated or striped appearance. The **sarcomere**, bounded by the dark Z lines, forms one complete contractile unit.

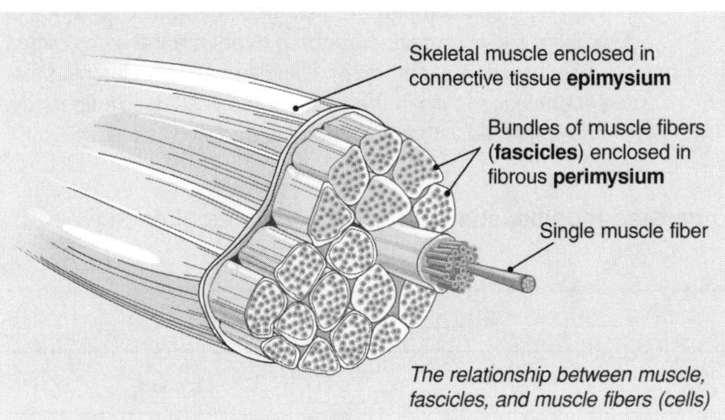
Skeletal muscle enclosed in connective tissue **epimysium**
Bundles of muscle fibers (**fascicles**) enclosed in fibrous **perimysium**
Single muscle fiber

The relationship between muscle, fascicles, and muscle fibers (cells)

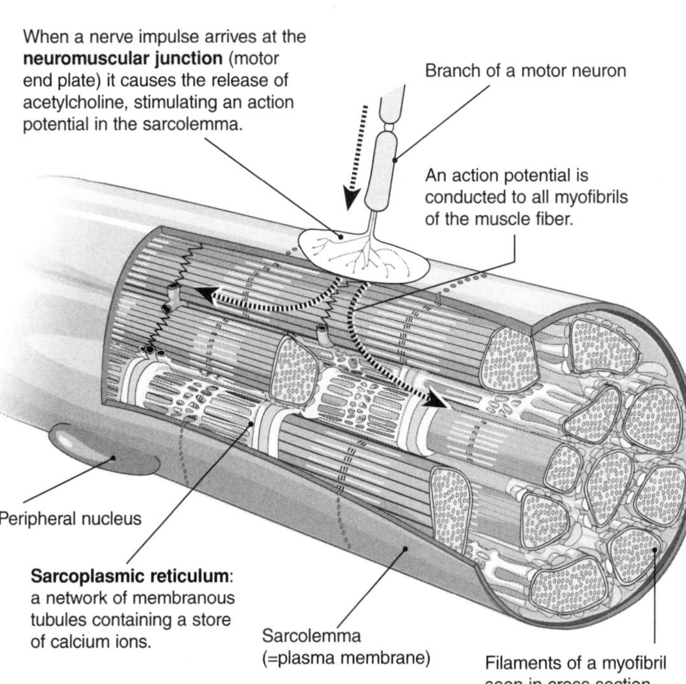
When a nerve impulse arrives at the **neuromuscular junction** (motor end plate) it causes the release of acetylcholine, stimulating an action potential in the sarcolemma.
Branch of a motor neuron
An action potential is conducted to all myofibrils of the muscle fiber.
Peripheral nucleus
Sarcoplasmic reticulum: a network of membranous tubules containing a store of calcium ions.
Sarcolemma (=plasma membrane)
Filaments of a myofibril seen in cross section

Longitudinal section of a sarcomere

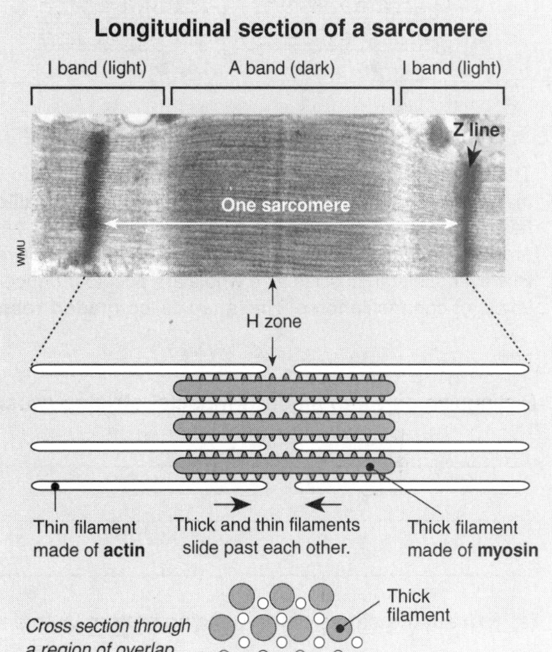
I band (light)
A band (dark)
I band (light)
Z line
One sarcomere
WMU
H zone
Thin filament made of **actin**
Thick and thin filaments slide past each other.
Thick filament made of **myosin**
Cross section through a region of overlap between thick and thin filaments.
Thick filament
Thin filament

The photograph of a sarcomere (above) shows the banding pattern arising as a result of the arrangement of thin and thick filaments. It is represented schematically in longitudinal section and cross section.

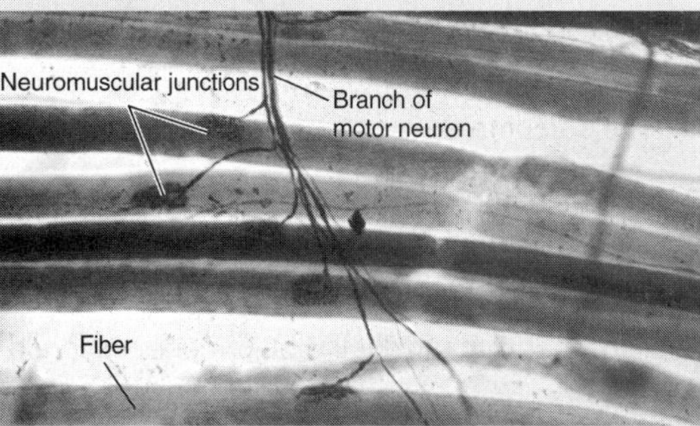
Neuromuscular junctions
Branch of motor neuron
Fiber

Above: Axon terminals of a motor neuron supplying a muscle. The branches of the axon terminate on the sarcolemma of a fiber at regions called the neuromuscular junction. Each fiber receives a branch of an axon, but one axon may supply many muscle fibers.

The Banding Pattern of Myofibrils

Within a myofibril, the thin filaments, held together by the **Z lines**, project in both directions. The arrival of an action potential sets in motion a series of events that cause the thick and thin filaments to slide past each other. This is called **contraction** and it results in shortening of the muscle fiber and is accompanied by a visible change in the appearance of the myofibril: the I band and the sarcomere shorten and H zone shortens or disappears (below).

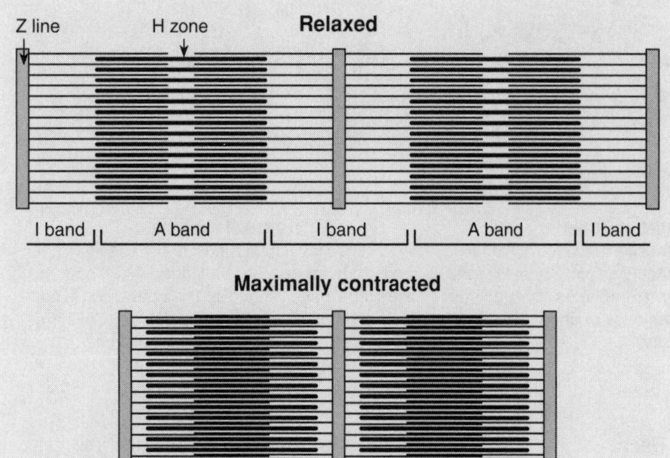

The response of a single muscle fiber to stimulation is to contract maximally or not at all; its response is referred to as the **all-or-none law** of muscle contraction. If the stimulus is not strong enough to produce an action potential, the muscle fiber will not respond. However skeletal muscles as a whole are able to produce varying levels of contractile force. These are called **graded responses**.

When Things Go Wrong

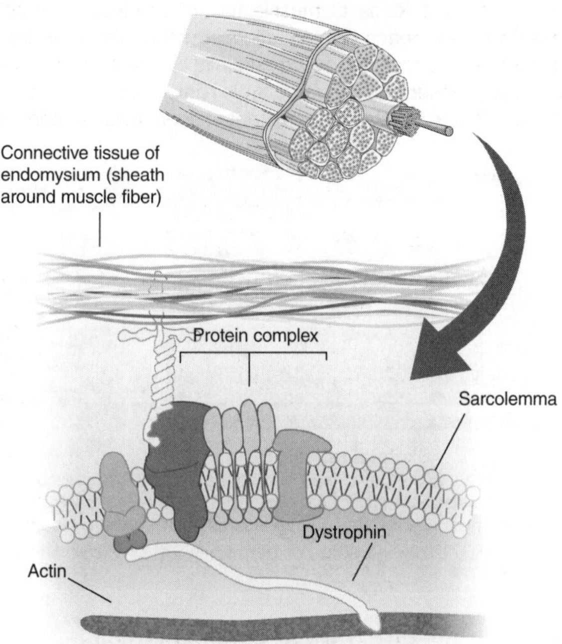

Duchenne's muscular dystrophy is an X-linked disorder caused by a mutation in the gene DMD, which codes for the protein **dystrophin**. The disease causes a rapid deterioration of muscle, eventually leading to loss of function and death. It is the most prevalent type of muscular dystrophy and affects only males. Dystrophin is an important structural component within muscle tissue and it connects muscles fibers to the extracellular matrix through a protein complex on the sarcolemma. The absence of dystrophin allows excess calcium to penetrate the sarcolemma (the fiber's plasma membrane). This damages the sarcolemma, and eventually results in the death of the cell. Muscle fibers die and are replaced with adipose and connective tissue.

1. Distinguish between **smooth muscle**, **striated muscle**, and **cardiac muscle**, summarizing the features of each type: ________________________

2. (a) Explain the cause of the banding pattern visible in striated muscle: ________________________

   (b) Explain the change in appearance of a myofibril during contraction with reference to the following:

   The I band: ________________________

   The H zone: ________________________

   The sarcomere: ________________________

3. Describe the purpose of the connective tissue sheaths surrounding the muscle and its fascicles: ________________________

4. Explain what is meant by the all-or-none response of a muscle fiber: ________________________

5. Explain why the inability to produce **dystrophin** leads to a loss of muscle function: ________________________

The Sliding Filament Theory

The previous activity described how muscle contraction is achieved by the thick and thin muscle filaments sliding past one another. This sliding is possible because of the structure and arrangement of the thick and thin filaments. The ends of the thick myosin filaments are studded with heads or **cross bridges** that can link to the thin filaments next to them. The thin filaments contain the protein actin, but also a regulatory protein complex. When the cross bridges of the thick filaments connect to the thin filaments, a shape change moves one filament past the other. Two things are necessary for cross bridge formation: calcium ions, which are released from the **sarcoplasmic reticulum** when the muscle receives an action potential, and ATP, which is hydrolyzed by ATPase enzymes on the myosin. When cross bridges attach and detach in sarcomeres throughout the muscle cell, the cell shortens. Although a muscle fiber responds to an action potential by contracting maximally, skeletal muscles as a whole can produce varying levels of contractile force. These **graded responses** are achieved by changing the frequency of stimulation (**frequency summation**) and by changing the number and size of motor units recruited (**multiple fiber summation**). Maximal contractions of a muscle are achieved when nerve impulses arrive at the muscle at a rapid rate and a large number of motor units are active at once.

The Sliding Filament Theory

Muscle contraction requires calcium ions (Ca^{2+}) and energy (in the form of ATP) in order for the thick and thin filaments to slide past each other. The steps are:

1. The binding sites on the **actin** molecule (to which myosin 'heads' will locate) are blocked by a complex of two protein molecules: tropomyosin and troponin.
2. Prior to muscle contraction, ATP binds to the heads of the myosin molecules, priming them in an erect high energy state. Arrival of an action potential causes a release of Ca^{2+} from the sarcoplasmic reticulum. The Ca^{2+} binds to the troponin and causes the blocking complex to move so that the myosin binding sites on the actin filament become exposed.
3. The heads of the cross-bridging myosin molecules attach to the binding sites on the actin filament. Release of energy from the hydrolysis of ATP accompanies the cross bridge formation.
4. The energy released from ATP hydrolysis causes a change in shape of the myosin **cross bridge**, resulting in a bending action (*the power stroke*). This causes the actin filaments to slide past the myosin filaments towards the centre of the sarcomere.
5. (Not illustrated). Fresh ATP attaches to the myosin molecules, releasing them from the binding sites and repriming them for a repeat movement. They become attached further along the actin chain as long as ATP and Ca^{2+} are available.

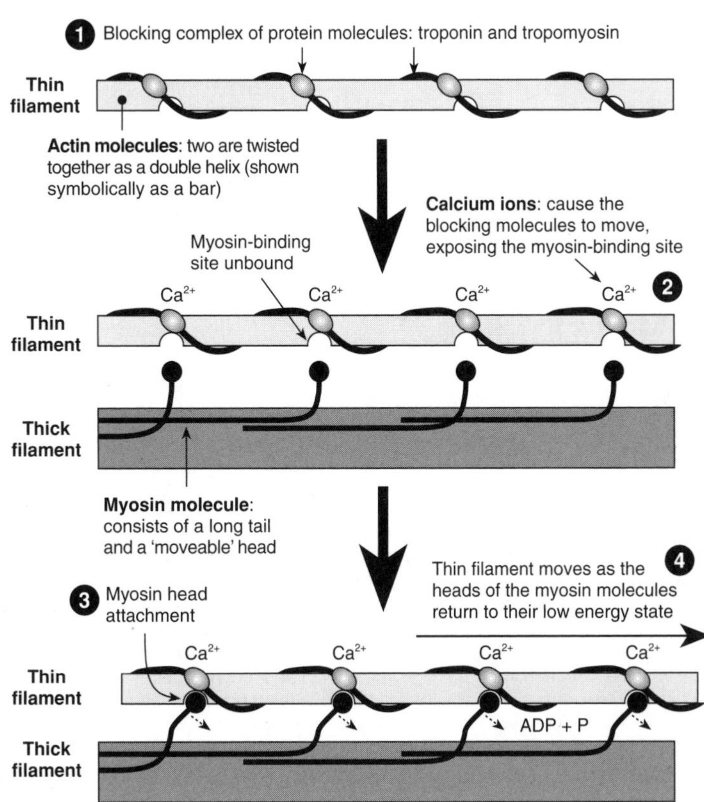

1. Match the following chemicals with their functional role in muscle movement (draw a line between matching pairs):

(a) Myosin	• Bind to the actin molecule in a way that prevents myosin head from forming a cross bridge
(b) Actin	• Supplies energy for the flexing of the myosin 'head' (power stroke)
(c) Calcium ions	• Has a moveable head that provides a power stroke when activated
(d) Troponin-tropomyosin	• Two protein molecules twisted in a helix shape that form the thin filament of a myofibril
(e) ATP	• Bind to the blocking molecules, causing them to move and expose the myosin binding site

2. Describe the two ways in which a muscle as a whole can produce contractions of varying force:

(a) ______________________________

____

(b) ______________________________

____

3. (a) Identify the two things necessary for cross bridge formation: ______________________________

(b) Explain where each of these comes from: ______________________________

____

Muscles and Movement

Related activities: Muscle Structure and Function
Web links: Muscle Cell Contraction, Sliding Filament

Muscle Innervation

Skeletal muscle contracts when a nerve impulse arrives at the neuromuscular junction and spreads to all fibers in the muscle through the system of T-tubules that run across the muscle cells. However, there are marked differences in how muscles are innervated in different animal phyla. Vertebrate muscle is innervated from a large number of neurons in the spinal cord. Each neuron may supply several muscle fibers but each fiber receives only one motor neuron. A motor neuron and the fibers it innervates is a called a **motor unit** and vertebrates achieve stronger contractions by activating more motor units. The motor neurons of vertebrates are only excitatory; they cannot inhibit contraction. In contrast, in arthropods, the nerve fibers from a relatively small number of neurons branch extensively to supply the entire muscle, and strength of contraction is controlled by which neurons (fast, slow, or inhibitory) are active. The contractile cells of a cnidarian are included below for comparison.

Making Muscles Work

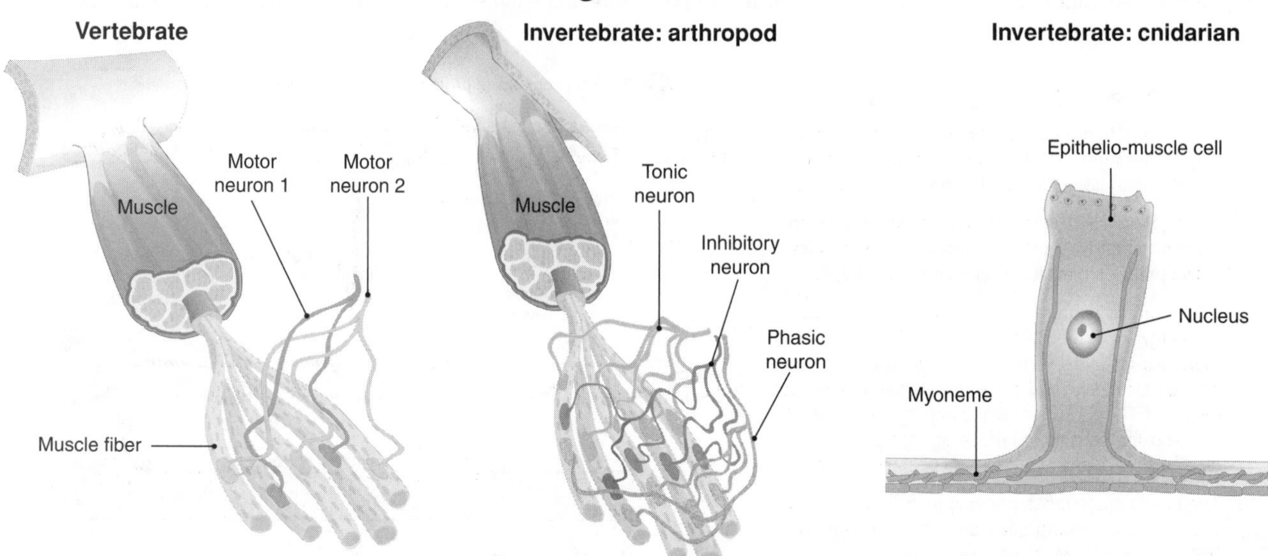

In vertebrates, each muscle fiber is controlled by a **motor neuron**. The contraction force exerted by the muscle is proportional to the number of fibers stimulated by motor neurons. This allows vertebrates to control the amount and speed of whole muscle contraction. A small number of fibers stimulated causes a small contraction.

Arthropod muscles contain only a small number of fibers. Invertebrates achieve muscular control by using use three distinct types of motor neuron. Each muscle fiber is innervated with **fast** (phasic), **slow** (tonic), and **inhibitory** neurons and each type of neuron elicits a difference response from the muscle fiber.

Hydra (anemone like animals), do not possess true muscle cells. Instead, movements are achieved using contractile cells called **epithelio-muscle** cells. These contain fibers called **myonemes**, which are able to perform simple contractile functions such as contracting the body to force out water from the mouth.

1. Explain how *Hydra* produces contractile movement, even though it lacks muscle tissue: ______

2. Compare and contrast muscle innervation in vertebrates and arthropods with reference to the following:

   (a) The number of motor neurons supplying each muscle fiber: ______

   (b) The type of neuron supplying each fiber: ______

   (c) How contraction strength is varied: ______

3. In vertebrate muscle, there are no inhibitory fibers. Explain how inhibition is achieved in vertebrates: ______

Related activities: *The Sliding Filament Theory*

Muscle Fatigue

Muscle fatigue refers to the decline in a muscle's ability to maintain force in a prolonged or repeated contraction. It is a normal result of vigorous exercise but the reasons for it are complex. During strenuous activity, the oxygen demand of the working muscle exceeds supply and the muscle metabolizes anaerobically. This leads to the metabolic changes (including lactic acid accumulation) that result in fatigue. The lactic acid (lactate) produced during muscle contraction diffuses into the bloodstream and is carried to the liver where it is reconverted to glucose via the **Cori cycle**. This requires energy from aerobic respiration and accounts for the **oxygen debt** which causes the continued heavy breathing after exercise has stopped. During moderate intensity exercise, the rate at which lactate enters the blood from the muscles is equal to its rate of removal by the liver. This steady-state situation is suitable for endurance training, because the lactic acid doesn't accumulate in the muscle tissue.

At Rest

- Muscles produce a surplus of ATP
- This extra energy is stored in CP (creatine phosphate) and glycogen

During Moderate Activity

- ATP requirements are met by the aerobic metabolism of glycogen and lipids

During Peak Activity

- Effort is limited by ATP. ATP production is ultimately limited by availability of oxygen.
- During short-term, intense activity, ATP is increased through anaerobic metabolism of glycogen (glycolysis).
- A by-product of this is lactic acid, which lowers tissue pH and affects cellular activity.
- When **fatigued**, the muscle can no longer contract fully.

Lactic Acid and Muscle Fatigue

Lactic acid is a by-product of ATP production through anaerobic metabolism when oxygen demand exceeds supply. Lactic acid accumulation in the muscle causes a fall in tissue pH and inhibits the activity of the key enzymes involved in ATP production. This decline in ATP supply limits muscular performance during peak activity (graph below). Together with the effects of ATP and **creatine phosphate** breakdown (accumulating phosphate (*Pi*) for example), lactic acid buildup also slows the release of calcium into the T tubules and affects the ion pumps responsible for moving calcium ions back into the sarcoplasmic reticulum. This contributes to fatigue because calcium is a key component in muscle contraction.

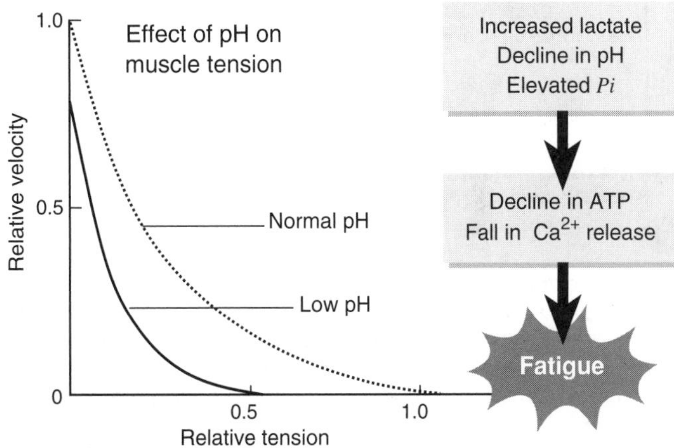

Short term maximal exertion (sprint)
- Lactic acid build-up lowers pH
- Depletion of creatine phosphate
- Buildup of phosphate (P_i) affects the sensitivity of the muscle to Ca^{2+}

Mixed aerobic and anaerobic (5 km race)
- Lactate accumulation in the muscle
- Build-up of ADP and P_i
- Decline in Ca^{2+} release affects the ability of the muscle to contract

Extended sub-maximal effort (marathon)
- Depletion of all energy stores (glycogen, lipids, amino acids) leads to a failure of Ca^{2+} release
- Repetitive overuse damages muscle fibers.

1. Explain the mechanism by which lactic acid accumulation leads to muscle fatigue: ____________________

____

2. Identify the two physiological changes in the muscle that ultimately result in a decline in muscle performance:

(a) ____________________ (b) ____________________

3. Suggest why the reasons for fatigue in a long distance race are different to those in a 100 m sprint: ____________________

____

____

Related activities: Muscle Structure and Function

Homeostasis During Exercise

Physical exercise places greater demands on the abilities of the body to maintain a steady state. Extra heat generated during exercise must be dissipated, oxygen demands increase, and there are more waste products produced. The body has an immediate response to exercise but also, over time, responds to the stress of repeated exercise (**training**) by adapting and improving its capacity for exercise and the efficiency with which it performs. This concept is illustrated below. Training causes tissue damage and depletes energy stores, but the body responds by repairing the damage, replenishing energy stores, and adjusting its responses in order to minimize the impact of exercise in the future. The maintenance of homeostasis during exercise is principally the job of the circulatory and respiratory systems, although the skin, kidneys, and liver are also important.

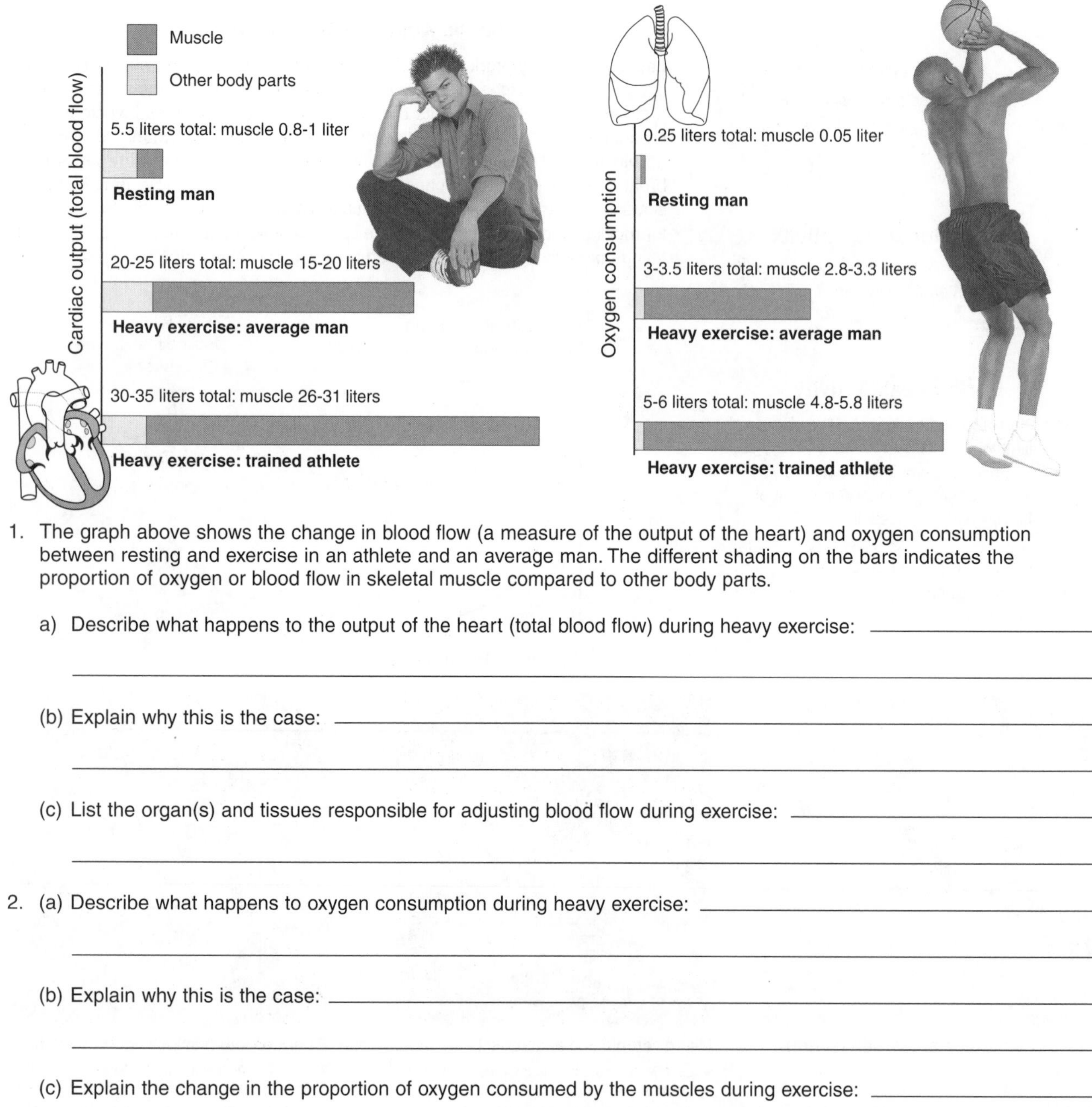

1. The graph above shows the change in blood flow (a measure of the output of the heart) and oxygen consumption between resting and exercise in an athlete and an average man. The different shading on the bars indicates the proportion of oxygen or blood flow in skeletal muscle compared to other body parts.

   a) Describe what happens to the output of the heart (total blood flow) during heavy exercise: ____________________

   (b) Explain why this is the case: ____________________

   (c) List the organ(s) and tissues responsible for adjusting blood flow during exercise: ____________________

2. (a) Describe what happens to oxygen consumption during heavy exercise: ____________________

   (b) Explain why this is the case: ____________________

   (c) Explain the change in the proportion of oxygen consumed by the muscles during exercise: ____________________

3. Explain the difference in oxygen consumption and blood flow between a trained athlete and an average man:

____

***Related activities:** Exercise and Blood Flow*

KEY TERMS Word Find

Use the clues below to find the relevant key terms in the WORD FIND grid

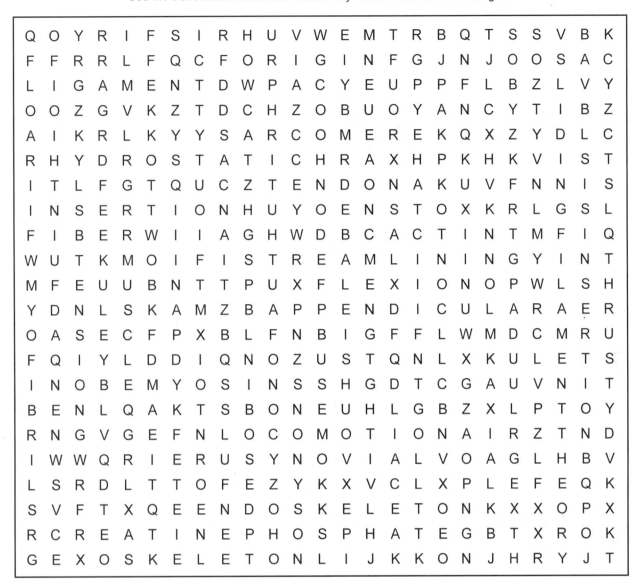

```
Q O Y R I F S I R H U V W E M T R B Q T S S V B K
F F R R L F Q C F O R I G I N F G J N J O O S A C
L I G A M E N T D W P A C Y E U P P F L B Z L V Y
O O Z G V K Z T D C H Z O B U O Y A N C Y T I B Z
A I K R L K Y Y S A R C O M E R E K Q X Z Y D L C
R H Y D R O S T A T I C H R A X H P K H K V I S T
I T L F G T Q U C Z T E N D O N A K U V F N N I S
I N S E R T I O N H U Y O E N S T O X K R L G S L
F I B E R W I I A G H W D B C A C T I N T M F I Q
W U T K M O I F I S T R E A M L I N I N G Y I N T
M F E U U B N T T P U X F L E X I O N O P W L S H
Y D N L S K A M Z B A P P E N D I C U L A R A E R
O A S E C F P X B L F N B I G F F L W M D C M R U
F Q I Y L D D I Q N O Z U S T Q N L X K U L E T S
I N O B E M Y O S I N S S H G D T C G A U V N I T
B E N L Q A K T S B O N E U H L G B Z X L P T O Y
R N G V G E F N L O C O M O T I O N A I R Z T N D
I W W Q R I E R U S Y N O V I A L V O A G L H B V
L S R D L T T O F E Z Y K X V C L X P L E F E Q K
S V F T X Q E E N D O S K E L E T O N K X X O P X
R C R E A T I N E P H O S P H A T E G B T X R O K
G E X O S K E L E T O N L I J K K O N J H R Y J T
```

Material composed mainly of calcium, phosphorus and sodium that forms the endoskeleton of vertebrates.

A model for muscle contraction in which filaments slide past one another.

The action of actin and myosin cross-bridge cycling creates this in muscle.

A connective tissue structure that connects bones to bones.

A condition in which the muscle has a reduced ability to generate force, usually as a result of energy supplies being exhausted.

Hard, waterproof structure outside the soft tissues of insects and other invertebrates.

Bending movement that decreases the angle between two body parts.

Tissue composed of contractile fibers.

A muscle cell is also called this.

The most common and most movable joint type in mammals, in which the articulating surfaces are surrounded by a capsule.

This protein forms the thick filaments in muscle myofibrils.

An upward force produced by objects less dense than the medium they are in.

The end of the muscle attaching to the bone that is not moved during a contraction.

The end of the muscle attaching to the freely moving bone of its joint.

Support structure that is found within the tissues of animals. Provides for the attachment of muscles and as a system of levers to manipulate objects.

The part of the human skeleton that includes the shoulder and pelvic girdles.

A phosphorylated molecule that acts as a rapidly available reserve of energy in skeletal muscle. (2 words: 8, 9)

Cylindrical organelles in muscle cells comprising bundles of actin and myosin filaments.

The act of self propulsion. Includes walking, swimming and flying.

This protein forms the thin filaments in muscle myofibrils.

Part of the muscle structure where filaments of myosin and actin overlap. The contraction of this structure is caused by the movement of these filaments past each other.

The part of the human skeleton that includes the cranium and spinal column.

Skeleton that relies on fluid pressure to support body parts and provide movement.

The name describing shortening of a muscle fiber.

A connective tissue structure that connects muscles to bone.

Forward motion produced by fins causing the backward displacement of water.

The process of reducing drag in order to use less effort to provide motion.

Force created in bony fish by the swim bladder and in sharks by the pectoral and caudal fins. In birds the force is created by the flapping and angling of the wings.

A straightening movement that increases the angle between body parts.

AP Topic 3B
IB SL/HL 6.6
AHL 11.4

The Next Generation

KEY CONCEPTS

- Almost all animals reproduce sexually at some stage in their life cycle.
- Different reproductive strategies are associated with internal or external fertilization and internal or external development of the embryo.
- Hormonal cycles govern all aspects of human reproduction.
- The placenta is a temporary organ to support the fetus during pregnancy in mammals.

KEY TERMS

aging (=senescence)
asexual reproduction
childhood
conception
contraception
copulation
corpus luteum
embryo
endometrium
estrogen
fertilization
fetus
gamete
gametogenesis
hermaphrodite
lactation
meiosis
menopause
menstrual cycle
mitosis
oogenesis
oviparous
ovoviviparous
ovulation
ovum (=egg)
oxytocin
parthenogenesis
parturition
placenta
pregnancy
progesterone
prolactin
puberty
sexual reproduction
sperm
spermatogenesis
testosterone
viviparous
zygote

OBJECTIVES

☐ 1. Use the **KEY TERMS** to help you understand and complete these objectives.

Principles of Reproduction

pages 229-232

☐ 2. Compare and contrast **sexual** and **asexual reproduction**.

☐ 3. Explain the role of **gamete formation** and **fertilization** in **sexual reproduction**. Describe differences between male gametes (**sperm**) and female gametes (**eggs**) in terms of size, the number produced, and motility.

☐ 4. Recall the role of **meiosis** in producing **gametes**. Explain how **fertilization** restores the diploid chromosome number in the **zygote**.

Reproductive Strategies

page 231-234

☐ 5. Distinguish between **internal** and **external fertilization**. Describe the characteristics, advantages, and disadvantages of each strategy.

☐ 6. Distinguish **copulation** from **fertilization** and identify the structures that have developed in males and females to enable internal fertilization.

☐ 7. Distinguish between **oviparous**, **ovoviviparous**, and **viviparous**, and describe examples of each type of reproductive strategy.

Human Reproduction & Development

page 235-248

☐ 8. Describe the structure and function of adult male and female reproductive systems. Describe the processes involved in **gametogenesis** (**oogenesis** in females and **spermatogenesis** in males) and where they occur.

☐ 9. Contrast gametogenesis in males and females.

☐ 10. Explain the features of the **menstrual cycle** including the development of the ovarian follicles and corpora lutea, the cyclical changes to the **endometrium**, and **menstruation**. Relate the changes in the menstrual cycle to the changes in the hormones regulating the cycle.

☐ 11. Describe **fertilization**, including the **acrosome reaction** and the **cortical reaction**. Describe and explain the main events in **embryonic development** between **implantation** and 5-8 weeks.

☐ 12. Describe the structure and function of the **placenta**, including reference to how it is maintained during the **pregnancy**.

☐ 13. Describe **birth** (parturition) and its control, including the role of **oxytocin** and **progesterone**.

☐ 14. Describe **lactation** and its control. Explain its importance to early nutrition.

☐ 15. Describe milestones in the growth and development of humans, including **infancy**, **childhood**, **puberty**, and **senescence**.

Periodicals: *listings for this chapter are on page 390*

Weblinks: *www.thebiozone.com/weblink/SB2-2603.html*

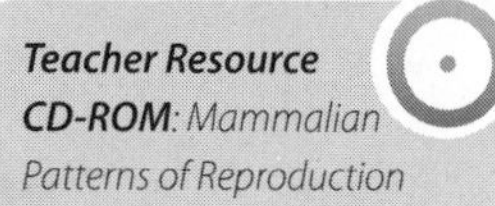

Teacher Resource CD-ROM: *Mammalian Patterns of Reproduction*

Asexual Reproduction

In most forms of asexual reproduction, the parent splits, fragments, or buds to produce offspring identical to itself. Parthenogenesis is a special type of asexual reproduction where unfertilized eggs give rise to clones. Asexually reproducing organisms do not need to find a mate, so the energy that might otherwise be used for sexual activity can be used for other things. Asexual reproduction is rapid, but all the offspring are genetically identical. If conditions change there is little ability to adapt.

Binary Fission

Binary fission is a method of asexual reproduction that occurs in prokaryotes (bacteria and cyanobacteria) and protists (e.g. ***Amoeba*** and ***Paramecium***). Binary fission occurs by division of a parent body into two, more or less equal, parts. The cell's DNA is replicated, followed by division of the nucleoplasm (in prokaryotes) or the cytoplasm (protists). The series (right) shows stages in the process of binary fission in *Amoeba*. The photograph below this series shows binary fission in *Paramecium*. The nucleus has divided and the cytoplasm is dividing. The arrows indicate where a constriction is developing in the cell.

Note that some life cycle stages of parasitic protozoans, such as the malarial parasite *Plasmodium*, undergo **multiple fission**. The nucleus divides repeatedly before the final division of the cytoplasm to produce many new cells. Repeated cycles of multiple fission produce large numbers of offspring very rapidly.

Amoeba **undergoing fission**

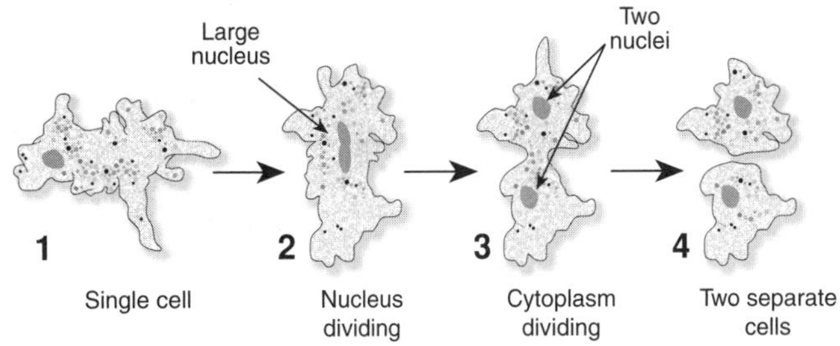

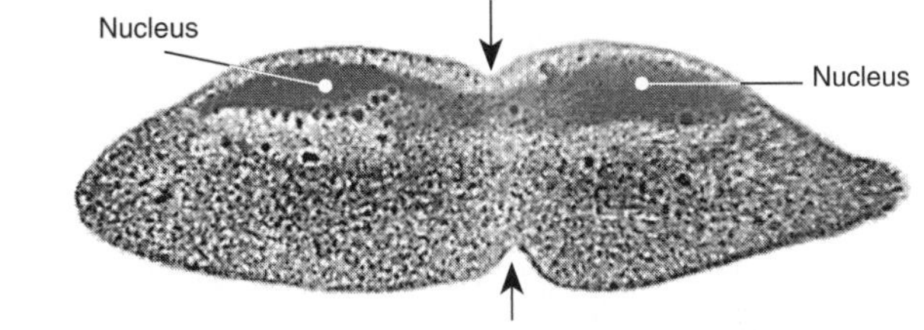

Paramecium (photo RCN)

Budding and Fragmentation

Sponges and most cnidarians (e.g. *Hydra*) can reproduce by **budding**. A small part of the parent body separates from the rest and develops into a new individual. This new individual may remain attached as part of the colony, or the budding offspring may constrict at its point of attachment (arrowed on the photograph) and eventually be released as an independent organism.

The photo (right) shows *Hydra* budding. The new individual is forming on the right of the animal. Cnidarians also undergo **fragmentation**. In this natural process, the organism spontaneously divides into fragments which then regenerate. Fragmentation also occurs in sponges and flatworms (platyhelminthes).

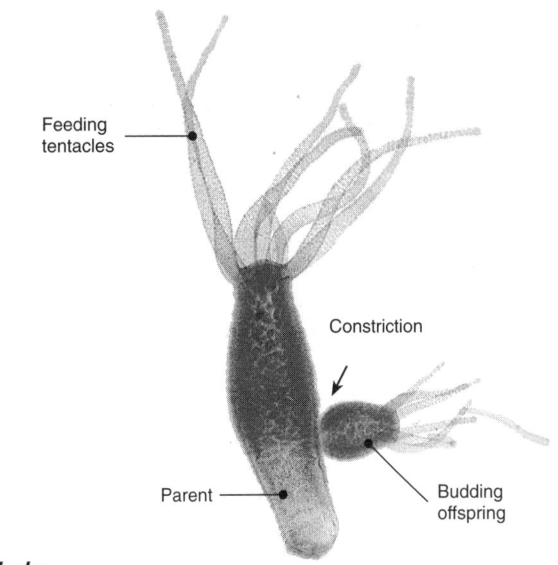

Hydra (photo EII)

1. Name the reproductive process occurring in the photograph of *Paramecium* above: ____________________

2. Name the reproductive process occurring in *Hydra*: ____________________

3. (a) Suggest why multiple fission produces offspring more rapidly than does simple binary fission: ____________________

____

____

____

(b) Explain the advantage of multiple fission to an intracellular parasite such as *Plasmodium*: ____________________

____

____

Alternating Asexual and Sexual Cycles of Reproduction

Some organisms combine several cycles of asexual reproduction by **parthenogenesis** (below, left) with periods when they reproduce sexually (producing gametes by meiosis which combine in fertilization). The parthenogenetic phase enables the rapid reproduction of a well adapted clone. The sexual phase is induced when the environment becomes unfavorable for the clone. The new generation, produced by sexual reproduction, may include some individuals that are better adapted to a new set of environmental conditions.

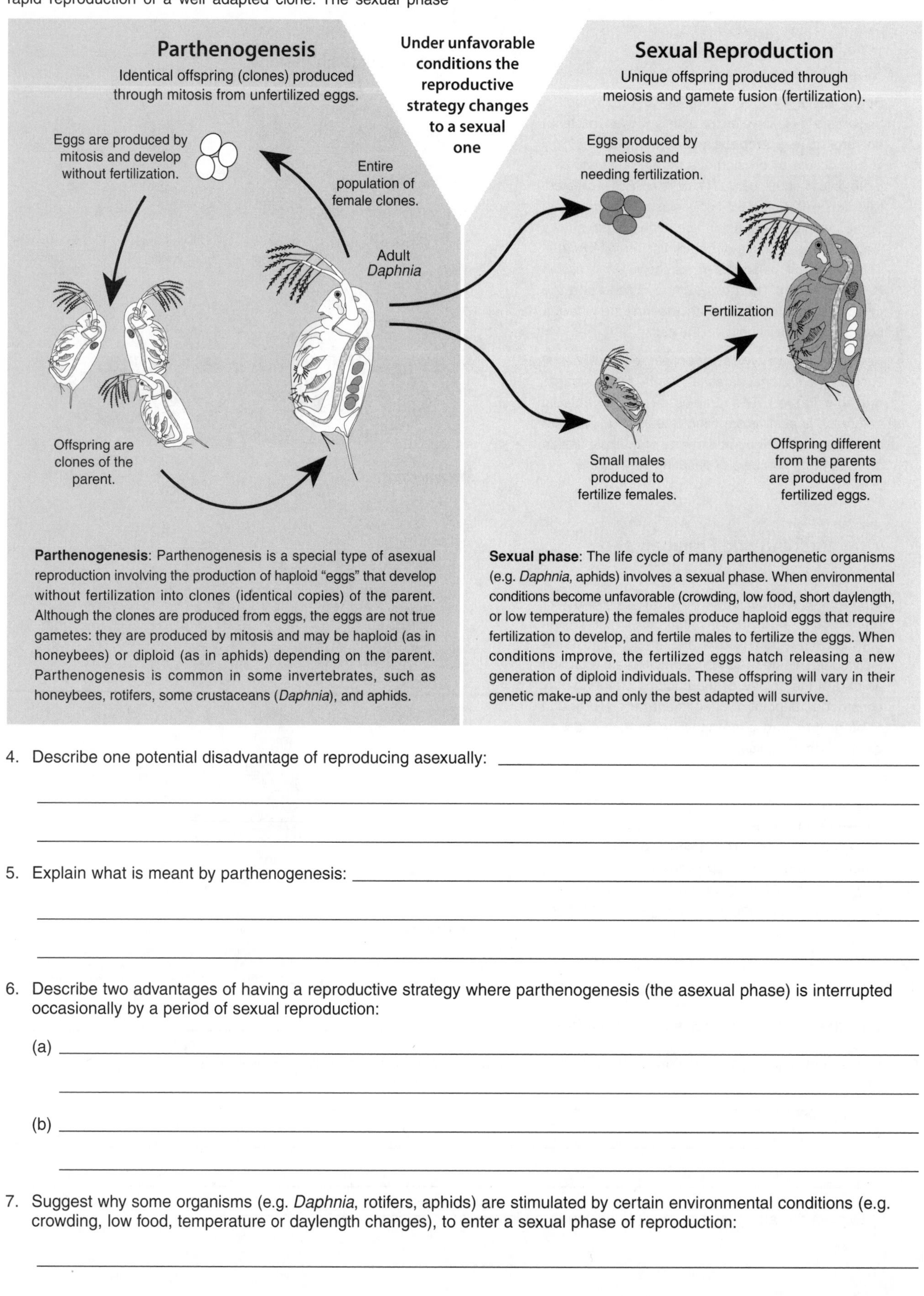

Parthenogenesis
Identical offspring (clones) produced through mitosis from unfertilized eggs.
Under unfavorable conditions the reproductive strategy changes to a sexual one
Sexual Reproduction
Unique offspring produced through meiosis and gamete fusion (fertilization).
Eggs are produced by mitosis and develop without fertilization.
Entire population of female clones.
Adult Daphnia
Offspring are clones of the parent.
Eggs produced by meiosis and needing fertilization.
Fertilization
Small males produced to fertilize females.
Offspring different from the parents are produced from fertilized eggs.

Parthenogenesis: Parthenogenesis is a special type of asexual reproduction involving the production of haploid "eggs" that develop without fertilization into clones (identical copies) of the parent. Although the clones are produced from eggs, the eggs are not true gametes: they are produced by mitosis and may be haploid (as in honeybees) or diploid (as in aphids) depending on the parent. Parthenogenesis is common in some invertebrates, such as honeybees, rotifers, some crustaceans (*Daphnia*), and aphids.

Sexual phase: The life cycle of many parthenogenetic organisms (e.g. *Daphnia*, aphids) involves a sexual phase. When environmental conditions become unfavorable (crowding, low food, short daylength, or low temperature) the females produce haploid eggs that require fertilization to develop, and fertile males to fertilize the eggs. When conditions improve, the fertilized eggs hatch releasing a new generation of diploid individuals. These offspring will vary in their genetic make-up and only the best adapted will survive.

4. Describe one potential disadvantage of reproducing asexually: ______________________

5. Explain what is meant by parthenogenesis: ______________________

6. Describe two advantages of having a reproductive strategy where parthenogenesis (the asexual phase) is interrupted occasionally by a period of sexual reproduction:

   (a) ______________________

   (b) ______________________

7. Suggest why some organisms (e.g. *Daphnia*, rotifers, aphids) are stimulated by certain environmental conditions (e.g. crowding, low food, temperature or daylength changes), to enter a sexual phase of reproduction:

____

Animal Sexual Reproduction

All types of sexual reproduction involve the production of **gametes** (sex cells), produced by special sex organs called **gonads**. Female gametes (**eggs**) and male gametes (**sperm**) come together in **fertilization**. Animal sexual reproduction follows one of three main patterns, determined by the location of fertilization and embryonic development. These patterns are: external fertilization and development; internal fertilization followed by external development; internal fertilization and development. **External fertilization** is found in many aquatic invertebrates and most fish, where eggs and sperm are released into the surrounding water. Male and female parents usually release their gametes (spawn) at the same time and place in order to increase the chances of successful fertilization. In other invertebrates, reptiles, sharks, birds, and mammals, sperm are transferred from the male to inside the female's genital tract during the act of **copulation**. This **internal fertilization** increases the chance that the gametes will meet successfully. In birds and most reptiles, one adaptation to life on land has been the evolution of the **amniote egg**: a structure that enables the embryo to complete its development outside the parent surrounded by a protective shell and nourished by a yolk sac. The pattern of internal development in mammals provides the most advantages for the embryo in terms of nourishment and protection during development.

Achieving Fertilization: The Mating Game

Many marine invertebrates release gametes into the sea. Fertilization and development are external to the parent. *Example: giant clam*

Insects often have elaborate courtship rituals. Fertilization is internal, but the eggs are laid and develop externally. *Example: dipteran flies*

In amphibians, a prolonged coupling, called amplexus, precedes gamete release and external fertilization. *Example: frogs*

In birds and reptiles gamete fertilization is internal but the eggs are laid (usually in nests) and develop externally. *Example: quail*

Mammals exhibit internal fertilization, a long period of internal development, and often prolonged parental care. *Example: African lions*

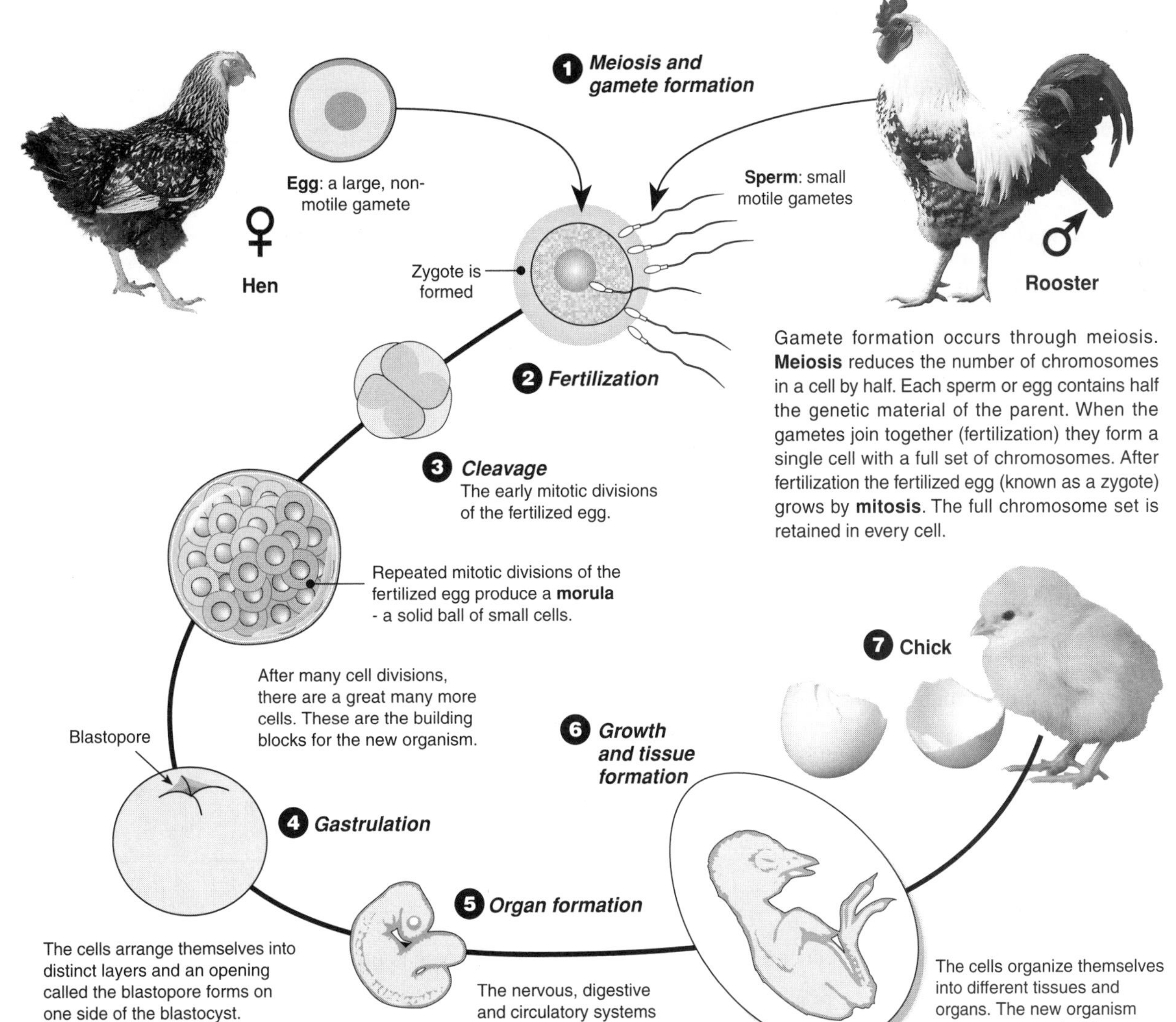

Gamete formation occurs through meiosis. **Meiosis** reduces the number of chromosomes in a cell by half. Each sperm or egg contains half the genetic material of the parent. When the gametes join together (fertilization) they form a single cell with a full set of chromosomes. After fertilization the fertilized egg (known as a zygote) grows by **mitosis**. The full chromosome set is retained in every cell.

Hermaphroditism

Most animals have separate sexes (individuals are either male or female). However, in some animals both sperm and eggs can be produced in the same individual. Such animals are known as **hermaphrodites**. In earthworms (below), flatworms, and some molluscs (e.g. land snails), both male and female organs are active in the same animal and there is typically a reciprocal transfer of sperm (each receives sperm from the other during copulation). In this type of hermaphroditism, there is no self fertilization: a mate is necessary for any fertilization to occur. However, some specialized hermaphroditic animals, such as parasitic tapeworms, are capable of self-fertilization.

The photo above shows two earthworms in a mating clasp. Each worm places its reproductive region (the clitellum) against the reproductive region of the other worm, and sperm is exchanged.

Courtship and mating in the snail *Cantareus aspersus* (formerly *Helix aspersa*, above). During an elaborate courtship (A), calcareous darts are fired from the genital opening (behind the tentacle) into the body of the partner. Mating (B) involves reciprocal transfer of sperm via a penis (P).

1. Describe one advantage of sexual reproduction: ______

2. Describe one potential disadvantage of sexual reproduction: ______

3. Compare and contrast the key differences between male and female gametes in relation to:

   (a) The size of gametes: ______

   (b) Number of gametes produced: ______

   (c) Motility of gametes: ______

4. Distinguish between **internal** fertilization and **external** fertilization, identifying advantages of each strategy:

5. (a) Name an animal group with internal fertilization but external development: ______

   (b) Name an animal group with internal fertilization and internal development: ______

   (c) Describe one benefit and one cost involved in providing for internal development of an embryo:

   Benefit: ______

   Cost: ______

6. Explain why each new individual produced from the fusion of the two gametes is unique: ______

Animal Reproductive Strategies

To reproduce sexually, animals must have systems to ensure that gametes meet and fertilization takes place. There is a huge range in the complexity of reproductive structures in animals: the least complex do not even have distinct gonads, whereas the most complex comprise numerous ducts, glands, and accessory structures to produce the gametes and protect the eggs and developing embryos. Diverse systems for reproduction have evolved amongst the invertebrates; some of the most complex are found in parasitic flatworms. Insects, which have separate sexes and highly developed reproductive systems, are illustrated in detail here. Insects frequently also have elaborate courtship and mating behaviors associated with successful transfer of gametes. Amongst the vertebrates, the basic structures of the reproductive system are relatively uniform, but there is huge variation in the strategies shown by different vertebrate groups. Biologists distinguish **oviparous** (egg laying) animals from, **viviparous** (live bearing) animals. A small number of vertebrates are also **ovoviviparous**: the young are born live but their nutrition inside the mother is derived from stores within the egg. This strategy is typical of non-mammal species that bear live young. Here the contrast is made between the reproductive strategies of amphibians, which rely heavily on water for their reproduction, and the strategy of birds, in which the evolution of a shelled egg has freed them from their reproductive dependence on water.

Reproductive Strategies of Frogs

In most frogs and toads, fertilization is external. This is achieved when the male clasps the female. Called **amplexus**, this clasping may last several hours or even days until the female lays her eggs. Not all frogs lay eggs in water. Some frogs have adopted novel behavior and physiological strategies to avoid their eggs becoming an easy meal. The examples here (right and below) illustrate the variety of solutions developed by different frog species.

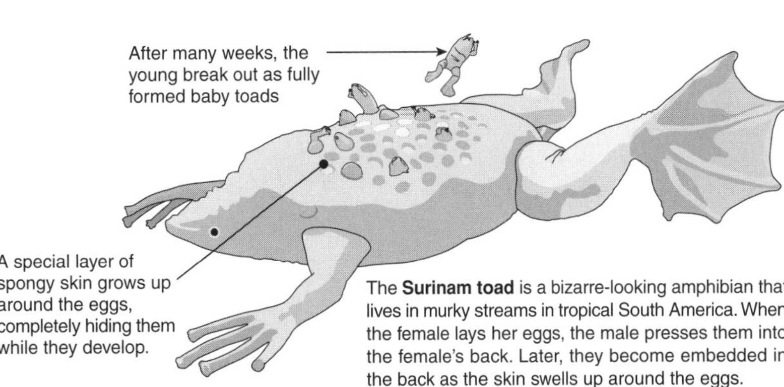

The **Surinam toad** is a bizarre-looking amphibian that lives in murky streams in tropical South America. When the female lays her eggs, the male presses them into the female's back. Later, they become embedded in the back as the skin swells up around the eggs.

Some frogs lay their eggs in a nest of foam either on land (attached to leaves) or floating on water. The foam not only hides the eggs from predators, but it keeps them moist and prevents them from drying out.

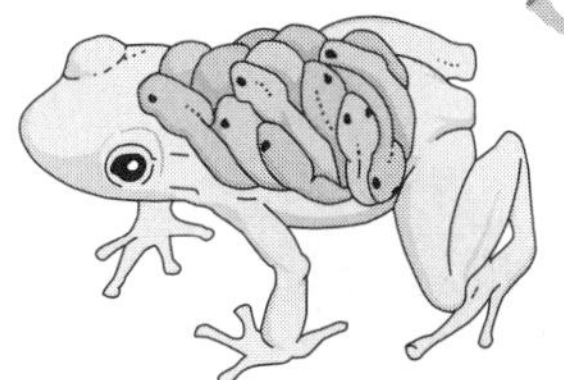

Some species of small frogs in both South America and Africa lay eggs that hatch into tadpoles on land. The tadpoles stick to the back of one of the parents with mouthparts modified to function as suckers. The parent carries the tadpoles to water.

Some frog species lay eggs on leaves or branches overhanging the water. In some of these species one of the parents remains with the eggs until they hatch. The tadpoles that emerge from the eggs drop into the water below to complete their development.

When a female midwife toad lays her string of eggs, the male winds them around his back legs. He carries the eggs for about a month, visiting puddles to keep them moist. When the eggs are ready to hatch he places them in a suitable pool.

The Structure and Physiology of a Bird's Egg

Both reptiles and birds (which evolved from reptilian ancestors) developed watertight shelled eggs, called **cleidoic eggs**. The egg is supplied with all the necessary food material as well as fats that yield water when metabolized. The shell enclosing the egg provides protection and reduces water loss, yet permits gas exchange. Waste materials from the developing chick embryo are stored in the egg.

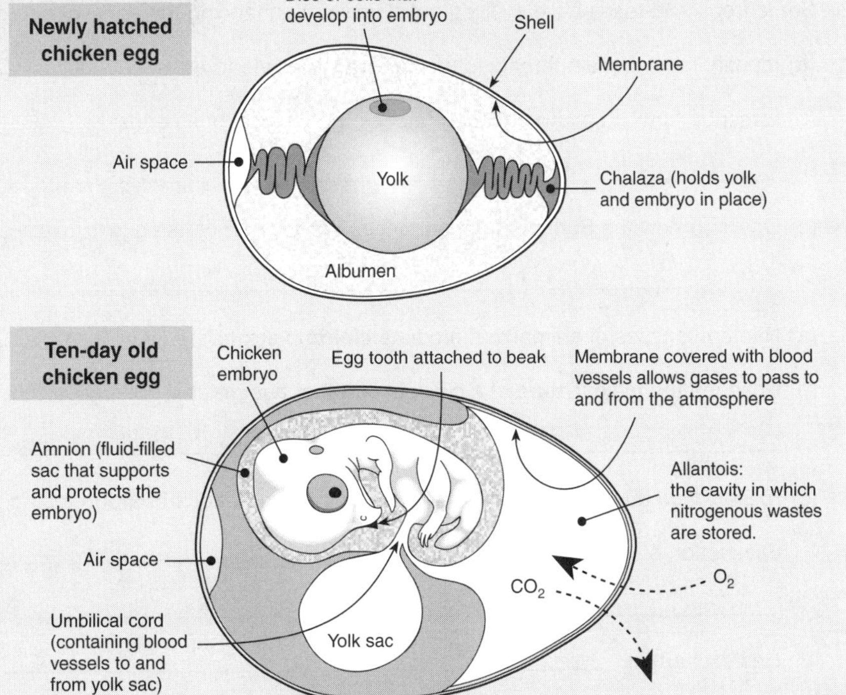

Reproductive Strategies of Insects

Insects reproduce sexually, mainly by internal fertilization followed by the production of yolk-filled eggs. A single pair of gonads is located in the abdomen. Most insects transfer sperm within small packets called **spermatophores**. Claspers at the end of the male's abdomen hold the female's abdomen during copulation. The terminal segments of the female's abdomen may form an **ovipositor**, often extendable or needle-like, with which they lay their eggs. Depending on the insect species, eggs may be buried in soil, animal dung, or rotting carcasses, injected into plant tissue or living hosts, or cemented to twigs or leaves.

Insect reproductive organs

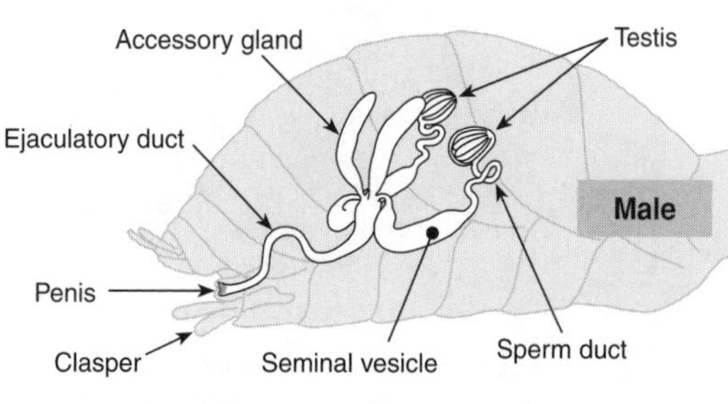

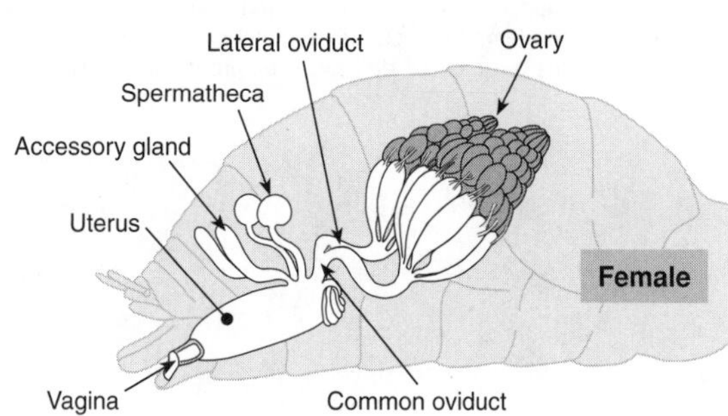

Mating in Damselflies

Species in the order Odonata (damselflies and dragonflies) are unique amongst the insects in that the male copulatory apparatus is situated on abdominal segments close to its thorax. However, the testes are located at the end of its abdomen so that, prior to mating, the male has to bend the abdomen round to transfer semen from the testes to the penis. The location of the male genitals also accounts for the unique "wheel position" adopted by members of this order (below).

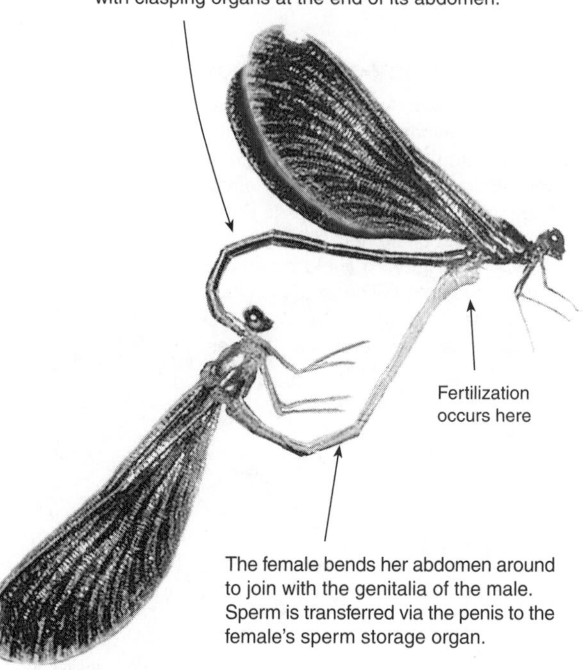

The male damselfly grips onto a twig with its front legs during mating while holding the female behind the head with clasping organs at the end of its abdomen.
Fertilization occurs here
The female bends her abdomen around to join with the genitalia of the male. Sperm is transferred via the penis to the female's sperm storage organ.

1. Giving examples, distinguish between **oviparous** and **viviparous** vertebrates: ___

2. Some frogs and toads have evolved novel ways of enhancing the survival of their eggs.

   (a) Identify the two main threats to the survival of frog/toad eggs: ___

   (b) Describe how the midwife toad enhances the survival of its eggs: ___

   (c) Describe how the Surinam toad enhances the survival of its eggs: ___

3. (a) Name two types of animal that produce **cleidoic eggs**: ___

   (b) Describe the main feature of a cleidoic egg that has made it so successful: ___

   (c) Explain how a cleidoic egg provides for the following needs of a developing embryo:

   Elimination of wastes: ___

   Gas exchange: ___

   Nutrition (food supply): ___

Female Reproductive System

The female reproductive system in mammals produces eggs, receives the penis and sperm during sexual intercourse, and houses and nourishes the young. Female reproductive systems in mammals are similar in their basic structure (uterus, ovaries etc.) but the shape of the uterus and the form of the placenta during pregnancy vary. The human system is described below.

Oogenesis

Oogenesis is the process by which mature ova (egg cells) are produced by the ovary. Oogonia are formed in the female embryo and undergo repeated mitotic divisions to form the primary oocyte. These remain in prophase of meiosis I throughout childhood. At this stage, all the eggs a female will ever have are present, but they remain in this resting phase until puberty. At puberty, meiosis resumes. Eggs are released, arrested in metaphase of meiosis II. This second division is only completed upon fertilization.

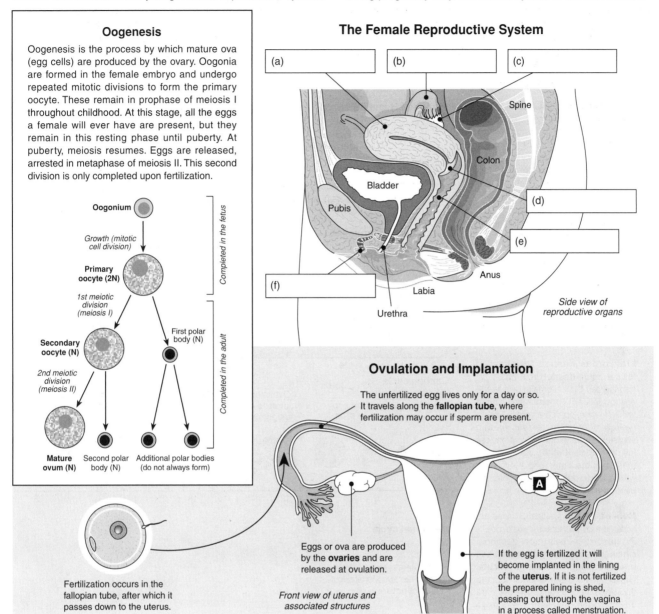

The Female Reproductive System

Side view of reproductive organs

Ovulation and Implantation

Fertilization occurs in the fallopian tube, after which it passes down to the uterus.

Front view of uterus and associated structures

1. The female human reproductive system and associated structures are illustrated above. Using the word list, identify the labeled parts. **Word list**: *ovary, uterus (womb), vagina, fallopian tube (oviduct), cervix, clitoris.*

2. In a few words or a short sentence, state the function of each of the structures labeled (a) - (d) in the above diagram:

   (a) ______

   (b) ______

   (c) ______

   (d) ______

3. (a) Name the organ labeled (**A**) in the diagram: ______

   (b) Name the event associated with this organ that occurs every month: ______

   (c) Name the process by which mature ova are produced: ______

4. (a) Name the stage in meiosis at which the oocyte is released from the ovary: ______

   (b) State when in the reproductive process meiosis II is completed: ______

Related activities: *Control of the Menstrual Cycle*
Web links: *Menstrual Cycle Animation, Ovarian & Uterine Cycle*

The Menstrual Cycle

In non-primate mammals the reproductive cycle is characterized by a **breeding season** and an **estrous cycle** (a period of greater sexual receptivity during which ovulation occurs). In contrast, humans and other primates are sexually receptive throughout the year and may mate at any time. Like all placental mammals, their uterine lining thickens in preparation for pregnancy. However, unlike other mammals, primates shed this lining as a discharge through the vagina if fertilization does not occur. This event, called **menstruation**, characterizes the human reproductive or **menstrual cycle**. In human females, the menstrual cycle starts from the first day of bleeding and lasts for about 28 days. It involves a predictable series of changes that occur in response to hormones. The cycle is divided into three phases (see below), defined by the events in each phase.

The Menstrual Cycle

Luteinizing hormone (LH) and follicle stimulating hormone (FSH): These hormones from the anterior pituitary have numerous effects. FSH stimulates the development of the ovarian follicles resulting in the release of estrogen. Estrogen levels peak, stimulating a surge in LH and triggering ovulation.

Hormone levels: Of the follicles that begin developing in response to FSH, usually only one (the Graafian follicle) becomes dominant. In the first half of the cycle, estrogen is secreted by this developing Graafian follicle. Later, the Graafian follicle develops into the corpus luteum (below right) which secretes large amounts of progesterone (and smaller amounts of estrogen).

The corpus luteum: The Graafian follicle continues to grow and then (around day 14) ruptures to release the egg (ovulation). LH causes the ruptured follicle to develop into a corpus luteum (yellow body). The corpus luteum secretes progesterone which promotes full development of the uterine lining, maintains the embryo in the first 12 weeks of pregnancy, and inhibits the development of more follicles.

Menstruation: If fertilization does not occur, the corpus luteum breaks down. Progesterone secretion declines, causing the uterine lining to be shed (menstruation). If fertilization occurs, high progesterone levels maintain the thickened uterine lining. The placenta develops and nourishes the embryo completely by 12 weeks.

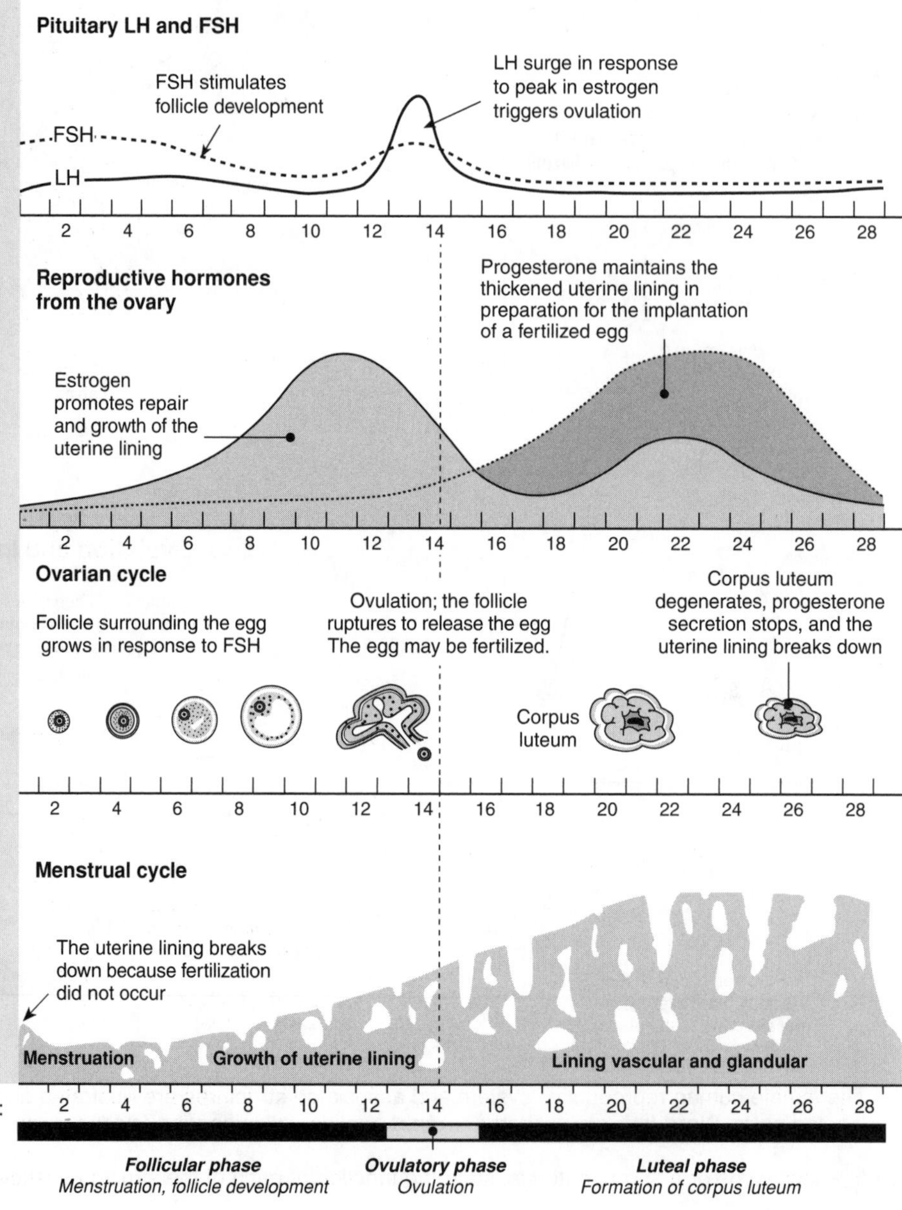

Pituitary LH and FSH
FSH stimulates follicle development
LH surge in response to peak in estrogen triggers ovulation
FSH
LH
2 4 6 8 10 12 14 16 18 20 22 24 26 28
Reproductive hormones from the ovary
Progesterone maintains the thickened uterine lining in preparation for the implantation of a fertilized egg
Estrogen promotes repair and growth of the uterine lining
2 4 6 8 10 12 14 16 18 20 22 24 26 28
Ovarian cycle
Follicle surrounding the egg grows in response to FSH
Ovulation; the follicle ruptures to release the egg The egg may be fertilized.
Corpus luteum degenerates, progesterone secretion stops, and the uterine lining breaks down
Corpus luteum
2 4 6 8 10 12 14 16 18 20 22 24 26 28
Menstrual cycle
The uterine lining breaks down because fertilization did not occur
Menstruation
Growth of uterine lining
Lining vascular and glandular
Day of the cycle:
2 4 6 8 10 12 14 16 18 20 22 24 26 28
Follicular phase
Menstruation, follicle development
Ovulatory phase
Ovulation
Luteal phase
Formation of corpus luteum

1. Name the hormone responsible for:

   (a) Follicle growth: ______________________ (b) Ovulation: ______________________

2. Each month, several ovarian follicles begin development, but only one (the Graafian follicle) develops fully:

   (a) Name the hormone secreted by the developing follicle: ______________________

   (b) State the role of this hormone during the follicular phase: ______________________

   (c) Suggest what happens to the follicles that do not continue developing: ______________________

3. (a) Identify the principal hormone secreted by the corpus luteum: ______________________

   (b) State the purpose of this hormone: ______________________

4. State the hormonal trigger for menstruation: ______________________

Related activities: Control of the Menstrual Cycle
Web links: The Menstrual Cycle Animation

Periodicals: Measuring female hormones in saliva

Control of the Menstrual Cycle

The female menstrual cycle is regulated by the interplay of several reproductive hormones. The main control centers for this regulation are the **hypothalamus** and the **anterior pituitary gland**. The hypothalamus secretes GnRH (gonadotrophin releasing hormone), a hormone that is essential for normal gonad function in males and females. GnRH is transported in blood vessels to the anterior pituitary where it brings about the release of two hormones: follicle stimulating hormone (FSH) and luteinizing hormone (LH). It is these two hormones that induce the cyclical changes in the ovary and uterus. Regulation of blood hormone levels during the menstrual cycle is achieved through **negative feedback** mechanisms. The exception to this is the mid cycle surge in LH (see previous page) which is induced by the rapid increase in estrogen secreted by the developing follicle.

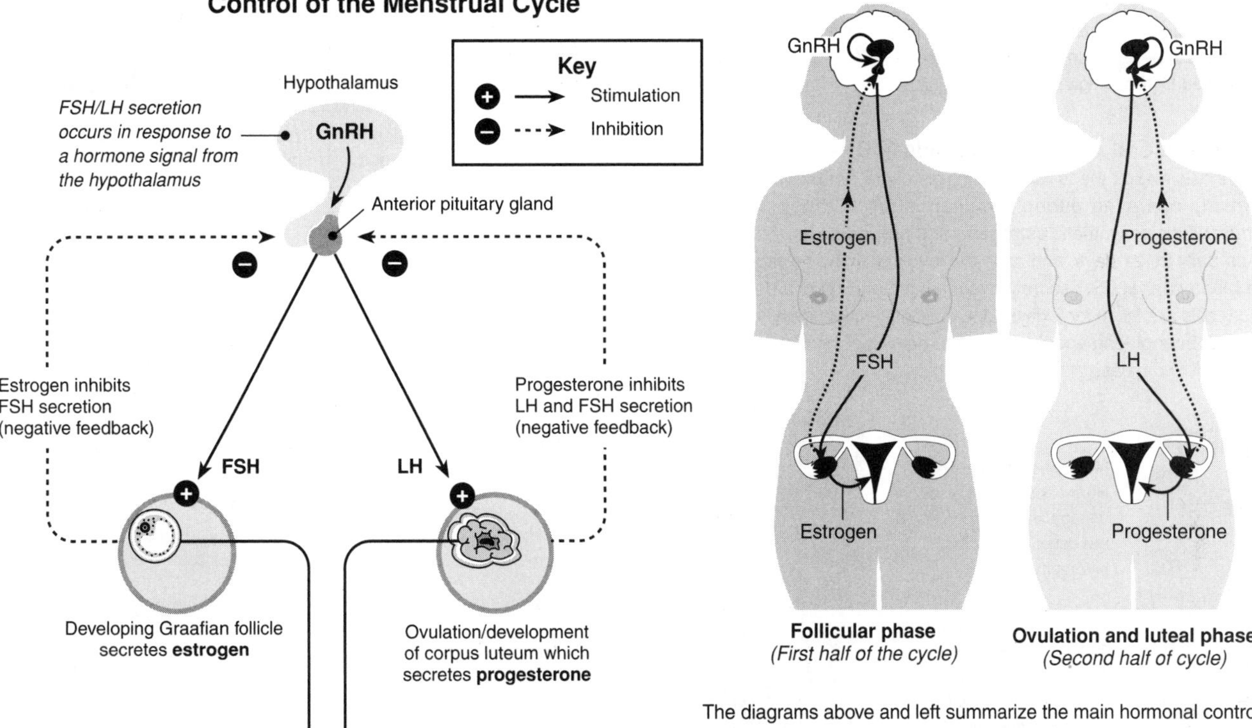

The diagrams above and left summarize the main hormonal controls during the two halves of the menstrual cycle. In the first half of the cycle, FSH stimulates follicle development in the ovary. The developing follicle secretes estrogen which acts on the uterus and, in the anterior pituitary, inhibits FSH secretion. In the second half of the cycle, LH induces ovulation and development of the corpus luteum. The corpus luteum secretes progesterone which acts on the uterus and also inhibits further secretion of LH (and also FSH).

1. Using the information above and on the previous page, complete the table below summarizing the role of hormones in the control of the menstrual cycle. To help you, some of the table has been completed:

Hormone	Site of secretion	Main effects and site of action during the menstrual cycle
GnRH		
		Stimulates the growth of ovarian follicles
LH		
		At high levels, stimulates LH surge. Promotes growth and repair of the uterine lining.
Progesterone		

2. Briefly explain the role of negative feedback in the control of hormone levels in the menstrual cycle:

3. **FSH** and **LH** (called ICSH or interstitial cell stimulating hormone in males) also play a central role in male reproduction. Refer to the activity *Male Reproductive System* and state how these two hormones are involved **in male reproduction**:

Contraception

Humans have many ways in which to manage their own reproduction. They may choose to prevent or assist fertilization of an egg by a sperm (conception). **Contraception** refers to the use of methods or devices that prevent conception. There are many contraceptive methods available including physical barriers (such as condoms) that prevent egg and sperm ever meeting. The most effective methods (excluding sterilization) involve chemical interference in the normal female cycle so that egg production is inhibited. This is done by way of **oral contraceptives** (below, left) or hormonal implants. If taken properly, oral contraceptives are almost 100% effective at preventing pregnancy. The placement of their action in the normal cycle of reproduction (from gametogenesis to pregnancy) is illustrated below. Other contraceptive methods are included for comparison.

Hormonal Contraception

The most common method by which to prevent conception using hormones is by using an oral contraceptive pill (OCP). These may be **combined OCPs**, or low dose mini pills.

Combined oral contraceptive pills (OCPs)

These pills exploit the feedback controls over hormone secretion normally operating during a menstrual cycle. They contain combinations of synthetic **estrogens** and **progesterone**. They are taken daily for 21 days, and raise the levels of these hormones in the blood so that FSH secretion is inhibited and no ova develop. Sugar pills are taken for 7 days; long enough to allow menstruation to occur but not long enough for ova to develop. Combined OCPs can be of two types:

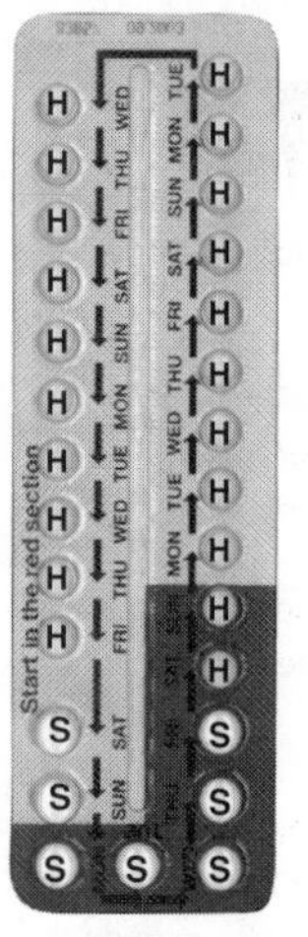

Monophasic pills (left): Hormones (**H**) are all at one dosage level. Sugar pills (**S**) are usually larger and differently colored.

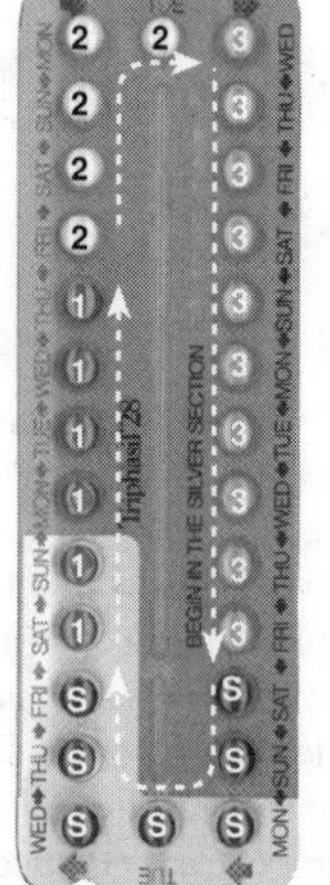

Triphasic pills (right): The hormone dosage increases in stages (**1**,**2**,**3**), mimicking the natural changes in a menstrual cycle.

Mini-pill (progesterone only)

The mini-pill contains 28 days of low dose progesterone; generally too low to prevent ovulation. The pill works by thickening the cervical mucus and preventing endometrial thickening. The mini-pill is less reliable than combined pills and must be taken at a regular time each day. However, it is safer for older women and those who are breastfeeding.

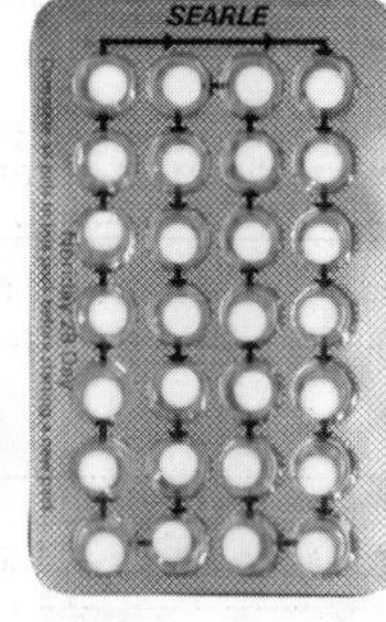

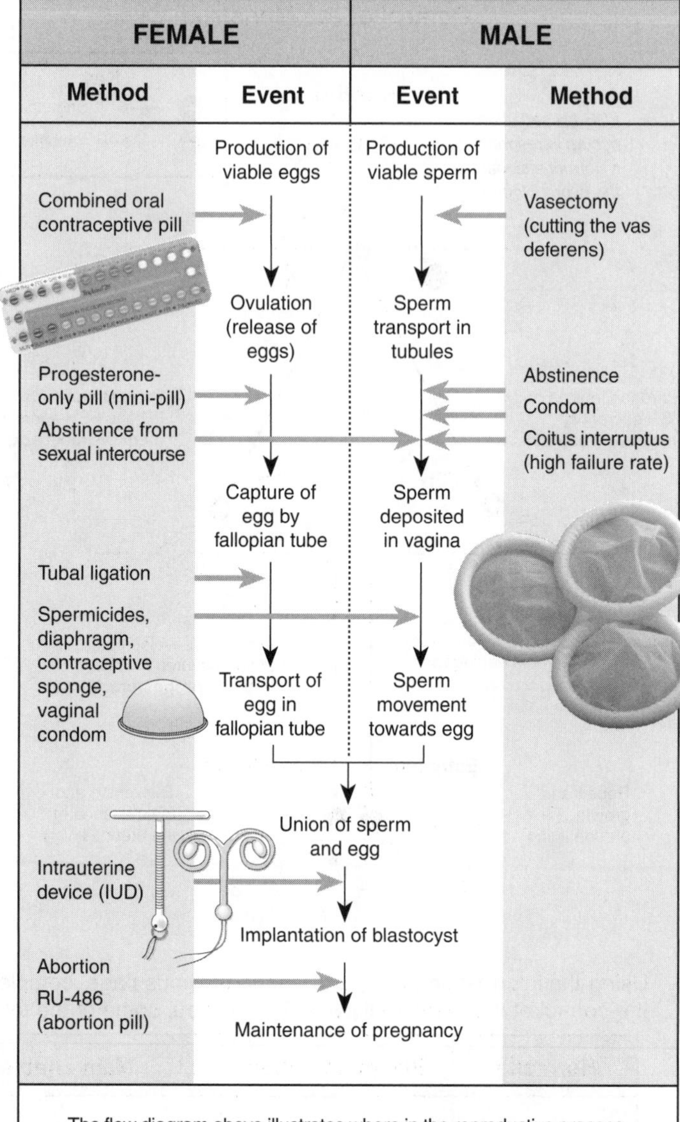

The flow diagram above illustrates where in the reproductive process, from gametogenesis to pregnancy, various contraceptive methods operate. Note the early action of hormonal contraceptives.

1. Explain briefly how the **combined oral contraceptive pill** acts as a contraceptive: ___________

____

____

2. Contrast the mode of action of OCPs with that of the mini-pill, giving reasons for the differences: ___________

____

____

3. Suggest why oral contraceptives offer such effective control over conception: ___________

____

____

Male Reproductive System

The reproductive role of the male is to produce the sperm and deliver them to the female. When a sperm combines with an egg, it contributes half the genetic material of the offspring and, in humans and other mammals, determines its sex. The reproductive structures in human males (shown below) are in many ways typical of other mammals.

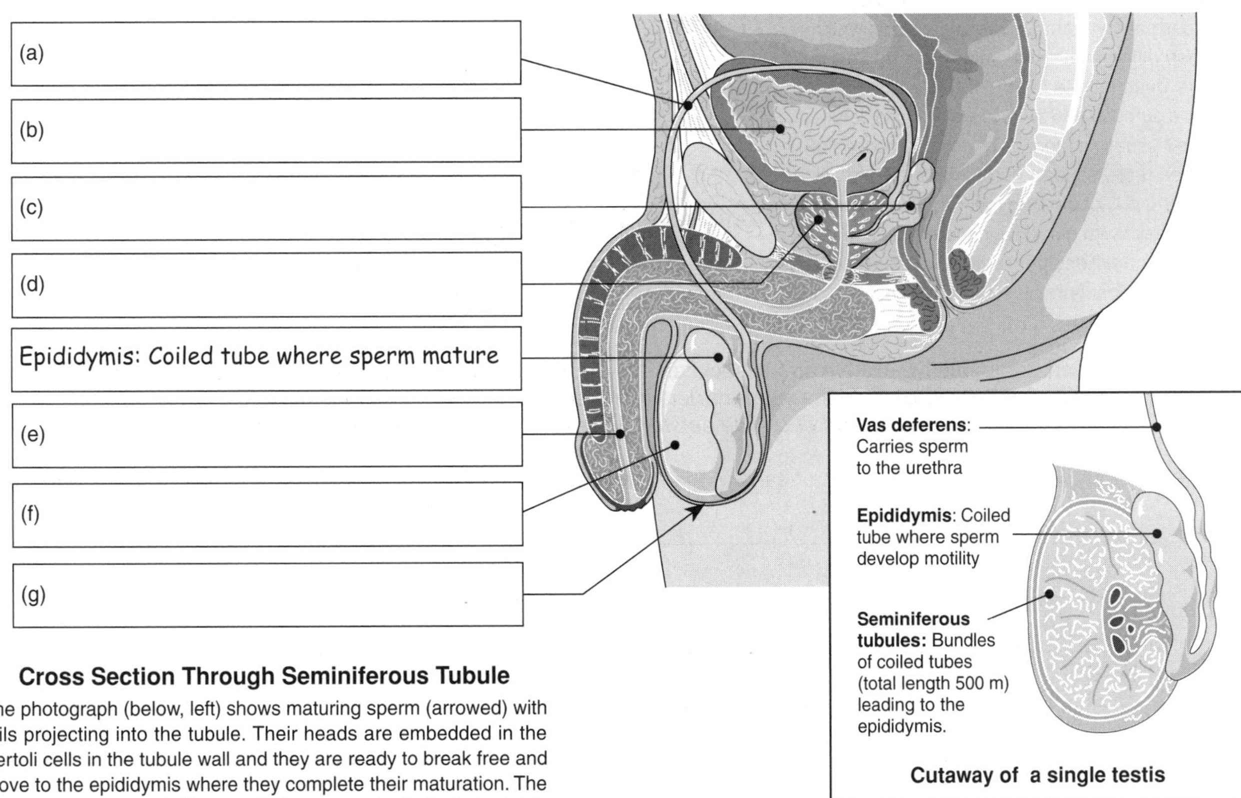

Cross Section Through Seminiferous Tubule

The photograph (below, left) shows maturing sperm (arrowed) with tails projecting into the tubule. Their heads are embedded in the Sertoli cells in the tubule wall and they are ready to break free and move to the epididymis where they complete their maturation. The same cross-section is illustrated diagrammatically (below, right).

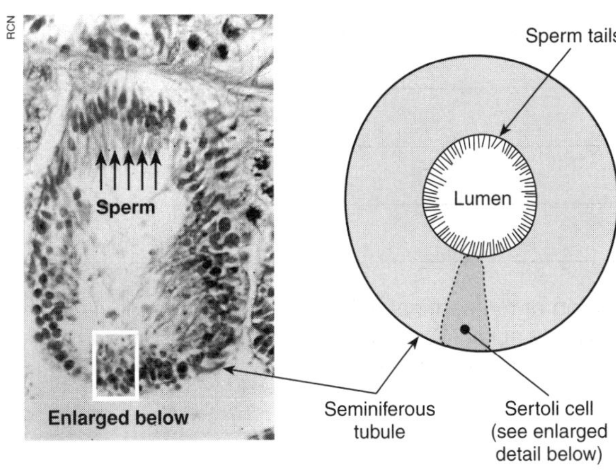

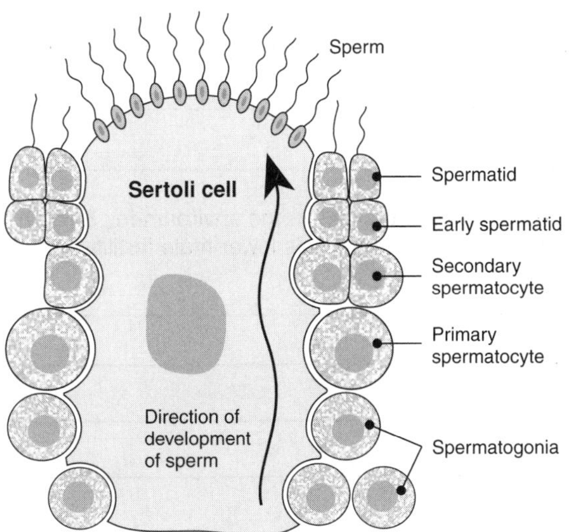

Spermatogenesis

Spermatogenesis is the process by which mature spermatozoa (sperm) are produced in the testis. In humans, they are produced at the rate of about 120 million per day. Spermatogenesis is regulated by the hormones **FSH** (from the anterior pituitary) and testosterone (secreted from the testes in response to **ICSH** (LH) from the anterior pituitary). Spermatogonia, in the outer layer of the seminiferous tubules, multiply throughout reproductive life. Some of them divide by meiosis into spermatocytes, which produce spermatids. These are transformed into mature sperm by the process of spermiogenesis in the seminiferous tubules of the testis. Full sperm motility is achieved in the epididymis.

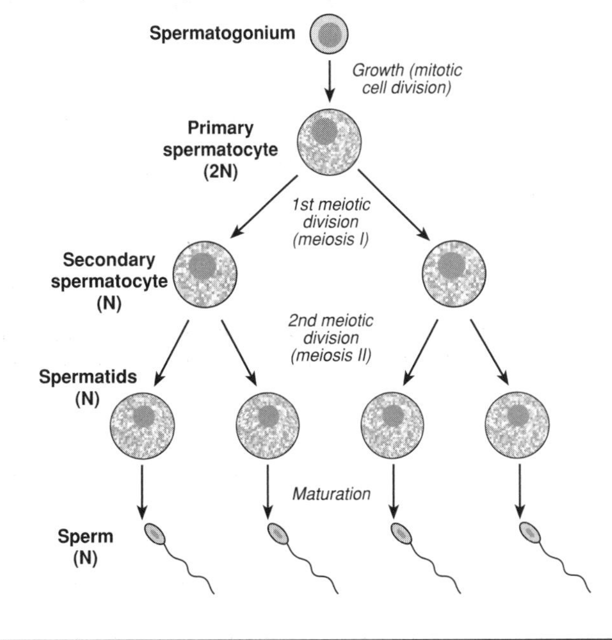

Sperm Structure

Mature spermatozoa (sperm) are produced by a process called spermatogenesis in the testes (see description of the process on the previous page). Meiotic division of spermatocytes produces spermatids which then differentiate into mature sperm. Sperm are quite simple in structure because their sole purpose is to swim to the egg and donate their genetic material. They are composed of three regions: headpiece, midpiece, and tail. Sperm do not live long (only about 48 hours), but they swim quickly and there are so many of them (millions per ejaculation) that some are able to reach the egg to fertilize it.

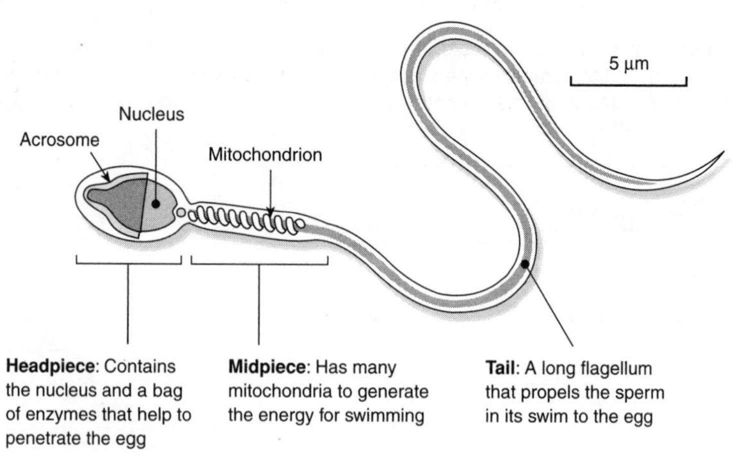
5 µm
Nucleus
Acrosome
Mitochondrion
Headpiece: Contains the nucleus and a bag of enzymes that help to penetrate the egg
Midpiece: Has many mitochondria to generate the energy for swimming
Tail: A long flagellum that propels the sperm in its swim to the egg

1. The male human reproductive system and associated structures are shown on the previous page. Using the following word list identify the labeled parts (write your answers in the spaces provided on the diagram).
   Word list: *bladder, scrotal sac, sperm duct (vas deferens), seminal vesicle, testis, urethra, prostate gland*

2. In a short sentence, state the function of each of the structures labeled (a)-(g) in the diagram on the previous page:

   (a) ______

   (b) ______

   (c) ______

   (d) ______

   (e) ______

   (f) ______

   (g) ______

3. (a) Name the process by which mature sperm are formed: ______

   (b) Name the hormones regulating this process: ______

   (c) State where most of this process occurs: ______

   (d) State where the process is completed: ______

4. The secretions of the prostate gland (which make up a large proportion of the seminal fluid produced in an ejaculation) are of alkaline pH, while the secretions of the vagina are normally slightly acidic. With this information, explain the role the prostate gland secretions have in maintaining the viability of sperm deposited in the vagina.

5. Each ejaculation of a healthy, fertile male contains 100-400 million sperm. Suggest why so many sperm are needed:

6. Recently, concern has been expressed about the level of synthetic estrogen-like chemicals in the environment. Explain the reason for this concern and discuss evidence in support of the claim that these chemicals lower male fertility:

Fertilization and Early Growth

When an egg cell is released from the ovary it is arrested in metaphase of meiosis II and is termed a secondary oocyte. **Fertilization** occurs when a sperm penetrates an egg cell at this stage and the sperm and egg nuclei unite to form the zygote. Fertilization is always regarded as time 0 in a period of gestation (pregnancy) and has five distinct stages (below). After fertilization, the zygote begins its **development** i.e. its growth and differentiation into a multicellular organism (see next page).

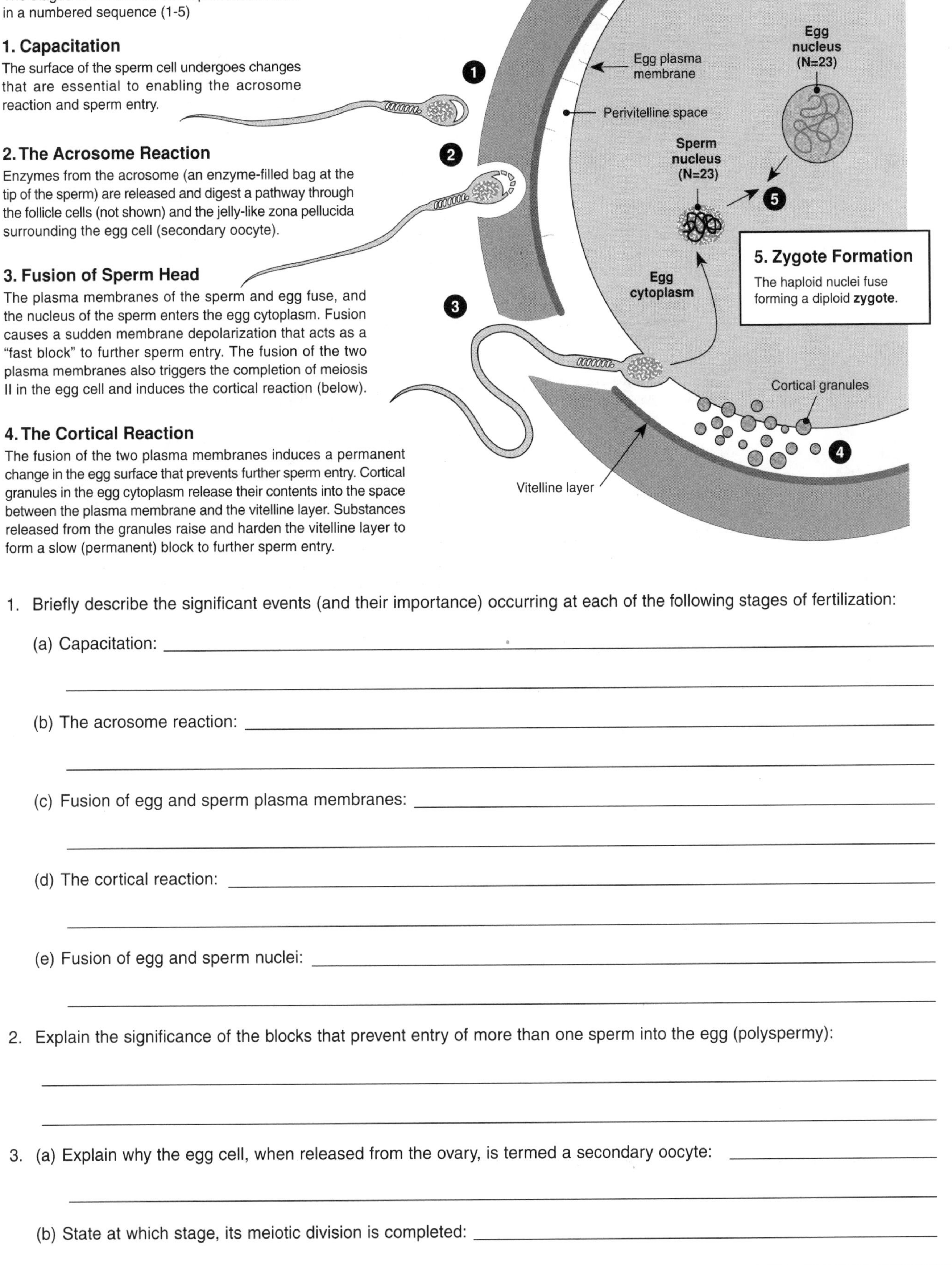

Fertilization (Time 0)

The stages in fertilization are represented below in a numbered sequence (1-5)

1. Capacitation

The surface of the sperm cell undergoes changes that are essential to enabling the acrosome reaction and sperm entry.

2. The Acrosome Reaction

Enzymes from the acrosome (an enzyme-filled bag at the tip of the sperm) are released and digest a pathway through the follicle cells (not shown) and the jelly-like zona pellucida surrounding the egg cell (secondary oocyte).

3. Fusion of Sperm Head

The plasma membranes of the sperm and egg fuse, and the nucleus of the sperm enters the egg cytoplasm. Fusion causes a sudden membrane depolarization that acts as a "fast block" to further sperm entry. The fusion of the two plasma membranes also triggers the completion of meiosis II in the egg cell and induces the cortical reaction (below).

4. The Cortical Reaction

The fusion of the two plasma membranes induces a permanent change in the egg surface that prevents further sperm entry. Cortical granules in the egg cytoplasm release their contents into the space between the plasma membrane and the vitelline layer. Substances released from the granules raise and harden the vitelline layer to form a slow (permanent) block to further sperm entry.

5. Zygote Formation

The haploid nuclei fuse forming a diploid **zygote**.

1. Briefly describe the significant events (and their importance) occurring at each of the following stages of fertilization:

   (a) Capacitation: ____________________

   (b) The acrosome reaction: ____________________

   (c) Fusion of egg and sperm plasma membranes: ____________________

   (d) The cortical reaction: ____________________

   (e) Fusion of egg and sperm nuclei: ____________________

2. Explain the significance of the blocks that prevent entry of more than one sperm into the egg (polyspermy):

   ____

3. (a) Explain why the egg cell, when released from the ovary, is termed a secondary oocyte: ____________________

   (b) State at which stage, its meiotic division is completed: ____________________

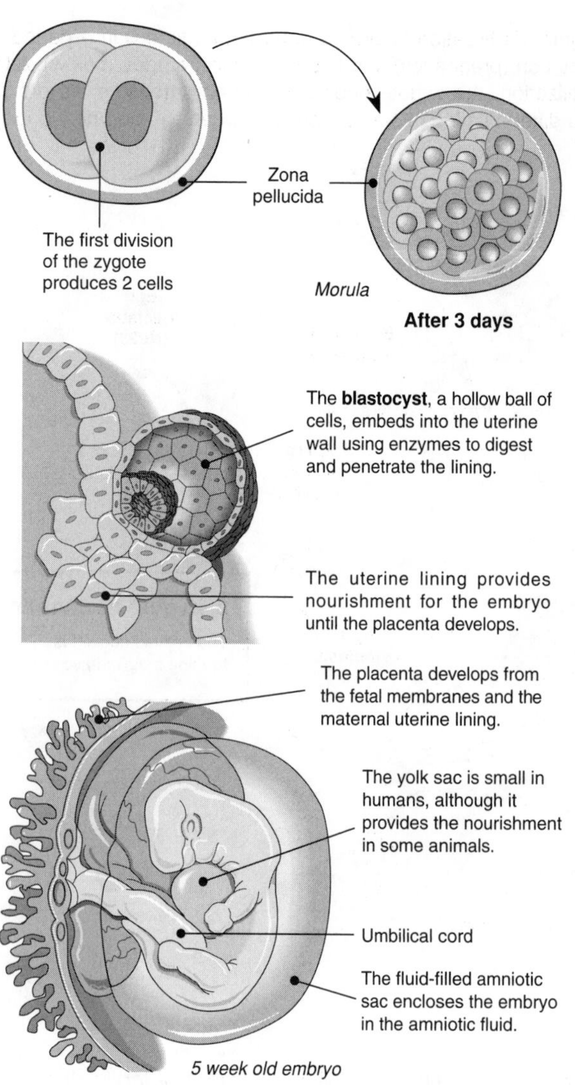

The first division
of the zygote
produces 2 cells
Zona
pellucida
Morula
After 3 days
The **blastocyst**, a hollow ball of cells, embeds into the uterine wall using enzymes to digest and penetrate the lining.
The uterine lining provides nourishment for the embryo until the placenta develops.
The placenta develops from the fetal membranes and the maternal uterine lining.
The yolk sac is small in humans, although it provides the nourishment in some animals.
Umbilical cord
The fluid-filled amniotic sac encloses the embryo in the amniotic fluid.
5 week old embryo

Early Growth and Development

Cleavage and Development of the Morula

Immediately after fertilization, rapid cell division takes place. These early cell divisions are called **cleavage** and they increase the number of cells, but not the size of the zygote. The first cleavage is completed after 36 hours, and each succeeding division takes less time. After 3 days, successive cleavages have produced a solid mass of cells called the **morula**, (left) which is still about the same size as the original zygote.

Implantation of the Blastocyst (after 6-8 days)

After several days in the uterus, the morula develops into the blastocyst. It makes contact with the uterine lining and pushes deeply into it, ensuring a close maternal-fetal contact. Blood vessels provide early nourishment as they are opened up by enzymes secreted by the blastocyst. The embryo produces **HCG** (human chorionic gonadotropin), which prevents degeneration of the corpus luteum and signals that the woman is pregnant.

Embryo at 5-8 Weeks

Five weeks after fertilization, the embryo is only 4-5 mm long, but already the central nervous system has developed and the heart is beating. The embryonic membranes have formed; the amnion encloses the embryo in a fluid-filled space, and the allanto-chorion forms the fetal portion of the placenta. From two months the embryo is called a fetus. It is still small (30-40 mm long), but the limbs are well formed and the bones are beginning to harden. The face has a flat, rather featureless appearance with the eyes far apart. Fetal movements have begun and brain development proceeds rapidly. The placenta is well developed, although not fully functional until 12 weeks. The umbilical cord, containing the fetal umbilical arteries and vein, connects fetus and mother.

4. State what contribution the sperm and egg cell make to each of the following:

   (a) The nucleus of the zygote: Sperm contribution: ________________ Egg contribution: ________________

   (b) The cytoplasm of the zygote: Sperm contribution: ________________ Egg contribution: ________________

5. Explain what is meant by cleavage and comment on its significance to the early development of the embryo:

6. (a) Explain the importance of implantation to the early nourishment of the embryo:

   (b) Identify the fetal tissues that contribute to the formation of the placenta:

   (c) Suggest a purpose of the amniotic sac and comment on its importance to the developing embryo:

   (d) Suggest why the heart is one of the very first structures to develop in the embryo:

7. State why the fetus is particularly prone to damage from drugs towards the end of the first trimester (2-3 months):

The Placenta

As soon as an embryo embeds in the uterine wall it begins to obtain nutrients from its mother and increase in size. At two months, when the major structures of the adult are established, it is called a fetus. It is entirely dependent on its mother for nutrients, oxygen, and elimination of wastes. The placenta is the specialized organ that performs this role, enabling exchange between fetal and maternal tissues, and allowing a prolonged period of fetal growth and development within the protection of the uterus. The placenta also has an endocrine role, producing hormones that enable the pregnancy to be maintained.

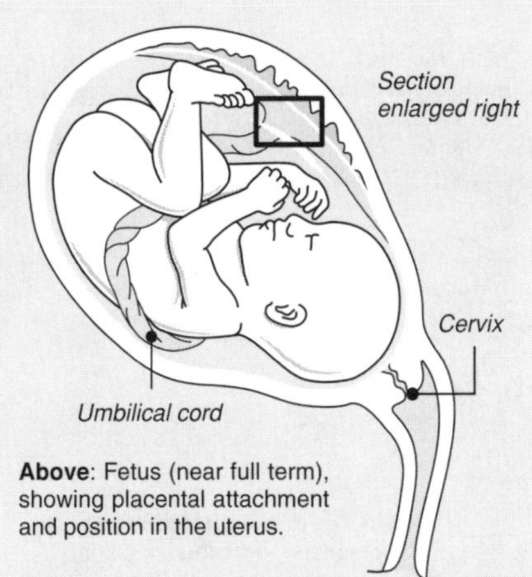

Above: Fetus (near full term), showing placental attachment and position in the uterus.

Below: Photograph shows a 14 week old fetus. Limbs are fully formed, many bones are beginning to ossify, and joints begin to form. Facial features are becoming more fully formed.

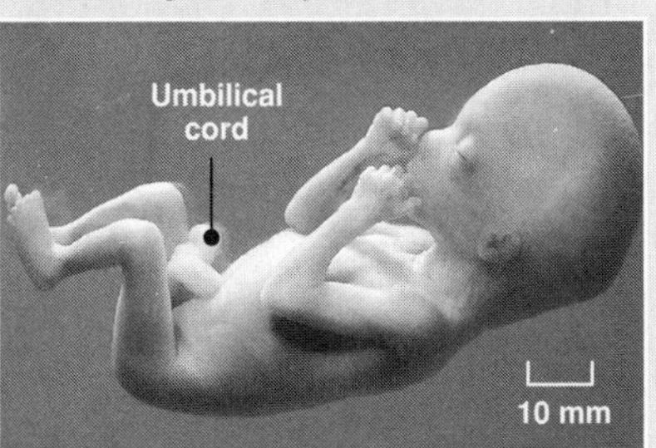

Schematic diagram showing part of the placenta in section

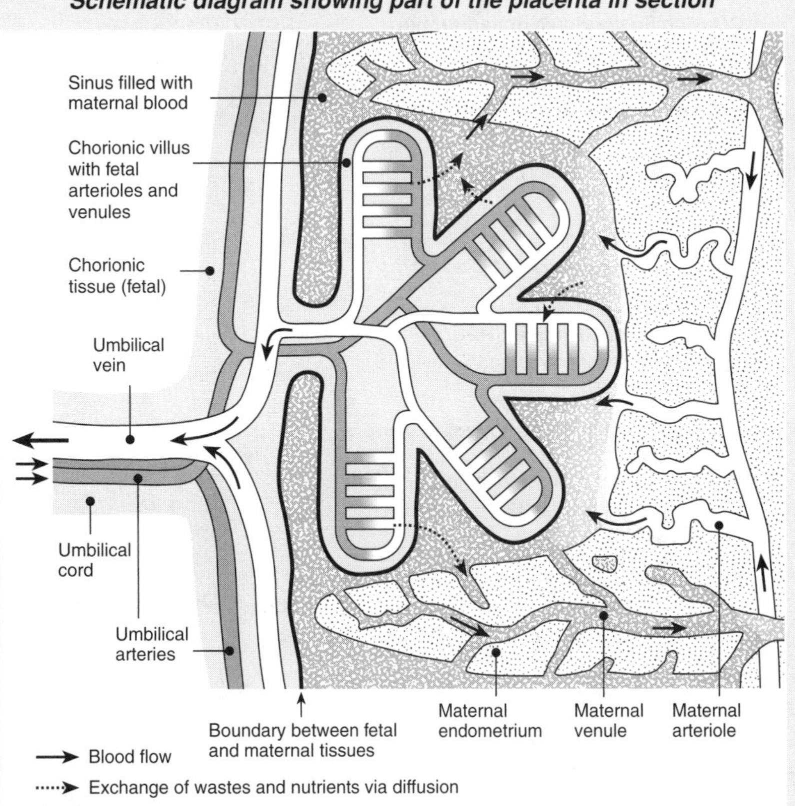

The placenta is a disc-like organ, about the size of a dinner plate and weighing about 1 kg. It develops when fingerlike projections of the fetal chorion (the chorionic villi) grow into the endometrium of the uterus. The villi contain the numerous capillaries connecting the fetal arteries and vein. They continue invading the maternal tissue until they are bathed in the maternal blood sinuses. The maternal and fetal blood vessels are in such close proximity that oxygen and nutrients can diffuse from the maternal blood into the capillaries of the villi. From the villi, the nutrients circulate in the umbilical vein, returning to the fetal heart. Carbon dioxide and other wastes leave the fetus through the umbilical arteries, pass into the capillaries of the villi, and diffuse into the maternal blood. Note that fetal blood and maternal blood do not mix: the exchanges occur via diffusion through thin walled capillaries.

1. In simple terms, describe the basic structure of the human placenta: ______

2. The umbilical cord contains the fetal arteries and vein. Describe the status of the blood in each type of fetal vessel:

   (a) Fetal arteries: Oxygenated and containing nutrients / Deoxygenated and containing nitrogenous wastes (delete one)

   (b) Fetal vein: Oxygenated and containing nutrients / Deoxygenated and containing nitrogenous wastes (delete one)

3. Teratogens are substances that may cause malformations in embryonic development (e.g. nicotine, alcohol):

   (a) Give a general explanation why substances ingested by the mother have the potential to be harmful to the fetus: ______

   (b) Explain why cigarette smoking is so harmful to fetal development: ______

Periodicals:
The Placenta

***Related activities**: Fertilization and Early Growth*

The Hormones of Pregnancy

Human reproductive physiology occurs in a cycle (the menstrual cycle) which follows a set pattern and is regulated by the interplay of several hormones. Control of hormone release is brought about through feedback mechanisms: the levels of the female reproductive hormones, estrogen and progesterone, regulate the secretion of the pituitary hormones that control the ovarian cycle (see earlier pages). Pregnancy interrupts this cycle and maintains the corpus luteum and the placenta as endocrine organs with the specific role of maintaining the developing fetus for the period of its development. During the last month of pregnancy the peptide hormone oxytocin induces the uterine contractions that will expel the baby from the uterus.

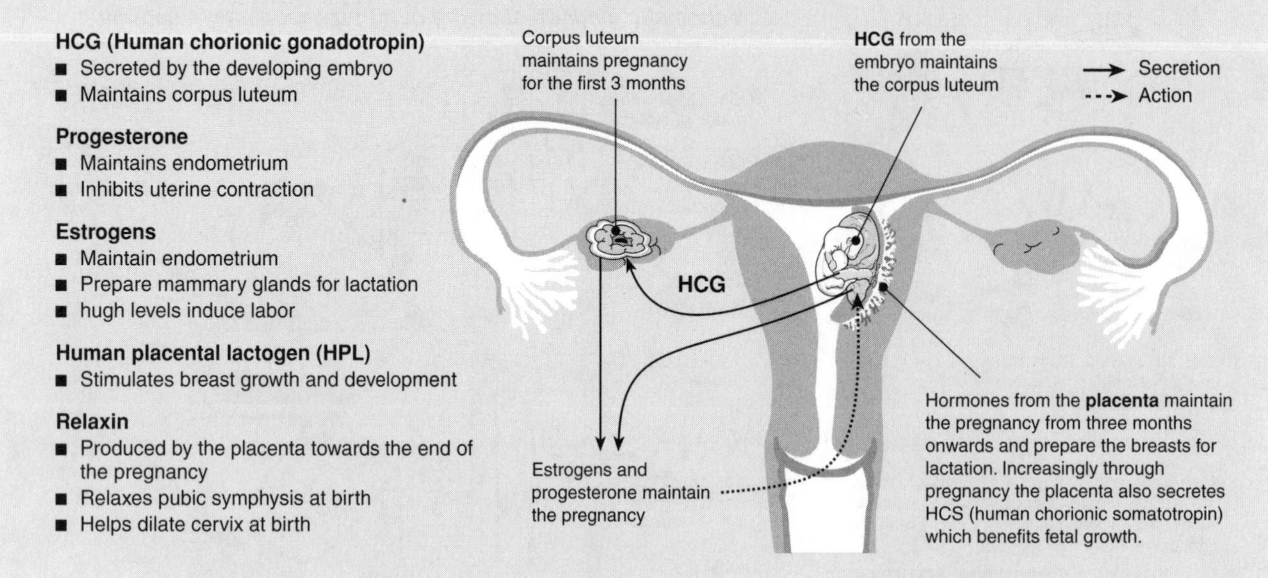

HCG (Human chorionic gonadotropin)
■ Secreted by the developing embryo
■ Maintains corpus luteum
Progesterone
■ Maintains endometrium
■ Inhibits uterine contraction
Estrogens
■ Maintain endometrium
■ Prepare mammary glands for lactation
■ hugh levels induce labor
Human placental lactogen (HPL)
■ Stimulates breast growth and development
Relaxin
■ Produced by the placenta towards the end of the pregnancy
■ Relaxes pubic symphysis at birth
■ Helps dilate cervix at birth
Corpus luteum maintains pregnancy for the first 3 months
HCG from the embryo maintains the corpus luteum
Secretion
Action
HCG
Estrogens and progesterone maintain the pregnancy
Hormones from the placenta maintain the pregnancy from three months onwards and prepare the breasts for lactation. Increasingly through pregnancy the placenta also secretes HCS (human chorionic somatotropin) which benefits fetal growth.

Hormonal Changes During Pregnancy, Birth, and Lactation

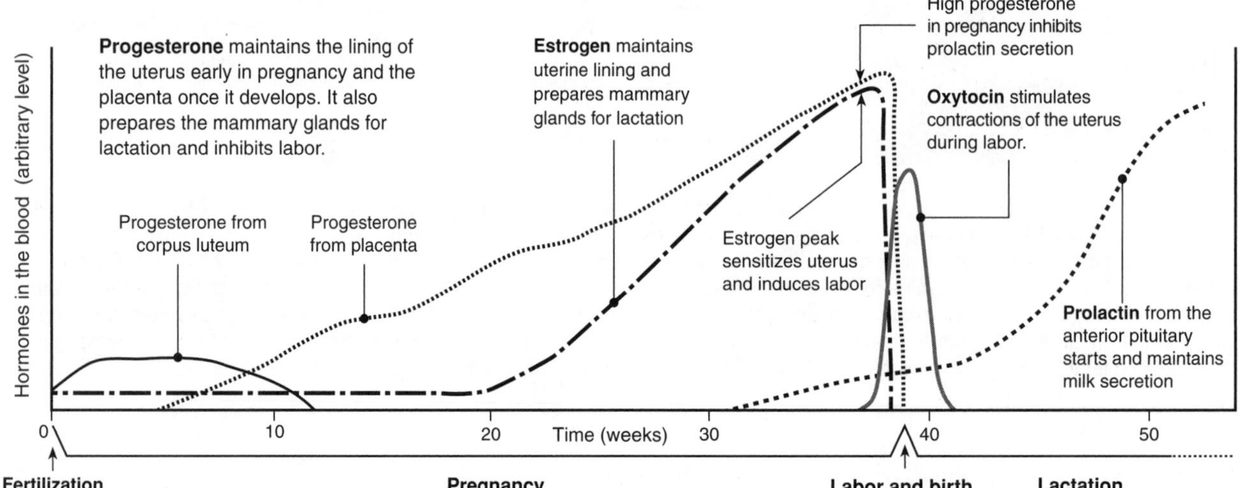

During the first 12-16 weeks pregnancy, the **corpus luteum** secretes enough progesterone to maintain the uterine lining and sustain the developing embryo. After this, the placenta takes over as the primary endocrine organ of pregnancy. **Progesterone** and **estrogen** from the placenta maintain the uterine lining, inhibit the development of further ova (eggs), and prepare the breast tissue for **lactation** (milk production). At the end of pregnancy, the placenta loses competency, progesterone levels fall, and high estrogen levels trigger the onset of labor. The estrogen peak coincides with an increase in oxytocin, which stimulates uterine contractions in a postive feedback loop: the contractions and the increasing pressure of the cervix from the infant stimulate release of more oxytocin, and more contractions and so on, until the infant exits the birth canal. After birth, the secretion of prolactin increases. Prolactin maintains lactation during the period of infant nursing.

1. (a) Explain why the corpus luteum is the main source of progesterone in early pregnancy:

   (b) Name the hormones responsible for maintaining pregnancy:

2. (a) Identify two hormones involved in **labor** and explain their roles:

   (b) Describe two physiological factors in initiating labor:

Related activities: *The Menstrual Cycle, Control of the Menstrual Cycle, Birth and Lactation, The Placenta*

Periodicals: *Pregnancy tests*

Birth and Lactation

A human pregnancy (the period of **gestation**) lasts, on average, about 38 weeks after fertilization. It ends in labor, the birth of the baby, and expulsion of the placenta. During pregnancy, progesterone maintains the placenta and inhibits contraction of the uterus. At the end of a pregnancy, increasing estrogen levels overcome the influence of progesterone and labor begins. Prostaglandins, factors released from the placenta, and the physiological state of the baby itself are also involved in triggering the actual timing of labor onset. Labor itself comprises three stages (below), and ends with the delivery of the placenta. After birth, the mother provides nutrition for the infant through **lactation**: the production and release of milk from mammary glands. Breast milk provides infants with a complete, easily digested food for the first 4-6 months of life. All breast milk contains maternal antibodies, which give the infant protection against infection while its own immune system develops.

Birth and the Stages of Labor

Stage 1: Dilation

Duration: 2-20 hours

The time between the onset of labor and complete opening (dilation) of the cervix. The amniotic sac may rupture at this stage, releasing its fluid. The hormone **oxytocin** stimulates the uterine contractions necessary to dilate the cervix and expel the baby. It is these uterine contractions that give the pain of labor, most of which is associated with this first stage.

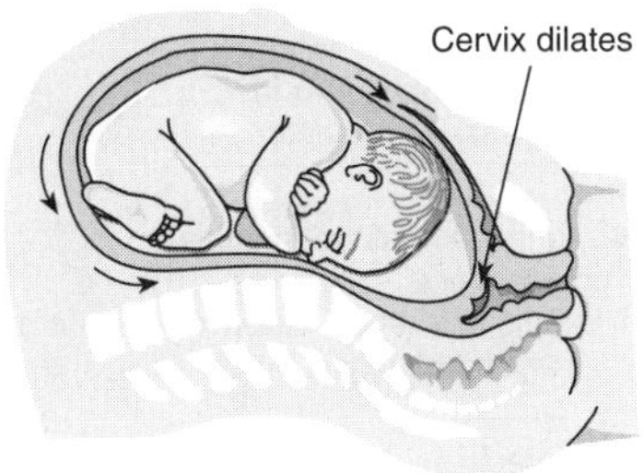

Stage 2: Expulsion

Duration: 2-100 minutes

The time from full dilation of the cervix to delivery. Strong, rhythmic contractions of the uterus pass in waves (arrows), and push the baby to the end of the vagina, where the head appears.

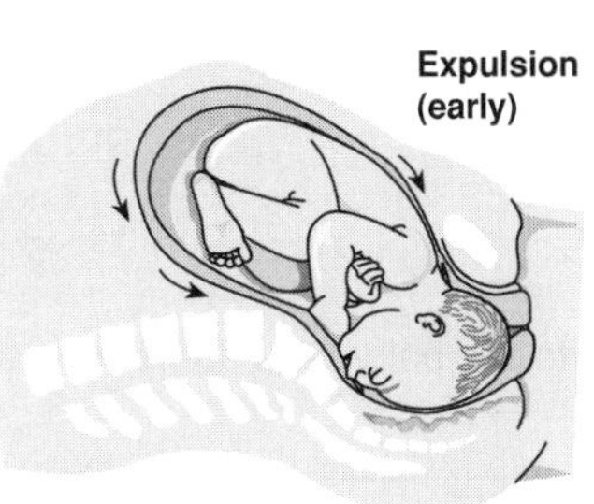

As labor progresses, the time between each contraction shortens. Once the head is delivered, the rest of the body usually follows very rapidly. Delivery completes stage 2.

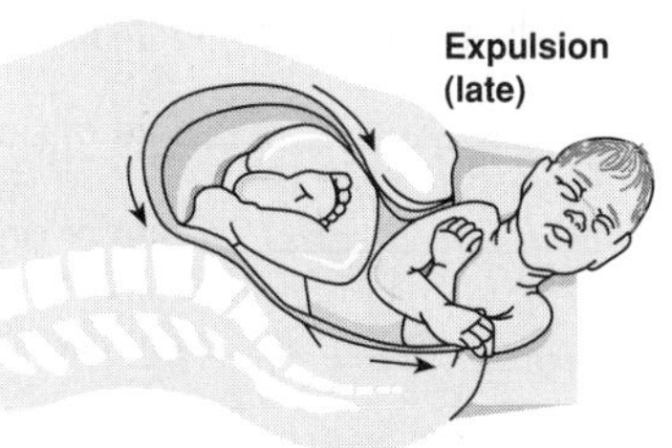

Stage 3: Delivery of placenta

Time: 5-45 minutes after delivery

The third or **placental stage**, refers to the expulsion of the placenta from the uterus. After the placenta is delivered, the placental blood vessels constrict to stop bleeding.

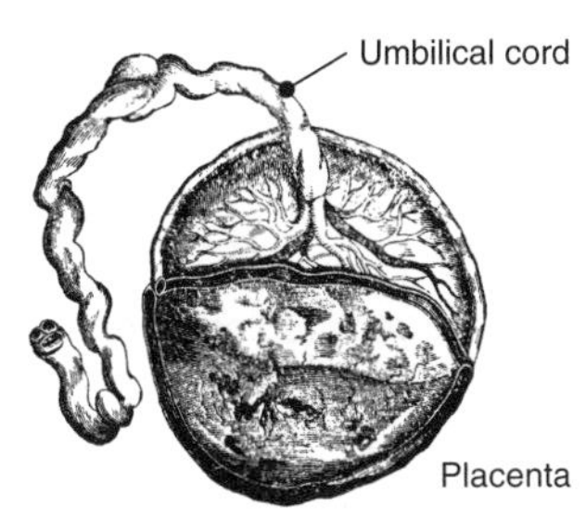

Delivery of the Baby: The End of Stage 2

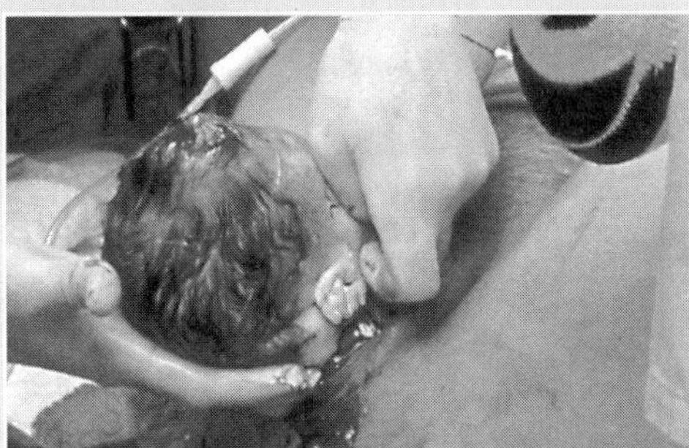

Delivery of the head. This baby is face forward. The more usual position for delivery is face to the back of the mother.

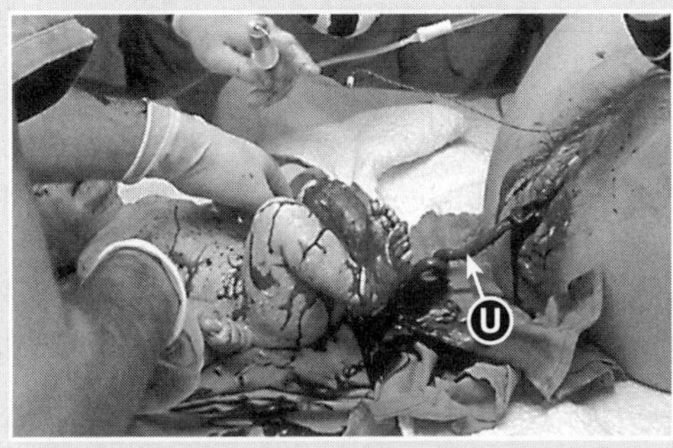

Full delivery of the baby. Note the umbilical cord (U), which supplies oxygen until the baby's breathing begins.

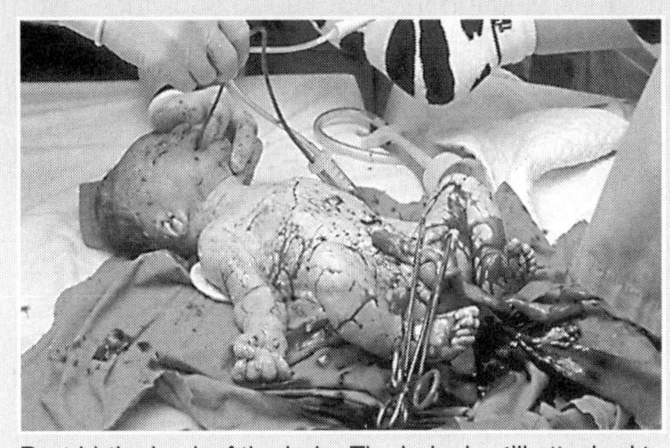

Post-birth check of the baby. The baby is still attached to the placenta and the airways are being cleared of mucus.

1. Name the three stages of birth, and briefly state the main events occurring in each stage:

   (a) Stage 1: ___

   (b) Stage 2: ___

   (c) Stage 3: ___

2. (a) Name the hormone responsible for triggering the onset of labor: ___

   (b) Describe two other factors that might influence the timing of labor onset: ___

Lactation and its Control

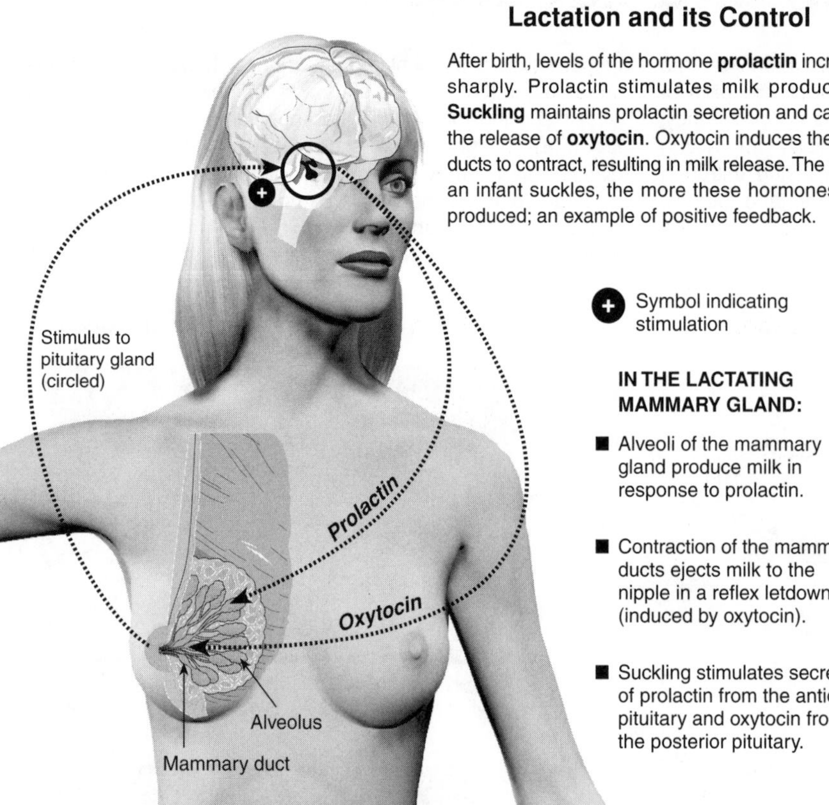

After birth, levels of the hormone **prolactin** increase sharply. Prolactin stimulates milk production. **Suckling** maintains prolactin secretion and causes the release of **oxytocin**. Oxytocin induces the milk ducts to contract, resulting in milk release. The more an infant suckles, the more these hormones are produced; an example of positive feedback.

+ Symbol indicating stimulation

IN THE LACTATING MAMMARY GLAND:

- Alveoli of the mammary gland produce milk in response to prolactin.
- Contraction of the mammary ducts ejects milk to the nipple in a reflex letdown (induced by oxytocin).
- Suckling stimulates secretion of prolactin from the antior pituitary and oxytocin from the posterior pituitary.

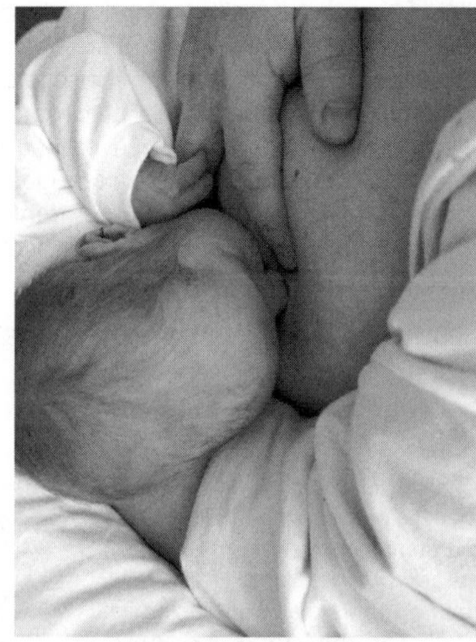

It is essential to establish breast feeding soon after birth, as this is when infants exhibit the strong reflexes that enable them to learn to suckle effectively. The first formed milk, colostrum, has very little sugar, virtually no fat, and is rich in maternal antibodies. Breast milk that is produced later has a higher fat content, and its composition varies as the nutritional needs of the infant change during growth.

3. Explain why the umbilical cord continues to supply blood to the baby for a short time after delivery: ____________________

4. For each of the following processes, state the primary controlling hormone and its site of production:

   (a) Uterine contraction during labor: Hormone: ______________ Site of production: ______________

   (b) Production of milk: Hormone: ______________ Site of production: ______________

   (c) Milk ejection in response to suckling: Hormone: ______________ Site of production: ______________

5. State which hormone inhibits prolactin secretion during pregnancy: ____________________

6. Describe two benefits of breast feeding to the health of the infant:

   (a) ____________________

   (b) ____________________

7. (a) Describe the nutritional differences between the first formed milk (colostrum) and the milk that is produced later:

   ____

   (b) Suggest a reason for these differences: ____________________

8. Explain why the nutritional composition of breast milk might change during a six-month period of breast feeding:

   ____

9. Infants exhibit marked growth spurts at six weeks and three months of age. At these times, their caloric (energy intake) requirements also increase sharply. With reference to what you know about the control of lactation, suggest how a breast-feeding mother could continue to provide for the increased energy requirements of her infant:

   ____

Sexual Development

Like many animals, humans differentiate into the male or female sex by the action of a combination of different hormones. The hormones testosterone (in males), and estrogen and progesterone (in females), are responsible for puberty (the onset of sexual maturity), the maintenance of gender differences, and the production of gametes. In females, estrogen and progesterone also regulate the menstrual cycle, and ensure the maintenance of pregnancy and nourishment of young.

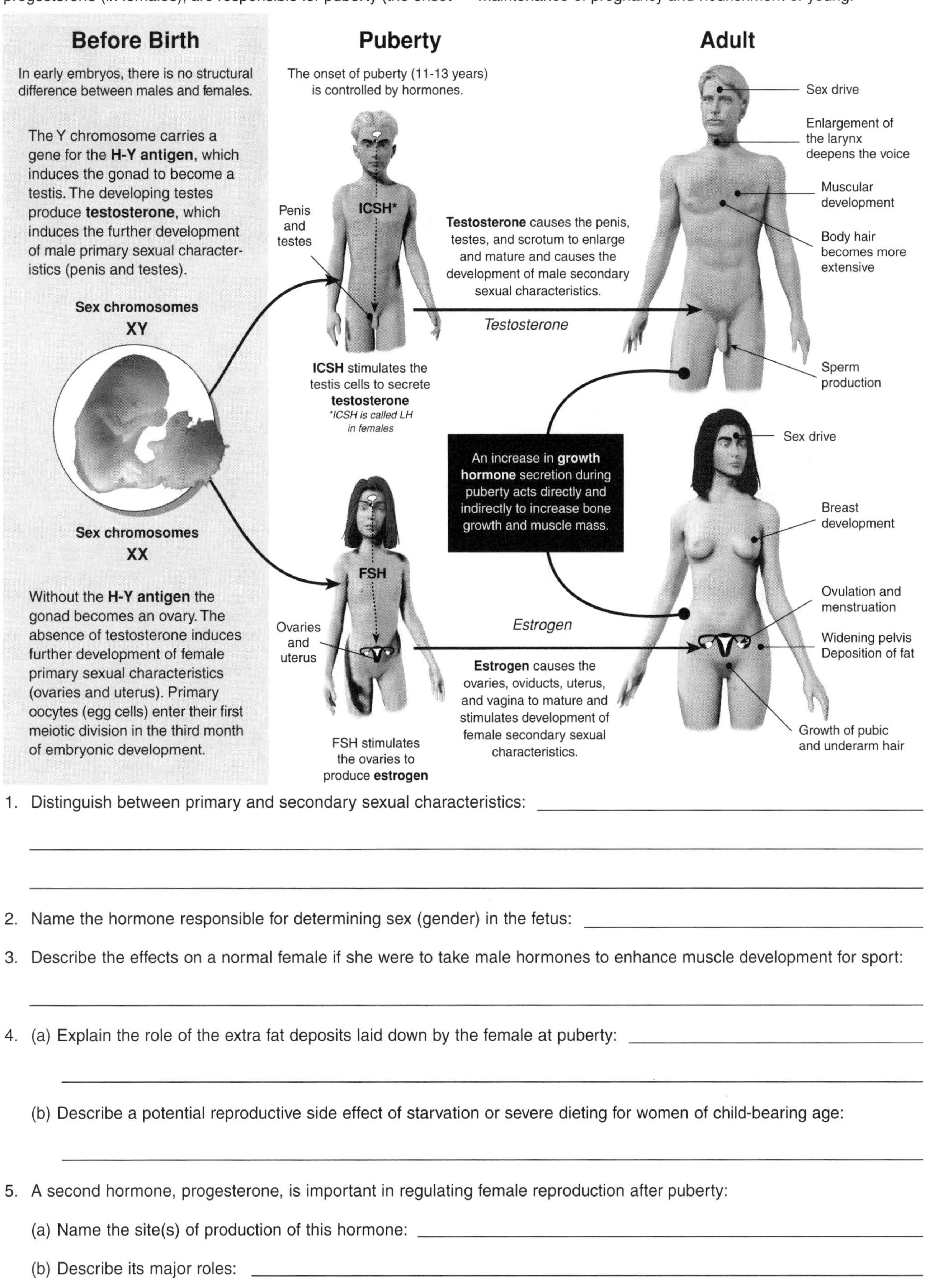

Before Birth
In early embryos, there is no structural difference between males and females.
The Y chromosome carries a gene for the H-Y antigen, which induces the gonad to become a testis. The developing testes produce testosterone, which induces the further development of male primary sexual characteristics (penis and testes).
Sex chromosomes XY
Sex chromosomes XX
Without the H-Y antigen the gonad becomes an ovary. The absence of testosterone induces further development of female primary sexual characteristics (ovaries and uterus). Primary oocytes (egg cells) enter their first meiotic division in the third month of embryonic development.
Puberty
The onset of puberty (11-13 years) is controlled by hormones.
ICSH*
Penis and testes
ICSH stimulates the testis cells to secrete testosterone
*ICSH is called LH in females
Testosterone causes the penis, testes, and scrotum to enlarge and mature and causes the development of male secondary sexual characteristics.
Testosterone
An increase in growth hormone secretion during puberty acts directly and indirectly to increase bone growth and muscle mass.
FSH
Ovaries and uterus
FSH stimulates the ovaries to produce estrogen
Estrogen
Estrogen causes the ovaries, oviducts, uterus, and vagina to mature and stimulates development of female secondary sexual characteristics.
Adult
Sex drive
Enlargement of the larynx deepens the voice
Muscular development
Body hair becomes more extensive
Sperm production
Sex drive
Breast development
Ovulation and menstruation
Widening pelvis
Deposition of fat
Growth of pubic and underarm hair

1. Distinguish between primary and secondary sexual characteristics: ______________________

2. Name the hormone responsible for determining sex (gender) in the fetus: ______________________

3. Describe the effects on a normal female if she were to take male hormones to enhance muscle development for sport:

4. (a) Explain the role of the extra fat deposits laid down by the female at puberty: ______________________

   (b) Describe a potential reproductive side effect of starvation or severe dieting for women of child-bearing age:

5. A second hormone, progesterone, is important in regulating female reproduction after puberty:

   (a) Name the site(s) of production of this hormone: ______________________

   (b) Describe its major roles: ______________________

The Next Generation

Growth and Development

Development describes the process of growing to maturity, from zygote to adult. After birth, development continues rapidly and is marked by specific stages recognized by the set of physical and cognitive skills present. Obvious physical changes include the elongation of the bones, increasing ossification of cartilage, and changes to the proportions of the body. These proportional changes are the result of **allometric growth** (differential growth rates) and occur concurrently with motor, intellectual, and emotional and social development. These changes lead the child to increasing independence and maturity.

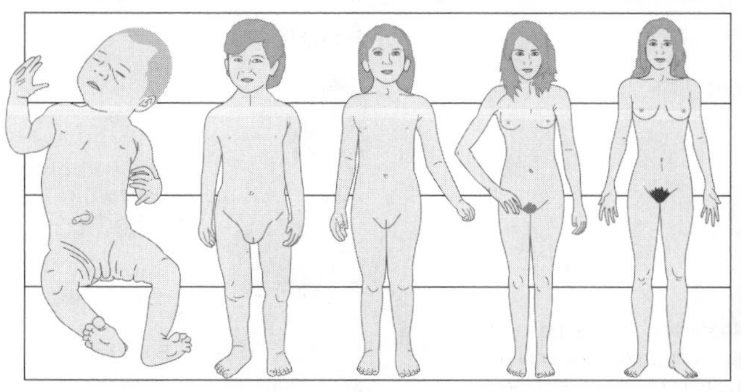

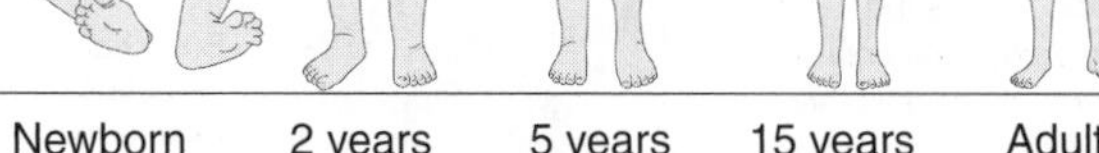

At birth the cranium is very large in comparison to the face and the skull makes up around one quarter of the infant's height. During early life, the face continues to grow outward, reducing the relative proportions of the cranium, while at adulthood the size of the skull in proportion to the body is much less.

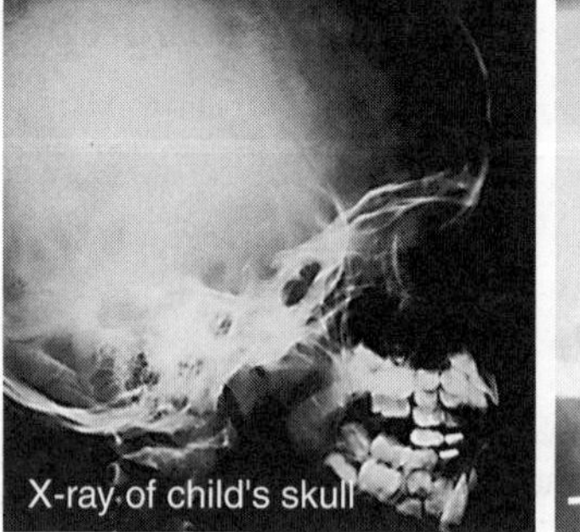

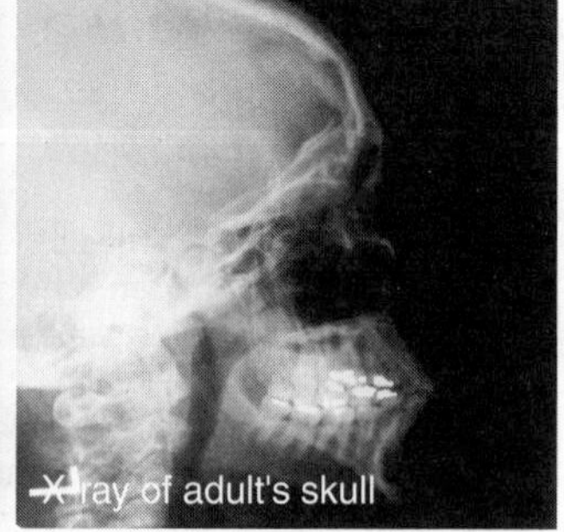

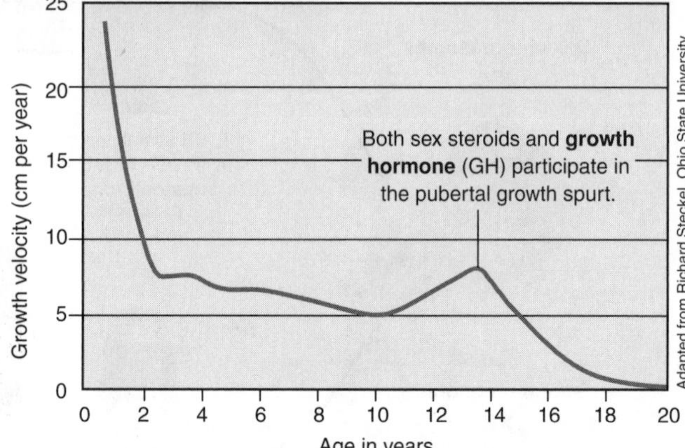

By 6 weeks old, a human baby is usually able to hold its head up if placed on its stomach. At 3 months the infant will exercise limbs aimlessly but by 5 months is able to grasp objects and sit up. The infant may be able to crawl by 8 months and walk by 12 months. It is more or less independent by two years and undergoes changes to adulthood at around 11 years of age (puberty).

Babies are effectively born premature so that they complete much of their early development in the first two to three years of life. The rate of growth declines slowly through childhood, but increases again to a peak in puberty (the growth spurt). By 20 years of age the cartilage in the long bones has been replaced by bone and growth stops.

1. Describe the most noticeable change in body proportion from birth to adulthood: __________

2. Describe the changes that occur in the first period of rapid growth in humans: __________

3. Describe the changes that occur in the second period of rapid growth in humans: __________

4. Answer the following questions with respect to the graph depicting growth rates 0-20 years (above, right):

   (a) Describe what happens to growth velocity in the first two years of an infant's life: __________

   (b) Describe what the graph infers about the rate of growth in the period before birth: __________

   (c) Identify the age range (in years) marking the pubertal growth spurt: __________

5. Relate the changes in physical development to the changes occurring in the mental development of an infant:

Related activities: *Sexual Development*

Periodicals:
Menopause-design fault or by design, Age-old story

KEY TERMS Mix and Match

INSTRUCTIONS: Test your vocabulary by matching each term to its definition, as identified by its preceding letter code.

ASEXUAL REPRODUCTION ....

CONTRACEPTION ....

CORPUS LUTEUM ....

EMBRYO ....

ENDOMETRIUM ....

(O)ESTROGEN ....

FERTILIZATION ....

GAMETOGENESIS ....

LACTATION ....

MEIOSIS ....

MENSTRUAL CYCLE ....

OOGENESIS ....

OVIPAROUS ....

OVOVIVIPAROUS ....

OVULATION ....

OVUM ....

OXYTOCIN ....

PARTHENOGENESIS ....

PLACENTA ....

PREGNANCY ....

PROGESTERONE ....

PROLACTIN ....

PUBERTY ....

SPERM ....

SPERMATOGENESIS ....

TESTOSTERONE ....

VIVIPAROUS ....

ZYGOTE ....

A The production and secretion of milk from the mammary glands.

B The principal male sex hormone.

C A mode of reproduction in which embryos develop inside eggs that are retained within the mother's body until they are ready to hatch.

D The inner membrane of the mammalian uterus.

E A steroid hormone, which functions as the primary female sex hormone.

F A term meaning to give birth to live young.

G The initial cell formed from the union of two gametes when a new organism is produced by means of sexual reproduction.

H A form of asexual reproduction in females, where growth and development of embryos occurs without fertilization by a male.

I The cycle of changes in reproductive physiology occurring in fertile female humans.

J A reduction division involved in the production of gametes.

K An organ, characteristic of most mammals, that enables exchanges (via the blood supply) of nutrient, gases, and wastes, between mother and fetus.

L A temporary endocrine structure in mammals, involved in the hormonal maintenance of the endometrium.

M The state of carrying offspring (a fetus), inside the uterus of a female.

N The fusion of sperm and egg. Also called conception.

O The prevention of fertilization of the ovum by sperm cells.

P The process by which gametes are produced.

Q The process in a female's menstrual cycle by which a mature ovarian follicle ruptures and releases an ovum.

R A peptide hormone with a primary role in lactation in mammals.

S A steroid hormone involved in the female menstrual cycle, pregnancy, and embryogenesis.

T Egg laying, with little or no other embryonic development within the mother.

U The male gamete.

V The creation of mature spermatozoa (sperm cells) in males.

W A hormone with a role in uterine contraction in labor and milk letdown in lactating females.

X An egg cell of female gamete.

Y The earliest stage of development, until about 8 weeks post-fertilization in humans.

Z The period of physical changes during which a child's body becomes a reproductively capable adult body

AA The creation of an ovum.

BB Reproduction which does not involve meiosis or fertilization.

Important in this section ...

- *Develop an understanding of basic plant structure*
- *Explain transport processes in angiosperms*
- *Explain how plants respond to the environment*
- *Describe the features of angiosperm reproduction*

Plant Growth Responses

Responses to environment	• Plant responses • Plant rhythms • Vernalization and dormancy
Tropisms	• Phototropism and the role of auxin • Gravitropism
Responses to day length	• Photoperiodism • Long day and short day plants • Phytochrome action
Other plant hormones	• General effects of auxins and ABA • Gibberellins • Cytokinins

Plant Structure

The structure of plants	• Plant diversity • The plant body • Roots, stems, leaves
Plant tissues	• Xylem and phloem • Organization of plant tissues • Adaptations to environment

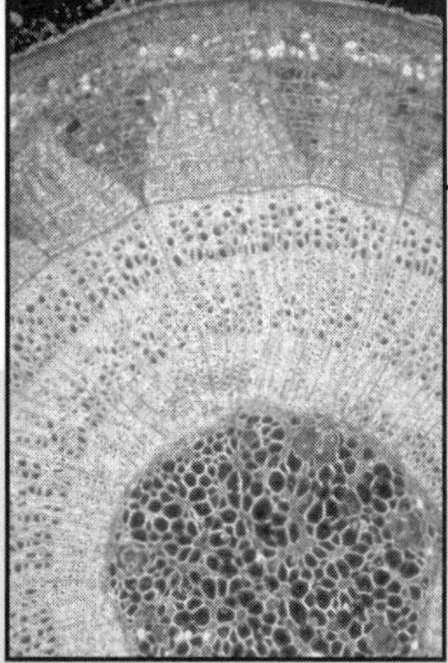

The basic plant body: roots, stems, and leaves, often shows extreme modifications for growth in different environments.

Plant responses are controlled by phytohormones which interact in complex ways to regulate growth and development.

Part 3

The Structure and Function of Plants

The structural and physiological adaptations of plants can be understood within the context of their environment.

Plants use plant growth substances to coordinate their activities appropriately in the environment.

Plant life cycles are characterized by alternation of generations. In angiosperms the sporophyte generation dominates.

The vascular tissues are connected throughout the plant body, acting to both support and transport.

Why reproduce?	• Successful reproduction = fitness • Sexual vs asexual reproduction
Diversity in plant reproduction	• Alternation of generations • Independence from water • The angiosperm plan
Angiosperm reproduction	• Wind pollination • Insect pollination • The double fertilization
Fruits and seeds	• The formation and role of fruits • Seed structure and germination • Seed dispersal

Plant Reproduction

Support in plants	• Support tissues • The interdependence of support and transport in plants
Root structure	• Diversity in root structure • Monocots vs dicots • Water and mineral uptake
Plant gas exchange	• Gas transfers in the leaf • Stomatal opening/closing • The effect of hormones
Plant transport	• Transpiration and water potential • Tension-cohesion hypothesis • Factors affecting transpiration rates • Translocation: sources and sinks

Plant Support and Transport

AP Topic 3B

IB AHL 9.1

Plant Structure and Growth Responses

KEY CONCEPTS

- Diversity in plant structural morphology reflects their adaptations to different environments.
- Dicot and monocot plants show distinct structural differences in the arrangements of their tissues.
- Plants grow at regions of meristematic tissue.
- Plant growth responses are regulated by phytohormones. The interaction of these creates differential responses in different tissues.

KEY TERMS

abscisic acid (ABA)
apical meristem
auxin
axillary bud
cytokinins
dicot
gibberelllins
gravitropism
hydrophyte
internode
lateral meristem
leaf
long day plants
meristem
mesophyll
monocot
nastic response
node
phloem
photoperiodism
phototropism
phytochrome
primary growth
root
secondary growth
short day plant
stem
tropism
vascular cambium
vascular bundle
xerophyte
xylem

OBJECTIVES

- ☐ 1. Use the **KEY TERMS** to help you understand and complete these objectives.

The Variety of Plants

pages 252-256, 261-262

- ☐ 2. Describe the pivotal role of plants as producers, in the global carbon budget, and as a source of valuable products.
- ☐ 3. Describe plant diversity as illustrated by the structural differences between **bryophytes**, **ferns**, **conifers**, and **angiosperms** (flowering plants).
- ☐ 4. Describe the external structure of a named **dicotyledonous plant**. Identify the location and function of root, stem, leaf, and axillary and terminal buds.

Angiosperm Leaves and Stems

pages 254, 257-260, 263-271

- ☐ 5. Use labeled **plan diagrams** to describe and compare the distribution of tissues in a dicot stem and leaf.
- ☐ 6. Explain the relationship between the distribution of tissues (e.g. phloem, xylem, parenchyma) in a leaf and the functions of these tissues. Include reference to: light absorption, gas exchange, support, water conservation, and transport. Describe the adaptations of leaves to maximize photosynthesis in different environments.
- ☐ 7. Describe differences in leaf and stem structure in **monocots** and dicots.
- ☐ 8. Describe some modifications of leaves, roots, or stems for different functions, e.g. for support, as food storage organs, or for food capture.
- ☐ 9. Describe adaptations in **xerophytes** and/or **hydrophytes**. In each case, explain how the adaptation enhances survival in the environment.
- ☐ 10. Describe the location and function of the **meristems** in dicots. Compare and contrast **primary growth** (apical growth) and **secondary growth** (lateral growth). Recognize that not all plants show secondary growth.

Plant Responses to Environment

pages 272-278

- ☐ 11. Describe environmental factors that are important to plants and summarize how plants respond to each and why.
- ☐ 12. Describe and explain **tropisms**, including **phototropism**. Describe the role of **auxins** in tropic responses and discuss the evidence for it.
- ☐ 13. Explain the role of **phytohormones** in regulating plant responses to environmental factors. Include reference to any of: **auxins**, **abscisic acid** (ABA), **cytokinins**, **ethylene**, and **gibberellins**.
- ☐ 14. Explain how flowering is controlled in short-day and long-day plants, including the role of **phytochrome** in plant responses to photoperiod.

Periodicals:
listings for this chapter are on page 390

Weblinks:
www.thebiozone.com/weblink/SB2-2603.html

Teacher Resource CD-ROM:
Separating the Wood From the Trees

The Variety of Plants

There are possibly 500,000 species of plant, from the smallest mosses and liverworts to the largest trees more than 100 metres tall and weighing over 100 tonnes. Plants are found in the driest deserts and submerged under the water. They can be parasitic or symbiotic. Some use animals as pollinators, others prey on them. In most classification schemes, there are ten plant divisions (a division is the botanical equivalent of phylum): the bryophytes or non-vascular plants form a division on their own and the tracheophytes or vascular plants make up the remaining nine divisions. The ferns (and their relatives) and the seed plants have developed the vascular tissue necessary to support a large size, and only the seed plants have developed true wood.

Aquatic Plants

Most higher aquatic plants occur in fresh water. Angiosperms are represented by a number of groups including the water lily (top) while ferns include the Nardoo water fern (above) found in Australia.

Orchids

Orchids are the largest family of angiosperms, occurring almost every habitat with most being found in the tropics. They are well known for the complex structural relationships they have evolved with their pollinators.

Desert Plants

Cacti inhabit the deserts of the Americas and are highly adapted for conserving water. Their leaves are reduced to needle-like spines and the stem is the main organ of photosynthesis. Similar adaptations are found in unrelated arid-adapted plants elsewhere.

Insectivorous Plants

Venus fly-traps, sundews, and pitcher plants are examples of plants that supplement their photosynthetic nutrition with the nitrogen from digesting small insects.

Epiphytes

Epiphytes have taken to growing high up in trees and use vegetation trapped in the forks of branches.

Parasitic Plants

Plants can be parasitic on other plants. Mistletoe (top) attaches to and draws sap from its host. New Zealand wood rose, *Dactylanthus* (above) infects the roots of other plants. It has no roots or leaves but still produces flowers, which are pollinated by bats.

Estuarine Plants

Mangroves are a large group of angiosperms that have adapted to living in tropical estuaries. They are able to tolerate constantly changing salt concentrations from the incoming tide and outgoing fresh water streams. Salt is extruded by the leaves and the roots grow up out of the mud.

Vines

Vines wrap themselves around a host for support and may smother and shade out the supporting tree. In the US, the invasive oriental bittersweet vine smothers native trees, while Australia has the aptly named strangler fig (above).

Giant Trees

Gymnosperms can grow to immense sizes and ages. Kauri can live for over 3000 years and grow over 50 m high. Some bristle cone pines in North America are over 4500 years old and giant sequoias can grow to 100 m high.

1. Discuss the adaptations that have enabled plants to occupy a wide variety of habitats:

Related activities: *Leaf and Stem Adaptations, Adaptations of Xerophytes Adaptations of Hydrophytes,*

The Importance of Plants

Via the process of photosynthesis, plants provide oxygen and are also the ultimate source of food and metabolic energy for nearly all animals. Besides foods (e.g. grains, fruits, and vegetables), plants also provide people with shelter, clothing, medicines, fuels, and the raw materials from which innumerable other products are made.

Plant tissues provide the energy for almost all heterotrophic life. Many plants produce delicious fruits in order to spread their seeds.

Plant tissues can be utilized to provide shelter in the form of framing, cladding, and roofing.

Many plants provide fibers for a range of materials including cotton (above), linen (from flax), and coir (from coconut husks).

Plant extracts, including rubber from rubber trees (above), can be utilized in many ways as an important manufacturing material.

Coal, petroleum, and natural gas are fossil fuels which were formed from the dead remains of plants and other organisms. Together with wood, they provide important sources of fuel.

Plants produce many beneficial and not so beneficial substances (e.g. the cannabis plant above). Over 25% of all modern medicines are derived from plant extracts.

1. Using examples, describe how plant species are used by people for each of the following:

   (a) Food: ______________________________

   (b) Fuel: ______________________________

   (c) Clothing: ______________________________

   (d) Building materials: ______________________________

   (e) Esthetic value: ______________________________

   (f) "Recreational" drugs: ______________________________

   (g) Therapeutic drugs (medicines): ______________________________

2. Outline three reasons why the destruction of native forests is of concern:

   (a) ______________________________

   (b) ______________________________

   (c) ______________________________

The General Structure of Plants

The support and transport systems in plants are closely linked; many of the same tissues are involved in both systems. Primitive plants (e.g. mosses and liverworts) are small and low growing, and have no need for support and transport systems. If a plant is to grow to any size, it must have ways to hold itself up against gravity and to move materials around its body. The body of a flowering plant has three parts: **roots** anchor the plant and absorb nutrients from the soil, **leaves** produce sugars by photosynthesis, and **stems** link the roots to the leaves and provide support for the leaves and reproductive structures. Vascular tissues (xylem and phloem) link all plant parts so that water, minerals, and manufactured food can be transported between different regions. All plants rely on fluid pressure within their cells (turgor) to give some support to their structure.

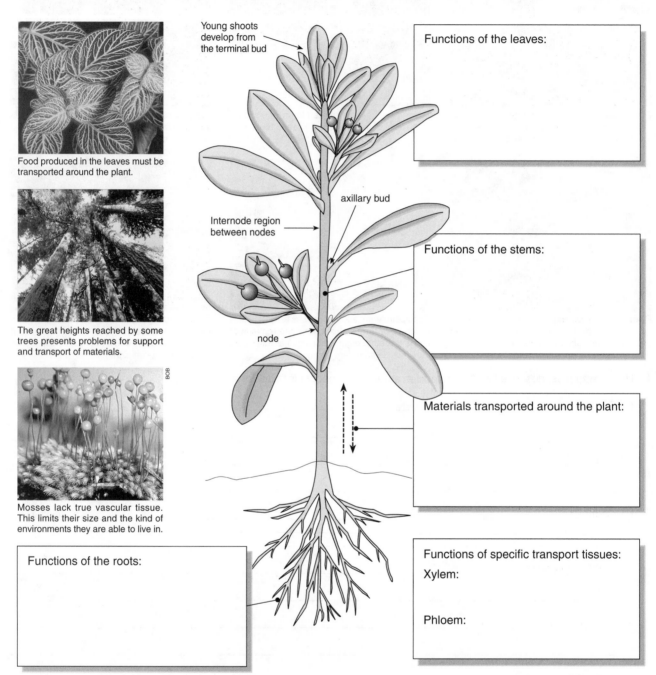

Food produced in the leaves must be transported around the plant.

The great heights reached by some trees presents problems for support and transport of materials.

BOB

Mosses lack true vascular tissue. This limits their size and the kind of environments they are able to live in.

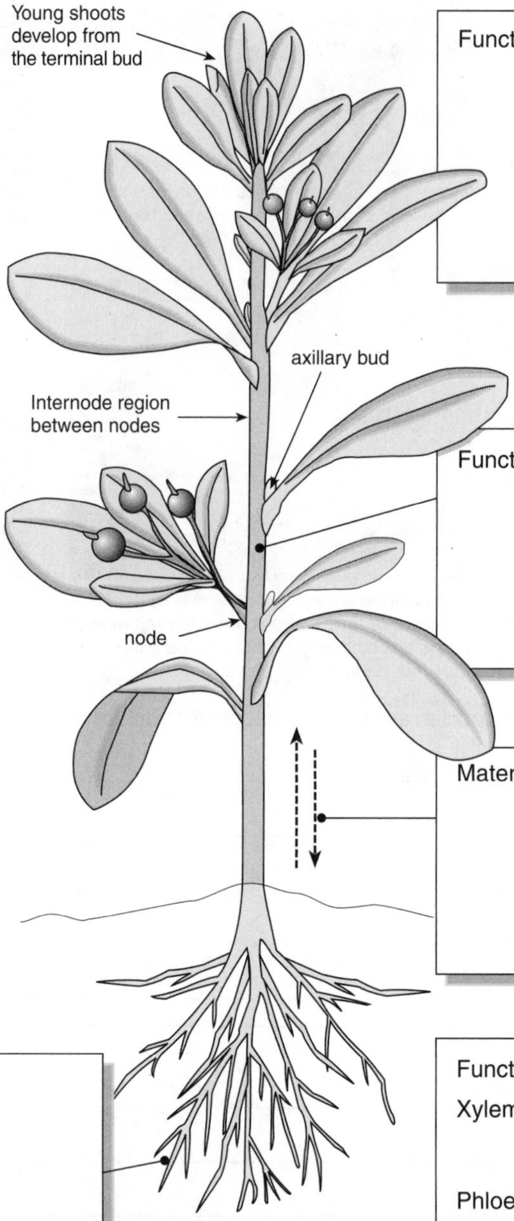

Young shoots develop from the terminal bud

axillary bud

Internode region between nodes

node

Functions of the leaves:

Functions of the stems:

Materials transported around the plant:

Functions of specific transport tissues:

Xylem:

Phloem:

Functions of the roots:

1. In the boxes provided in the diagram above:
   (a) Describe the main functions of the leaves, roots and stems (remember that the leaves themselves have leaf veins).
   (b) List the materials that are transported around the plant body.
   (c) Describe the functions of the transport tissues: xylem and phloem.

2. Name the solvent for all the materials that are transported around the plant: ____________________

3. Identify the processes involved in the transport of sap in the following tissues:

   (a) The xylem: ____________________

   (b) The phloem: ____________________

***Related activities:** Xylem, Phloem, Uptake in the Root, Translocation*

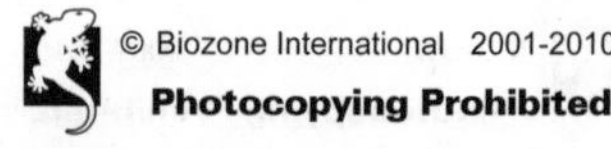

Diversity in Leaf Structure

The main function of leaves is to collect the radiant energy from the sun and convert it into a form that can be used by the plant. The sugars produced from **photosynthesis** are used as a source of chemical energy to do cellular work, and as building materials for growing new tissues. The structure of the leaf has evolved to cope with life on land, while still allowing efficient absorption of energy from the sun and carbon dioxide from the air. This must be achieved without excessive water loss. The most primitive plants have leaves without **vascular tissue** (veins), and stomata that are either non-existent or very simple. Some leaves are able to control the amount of carbon dioxide entering and the amount of water leaving the plant. Mosses and liverworts are the simplest plants to exhibit a **cuticle** which prevents water loss from the above ground parts. Regardless of their varying forms, foliage leaves of vascular plants comprise epidermal, mesophyll (the packing tissue of the leaf), and vascular tissues.

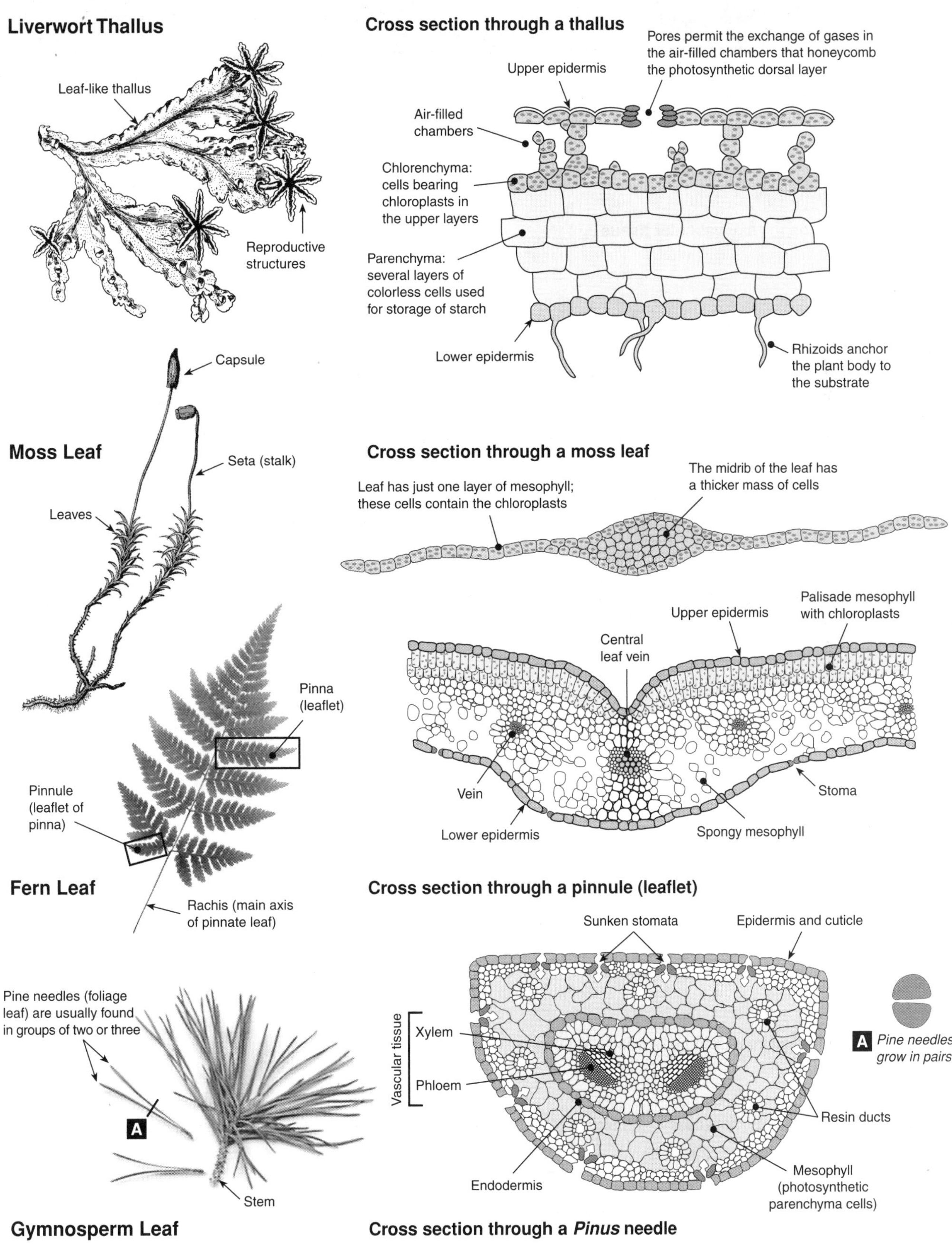

Related activities: *Angiosperm Leaf structure, Leaf and Stem Adaptations*
Web links: *Photographic Atlas of Plant Anatomy, Leaf Tissues*

This liverwort (above) shows the familiar thalloid form of the gametophyte plant body. Liverworts are reliant on the availability of free water for reproduction because the sperm must swim through a continuous film of water from the antheridia where they are produced to the archegonium where the eggs are held.

The spores of the most familiar order of ferns (the Filicales) give rise to tiny free-living **gametophytes** (above), which are restricted moist environments. A film of water is required for the sperm to swim to the eggs, but once the young sporophyte becomes established, the structure of its leaves enables it to be relatively tolerant of drier conditions.

The conspicuous conifer plants we see are representatives of the sporophyte generation. The conifer gametophytes borne on cones (above) and conifers are freed of a reproductive dependence on moist environments. The leaves, which are usually needle-like, show adaptations such as sunken stomata, which help reduce water losses.

1. (a) Describe the roles of **vascular tissue**: ______

   (b) List the plants on the previous page that lack true vascular tissue: ______

2. (a) Describe the purpose of the waxy cuticle that coats the leaf surface: ______

   (b) Explain why the leaf epidermis is transparent: ______

3. Explain why a leaf on a moss plant does not require a stoma or pore to allow gas exchange: ______

4. Describe where the photosynthetic tissue is located in:

   (a) A liverwort: ______

   (b) A moss: ______

   (c) A fern: ______

   (d) A gymnosperm: ______

5. (a) Describe a difference in external appearance between the leaves of gymnosperms and angiosperms (opposite):

   ______

   (b) Describe a distinctive feature of a fern leaf (frond): ______

6. Describe and explain differences in the structure of the "leafy" moss gametophyte and the leaves of true vascular plants:

   ______

Angiosperm Leaf Structure

Variations in the structure of angiosperm leaves are, to a great extent, related to habitat. Although angiosperms are able (to some extent) to control water losses through their stomata, the availability of water is still one of the most important factors affecting the structure and form of leaves. Plants can be categorized according to their water requirements or adaptations as mesophytes, hydrophytes, or xerophytes, but leaves often exhibit a combination of features characteristic of these three broad ecological types. Descriptions based on the structure of a rather generalized monocot (grass) leaf, and a similarly generalized dicot leaf are described below. Adaptation to habitat invariably results in deviations from these generalizations.

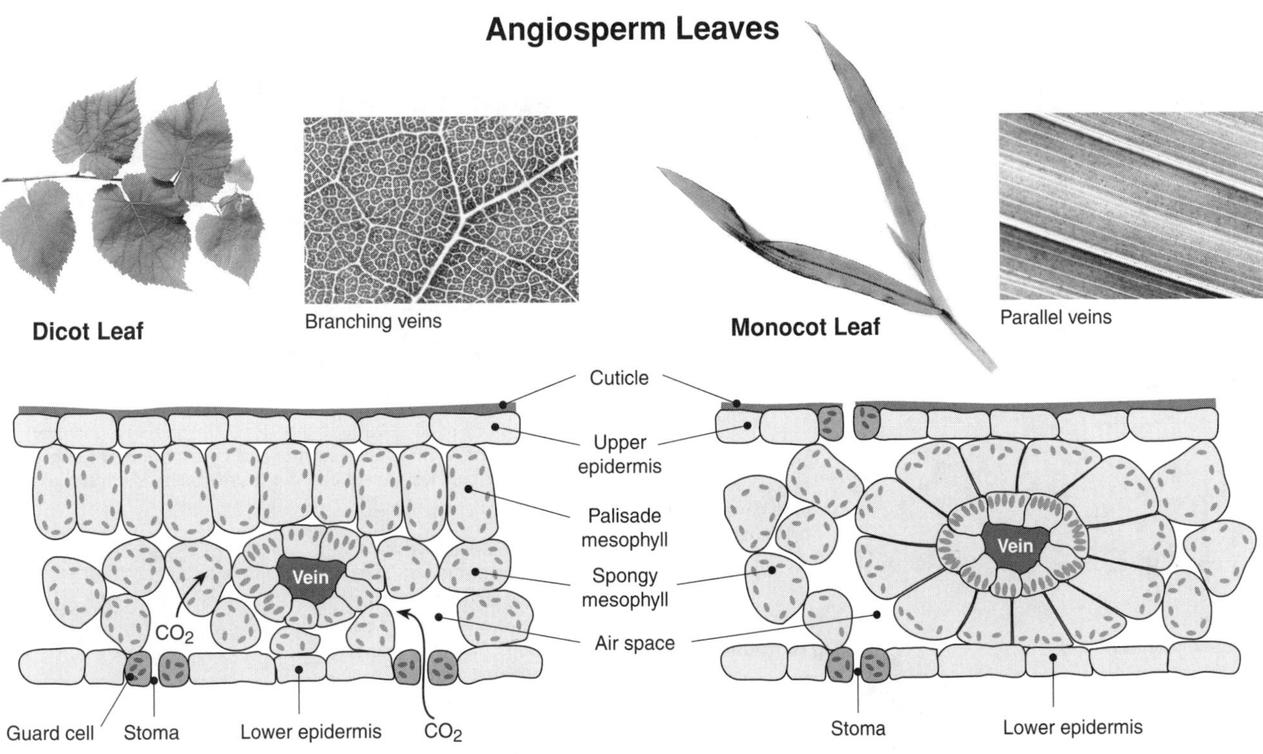

1. Explain how, on the basis of external morphology, you could distinguish a monocot and dicot leaf:

2. The internal arrangement of cells and tissues in typical monocot and dicot leaves are illustrated above. Describe two ways in which the internal leaf structure of the dicot differs from that of the monocot:

   (a)

   (b)

3. How might you expect the cuticle of an arid-adapted plant to differ from that of a mesophyte?

4. (a) Identify the region of a dicot leaf where most of the chloroplasts are found:

   (b) Name the important process that occurs in the chloroplasts:

5. (a) Explain the purpose of the air spaces in the leaf tissue:

   (b) Describe how gases enter and leave the leaf tissue:

   (c) Explain why leaves are usually flat and often broad:

Related activities: *Leaf and Stem Adaptations*
Web links: *Leaf Structure*

Xylem

Xylem is the principal **water conducting tissue** in vascular plants. It is also involved in conducting dissolved minerals, in food storage, and in supporting the plant body. As in animals, tissues in plants are groupings of different cell types that work together for a common function. Xylem is a **complex tissue**. In angiosperms, it is composed of five cell types: tracheids, vessels, xylem parenchyma, sclereids (short sclerenchyma cells), and fibers. The tracheids and vessel elements form the bulk of the tissue. They are heavily strengthened and are the conducting cells of the xylem. Parenchyma cells are involved in storage, while fibers and sclereids provide support. When mature, xylem is dead.

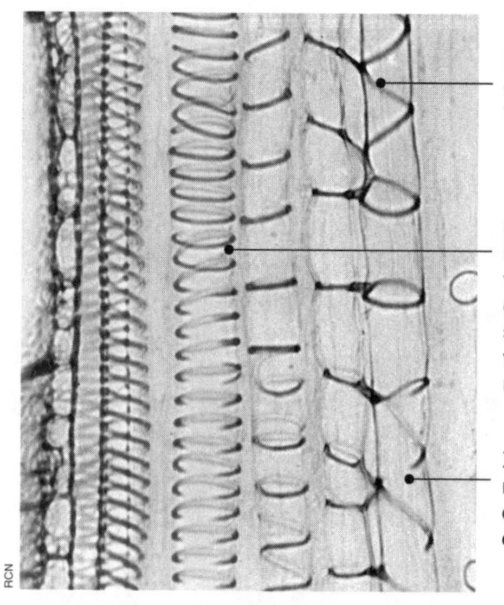

Xylem vessels form continuous tubes throughout the plant.

Spiral thickening of **lignin** around the walls of the vessel elements give extra strength allowing the vessels to remain rigid and upright.

Xylem is dead when mature. Note how the cells have lost their cytoplasm.

Vessel element

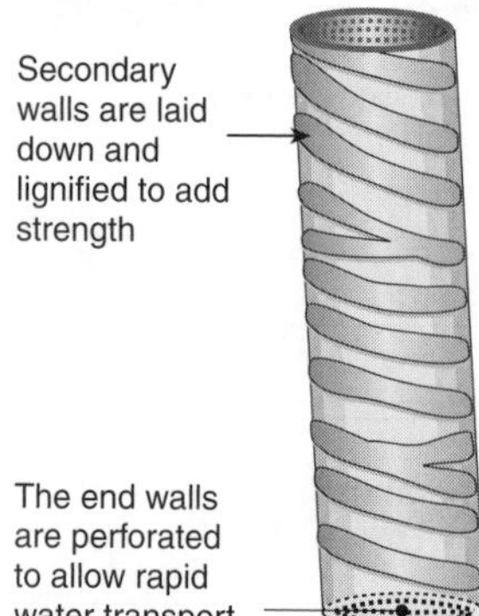

Secondary walls are laid down and lignified to add strength

The end walls are perforated to allow rapid water transport

Tip of tracheid

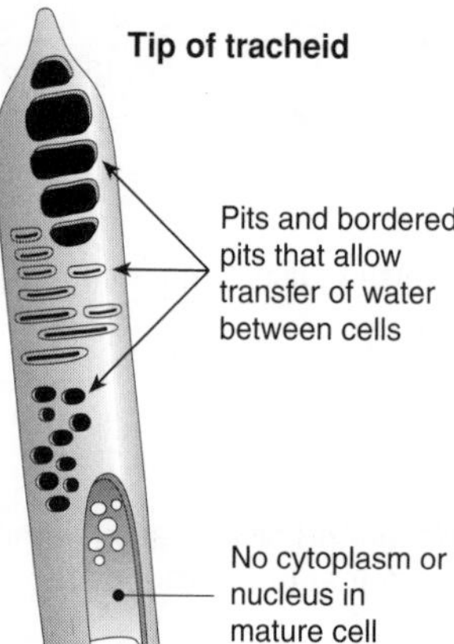

Pits and bordered pits that allow transfer of water between cells

No cytoplasm or nucleus in mature cell

Vessel elements and tracheids are the two conducting cells types in xylem. Tracheids are long, tapering hollow cells. Water passes from one tracheid to another through thin regions in the wall called **pits**. Vessel elements have pits, but the end walls are also perforated and water flows unimpeded through the stacked elements.

The Structure of Xylem Tissue

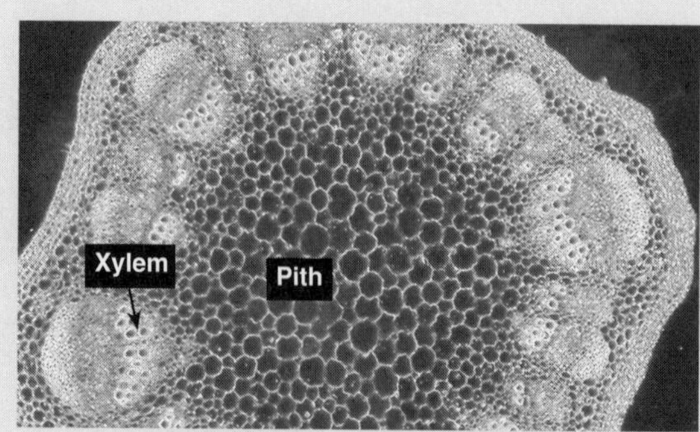

This cross section through the stem, *Helianthus* (sunflower) shows the central pith, surrounded by a peripheral ring of vascular bundles. Note the xylem vessels with their thick walls.

Fibers are a type of sclerenchyma cell. They are associated with vascular tissues and usually occur in groups. The cells are very elongated and taper to a point and the cell walls are heavily thickened. Fibers give mechanical support to tissues, providing both strength and elasticity.

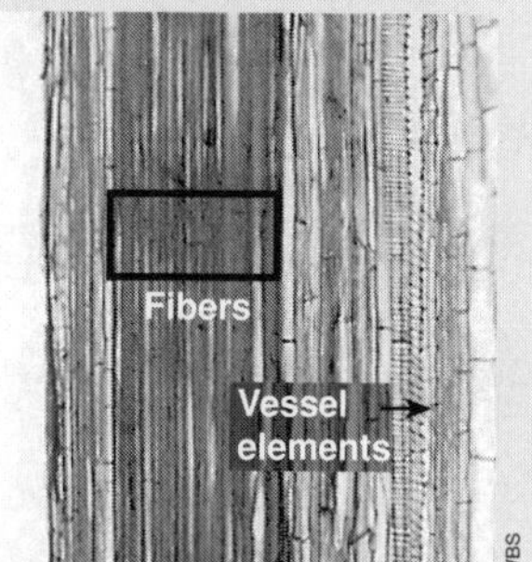

Vessel elements are found only in the xylem of angiosperms. They are large diameter cells that offer very low resistance to water flow. The possession of vessels (stacks of vessel elements) provides angiosperms with a major advantage over gymnosperms and ferns as they allow for very rapid water uptake and transport.

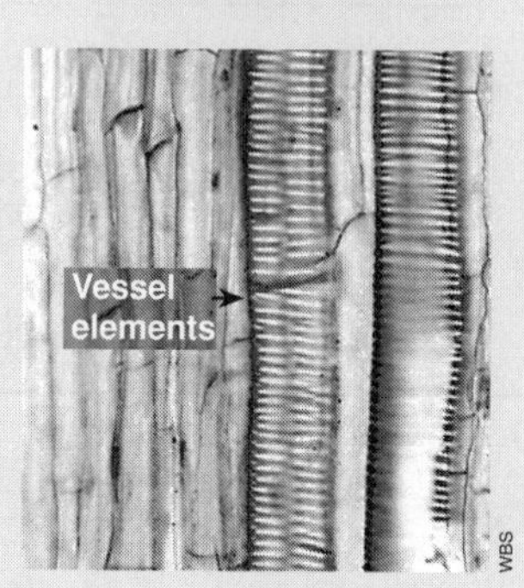

1. Describe the function of **xylem**: ______________________________

____

2. Identify the four main cell types in xylem and explain their role in the tissue:

(a) ______________________________

(b) ______________________________

(c) ______________________________

(d) ______________________________

3. Describe one way in which xylem is strengthened in a mature plant: ______________________________

4. Describe a feature of vessel elements that increases their efficiency of function: ______________________________

____

Phloem

Like xylem, **phloem** is a complex tissue, comprising a variable number of cell types. Phloem is the principal **food (sugar) conducting tissue** in vascular plants, transporting dissolved sugars around the plant. The bulk of phloem tissue comprises the **sieve tubes** (sieve tube members and sieve cells) and their companion cells. The sieve tubes are the principal conducting cells in phloem and are closely associated with the **companion cells** (modified parenchyma cells) with which they share a mutually dependent relationship. Other parenchyma cells, concerned with storage, occur in phloem, and strengthening fibers and sclereids (short sclerenchyma cells) may also be present. Unlike xylem, phloem is alive when mature.

LS through a sieve tube end plate

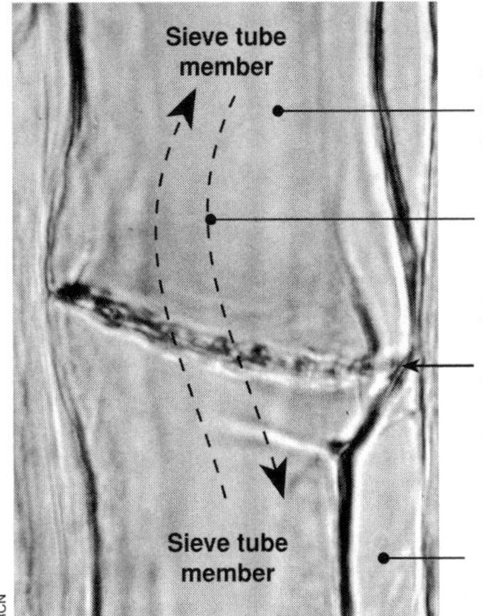

The sieve tube members lose most of their organelles but are still alive when mature

Sugar solution flows in both directions

Sieve tube end plate
Tiny holes (arrowed in the photograph below) perforate the sieve tube elements allowing the sugar solution to pass through.

Companion cell: a cell adjacent to the sieve tube member, responsible for keeping it alive

TS through a sieve tube end plate

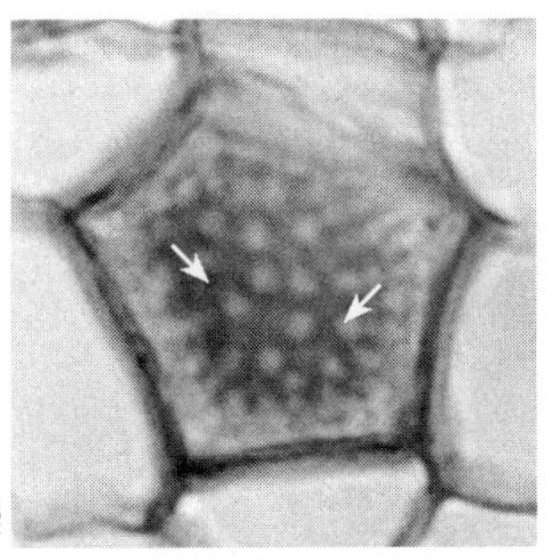

Adjacent sieve tube members are connected through **sieve plates** through which phloem sap flows.

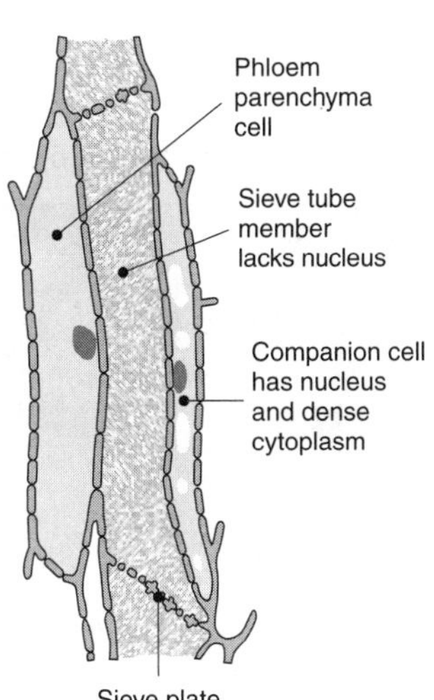

The Structure of Phloem Tissue

Phloem is alive at maturity and functions in the transport of sugars and minerals around the plant. Like xylem, it forms part of the structural vascular tissue of plants.

Fibers are associated with phloem as they are in xylem. Here they are seen in cross section where you can see the extremely thick cell walls and the way the fibers are clustered in groups. See the previous page for a view of fibers in longitudinal section.

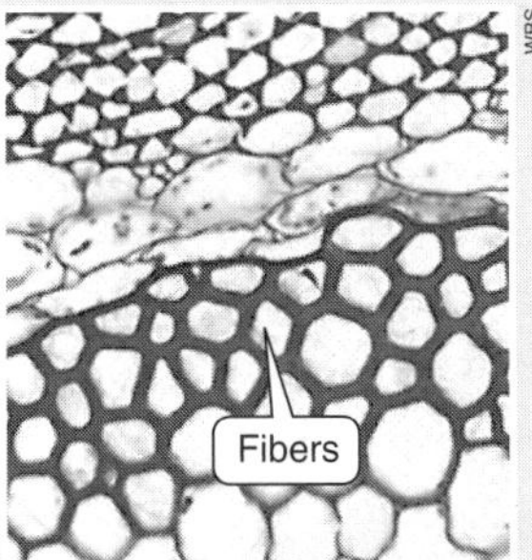

In this cross section through a buttercup root, the smaller companion cells can be seen lying alongside the sieve tube members. It is the sieve tube members that, end on end, produce the **sieve tubes**. They are the conducting tissue of phloem.

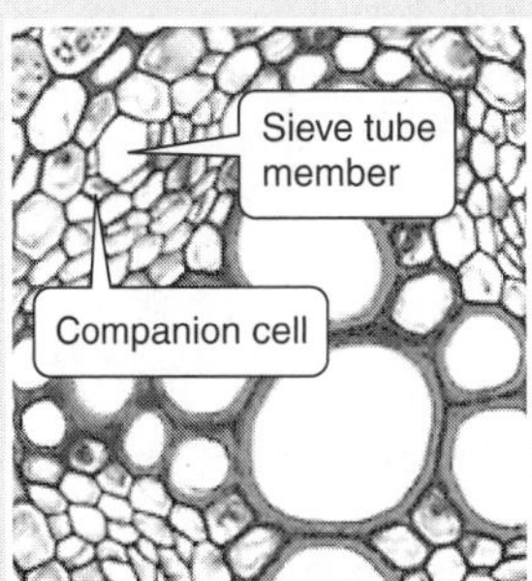

In this longitudinal section of a buttercup root, each sieve tube member has a thin **companion cell** associated with it. Companion cells retain their nucleus and control the metabolism of the sieve tube member next to them. They also have a role in the loading and unloading of sugar into the phloem.

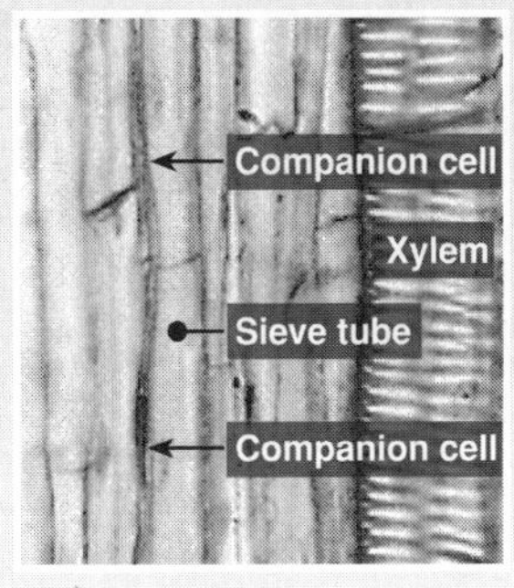

1. Describe the function of **phloem**: __________

2. Describe two differences between xylem and phloem: __________

3. Explain the purpose of the **sieve plate** at the ends of each sieve tube member: __________

4. (a) Name the conducting cell type in phloem: __________

   (b) Explain two roles of the companion cell in phloem: __________

5. State the purpose of the phloem parenchyma cells: __________

6. Identify a type of cell that provides strengthening in phloem: __________

Related activities: *Translocation*
Web links: *Photographic Atlas of Plant Anatomy*

Angiosperm Stem Structure

There is considerable variation in the primary structure of stems in seed plants, but there are three basic organizations. In some dicots, the vascular tissues form a continuous cylinder within the ground tissues, while in others the primary vascular tissues develop as discrete orderly cylinders separated by ground tissue (as depicted below left). In monocots and some herbaceous dicots, (below, right) the vascular bundles are scattered through the ground tissue. The arrangement of the vascular bundles indicates the extent to which secondary growth is possible. Where the vascular tissue forms a single cylinder or an orderly ring of vascular bundles, cambium develops within the vascular tissue. Where the vascular bundles are scattered in a disorderly fashion, vascular cambium does not develop and the bundles lose their potential for further growth.

Dicot Stem Structure

In the stems of many dicots, the **vascular bundles**, containing xylem (to the inside) and phloem (to the outside), are arranged in an orderly way around the stem. The vascular cambium lies between the phloem and the xylem. The cells of the vascular cambium divide to produce the thickening of the stem.

Monocot Stem Structure

The **vascular bundles** in a monocot stem are scattered randomly through the ground tissue. There is no distinction between inner pith and outer cortex as in dicots; there is only ground tissue, which is made up of large parenchyma (packing) cells.

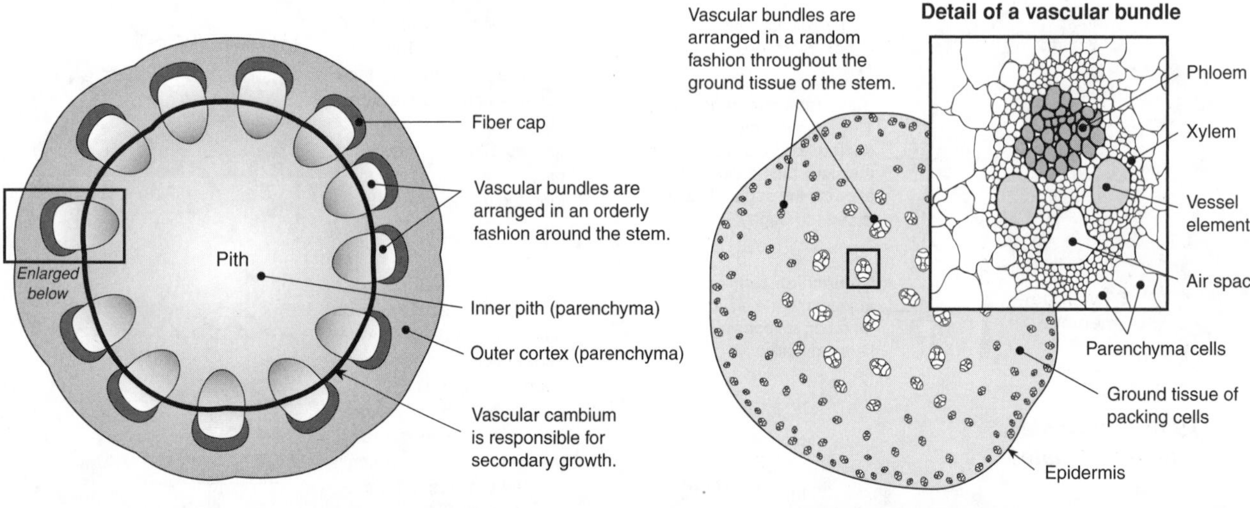

Cross section through a typical dicot stem

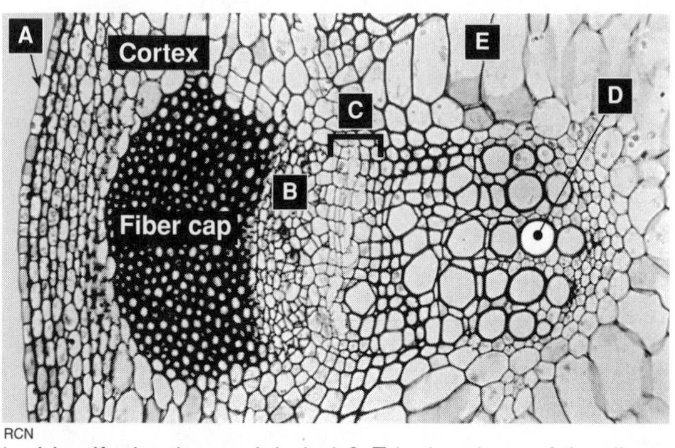

RCN

Cross section through a corn stem

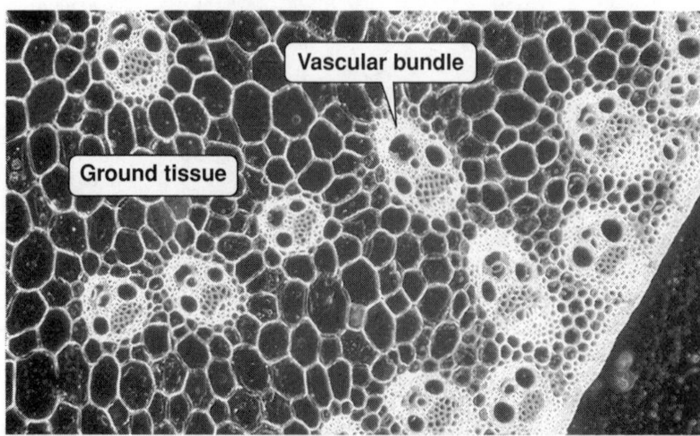

1. Identify the tissues labeled **A-E** in the photo of the dicot stem above. For each, describe the function of the tissue:

   (a) ___

   (b) ___

   (c) ___

   (d) ___

   (e) ___

2. (a) Contrast the arrangement of the vascular tissues in the stems of monocots and dicots: ___

   ___

   ___

   (b) Relate these differences to differences in growth (and stature) achieved by each broad taxon: ___

   ___

Related activities: Diversity in Stem Structure, Leaf and Stem Adaptations
Web links: Photographic Atlas of Plant Anatomy

Diversity in Stem Structure

Liverworts, the simplest of all living plants, lack specialized conducting tissue, and do not have stems. Stems and roots contain the **vascular tissues**, which are required to move materials around the plant, from where they are made or absorbed from, to where they are needed. "Vascular" simply means veins, therefore vascular plants are those with veins as a system of transport around the plant. The arrangement of the vascular tissue in a plant varies depending on the type of plant and whether the tissue is located in the stem or the root. The growth that leads to the young, flexible stem is called **primary growth**. The increase in the girth (diameter) of the stem is the result of **secondary growth** and is caused by the production of wood. All plants have primary growth but only some plants show secondary growth.

Moss Stems

True mosses may range in size from as small as 0.5 mm to up to 500 mm in length. The stems of the gametophyte and sporophyte have a central strand of water-conducting cells known as hydroids. These cells are long, and the end walls are thin, making them a preferential pathway for water.

Fern Stems

Ferns have a wide variety of stem arrangements. The rachis (frond) of a small fern may have a stem cross-section that appears to have discrete vascular bundles. The larger stems of tree ferns (e.g. *Dicksonia*) have rings of phloem and xylem. These may be interrupted by what are called 'leaf gaps' where part of the vascular tissue is diverted towards the leaf (frond).

Gymnosperm Stems

Gymnosperm stem structure is illustrated below in a cross-section of a young pine seedling (below right) and an older woody trunk. A stem in its first year of growth forms primary xylem and phloem on each side of the cambium layer. Further growth in following years undergoes what is called **secondary growth**. Note the annual rings and the change in cell size at the end of the season's growth.

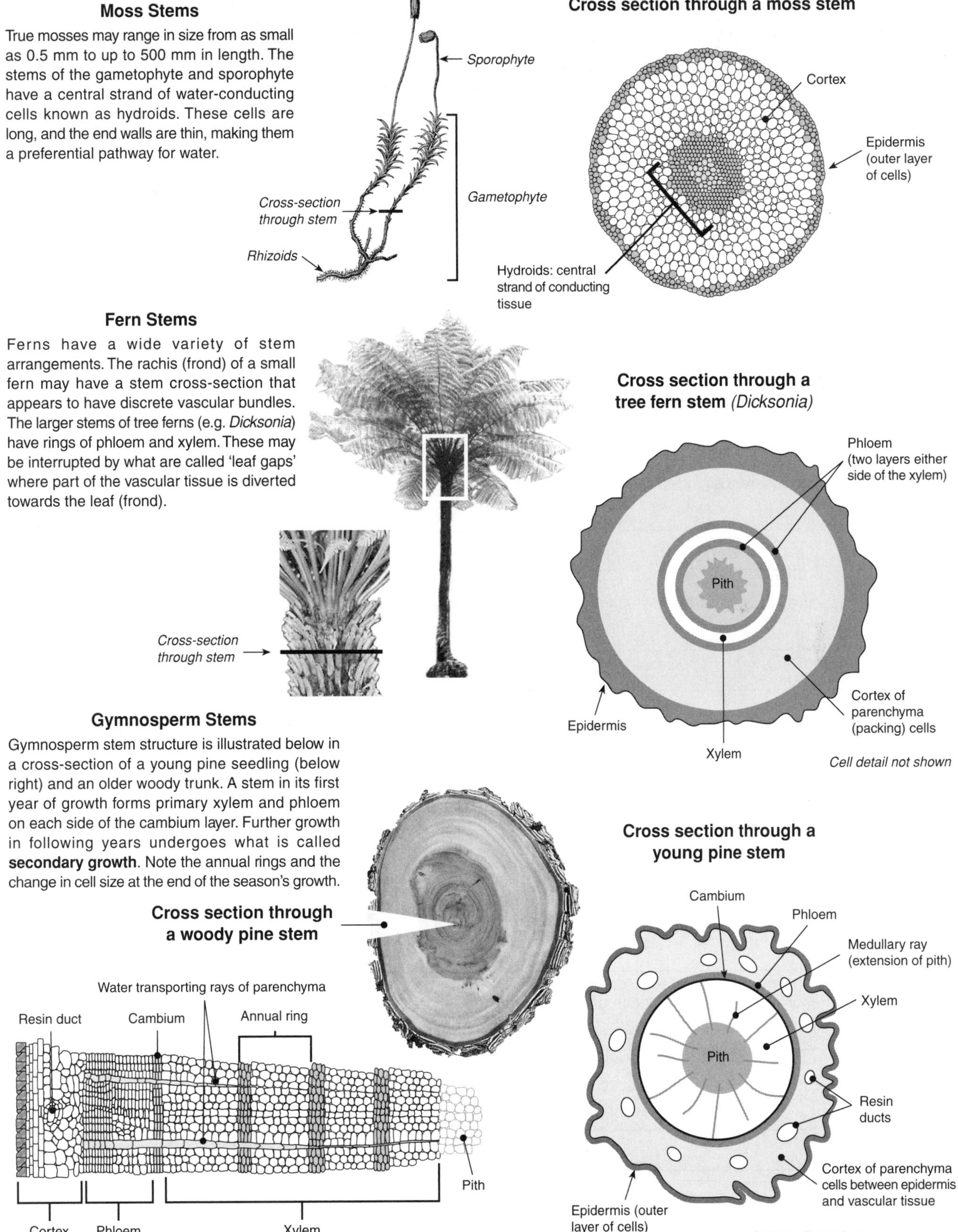

In this photograph, the leafy gametophyte moss body is contrasted against the stalked sporophytes. Liverworts and mosses lack true vascular tissue and this restricts the size to which they can grow, although the giant moss, *Dawsonia superba*, which is found throughout New Zealand, can grow up to 500 mm.

The evolution of vascular tissue was important in the diversification of plants, enabling them to achieve larger sizes and occupy a diverse range of habitats. Today, ferns occupy a wide range of habitats, including more marginal habitats such as the crevices of rock faces, where the success of flowering plants may be limited.

Conifers are vascular plants and the arrangement of their stem tissues is not unlike that of some dicots. The ability of their vascular tissue to produce lateral growth (secondary thickening) enables them to grow to a size. Some, like these coastal redwoods (*Sequoia*) in California are among the largest trees in the world.

1. (a) Describe the structural and functional differences between the stem tissues of a moss and a vascular plant:

   (b) Explain these differences in terms of the structural diversity of these plant taxa:

2. Describe how the structure of the fern stem (previous page) differs from the stems of gymnosperms and angiosperms:

3. (a) Describe a similarity between the organization of vascular tissues in gymnosperms and most dicot angiosperms:

   (b) Relate this to the structural diversity of these taxa:

4. Discuss the importance that the evolution of vascular tissue has had to the diversification of plant taxa:

Primary and Secondary Growth in Dicots

Two types of growth can contribute to an increase in the size of a plant. **Primary growth**, which occurs in the **apical meristem** of the shoot bud, increases the length (height) of a plant. **Secondary growth** increases plant girth and occurs in the **secondary** (lateral) **meristem** in the plant stem (or trunk). All plants show primary growth but only some dicots and gymnosperms show secondary growth. Secondary growth leads to the development of **wood**, supporting the plant as it grows taller.

Primary Growth

Primary growth occurs at the **apical meristem** and produces leaf and shoot buds. Three types of **primary meristem** are produced from the apical meristem: **procambium**, **protoderm**, and **ground meristem**.

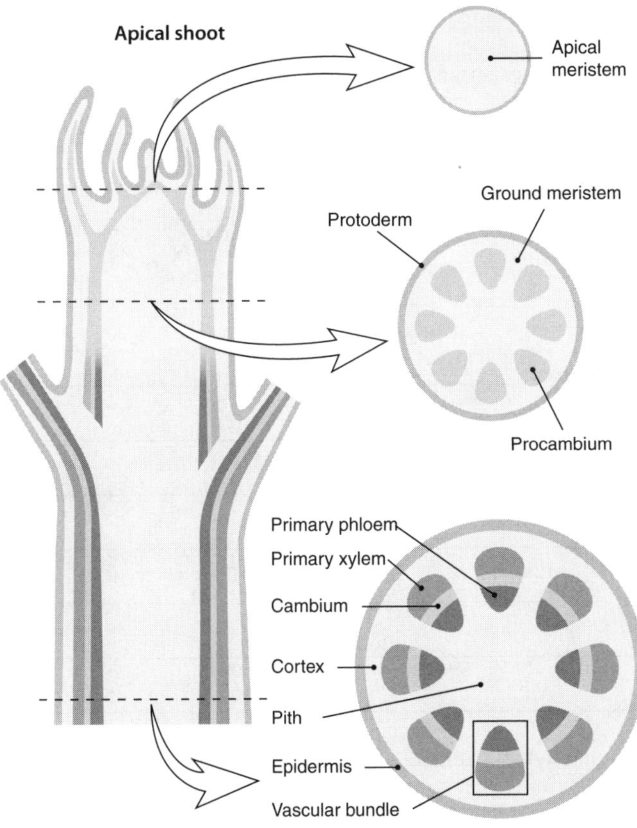

Primary Meristems

Procambium: In dicots, the procambium forms vascular bundles that are found in a ring near the epidermis and surrounded by cortex. As the procambium divides, the cells on the inside become **primary xylem** and those on the outside become **primary phloem**.

Protoderm: The outer layer of cells produced by the apical meristem become protoderm. When these mature, they form epidermis, including guard cells, cuticle, and trichomes (hair-like cells).

Ground meristem: This forms from the central cells of the apical meristem. These cells then differentiate into parenchyma cells forming pith (cells found in the central core of the stem) and cortex (the ring of cells found just inside the epidermis).

Secondary Growth

Growth in diameter occurs in the lateral meristems (**vascular cambium**). The vascular cambium comprises two types of dividing cell. The **fusiform initials** are orientated vertically and become either secondary phloem or xylem. The **ray initials** are orientated horizontally and form ray cells which join xylem and phloem transport systems.

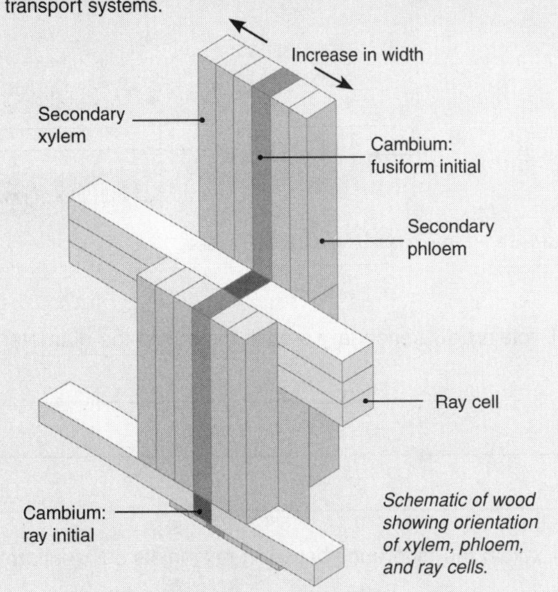

Schematic of wood showing orientation of xylem, phloem, and ray cells.

What is Wood?

Wood is composed mainly of **secondary xylem**. The phloem forms a thin layer just inside the outer periderm and along with the periderm forms the **bark**.

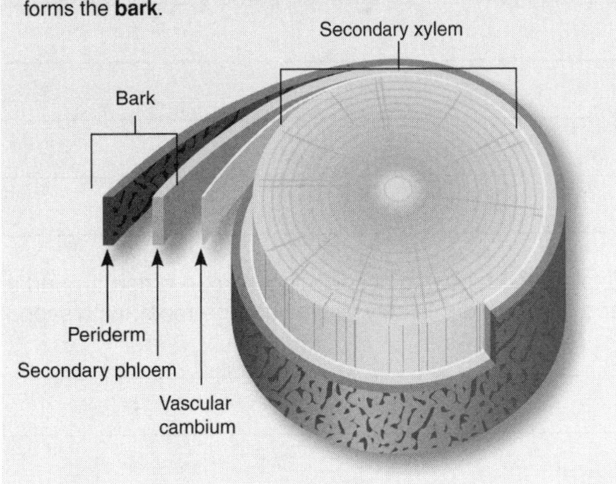

1. Discuss the structure and formation of primary tissues in dicot plants and contrast this with the structure and formation of the secondary tissues:

***Related activities**: Separating the Wood from the Trees*
***Web links**: Web links: Photographic Atlas of Plant Anatomy*

Tissues Generated by the Meristem

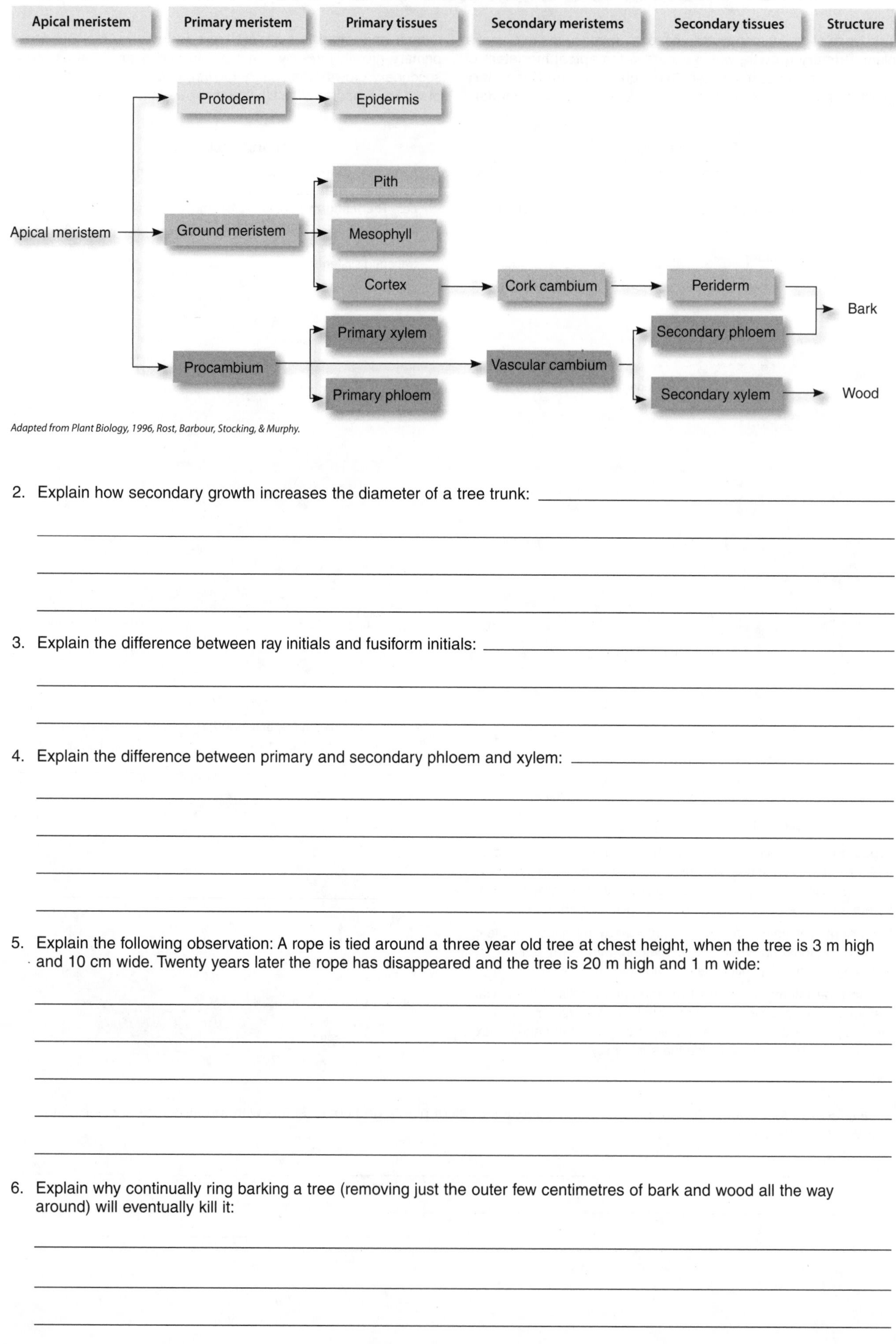

Adapted from Plant Biology, 1996, Rost, Barbour, Stocking, & Murphy.

2. Explain how secondary growth increases the diameter of a tree trunk:

3. Explain the difference between ray initials and fusiform initials:

4. Explain the difference between primary and secondary phloem and xylem:

5. Explain the following observation: A rope is tied around a three year old tree at chest height, when the tree is 3 m high and 10 cm wide. Twenty years later the rope has disappeared and the tree is 20 m high and 1 m wide:

6. Explain why continually ring barking a tree (removing just the outer few centimetres of bark and wood all the way around) will eventually kill it:

Leaf and Stem Adaptations

In order to photosynthesize, plants must obtain a regular supply of carbon dioxide (CO_2) gas; the raw material for the production of carbohydrate. In green plants, the systems for gas exchange and photosynthesis are linked; without a regular supply of CO_2, photosynthesis ceases. The leaf, as the primary photosynthetic organ, is adapted to maximize light capture and facilitate the entry of CO_2, while minimizing water loss. There are various ways in which plant leaves are adapted to do this. The ultimate structure of the leaf reflects the environment of the leaf (sun or shade, terrestrial or aquatic), its resistance to water loss, and the importance of the leaf relative to other parts of the plant that may be photosynthetic, such as the stem.

Sun Plant

A **sun leaf**, when exposed to high light intensities, can absorb much of the light available to the cells.

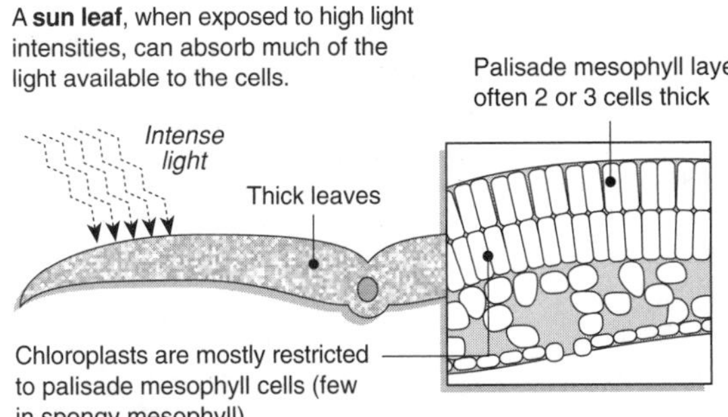
Palisade mesophyll layer often 2 or 3 cells thick
Intense light
Thick leaves
Chloroplasts are mostly restricted to palisade mesophyll cells (few in spongy mesophyll).

Sun leaves

Sun plants are adapted for growth in full sunlight. They have higher levels of respiration but can produce sugars at rates high enough to compensate for this. Sun plants include many weed species found on open exposed grassland. They expend more energy on the construction and maintenance of thicker leaves than do shade plants. The benefit of this investment is that they can absorb the higher light intensities available and grow rapidly.

Shade Plant

A **shade leaf** can absorb the light available at lower light intensities. If exposed to high light, most would pass through.

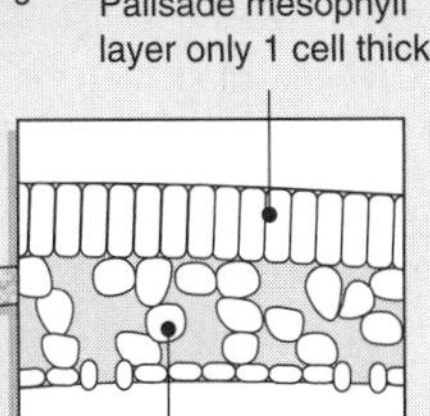
Palisade mesophyll layer only 1 cell thick
Low light intensity
Thin leaves
Chloroplasts occur throughout the mesophyll (as many in the spongy as palisade mesophyll).

Shade leaves

Shade plants typically grow in forested areas, partly shaded by the canopy of larger trees. They have lower rates of respiration than sun plants, mainly because they build thinner leaves. The fewer number of cells need less energy for their production and maintenance. In competition with sun plants, they are disadvantaged by lower rates of sugar production, but in low light environments this is offset by their lower respiration rates.

1. (a) Identify the structures in leaves that facilitate gas exchange: ______________________

   (b) Explain their critical role in plant nutrition: ______________________

   ____

   ____

2. (a) State which type of plant (sun or shade adapted) has the highest level of respiration: ______________

   (b) Explain how the plant compensates for the higher level of respiration: ______________________

   ____

3. Discuss the adaptations of leaves in **sun** and **shade plants**: ______________________

   ____

   ____

   ____

   ____

   ____

   ____

   ____

Related activities: *Diversity in Leaf Structure, Diversity in Stem Structure, Modifications in Plants*

Adaptations for Photosynthesis and Gas Exchange in Plants

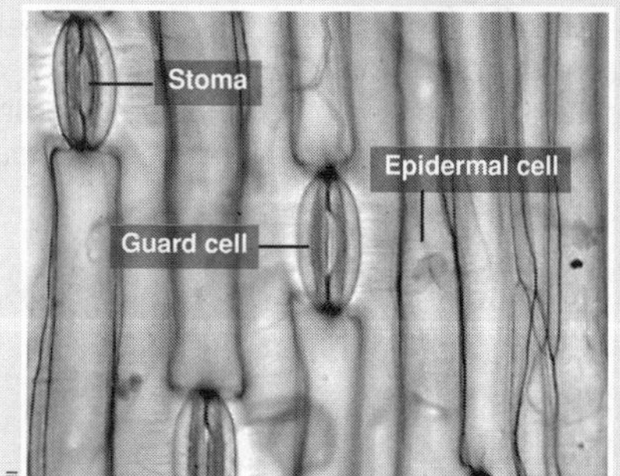

Surface view of stomata on a monocot leaf (grass). The parallel arrangement of stomata is a typical feature of monocot leaves. Grass leaves show properties of **xerophytes**, with several water conserving features (see right).

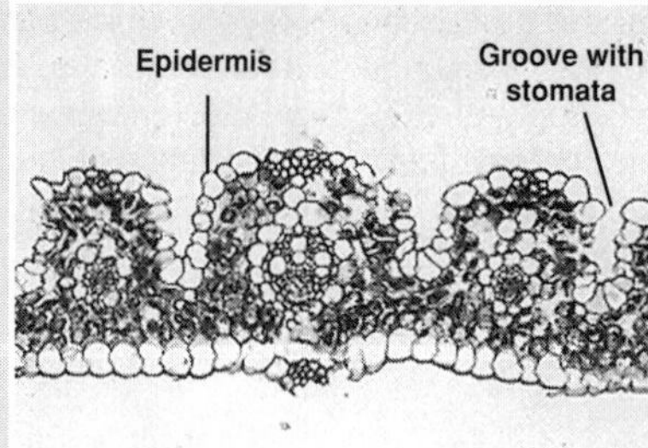

Cross section through a grass leaf showing the stomata housed in grooves. When the leaf begins to dehydrate, it may fold up, closing the grooves and thus preventing or reducing water loss through the stomata.

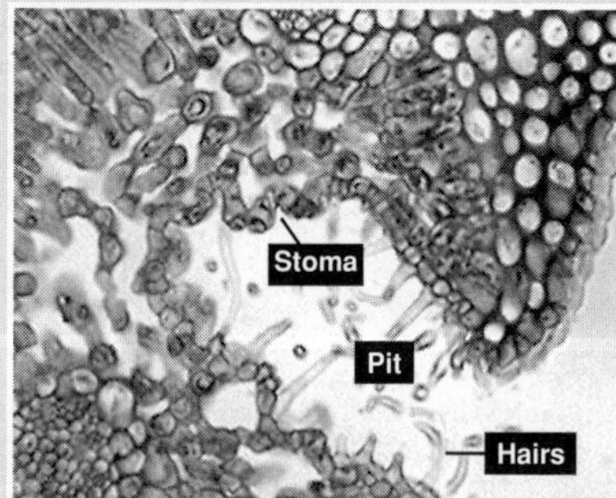

Oleander (above) is a xerophyte that displays many water conserving features. The stomata are found at the bottom of pits on the underside of the leaf. The pits restrict water loss to a greater extent than they reduce CO_2 uptake.

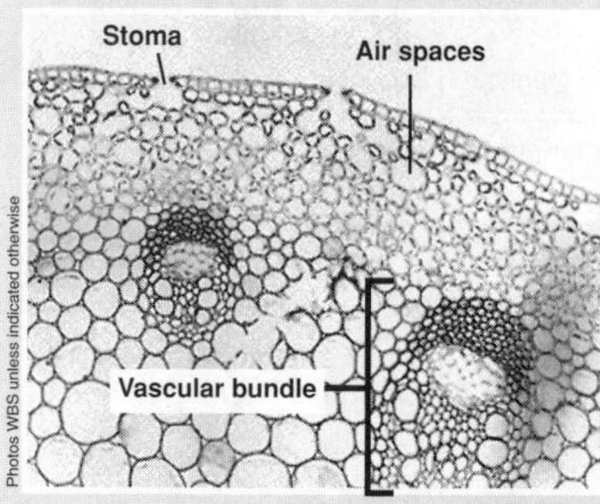

Some plants (e.g. buttercup above) have photosynthetic stems, and CO_2 enters freely into the stem tissue through stomata in the epidermis. The air spaces in the cortex are more typical of leaf mesophyll than stem cortex.

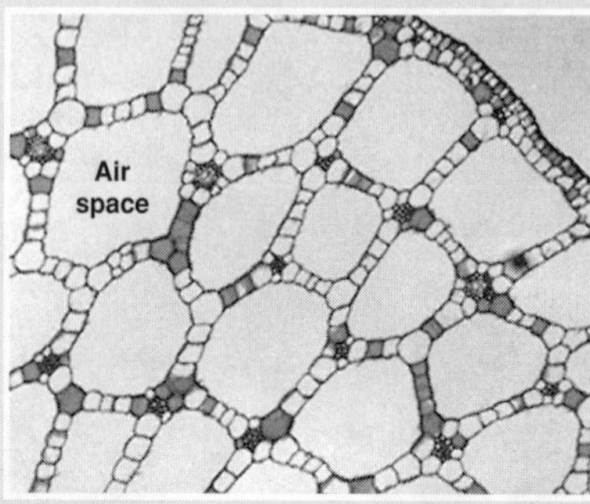

Hydrophytes, such as *Potamogeton*, above, have stems with massive air spaces. The presence of air in the stem means that they remain floating in the zone of light availability and photosynthesis is not compromised.

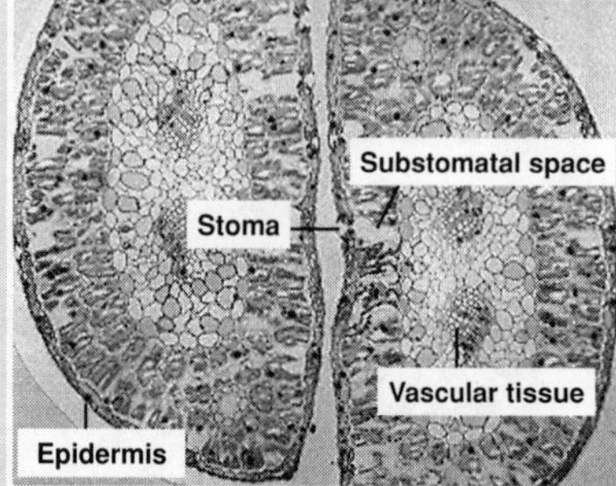

This transverse view of the twin leaves of a two-needle pine shows the sunken stomata and substomatal spaces. This adaptation for arid conditions reduces water loss by creating a region of high humidity around the stoma.

4. Describe two adaptations in plants for reducing water loss while maintaining entry of gas into the leaf:

   (a) ______________________________

   ____

   (b) ______________________________

   ____

5. Describe two adaptations of photosynthetic stems that are not present in non-photosynthetic stems, and explain the reasons for these:

   (a) ______________________________

   ____

   (b) ______________________________

   ____

6. The example of a photosynthetic stem above is from a buttercup, a plant in which the leaves are still the primary organs of photosynthesis.

   (a) Identify an example of the plant where the stem is the **only** photosynthetic organ: ______________

   (b) Describe the structure of the leaves in your example and suggest a reason for their particular structure:

   ____

   ____

7. Describe one role of the air spaces in the stems of *Potamogeton* related to maintaining photosynthesis: ______________

   ____

Modifications in Plants

Various parts of the plant body may be modified for a specific role. Some **biennial plants**, e.g. carrots, store carbohydrates in fleshy **storage roots** during their first year of growth and use this store the following year to fuel the development of flowers, fruits, and seeds. The specialized **buttress** roots and **aerial** roots of some large tropical tree species (e.g. the banyan tree) provide support in thin tropical soils. Mangroves also have specialized aerial roots, called pneumatophores, which enable gas exchange in the water-logged substrate. In some **epiphytes**, e.g. the orchids, the aerial roots may be photosynthetic. **Parasitic** plants, such as mistletoe, produce rootlike organs that penetrate and parasitize the host's tissues. The stems of some plants also function as storage or photosynthetic organs. Horizontal underground stems, or **rhizomes**, can become swollen to provide a food store in the same way as root tubers. Similarly, **corms** are upright underground stems that become thickened with stored food. **Bulbs** are large buds with thick non-photosynthetic, food storage leaves clustered on short stems, e.g. onions. Cacti are a familiar example of stem and leaf modification. The fleshy green stems store water and photosynthesize while the leaves are modified into defensive spines. The tendrils of legumes and the traps of insectivorous plants are other familiar leaf modifications.

Modifications of Plant Parts for Storage

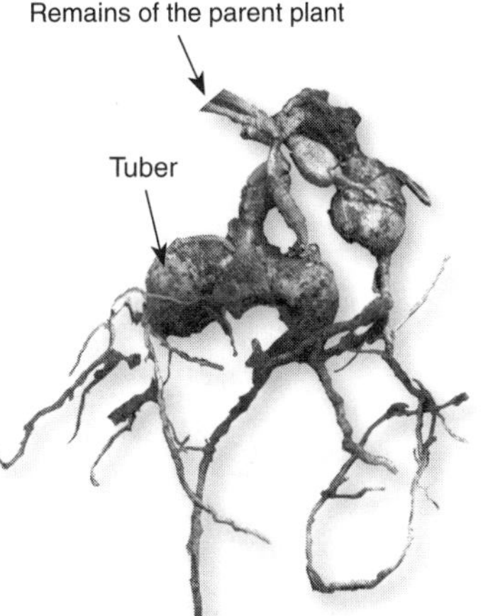

Dahlia root tuber

Root tubers, e.g. dahlias (above), lack terminal and lateral buds. Both stem and root tubers can give rise to new individuals, thereby providing a means of vegetative propagation.

Tubers are the swollen part of an underground stem or root, usually modified for storing food. The potato (below) is a **stem tuber**, as indicated by the presence of terminal and lateral buds.

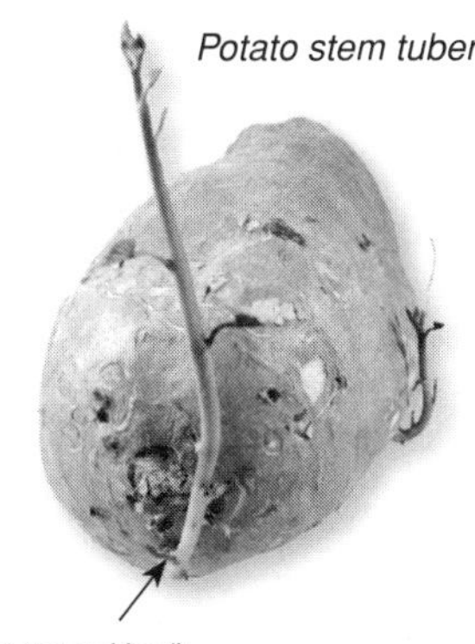

Potato stem tuber

Bulbs

A true bulb is really just a typical shoot compressed into a shortened form. Fleshy storage leaves are attached to a stem plate and form concentric circles around the growing tip. New roots form from the lower part of the stem.

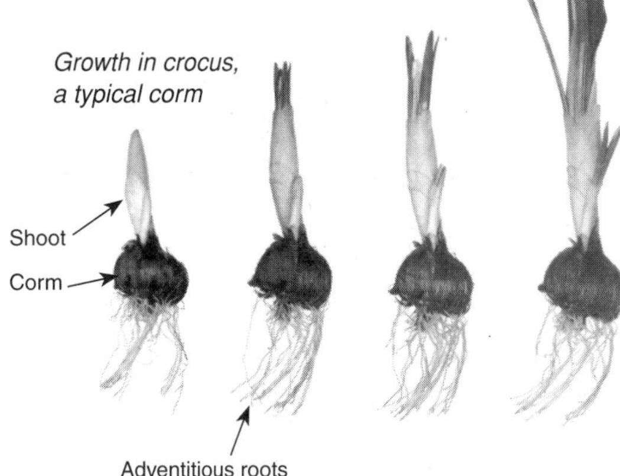

Growth in crocus, a typical corm

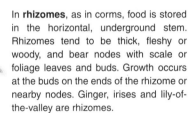

In a **corm**, food is stored in stem tissue. Corms look like bulbs, but if you cut a corm in half you see a mass of homogenous tissue rather than concentric rings of fleshy leaves as in a bulb. Cyclamen, gladiolus, and crocus (above) are corms.

Underground stem containing stored food

Iris rhizome

In **rhizomes**, as in corms, food is stored in the horizontal, underground stem. Rhizomes tend to be thick, fleshy or woody, and bear nodes with scale or foliage leaves and buds. Growth occurs at the buds on the ends of the rhizome or nearby nodes. Ginger, irises and lily-of-the-valley are rhizomes.

Supportive and Breathing Roots

Many plants have specialized roots growing from the stem into the soil. These **prop roots** are seen in mangroves (above) and corn. They provide stability for the plant in the substrate.

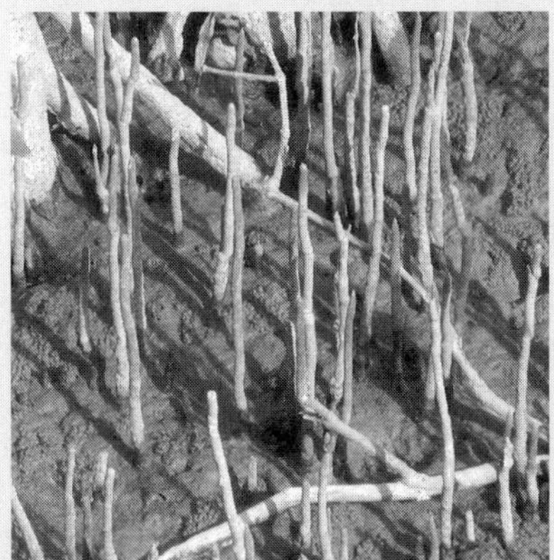

Pneumatophores are the specialized 'breathing roots' of some types of mangroves. They grow up into the air and absorb oxygen-rich air via surface openings in the wood called lenticels.

The tropical **banyan tree** sends down rope-like aerial roots from its branches. These anchor in the soil and become very thick, forming massive columns that support the heavy branches.

Related activities: *Adaptations of Xerophytes, Adaptations of Hydrophytes*
Web links: *Types of Roots*

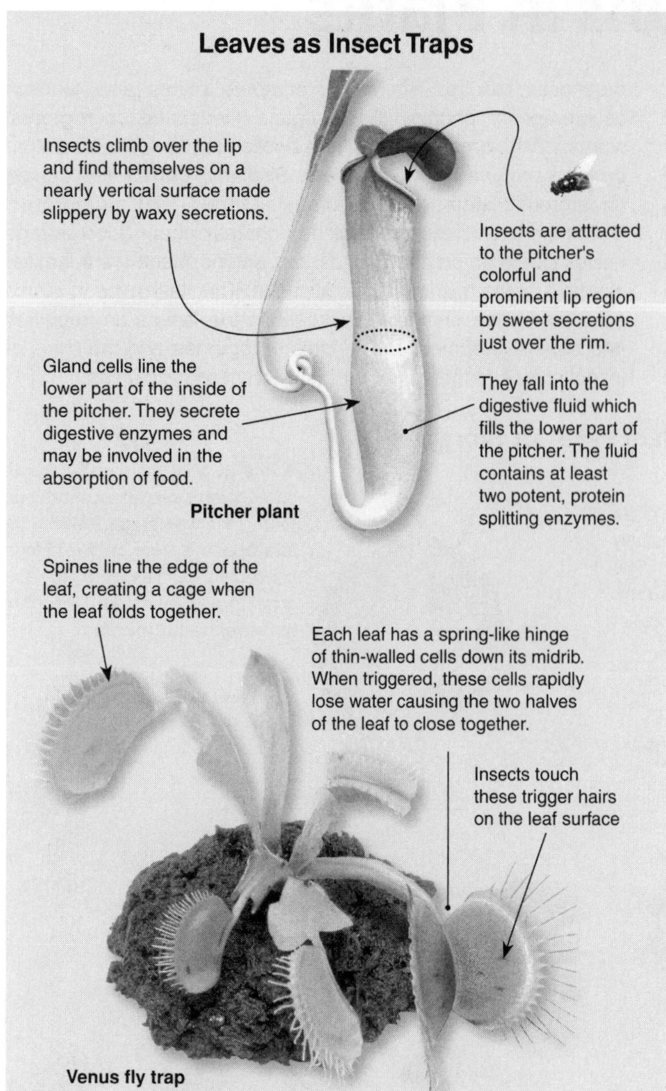

Leaves as Insect Traps
Insects climb over the lip and find themselves on a nearly vertical surface made slippery by waxy secretions.
Insects are attracted to the pitcher's colorful and prominent lip region by sweet secretions just over the rim.
Gland cells line the lower part of the inside of the pitcher. They secrete digestive enzymes and may be involved in the absorption of food.
They fall into the digestive fluid which fills the lower part of the pitcher. The fluid contains at least two potent, protein splitting enzymes.
Pitcher plant
Spines line the edge of the leaf, creating a cage when the leaf folds together.
Each leaf has a spring-like hinge of thin-walled cells down its midrib. When triggered, these cells rapidly lose water causing the two halves of the leaf to close together.
Insects touch these trigger hairs on the leaf surface
Venus fly trap

Many arid-adapted plants, like this aloe, have succulent leaves that are photosynthetic and also modified for internal storage of water.

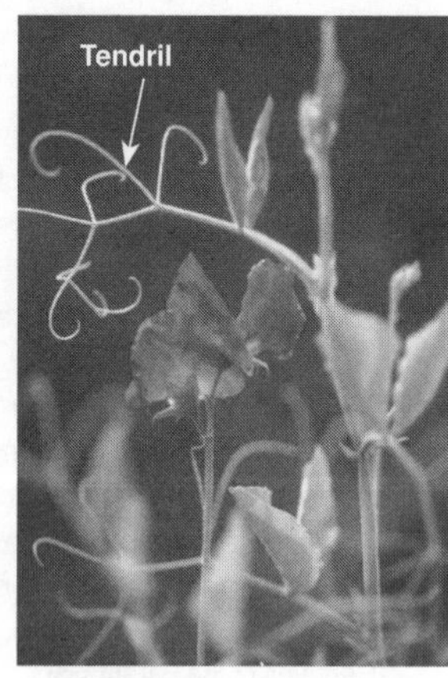

A tendril is a thread-like structure leaf modification that helps a plant to climb over other plants or objects to gain access to light.

In temperate climates, the buds of woody plants (e.g. **hickory**) are protected over winter by modified leaves called bud scales. The waxy scales prevent desiccation and insulate the bud against the cold.

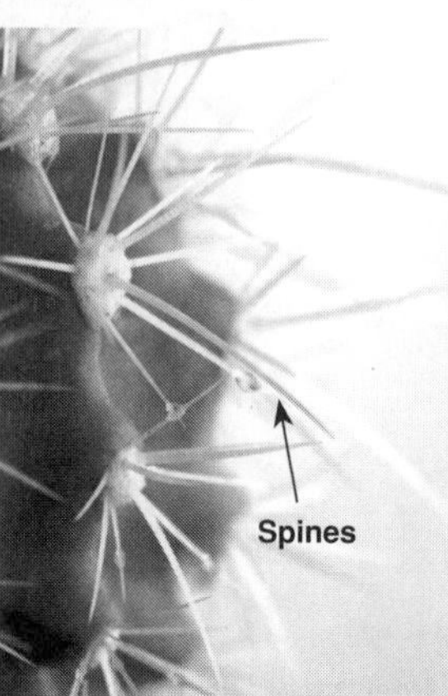

Many cacti have leaves reduced to sharp non-photosynthetic spines, which act to deter browsers. In the example above, the spines (leaves) grow from shortened shoots that arise from the photosynthetic stem.

1. For each of the following, identify the plant part that has been modified, and describe the modification and its purpose. Give an example in each case. The first one has been completed for you:

   (a) Bulb: Leaves are modified for food storage in the dormant plant. The leaf bases are fleshy and tightly packed together on a shortened stem. Example: onion, garlic, tulip, lily.

   (b) Corm: ______________________________

   (c) Bud scales: ______________________________

   (d) Tendrils: ______________________________

   (e) Venus flytrap 'trap': ______________________________

2. Discuss the role of aerial roots in named examples, explaining how they benefit the plant in each case:

____

Adaptations of Xerophytes

Plants adapted to dry conditions are called **xerophytes** and they show structural (xeromorphic) and physiological adaptations for water conservation. These typically include small, hard leaves, and epidermis with a thick cuticle, sunken stomata, succulence, and permanent or temporary absence of leaves. Xerophytes may live in humid environments, provided that their roots are in dry microenvironments (e.g. the roots of epiphytic plants that grow on tree trunks or branches). The nature of the growing environment is important in many other situations too. **Halophytes** (salt tolerant plants) and alpine species may also show xeromorphic features in response to the scarcity of obtainable water and high transpirational losses in these environments.

Leaves modified into spines or hairs to reduce water loss. Light colored spines reflect solar radiation.

Squat, rounded shape reduces surface area. The surface tissues of many cacti are tolerant of temperatures in excess of 50°C.

Shallow, but extensive fibrous root system.

Stem becomes the major photosynthetic organ, plus a reservoir for water storage.

Water table low

Dry Desert Plant

Desert plants, such as cacti, must cope with low or sporadic rainfall and high transpiration rates. A number of structural adaptations (diagram left) reduce water losses, and enable them to access and store available water. Adaptations such as waxy leaves also reduce water loss and, in many desert plants, germination is triggered only by a certain quantity of rainfall.

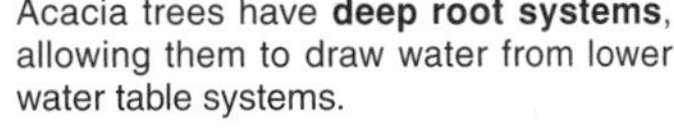

Acacia trees have **deep root systems**, allowing them to draw water from lower water table systems.

The outer surface of many succulents are coated in fine hairs, which traps air close to the surface reducing transpiration rate.

Seaweeds, which are protists, not plants, tolerate drying between tides even though they have no xeromorphic features.

A waxy coating of **suberin** on mangrove roots excludes 97% of salt from the water.

Ocean Margin Plant

Land plants that colonize the shoreline must have adaptations to obtain water from their saline environment while maintaining their osmotic balance. In addition, the shoreline is often a windy environment, so they frequently show xeromorphic adaptations that enable them to reduce transpirational water losses.

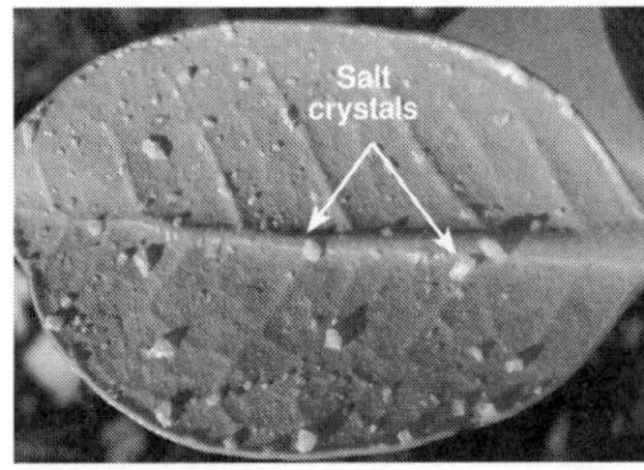

To maintain osmotic balance, mangroves can secrete absorbed salt as salt crystals (above), or accumulate salt in old leaves which are subsequently shed.

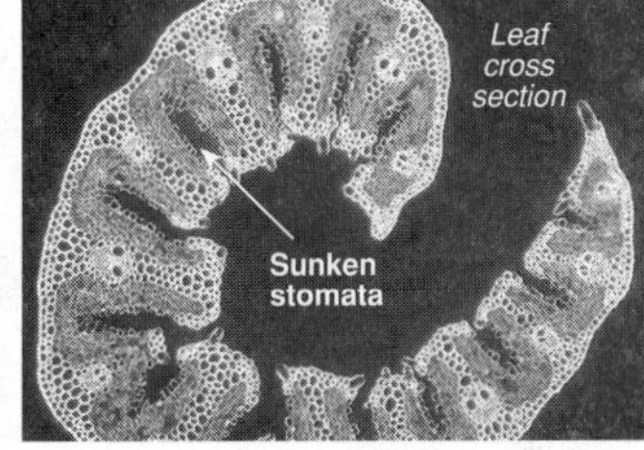

Grasses found on shoreline coasts (where it is often windy), curl their leaves and have sunken stomata to reduce water loss by transpiration.

Methods of water conservation in various plant species

Adaptation for water conservation	Effect of adaptation	Example
Thick, waxy cuticle to stems and leaves	Reduces water loss through the cuticle.	*Pinus* sp. ivy (*Hedera*), sea holly (*Eryngium*), prickly pear (*Opuntia*).
Reduced number of stomata	Reduces the number of pores through which water loss can occur.	Prickly pear (*Opuntia*), *Nerium* sp.
Stomata sunken in pits, grooves, or depressions Leaf surface covered with fine hairs Massing of leaves into a rosette at ground level	Moist air is trapped close to the area of water loss, reducing the diffusion gradient and therefore the rate of water loss.	**Sunken stomata**: *Pinus* sp., *Hakea* sp. **Hairy leaves**: lamb's ear. **Leaf rosettes**: dandelion (*Taraxacum*), daisy.
Stomata closed during the light, open at night	CAM metabolism: CO_2 is fixed during the night, water loss in the day is minimized.	**CAM plants**, e.g. American aloe, pineapple, *Kalanchoe, Yucca.*
Leaves reduced to scales, stem photosynthetic Leaves curled, rolled, or folded when flaccid	Reduction in surface area from which transpiration can occur.	**Leaf scales**: broom (*Cytisus*). **Rolled leaf**: marram grass (*Ammophila*), *Erica* sp.
Fleshy or succulent stems Fleshy or succulent leaves	When readily available, water is stored in the tissues for times of low availability.	**Fleshy stems**: *Opuntia*, candle plant (*Kleinia*). **Fleshy leaves**: *Bryophyllum.*
Deep root system below the water table	Roots tap into the lower water table.	Acacias, oleander.
Shallow root system absorbing surface moisture	Roots absorb overnight condensation.	Most cacti

Adaptations in halophytes and drought tolerant plants

Ice plant (*Carpobrotus*): The leaves of many desert and beach dwelling plants are fleshy or succulent. The leaves are triangular in cross section and crammed with water storage cells. The water is stored after rain for use in dry periods. The shallow root system is able to take up water from the soil surface, taking advantage of any overnight condensation.

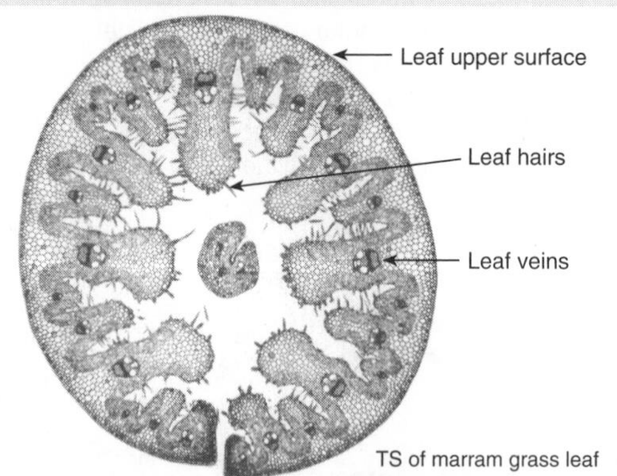

Marram grass (*Ammophila*): The long, wiry leaf blades of this beach grass are curled downwards with the stomata on the inside. This protects them against drying out by providing a moist microclimate around the stomata. Plants adapted to high altitude often have similar adaptations.

Ball cactus (*Echinocactus grusonii*): In cacti, the leaves are modified into long, thin spines which project outward from the thick fleshy stem (see close-up above right). This reduces the surface area over which water loss can occur. The stem takes over the role of producing the food for the plant and also stores water during rainy periods for use during drought. As in succulents like ice plant, the root system in cacti is shallow to take advantage of surface water appearing as a result of overnight condensation.

1. Explain the purpose of **xeromorphic** adaptations: ______________________

2. Describe three xeromorphic adaptations of plants:

   (a) ______________________

   (b) ______________________

   (c) ______________________

3. Describe a physiological mechanism by which plants can reduce water loss during the daylight hours:

4. Explain why creating a moist microenvironment around the areas of water loss reduces transpiration rate:

5. Explain why seashore plants (halophytes) exhibit many desert-dwelling adaptations: ______________________

Adaptations of Hydrophytes

Hydrophytes are a group of plants which have adapted to living either partially or fully submerged in water. Survival in water poses different problems to those faced by terrestrial plants. Hydrophytes have a reduced root system, a feature that is often related to the relatively high concentration of nutrients in the sediment and the plant's ability to remove nitrogen and phosphorus directly from the water. The leaves of submerged plants are thin to increase the surface area of photosynthetic tissue and reduce internal shading. Hydrophytes typically have no cuticle (waterproof covering) or the cuticle is very thin. This enables the plant ability to absorb minerals and gases directly from the water. In addition, being supported by the water, they require very little in the way of structural support tissue.

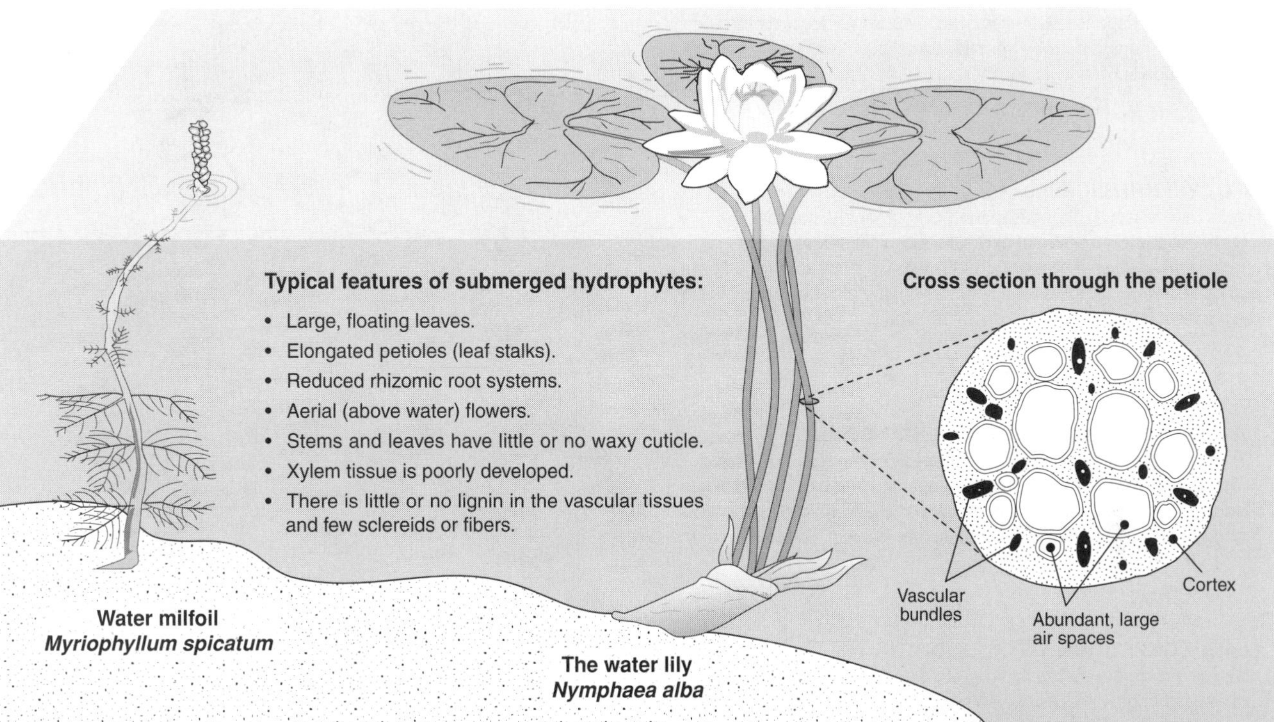

Typical features of submerged hydrophytes:
• Large, floating leaves.
• Elongated petioles (leaf stalks).
• Reduced rhizomic root systems.
• Aerial (above water) flowers.
• Stems and leaves have little or no waxy cuticle.
• Xylem tissue is poorly developed.
• There is little or no lignin in the vascular tissues and few sclereids or fibers.
Cross section through the petiole
Vascular bundles
Abundant, large air spaces
Cortex
Water milfoil
Myriophyllum spicatum
The water lily
Nymphaea alba

Myriophyllum's submerged leaves are well spaced and taper towards the surface to assist with gas exchange and distribution of sunlight.

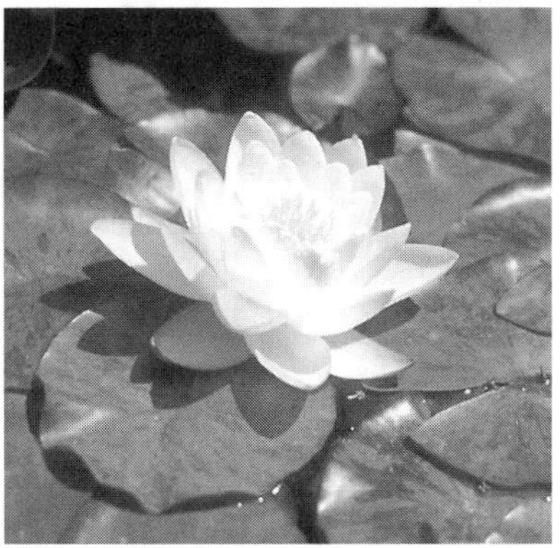

The floating leaves of water lilies (*Nymphaea*) have a high density of stomata on the upper leaf surface so they are not blocked by water.

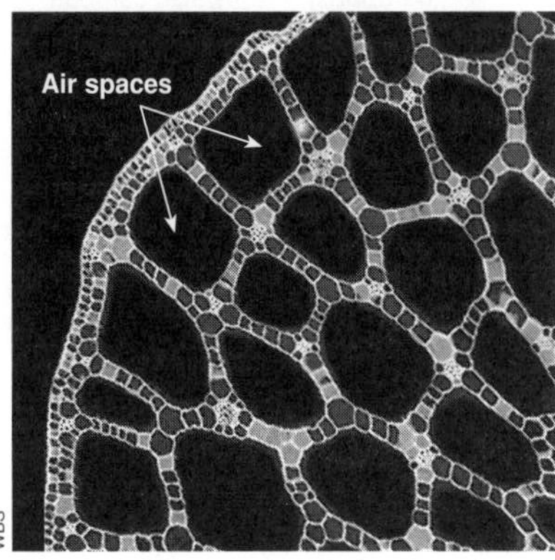

Cross section through oxygen weed (*Potamogeton*), showing the large air spaces which assist with flotation and gas exchange.

1. Explain how the following adaptations assist hydrophytes to survive in an aquatic environment:

   (a) Large air spaces within the plant's tissues: ______________________

   (b) Thin cuticle: ______________________

   (c) High stomatal densities on the upper leaf surface: ______________________

2. Explain why hydrophytic plants have retained an aerial (above water) flowering system: ______________________

Plant Responses

Even though most plants are firmly rooted in the ground, they are still capable of responding and making adjustments to changes in their external environment. This ability is manifested chiefly in changing patterns of growth. These responses may involve relatively sudden physiological changes, as occurs in flowering, or a steady growth response, such as a **tropism**. Other responses made by plants include nastic movements, circadian rhythms, photoperiodism, dormancy, and vernalization.

TROPISMS
Tropisms are growth responses made by plants to directional external stimuli, where the direction of the stimulus determines the direction of the growth response. A tropism may be positive (towards the stimulus), or negative (away from the stimulus). Common stimuli for plants include light, gravity, touch, and chemicals.

LIFE CYCLE RESPONSES
Plants use seasonal changes in the environment as cues for the commencement or ending of particular life cycle stages. Such changes are mediated by **plant growth factors**, such as phytochrome and gibberellin. Examples include flowering and other **photoperiodic responses**, dormancy and germination, and leaf fall.

RAPID RESPONSES TO ENVIRONMENTAL STIMULI
Plants are capable of quite rapid responses. Examples include the closing of **stomata** in response to water loss, opening and closing of flowers in response to temperature (photo, below right), and **nastic responses** (photos, below left). These responses often follow a circadian rhythm.

PLANT COMPETITION AND ALLELOPATHY
Although plants are rooted in the ground, they can still compete with other plants to gain access to resources. Some plants produce chemicals that inhibit the growth of neighboring plants. Such chemical inhibition is called **allelopathy**. Plants also compete for light and may grow aggressively to shade out slower growing competitors.

PLANT RESPONSES TO HERBIVORY
Many plant species have responded to grazing or browsing pressure with evolutionary adaptations enabling them to survive constant cropping. Examples include rapid growth to counteract the constant loss of biomass (grasses), sharp spines or thorns to deter browsers (acacias, cacti), or toxins in the leaf tissues (eucalyptus).

Shoots are **positively phototropic** and grow toward the light.

Roots are positively gravitropic and grow towards the Earth's gravitational pull.

RCN

Some plants, such as *Mimosa* (above), are capable of **nastic responses**. These are relatively rapid, reversible movements, such as leaf closure in response to touch. Unlike tropisms, nastic responses are independent of stimulus direction.

The growth of tendrils around a support is a response to a mechanical stimulus, and is called thigmomorphogenesis.

RA

The opening and closing of this tulip flower is temperature dependent. The flowers close when it is cool at night.

1. Identify the stimuli to which plants typically respond: ______________________

2. Explain how plants benefit by responding appropriately to the environment: ______________________

____

____

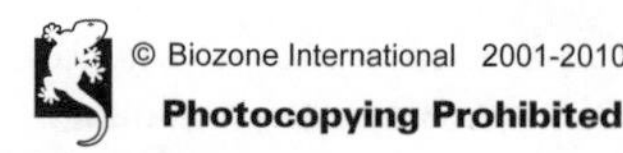

Investigating Phototropism

Phototropism in plants was linked to a growth promoting substance in the 1920s. A number of classic experiments, investigating phototropic responses in severed coleoptiles, gave evidence for the hypothesis that **auxin** was responsible for the tropic responses of stems. Auxins promote cell elongation. Stem curvature in response to light can therefore result from the differential distribution of auxin either side of a stem. However, the mechanisms of hormone action in plants are still not well understood. Auxins increase cell elongation only over a certain concentration range and, at certain levels, auxins stop inducing elongation and begin to inhibit it. Note that there is *some* experimental evidence to contradict the original auxin hypothesis and the early experiments have been criticized for oversimplifying the real situation. Some of the early experiments investigating the phototropic response, and the role of hormone(s) in controlling it, are outlined below.

1. **Directional light**: A pot plant is exposed to direct sunlight near a window and as it grows, the shoot tip turns in the direction of the sun. If the plant was rotated, it adjusted by growing towards the sun in the new direction.

   (a) Name the hormone that regulates this growth response:

   ___

   (b) Give the full name of this growth response:

   ___

   (c) State how the cells behave to cause this change in shoot direction at:

   Point A: ___

   Point B: ___

   (d) State which side (A or B) would have the highest concentration of hormone:

   ___

   (e) Draw a diagram of the cells as they appear across the stem from point A to B (in the rectangle on the right).

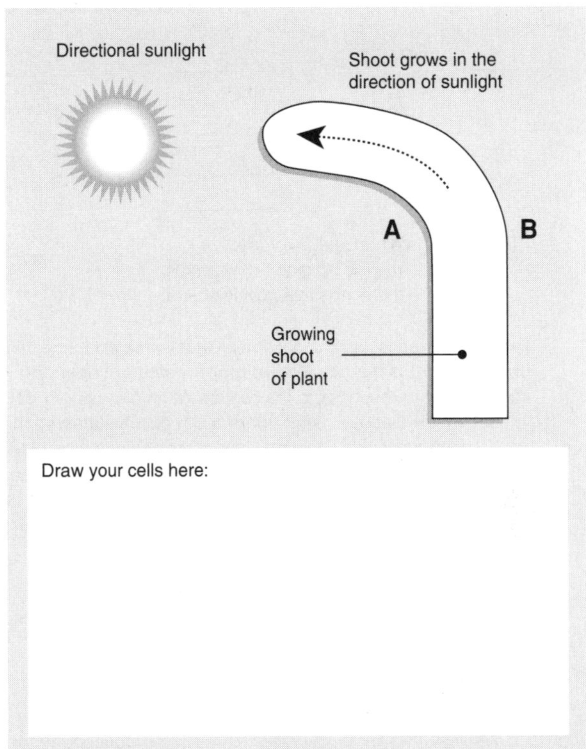

2. **Light excluded from shoot tip**: With a foil cap placed over the top of the shoot tip, light is prevented from reaching it. Under these conditions, the direction of growth does not change towards the light source, but continues straight up. State the conclusion you could make about the source and activity of the hormone that controls the growth response:

   ___

   ___

   ___

   ___

   ___

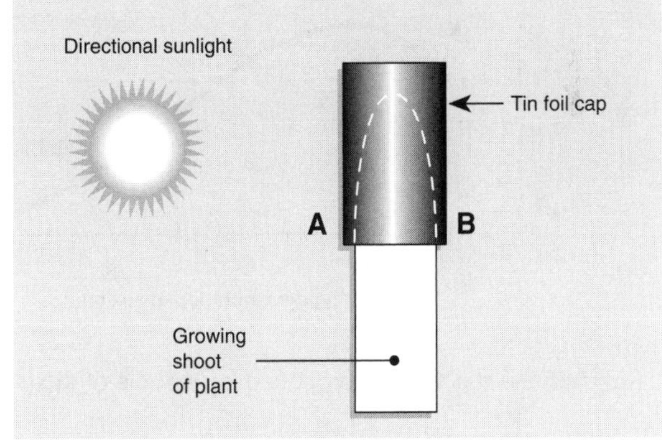

3. **Cutting into the transport system**: Two identical plants were placed side-by-side and subjected to the same directional light source. Razor blades were cut half-way into the stem, interfering with the transport system of the stem. Plant A had the cut on the same side as the light source, while Plant B was cut on the shaded side. Predict the growth responses of:

   Plant A: ___

   ___

   Plant B: ___

   ___

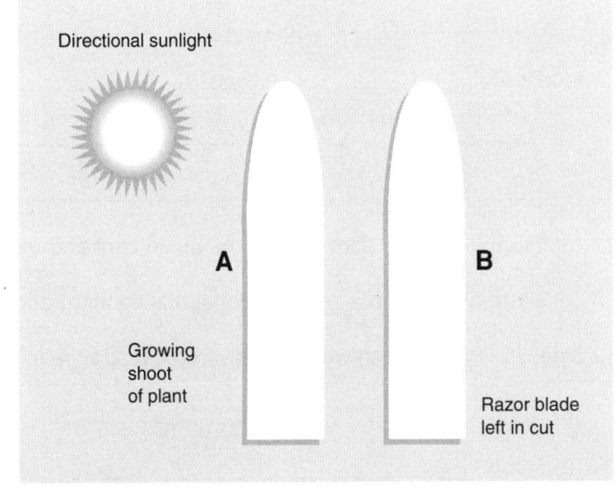

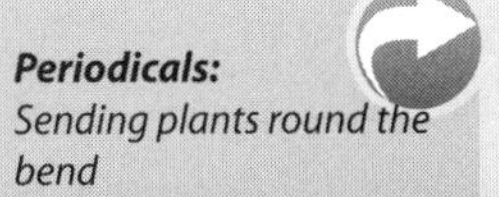
Periodicals:
Sending plants round the bend

Related activities: *Plant Responses, Auxins, Gibberellins and ABA*

Investigating Gravitropism

Although the response of shoots and roots to gravity is well known, the mechanism behind it is not at all well understood. The importance of auxin as a plant growth regulator, as well as its widespread occurrence in plants, led to it being proposed as the primary regulator in the gravitropic response. The basis of auxin's proposed role in **gravitropism** is outlined below. The mechanism is appealing in its simplicity but, as noted below, has been widely criticized, and there is not a great deal of evidence to support it. Many of the early plant growth experiments (including those on phototropism) involved the use of coleoptiles. Their use has been criticized because the coleoptile (the sheath surrounding the young shoot of grasses) is a specialized and short-lived structure and is probably not representative of plant tissues generally.

Auxins and Gravitropic Responses

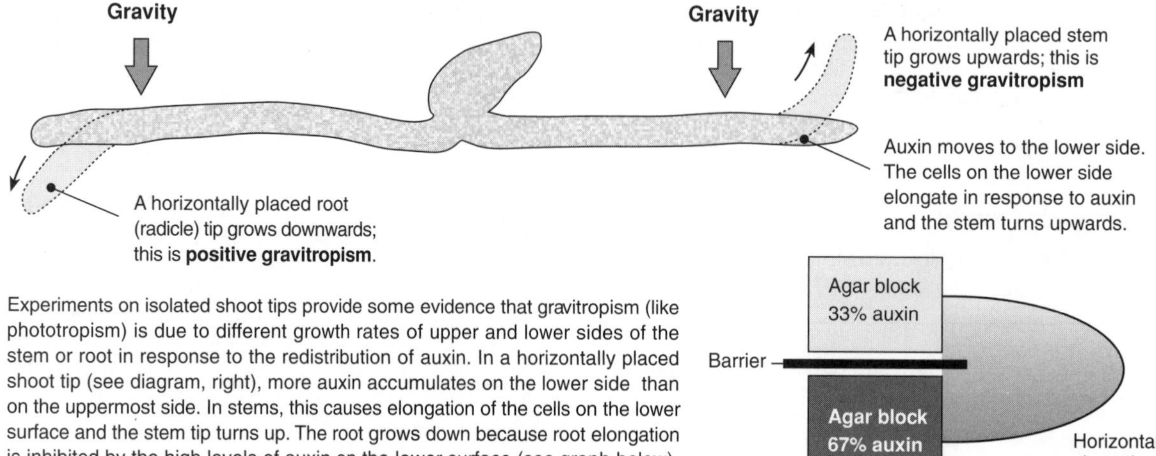

Experiments on isolated shoot tips provide some evidence that gravitropism (like phototropism) is due to different growth rates of upper and lower sides of the stem or root in response to the redistribution of auxin. In a horizontally placed shoot tip (see diagram, right), more auxin accumulates on the lower side than on the uppermost side. In stems, this causes elongation of the cells on the lower surface and the stem tip turns up. The root grows down because root elongation is inhibited by the high levels of auxin on the lower surface (see graph below).

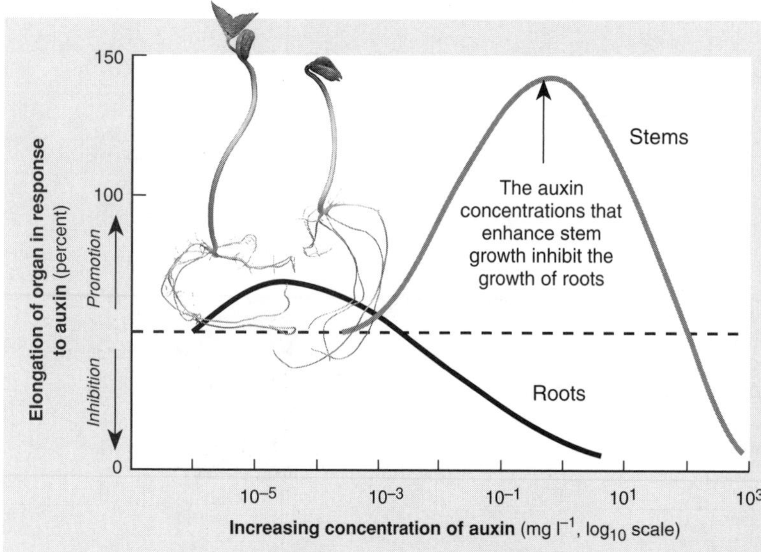

Auxin Concentration and Root Growth

In a horizontally placed seedling, auxin moves to the lower side of the organ in both the stem and root. Whereas the stem tip grows upwards, the root tip responds by growing down. Root elongation is inhibited by the same level of auxin that stimulates stem growth (see graph left). The higher auxin levels on the lower surface cause growth inhibition there. The most elongated cells are then on the upper surface and the root turns down. This simple auxin explanation for the gravitropic response has been much criticized: the concentrations of auxins measured in the upper and lower surfaces of horizontal stems and roots are too small to account for the growth movements observed. Alternative explanations suggest that growth inhibitors are also somehow involved in the gravitropic response.

1. Explain the mechanism proposed for the role of auxin in the gravitropic response in:

   (a) Shoots (stems): ______________________________

   (b) Roots: ______________________________

2. (a) From the graph above, state the auxin concentration at which root growth becomes inhibited: ______________

   (b) State the response of the stem at this concentration: ______________

3. Briefly state a reason why the gravitropic response in stems or roots is important to the survival of a seedling:

   (a) Stems: ______________________________

   (b) Roots: ______________________________

Related activities: *Investigating Phototropism, Auxins, Gibberellins and ABA*

Auxins, Gibberellins, and ABA

Auxin was the first substance to be identified as a plant hormone. Charles Darwin and his son Francis were first to recognize its role in stimulating cell elongation, while Frits W. Went isolated this growth-regulating substance, which he called auxin. **Indole-acetic acid** (IAA) is the only known naturally occuring auxin. Shortly after its discovery it was found to have a role in suppressing the growth of lateral buds. This inhibitory influence of a shoot tip or apical bud on the lateral buds is called **apical dominance**. Two Japanese scientists isolated **gibberellin** in 1934, eight years after the isolation of auxin, and more than 78 gibberellins have now been identified. Gibberellins are involved in stem and leaf elongation, as well as breaking dormancy in seeds. Specifically, they stimulate cell division and cell elongation, allowing stems to 'bolt' and the root to penetrate the testa in germination. During the 1960s, Frederick T. Addicott discovered a substance apparently capable of accelerating **abscission** in leaves and fruit (which he called abscisin), and which is now called **abscisic acid** (ABA). Although it now seems that ABA has very little to do with leaf abscission, it is a growth inhibitor and also stimulates the closing of stomata in most plant species. It is also involved in preventing premature germination and development of seeds.

Auxins, Gibberellins, and Abscisic Acid (ABA) and Plant Responses

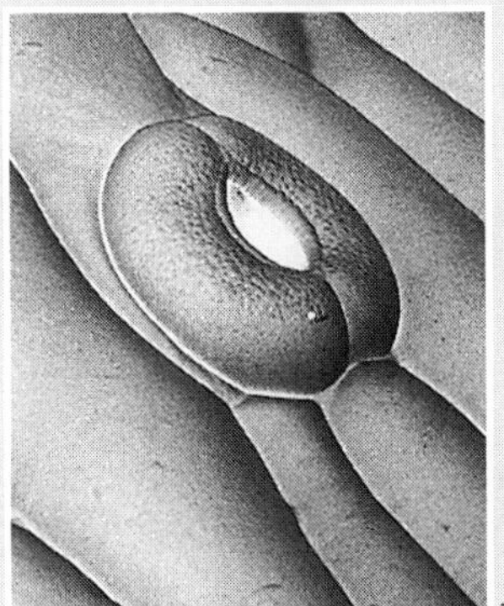

ABA stimulates the closing of stomata in most plant species. Its synthesis is stimulated by water deficiency (water stress).

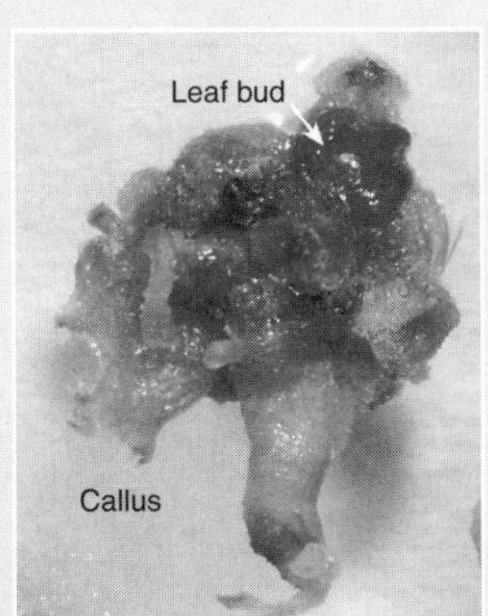

Cytokinins promote the development of leaf buds in calluses (above). In contrast, increased auxin levels will promote the development of roots.

Gibberellins are responsible for breaking dormancy in seeds and promote the growth of the embryo and emergence of the seedling.

ABA promotes seed dormancy. It is concentrated in senescent leaves, but it is probably not involved in leaf abscission except in a few species.

Gibberellins cause stem and leaf elongation by stimulating cell division and cell elongation. They are responsible for **bolting** in brassicas.

Gibberellins are used to hasten seed germination and ensure germination uniformity in the production of barley malt in brewing.

ABA is produced in ripe fruit and induces fruit fall. The effects of ABA are generally opposite to those of cytokinins.

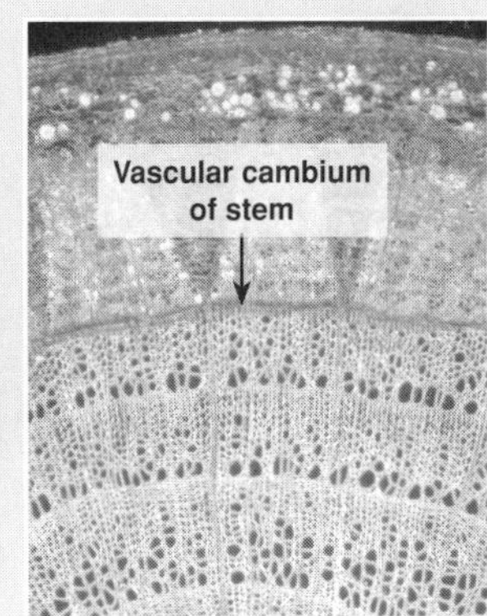

Auxins promote the activity of the vascular cambium (above), stem length, differentiation of tissues, and **apical dominance**.

1. Describe the role of **auxins** in apical dominance: ______

2. Describe the role of **gibberellins** in stem elongation and in the germination of grasses such as barley: ______

3. Describe the role of abscisic acid in closure of stomata: ______

Plant Rhythms

Plants are capable of marked physiological responses to a wide range of environmental variables. Some plant activities, such as the daily opening of flowers, follow **daily rhythms**, others are seasonal. Gardeners are well aware of the seasonal effects of temperature on plant growth and development. Woody plants are able to survive freezing temperatures because of metabolic changes that occur in the plant between summer and winter. This process of **acclimation** involves the production of thicker cell walls and accumulation of growth inhibitors in the plant tissues. Cold hardiness is genetically determined but can be influenced by horticultural practices that mimic seasonal changes. Alternation of periods of growth with periods of **dormancy** allows the plant to survive water shortages and extremes of hot or cold. When dormant, growth will not resume until the right combination of environmental cues are met. Such cues include exposure to cold, dryness, and suitable photoperiod. **Annuals** produce dormant seeds, whereas in **biennials** and **perennials** the shoots may die back while the overwintering structures become dormant. Low temperature stimulation of flowering (**vernalization**) and seed germination (**stratification**) are common in many species. These temperature responses are usually associated with increased activity in plant growth regulators such as gibberellic acid.

Dormancy is a condition of arrested growth in which the entire plant, or its seeds or buds, do not renew growth without certain environmental cues.

In many plants, after the seeds have taken up water, exposure to a period of low temperature (5°C) will break dormancy. This is called **stratification**.

Bud burst and flowering follow exposure to a cold period in many plants, including bulbs and many perennials. This process is called **vernalization**.

Daily Rhythm in Tulips

Many flowers, including tulips, show **sleep movements**. In many species, these are triggered by daylength, but in tulips the environmental cue is temperature. This series of photographs show the sleep movements of a single **tulip** flower over one 12 hour period during spring

7.00 am

9.30 am

11.00 am

5.00 pm

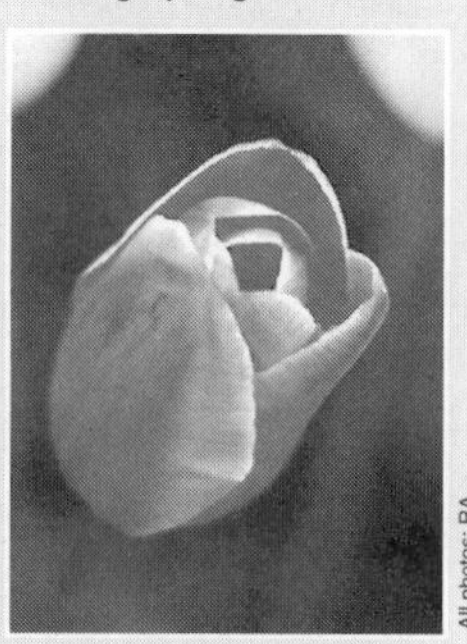

7.00 pm

All photos: RA

1. Describe two physiological responses of plants to seasonal changes in temperature:

   (a) ___

   (b) ___

2. (a) Explain the adaptive value of **stratification** in plants: ___

   (b) State whether this response is endogenous or exogenous: ___

3. Suggest why evergreen trees have a growth advantage in regions with a very short growing season: ___

4. (a) Describe the sleep movements of tulips in response to temperature: ___

   (b) Explain the adaptive value of these movements: ___

Photoperiodism in Plants

Photoperiodism is the response of a plant to the relative lengths of daylight and darkness. Flowering is a photoperiodic activity; individuals of a single species will all flower at much the same time, even though their germination and maturation dates may vary. The exact onset of flowering varies depending on whether the plant is a short-day or long-day type (see next page). Photoperiodic activities are controlled through the action of a pigment called **phytochrome**. Phytochrome acts as a signal for some biological clocks in plants and is also involved in other light initiated responses, such as germination, shoot growth, and chlorophyll synthesis. Plants do not grow at the same rate all of the time. In temperate regions, many perennial and biennial plants begin to shut down growth as autumn approaches. During unfavorable seasons, they limit their growth or cease to grow altogether. This condition of arrested growth is called **dormancy**, and it enables plants to survive periods of water scarcity or low temperature. The plant's buds will not resume growth until there is a convergence of precise environmental cues in early spring. Short days and long, cold nights (as well as dry, nitrogen deficient soils) are strong cues for dormancy. Temperature and daylength change seasonally in most parts of the world, so changes in these variables also influence many plant responses, including germination and flowering. In many plants, flowering is triggered only after a specific period of exposure to low winter temperatures. As described in the previous activity, this low-temperature stimulation of flowering is called **vernalization**.

Photoperiodism

Photoperiodism is based on a system that monitors the day/night cycle. The photoreceptor involved in this, and a number of other light-initiated plant responses, is a blue-green pigment called **phytochrome**. Phytochrome is universal in vascular plants and has two forms: active and inactive. On absorbing light, it readily converts from the inactive form (**P_r**) to the active form (**P_{fr}**). **P_{fr}** predominates in daylight, but reverts spontaneously back to the inactive form in the dark. The plant measures daylength (or rather night length) by the amount of phytochrome in each form.

Summary of phytochrome related activities in plants

Process	Effect of daylight	Effect of darkness
Conversion of phytochrome	Promotes $P_r \rightarrow P_{fr}$	Promotes $P_{fr} \rightarrow P_r$
Seed germination	Promotes	Inhibits
Leaf growth	Promotes	Inhibits
Flowering: long day plants	Promotes	Inhibits
Flowering: short day plants	Inhibits	Promotes
Chlorophyll synthesis	Promotes	Inhibits

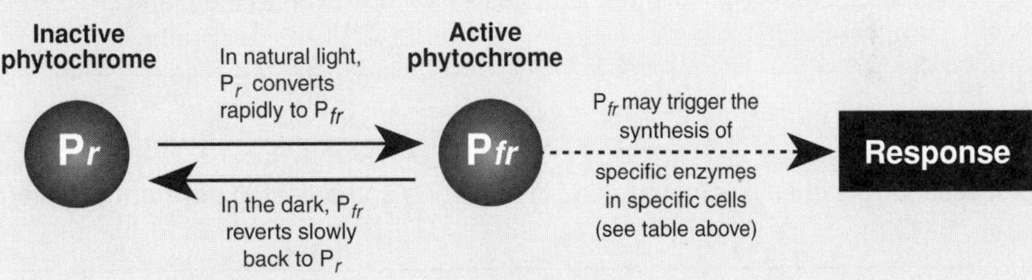

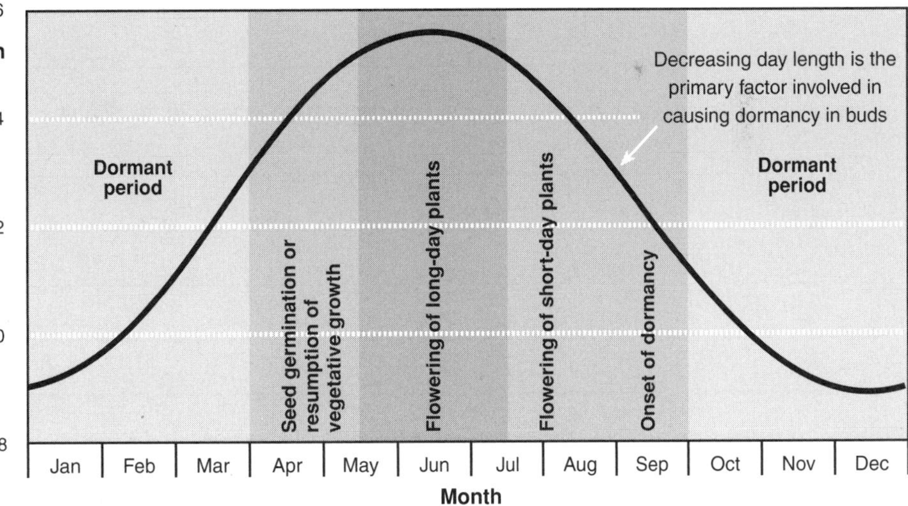

Day length and life cycle in plants (Northern Hemisphere)
The cycle of active growth and dormancy shown by temperate plants is correlated with the number of daylight hours each day (right). In the southern hemisphere, the pattern is similar, but is six months out of phase. The duration of the periods may also vary on islands and in coastal regions because of the moderating effect of nearby oceans.

1. Describe two plant responses, initiated by exposure to light, which are thought to involve the action of phytochrome:

(a) ______________________________

(b) ______________________________

2. Discuss the role of phytochrome in a plant's ability to measure daylength: ______________________________

____

____

____

____

Photoperiodism in Plants

An experiment was carried out to determine the environmental cue that triggers flowering in 'long-day' and 'short-day' plants. The diagram below shows 3 different light regimes to which a variety of long-day and short-day plants were exposed.

Long-day plants

When subjected to the light regimes on the right, the 'long-day' plants below flowered as indicated:

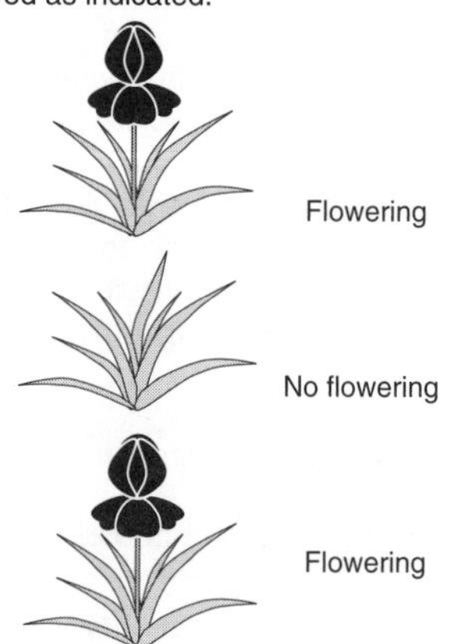

Examples: *lettuce, clover, delphinium, gladiolus, beetscorn, coreopsis*

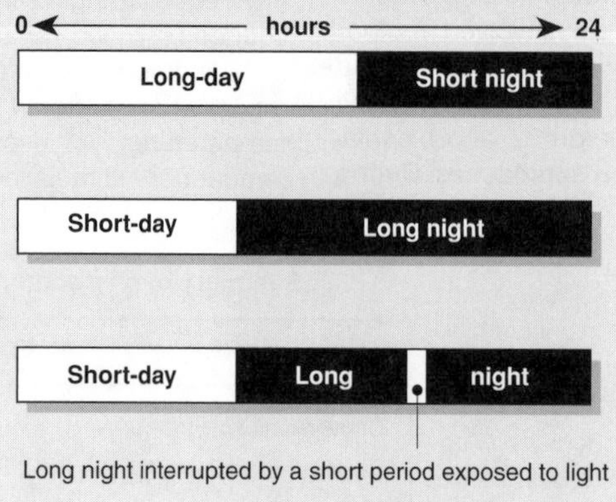

Long night interrupted by a short period exposed to light

Short-day plants

When subjected to the light regimes on the left, the 'short-day' plants below flowered as indicated:

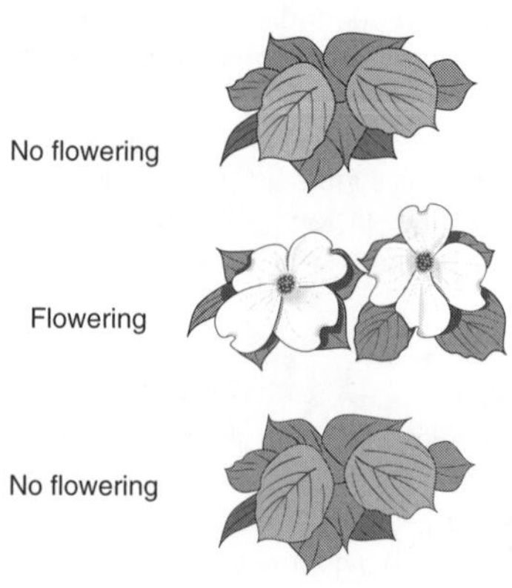

Examples: *potatoes, asters, dahlias, cosmos, chrysanthemums, pointsettias*

3. (a) Identify the environmental cue that synchronizes flowering in plants: ________________

   (b) Describe one biological advantage of this synchronization to the plants: ________________

4. Discuss the role of environmental cues in triggering and breaking **dormancy** in plants:

5. Discuss the adaptive value of **dormancy** and **vernalization** in temperate climates:

6. Study the three light regimes above and the responses of short-day and long-day flowering plants to that light. From this observation, describe the most important factor controlling the onset of flowering in:

   (a) Short-day plants: ________________

   (b) Long-day plants: ________________

7. Using information from the experiment described above, discuss the evidence for the statement "*Short-day plants are really better described as long-night plants.*"

KEY TERMS Crossword

Complete the crossword below, which will test your understanding of key terms in this chapter and their meanings

Clues Across

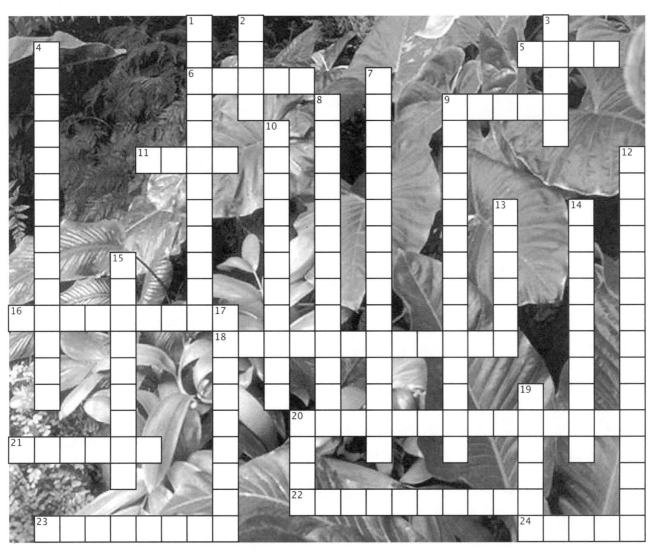

5. Part of the plant which collects sunlight and contains cells that carry out photosynthesis.
6. Cylinder of vascular tissue consisting of the xylem, phloem and the pericycle surrounded by the endodermis.
9. Plant hormone responsible for apical dominance.
11. Class of bryophyte. Normally grow as erect stems that may reach 50cm in larger specimens.
16. The region on a flowering plant between nodes.
18. A directional growth response in plants to light.
20. The physiological response to light.
21. A group of angiosperms with two cotyledons in the seed.
22. A plant that is adapted to growth partially or wholly submerged in water.
23. Group of angiosperms that produce only one cotyledon in the seed.
24. Vascular tissue that conducts water and minerals salts from the roots to the rest of the plant.

Clues Down

1. A plant hormone involved in plant development, including bud dormancy.
2. A plant structure that supports the leaves in the light and provides a place for flowers and fruit.
3. A group of plants that reproduce by spores. They develop vascular tissue but not wood.
4. A grouping of tissues that contains phloem, xylem, and support and strengthening tissues. (2 words: 8, 6).
7. Growth that gives rise to increase in diameter (2 words: 9, 6).
8. Growth that leads to increase in length. (2 words: 7, 6).
9. The undifferentiated tissue found in the buds and growing tips of roots in plants. (2 words: 6, 8).
10. A photoreceptor in plants that measures day length.
12. A layer of tissue that is the source of cells for secondary growth. (2 words: 8, 7).
13. Vascular tissue that conducts foods (e.g. sugars and proteins) through the plant.
14. The innermost layer of the cortex surrounding the stele.
15. A plant with adaptations to arid conditions.
17. The outer most layer of cells. In plants it is one cell layer thick and usually covered with a protective cuticle.
19. Parenchyma tissue surrounding the vascular cylinder.
20. Core of parenchyma cells within the vascular cylinder.

AP Topic 3B

IB AHL 9.2

Plant Support and Transport

KEY CONCEPTS

- Plant roots are specialized to provide a high surface area for water and mineral uptake.
- Gases enter and leave the plant mainly through stomata, but water loss is a consequence of this.
- Plants have adaptations to reduce water loss.
- Transpiration drives water uptake in plants.
- Translocation moves carbohydrate around the plant from sources to sinks.

KEY TERMS

bulk (=mass) flow
capillary action
Casparian strip
companion cell
cortex
endodermis
flaccid
guard cells
mass flow hypothesis
mineral
osmosis
phloem
root hair
root pressure
sieve tube cell
sink
source
stomata
tension-cohesion hypothesis
tracheid
translocation
transpiration
transpiration pull
turgid
vessel
water potential
xylem

OBJECTIVES

□ 1. Use the **KEY TERMS** to help you understand and complete these objectives.

Angiosperm Roots pages 281-283

□ 2. Describe features of the **root system** in plants that facilitate the uptake of water and mineral ions, including reference to the structure and role of root hairs and different patterns of root growth (e.g. fibrous root systems).

□ 3. Recognize that transport tissues in plants also have a role in support. Describe the roles of **cell turgor**, transport tissues (especially **xylem**), and cellulose in supporting the plant.

□ 4. Describe the structure of a typical dicot root, including reference to the location of the vascular tissue (**stele**), **pith**, **cortex**, and **endodermis**.

Plant Transport Processes pages 258-259, 269-270, 275, 284-292

□ 5. Describe the mechanism and pathways for water uptake in plant roots.

□ 6. Describe passive and active uptake of minerals in plant roots. Identify the role of some of the mineral ions important to plants, including: nitrate (NO^{3-}), phosphate (PO_4^{3-}), and magnesium (Mg^{2+}) ions.

□ 7. Describe **transpiration** in a flowering plant, explaining how water moves through the plant. Include reference to the roles of xylem, **cohesion-tension**, **transpiration pull** and **evaporation**, and **root pressure**.

□ 8. Describe the role of **stomata** and explain the movement of gases into and out of the spongy mesophyll of the leaf. Recognize transpiration as a necessary consequence of gas exchange in plants. Describe the role of **lenticels** in gas exchange in woody plants.

□ 9. Describe how guard cells regulate transpirational losses by opening and closing the stomata and explain the role of **abscisic acid** in this.

□ 10. Investigate and explain the effect of humidity, light, air movement, temperature, and water availability on **transpiration rate**.

□ 11. Describe the adaptations of **xerophytes** that help to reduce water loss.

□ 12. Describe and explain **translocation** in the **phloem**, identifying **sources** and **sinks** in sucrose transport. Evaluate the evidence for and against the **mass flow** (pressure-flow) **hypothesis** for the mechanism of translocation.

Periodicals:
listings for this chapter are on page 390

Weblinks:
www.thebiozone.com/weblink/SB2-2603.html

Support in Plants

Plants support themselves in their environment and maintain the positions that enable them to carry out essential processes. All plants are provided some support by **cell turgor**. For very small plants, this is sufficient. Terrestrial vascular plants have strengthening tissues that may be hardened with lignin, and many also produce secondary growth (wood). For aquatic plants the water provides support, and adaptations are primarily to maintain the plant in the photic zone and to remain anchored.

Aquatic Environment

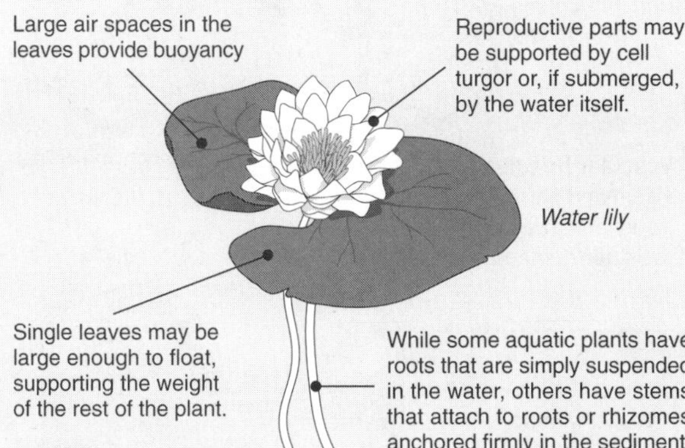

Large air spaces in the leaves provide buoyancy
Reproductive parts may be supported by cell turgor or, if submerged, by the water itself.
Water lily
Single leaves may be large enough to float, supporting the weight of the rest of the plant.
While some aquatic plants have roots that are simply suspended in the water, others have stems that attach to roots or rhizomes anchored firmly in the sediment.

Some aquatic plants, like **water hyacinth**, have swollen petioles that act as floats. Many form floating mats which block water ways and are serious weeds e.g. *Salvinia* and alligator weed.

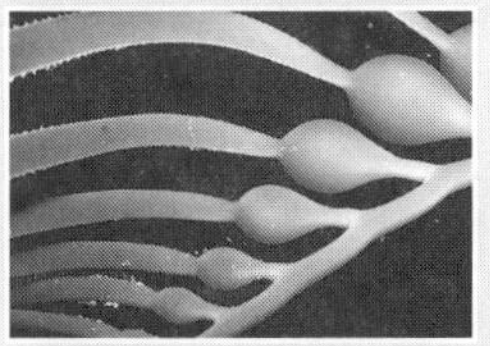

Marine and freshwater **algae** are not plants but have plant like qualities (e.g. chlorophyll). They lack vascular tissue and are supported by the water. Buoyancy may be assisted by airfilled floats or projections of the cell wall, which increase surface area (as in the case of diatoms).

Many floating or semi-aquatic plants, such as **water lilies**, have expanded leaves that provide a large surface area for flotation and photosynthesis. Such floating leaves support the submerged parts of the plant. These plants have roots that are attached to the bottom sediment.

Terrestrial Environment

Vascular plants with secondary thickening (woody tissue) can reach an enormous size. The Australian *Eucalyptus regnans* is reputed to grow to over 140 m in height. Their structural tissues (e.g. **xylem** and **wood**) provide support against gravity. Leaves and reproductive parts are supported by **cell turgor**.

Roots anchor the plant in the ground, forming a stable base for growth. The roots of large hard-wood trees can form **buttresses**, providing extra support in poor soils.

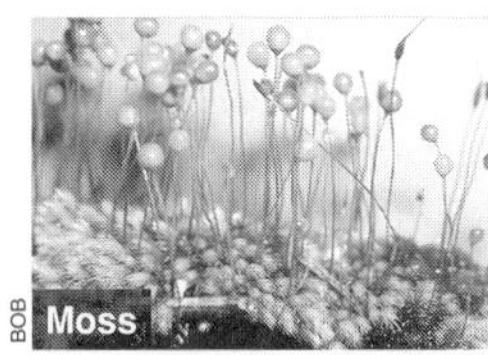

Bryophytes (mosses and liverworts) lack any vascular tissue: the plant body is supported by **cell turgor**. As a consequence, they are small and their upward growth is restricted. Although there are no true roots, filamentous rhizoids anchor the plant in the ground.

Liverworts are simpler in structure than mosses. The gametophyte (the main plant body) is a flattened structure that may be a lobed thallus or leaf-like, depending on the species. The plant lies flat against the substrate, rising just a centimeter or two above the ground.

Ferns are tracheophytes (vascular plants). They have well developed vascular tissues that provide support and allow transport of nutrients and water around the plant. Because of this, they are able to grow to considerable heights (e.g. tree ferns).

The **Lord Howe Island fig** has ten or more trunks that develop and form aerial roots. These grow downwards and once anchored in the ground, they provide extra support for the heavy weight of the trunks, allowing them to cover a wide area. Without such support, the trunks would collapse.

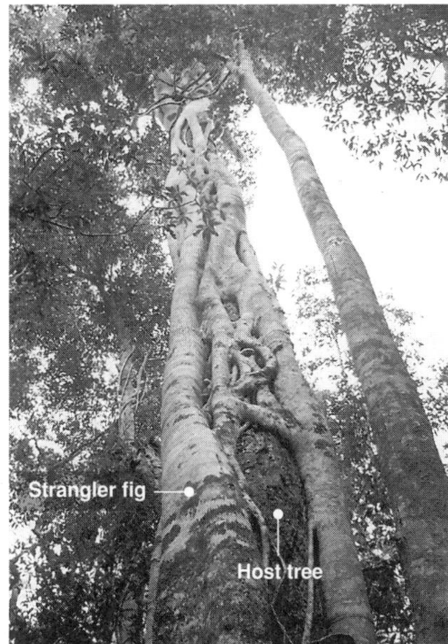

Strangler figs begin life high in the forest canopy as epiphytes. They develop roots that grow towards the forest floor. Once rooted in the soil they grow rapidly, embracing the trunk of a host tree with roots and shading it. As the host tree increases in girth, the fig cuts off the sap supply to its roots, killing it.

Mangroves grow on mudflat shorelines. The root system cannot penetrate far into the mud due to the lack of oxygen. Support is provided by roots that are sent out in all directions just below the surface. Pneumatophores or breathing roots, (seen above) arise from these shallow lateral roots.

Related activities: *Diversity in Stem Structure, Modifications in Plants*

Support in Woody Plants

Secondary xylem which consists of massed xylem vessels and fibers, makes up the bulk of the stem as **wood**, providing considerable strength.

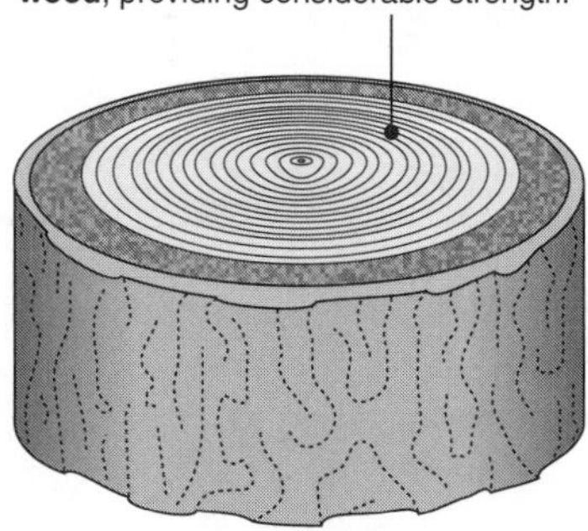

The strengthening of cells with lignin gives support to stems in all vascular plants. Lignin, together with cell turgor, is particularly important in non-woody (herbaceous) plants.

Support in Herbaceous Plants

Turgor pressure inside the parenchyma cells provides a strong inflating force that pushes against the epidermal layer.

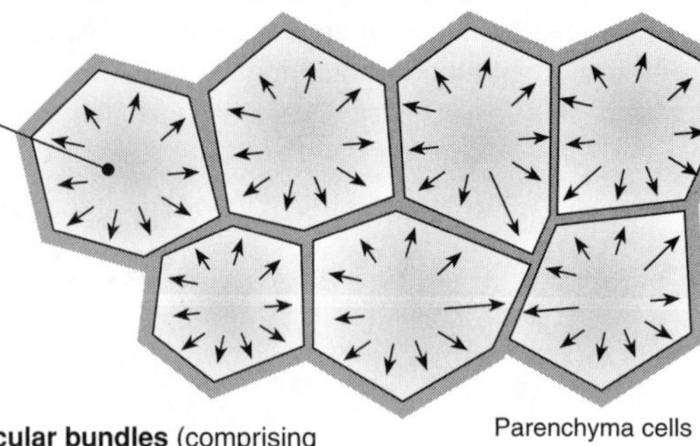

Parenchyma cells

Vascular bundles (comprising xylem and phloem) enhance the ability of herbaceous stems to resist tension and compression.

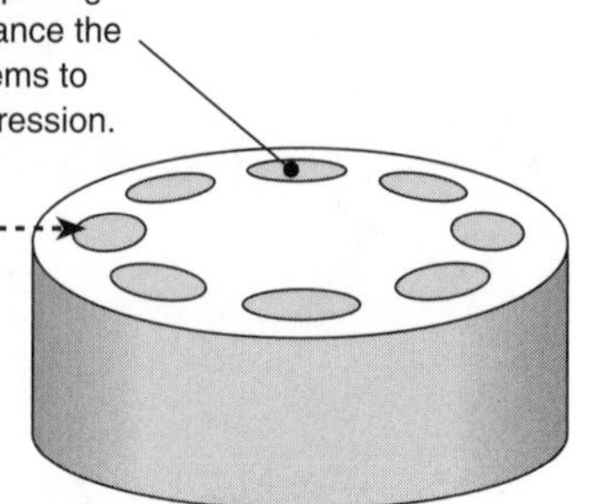

Xylem vessel with spiral thickening produced as a result of **lignin** deposition.

1. Contrast the main problems experienced by aquatic and terrestrial plants in supporting themselves:

2. Describe how the following are achieved in the aquatic protists and plants named below:

   (a) Maintaining a stationary position in seaweeds:

   (b) Keeping the fronds of kelp near the surface:

   (c) Keeping water lily pads floating on the surface:

3. Describe the function of **buttresses** on the trunks of large rainforest hardwood trees:

4. Explain the role of the following in providing support for vascular plants:

   (a) Lignin:

   (b) Turgor pressure:

   (c) Vascular bundles:

   (d) Secondary xylem:

5. Describe how strangler fig trees overcome support problems during the early stage of their development:

Angiosperm Root Structure

Roots are essential plant organs. They anchor the plant in the ground, absorb water and minerals from the soil, and transport these materials to other parts of the plant body. Roots may also act as storage organs, storing excess carbohydrate reserves until they are required by the plant. Roots are covered in an epidermis but, unlike the epidermis of leaves, the root epidermis has only a thin cuticle that presents no barrier to water entry. Young roots are also covered with **root hairs** (see below). Much of a root comprises a cortex of parenchyma cells. The air spaces between the cells are essential for aeration of the root tissue. Minerals and water must move from the soil into the xylem before they can be transported around the plant. Compared with stems, roots are relatively simple and uniform in structure. The structure of monocot and dicot roots is compared in the photographs below.

The Structure of a Dicot Root

These photographs (left and below) show cross sections through a young dicot root (i.e. primary tissues). In the photograph to the left, note the large area of the root occupied by the cortex. The parenchyma (packing) cells of the cortex store starch and other substances. The air spaces between the cells are essential for aeration of the root tissue, which is non-photosynthetic. The vascular tissue, xylem (X) and phloem (P) forms a central cylinder through the root and is surrounded by the **pericycle**, a ring of cells from which lateral roots arise. The primary xylem of dicot roots forms a star shape in the center of the vascular cylinder with usually 3 or 4 points. Unlike monocots, there is no central pith of parenchyma cells.

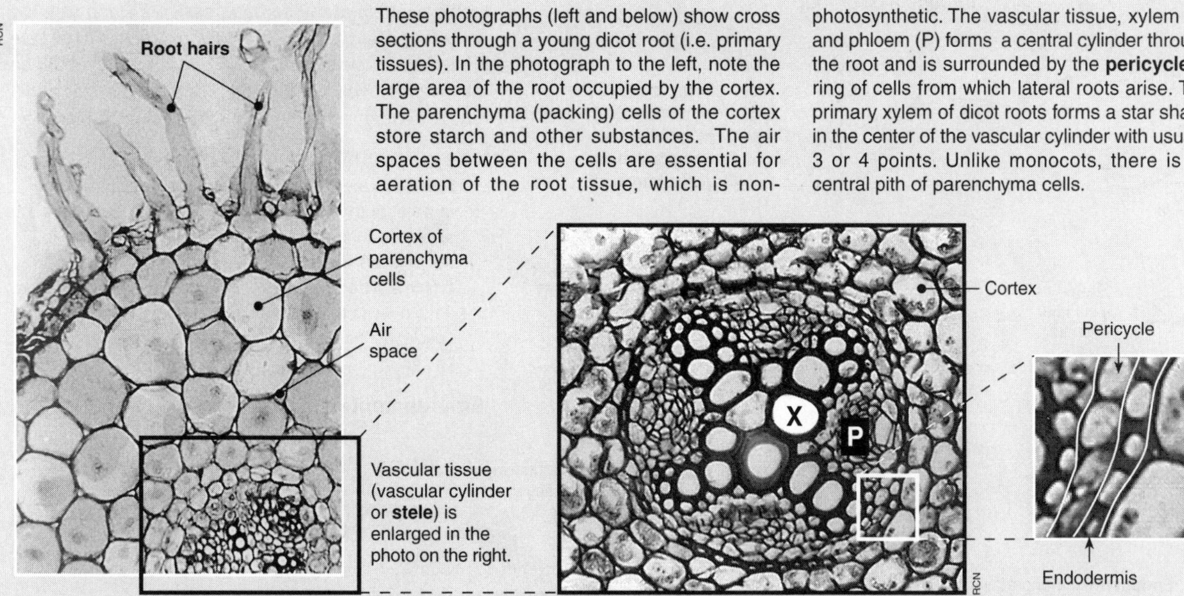

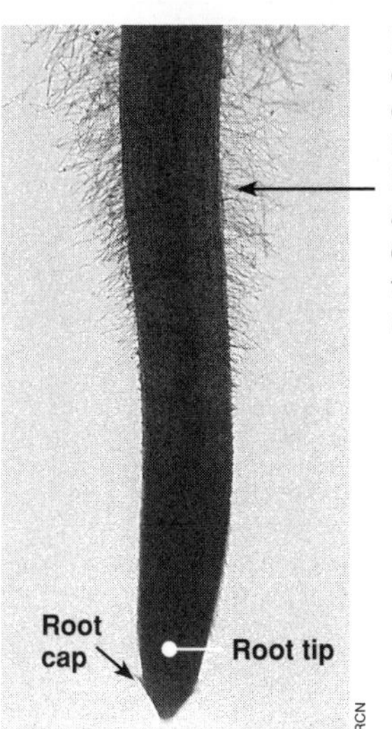

Root hairs are located behind the region of cell elongation in the root tip. They are single celled extensions of the epidermal cells that increase the surface area for absorption. Individual root hairs are short lived, but they are produced continually. The root tip is covered by a slimy **root cap**. This protects the dividing cells of the tip and aids the root's movement through the soil.

The Structure of a Monocot Root

Monocot roots (right) vary from dicot roots in several ways. As in dicots, there is a large cortex but the **endodermis** is very prominent and heavily thickened. The stele (ring of vascular tissue) is large compared with the size of the root and there are many xylem points. There is a central pith inside the vascular tissue that is absent in dicot roots.

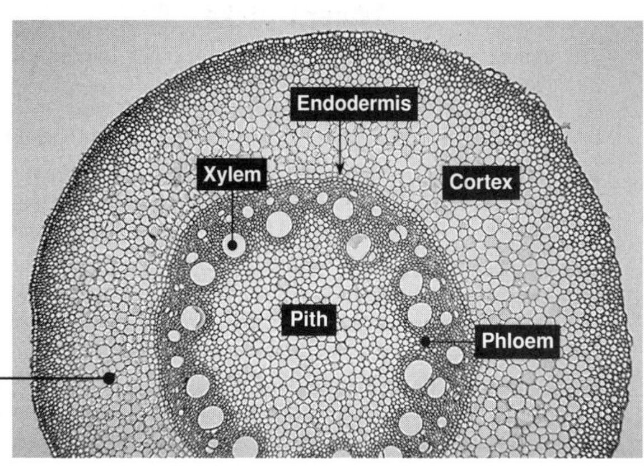

Cross section through an old root of corn (Zea mays), a typical monocot.

1. Explain the purpose of the root hairs: ______

2. Explain why the root tip is covered by a cap of cells: ______

3. Describe two features of internal anatomy that distinguish monocot and dicot roots:

(a) ______

(b) ______

4. Describe one feature that monocot and dicot roots have in common: ______

5. Describe the role of the parenchyma cells of the cortex: ______

Related activities: *Uptake in the Root*
Web links: *Photographic Atlas of Plant Anatomy*

Uptake in the Root

Plants need to take up water and minerals constantly to compensate for water losses and obtain the materials they need to manufacture food. The uptake of water and minerals is mostly restricted to the younger, most recently formed cells of the roots and the root hairs. Some water moves through the plant via the plasmodesmata of the cells (the **symplastic route**), but most passes through the free spaces between cell walls (the **apoplast**). Water uptake is assisted by root pressure, which arises because the soil and root tissue has a higher water potential than other plant tissues. Uptake by the roots is largely a passive process, although some mineral uptake involves active transport. Note: An alternative version of this activity, without reference to water potential, is available as a web link (see below) or on the Teacher Resource CD-ROM.

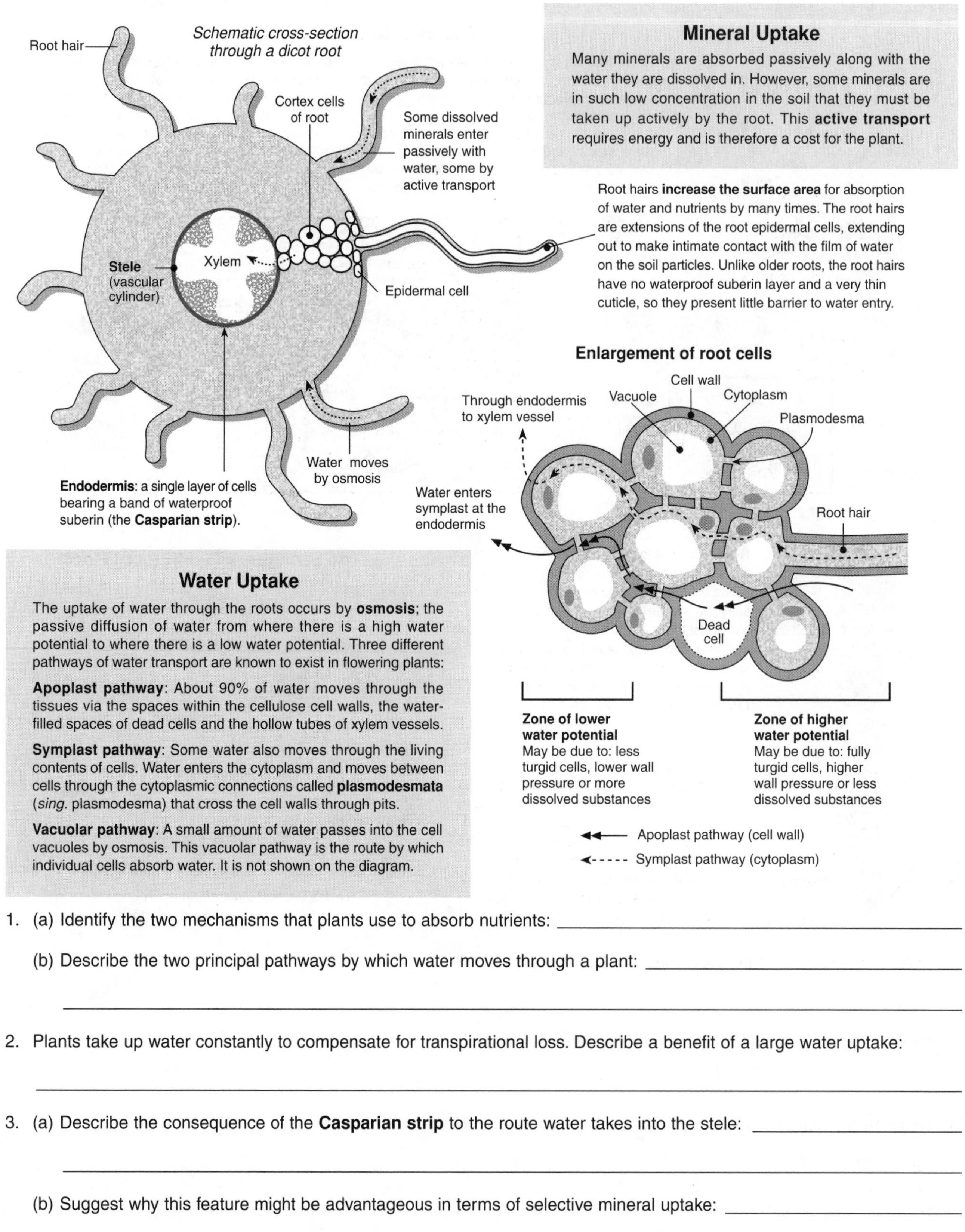

Schematic cross-section through a dicot root
Root hair
Cortex cells of root
Some dissolved minerals enter passively with water, some by active transport
Stele (vascular cylinder)
Xylem
Epidermal cell
Water moves by osmosis
Endodermis: a single layer of cells bearing a band of waterproof suberin (the **Casparian strip**).
Root hairs **increase the surface area** for absorption of water and nutrients by many times. The root hairs are extensions of the root epidermal cells, extending out to make intimate contact with the film of water on the soil particles. Unlike older roots, the root hairs have no waterproof suberin layer and a very thin cuticle, so they present little barrier to water entry.
Enlargement of root cells
Through endodermis to xylem vessel
Vacuole
Cell wall
Cytoplasm
Plasmodesma
Water enters symplast at the endodermis
Root hair
Dead cell
Zone of lower water potential
May be due to: less turgid cells, lower wall pressure or more dissolved substances
Zone of higher water potential
May be due to: fully turgid cells, higher wall pressure or less dissolved substances
Apoplast pathway (cell wall)
Symplast pathway (cytoplasm)

Mineral Uptake

Many minerals are absorbed passively along with the water they are dissolved in. However, some minerals are in such low concentration in the soil that they must be taken up actively by the root. This **active transport** requires energy and is therefore a cost for the plant.

Water Uptake

The uptake of water through the roots occurs by **osmosis**; the passive diffusion of water from where there is a high water potential to where there is a low water potential. Three different pathways of water transport are known to exist in flowering plants:

Apoplast pathway: About 90% of water moves through the tissues via the spaces within the cellulose cell walls, the water-filled spaces of dead cells and the hollow tubes of xylem vessels.

Symplast pathway: Some water also moves through the living contents of cells. Water enters the cytoplasm and moves between cells through the cytoplasmic connections called **plasmodesmata** (*sing.* plasmodesma) that cross the cell walls through pits.

Vacuolar pathway: A small amount of water passes into the cell vacuoles by osmosis. This vacuolar pathway is the route by which individual cells absorb water. It is not shown on the diagram.

1. (a) Identify the two mechanisms that plants use to absorb nutrients: ______

   (b) Describe the two principal pathways by which water moves through a plant: ______

2. Plants take up water constantly to compensate for transpirational loss. Describe a benefit of a large water uptake:

   ______

3. (a) Describe the consequence of the **Casparian strip** to the route water takes into the stele: ______

   (b) Suggest why this feature might be advantageous in terms of selective mineral uptake: ______

Plant Mineral Requirements

Plants normally obtain minerals from the soil. The availability of mineral ions to plant roots depends on soil texture, since this affects the permeability of the soil to air and water. Mineral ions may be available to the plant in the soil water, adsorbed on to clay particles, or via release from humus and soil weathering. **Macronutrients** (e.g. nitrogen, sulfur, phosphorus) are required in large amounts for building basic constituents such as proteins. **Trace elements** (e.g. manganese, copper, and zinc) are required in small amounts. Many are components of, or activators for, enzymes. After being absorbed, mineral ions diffuse into the endodermis and may diffuse, or be actively transported, to the xylem for transport to other regions of the plant.

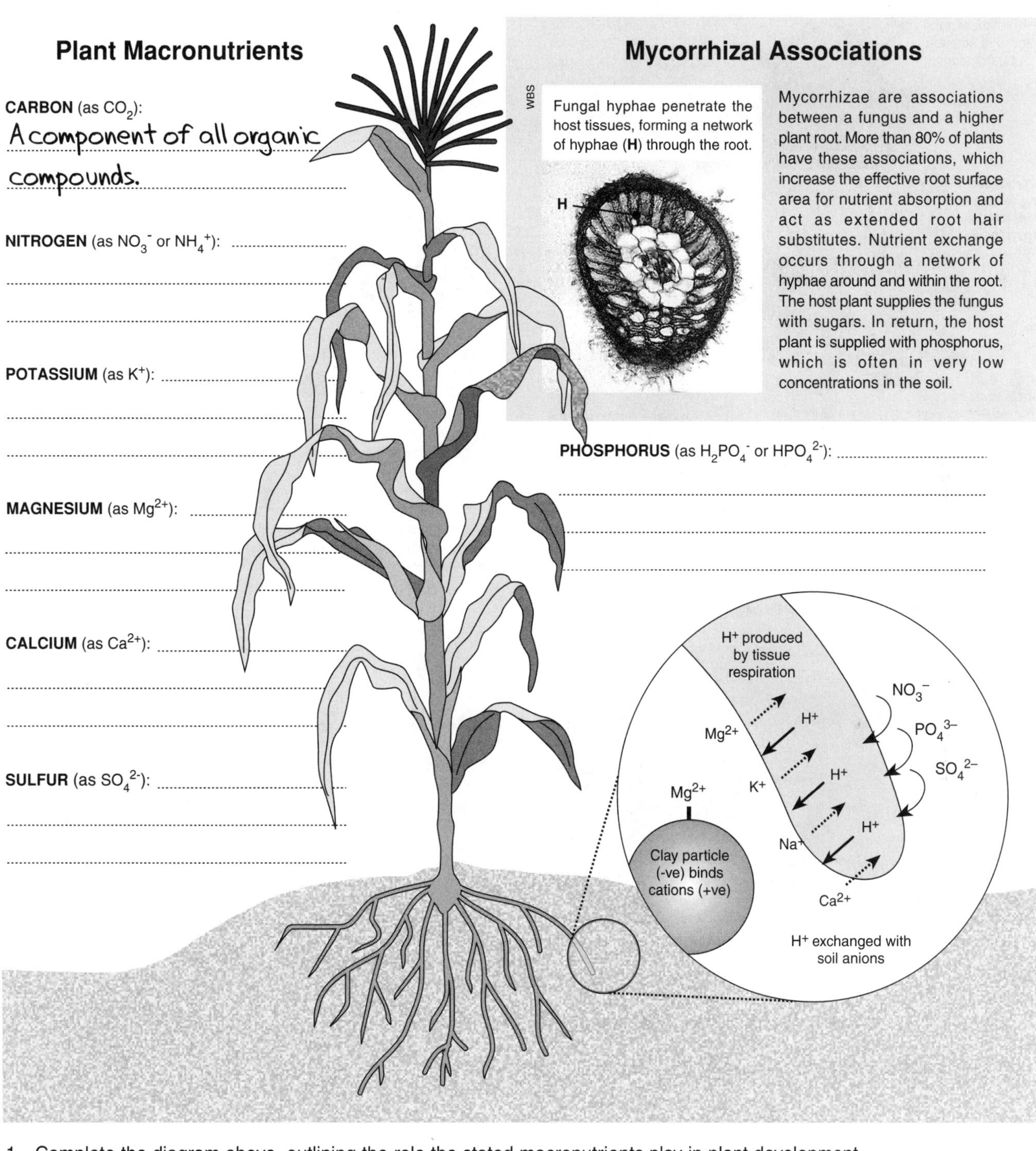

1. Complete the diagram above, outlining the role the stated macronutrients play in plant development.

2. Briefly describe the ways in which minerals are available to plants:

3. (a) Describe the role of mycorrhizal associations in plant nutrition:

   (b) Suggest why plants without mycorrhizal associations might show poor growth:

Related activities: Uptake in the Root
Web links: Mycorrhizas

Gas Exchange in Plants

Respiring tissues require oxygen, and the photosynthetic tissues of plants also require carbon dioxide in order to produce the sugars needed for their growth and maintenance. The principal gas exchange organs in plants are the leaves, and sometimes the stems. In most plants, the exchange of gases directly across the leaf surface is prevented by the waterproof, waxy cuticle layer. Instead, access to the respiring cells is by means of **stomata**, which are tiny pores in the leaf surface. The plant has to balance its need for carbon dioxide (keeping stomata open) against its need to reduce water loss (stomata closed).

Terrestrial Environment

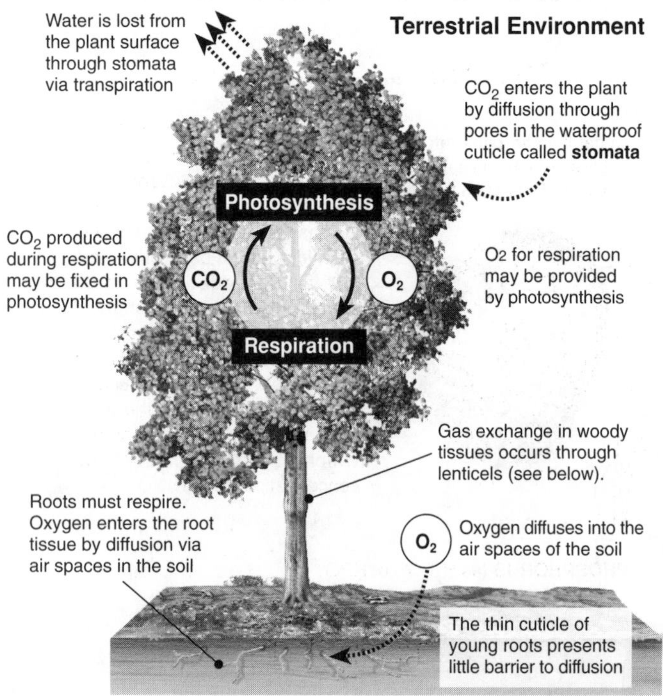

Most gas exchange in plants occurs through the leaves, but some also occurs through the stems and the roots. The shape and structure of leaves (very thin with a high surface area) assists gas exchange by diffusion.

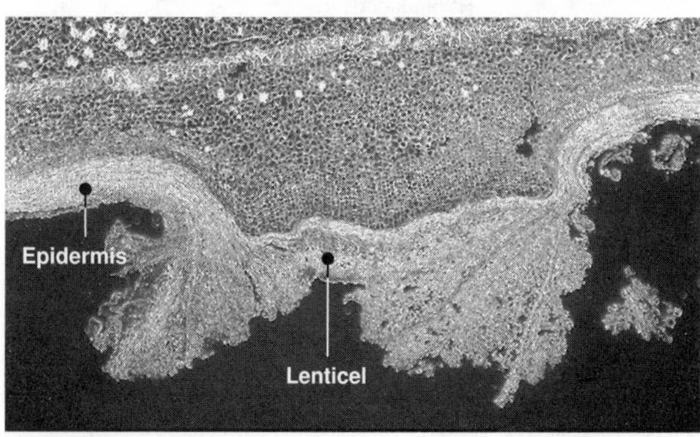

In woody plants, the wood prevents gas exchange. A lenticel is a small area in the bark where the loosely arranged cells allow entry and exit of gases into the stem tissue underneath.

Aquatic Environment

The aquatic environment presents special problems for plants. Water loss is not a problem, but CO_2 availability is often very limited because most of the dissolved CO_2 is present in the form of bicarbonate ions, which is not directly available to plants. Maximizing uptake of gaseous CO_2 by reducing barriers to diffusion is therefore important.

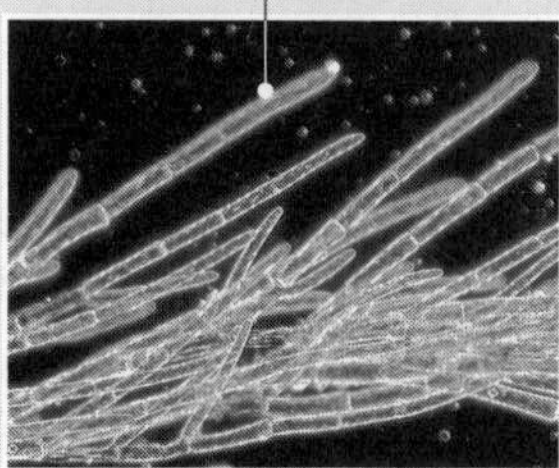
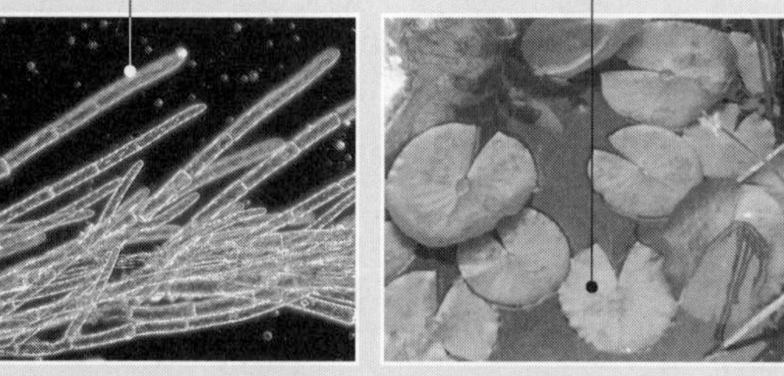

Algae lack stomata but achieve adequate gas exchange through simple diffusion into the cells.

Floating leaves, such as the water lilies above, generally lack stomata on their lower surface.

With the exception of liverworts, all terrestrial plants and most aquatic plants have stomata to provide for gas exchange. CO_2 uptake is aided in submerged plants because they have little or no cuticle to form a barrier to diffusion of gases. The few submerged aquatics that lack stomata altogether rely only on diffusion through the epidermis. Most aquatic plants also have air spaces in their spongy tissues (which also assist buoyancy).

Transitional Environment

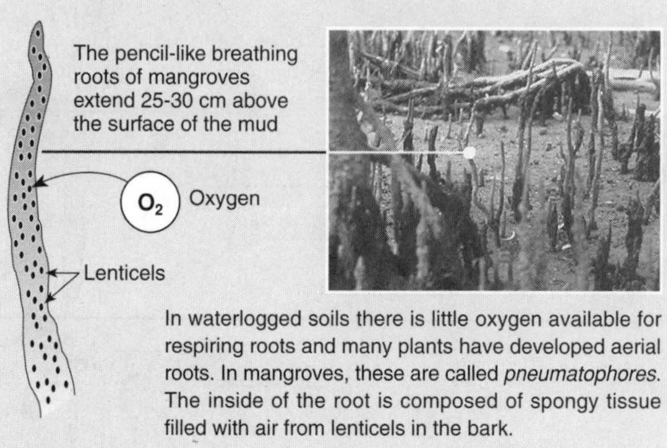

In waterlogged soils there is little oxygen available for respiring roots and many plants have developed aerial roots. In mangroves, these are called *pneumatophores*. The inside of the root is composed of spongy tissue filled with air from lenticels in the bark.

1. Name the gas produced by cellular respiration that is also a raw material for photosynthesis: ________

2. Describe the role of lenticels in plant gas exchange: ________

3. Identify two properties of leaves that assist gas exchange: ________

4. With respect to gas exchange and water balance, describe the most important considerations for:

   (a) Terrestrial plants: ________

   (b) Aquatic plants: ________

5. Describe an adaptation for gas exchange in the following plants:

   (a) A submerged aquatic angiosperm: ________

   (b) A mangrove in a salty mudflat: ________

Gas Exchange and Stomata

The leaf epidermis of angiosperms is covered with tiny pores, called **stomata**. Angiosperms have many air spaces between the cells of the stems, leaves, and roots. These air spaces are continuous and gases are able to move freely through them and into the plant's cells via the stomata. Each stoma is bounded by two **guard cells**, which together regulate the entry and exit of gases and water vapor. Although stomata permit gas exchange between the air and the photosynthetic cells inside the leaf, they are also the major routes for water loss through transpiration. Note: An alternative version of this activity, without reference to water potential, is available as a web link (see below) or on the Teacher Resource CD-ROM.

Gas Exchanges and the Function of Stomata

Gases enter and leave the leaf by way of stomata. Inside the leaf (as illustrated by a dicot, right), the large air spaces and loose arrangement of the spongy mesophyll facilitate the diffusion of gases and provide a large surface area for gas exchanges.

Respiring plant cells use oxygen (O_2) and produce carbon dioxide (CO_2). These gases move in and out of the plant and through the air spaces by diffusion.

When the plant is photosynthesizing, the situation is more complex. Overall there is a net consumption of CO_2 and a net production of oxygen. The fixation of CO_2 maintains a gradient in CO_2 concentration between the inside of the leaf and the atmosphere. Oxygen is produced in excess of respiratory needs and diffuses out of the leaf. These **net** exchanges are indicated by the arrows on the diagram.

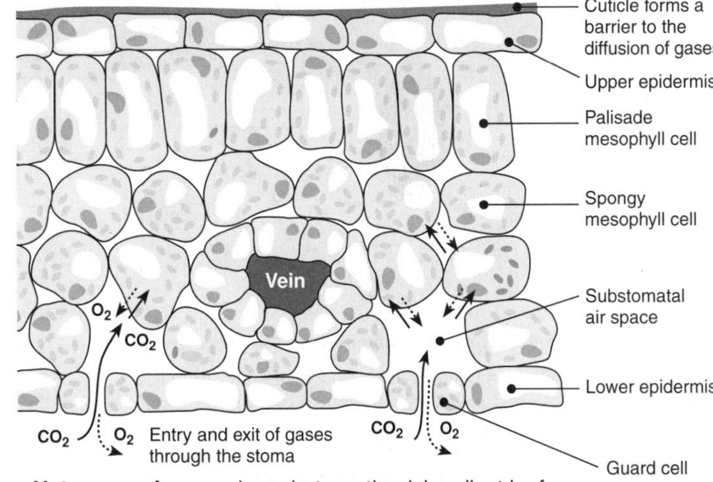

Net gas exchanges in a photosynthesizing dicot leaf

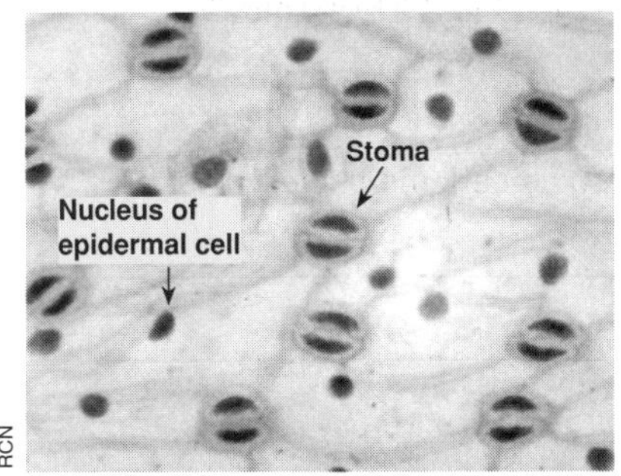

A surface view of the leaf epidermis of a dicot (above) illustrating the density and scattered arrangement of stomata. In dicots, stomata are usually present only on the lower leaf surface.

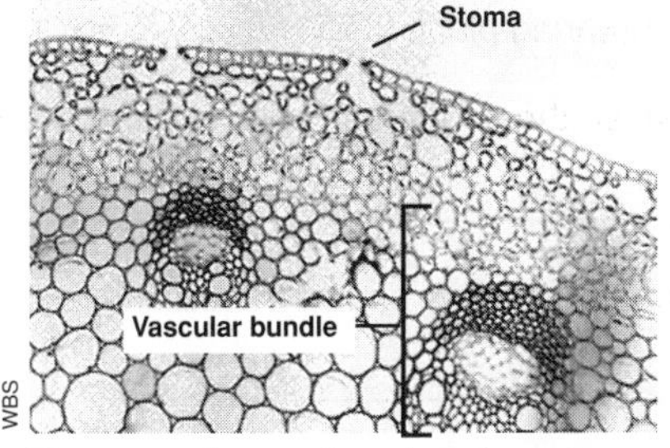

The stems of some plants (e.g. the buttercup above) are photosynthetic. Gas exchange between the stem tissues and the environment occurs through stomata in the outer epidermis.

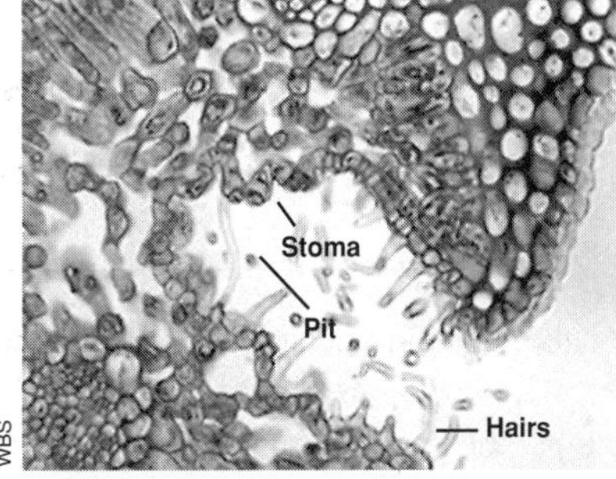

Oleander (above) is a xerophyte with many water conserving features. The stomata are in pits on the leaf underside. The pits restrict water loss to a greater extent than they reduce CO_2 uptake.

The cycle of opening and closing of stomata

The opening and closing of stomata shows a daily cycle that is largely determined by the hours of light and dark.

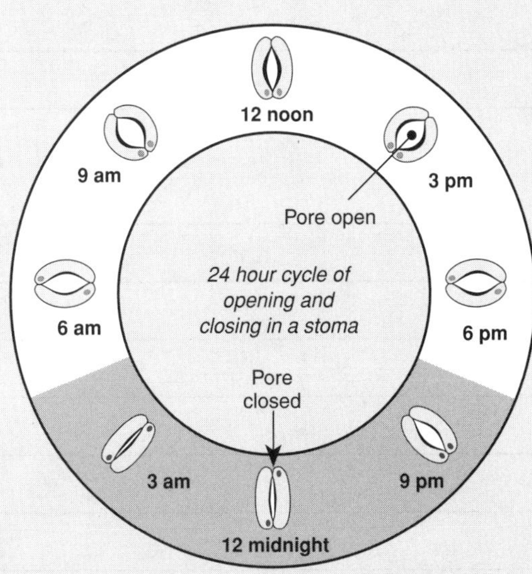

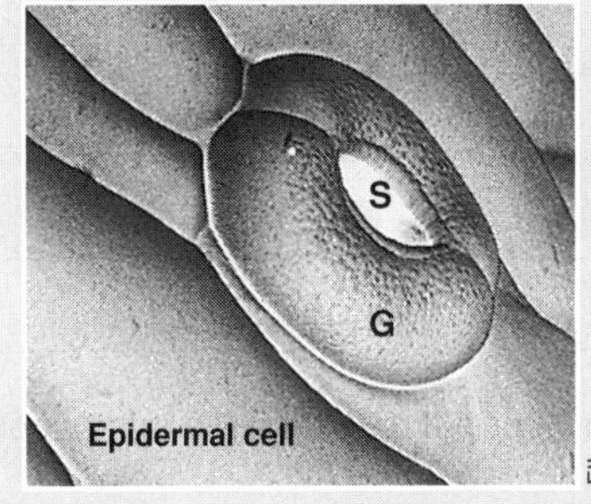

The image left shows a scanning electron micrograph (SEM) of a single stoma from the leaf epidermis of a dicot.

Note the guard cells (G), which are swollen tight and open the pore (S) to allow gas exchange between the leaf tissue and the environment.

Factors influencing stomatal opening

Stomata	Guard cells	Daylight	CO_2	Soil water
Open	Turgid	Light	Low	High
Closed	Flaccid	Dark	High	Low

The opening and closing of stomata depends on environmental factors, the most important being light, carbon dioxide concentration in the leaf tissue, and water supply. Stomata tend to open during daylight in response to light, and close at night (left and above). Low CO_2 levels also promote stomatal opening. Conditions that induce water stress cause the stomata to close, regardless of light or CO_2 level.

Related activities: *Gas Exchange in Plants, Transpiration*
Web links: *Opening and Closing of a Stoma, Gas Exchange and Stomata*

The guard cells on each side of a stoma control the diameter of the pore by changing shape. When the guard cells take up water by osmosis they swell and become turgid, making the pore wider. When the guard cells lose water, they become flaccid, and the pore closes up. By this mechanism a plant can control the amount of gas entering, or water leaving, the plant. The changes in turgor pressure that open and close the pore result mainly from the reversible uptake and loss of potassium ions (and thus water) by the guard cells.

Stomatal Pore Open

K^+ enters the guard cells from the epidermal cells (active transport coupled to a proton pump).

Water follows K^+ by osmosis.

Guard cell swells and becomes turgid.

Thickened ventral wall

Pore opens

Nucleus of guard cell

H_2O K^+

ψguard cell < ψepidermal cell: water enters the guard cells

Stomata open when the guard cells actively take up K^+ from the neighboring epidermal cells. The ion uptake causes the water potential (ψ) to become more negative in the guard cells. As a consequence, water is taken up by the cells and they swell and become turgid. The walls of the guard cells are thickened more on the inside surface (the ventral wall) than the outside wall, so that when the cells swell they buckle outward, opening the pore.

Stomatal Pore Closed

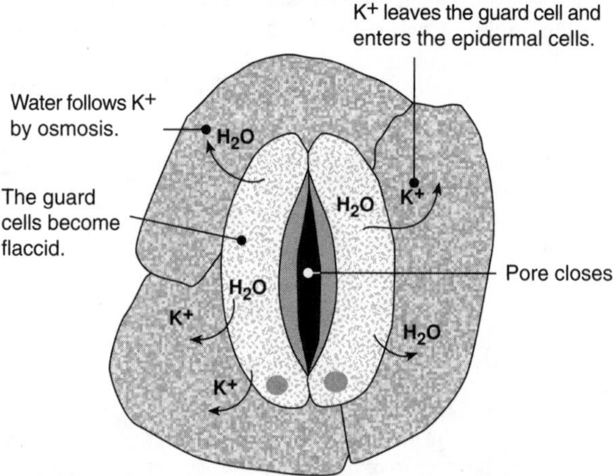

ψepidermal cell < ψguard cell: water leaves the guard cells

Stomata close when K^+ leaves the guard cells. The loss causes the water potential (ψ) to become less negative in the guard cells, and more negative in the epidermal cells. As a consequence, water is lost by osmosis and the cells sag together and close the pore. The K^+ movements in and out of the guard cells are thought to be triggered by blue-light receptors in the plasma membrane, which activate the active transport mechanisms involved.

1. With respect to a mesophytic, terrestrial flowering plant:

   (a) Describe the **net** gas exchanges between the air and the cells of the mesophyll in the dark (no photosynthesis): ____________________

   (b) Explain how this situation changes when a plant is photosynthesizing: ____________________

2. Identify two ways in which the continuous air spaces through the plant facilitate gas exchange:

   (a) ____________________

   (b) ____________________

3. Briefly outline the role of stomata in gas exchange in an angiosperm: ____________________

4. Summarize the mechanism by which the guard cells bring about:

   (a) Stomatal opening: ____________________

   (b) Stomatal closure: ____________________

Transpiration

Plants lose water all the time, despite the adaptations they have to help prevent it (e.g. waxy leaf cuticle). Approximately 99% of the water a plant absorbs from the soil is lost by evaporation from the leaves and stem. This loss, mostly through stomata, is called **transpiration** and the flow of water through the plant is called the **transpiration stream**. Plants rely on a gradient in water potential (ψ) to move water through their cells. Water flows passively from soil to air along a gradient of decreasing water potential. The gradient in water potential is the driving force in the ascent of water up a plant. A number of processes contribute to water movement up the plant: transpiration pull, cohesion, and root pressure. Transpiration may seem to be a wasteful process, but it has benefits. Evaporation cools the plant and the transpiration stream helps the plant to maintain an adequate mineral uptake, as many essential minerals occur in low concentrations in the soil. Note: An alternative version of this activity, without reference to water potential, is available as a web link (see below) or on the Teacher Resource CD-ROM.

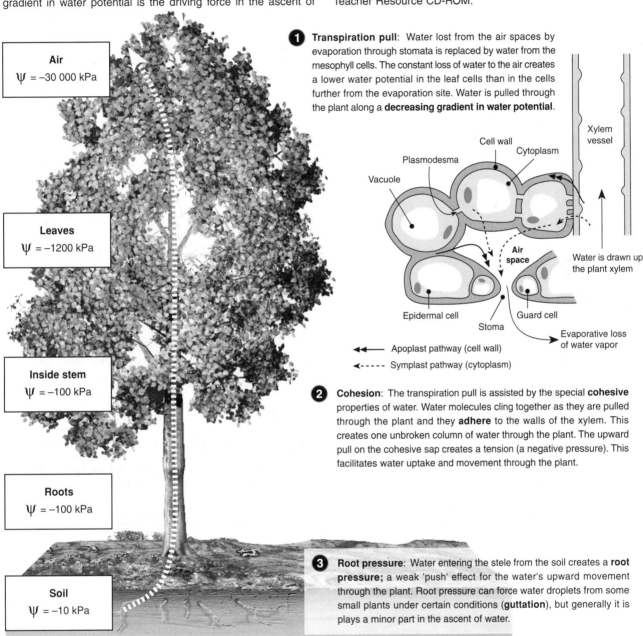

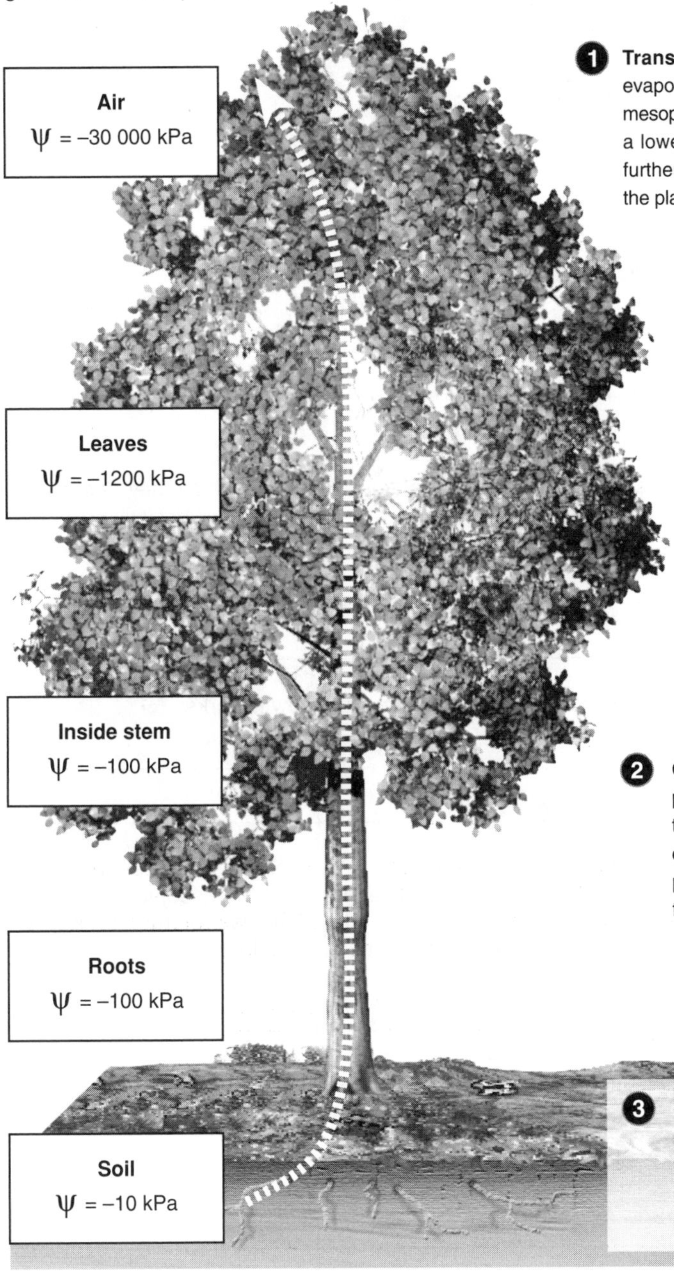

Air ψ = –30 000 kPa

Leaves ψ = –1200 kPa

Inside stem ψ = –100 kPa

Roots ψ = –100 kPa

Soil ψ = –10 kPa

1 Transpiration pull: Water lost from the air spaces by evaporation through stomata is replaced by water from the mesophyll cells. The constant loss of water to the air creates a lower water potential in the leaf cells than in the cells further from the evaporation site. Water is pulled through the plant along a **decreasing gradient in water potential**.

Xylem vessel
Cell wall
Cytoplasm
Plasmodesma
Vacuole
Air space
Water is drawn up the plant xylem
Epidermal cell
Guard cell
Stoma
Evaporative loss of water vapor
Apoplast pathway (cell wall)
Symplast pathway (cytoplasm)

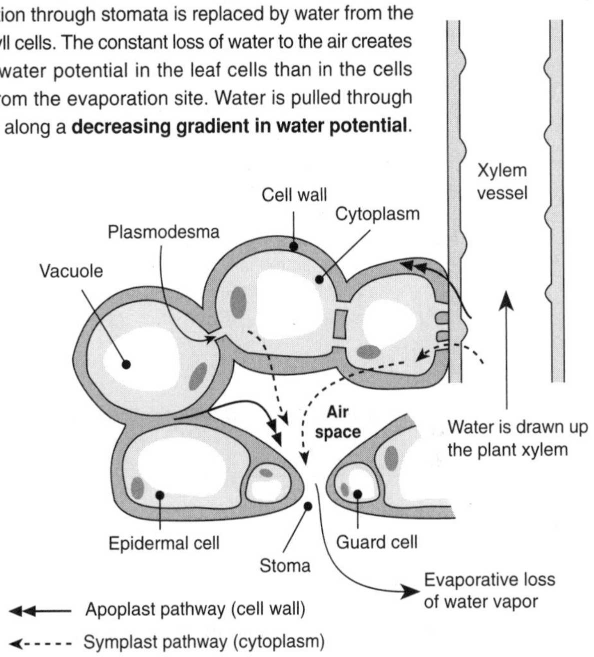

2 Cohesion: The transpiration pull is assisted by the special **cohesive** properties of water. Water molecules cling together as they are pulled through the plant and they **adhere** to the walls of the xylem. This creates one unbroken column of water through the plant. The upward pull on the cohesive sap creates a tension (a negative pressure). This facilitates water uptake and movement through the plant.

3 Root pressure: Water entering the stele from the soil creates a **root pressure;** a weak 'push' effect for the water's upward movement through the plant. Root pressure can force water droplets from some small plants under certain conditions (**guttation**), but generally it is plays a minor part in the ascent of water.

1. (a) Plants constantly lose water by transpiration. Explain how plants compensate for this: ______

   (b) Describe one benefit of the transpiration stream for a plant: ______

2. Briefly describe three processes that assist the transport of water from the roots of the plant upward:

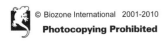

   (a) ______

   (b) ______

   (c) ______

The Potometer

A potometer is a simple instrument for investigating transpiration rate (water loss per unit time). The equipment is simple and easy to obtain. A basic potometer, such as the one shown right, can easily be moved around so that transpiration rate can be measured under different environmental conditions.

Some of the physical conditions investigated are:

- Humidity or vapor pressure (high or low)
- Temperature (high or low)
- Air movement (still or windy)
- Light level (high or low)
- Water supply

It is also possible to compare the transpiration rates of plants with different adaptations e.g. comparing transpiration rates in plants with rolled leaves vs rates in plants with broad leaves. If possible, experiments like these should be conducted simultaneously using replicate equipment. If conducted sequentially, care should be taken to keep the environmental conditions the same for all plants used.

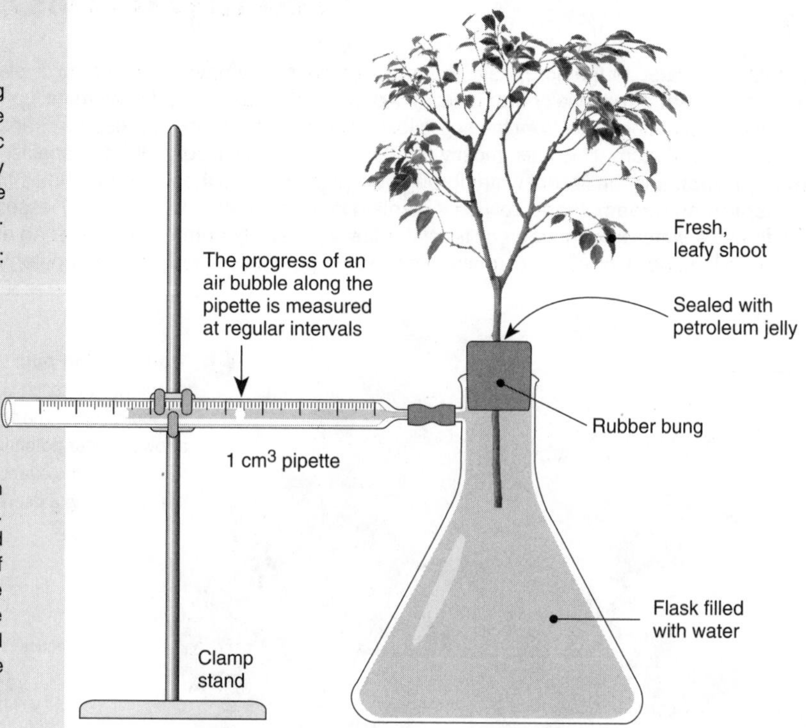

3. Describe three environmental conditions that increase the rate of transpiration in plants, and explain how they operate:

   (a) ______

   (b) ______

   (c) ______

4. The **potometer** (above) is an instrument used to measure transpiration rate. Briefly explain how it works:

   ______

   ______

5. An experiment was conducted on transpiration from a hydrangea shoot in a potometer. The experiment was set up and the plant left to stabilize (environmental conditions: still air, light shade, 20°C). The plant was then subjected to different environmental conditions and the water loss was measured each hour. Finally, the plant was returned to original conditions, allowed to stabilize and transpiration rate measured again. The data are presented below:

Experimental conditions	**Temperature** (°C)	**Humidity** (%)	**Transpiration** (gh^{-1})
(a) Still air, light shade, 20°C	18	70	1.20
(b) Moving air, light shade, 20°C	18	70	1.60
(c) Still air, bright sunlight, 23°C	18	70	3.75
(d) Still air and dark, moist chamber, 19.5°C	18	100	0.05

   (a) Name the control in this experiment: ______

   ______

   (b) Identify the factors that increased transpiration rate, explaining how each has its effect: ______

   ______

   ______

   (c) Suggest a possible reason why the plant had such a low transpiration rate in humid, dark conditions:

   ______

   ______

Translocation

Phloem transports the organic products of photosynthesis (sugars) through the plant in a process called **translocation**. In angiosperms, the sugar moves through the sieve elements, which are arranged end-to-end and perforated with sieve plates. Apart from water, phloem sap comprises mainly sucrose (up to 30%). It may also contain minerals, hormones, and amino acids, in transit around the plant. Movement of sap in the phloem is from a **source** (a plant organ where sugar is made or mobilized) to a **sink** (a plant organ where sugar is stored or used). Loading sucrose into the phloem at a source involves energy expenditure; it is slowed or stopped by high temperatures or respiratory inhibitors. In some plants, unloading the sucrose at the sinks also requires energy, although in others, diffusion alone is sufficient to move sucrose from the phloem into the cells of the sink organ. A version of this activity without reference to water potential is available as a web link (see below) or on the Teacher Resource CD-ROM.

Transport in the Phloem by Pressure-Flow

Phloem sap moves from source (region where sugar is produced or mobilised) to sink (region where sugar is used or stored) at rates as great as 100 m h^{-1}: too fast to be accounted for by cytoplasmic streaming. The most acceptable model for phloem movement is the **pressure-flow** (bulk flow) hypothesis. Phloem sap moves by bulk flow, which creates a pressure (hence the term "pressure-flow"). The key elements in this model are outlined below and in steps 1-4 right. For simplicity, the cells that lie between the source or sink cells and the phloem sieve-tube have been omitted.

1. Loading sugar into the phloem from a source (e.g. leaf cell) increases the solute concentration (decreases the water potential, ψ) inside the sieve-tube cells. This causes the sieve-tubes to take up water from the surrounding tissues by osmosis.
2. The water absorption creates a hydrostatic pressure that forces the sap to move along the tube (bulk flow), just as pressure pushes water through a hose.
3. The gradient of pressure in the sieve tube is reinforced by the active unloading of sugar and consequent loss of water by osmosis at the sink (e.g. root cell).
4. Xylem recycles the water from sink to source.

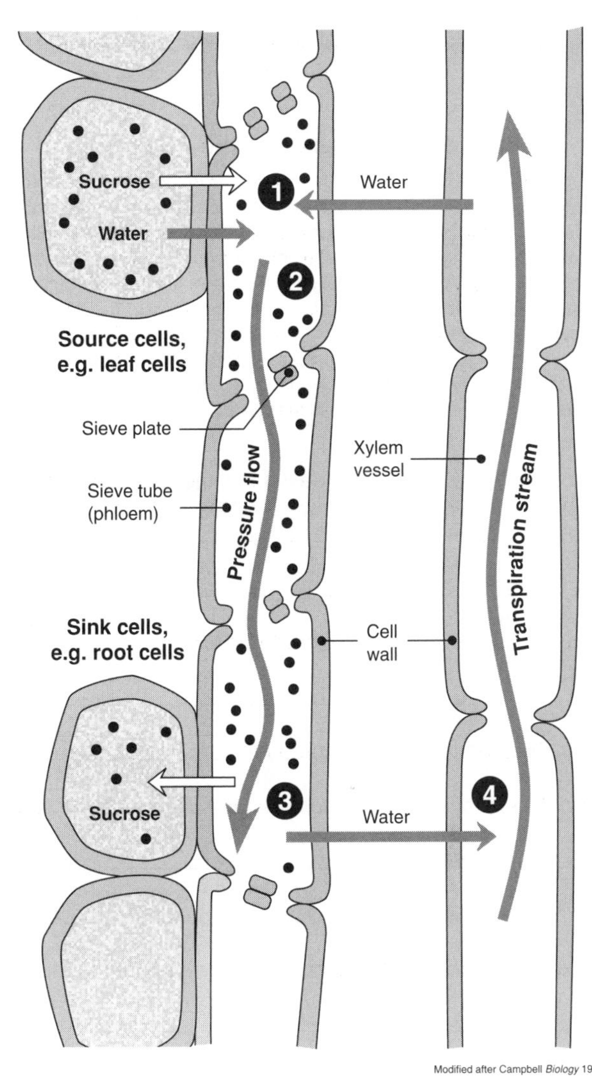

Modified after Campbell *Biology* 1993

Measuring Phloem Flow

Experiments investigating flow of phloem often use aphids. Aphids feed on phloem sap (left) and act as natural **phloem probes**. When the mouthparts (stylet) of an aphid penetrate a sieve-tube cell, the pressure in the sieve-tube force-feeds the aphid. While the aphid feeds, it can be severed from its stylet, which remains in place in the phloem. The stylet serves as a tiny tap that exudes sap. Using different aphids, the rate of flow of this sap can be measured at different locations on the plant.

1. (a) Explain what is meant by '**source to sink**' flow in phloem transport: ______

   (b) Name the usual **source** and **sink** in a growing plant:

   Source: ______ Sink: ______

   (c) Name another possible **source** region in the plant and state when it might be important: ______

   (d) Name another possible **sink** region in the plant and state when it might be important: ______

2. Explain why energy is required for translocation and where it is used: ______

Periodicals: *High tension*

Related activities: *Xylem, Phloem, Modifications in Plants*
Web links: *Translocation, Sucrose Transport*

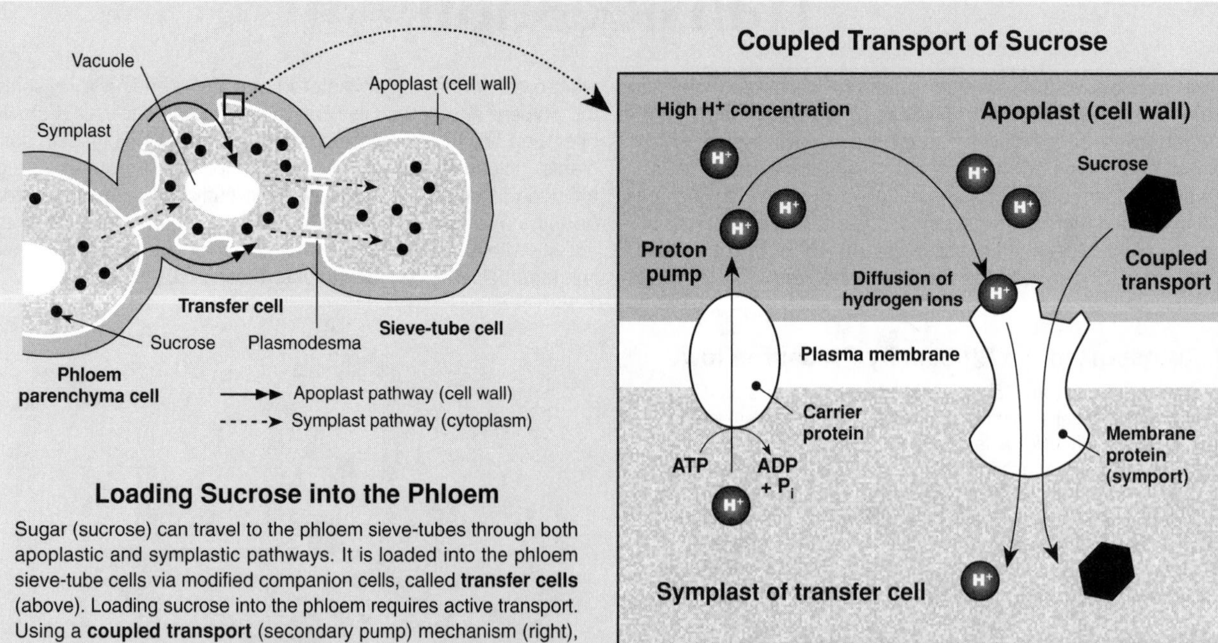

Loading Sucrose into the Phloem

Sugar (sucrose) can travel to the phloem sieve-tubes through both apoplastic and symplastic pathways. It is loaded into the phloem sieve-tube cells via modified companion cells, called **transfer cells** (above). Loading sucrose into the phloem requires active transport. Using a **coupled transport** (secondary pump) mechanism (right), transfer cells expend energy to accumulate the sucrose. The sucrose then passes into the sieve tube through plasmodesmata. The transfer cells have wall ingrowths that increase surface area for the transport of solutes. Using this mechanism, some plants can accumulate sucrose in the phloem to 2-3 times the concentration in the mesophyll.

Above: Proton pumps generate a hydrogen ion gradient across the membrane of the transfer cell. This process requires expenditure of energy. The gradient is then used to drive the transport of sucrose, by coupling the sucrose transport to the diffusion of H^+ back into the cell.

3. In your own words, describe what is meant by the following:

   (a) Translocation: ______________________________

   ____

   (b) Pressure-flow movement of phloem: ______________________________

   ____

   (c) Coupled transport of sucrose: ______________________________

4. Briefly explain why water follows the sucrose as the sucrose is loaded into the phloem sieve-tube cell:

   ____

   ____

5. Explain the role of the companion (transfer) cell in the loading of sucrose into the phloem: ______________________________

   ____

   ____

6. Contrast the composition of phloem sap and xylem sap (see the activities on xylem and phloem if you need help):

   ____

   ____

7. Explain why it is necessary for phloem to be alive to be functional, whereas xylem can function as a dead tissue:

   ____

   ____

8. The sieve plate represents a significant barrier to effective mass flow of phloem sap. Suggest why the presence of the sieve plate is often cited as evidence against the pressure-flow model for phloem transport:

   ____

   ____

   ____

KEY TERMS Mix and Match

INSTRUCTIONS: Test your vocab by matching each term to its correct definition, as identified by its preceding letter code.

- BULK FLOW ....
- CAPILLARY ACTION ....
- CARBON DIOXIDE ....
- CASPARIAN STRIP ....
- COHESION-TENSION HYPOTHESIS ....
- COMPANION CELLS ....
- CORTEX ....
- ENDODERMIS ....
- FLACCID ....
- GUARD CELLS ....
- MASS FLOW HYPOTHESIS ....
- MINERAL ....
- OSMOSIS ....
- PHLOEM ....
- ROOT HAIR ....
- ROOT PRESSURE ....
- SIEVE TUBE CELLS ....
- STOMATA ....
- TRACHEID ....
- TRANSLOCATION ....
- TRANSPIRATION ....
- TRANSPIRATION PULL ....
- TURGID ....
- VESSEL ....
- WATER POTENTIAL ....
- XYLEM ....

A Without turgor; limp.

B Water conducting cells in the xylem of angiosperms but absent from most gymnosperms.

C The passive movement of water molecules across a partially permeable membrane down a water potential gradient.

D **The** outer layer of a plant stem or root, bounded on the outside by the epidermis and on the inside by the endodermis.

E One proposed mechanism for the movement of sugars from source to sink in the phloem.

F A measure of the potential energy of water per unit volume relative to pure water. It quantifies the tendency of water to move from one area to another.

G Complex plant tissue specialized for the transport of water and dissolved mineral ions.

H Elongated cells in phloem for transporting carbohydrate (sugar).

I Living cells in close association with the sieve-tube members in phloem.

J Specialized cells, which occur in pairs and which regulate movement of gases and water vapor through the stomata.

K Pores in the leaf surface through which gases can move.

L The tendency of fluids in narrow tubes to move upwards, against the pull of gravity.

M The hypothesis for the movement of water through the plant based on transpiration pull and the cohesive properties of water.

N This gas is required for photosynthesis and enters the plant by diffusion.

O Movement of water and solutes together as a single mass due to a pressure gradient.

P A thin layer of parenchyma tissue found just outside the vascular cylinder in roots, which helps to regulate the passive movements of water and ions,.

Q Vascular tissue that conducts sugars through the plant. Characterized by the presence of sieve tubes.

R The transport of materials within a plant by the phloem.

S Elongated cells in the xylem that transport water and mineral salts.

T One of the many chemical elements required by living organisms, other than the four elements carbon, hydrogen, nitrogen, and oxygen present in common organic molecules.

U A force resulting from the evaporation of water from the surfaces of cells in the interior of the leaves.

V A term meaning swollen or tight.

W Osmotic pressure within the cells of a root system that aids the movement of water through the plant. It occurs when the soil moisture level is high during the night or when transpiration is low during the day.

X **A** tubular outgrowth of an epidermal cell on a plant root.

Y A band of waterproof material in the radial and transverse walls of the endodermis, which blocks the passive movement of materials, such as water and solutes into the stele.

Z The loss of water vapor by plants, mainly from leaves via the stomata.

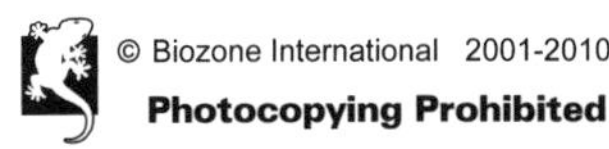

AP Topic 3B

IB AHL 9.3

Plant Reproduction

KEY CONCEPTS

- Alternation of generations is a characteristic feature of plant life cycles.
- The sporophyte generation dominates the life cycle of angiosperms.
- Diversity in flower structure reflects the diversity in methods of pollination.
- Fruit diversity reflects the different ways in which plants spread their seeds.
- The seed houses a dormant embryonic plant.

KEY TERMS

anemophily
anther
calyx
carpel
cotyledon
cross-pollination
double fertilization
egg
endosperm
entomophily
fertilization
filament
floret
flower
fruit
germination
inflorescence
meiosis
micropyle
mitosis
ovary
ovule
petal
plumule
pollen
pollen tube
pollination
radicle
seed
sepal
sporophyte
stamen
stigma
style
testa
zygote

Periodicals: *listings for this chapter are on page 390*

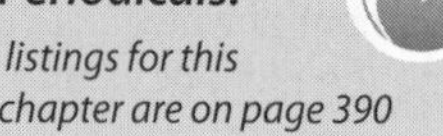

Weblinks: *www.thebiozone.com/weblink/SB2-2603.html*

OBJECTIVES

□ 1. Use the **KEY TERMS** to help you understand and complete these objectives.

Plant Life Cycles

pages 267-268, 295-298

□ 2. Describe **alternation of generations** as a distinctive feature of plants. Explain the significance of the **haploid** (**gametophyte**) and **diploid** (**sporophyte**) generations and their relative dominance in different phyla.

□ 3. Describe alternation of generations in representative plant phyla.

□ 4. Describe the diversity of mechanisms for vegetative propagation in angiosperms and explain its adaptive value as a reproductive strategy.

Angiosperm Reproduction

pages 277-278, 297-305

□ 5. Recall the dominance of the sporophyte generation in the life cycle of an angiosperm. Describe the relationship between the adult plants, **egg** and **pollen**, **zygote**, and **embryo** in the angiosperm life cycle. Indicate where **mitosis**, **meiosis**, and **fertilization** occur.

□ 6. Describe and explain the structure and function of reproductive structures (**flowers** and associated structures) in insect-pollinated and/or wind-pollinated angiosperms.

□ 7. Explain how flowering is controlled by the action of phytochrome.

□ 8. Describe **pollination** and the events leading to **fertilization** in an angiosperm. Include reference to the development of the **pollen tube**, and the significance of the **double fertilization**.

□ 9. Explain the role of **cross pollination** in angiosperm reproduction. Describe and explain diversity in the mechanisms for achieving cross pollination.

□ 10. Explain the purpose of the **fruit** in the angiosperm life cycle. Describe stages in fruit development. Explain the relationship between the type of fruit produced by a plant and its method of **seed dispersal**.

□ 11. Explain the purpose of a **seed**. Describe the internal and external structure of a seed in a monocot and/or a dicot (e.g. maize and/or bean).

□ 12. Describe and explain diversity in methods of seed dispersal in seed-producing plants (gymnosperms and angiosperms).

□ 13. Describe and explain **germination** in a starchy seed. Describe the metabolic processes involved during germination, explaining how the food store is mobilized to supply the growing plant.

Alternation of Generations

The life cycles of all plants include a gametophyte generation (haploid or N phase) and a sporophyte generation (diploid or 2N phase). The two generations alternate, each giving rise to the other (commonly termed **alternation of generations**). The two plant forms (sporophyte and gametophyte) are named for the type of reproductive cells they produce. Gametophytes produce gametes by mitosis, whereas sporophytes produce spores by meiosis. Spores develop directly into organisms. Gametes (egg and sperm) unite during fertilization to form a zygote which gives rise to an organism.

Algae

The sea lettuce *Ulva*, a green alga, has a life cycle alternating between two generations that seem to be the same. Although they appear identical, the cells of the sporophyte generation contain **26** chromosomes, while the gametophyte cells contain only **13** chromosomes. The sporophyte produces spores that settle on rock surfaces to grow into male and female gametophytes. These in turn mature and release gametes into the water. In fertilization, two gametes fuse and grow into a sporophyte.

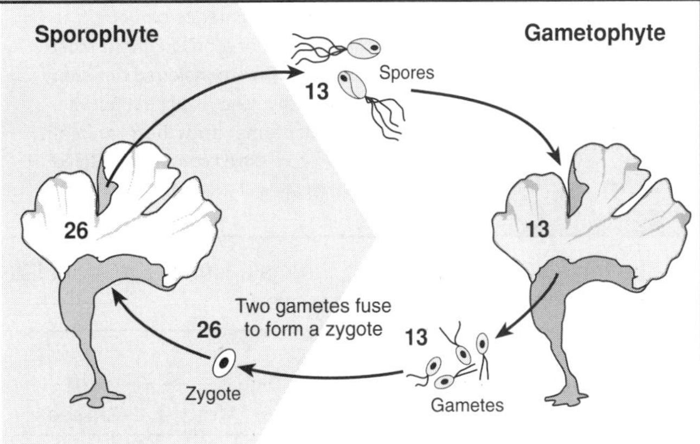

Moss

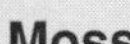

Most moss species have separate male and female gametophytes. A sperm swims through a film of moisture to the female gametophyte (archegonium) and fertilizes an egg. The resulting zygote develops into an embryonic sporophyte on top of the female gametophyte. It grows a long stalk that has a spore-producing capsule (sporangium) at the tip. When the sporangium matures, it bursts, scattering the spores. Those landing on moist soil will germinate to form a new gametophyte.

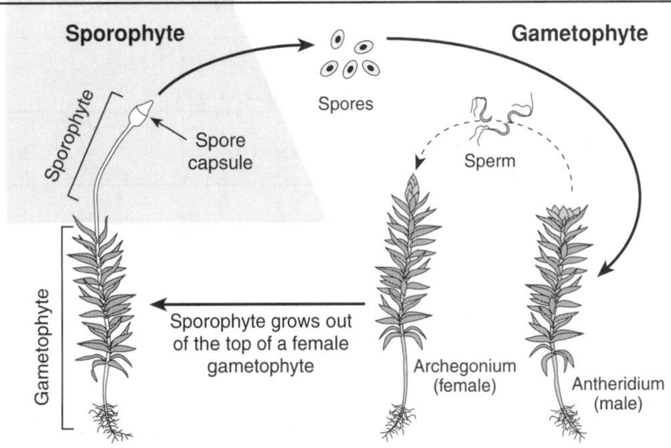

Fern

Ferns are sporophytes. Spores are formed on the back of the fronds and are released into the air to be dispersed by the wind. If a spore lands on moist soil, it will germinate into a very small **prothallus**. On this flat, heart-shaped gametophyte, male and female organs develop, but at different times. As with mosses, fern sperm use flagella to swim through moisture to the female cells to fertilize them. A new sporophyte grows out of the female organ (archegonium) and matures into a fern.

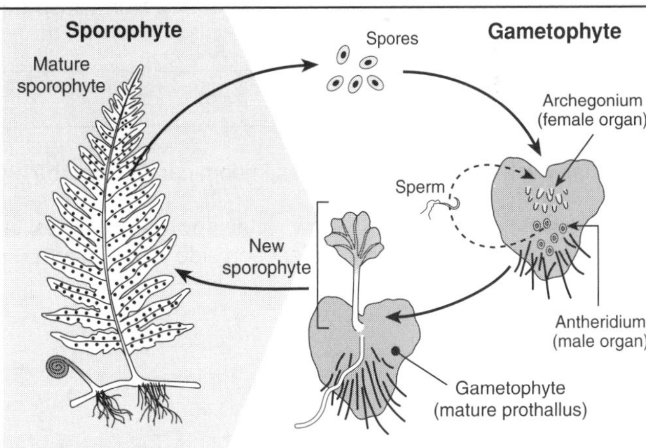

Gymnosperm

Trees are sporophytes. Most gymnosperm species produce both pollen cones and ovulate (female) cones. Male cones produce hundreds of pollen grains (male gametophytes). A female cone contains many scales, each with two ovules. Wind blown pollen falls on the female cone and is drawn into the ovule through a tiny opening called the micropyle. The pollen grain germinates in the ovule and grows a pollen tube that seeks out the female gametophyte and fertilizes the egg cell within it.

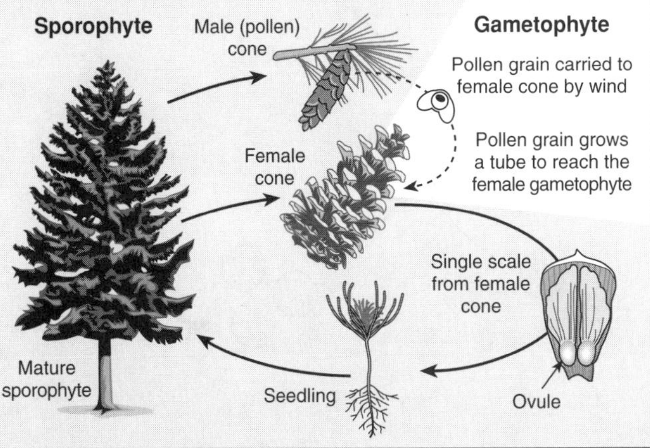

Related activities: *Angiosperm Reproduction*
Web links: *Fern Life Cycle*

Angiosperm

Angiosperms, or flowering plants, are the most successful plants on Earth today. They reproduce sexually by forming flowers, fruits, and seeds. The sporophyte generation is clearly dominant, and the gametophyte generation is reduced in size to just a small number of cells (there are no archegonia or antheridia). Anthers develop an immature gametophyte in the form of a pollen grain. Most often, pollen is transferred between flowers by wind or animal activity; angiosperms and their animal pollinators exhibit coevolution. Some plants are self-pollinating.

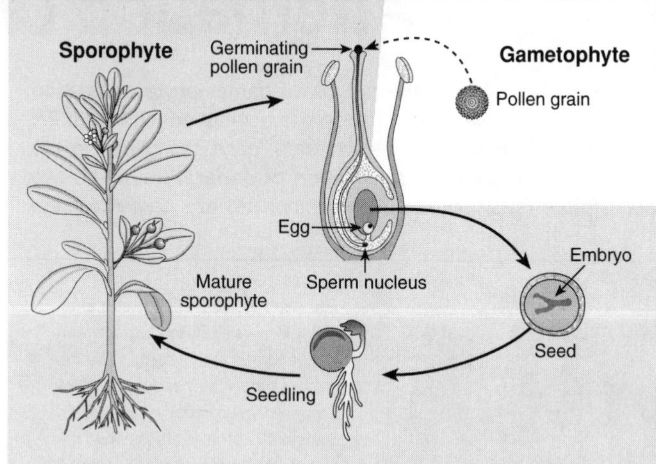

1. The table below summarizes the main features of plant life cycles. Using brief explanations, complete the table (the first example has been completed for you):

Life cycle feature	Taxonomic Group				
	Algae	Mosses	Ferns	Gymnosperms	Angiosperms
Dominant generation	Gametophyte				
Alternate generation	Sporophyte (zygospore)				
Movement of sperm	Needs water				
Gametophyte reliance on sporophyte	None				
Sporophyte reliance on gametophyte	None				
Ecological niche with respect to reproduction	Water needed				

2. One of the principal trends evident in plant life cycles is the increasing independence on water for reproduction. Explain what feature of the male gamete (sperm or pollen) illustrates this:

____

____

3. State which generation increases in dominance from the algae to the angiosperms: ____________________

4. In the schematic diagram below, label: **spores**, **gametes**, and **zygote**. Beside each, indicate the chromosome state: haploid (**N**) or diploid (**2N**). Label each side of the diagram (gray and white) with haploid or diploid as appropriate:

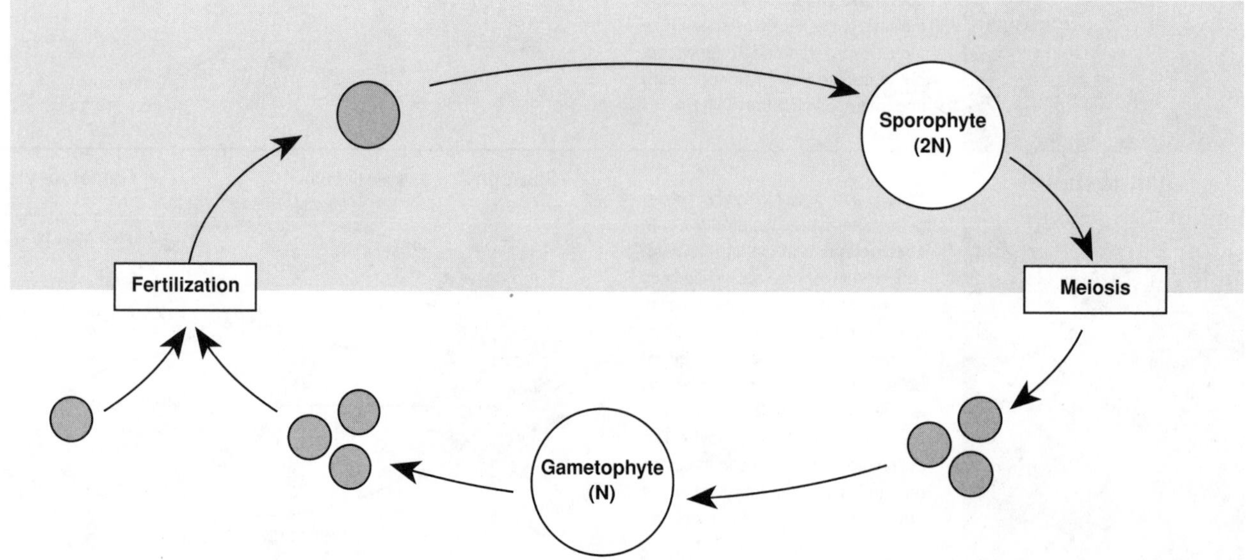

Angiosperm Reproduction

The primary method of reproduction for flowering plants is by seeds, which develop after fertilization of the female parts of the flower, and contain the protected plant embryo together with a store of food. The typical life cycle of a flowering plant such as the bean (below) involves the formation of gametes (egg and sperm) from the haploid gametophytes, the fertilization of the egg by a sperm cell to form the zygote, the production of fruit around the seed, and the germination of the seed and its growth by mitosis. The eggs and sperm are housed within the female and male gametophytes (embryo sac and pollen grain respectively). Each mature pollen grain contains two sperm nuclei and each mature embryo sac contains the egg and two polar nuclei. The leafy plant bearing the flowers represents the sporophyte generation of the life cycle.

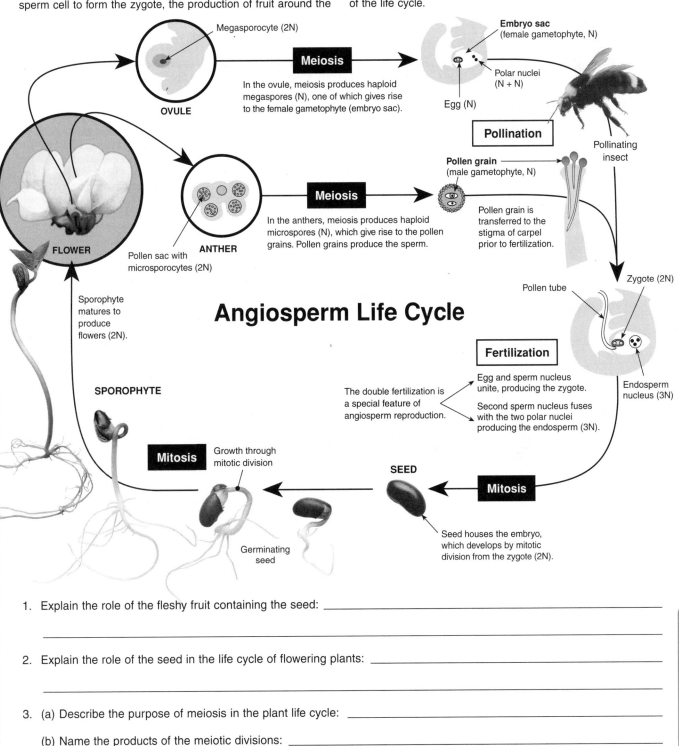

1. Explain the role of the fleshy fruit containing the seed: ____________________

____

2. Explain the role of the seed in the life cycle of flowering plants: ____________________

____

3. (a) Describe the purpose of meiosis in the plant life cycle: ____________________

   (b) Name the products of the meiotic divisions: ____________________

   (c) Explain the advantage of seed production through sexual reproduction over vegetative propagation:

   ____

   ____

4. Explain the purpose of mitosis in the plant life cycle: ____________________

____

Related activities: *Alternation of Generations, Seed Structure and Germination*
Web links: *Cycle of an Angiosperm, Plant reproduction Animation*

A Most Accomplished Traveler

The origin of the coconut (*Cocos nucifera*) is one of botany's mysteries. It is so extensively cultivated and so widespread in the wild, determining its origin and dispersal around the globe is extremely difficult. Suggestions have been made that the coconut originated on the coastline of the Gondwanan continent, and spread to volcanic islands where competition had been eliminated by volcanic activity. It has also been suggested that the coconut originated in South Asia or South America. Fossils show that it has been wide spread for some time, with the oldest known fossils of coconut-like palm trees found in Bangladesh and fossils from New Zealand showing it was established there some 15 million years ago.

Coconuts are a single seeded fruit (a type known as a drupe and not, in fact, a nut at all), most commonly seen as the seed with the fibrous husk removed. The coconut fruit possesses a number of features that have allowed it to spread throughout the tropics. The fibrous husk allows it to float and keeps out the seawater. Its oval shape is very stable and allows it to ride high in the water. The seed is the largest of any plant except the coco-de-mer (*Lodoicea maldivica*) and it has a thin but tough shell. The endosperm takes up only a small lining inside the seed, leaving a hollow that is filled with liquid. As the seed matures on its voyage across the sea, the liquid is absorbed. This adds to the buoyancy of the fruit. Additionally, the seed takes a long time to germinate, from 30 to 220 days. The seed is never dormant, as this is not necessary in an equable tropical climate, so the long germination period is an adaptation to extended periods of ocean-going travel between islands. Coconuts can still be viable after travelling for as long as 200 days and covering up to 4000 km.

Above: Coconut palms fringing a beach, Thailand

Right: Coconut germinating on black sand on the island of Hawaii

Wiki: Wmpearl

Before the arrival of humans in the Pacific, coconuts were already widespread, but because of its valuable features, it has been even more widely dispersed by humans, both in prehistory and in modern times. Not only is it a portable, storable food and water source conveniently sealed in a hard shell, but the husk fibers can be used for making ropes, bedding, and many other products. Coconut oils, flesh, and fibers are still extensively used today.

The only tropical coastlines the coconut failed to reach were those of the Atlantic and Caribbean as these bodies of water do not mix with the Pacific or Indian Oceans except in polar regions. However, the arrival of Europeans around 1500 AD, soon saw the coconut cross the isthmus of Central America to colonise new coastlines. It is now common in every tropical and subtropical region on Earth.

1. Explain why the origin of the coconut palm is difficult to determine: ______

2. Describe the features of the coconut that allowed it to spread across the Pacific and Indian Oceans: ______

3. Explain why humans found the coconut so useful: ______

4. Explain why the coconut never established in the Atlantic before humans introduced it. ______

Related activities: *Seed Dispersal*

Wind Pollinated Flowers

The flowers of wind pollinated plants are not as brightly coloured or as large as those of insect pollinated plants. In general, their flowers are small and may lack petals altogether, while the anthers and stigma are large and hang clear of the surrounding structures. Because wind pollen is highly inefficient, with most pollen falling to the ground within one hundred metres, the anthers also produce large amounts of pollen to ensure cross pollination with other plants. Only around 10% of flowering plants are wind pollinated, but this 10% includes the grasses, arguably one of the most successful of all the angiosperm groups.

The Structure of Wind Pollinated Flowers

Awn

Lemma

Palea

Rachilla

Bract (glume)

Oat spikelets

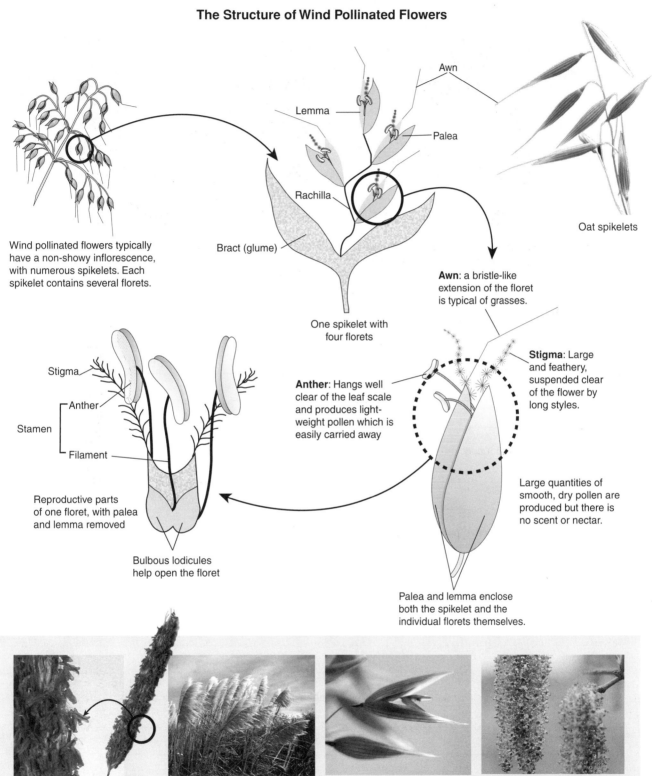

Wind pollinated flowers typically have a non-showy inflorescence, with numerous spikelets. Each spikelet contains several florets.

One spikelet with four florets

Awn: a bristle-like extension of the floret is typical of grasses.

Stigma

Anther

Stamen

Filament

Anther: Hangs well clear of the leaf scale and produces light-weight pollen which is easily carried away

Stigma: Large and feathery, suspended clear of the flower by long styles.

Large quantities of smooth, dry pollen are produced but there is no scent or nectar.

Reproductive parts of one floret, with palea and lemma removed

Bulbous lodicules help open the floret

Palea and lemma enclose both the spikelet and the individual florets themselves.

As with most wind pollinated plants, grass flowers are small, but are produced in large numbers. The arrow in the photograph above points to the anthers of a grass plant, which are held above the ground on a stalk.

Grasses grow in stands that can cover hundreds of km^2. This gives a much greater chance of pollen reaching the stigma of a different plant. Pampas grass (above) grows to 4 m and produces huge plumes that hold the flowers up in the wind.

Cereal plants, being grasses, are pollinated by wind. Oats produce inflorescences as spikelets which contain smaller florets which hold the anthers and stigma. The photo above clearly shows the bract opening to reveal the florets inside.

The flowers of wind-pollinated trees are small, often with green petals, and very numerous. Wind pollinated trees are common in temperate regions where wind movements through deciduous trees easily disperses the pollen.

Related activities*: Insect Pollinated Flowers, Pollination and Fertilization*

1. Describe three differences between the insect (left) and wind (right) pollinated flowers shown in the photographs below:

(a)

(b)

(c)

2. Using the photographs and your answers above to help you, contrast wind and insect pollinated flowers with respect to each of the following characteristics. For each, give reasons for the differences observed:

(a) Appearance of the flowers:

(b) Production of scent and nectar:

(c) Amount of pollen produced:

(d) Position of the reproductive parts (stigma, stamens):

3. Describe two adaptations of wind pollinated flowers:

(a)

(b)

4. Contrast the efficiency of wind and animals as pollinating agents, giving a reason for your answer:

5. Describe the main the advantage of **cross pollination** and discuss the ways in which plants can ensure this occurs:

Insect Pollinated Flowers

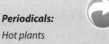

Flowering plants (**angiosperms**) are highly successful organisms. The egg cell is retained within the flower of the parent plant and the male gametes (contained in the **pollen**) must be transferred to it by **pollination** in order for fertilization to occur. Most angiosperms are **monoecious**, with male and female parts on the same plant. Some of these plants will self-pollinate, but most have mechanisms that make this difficult or impossible. The female and male parts may be physically separated in the flower, or they may mature at different times. **Dioecious plants** avoid this problem by carrying the male and female flowers on separate plants. Flowers are pollinated in three different ways (animal, wind or water) and their structures differ accordingly. Of the animal pollinators, insects provide the greatest effectiveness of pollination as well as the most specialised pollination. Flowers attract insects with brightly coloured petals, scent, and offers of food such as nectar and pollen.

Cross Section of an Insect Pollinated Flower

Stigma: The receptive part of the carpel. Pollen grains will germinate only if they land here.

Style: The structure that supports the stigma.

Ovary: The base of the carpel where the ovules develop.

Ovules: These are eggs and once fertilised, become the seeds. The ovule skin becomes the seed coat or testa

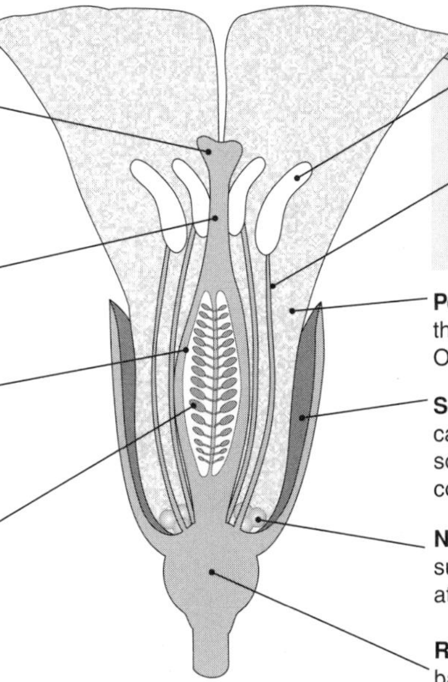

Anther: Top portion of the stamen, the male organ of reproduction.

Filament: The slender stalk of the stamen that supports the anther.

Petals: Collectively, these form the corolla. Often brightly coloured

Sepals: Together form the calyx. Usually green, but sometimes the same colour as the petals

Nectary: Plants produce a sugary liquid called nectar to attract insects to the flower.

Receptacle: The swollen base of the flower. Sometimes it forms the succulent tissue of the fruit.

An entire female part is the carpel. There may be one or more carpels per flower.

Magnolias are an ancient plant group, with very generalised flowers that are accessible to their beetle pollinators.

Orchids are well known for the many structural variations in their flowers, which are often highly specialised. They frequently have only one specific insect pollinator; a relationship that has arisen through coevolution.

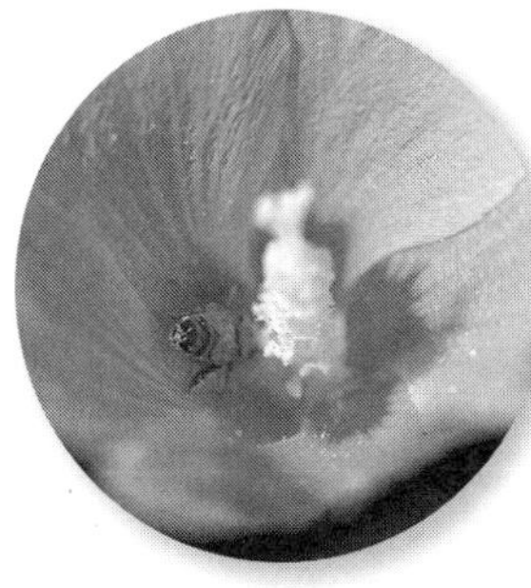

The petals of flowers guide insects towards the pollen or nectar at the centre of the flower using various colours and lines known as nectar guides. In this way, wandering insects are enticed into entering the flower and transfer pollen in the most efficient way.

Bees and many other insects are able to detect ultraviolet light. Many flowers contain pigments that reflect UV producing a specific pattern visible to insect but not to other animals. In this way, plants can use their flowers to specifically attract preferred insect pollinators.

1. Describe the difference between moecious and dioecious plants: ______________________

2. Describe the difference between the stigma and the anther: ______________________

3. Discuss how flowers are used to attract specific insect pollinators to a plant: ______________________

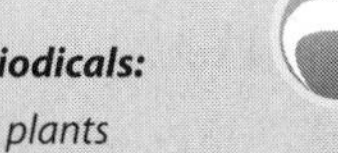

Pollination and Fertilization

Before the egg and sperm can fuse in fertilization, the pollen (which contains the male gametes) must be transferred from the male anthers to the female stigma in **pollination**. Plants rarely self-pollinate, although they can be made to do so. Most often the stigma of one plant receives pollen from other plants in **cross-pollination**. After pollination, the sperm nuclei can enter the ovule and fertilization can occur. In angiosperms, there is a double fertilization: one to produce the embryo and the other to produce the endosperm nucleus. The endosperm nucleus gives rise to the endosperm: the food store for the embryonic plant.

Growth of the pollen tube and double fertilization

Pollen grains are immature male gametophytes, formed by mitosis of haploid microspores within the pollen sac. Pollination is the actual transfer of the pollen from the stamens to the stigma. Pollen grains cannot move independently. They are usually carried by wind (**anemophily**) or animals (**entomophily**). After landing on the sticky stigma, the pollen grain is able to complete development, germinating and growing a pollen tube that extends down to the ovary. Directed by chemicals (usually calcium), the pollen tube enters the ovule through the **micropyle**, a small gap in the ovule. A **double fertilization** takes place. One sperm nucleus fuses with the egg to form the zygote. A second sperm nucleus fuses with the two polar nuclei within the embryo sac to produce the endosperm tissue (3N). There are usually many ovules in an ovary, therefore many pollen grains (and fertilizations) are needed before the entire ovary can develop.

Different pollens are variable in shape and pattern, and genera can be easily distinguished on the basis of their distinctive pollen. This feature is exploited in the relatively new field of forensic botany; the tracing of a crime through botanical evidence. The species specific nature of pollen ensures that only genetically compatible plants will be fertilized. Some species, such as *Primula*, produce two pollen types, and this assists in cross pollination between different flower types.

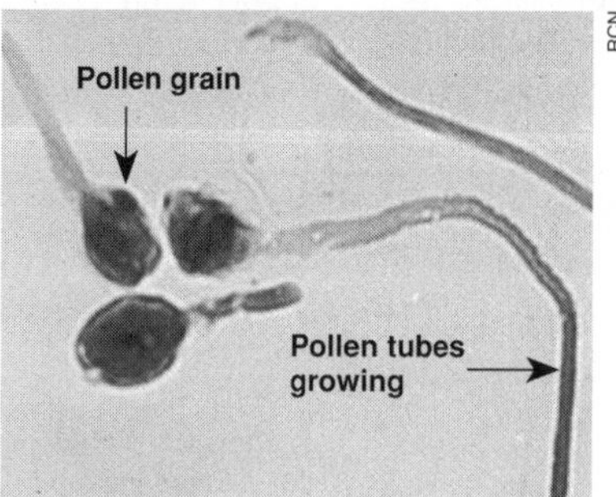

Germinating pollen grains

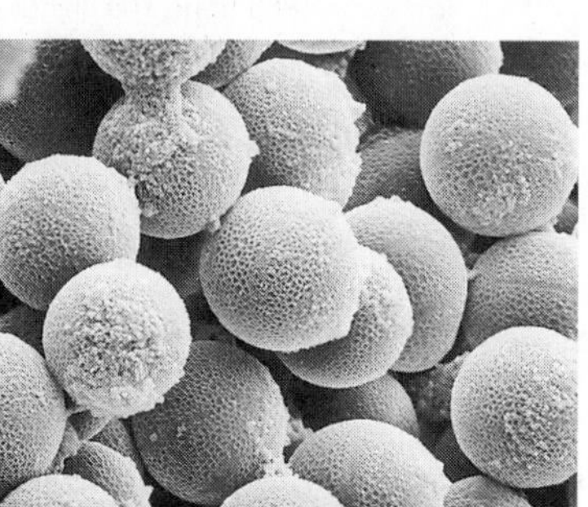

SEM: *Primula* (primrose) pollen

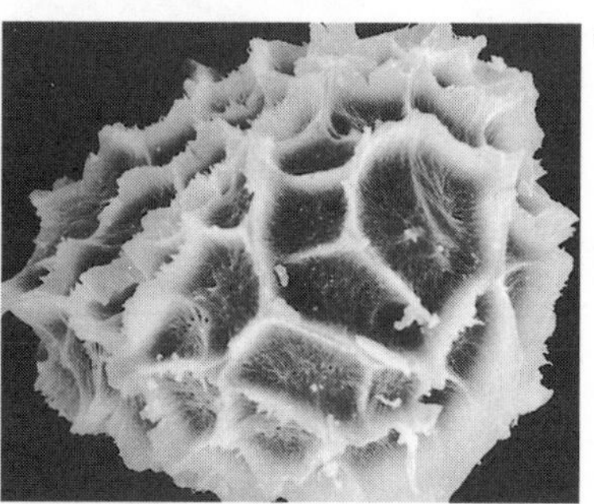

SEM: Dandelion pollen

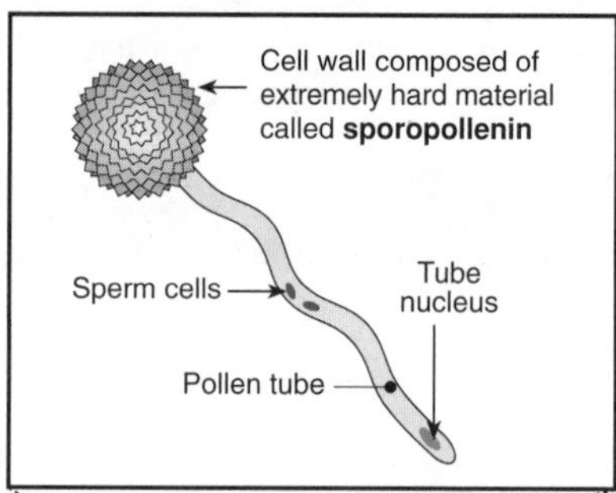

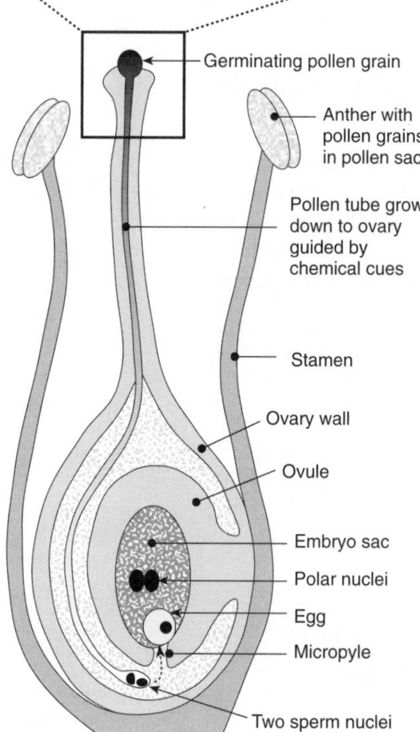

1. Distinguish clearly between **pollination** and **fertilization**:

2. Describe the role of the double fertilization in angiosperm reproduction:

3. Name the main chemical responsible for pollen tube growth:

4. Suggest a reason for the great variability seen in the structure of pollen grains:

5. Pollen can be used as an indicator of past climates and vegetation. Give two reasons why pollen is well suited to this use:

   (a)

   (b)

Fruits

A **fruit** is a mature, ripened ovary, although other plant parts, in addition to the ovary, may contribute to the fleshy parts of what we call the fruit. As a seed develops, the ovary wall around it enlarges and changes to become the fruit wall or **pericarp**. The pericarp has three regions: the outer exocarp, central mesocarp, and inner endocarp. Fruits may open to release the seeds or they may retain the seeds and be dispersed whole. They are classified according to the number of ovaries involved in their formation and the nature of the fruit wall (dry or fleshy). Succulent fruits are usually dispersed by animals and dry fruits by wind, water, or mechanical means. Fruits occur only in angiosperms. Their development has been a central feature of angiosperm evolution.

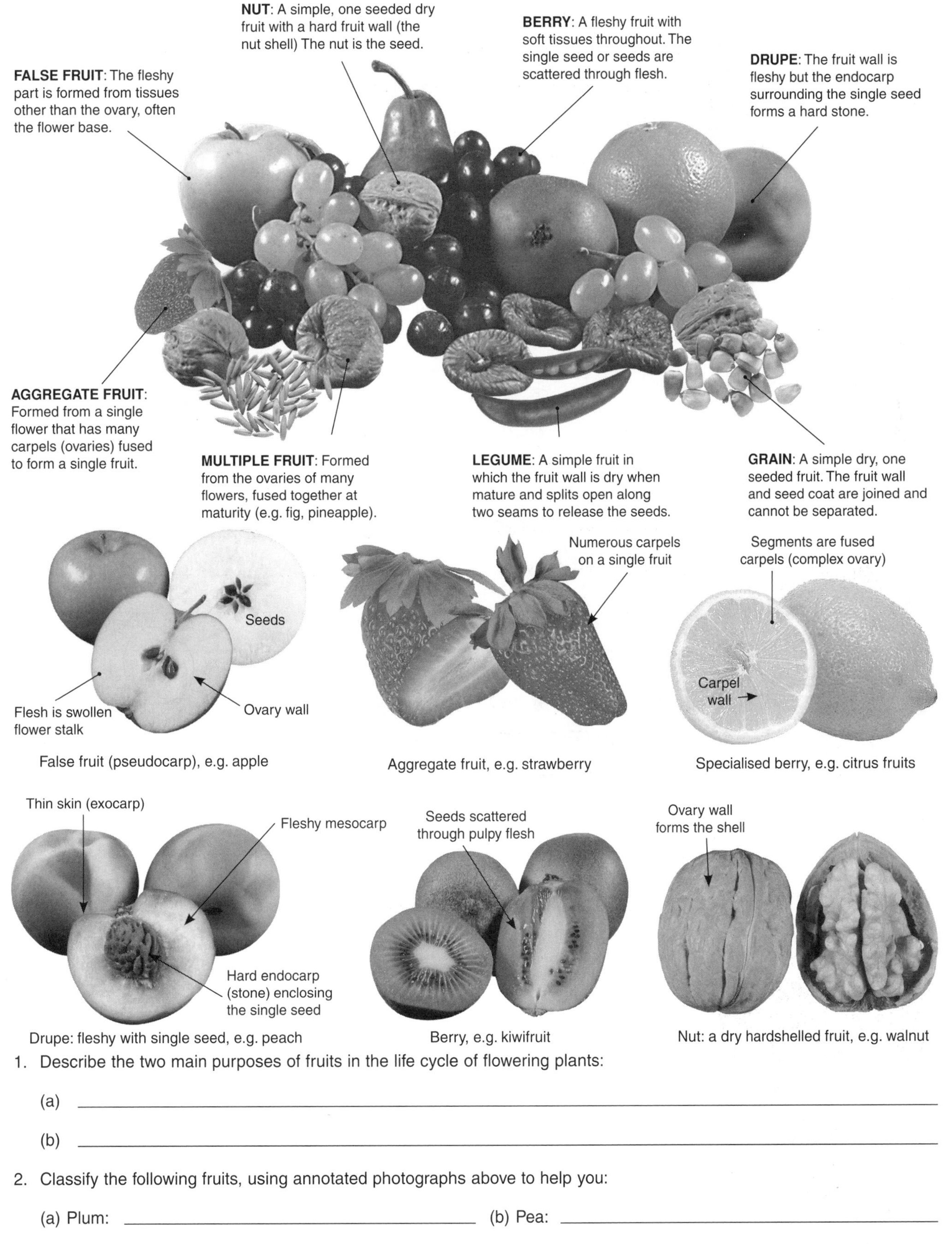
NUT: A simple, one seeded dry fruit with a hard fruit wall (the nut shell) The nut is the seed.
BERRY: A fleshy fruit with soft tissues throughout. The single seed or seeds are scattered through flesh.
FALSE FRUIT: The fleshy part is formed from tissues other than the ovary, often the flower base.
DRUPE: The fruit wall is fleshy but the endocarp surrounding the single seed forms a hard stone.
AGGREGATE FRUIT: Formed from a single flower that has many carpels (ovaries) fused to form a single fruit.
MULTIPLE FRUIT: Formed from the ovaries of many flowers, fused together at maturity (e.g. fig, pineapple).
LEGUME: A simple fruit in which the fruit wall is dry when mature and splits open along two seams to release the seeds.
GRAIN: A simple dry, one seeded fruit. The fruit wall and seed coat are joined and cannot be separated.
Seeds
Flesh is swollen flower stalk
Ovary wall
Numerous carpels on a single fruit
Segments are fused carpels (complex ovary)
Carpel wall
Thin skin (exocarp)
Fleshy mesocarp
Hard endocarp (stone) enclosing the single seed
Seeds scattered through pulpy flesh
Ovary wall forms the shell

False fruit (pseudocarp), e.g. apple

Aggregate fruit, e.g. strawberry

Specialised berry, e.g. citrus fruits

Drupe: fleshy with single seed, e.g. peach

Berry, e.g. kiwifruit

Nut: a dry hardshelled fruit, e.g. walnut

1. Describe the two main purposes of fruits in the life cycle of flowering plants:

   (a) ______________________________

   (b) ______________________________

2. Classify the following fruits, using annotated photographs above to help you:

   (a) Plum: ______________ (b) Pea: ______________

   (c) Raspberry: ______________ (d) Watermelon: ______________

Seed Dispersal

Flowering plants have evolved many ways to ensure that their seeds are dispersed. This has given them greater opportunities to expand their range. If a seed is carried into an area suitable for its germination, it will become established there. In some cases the seed itself is the agent of dispersal, but often it is the fruit. The chief agents of seed dispersal are wind, water, and animals. Many seeds are readily dispersed by water, even when they lack special buoyancy mechanisms. Wind also spreads the seeds of many plants. Such seeds have wing-like or feathery structures that catch the air currents and carry the seeds long distances. Plants that rely on animals to spread their seeds may have hooks or barbs that catch the animal hair, sticky secretions that adhere to the skin or hair, or fleshy fruits that are eaten leaving the seed to be deposited in faeces some distance from the parent plant. Other dispersal mechanisms rely on explosive discharge or shaking from pods or capsules (e.g. legumes, poppy).

For each of the examples below, describe the method of dispersal and the adaptive features associated with the method:

1. **Dandelion** seeds are held in a puff-like cluster:
   (a) Dispersal mechanism: ______
   (b) Adaptive features: ______

IMSI

2. **Acorns** are heavy fruits in which the fleshy seeds are encased in a resistant husk:
   (a) Dispersal mechanism: ______
   (b) Adaptive features: ______

3. **Coconuts** are heavy buoyant fruits with a thick husk:
   (a) Dispersal mechanism: ______
   (b) Adaptive features: ______

4. **Maple** fruits are winged, two-seeded samaras:
   (a) Dispersal mechanism: ______
   (b) Adaptive features: ______

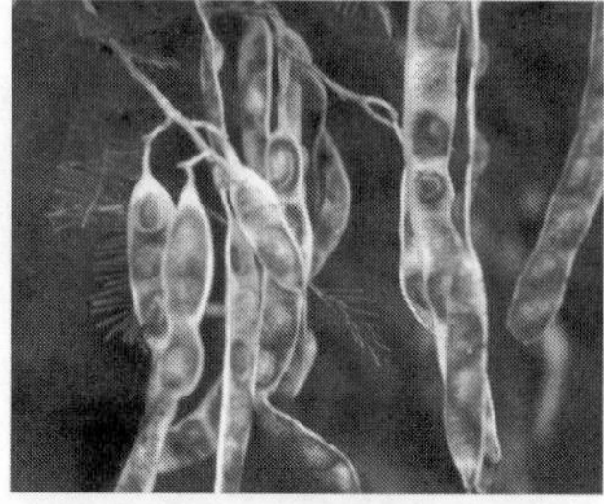

5. **Wattle** (*Acacia* spp.) seeds are enclosed in pods. A fleshy strip surrounds each seed:
   (a) Dispersal mechanism: ______
   (b) Adaptive features: ______

Karori Wildlife Sanctuary

6. **New Zealand flax** (*Phormium* spp.) produces seeds in pods:
   (a) Dispersal mechanism: ______
   (b) Adaptive features: ______

Seed Structure and Germination

After fertilization has occurred, the ovary develops into the fruit and the ovules within the ovary become the **seeds**. Recall that in plants there is double fertilization. One sperm fertilizes the egg to form the embryo, but another sperm combines with the diploid endosperm nucleus to form a large triploid cell which gives rise to the endosperm. The development of the endosperm is important and begins before embryonic development in order to produce a nutrient store for the young plant. A seed is an entire reproductive unit, housing the embryonic plant in a state of dormancy. During the last stages of maturing, the seed dehydrates until its water content is only 5-15% of its weight. The embryo stops growing and remains **dormant** until the seed germinates. At germination, the food store is mobilized to provide the nutrients for plant growth and development.

Dicot seeds: soy (above) cashew (below) There are two fleshy cotyledons. These store food that was absorbed from the endosperm.

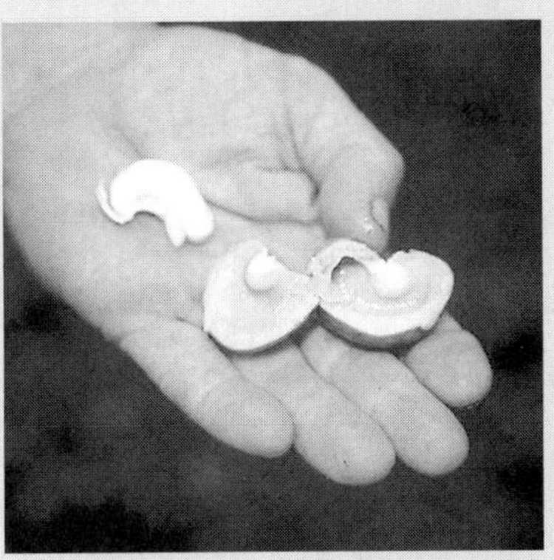

Germination requires rehydration of the seed and reactivation of the metabolism. The seed absorbs water through the seed coat (testa) and micropyle. As the dry substances in the seed tissue take up water, the cells expand, metabolism is reactivated, and embryonic growth begins. Activation begins with the release of gibberellin (GA) from the embryo. GA enhances cell elongation, making it possible for the root to penetrate the testa. It also stimulates the synthesis of enzymes, which hydrolyze the starch to produce sugars. The mobilized food stores are then delivered to the developing roots and shoots.

Seed Structure and Formation

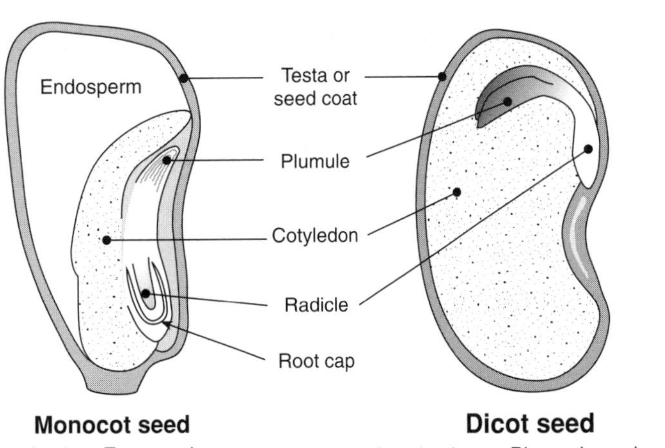

Monocot seed (maize: *Zea mays*)

Dicot seed (garden bean: *Phaseolus vulgaris*)

Every seed contains an embryo comprising a rudimentary shoot (plumule), root (radicle), and one or two cotyledons (seed leaves). The embryo and its food supply are encased in a tough, protective seed coat or **testa**. In monocots, the endosperm provides the food supply, whereas in most dicot seeds, the nutrients from the endosperm are transferred to the large, fleshy cotyledons.

Germination in a Dicot Seed
(garden bean: *Phaseolus vulgaris*)

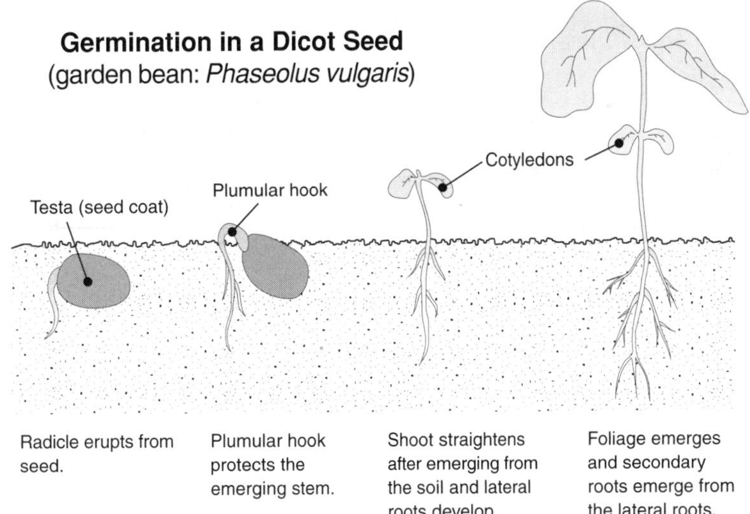

Radicle erupts from seed.

Plumular hook protects the emerging stem.

Shoot straightens after emerging from the soil and lateral roots develop.

Foliage emerges and secondary roots emerge from the lateral roots.

1. Explain the purpose of a **seed**: ______________________________

____

2. (a) State the function of the endosperm in angiosperms: ______________________________

(b) State how the endosperm is derived: ______________________________

3. Explain the role of the testa in seeds: ______________________________

4. Explain why the seed requires a food store: ______________________________

____

5. Explain why stored seeds must be kept dry: ______________________________

Related activities: *Pollination and Fertilization, Fruits*

Events in Germination

Seed germination refers to the beginning of seed growth. It involves a process of rehydration of the seed and reactivation of normal metabolism. The development of the **endosperm** begins before embryo development in order to produce a nutrient-rich store for the young plant. In dicots (e.g. beans), the endosperm is transferred to the two fleshy cotyledons during seed development and may disappear altogether. However, in moncots (grasses and grains) the endosperm remains as the main food source for the embryo. The single cotyledon is called the **scutellum** and the shoot (plumule) is sheathed in a **coleoptile**.

Metabolic Events in Germination

Germination begins when a mature seed begins to take up water through the micropyle and testa (**imbibition**). Imbibition causes the seed to swell and the testa to split. The food stored in the seed is hydrolyzed to produce substrates for respiration (e.g. glucose). Germinating seeds have a high oxygen requirement and respire rapidly. Water is essential to the germination process: it enables expansion of the growing cells and activates the enzymes needed for germination. Some of these enzymes are produced in response to the release of gibberellic acid (GA) from the growing embryo. Water is also required for the hydrolysis of stored starch and for the translocation of the mobilized food to the sites of growth.

The series **A** - **E** below shows the conditions within a **monocot seed** at the times that appear directly above on the graph. (**A**) GA is released from the embryo. (**B**) Digestive enzymes (filled black dots) produced by aleurone cells. (**C**) Enzymes mobilize food in the endosperm, releasing soluble nutrients (open dots). (**D**) Embryo's cotyledon absorbs food and delivers it to the shoot and root. (**E**) The first leaves are photosynthesizing by the time food reserves have been depleted.

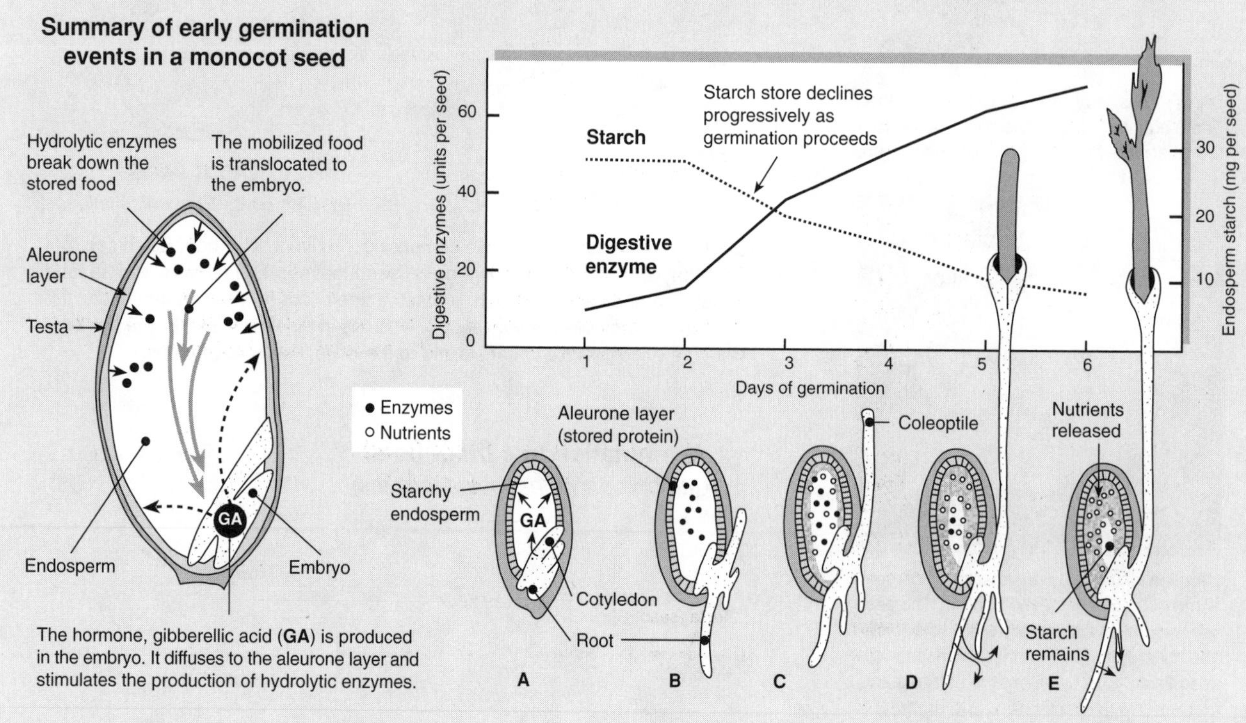

1. Identify the two processes involved in seed germination and describe the purpose of each:

(a) ____________________

____

(b) ____________________

____

2. Explain the role of the following in germination of a typical monocot seed:

(a) Water: ____________________

____

(b) Oxygen: ____________________

____

(c) Gibberellic acid: ____________________

____

Related activities: *Seed Structure and Germination*

Important in this section ...

- *Develop an understanding of ecological processes*
- *Describe and explain influences on population dynamics and community diversity*
- *Develop an understanding of socio-scientific issues*

Habitat and Distribution

Biomes	• Classification of the Earth's biomes • The effect of climate and altitude
Environmental gradients	• Gradients in abiotic factors • Distributional patterns in communities • Quantifying community patterns
Habitat	• Habitat and tolerance range • Microclimates and microhabitats

Population Ecology

Features of populations	• Features of populations • Population density and distribution • Sampling populations
Population dynamics	• Population regulation • Population growth • Population cycles
Survivorship	• Population age structure • Life tables and age specific survival • r and K selection
Human populations	• Age structures in human populations • Global demography

Gradients in physical factors contribute to ecosystem and community diversity.

Population fluctuations are determined by factors affecting age structure and survival.

Part 4

Ecology and the Biosphere

Abiotic factors and biotic interactions influence community diversity and the growth and sustainability of populations.

Ecosystems are open systems through which energy flows and nutrients are cycled.

Humans can intervene in natural cycles.

The predominant interactions between species are associated in some way with the acquisition of food.

Human activities influence the changes and cycles occurring naturally within ecosystems.

Ecological niche	• The concept of the niche • Niche and adaptation • Constraints on niche exploitation
Energy transfers	• Trophic groups • Food chains and food webs
Interactions between species	• Competition • Exploitative relationships • Mutualism • Key species and community structure
Ecological succession	• Primary succession • Secondary succession

Community Ecology

Quantifying energy flow	• Trophic efficiencies • Ecological pyramids • The productivity of ecosystems
Nutrient cycles and human activity	• The carbon cycle • Global warming • Microorganisms ad the nitrogen cycle • Nitrogen pollution
Biodiversity	• The importance of biodiversity • Human impacts on biodiversity • The effects of global warming • Conservation vs preservation

Ecosystem Ecology & Human Impact

AP Topic 3C

IB SL/HL 5.1

Habitat and Distribution

KEY CONCEPTS

- The world's biomes are distributed according to broad zones of climate, altitude, and latitude.
- Gradients in physical factors in an environment contribute to community diversity.
- Zonation and stratification are common types of distributional variation in communities.
- Organisms preferentially occupy habitats (and microhabitats) best suited to their requirements.

KEY TERMS

abiotic (=physical) factor
altitude
belt transect
biome
biotic factor
environmental gradient
habitat
latitude
line transect
limiting factor
microclimate
microhabitat
physical (=abiotic) factor
quadrat
stratification
tolerance range
transect
zonation

OBJECTIVES

☐ 1. Use the **KEY TERMS** to help you understand and complete these objectives.

Biomes and the Influence of Climate

pages 309, 344

☐ 2. Define the terms **biosphere** and **biome**. Recognize and describe the main features of the major biomes and explain how they are classified.

☐ 3. Explain the effect of **latitude**, **altitude**, and **rainfall** in determining the distribution of world biomes.

Physical Gradients and Community Patterns

pages 310-313, 317-321

☐ 4. Describe the **biotic** and **abiotic factors** influencing plant and animal distribution in an environment.

☐ 5. Explain how gradients in physical factors can occur over relatively short distances, e.g. on a rocky shore, or in a forest, desert, or lake. Explain the role of these **environmental gradients** in species distribution and in contributing to community diversity.

☐ 6. Describe the features of the two most common types of distributional variation in communities: zonation and stratification. Explain how these patterns arise and how they increase the amount of community diversity.
(a) **Zonation**: Describe and explain zonation in an intertidal community or altitudinal zonation in a forest community.
(b) **Stratification**: Describe and explain vertical stratification in a rainforest.

☐ 7. Describe and explain techniques used to quantify distributional variation in communities, e.g. **belt transects** and **line transects** using point sampling or **quadrats**. Describe advantages and limitations of different methods with respect to sampling time, cost, and suitability to habitat.

Habitat

pages 314-316, 326-327

☐ 8. Define the term **habitat** and explain how an organism's **tolerance range** determines, in part, where it can live. Explain how **microclimates** develop and recognize the role of microclimates and **limiting factors** in species distribution and community diversity.

☐ 9. Describe methods to measure abiotic factors in a habitat. Include reference to the following (as appropriate): pH, light, temperature, dissolved oxygen, current speed, total dissolved solids, and conductivity.

☐ 10. Describe the habitat for a named organism. Recognize that the same species may occupy a wide habitat range but that some of these habitats will be more favorable than others.

Periodicals:
listings for this chapter are on page 391

Weblinks:
www.thebiozone.com/weblink/SB2-2603.html

Teacher Resource CD-ROM:
Ecological Sampling

Biomes

The Earth's **biomes** are the largest geographically based biotic communities that can be conveniently recognized. These are large areas where the vegetation type shares a particular suite of physical requirements. Biomes have characteristic features, but the boundaries between them are not distinct. The same biome may occur in widely separated regions of the world wherever the climatic and soil conditions are similar. Terrestrial biomes are recognized for all the major climatic regions of the world. They are classified by their predominant vegetation type.

Earth's Climate and Biomes

Biomes are closely related to the major air cells that circle the Earth and are reflected in the Northern and Southern Hemispheres.

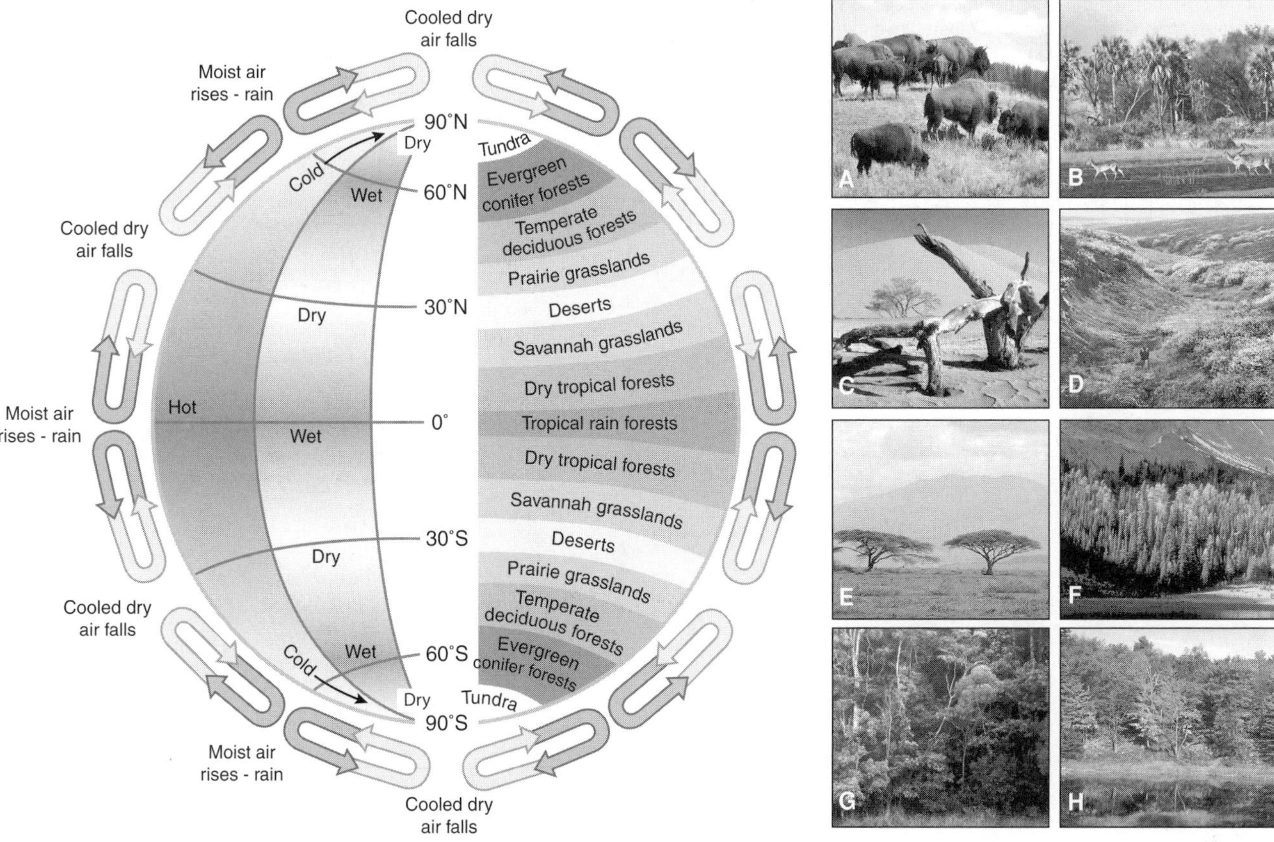

Biomes and Landscapes

Climate is heavily modified by the landscape. Where there are large mountain ranges, wind is deflected upwards causing rain on the windward side and a **rain shadow** on the leeward side. The biome that results from this is considerably different from the one that may have appeared with no wind deflection. Large expanses of ocean and flat land also change the climate by modifying air temperatures and the amount of rainfall.

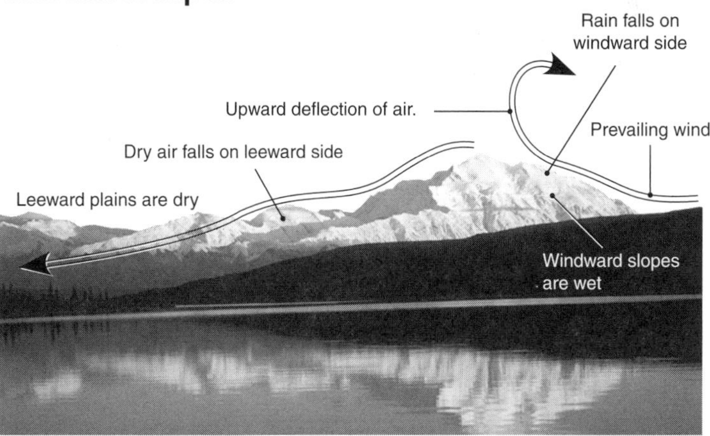

1. Match the lettered biome images with the appropriate biome:

(a) Tundra: ______________________

(b) Temperate deciduous forest: ______________________

(c) Deserts: ______________________

(d) Dry tropical forests: ______________________

(e) Evergreen conifer forest: ______________________

(f) Prairie grasslands: ______________________

(g) Savannah grasslands: ______________________

(h) Tropical rain forests: ______________________

2. Identify which abiotic factor(s) limit the extent of temperate deciduous forests: ______________________

____

Physical Factors and Gradients

Gradients in abiotic factors are found in almost every environment; they influence habitats and **microclimates**, and determine patterns of species distribution. This activity, covering the next four pages, examines the physical gradients and microclimates that might typically be found in four very different environments. Note that **dataloggers** (pictured right), are being increasingly used to gather such data.

A Desert Environment

Desert environments experience extremes in temperature and humidity, but they are not uniform with respect to these factors. This diagram illustrates hypothetical values for temperature and humidity for some of the microclimates found in a desert environment at midday.

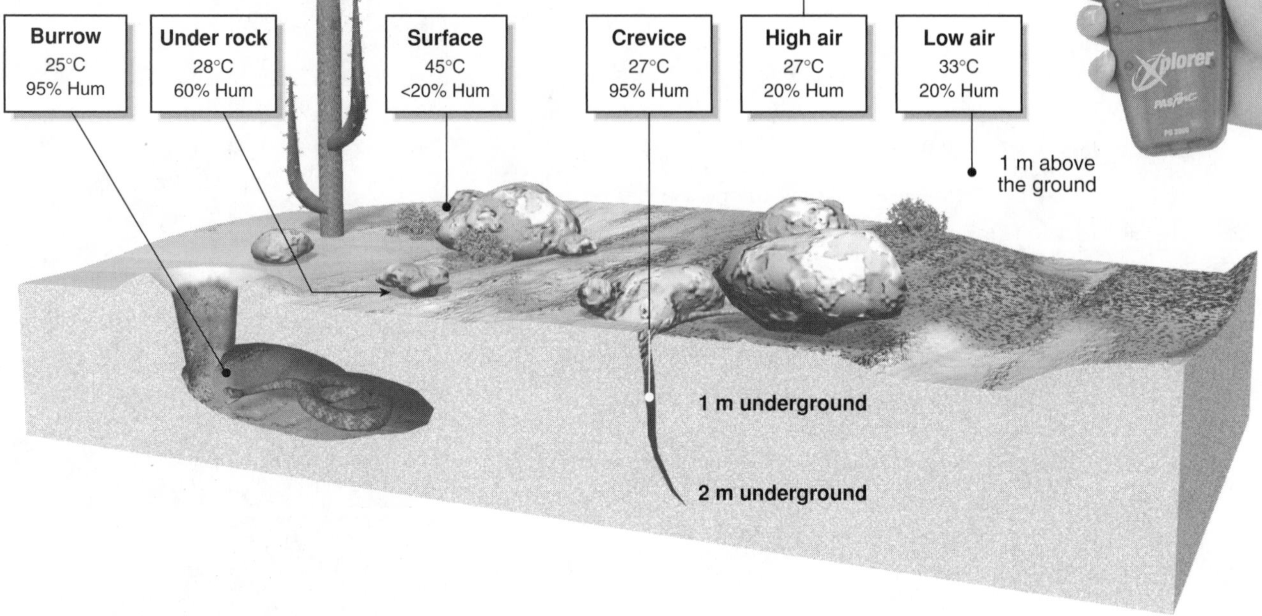

1. Distinguish between **climate** and **microclimate**: ______

2. Study the diagram above and describe the general conditions where high humidity is found: ______

3. Identify the three microclimates that a land animal might exploit to avoid the extreme high temperatures of midday: ______

4. Describe the likely consequences for an animal that was unable to find a suitable microclimate to escape midday sun: ______

5. Describe the advantage of high humidity to the survival of most land animals: ______

6. Describe the likely changes to the temperature and relative humidity that occur during the night: ______

Related activities: *Sampling Populations*
Web links: *Tide Pool Ecology*

Physical Factors in a Tropical Rainforest

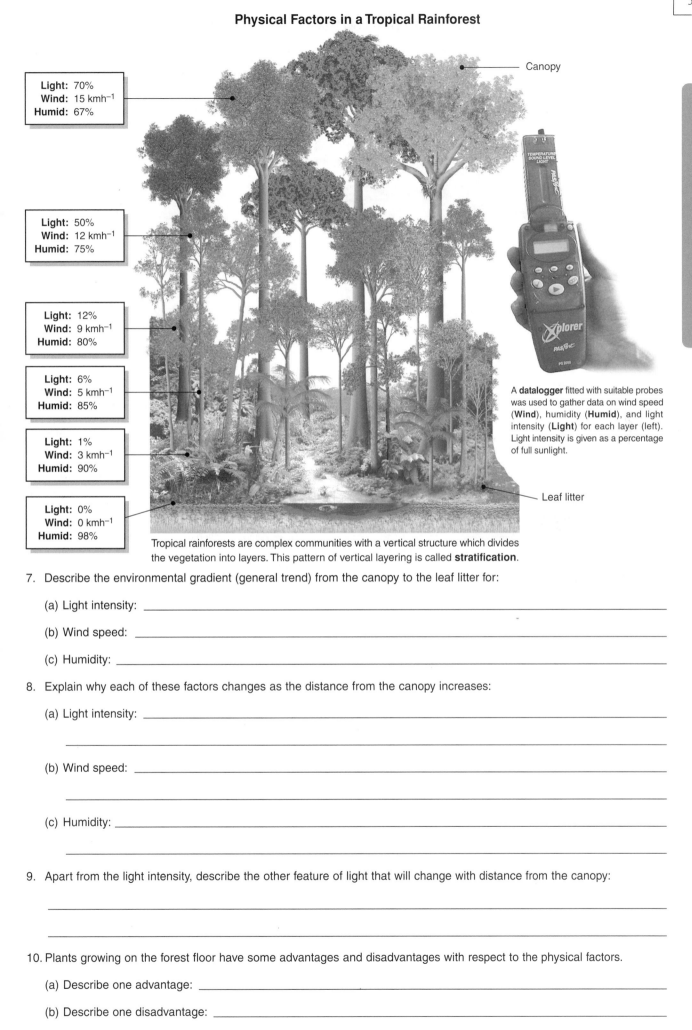

Canopy

Light: 70%
Wind: 15 kmh^{-1}
Humid: 67%

Light: 50%
Wind: 12 kmh^{-1}
Humid: 75%

Light: 12%
Wind: 9 kmh^{-1}
Humid: 80%

Light: 6%
Wind: 5 kmh^{-1}
Humid: 85%

Light: 1%
Wind: 3 kmh^{-1}
Humid: 90%

Light: 0%
Wind: 0 kmh^{-1}
Humid: 98%

Leaf litter

A **datalogger** fitted with suitable probes was used to gather data on wind speed (**Wind**), humidity (**Humid**), and light intensity (**Light**) for each layer (left). Light intensity is given as a percentage of full sunlight.

Tropical rainforests are complex communities with a vertical structure which divides the vegetation into layers. This pattern of vertical layering is called **stratification**.

7. Describe the environmental gradient (general trend) from the canopy to the leaf litter for:

   (a) Light intensity: ______________________

   (b) Wind speed: ______________________

   (c) Humidity: ______________________

8. Explain why each of these factors changes as the distance from the canopy increases:

   (a) Light intensity: ______________________

   (b) Wind speed: ______________________

   (c) Humidity: ______________________

9. Apart from the light intensity, describe the other feature of light that will change with distance from the canopy:

   ____

10. Plants growing on the forest floor have some advantages and disadvantages with respect to the physical factors.

    (a) Describe one advantage: ______________________

    (b) Describe one disadvantage: ______________________

Physical Factors at Low Tide on a Rock Platform

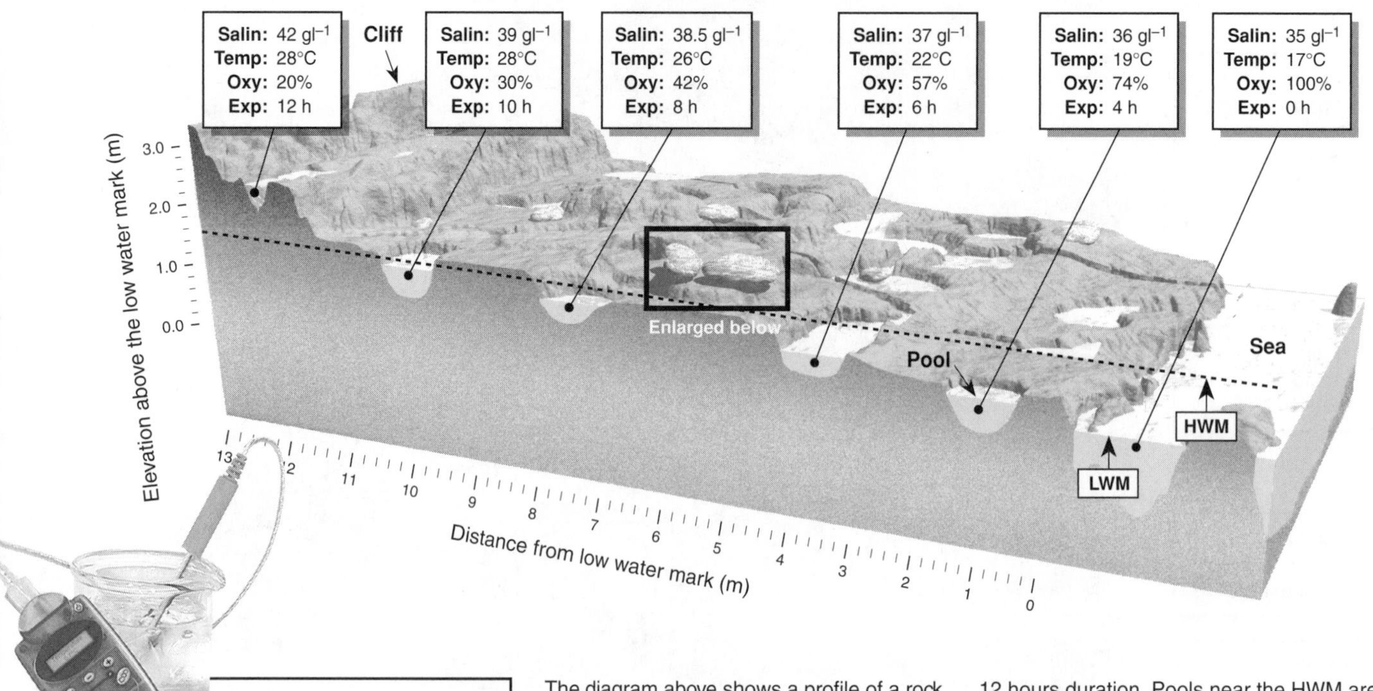

Boulders

A

B

C

The diagram above shows a profile of a rock platform at low tide. The **high water mark** (HWM) shown here is the average height the spring tide rises to. In reality, the high tide level will vary with the phases of the moon (i.e. spring tides and neap tides). The **low water mark** (LWM) is an average level subject to the same variations due to the lunar cycle. The rock pools vary in size, depth, and position on the platform. They are isolated at different elevations, trapping water from the ocean for time periods that may be brief or up to 10 – 12 hours duration. Pools near the HWM are exposed for longer periods of time than those near the LWM. The difference in exposure times results in some of the physical factors exhibiting a **gradient**; the factor's value gradually changes over distance. Physical factors sampled in the pools include salinity, or the amount of dissolved salts (g) per liter (**Salin**), temperature (**Temp**), dissolved oxygen compared to that of open ocean water (**Oxy**), and exposure, or the amount of time isolated from the ocean water (**Exp**).

11. Describe the environmental gradient (general trend) from the low water mark (LWM) to the high water mark (HWM) for:

(a) Salinity: ______________________

(b) Temperature: ______________________

(c) Dissolved oxygen: ______________________

(d) Exposure: ______________________

12. Rock pools above the normal high water mark (HWM), such as the uppermost pool in the diagram above, can have wide extremes of salinity. Explain the conditions under which these pools might have either:

(a) Very low salinity: ______________________

(b) Very high salinity: ______________________

13. (a) The inset diagram (above, left) is an enlarged view of two boulders on the rock platform. Describe how the physical factors listed below might differ at each of the labeled points **A**, **B**, and **C**:

Mechanical force of wave action: ______________________

Surface temperature when exposed: ______________________

(b) State the term given to these localized variations in physical conditions: ______________________

Physical Factors in an Oxbow Lake in Summer

Oxbow lakes are formed from old river meanders which have been cut off and become isolated from the main channel following a change of the river's course. They are commonly shallow (about 2-4 m deep) but may be deep enough to develop temporary, but relatively stable, temperature gradients from top to bottom (below). Small lakes are relatively closed systems and events in them are independent of those in other nearby lakes, where quite different water quality may be found. The physical factors are not constant throughout the water in the lake. Surface water and water near the lake margins can have quite different values for factors such as water temperature (**Temp**), dissolved oxygen (**Oxygen**) measured in milligrams per liter ($mg\,l^{-1}$), and light penetration (**Light**), indicated here as a percentage of the light striking the surface.

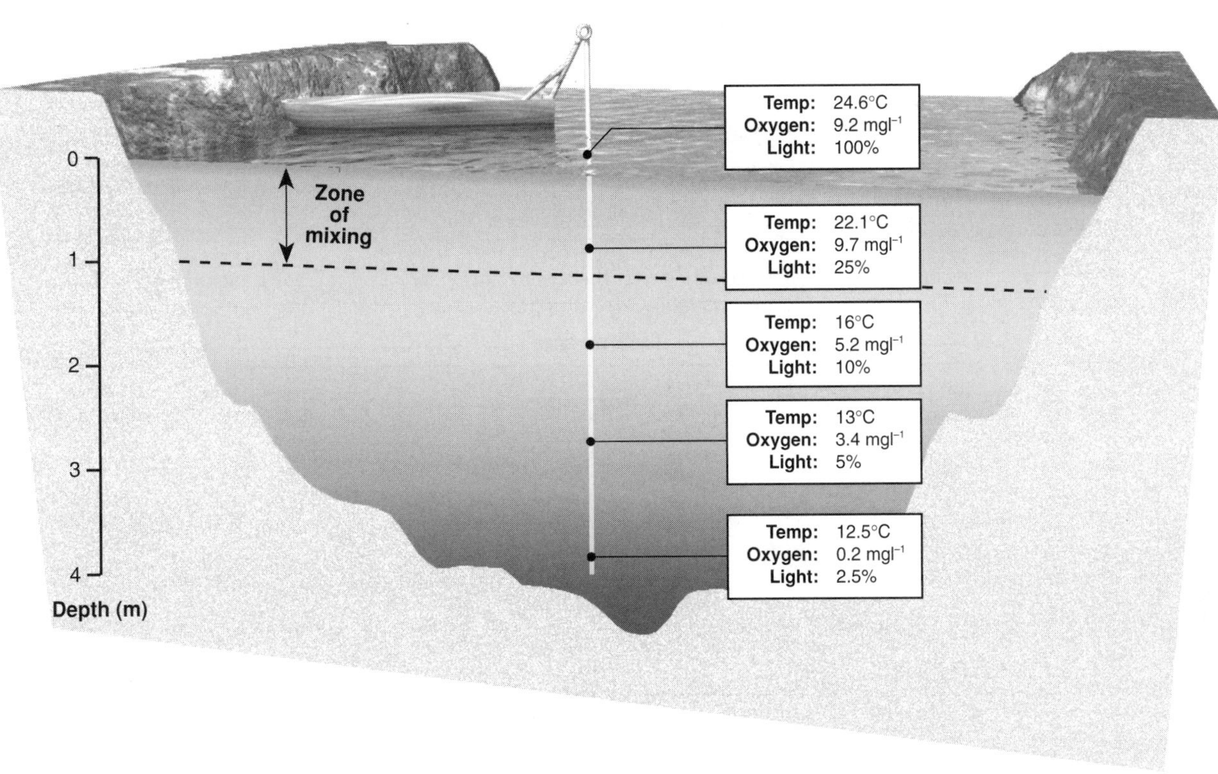

14. With respect to the diagram above, describe the environmental gradient (general trend) from surface to lake bottom for:

(a) Water temperature: ____________________

(b) Dissolved oxygen: ____________________

(c) Light penetration: ____________________

15. During the summer months, the warm surface waters are mixed by gentle wind action. Deeper cool waters are isolated from this surface water. This sudden change in the temperature profile is called a **thermocline** which itself is a further barrier to the mixing of shallow and deeper water.

(a) Explain the effect of the thermocline on the dissolved oxygen at the bottom of the lake: ____________________

(b) Explain what causes the oxygen level to drop to the low level: ____________________

16. Many of these shallow lakes can undergo great changes in their salinity (sodium, magnesium, and calcium chlorides):

(a) Name an event that could suddenly reduce the salinity of a small lake: ____________________

(b) Name a process that can gradually increase the salinity of a small lake: ____________________

17. Describe the general effect of physical gradients on the distribution of organisms in habitats: ____________________

Habitat

The environment in which a species population (or an individual organism) lives (including all the physical and biotic factors) is termed its **habitat**. Within a prescribed habitat, each species population has a range of tolerance to variations in its physical and chemical environment. Within the population, individuals will have slightly different tolerance ranges based on small differences in genetic make-up, age, and health. The wider an organism's tolerance range for a given abiotic factor (e.g. temperature or salinity), the more likely it is that the organism will be able to survive variations in that factor. Species **dispersal** is also strongly influenced by **tolerance range**. The wider the tolerance range of a species, the more widely dispersed the organism is likely to be. As well as a tolerance range, organisms have a narrower **optimum range** within which they function best. This may vary from one stage of an organism's development to another or from one season to another. Every species has its own optimum range. Organisms will usually be most abundant where the abiotic factors are closest to the optimum range.

Habitat Occupation and Tolerance Range

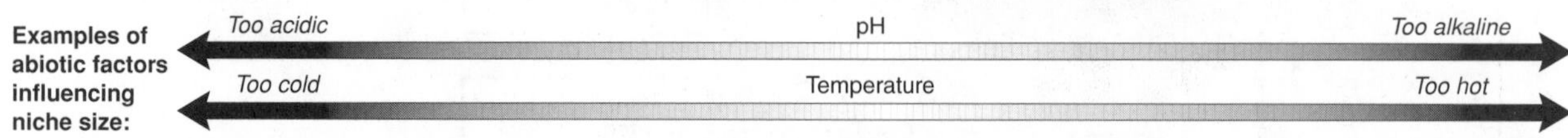

The law of tolerances states that *for each abiotic factor, a species population (or organism) has a tolerance range within which it can survive. Toward the extremes of this range, that abiotic factor tends to limit the organism s ability to survive* .

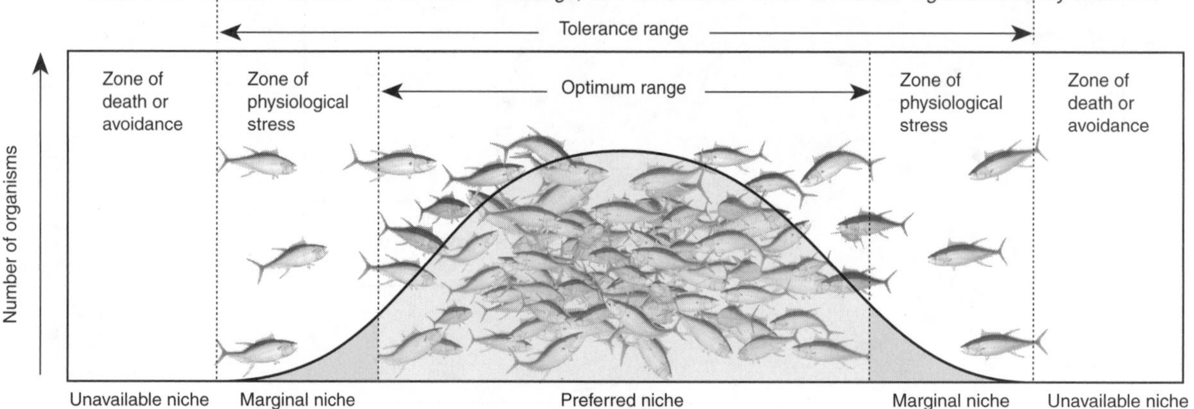

The Scale of Available Habitats

A habitat may be vast and relatively homogeneous, as is the open ocean. Barracuda (above) occur around reefs and in the open ocean where they are aggressive predators.

For non-mobile organisms, such as the fungus above, a suitable habitat may be defined by the particular environment in a relatively tiny area, such as on this decaying log.

For microbial organisms, such as the bacteria and protozoans of the ruminant gut, the habitat is defined by the chemical environment within the rumen (R) of the host animal, in this case, a cow.

1. Explain how an organism's habitat occupation relates to its tolerance range: ________________

____

____

2. (a) Identify the range in the diagram above in which most of the species population is found. Explain why this is the case:

____

____

(b) Describe the greatest constraints on an organism's growth and reproduction within this range: ________________

____

3. Describe some probable stresses on an organism forced into a marginal niche: ________________

____

____

Related activities: *Ecological Niche*

Dingo Habitats

An organism's habitat is not always of a single type. Some animals range over a variety of habitats, partly in order to obtain different resources from different habitats, and sometimes simply because they are forced into marginal habitats by competition. Dingoes are found throughout Australia, in ecosystems as diverse as the tropical rainforests of the north to the arid deserts of the Center. Within each of these ecosystems, they may frequent several habitats or **microhabitats**. The information below shows how five dingo packs exploit a variety of habitats at one location in Australia. The table on the following page shows how dingoes are widespread in their distribution and are found living in a variety of ecosystems.

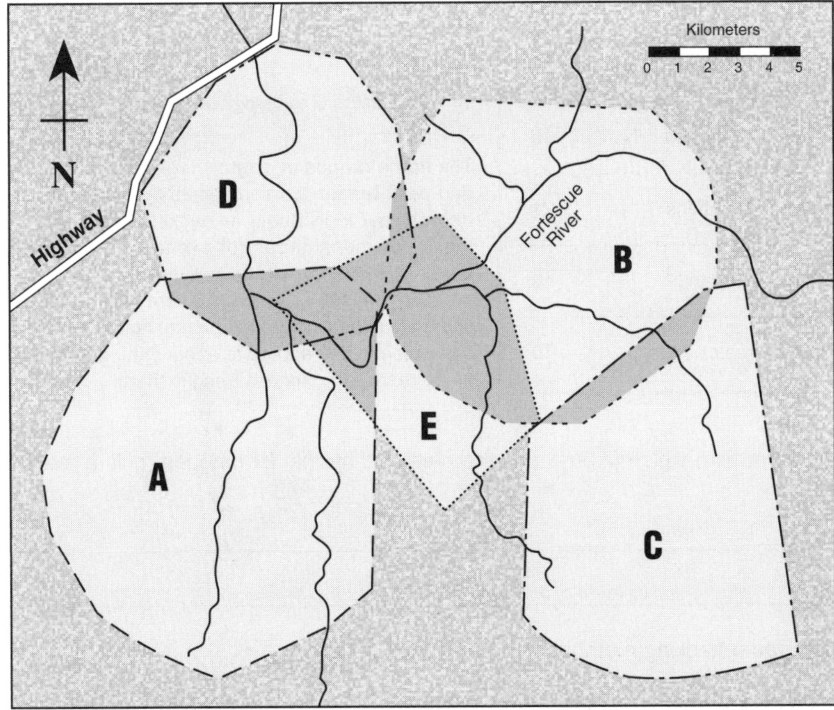

The map on the left shows the territories of five stable dingo packs (A to E) in the Fortescue River region in north-west Australia. The territories were determined by 4194 independent radio-tracking locations over 4 years. The size and nature of each territory, together with the makeup of each pack is given in the table below. The major prey of the dingoes in this region are large kangaroos (red kangaroos and euros).

Adapted from: Corbett, L. *The dingo in Australia and Asia*, 1995. University of NSW Press, after (original source) Thomson, P.C. 1992. *The behavioral ecology of dingoes in north-west Australia. IV. Social and spatial organization, and movements. Wildlife Research* 19: 543-563.

Dingo pack name	Territory area (km^2)	Pack size	Dingo density	Index of kangaroo abundance (%)	Habitat types (%) and habitat usage (%) in each territory							
					Riverine		Stony		Floodplain		Hills	
Pack A	113	12	10.6	15.9	10	(49)	1	(2)	21	(6)	69	(44)
Pack B	94	12		8.5	14	(43)	9	(10)	38	(25)	39	(23)
Pack C	86	3		3.9	2	(3)	0	(0)	63	(94)	35	(3)
Pack D	63	6		12.3	12	(35)	5	(8)	46	(20)	37	(37)
Pack E	45	10	22.2	8.4	14	(31)	6	(4)	39	(18)	42	(47)

Dingo density: mean number of dingoes per 100 km^2 (calculated by you)

Index of kangaroo abundance: Percentage of observations of kangaroos per observations of dingoes

Habitat type: Portion of the territory with this kind of habitat

Habitat usage (in brackets): Percentage of time spent by the pack in this habitat

1. Calculate the density of each of the dingo packs at the Fortescue River site above (two have been done for you). Remember that to determine the density, you carry out the following calculation:

Density = pack size ÷ territory area x 100 (to give the mean number per 100 km^2)

2. Name the dominant habitat (or habitats) for each territory in the table above (e.g. riverine, stony, floodplain, hills):

(a) Pack A: ______

(b) Pack B: ______

(c) Pack C: ______

(d) Pack D: ______

(e) Pack E: ______

Dingo home range size in contrasting ecosystems

Location (study site)	Ecosystem	Range (km^2)
1 Fortescue River, North-west Australia	Semi-arid, coastal plains and hills	77
2 Simpson Desert, Central Australia	Arid, gibber (stony) and sandy desert	67
3 Kapalga, Kakadu N.P., North Australia	Tropical, coastal wetlands and forests	39
4 Harts Ranges, Central Australia	Semi-arid, river catchment and hills	25
5 Kosciusko N.P., South-east Australia	Moist, cool forested mountains	21
6 Georges Creek N.R., East Australia	Moist, cool forested tablelands	18
7 Nadgee N.R., South-east Australia	Moist, cool coastal forests	10

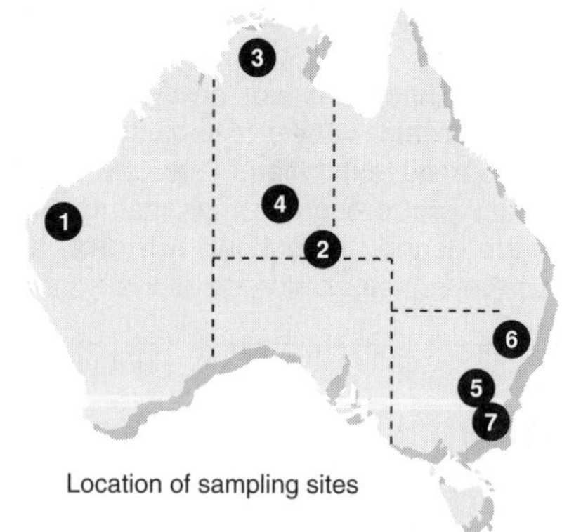

Location of sampling sites

The **home ranges** of neighboring dingo individuals and pack **territories** sometimes overlap to some degree, but individuals or packs avoid being in these communal areas (of overlap) at the same time. This overlap occurs especially during the breeding season or in areas where resources are shared (e.g. hunting grounds, water holes). Different ecosystem types appear to affect the extent of the home ranges for dingoes living in them.

3. Study the table on the previous page and determine which (if any) was the preferred habitat for dingoes (give a reason for your answer):

4. The dingoes at this site were studied using radio-tracking methods.

   (a) Explain how radio-tracking can be used to determine the movements of dingoes:

   (b) State how many independent tracking locations were recorded during this study:

   (c) State how long a period of time the study was run for:

   (d) Explain why so many records were needed over such a long period of time:

5. From the table on the previous page, state whether the relative kangaroo abundance (the major prey of the dingo) affects the density that a given territory can support:

6. Study the table at the top of this page which shows the results of an investigation into the sizes of home ranges in dingo populations from different ecosystems.

   (a) Describe the feature of the ecosystems that appears to affect the size of home ranges:

   (b) Explain how this feature might affect how diverse (varied) the habitats are within the ecosystem:

Community Change With Altitude

The Kosciusko National Park lies on the border between Victoria and New South Wales. In 1959, a researcher by the name of A.B. Costin carried out a detailed sampling of a transect between Berridale and the summit of Mt. Kosciusko. The map on the right shows the transect as a dotted line representing a distance of some 50-60 km.

The distribution of the various plant species up the slope of Mount Kosciusko is affected by changes in the physical conditions with increasing altitude. Below are two diagrams, one showing the profile of the transect showing changes in vegetation and soil types with increasing altitude. The diagram below the profile shows the changes in temperature and rainfall (precipitation) with altitude.

The low altitude soil around Berridale has low levels of organic matter supporting dry tussock grassland vegetation. The high altitude alpine soils are rich in organic matter, largely due to slow decay rates.

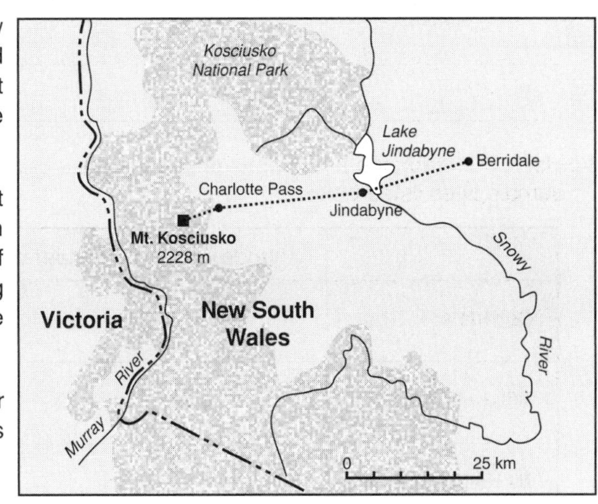

Profile of Mount Kosciusko

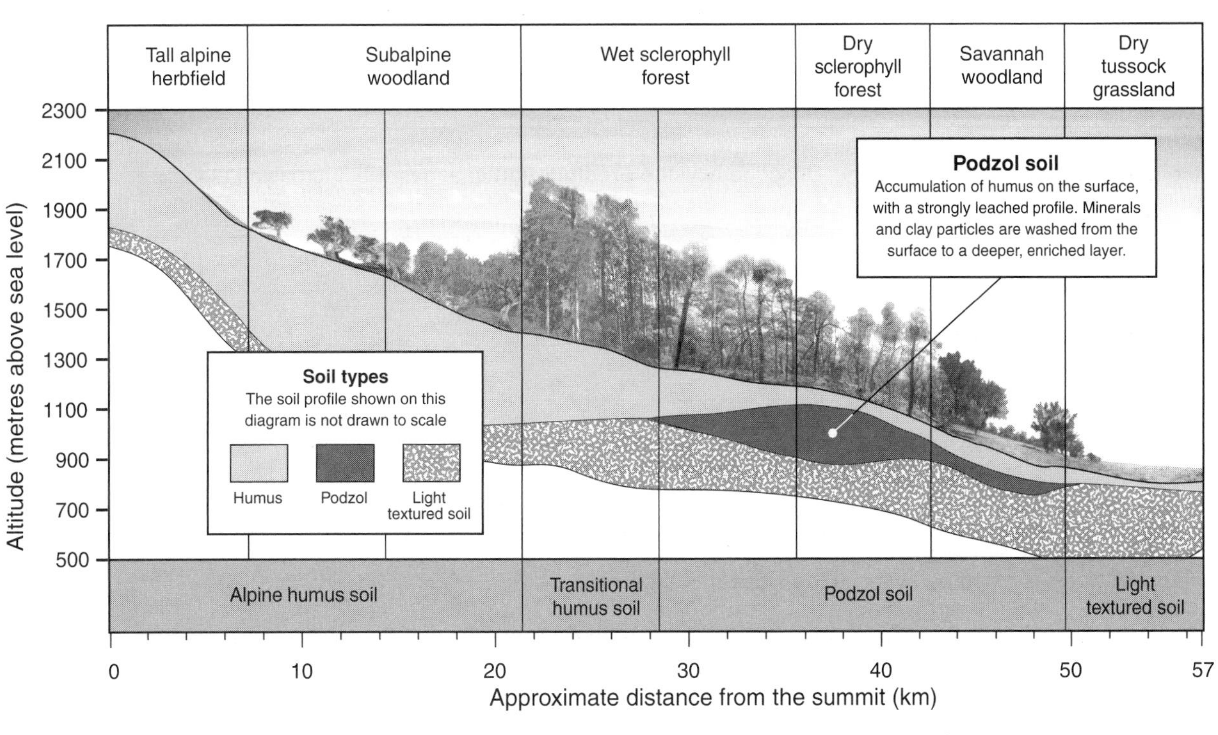

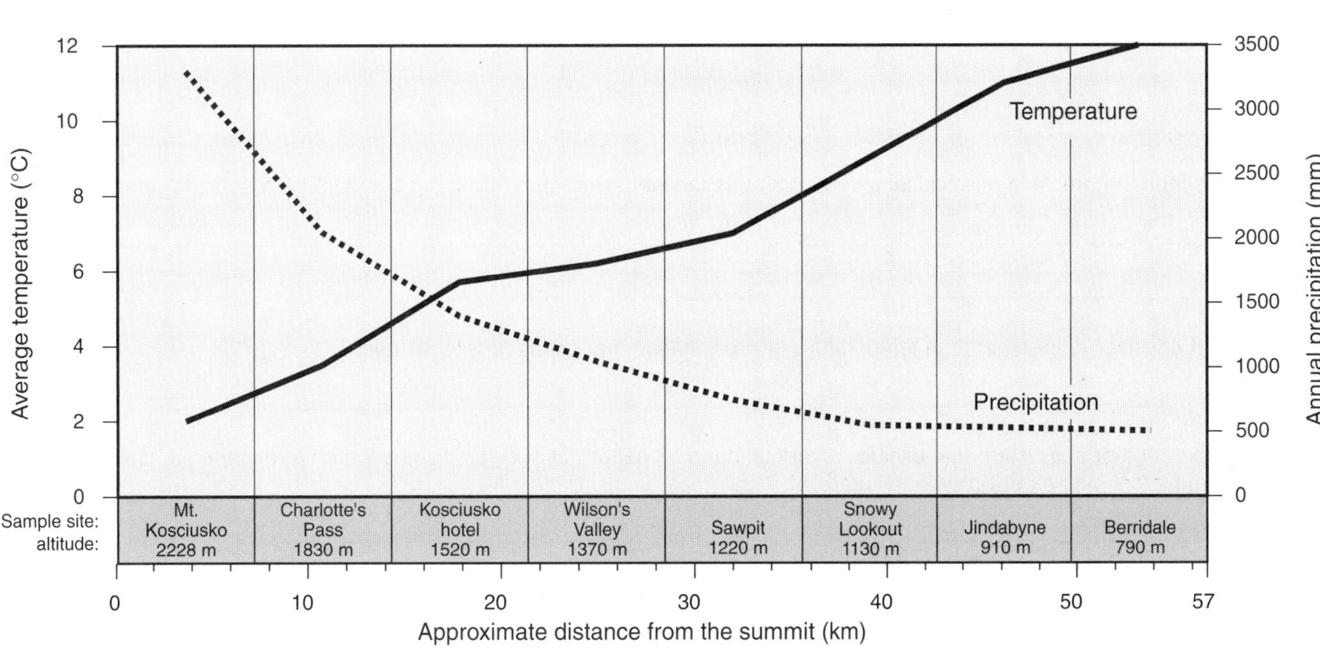

Related activities: *Shoreline Zonation*

1. Calculate the vertical distance (change in altitude) in metres, between Berridale and Mount Kosciusko: ______________

2. Name the three physical factors that are illustrated on the diagrams on the previous page:

3. Using the diagrams and graphs on the previous page, describe the following physical measurements for the three sample sites listed below:

	Altitude (m)	Temperature (°C)	Precipitation (mm)	Soil type
Berridale:				
Wilson's Valley:				
Mt Kosciusko:				

4. Study the graph of temperature vs altitude. Describe how the **temperature** changes with increasing altitude:

5. Study the graph of precipitation vs altitude. Describe how the **precipitation** changes with increasing altitude:

6. Suggest a reason why the leaf litter is slow to decay in the alpine soil: ______________

7. The different vegetation types are distributed on the slopes of Mt. Kosciusko in a banded pattern from low altitude to the summit. Give the name of this kind of distribution pattern:

8. Wet sclerophyll forest is found part way up the slope of Mt. Kosciusko.

   (a) Study the profile on the previous page and determine the **altitude range** for wet sclerophyll forest (in metres):

   (b) Describe the probable physical factor that prevents the wet sclerophyll forest not being found at a lower altitude:

   (c) Describe the probable physical factor that prevents the wet sclerophyll forest not being found at a higher altitude:

9. Name a physical factor other than temperature or precipitation that changes with altitude:

10. Describe another ecosystem that exhibits a vertical banding pattern of species distribution as a response to changing physical factors over an environmental gradient:

Transect Sampling

A **transect** is a line placed across a community of organisms. Transects are usually carried out to provide information on the **distribution** of species in the community. This is of particular value in situations where environmental factors that change over the sampled distance. This change is called an **environmental gradient** (e.g. up a mountain or across a seashore). The usual practice for small transects is to stretch a string between two markers. The string is marked off in measured distance intervals, and the species at each marked point are noted. The sampling points along the transect may also be used for the siting of quadrats, so that changes in density and community composition can be recorded. Belt transects are essentially a form of continuous quadrat sampling. They provide more information on community composition but can be difficult to carry out. Some transects provide information on the vertical, as well as horizontal, distribution of species (e.g. tree canopies in a forest).

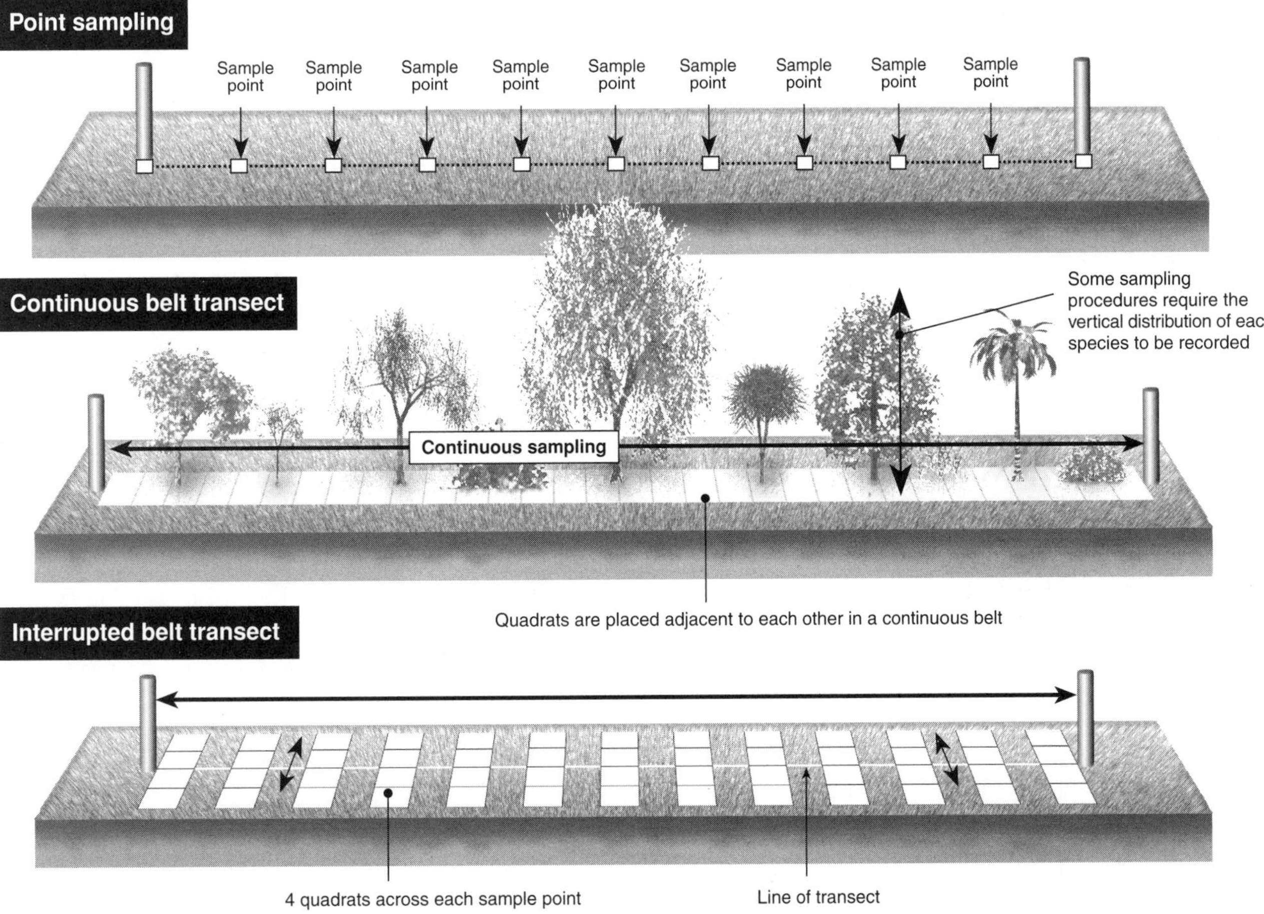

1. Belt transect sampling uses quadrats placed along a line at marked intervals. In contrast, point sampling transects record only the species that are touched or covered by the line at the marked points.

   (a) Describe one disadvantage of belt transects: ______________________________

   (b) Explain why line transects may give an unrealistic sample of the community in question: ______________________________

   (c) Explain how belt transects overcome this problem: ______________________________

   (d) Describe a situation where the use of transects to sample the community would be inappropriate: ______________________________

2. Explain how you could test whether or not a transect sampling interval was sufficient to accurately sample a community:

____

Related activities: *Density and Distribution, Physical Factors and Gradients*

Kite graphs are an ideal way in which to present distributional data from a belt transect (e.g. abundance or percentage cover along an environmental gradient). Usually, they involve plots for more than one species. This makes them good for highlighting probable differences in habitat preference between species. Kite graphs may also be used to show changes in distribution with time (e.g. with daily or seasonal cycles).

3. The data on the right were collected from a rocky shore field trip. Periwinkles from four common species of the genus *Littorina* were sampled in a continuous belt transect from the low water mark, to a height of 10 m above that level. The number of each of the four species in a 1 m^2 quadrat was recorded.

   Plot a **kite graph** of the data for all four species on the grid below. Be sure to choose a scale that takes account of the maximum number found at any one point and allows you to include all the species on the one plot. Include the scale on the diagram so that the number at each point on the kite can be calculated.

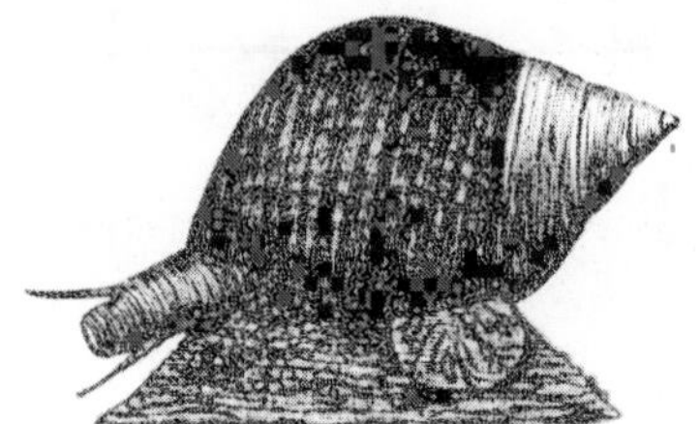

Field data notebook

Numbers of periwinkles (4 common species) showing vertical distribution on a rocky shore

	Periwinkle species:			
Height above low water (m)	*L. littorea*	*L. saxatalis*	*L. neritoides*	*L. littoralis*
0-1	0	0	0	0
1-2	1	0	0	3
2-3	3	0	0	17
3-4	9	3	0	12
4-5	15	12	0	1
5-6	5	24	0	0
6-7	2	9	2	0
7-8	0	2	11	0
8-9	0	0	47	0
9-10	0	0	59	0

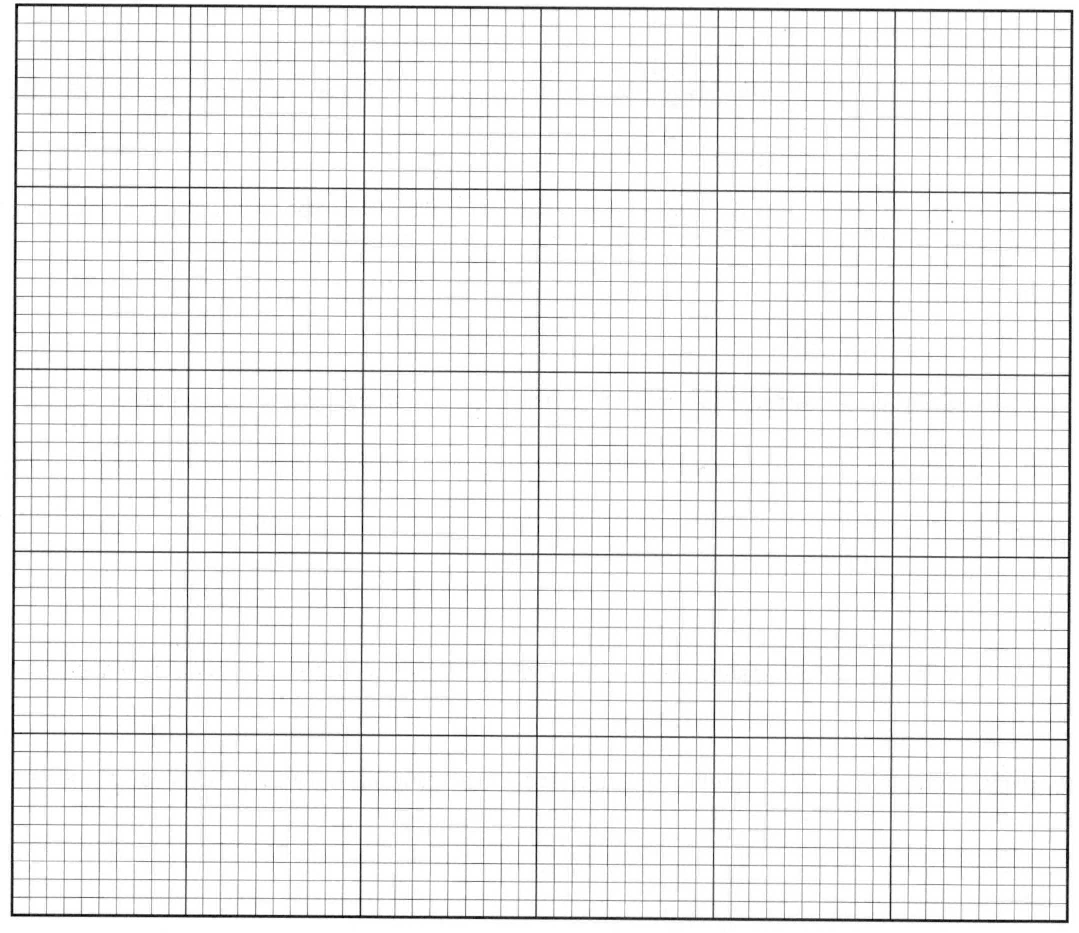

Shoreline Zonation

Zonation refers to the division of an ecosystem into distinct zones that experience similar abiotic conditions. In a more global sense, zonation may also refer to the broad distribution of vegetation according to latitude and altitude. Zonation is particularly clear on a rocky seashore, where assemblages of different species form a banding pattern approximately parallel to the waterline. This effect is marked in temperate regions where the prevailing weather comes from the same general direction. Exposed shores show the clearest zonation. On sheltered rocky shores there is considerable species overlap and it is only on the upper shore that distinct zones are evident. Rocky shores exist where wave action prevents the deposition of much sediment. The rock forms a stable platform for the secure attachment of organisms such as large seaweeds and barnacles. Sandy shores are less stable than rocky shores and the organisms found there are adapted to the more mobile substrate.

Tide pools on a rocky shore, Santa Cruz, California

Seashore Zonation Patterns

The zonation of species distribution according to an environmental gradient is well shown on rocky shores. Variations in low and high tide affect zonation, and in areas with little tidal variation, zonation is restricted. High on the shore, some organisms may be submerged only at spring high tide. Low on the shore, others may be exposed only at spring low tide. There is a gradation in extent of exposure and the physical conditions associated with this. Zonation patterns generally reflect the vertical movement of seawater. Sheer rocks can show marked zonation as a result of tidal changes with little or no horizontal shift in species distribution. The profiles below, show generalized zonation patterns on an exposed rocky shore (left) with an exposed sandy shore for comparison (right). **SLT** = Spring low tide mark, **MLT** = Mean low tide mark, **MHT** = Mean high tide mark, **SHT** = Spring high tide mark.

Key to species

1. Lichen: sea ivory
2. Small periwinkle *Littorina neritoides*
3. Lichen *Verrucaria maura*
4. Rough periwinkle *Littorina saxatilis*
5. Common limpet *Patella vulgaris*
6. Laver *Porphyra*
7. Spiral wrack *Fucus spiralis*
8. Australian barnacle
9. Common mussel *Mytilus edulis*
10. Common whelk *Buccinum undatum*
11. Grey topshell *Gibbula cineraria*
12. Carrageen (Irish moss) *Chondrus crispus*
13. Thongweed *Himanthalia elongata*
14. Toothed wrack *Fucus serratus*
15. Dabberlocks *Alaria esculenta*
16. Common sandhopper
17. Sandhopper *Bathyporeia pelagica*
18. Common cockle *Cerastoderma edule*
19. Lugworm *Arenicola marina*
20. Sting winkle *Ocinebra erinacea*
21. Common necklace shell *Natica alderi*
22. Rayed trough shell *Mactra corallina*
23. Sand mason worm *Lanice conchilega*
24. Sea anemone *Halcampa*
25. Pod razor shell *Ensis siliqua*
26. Sea potato *Echinocardium* (a heart urchin)

Note: Where several species are indicated within a single zonal band, they occupy the entire zone, not just the position where their number appears.

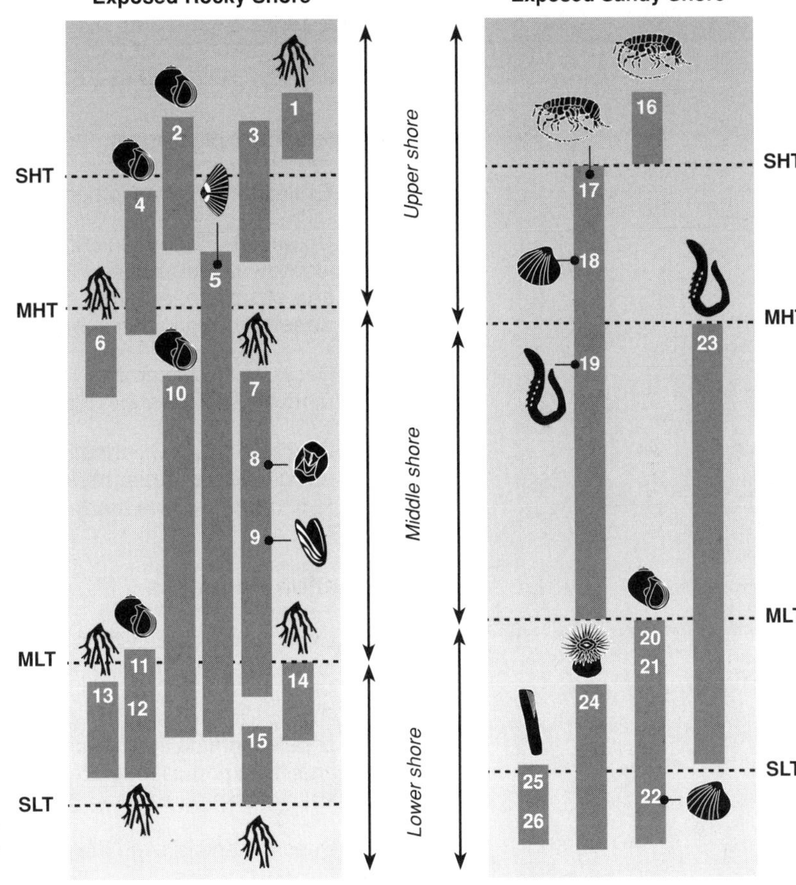

1. (a) Suggest why the time of exposure above water is a major factor controlling species distribution on a rocky shore:

   (b) Identify two other abiotic factors that might influence species distribution on a rocky shore:

   (c) Identify two biotic factors that might influence species distribution on a rocky shore:

2. Describe the zonation pattern on a rocky shore:

Related activities: *Physical Factors and Gradients*
Web links: *Tide Pool Ecology*

AP Topic 3C

IB SL/HL 5.3

Population Ecology

KEY CONCEPTS

- Population specific characteristics include age structure, natality, and mortality.
- We can quantify the attributes of populations through sampling.
- Population growth is regulated by density dependent and density independent factors.
- Certain demographic trends in human populations are associated with changes in economic development.

KEY TERMS

abundance
age structure
carrying capacity (K)
community
demography
density
density dependent factor
density independent factor
distribution
environmental resistance
exponential growth
intrinsic rate of natural increase (*r*)
K selection
life tables
logistic growth
mark and recapture
migration rate
mortality
natality
population
population growth
population sampling
population size
quadrat
r selection
survivorship curve

OBJECTIVES

☐ 1. Use the **KEY TERMS** to help you understand and complete these objectives.

Features of Populations

pages 324-329, 339-340

☐ 2. Explain the difference between a **population** and a **community**.

☐ 3. Understand that populations are dynamic and exhibit attributes not shown by individuals themselves. Describe population specific characteristics including **age structure**, **natality**, **mortality**, and **distribution**. Distinguish population **density** and population **size**.

☐ 4. Describe the distribution patterns of organisms within their range: uniform, random, and clumped. Describe factors governing each type of distribution.

☐ 5. Describe how the attributes of populations can be assessed quantitatively using **population sampling**. Describe methods for population sampling (e.g. **quadrats** and **mark and recapture**).

Population Dynamics

pages 323, 330-338, 341, 355-359

☐ 6. Explain how population size can be affected by births, deaths, and migration and express the relationship in an equation.

☐ 7. Recognize the value of **life tables** in providing information of patterns of population natality and mortality. Explain the role of **survivorship curves** in analyzing populations. Describe and explain the features of type I, II, and III survivorship curves and give examples.

☐ 8. Describe and explain the typical features of *r* and K selection.

☐ 9. Describe how the trends in population change can be shown in a population growth curve of population numbers (Y axis) against time (X axis).

☐ 10. Describe factors affecting final population size. Include reference to **carrying capacity** (K), **environmental resistance**, **density dependent factors**, and **density independent factors**.

☐ 11. Describe **exponential** and **logistic growth**. Explain patterns of population growth in colonizing, stable, and declining populations.

☐ 12. Describe and explain population cycles in interacting species (e.g. snowshoe hare and Canada lynx) in which the fluctuations in population growth in one species lag behind those of the other, usually in a predictable way.

☐ 13. Describe and explain demographic trends in human populations.

Periodicals:
listings for this chapter are on page 391

Weblinks:
www.thebiozone.com/weblink/SB2-2603.html

Teacher Resource CD-ROM:
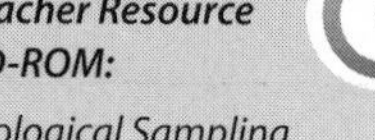
Ecological Sampling

The Rise and Fall of Human Populations

Human populations are subject to rises and collapses in the same way as natural animal populations. Throughout history there have been a number of peaks of human civilisation followed by collapse. These collapses have been triggered by various events but can generally be attributed either to the spread of disease or to the collapse of a food source (normally agriculture). Examples can be traced right back to the origins of humans.

Mitochondrial DNA analyses show that the human population may have been on the brink of extinction with only around 10,000 individuals alive 150,000 years ago. The population remained low for virtually the whole of human prehistory. When the first towns and cities were being built, around 10,000 years ago, the human population had reached barely 5 million. By around 700 AD, the human population had reached 150 million, and the first very large cities were developing. One such city was the Mayan city of Tikal.

TIKAL: At its peak around 800 AD, Tikal and the surrounding area, was inhabited by over 400,000 people. Extensive fields were used to cultivate crops and the total area of the city and its satellite towns and fields may have reached over 250 km^2. Eventually the carrying capacity of the tropical, nutrient-poor land was overextended and people began to starve. By 900 AD the city had been deserted and the surrounding area abandoned.

EASTER ISLAND: Similar events happened elsewhere. Easter Island is located 3,000 km from South America and 2,000 km from the nearest occupied land (the tiny, isolated Pitcairn Island). Easter Island has a mild climate and fertile volcanic soil, but when Europeans discovered it in the 1700s, it was covered in dry grassland, lacking trees or any vegetation above 1m high. Around 2,000 people survived on the island by subsistence farming, yet all around stood huge stone statues, some 30 m tall and weighing over 200 tonnes. Clearly a much larger more advanced society had been living on the island at some time in the past. Archaeological studies have found that populations reached 20,000 people prior to 1500AD. Exhaustion of the island's resources by the population was followed by war and civil unrest and the population fell to the subsistence levels found in the 1700s.

EUROPE: Despite isolated events, the world population continued to grow so that by 1350 AD it had reached around 450 million. As a result of the continued rise of urban populations, often living in squalid conditions, disease spread rapidly. The bubonic plague, which swept through Europe at this time, reduced its population almost by half, and reduced the world's population to 350 million. Despite further outbreaks of plague and the huge death tolls of various wars, the human population had reached 2.5 billion by 1950. By 1990 it was 5 billion and today it is around 6.5 billion. In slightly less than 60 years the human population has grown almost twice as much as it did in the whole of human history up until 1950. Much of this growth can be attributed to major advances in agriculture and medicine. However, signs are appearing that the human population are approaching maximum sustainable levels. Annual crop yields have ceased increasing and many common illnesses are becoming more difficult to treat. The rapid spread of modern pandemics, such as H1N1 swine flu, illustrates the vulnerability of modern human populations. Could it be, perhaps, that another great reduction in the human population is imminent?

1. Describe the general trend of human population growth over the last 100,000 years: ______

2. Explain why the human population has grown at such a increased rate in the last 60 years: ______

3. Discuss similarities between the events at Tikal and on the Easter Islands and how they can help us plan for the future: ______

Features of Populations

Populations have a number of attributes that may be of interest. Usually, biologists wish to determine **population size** (the total number of organisms in the population). It is also useful to know the **population density** (the number of organisms per unit area). The density of a population is often a reflection of the **carrying capacity** of the environment, i.e. how many organisms a particular environment can support. Populations also have structure; particular ratios of different ages and sexes. These data enable us to determine whether the population is declining or increasing in size. We can also look at the **distribution** of organisms within their environment and so determine what particular aspects of the habitat are favored over others. One way to retrieve information from populations is to **sample** them. Sampling involves collecting data about features of the population from samples of that population (since populations are usually too large to examine in total). Sampling can be done directly through a number of sampling methods or indirectly (e.g. monitoring calls, looking for droppings or other signs). Some of the population attributes that we can measure or calculate are illustrated on the diagram below.

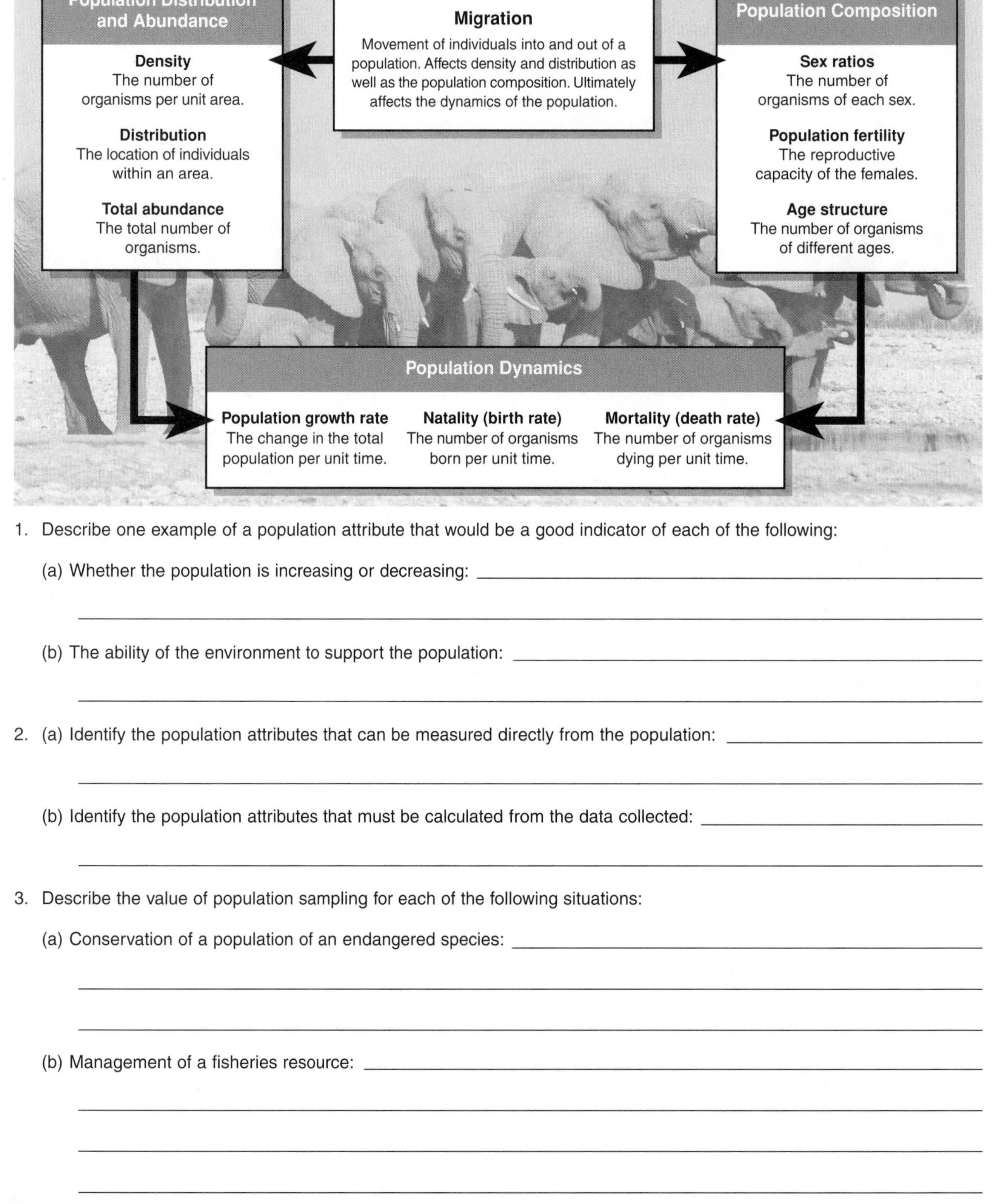

Population Distribution and Abundance
Density
The number of organisms per unit area.
Distribution
The location of individuals within an area.
Total abundance
The total number of organisms.
Migration
Movement of individuals into and out of a population. Affects density and distribution as well as the population composition. Ultimately affects the dynamics of the population.
Population Composition
Sex ratios
The number of organisms of each sex.
Population fertility
The reproductive capacity of the females.
Age structure
The number of organisms of different ages.
Population Dynamics
Population growth rate
The change in the total population per unit time.
Natality (birth rate)
The number of organisms born per unit time.
Mortality (death rate)
The number of organisms dying per unit time.

1. Describe one example of a population attribute that would be a good indicator of each of the following:

   (a) Whether the population is increasing or decreasing: __________

   (b) The ability of the environment to support the population: __________

2. (a) Identify the population attributes that can be measured directly from the population: __________

   (b) Identify the population attributes that must be calculated from the data collected: __________

3. Describe the value of population sampling for each of the following situations:

   (a) Conservation of a population of an endangered species: __________

   (b) Management of a fisheries resource: __________

Related activities: Density and Distribution, Population Age Structure

Sampling Populations

In most ecological studies, it is not possible to measure or count all the members of a population. Instead, information is obtained through **sampling** in a manner that provides a fair (unbiased) representation of the organisms present and their distribution. This is usually achieved through **random sampling**, a technique in which every possible sample of a given size has the same chance of selection. Most practical exercises in community ecology involve the collection or census of living organisms, with a view to identifying the species and quantifying their abundance and other population features of interest. Sampling techniques must be appropriate to the community being studied and the information you wish to obtain. Any field study must also consider the time and equipment available, the organisms involved, and the impact of the sampling method on the environment. Often indicator species and **species diversity indices** are used as a way of quantifying biodiversity and ecosystem "health". Such indicators can be particularly useful when monitoring ecosystem change and looking for causative factors in species loss.

Quantifying the Diversity of Ecosystems

The methods we use to sample communities and their constituent populations must be appropriate to the ecosystem being investigated. Communities in which the populations are at low density and have a random or clumped distribution will require a different sampling strategy to those where the populations are uniformly distributed and at higher density. There are many sampling options, each with advantages and drawbacks for particular communities. How would you assess aspects (e.g. species richness, abundance, or distribution) of the reef community above?

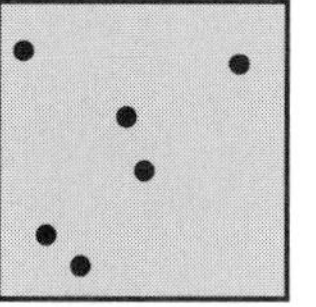
Random point sampling

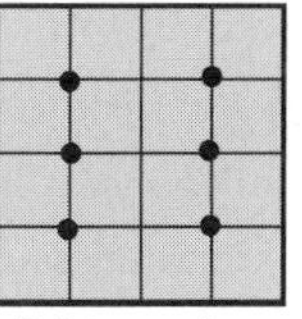
Point sampling: systematic grid

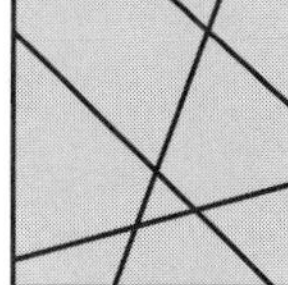
Line and belt transects

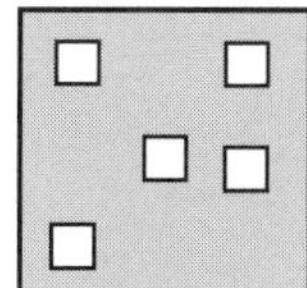
Random quadrats

Photo: www.coastal planning.net

Marine ecologists use quadrat sampling to estimate biodiversity prior to works such as dredging.

RA

Line transects are appropriate to estimate biodiversity along an environmental gradient.

Tagging has been used for more than 30 years to follow the migration of monarch butterflies. The photograph here depicts an older tagging method, which has largely been replaced by a tag on the underside of the hindwing (inset). The newer method results in better survival and recapture rates and interferes less with flight.

Which Sampling Method?

Field biologists take a number of factors into consideration when deciding on a sampling method for a chosen population or community. The benefits and drawbacks of some common methods are outlined below:

Point sampling is time efficient and good for determining species abundance and community composition. However, organisms in low abundance may be missed.

Transects are well suited to determining changes in community composition along an environmental gradient but can be time consuming to do well.

Quadrats are also good for assessments of community diversity and compostion but are largely restricted to plants and immobile animals. Quadrat size must also be appropriate for the organisms being sampled.

Mark and recapture is useful for highly mobile species which are otherwise difficult to record. However, it is time consuming to do well. **Radiotracking** (below) offers an alternative to mark and recapture and is now widely used in conservation to study the movements of both threatened species and pests.

Radio-tracking technology is now widely used in many aspects of conservation work. Radio-tracking has been used to study the movements and habitat use of a diverse range of animal species. The information allows conservation organisations to develop better strategies for managing these species in the wild or follow the progress of reintroduced captive-bred animals.

PH

Australian brush-tailed possums are a major pest in New Zealand where they damage native forests, and prey on native birds. Radio-tracking is used on possums in critical conservation areas in New Zealand to determine dispersal rates, distribution, and habitat use, so that pest control can be implemented more effectively.

Periodicals: *Bowels of the beast*

Related activities: *Quadrat Sampling, Transect Sampling, Mark and Recapture Sampling* ***Web links:*** *Ecological Sampling Methods*

Using Dataloggers in Field Studies

Usually, when we collect information about populations in the field, we also collect information about the physical environment. This provides important information about the local habitat and can be useful in assessing habitat preference. With the advent of **dataloggers**, collecting this information is straightforward.

Dataloggers are electronic instruments that record measurements over time. They are equipped with a microprocessor, data storage facility, and sensor. Different sensors are used to measure a range of variables in water or air. The datalogger is connected to a computer, and software is used to set the limits of operation (e.g. the sampling interval) and initiate the logger. The logger is then disconnected and used remotely to record and store data. When reconnected to the computer, the data are downloaded, viewed, and plotted. Dataloggers make data collection quick and accurate, and they enable prompt data analysis.

Datalogger photos courtesy of PASCO

Dataloggers fitted with sensors are portable and easy to use in a wide range of aquatic (left) and terrestrial (right) environments. Different variables can be measured by changing the sensor attached to the logger.

1. Explain why we **sample** populations: ______

2. Describe a sampling technique that would be appropriate for determining each of the following:

   (a) The percentage cover of a plant species in pasture: ______

   (b) The density and age structure of a plankton population: ______

   (c) Change in community composition from low to high altitude on a mountain: ______

3. Explain why it is common practice to also collect information about the physical environment when sampling populations:

Quadrat 5
Quadrat 4
Quadrat 3
Quadrat 2
Quadrat 1

0 50 100

Percentage cover

Red stem moss
Fern moss
Snake moss
Star moss
Eye brow moss
Broad leaved star moss
Tree moss
Lichens (various species)

QUADRAT	1	2	3	4	5
Height / m	0.4	0.8	1.2	1.6	2.0
Light / arbitrary units	40	56	68	72	72
Humidity / percent	99	88	80	76	78
Temperature / °C	12.1	12.2	13	14.3	14.2

Lichen

Moss

4. The figure (above) shows the changes in vegetation cover along a 2 m vertical transect up the trunk of an oak tree (*Quercus*). Changes in the physical factors light, humidity, and temperature along the same transect were also recorded. From what you know about the ecology of mosses and lichens, account for the observed vegetation distribution:

Density and Distribution

Distribution and density are two interrelated properties of populations. Population density is the number of individuals per unit area (for land organisms) or volume (for aquatic organisms). Careful observation and precise mapping can determine the distribution patterns for a species. The three basic distribution patterns are: random, clumped and uniform. In the diagram below, the circles represent individuals of the same species. It can also represent populations of different species.

Low Density

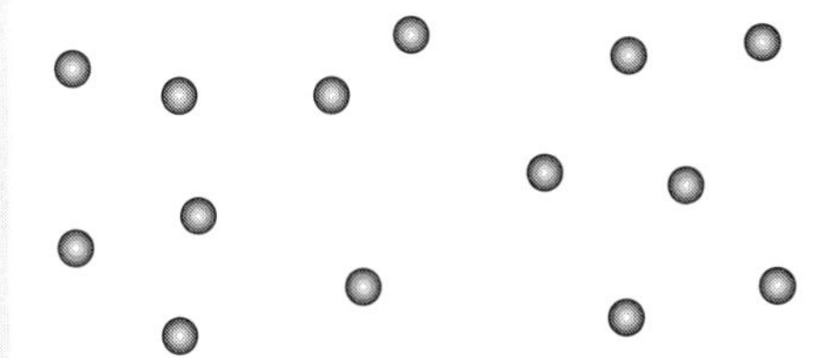

In low density populations, individuals are spaced well apart. There are only a few individuals per unit area or volume (e.g. highly territorial, solitary mammal species).

High Density

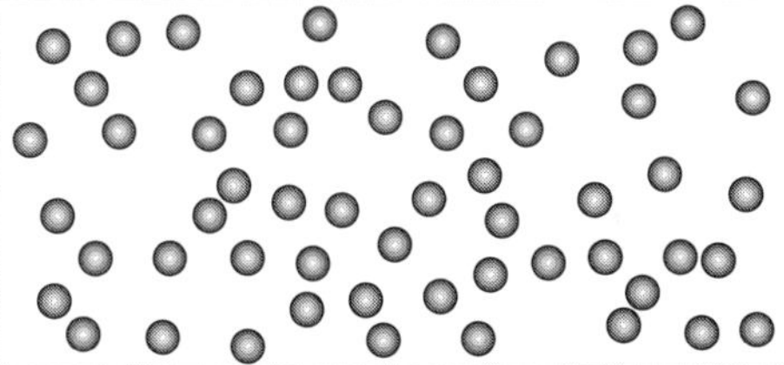

In high density populations, individuals are crowded together. There are many individuals per unit area or volume (e.g. colonial organisms, such as many corals).

Tigers are solitary animals, found at low densities.

Termites form well organized, high density colonies.

Random Distribution

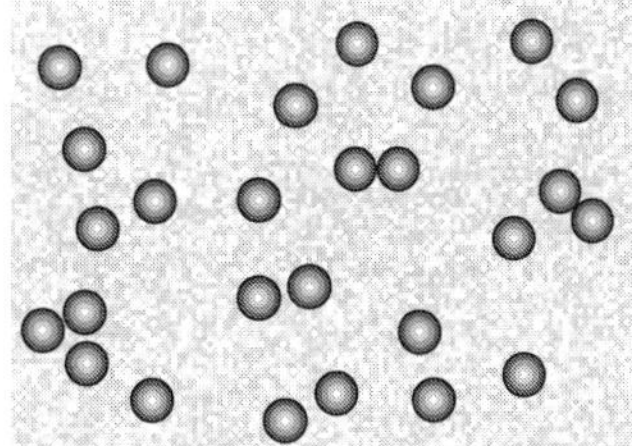

Random distributions occur when the spacing between individuals is irregular. The presence of one individual does not directly affect the location of any other individual. Random distributions are uncommon in animals but are often seen in plants.

Clumped Distribution

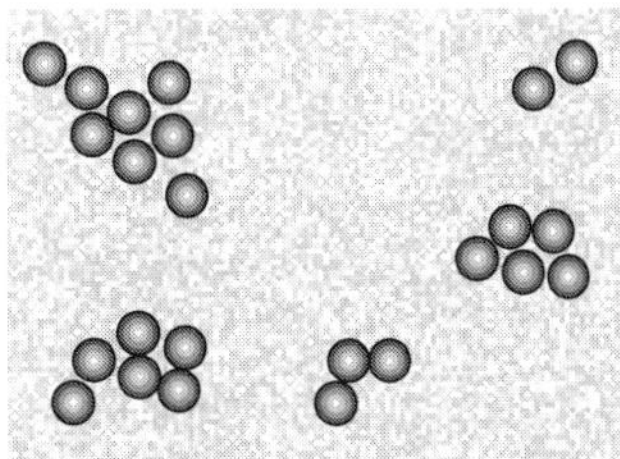

Clumped distributions occur when individuals are grouped in patches (sometimes around a resource). The presence of one individual increases the probability of finding another close by. Such distributions occur in herding and highly social species.

Uniform Distribution

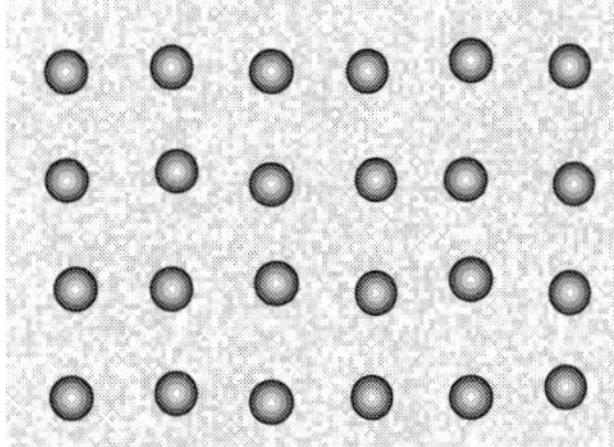

Regular distribution patterns occur when individuals are evenly spaced within the area. The presence of one individual decreases the probability of finding another individual very close by. The penguins illustrated above are also at a high density.

1. Describe why some organisms may exhibit a clumped distribution pattern because of:

   (a) Resources in the environment: ______

   (b) A group social behavior: ______

2. Describe a social behavior found in some animals that may encourage a uniform distribution: ______

3. Describe the type of environment that would encourage uniform distribution: ______

4. Describe an example of each of the following types of distribution pattern:

   (a) Clumped: ______

   (b) Random (more or less): ______

   (c) Uniform (more or less): ______

Quadrat Sampling

Quadrat sampling is a method by which organisms in a certain proportion (sample) of the habitat are counted directly. As with all sampling methods, it is used to estimate population parameters when the organisms present are too numerous to count in total. It can be used to estimate population **abundance** (number), **density**, **frequency of occurrence**, and **distribution**. Quadrats may be used without a transect when studying a relatively uniform habitat. In this case, the quadrat positions are chosen randomly using a random number table.

The general procedure is to count all the individuals (or estimate their percentage cover) in a number of quadrats of known size and to use this information to work out the abundance or percentage cover value for the whole area. The number of quadrats used and their size should be appropriate to the type of organism involved (e.g. grass vs tree).

Quadrat

Area being sampled

$$\text{Estimated average density} = \frac{\text{Total number of individuals counted}}{\text{Number of quadrats X area of each quadrat}}$$

Guidelines for Quadrat Use:

1. The **area of each quadrat** must be known exactly and ideally quadrats should be the same shape. The quadrat does not have to be square (it may be rectangular, hexagonal etc.).
2. **Enough quadrat samples** must be taken to provide results that are representative of the total population.
3. The **population of each quadrat** must be known exactly. Species must be distinguishable from each other, even if they have to be identified at a later date. It has to be decided beforehand what the count procedure will be and how organisms over the quadrat boundary will be counted.
4. The size of the quadrat should be appropriate to the organisms and habitat, e.g. a large size quadrat for trees.
5. The quadrats must be **representative of the whole area**. This is usually achieved by **random sampling** (right).

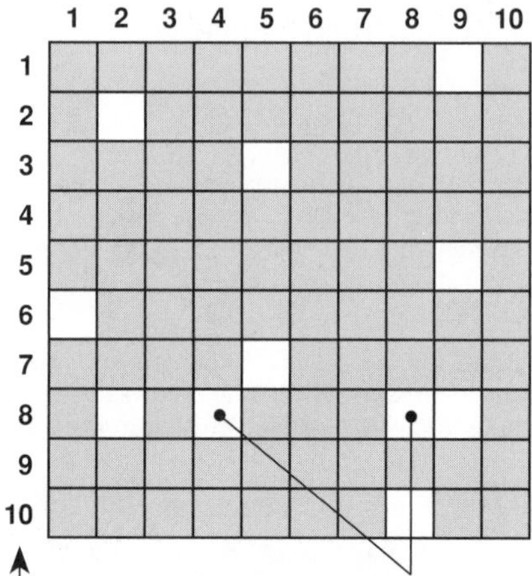

The area to be sampled is divided up into a grid pattern with indexed coordinates

Quadrats are applied to the predetermined grid on a random basis. This can be achieved by using a random number table.

Sampling a centipede population

A researcher by the name of Lloyd (1967) carried out a sampling of centipedes in Wytham Woods, near Oxford in England. A total of 37 hexagon–shaped quadrats were used, with a diameter of 30 cm (see diagram on right). These were arranged in a pattern so that they were all touching each other. Use the data in the diagram to answer the following questions:

1. Determine the average number of centipedes captured per quadrat:

____

2. Calculate the estimated average density of centipedes per square meter (remember that each quadrat is 0.08 square meters in area):

____

3. Looking at the data for individual quadrats, describe in general terms the distribution of the centipedes in the sample area:

____

4. Describe one factor that might account for the distribution pattern:

____

____

____

0 2 0 0
1 0 2 0 2
0 0 0 0 0 2
0 2 0 1 1 1 2
1 1 0 0 3 1
0 3 1 2 0
1 0 1 0

Each quadrat was a hexagon with a diameter of 30 cm and an area of 0.08 square meters.

The number in each hexagon indicates how many centipedes were caught in that quadrat.

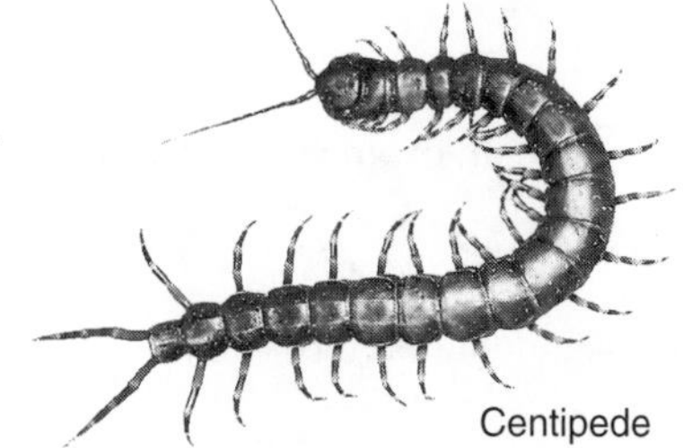

Centipede

Related activities: *Density and Distribution* ***Web links:*** *Investigating Marine Life, Using Quadrats to Sample, Sampling a Leaf Litter Population*

Quadrat-Based Estimates

The simplest description of a plant community in a habitat is a list of the species that are present. This qualitative assessment of the community has the limitation of not providing any information about the **relative abundance** of the species present. Quick estimates can be made using **abundance scales**, such as the ACFOR scale described below. Estimates of percentage cover provide similar information. These methods require the use of **quadrats**. Quadrats are used extensively in plant ecology. This activity outlines some of the common considerations when using quadrats to sample plant communities.

What Size Quadrat?

Quadrats are usually square, and cover 0.25 m^2 (0.5 m x 0.5 m) or 1 m^2, but they can be of any size or shape, even a single point. The quadrats used to sample plant communities are often 0.25 m^2. This size is ideal for low-growing vegetation, but quadrat size needs to be adjusted to habitat type. The quadrat must be large enough to be representative of the community, but not so large as to take a very long time to use.

A quadrat covering an area of 0.25 m^2 is suitable for most low growing plant communities, such as this alpine meadow, fields, and grasslands.

Larger quadrats (e.g. 1 m^2) are needed for communities with shrubs and trees. Quadrats as large as 4 m x 4 m may be needed in woodlands.

Small quadrats (0.01 m^2 or 100 mm x 100 mm) are appropriate for lichens and mosses on rock faces and tree trunks.

How Many Quadrats?

As well as deciding on a suitable quadrat size, the other consideration is how many quadrats to take (the sample size). In species-poor or very homogeneous habitats, a small number of quadrats will be sufficient. In species-rich or heterogeneous habitats, more quadrats will be needed to ensure that all species are represented adequately.

Determining the number of quadrats needed

- Plot the cumulative number of species recorded (on the *y* axis) against the number of quadrats already taken (on the *x* axis).
- The point at which the curve levels off indicates the suitable number of quadrats required.

Fewer quadrats are needed in species-poor or very uniform habitats, such as this bluebell woodland.

Describing Vegetation

Density (number of individuals per unit area) is a useful measure of abundance for animal populations, but can be problematic in plant communities where it can be difficult to determine where one plant ends and another begins. For this reason, plant abundance is often assessed using **percentage cover**. Here, the percentage of each quadrat covered by each species is recorded, either as a numerical value or using an abundance scale such as the ACFOR scale.

The ACFOR Abundance Scale

A = Abundant (30% +)
C = Common (20-29%)
F = Frequent (10-19%)
O = Occasional (5-9%)
R = Rare (1-4%)

The AFCOR scale could be used to assess the abundance of species in this wildflower meadow. Abundance scales are subjective, but it is not difficult to determine which abundance category each species falls into.

1. Describe one difference between the methods used to assess species abundance in plant and in animal communities:

2. Identify the main consideration when determining appropriate quadrat size:

3. Identify the main consideration when determining number of quadrats:

4. Explain two main disadvantages of using the ACFOR abundance scale to record information about a plant community:

(a)

(b)

Population Regulation

Very few species show continued exponential growth. Population size is regulated by factors that limit population growth. The diagram below illustrates how population size can be regulated by environmental factors. **Density independent factors** may affect all individuals in a population equally. Some, however, may be better able to adjust to them. **Density dependent factors** have a greater affect when the population density is higher. They become less important when the population density is low.

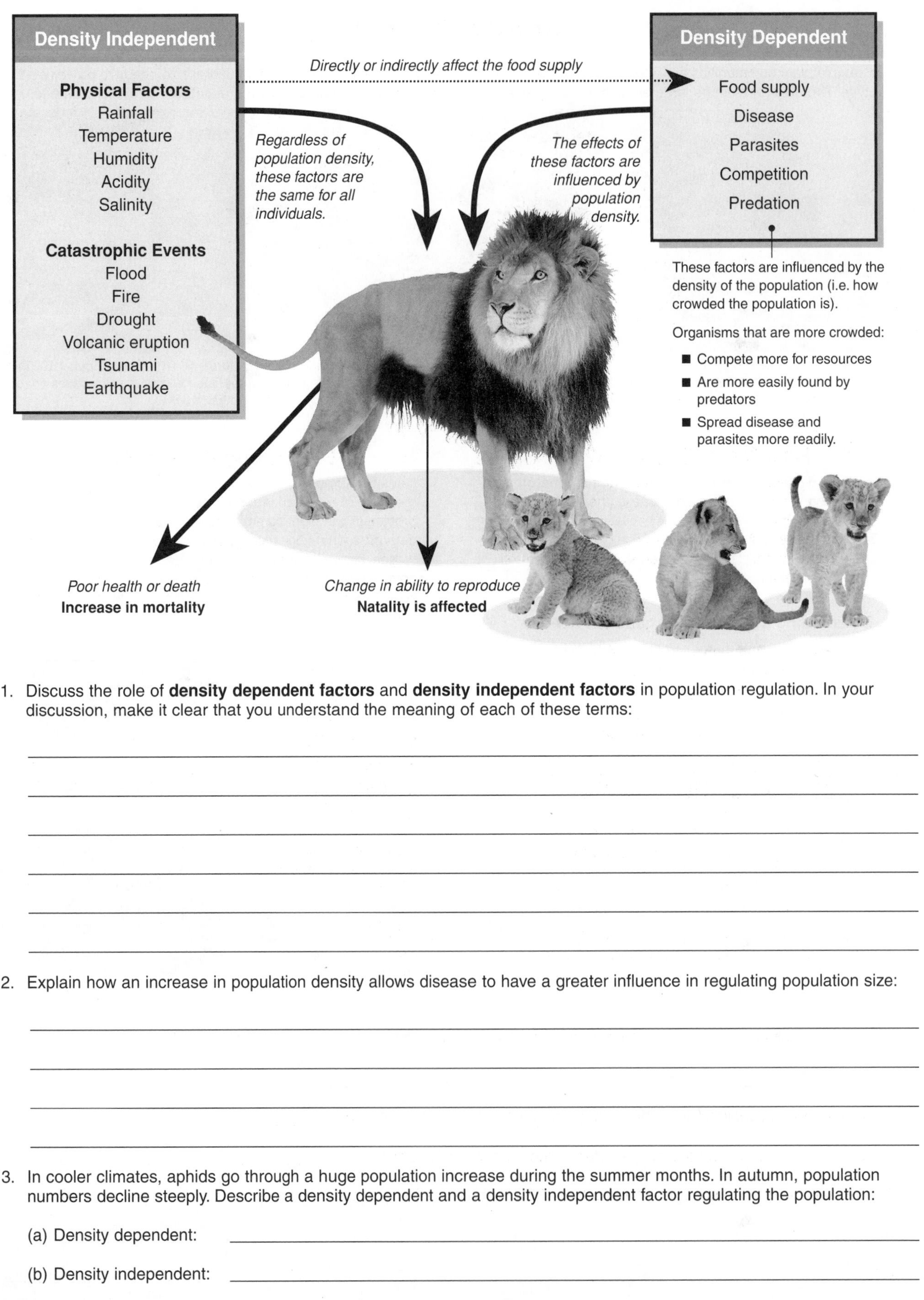
Density Independent
Physical Factors
Rainfall
Temperature
Humidity
Acidity
Salinity
Catastrophic Events
Flood
Fire
Drought
Volcanic eruption
Tsunami
Earthquake
Directly or indirectly affect the food supply
Regardless of population density, these factors are the same for all individuals.
The effects of these factors are influenced by population density.
Density Dependent
Food supply
Disease
Parasites
Competition
Predation
These factors are influenced by the density of the population (i.e. how crowded the population is).
Organisms that are more crowded:
Compete more for resources
Are more easily found by predators
Spread disease and parasites more readily.
Poor health or death
Increase in mortality
Change in ability to reproduce
Natality is affected

1. Discuss the role of **density dependent factors** and **density independent factors** in population regulation. In your discussion, make it clear that you understand the meaning of each of these terms:

2. Explain how an increase in population density allows disease to have a greater influence in regulating population size:

3. In cooler climates, aphids go through a huge population increase during the summer months. In autumn, population numbers decline steeply. Describe a density dependent and a density independent factor regulating the population:

   (a) Density dependent:

   (b) Density independent:

Related activities: *Density and Distribution, r and K Selection*

Population Growth

Organisms do not generally live alone. A **population** is a group of organisms of the same species living together in one geographical area. This area may be difficult to define as populations may comprise widely dispersed individuals that come together only infrequently (e.g. for mating). The number of individuals comprising a population may also fluctuate considerably over time. These changes make populations dynamic: populations gain individuals through births or immigration, and lose individuals through deaths and emigration. For a population in **equilibrium**, these factors balance out and there is no net change in the population abundance. When losses exceed gains, the population declines.

Births, deaths, immigrations (movements into the population) and emigrations (movements out of the population) are events that determine the numbers of individuals in a population. Population growth depends on the number of individuals added to the population from births and immigration, minus the number lost through deaths and emigration. This is expressed as:

Population growth =

Births – Deaths + Immigration – Emigration
(B) (D) (I) (E)

The difference between immigration and emigration gives net migration. Ecologists usually measure the **rate** of these events. These rates are influenced by environmental factors (see below) and by the characteristics of the organisms themselves. Rates in population studies are commonly expressed in one of two ways:

- Numbers per unit time, e.g. 20,150 live births per year.
- Per capita rate (number per head of population), e.g. 122 live births per 1000 individuals per year (12.2%).

Limiting Factors

Population size is also affected by limiting factors; factors or resources that control a process such as organism growth, or population growth or distribution. Examples include availability of food, predation pressure, or available habitat.

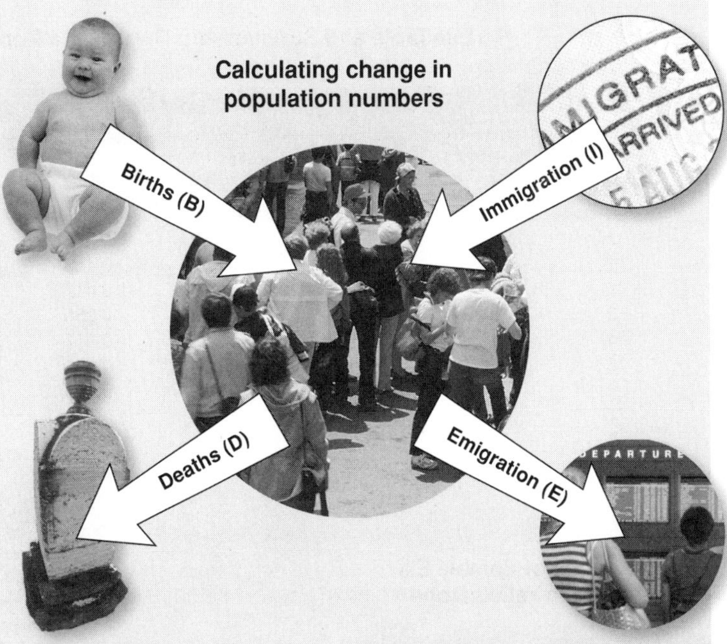

Human populations often appear exempt from limiting factors as technology and efficiency solve many food and shelter problems. However, as the last arable land is used and agriculture reaches its limits of efficiency, it is estimated that the human population may peak at around 10 billion by 2050.

1. Define the following terms used to describe changes in population numbers:

   (a) Death rate (mortality): ______

   (b) Birth rate (natality): ______

   (c) Net migration rate: ______

2. Explain how the concept of limiting factors applies to population biology: ______

3. Using the terms, B, D, I, and E (above), construct equations to express the following (the first is completed for you):

   (a) A population in equilibrium: B + I = D + E

   (b) A declining population: ______

   (c) An increasing population: ______

4. A population started with a total number of 100 individuals. Over the following year, population data were collected. Calculate birth rates, death rates, net migration rate, and rate of population change for the data below (as percentages):

   (a) Births = 14: Birth rate = ______ (b) Net migration = +2: Net migration rate = ______

   (c) Deaths = 20: Death rate = ______ (d) Rate of population change = ______

   (e) State whether the population is increasing or declining: ______

5. The human population is around 6.7 billion. Describe and explain two limiting factors for population growth in humans:

______

Life Tables and Survivorship

Life tables, such as those shown below, provide a summary of mortality for a population (usually for a group of individuals of the same age or **cohort**). The basic data are just the number of individuals remaining alive at successive sampling times (the **survivorship** or **l*x***). Life tables are an important tool when analyzing changes in populations over time. They can tell us the ages at which most mortality occurs in a population and can also provide information about life span and population age structure. Biologists use the l*x* column of a basic life table to derive a survivorship curve. Survivorship curves are standardized as the number of survivors per 1000 individuals so that populations of different types can be easily compared.

Life Table and Survivorship Curve for a Population of the Barnacle *Balanus*

Age in years (x)	No. alive each year (N_x)	Proportion surviving at the start of age x (l_x)	Proportion dying between x and $x+1$ (d_x)	Mortality (q_x)
0	142	1.000	0.563	0.563
1	62	0.437	0.198	0.452
2	34	0.239	0.098	0.412
3	20	0.141	0.035	0.250
4	15	0.106	0.028	0.267
5	11	0.078	0.036	0.454
6	6	0.042	0.028	0.667
7	2	0.014	0.0	0.000
8	2	0.014	0.014	1.000
9	0	0.0	0.0	–

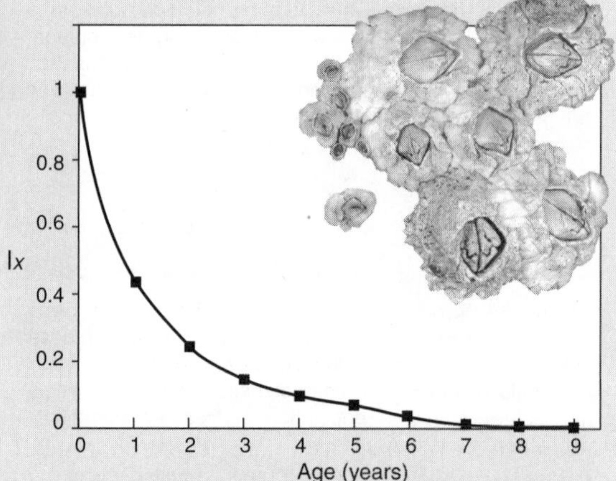

Life Table for Female Elk, Northern Yellowstone

x	l_x	d_x	q_x
0	1000	323	.323
1	677	13	.019
2	664	2	.003
3	662	2	.003
4	660	4	.006
5	656	4	.006
6	652	9	.014
7	643	3	.005
8	640	3	.005
9	637	9	.014
10	628	7	.001
11	621	12	.019
12	609	13	.021
13	596	41	.069
14	555	34	.061
15	521	20	.038
16	501	59	.118
17	442	75	.170
18	367	93	.253
19	274	82	.299
20	192	57	.297
21+	135	135	1.000

Survivorship Curve for Female Elk of Northern Yellowstone National Park

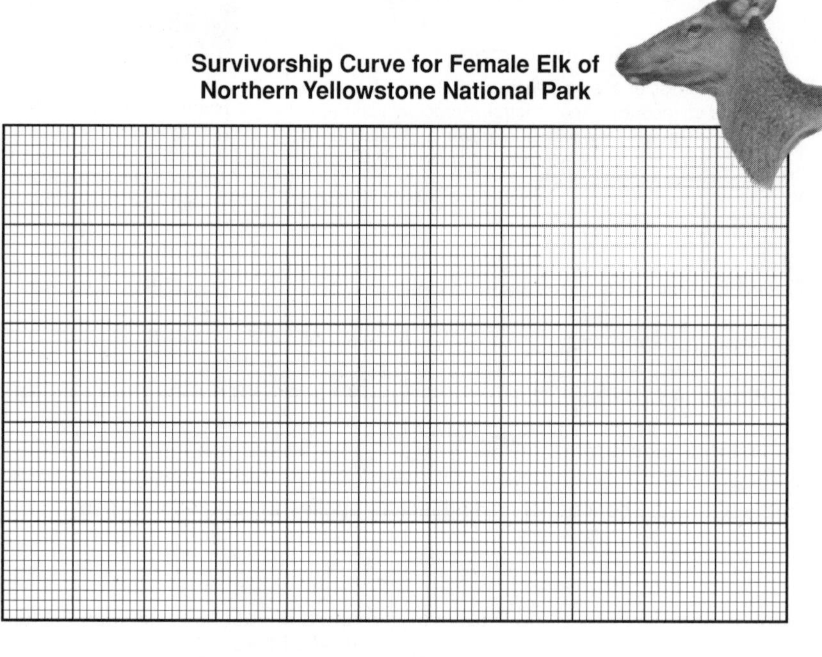

1. (a) In the example of the barnacle *Balanus* above, state when most of the group die: ________________

   (b) Identify the type of survivorship curve is represented by these data (see opposite): ________________

2. (a) Using the grid, plot a survivorship curve for elk hinds (above) based on the life table data provided:

   (b) Describe the survivorship curve for these large mammals: ________________

3. Explain how a biologist might use life table data to manage an endangered population: ________________

Survivorship Curves

The survivorship curve depicts age-specific mortality. It is obtained by plotting the number of individuals of a particular cohort against time. Survivorship curves are standardized to start at 1000 and, as the population ages, the number of survivors progressively declines. The shape of a survivorship curve thus shows graphically at which life stages the highest mortality occurs. Survivorship curves in many populations fall into one of three hypothetical patterns (below). Wherever the curve becomes steep, there is an increase in mortality. The convex Type I curve is typical of populations whose individuals tend to live out their physiological life span. Such populations usually produce fewer young and show some degree of parental care. Organisms that suffer high losses of the early life stages (a Type III curve) compensate by producing vast numbers of offspring. These curves are conceptual models only, against which real life curves can be compared. Many species exhibit a mix of two of the three basic types. Some birds have a high chick mortality (Type III) but adult mortality is fairly constant (Type II). Some invertebrates (e.g. crabs) have high mortality only when molting and show a stepped curve.

Hypothetical Survivorship Curves

Type I

Late loss survivorship curve

Mortality (death rate) is very low in the infant and juvenile years, and throughout most of adult life. Mortality increases rapidly in old age. **Examples**: Humans (in developed countries) and many other large mammals (e.g. big cats, elephants).

Type II

Constant loss survivorship curve

Mortality is relatively constant through all life stages (no one age is more susceptible than another). **Examples**: Some invertebrates such as *Hydra*, some birds, some annual plants, some lizards, and many rodents.

Type III

Early loss survivorship curve

Mortality is very high during early life stages, followed by a very low death rate for the few individuals reaching adulthood. **Examples**: Many fish (not mouth brooders) and most marine invertebrates (e.g. oysters, barnacles).

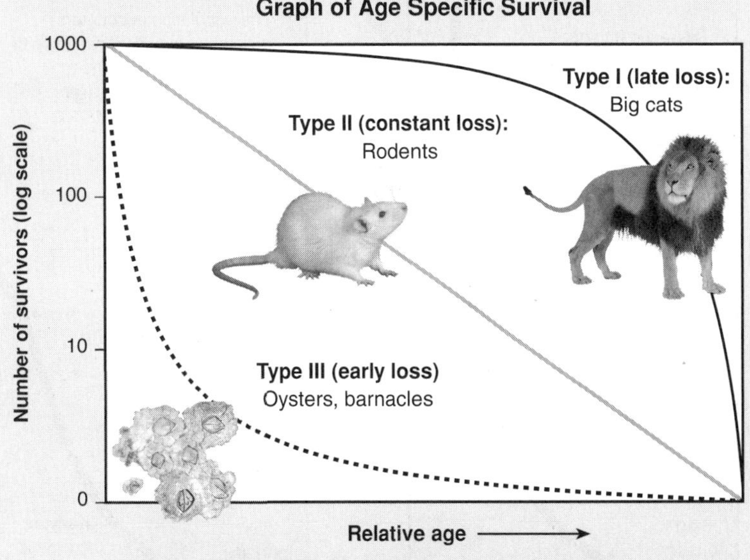

Three basic types of survivorship curves and representative organisms for each type. The vertical axis may be scaled arithmetically or logarithmically.

Elephants have a close matriarchal society and a long period of parental care. Elephants are long-lived and females usually produce just one calf.

Rodents are well known for their large litters and prolific breeding capacity. Individuals are lost from the population at a more or less constant rate.

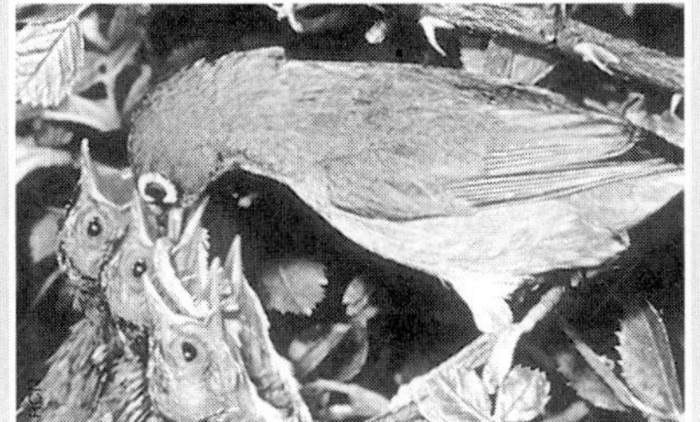

Despite vigilant parental care, many birds suffer high juvenile losses (Type III). For those surviving to adulthood, deaths occur at a constant rate.

1. Explain why human populations might not necessarily show a Type I curve: ______________________

____

2. Explain how organisms with a Type III survivorship compensate for the high mortality during early life stages:

____

3. Describe the features of a species with a Type I survivorship that aid in high juvenile survival: ______________________

____

4. Discuss the following statement: "There is no standard survivorship curve for a given species; the curve depicts the nature of a population at a particular time and place and under certain environmental conditions.":

____

____

____

Population Growth Curves

Populations becoming established in a new area for the first time are often termed **colonizing populations** (below, left). They may undergo a rapid **exponential** (logarithmic) increase in numbers as there are plenty of resources to allow a high birth rate, while the death rate is often low. Exponential growth produces a J-shaped growth curve that rises steeply as more and more individuals contribute to the population increase. If the resources of the new habitat were endless (inexhaustible) then the population would continue to increase at an **exponential** rate. However, this rarely happens in natural populations. Initially, growth may be exponential (or nearly so), but as the population grows, its increase will slow and it will stabilize at a level that can be supported by the environment (called the carrying capacity or K). This type of growth is called sigmoidal and produces the **logistic growth curve** (below, right). **Established populations** will fluctuate about K, often in a regular way (gray area on the graph below, right). Some species will have populations that vary little from this stable condition, while others may oscillate wildly.

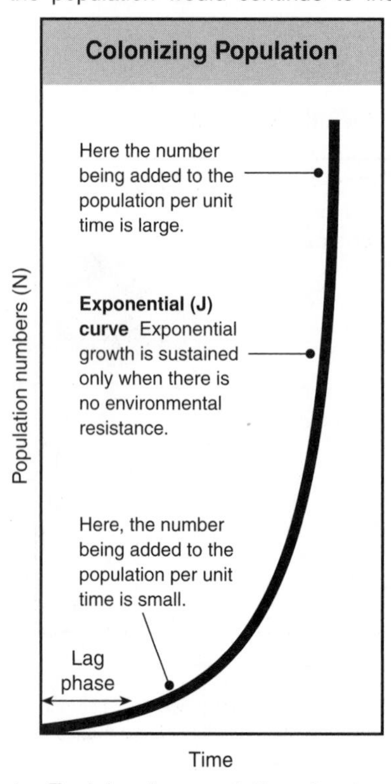
Colonizing Population
Here the number being added to the population per unit time is large.
Exponential (J) curve Exponential growth is sustained only when there is no environmental resistance.
Here, the number being added to the population per unit time is small.
Lag phase
Population numbers (N)
Time

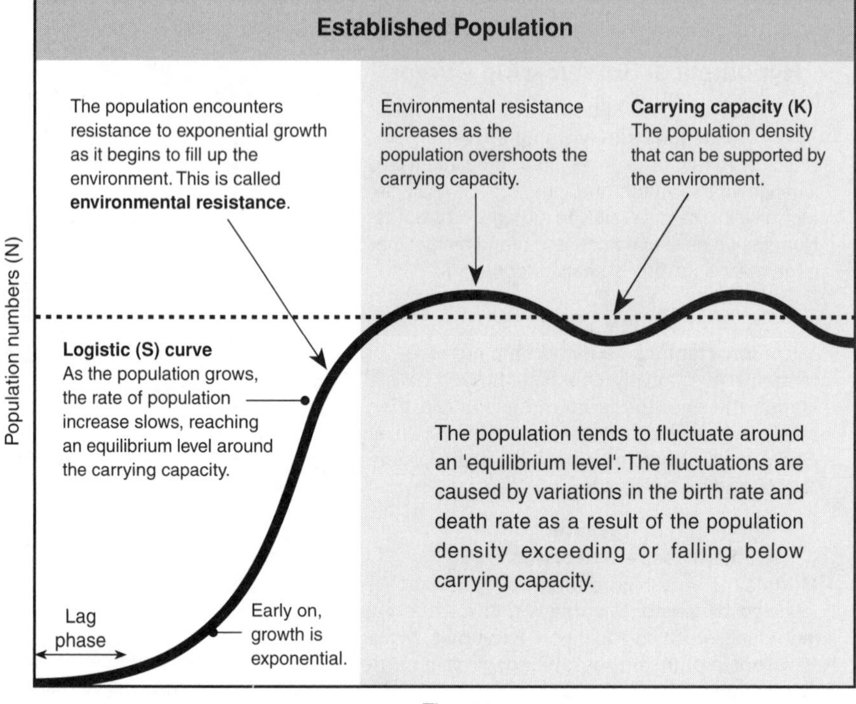
Established Population
The population encounters resistance to exponential growth as it begins to fill up the environment. This is called environmental resistance.
Environmental resistance increases as the population overshoots the carrying capacity.
Carrying capacity (K) The population density that can be supported by the environment.
Logistic (S) curve As the population grows, the rate of population increase slows, reaching an equilibrium level around the carrying capacity.
The population tends to fluctuate around an 'equilibrium level'. The fluctuations are caused by variations in the birth rate and death rate as a result of the population density exceeding or falling below carrying capacity.
Lag phase
Early on, growth is exponential.
Population numbers (N)
Time

1. Explain why populations tend not to continue to increase exponentially in an environment: ____________________

2. Explain what is meant by environmental resistance: ____________________

3. (a) Explain what is meant by carrying capacity: ____________________

   (b) Explain the importance of **carrying capacity** to the growth and maintenance of population numbers: ____________________

4. Species that expand into a new area, such as rabbits did in areas of Australia, typically show a period of rapid population growth followed by a slowing of population growth as density dependent factors become more important and the population settles around a level that can be supported by the carrying capacity of the environment.

   (a) Explain why a newly introduced consumer (e.g. rabbit) would initially exhibit a period of exponential population growth:

   (b) Describe a likely outcome for a rabbit population after the initial rapid increase had slowed: ____________________

5. Describe the effect that introduced grazing species might have on the carrying capacity of the environment:

Human Demography

Human populations through time have undergone demographic shifts related to societal changes and economic development. The demographic transition model (DTM) was developed in 1929 to explain the transformation of countries from high birth rates and high death rates to low birth rates and low death rates as part of their economic development from a pre-industrial to an industrialized economy. The transition involves four stages, or possibly five (with some nations, including England, being recognized as moving beyond stage four). Each stage of the transition reflects the changes in birth and death rates observed in human societies over the last 200 years. Most developed countries are beyond stage three of the model; the majority of developing countries are in stage two or stage three. The model was based on the changes seen in Europe, so these countries follow the DTM relatively well. Many developing countries have moved into stage three. The exceptions include some poor countries, mainly in sub-Saharan Africa and some Middle Eastern countries, which are poor or affected by government policy or civil strife.

Stage one: Birth and death rates balanced but high as a result of starvation and disease.

Stage two: Improvement in food supplies and public health result in reduced death rates.

Stage three moves the population towards stability through a decline in the birth rate.

Stage four: Birth and death rates are both low and the total population is high and stable.

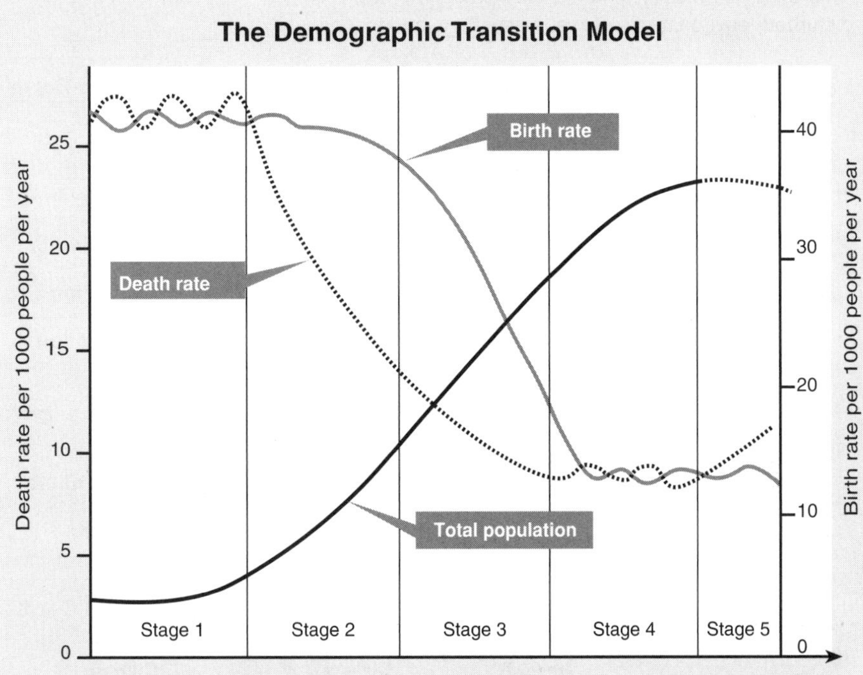

Stage one (pre-modern): A balance between birth and death rates as was true of all populations until the late 18th Century. Children are important contributors to the household economy. Losses as a result of starvation and disease are high. Stage 1 is sometimes called the "High Stationary Stage" of population growth (high birth and death rates and stationary total population numbers).

Stage two (early expanding): Rapid population expansion as a result of a decline in death rates. The changes leading to this stage in Europe were initiated in the Agricultural Revolution of the 18th century but have been more rapid in developing countries since then. Stage two is associated with more reliable food supplies and improvements in public health.

Stage three (late expanding): The population moves towards stability through a decline in the birth rate. This stage is associated with increasing urbanisation and a decreased reliance on children as a source of family wealth. Family planning in nations such as Malaysia (photo left) has been instrumental in their move to stage three.

Stage four (post-industrial): Birth and death rates are both low and the total population is high and stable. The population ages and in some cases the fertility rate falls below replacement.

Stage five (declining): Proposed by some theorists as representing countries that have undergone the economic transition from manufacturing based industries into service and information based industries and the population reproduces well below replacement levels. Countries in stage five include the United Kingdom (the earliest nation recognised as reaching Stage Five) and Germany.

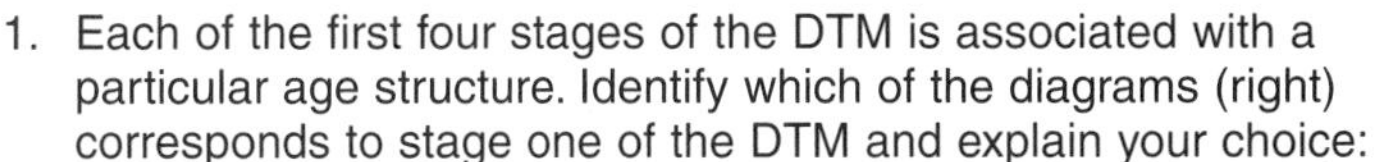

1. Each of the first four stages of the DTM is associated with a particular age structure. Identify which of the diagrams (right) corresponds to stage one of the DTM and explain your choice:

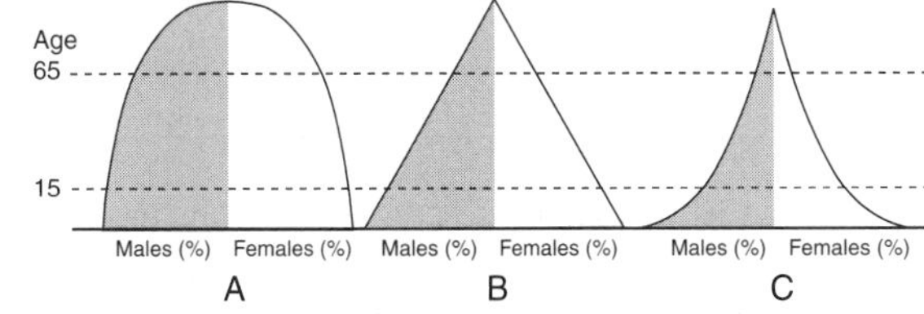

2. Suggest why it might become less important to have a large number of children in more economically developed nations:

r and K Selection

Two parameters govern the logistic growth of populations: the intrinsic rate of natural increase or biotic potential (this is the maximum reproductive potential of an organism, symbolized by an italicized ***r***), and the carrying capacity (saturation density) of the environment (represented by the letter **K**). Species can be characterized by the relative importance of *r* and K in their life cycles. Species with a high intrinsic capacity for population increase are called ***r*-selected species**, and include algae, bacteria, rodents, many insects, and most annual plants. These species show life history features associated with rapid growth in disturbed environments. To survive, they must continually invade new areas to compensate for being replaced by more competitive species. In contrast, **K-selected** species, which include most large mammals, birds of prey, and large, long-lived plants, exist near the carrying capacity of their environments and are pushed in competitive environments to use resources more efficiently. These species have fewer offspring and longer lives, and put their energy into nuturing their young to reproductive age. Most organisms have reproductive patterns between these two extremes. Both *r*-selected species (crops) and K-selected species (livestock) are found in agriculture.

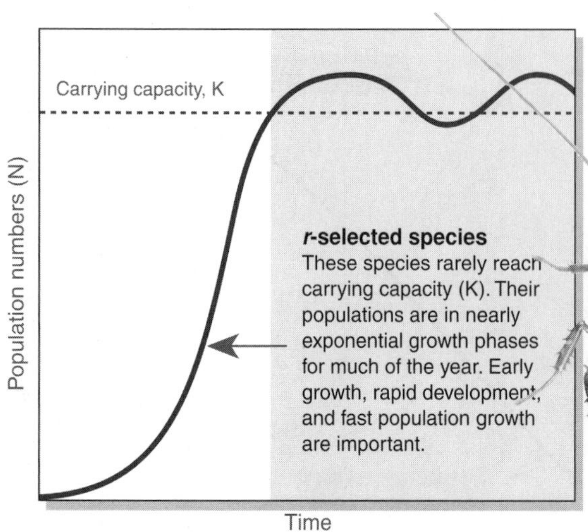

Carrying capacity, K
Population numbers (N)
Time
r-selected species
These species rarely reach carrying capacity (K). Their populations are in nearly exponential growth phases for much of the year. Early growth, rapid development, and fast population growth are important.

Correlates of *r*-Selected Species	
Climate	Variable and/or unpredictable
Mortality	Density independent
Survivorship	Often Type III (early loss)
Population size	Fluctuates wildly. Often below K.
Competition	Variable, often lax. Generalist niche.
Selection favors	Rapid development, high r_m, early reproduction, small body size, single reproduction (annual).
Length of life	Short (usually less than one year)
Leads to:	Productivity

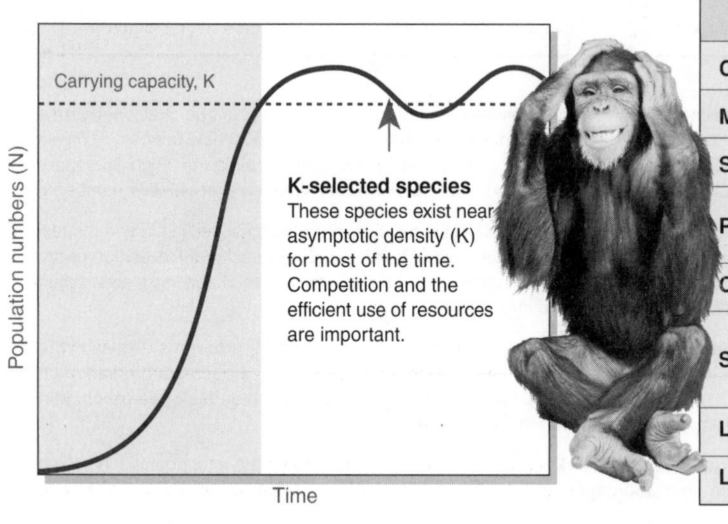

Carrying capacity, K
Population numbers (N)
Time
K-selected species
These species exist near asymptotic density (K) for most of the time. Competition and the efficient use of resources are important.

Correlates of K-Selected Species	
Climate	Fairly constant and/or predictable
Mortality	Density dependent
Survivorship	Types I or II (late or constant loss)
Population size	Fairly constant in time. Near equilibrium with the environment.
Competition	Usually keen. Specialist niche.
Selection favors	Slower development, larger body size, greater competitive ability, delayed reproduction, repeated reproduction.
Length of life	Longer (greater than one year)
Leads to:	Efficiency

1. Explain the significance of the *r* and the K notation when referring to ***r*** and **K selection**:

2. Giving an example, explain why *r*-selected species tend to be **opportunists**:

3. Explain why K-selected species are also called **competitor species**:

4. Suggest why many K-selected species are often vulnerable to extinction:

Related activities: *Life Tables and Survivorship, Population Growth Curves*

Population Age Structure

The **age structure** of a population refers to the relative proportion of individuals in each age group in the population. The age structure of populations can be categorized according to specific age categories (such as years or months), but also by other measures such as life stage (egg, larvae, pupae, instars), of size class (height or diameter in plants). Population growth is strongly influenced by age structure; a population with a high proportion of reproductive and prereproductive aged individuals has a much greater potential for population growth than one that is dominated by older individuals. The ratio of young to adults in a relatively stable population of most mammals and birds is approximately 2:1 (below, left). Growing populations in general are characterized by a large and increasing number of young, whereas a population in decline typically has a decreasing number of young. Population age structures are commonly represented as pyramids, in which the proportions of individuals in each age/size class are plotted with the youngest individuals at the pyramid's base. The number of individuals moving from one age class to the next influences the age structure of the population from year to year. The loss of an age class (e.g. through overharvesting) can profoundly influence a population's viability and can even lead to population collapse.

Age Structures in Animal Populations

These theoretical age pyramids, which are especially applicable to birds and mammals, show how growing populations are characterized by a high ratio of young (white bar) to adult age classes (gray bars). Aging populations with poor production are typically dominated by older individuals.

4
8
12
76

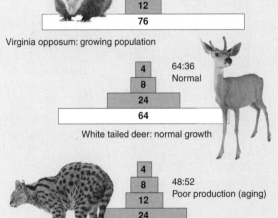

76 young : 24 adults
Rapidly growing population

Virginia opposum: growing population

4
8
24
64

64:36
Normal

White tailed deer: normal growth

4
8
12
24
48

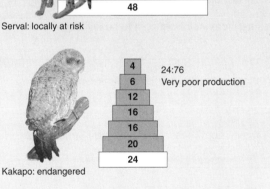

48:52
Poor production (aging)

Serval: locally at risk

4
6
12
16
16
20
24

24:76
Very poor production

Kakapo: endangered

Age Structures in Human Populations

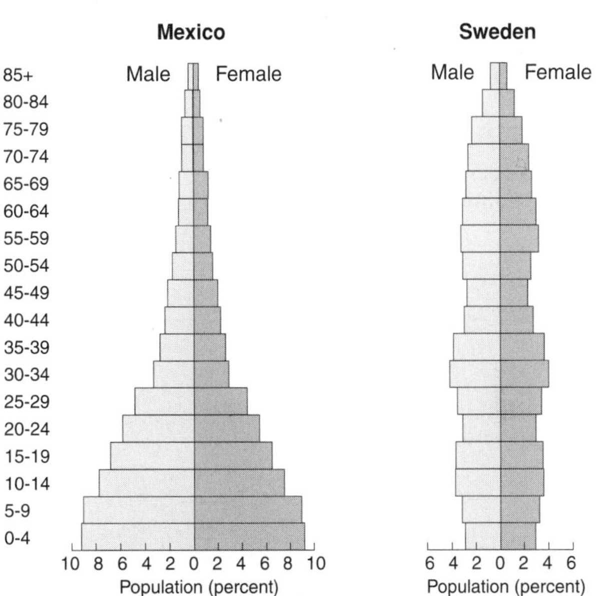

Most of the growth in human populations in recent years has occurred in the developing countries in Africa, Asia, and Central and South America. This is reflected in their age structure; a large proportion of the population comprises individuals younger than 15 years (age pyramid above, left). Even if each has fewer children, the population will continue to increase for many years. The stable age structure of Sweden is shown for comparison.

1. For the theoretical age pyramids above left:

   (a) State the approximate ratio of young to adults in a rapidly increasing population: ____________________

   (b) Suggest why changes in population age structure alone are not necessarily a reliable predictor of population trends:

   ____

   ____

2. Explain why the population of Mexico is likely to continue to increase rapidly even if the rate of population growth slows:

   ____

   ____

Analyzes of the age structure of populations can assist in their management because it can indicate where most population mortality occurs and whether or not reproductive individuals are being replaced. The age structure of plant and animal populations can be examined; a common method is through an analysis of size which is often related to age in a predictable way.

Managed Fisheries

The graphs below illustrate the age structure of a hypothetical fish population under different fishing pressures. The age structure of the population is determined by analyzing the fish catch to determine the frequency of fish in each size (age) class.

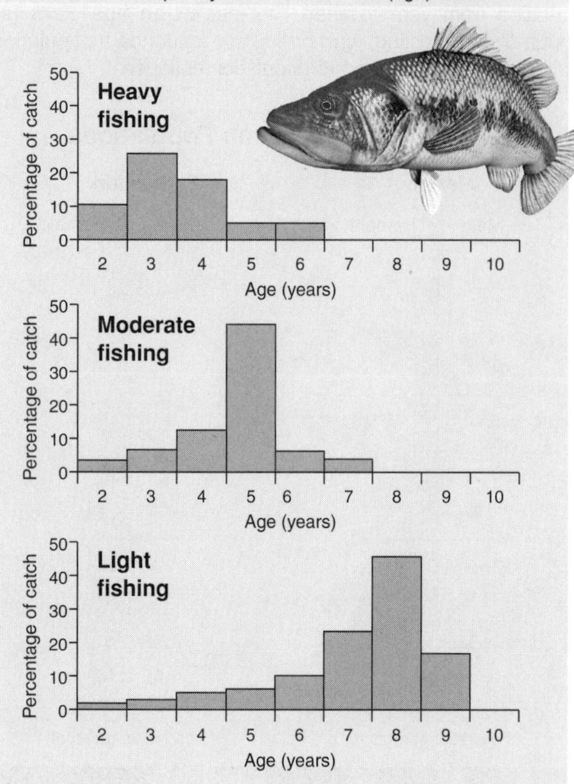

Thatch Palm Populations on Lord Howe Island

Lord Howe Island is a narrow sliver of land approximately 770 km northeast of Sydney. The age structure of populations of the thatch palm *Howea forsteriana* was determined at three locations on the island: the golf course, Gray Face and Far Flats. The height of the stem was used as an indication of age. The differences in age structure between the three sites are mainly due to the extent of grazing at each site.

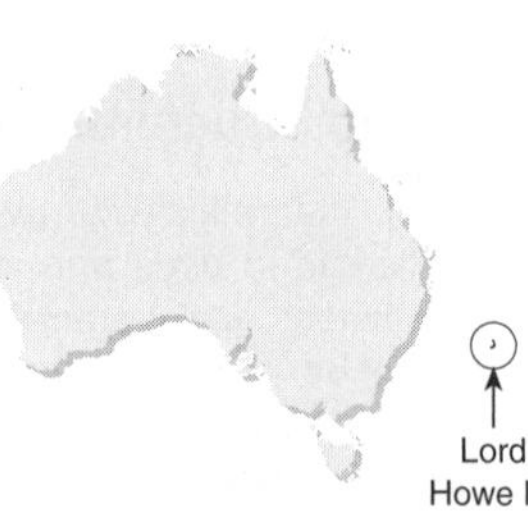

Golf Course

Far Flats

Gray Face

% of population

Stem height (m)

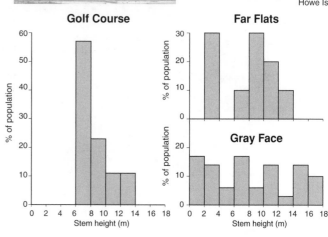

3. For the managed fish population above left:

   (a) Name the general factor that changes the age structure of this fish population: __________

   (b) Describe how the age structure changes when the fishing pressure increases from light to heavy levels:

   ____

4. State the most common age class for each of the above fish populations with different fishing pressures:

   (a) Heavy: __________ (b) Moderate: __________ (c) Light: __________

5. Determine which of the three sites sampled on Lord Howe Island (above, right) best reflects the age structure of:

   (a) An ungrazed population: __________

   Reason for your answer: __________

   (b) A heavily grazed and mown population: __________

   Reason for your answer: __________

6. Describe the likely long term prospects for the population at the golf course: __________

   ____

7. Describe a potential problem with using size to estimate age: __________

   ____

8. Explain why a knowledge of age structure could be important in managing a resource: __________

   ____

Mark and Recapture Sampling

The mark and recapture method of estimating population size is used in the study of animal populations where individuals are highly mobile. It is of no value where animals do not move or move very little. The number of animals caught in each sample must be large enough to be valid. The technique is outlined in the diagram below.

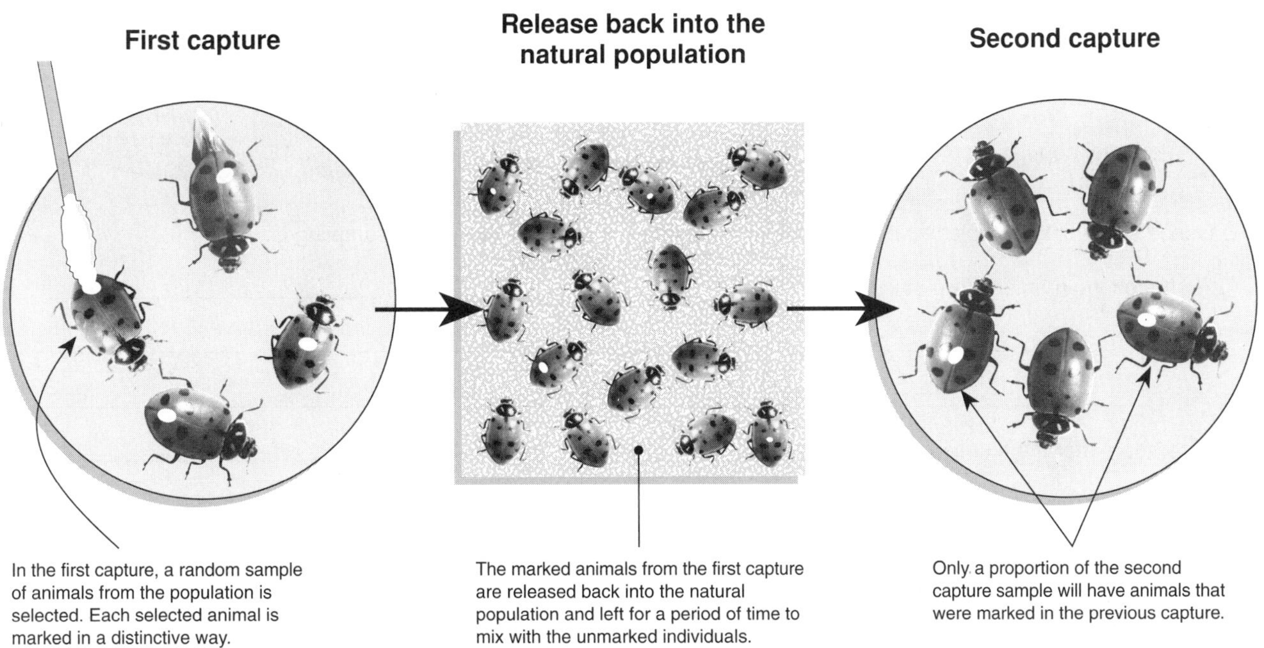

First capture
Release back into the natural population
Second capture
In the first capture, a random sample of animals from the population is selected. Each selected animal is marked in a distinctive way.
The marked animals from the first capture are released back into the natural population and left for a period of time to mix with the unmarked individuals.
Only a proportion of the second capture sample will have animals that were marked in the previous capture.

The Lincoln Index

$$\text{Total population} = \frac{\text{No. of animals in 1st sample (all marked)} \times \text{Total no. of animals in 2nd sample}}{\text{Number of marked animals in the second sample (recaptured)}}$$

The mark and recapture technique comprises a number of simple steps:

1. The population is sampled by capturing as many of the individuals as possible and practical.
2. Each animal is marked in a way to distinguish it from unmarked animals (unique mark for each individual not required).
3. Return the animals to their habitat and leave them for a long enough period for complete mixing with the rest of the population to take place.
4. Take another sample of the population (this does not need to be the same sample size as the first sample, but it does have to be large enough to be valid).
5. Determine the numbers of marked to unmarked animals in this second sample. Use the equation above to estimate the size of the overall population.

1. For this exercise you will need several boxes of matches and a pen. Work in a group of 2-3 students to 'sample' the population of matches in the full box by using the mark and recapture method. Each match will represent one animal.

   (a) Take out 10 matches from the box and mark them on 4 sides with a pen so that you will be able to recognize them from the other unmarked matches later.
   (b) Return the marked matches to the box and shake the box to mix the matches.
   (c) Take a sample of 20 matches from the same box and record the number of marked matches and unmarked matches.
   (d) Determine the total population size by using the equation above.
   (e) Repeat the sampling 4 more times (steps b-d above) and record your results:

	Sample 1	Sample 2	Sample 3	Sample 4	Sample 5
Estimated Population					

   (f) Count the actual number of matches in the matchbox : ____________________

   (g) Compare the actual number to your estimates and state by how much it differs: ____________________

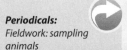

2. In 1919 a researcher by the name of Dahl wanted to estimate the number of trout in a Norwegian lake. The trout were subject to fishing so it was important to know how big the population was in order to manage the fish stock. He captured and marked 109 trout in his first sample. A few days later, he caught 177 trout in his second sample, of which 57 were marked. Use the **Lincoln index** (on the previous page) to estimate the total population size:

Size of 1st sample: ______________________

Size of 2nd sample: ______________________

No. marked in 2nd sample: ______________________

Estimated total population: ______________________

3. Describe some of the problems with the mark and recapture method if the second sampling is:

(a) Left too long a time before being repeated: ______________________

(b) Too soon after the first sampling: ______________________

4. Describe two important assumptions being made in this method of sampling, that would cause the method to fail if they were not true:

(a) ______________________

(b) ______________________

5. Some types of animal would be unsuitable for this method of population estimation (i.e. the method would not work).

(a) Name an animal for which this method of sampling would not be effective: ______________________

(b) Explain your answer above: ______________________

6. Describe three methods for marking animals for mark and recapture sampling. Take into account the possibility of animals shedding their skin, or being difficult to get close to again:

(a) ______________________

(b) ______________________

(c) ______________________

7. Scientists in the UK and Canada have, at various times since the 1950s, been involved in computerized tagging programs for Northern cod (a species once abundant in Northern Hemisphere waters but now severely depleted). Describe the type of information that could be obtained through such tagging programs:

____

Population Cycles

Some mammals, particularly in highly seasonal environments, exhibit regular cycles in their population numbers. Snowshoe hares in Canada exhibit such a cycle of population fluctuation that has a periodicity of 9–11 years. Populations of lynx in the area show a similar periodicity. Contrary to early suggestions that the lynx controlled the size of the hare population, it is now known that the fluctuations in the hare population are governed by other factors, probably the availability of palatable grasses. The fluctuations in the lynx numbers however, do appear to be the result of fluctuations in the numbers of hares (their principal food item). This is true of most **vertebrate** predator-prey systems: predators do not usually control prey populations, which tend to be regulated by other factors such as food availability and climatic factors. Most predators have more than one prey species, although one species may be preferred. Characteristically, when one prey species becomes scarce, a predator will "switch" to another available prey item. Where one prey species is the principal food item and there is limited opportunity for prey switching, fluctuations in the prey population may closely govern predator cycles.

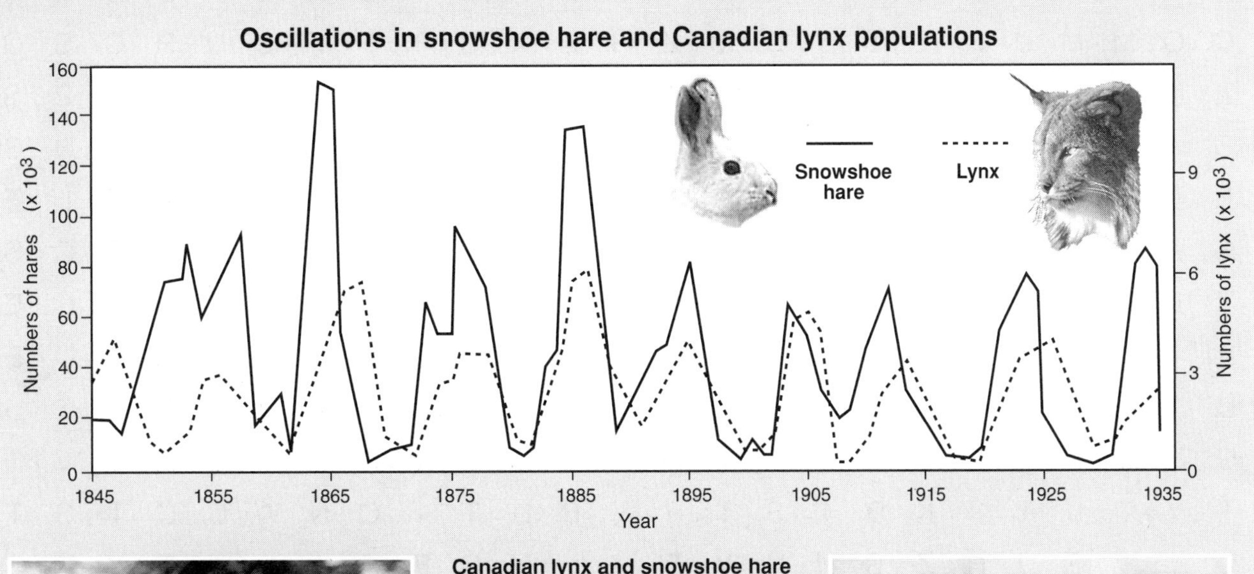

Canadian lynx and snowshoe hare

Regular trapping records of Canadian lynx (left) over a 90 year period revealed a cycle of population increase and decrease that was repeated every 10 years or so. The oscillations in lynx numbers closely matched those of the snowshoe hare (right), their principal prey item. There is little opportunity for prey switching in this system and the lynx are very dependent on the hares for food. Consequently, the oscillations in the two populations have a similar periodicity, with the lynx numbers lagging slightly behind those of the hare.

1. (a) From the graph above, determine the lag time between the population peaks of the hares and the lynx:

   (b) Explain why there is this time lag between the increase in the hare population and the response of the lynx:

2. Suggest why the lynx populations appear to be so dependent on the fluctuations on the hare:

3. (a) In terms of birth and death rates, explain how the availability of palatable food might regulate the numbers of hares:

   (b) Explain how a decline in available palatable food might affect their ability to withstand predation pressure:

Related activities: *Predator-Prey Strategies, Population Growth*

KEY TERMS Word Find

Use the clues below to find the relevant key terms in the WORD FIND grid

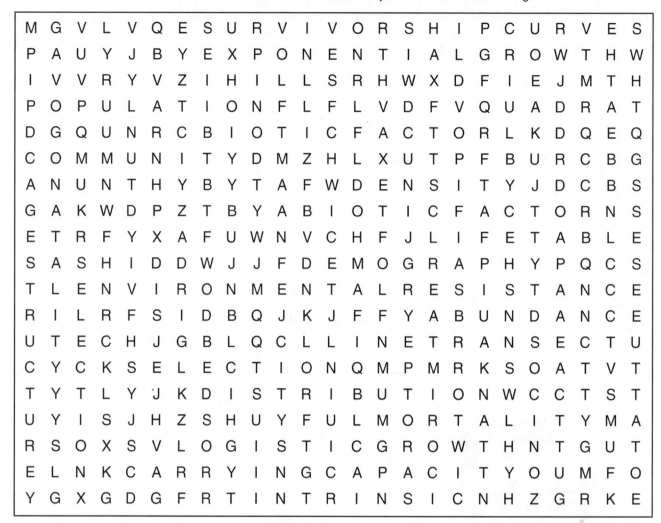

M	G	V	L	V	Q	E	S	U	R	V	I	V	O	R	S	H	I	P	C	U	R	V	E	S
P	A	U	Y	J	B	Y	E	X	P	O	N	E	N	T	I	A	L	G	R	O	W	T	H	W
I	V	V	R	Y	V	Z	I	H	I	L	L	S	R	H	W	X	D	F	I	E	J	M	T	H
P	O	P	U	L	A	T	I	O	N	F	L	F	L	V	D	F	V	Q	U	A	D	R	A	T
D	G	Q	U	N	R	C	B	I	O	T	I	C	F	A	C	T	O	R	L	K	D	Q	E	Q
C	O	M	M	U	N	I	T	Y	D	M	Z	H	L	X	U	T	P	F	B	U	R	C	B	G
A	N	U	N	T	H	Y	B	Y	T	A	F	W	D	E	N	S	I	T	Y	J	D	C	B	S
G	A	K	W	D	P	Z	T	B	Y	A	B	I	O	T	I	C	F	A	C	T	O	R	N	S
E	T	R	F	Y	X	A	F	U	W	N	V	C	H	F	J	L	I	F	E	T	A	B	L	E
S	A	S	H	I	D	D	W	J	J	F	D	E	M	O	G	R	A	P	H	Y	P	Q	C	S
T	L	E	N	V	I	R	O	N	M	E	N	T	A	L	R	E	S	I	S	T	A	N	C	E
R	I	L	R	F	S	I	D	B	Q	J	K	J	F	F	Y	A	B	U	N	D	A	N	C	E
U	T	E	C	H	J	G	B	L	Q	C	L	L	I	N	E	T	R	A	N	S	E	C	T	U
C	Y	C	K	S	E	L	E	C	T	I	O	N	Q	M	P	M	R	K	S	O	A	T	V	T
T	Y	T	L	Y	J	K	D	I	S	T	R	I	B	U	T	I	O	N	W	C	C	T	S	T
U	Y	I	S	J	H	Z	S	H	U	Y	F	U	L	M	O	R	T	A	L	I	T	Y	M	A
R	S	O	X	S	V	L	O	G	I	S	T	I	C	G	R	O	W	T	H	N	T	G	U	T
E	L	N	K	C	A	R	R	Y	I	N	G	C	A	P	A	C	I	T	Y	O	U	M	F	O
Y	G	X	G	D	G	F	R	T	I	N	T	R	I	N	S	I	C	N	H	Z	G	R	K	E

Two words describing the proportions of individuals in each class in a population (3, 9)

The relative representation of a species in a particular ecosystem.

A line across a habitat along which organisms are sampled at set intervals to determine changes in community composition.

A naturally occurring group of different species living together and interacting within the same general area.

The occurrences of a species over the total area in which it lives.

The statistical study of human population dynamics.

The number of individuals of a species per unit area.

The population size of a species that the environment can sustain indefinitely, denoted K in the equation describing population growth.

A curve of age-specific survival.

Population growth (change in numbers with time) described by an S-shaped or sigmoidal curve.

Any contribution to the environment by a living organism (2 words: 6, 6).

Any factor in an ecosystem that results from the physical environment is called this (2 words: 7, 6).

Number of individuals dying per unit time; death rate.

Number of individuals being born per unit time; birth rate.

A collection of a particular species in a certain area.

Selection for traits that favor rapid growth in an unsaturated environment.

A measured and marked shape, often a square, used to isolate an area within which the organisms are counted and/or measured.

A table of age specific mortality.

Selection for traits that favor slower growth and reproduction and competitive ability.

An organism's biotic potential is termed its _ _ _ _ _ _ _ _ _ rate of natural increase.

Also called geometric growth; an increase in number or size, at a constantly growing rate.

The effect of physical and biological factors in preventing a species reaching its biotic potential (reproducing at its maximum rate).

AP Topic 3C

IB SL/HL 5.1

Community Ecology

KEY CONCEPTS

- Energy in ecosystems is transferred through food chains.
- The decline in the energy available to each successive trophic level limits the number of feeding links in ecosystems.
- Species interact in ways that can be neutral, beneficial, or harmful to one or more parties.
- Ecological succession is a natural process by which an ecosystem changes over time.

KEY TERMS

adaptation
allelopathy
climax community
commensalism
competition
consumer
decomposer
detritivore
ecological succession
ecosystem stability
exploitation
food chain
food web
Gause's competitive exclusion principle
herbivory
interspecific competition
intraspecific competition
keystone species
mutualism
niche differentiation
parasite / parasitism
predator / predation
prey
primary consumer
primary succession
producer
saprotroph
secondary consumer
secondary succession
symbiosis
ten percent law
tertiary consumer
trophic level

□ 1. Use the **KEY TERMS** to help you understand and complete these objectives.

The Concept of the Niche

pages 350-352, 355-356

□ 2. Explain the concept of the **ecological niche** (niche).

□ 3. Distinguish between the **fundamental** and the **realized niche**. Describe and explain factors affecting **niche breadth**.

□ 4. Describe **physiological**, **structural**, and **behavioral adaptations** for survival in a given niche. Explain that these are the result of evolutionary changes to the species, but not to individuals within their own lifetimes.

Trophic Relationships

pages 344-349, also see 369-372

□ 5. Explain the relationship between **ecosystem**, **community**, and **population**. Recall the difference between **biotic** and **abiotic factors**. Identify the source of energy for almost all communities and comment on exceptions to this.

□ 6. Distinguish between **producers**, **consumers**, **detritivores**, and **saprotrophs** (saprophytes or decomposers). Describe the role of each of these **trophic groups** in energy transfer and nutrient cycling.

□ 7. Describe how energy is transferred between **trophic levels** in **food chains** and **food webs**. Comment on the efficiency of energy transfers.

□ 8. Construct **food chains** and a **food web** for a named community. Assign the organisms to trophic levels.

Community Interactions

pages 210, 353-365

□ 9. Recall how communities comprise different populations living together. Explain how biotic interactions contribute to **ecosystem stability** and comment on the significance of **keystone species** to ecosystem function.

□ 10. Describe and explain **interspecific** interactions in communities, including **commensalism**, **competition**, **mutualism**, **amensalism**, **exploitation** (parasitism, predation, herbivory), and **allelopathy**.

□ 11. In more detail than #10, explain the role of **interspecific competition** in constraining niche breadth in species. Describe consequences interspecific competition, e.g. **niche differentiation** or **competitive exclusion**.

□ 12. Describe and explain **intraspecific competition** in populations with and without a social structure.

□ 13. Explain the role of competition in limiting population size.

□ 14. Describe and explain community changes in an **ecological succession**. Distinguish between **primary** and **secondary succession**. Appreciate that successional changes do not always follow an idealized progression.

Periodicals:

listings for this chapter are on page 391

Weblinks:

www.thebiozone.com/weblink/SB2-2603.html

Teacher Resource CD-ROM: Production and Trophic Efficiency

Components of an Ecosystem

The concept of the ecosystem was developed to describe the way groups of organisms are predictably found together in their physical environment. A community comprises all the organisms within an ecosystem. Both physical (abiotic) and biotic factors affect the organisms in a community, influencing their distribution and their survival, growth, and reproduction.

The Biosphere

The **biosphere**, which contains all the Earth's living organisms, amounts to a narrow belt around the Earth extending from the bottom of the oceans to the upper atmosphere. Broad scale life-zones or **biomes** are evident within the biosphere, characterized according to the predominant vegetation. Within these biomes, **ecosystems** form natural units comprising the non-living, physical environment (the soil, atmosphere, and water) and the **community** (all the organisms living in a particular area).

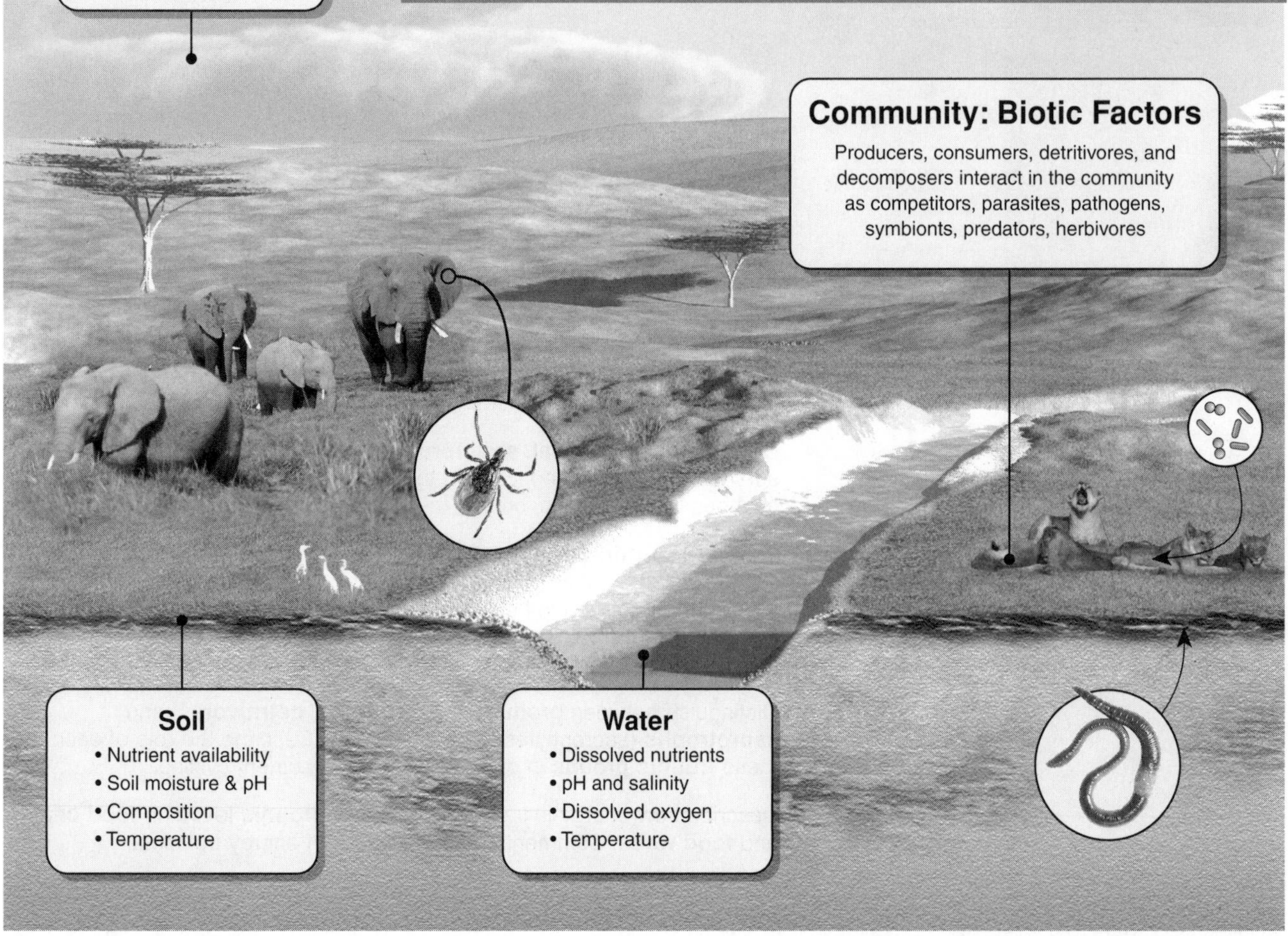

Physical Environment

Atmosphere
- Wind speed & direction
- Humidity
- Light intensity & quality
- Precipitation
- Air temperature

Community: Biotic Factors
Producers, consumers, detritivores, and decomposers interact in the community as competitors, parasites, pathogens, symbionts, predators, herbivores

Soil
- Nutrient availability
- Soil moisture & pH
- Composition
- Temperature

Water
- Dissolved nutrients
- pH and salinity
- Dissolved oxygen
- Temperature

1. Distinguish clearly between a community and an ecosystem: ______________________

2. Distinguish between biotic and abiotic factors: ______________________

3. Use one or more of the following terms to describe each of the features of a rainforest listed below:
Terms: *population, community, ecosystem, physical factor.*

(a) All the green tree frogs present: ____________ (c) All the organisms present: ____________

(b) The entire forest: ____________ (d) The humidity: ____________

Food Chains

Every ecosystem has a trophic structure: a hierarchy of feeding relationships that determines the pathways for energy flow and nutrient cycling. Species are assigned to trophic levels on the basis of their sources of nutrition. The first trophic level (**producers**), ultimately supports all other levels. The consumers are those that rely on producers for their energy. Consumers are ranked according to the trophic level they occupy (first order, second order, etc.). The sequence of organisms, each of which is a source of food for the next, is called a **food chain**. Food chains commonly have four links but seldom more than six. Those organisms whose food is obtained through the same number of links belong to the same trophic level. Note that some consumers (particularly top carnivores and omnivores) may feed at several different trophic levels, and many primary consumers eat many plant species. The different food chains in an ecosystem therefore tend to form complex webs of interactions (food webs).

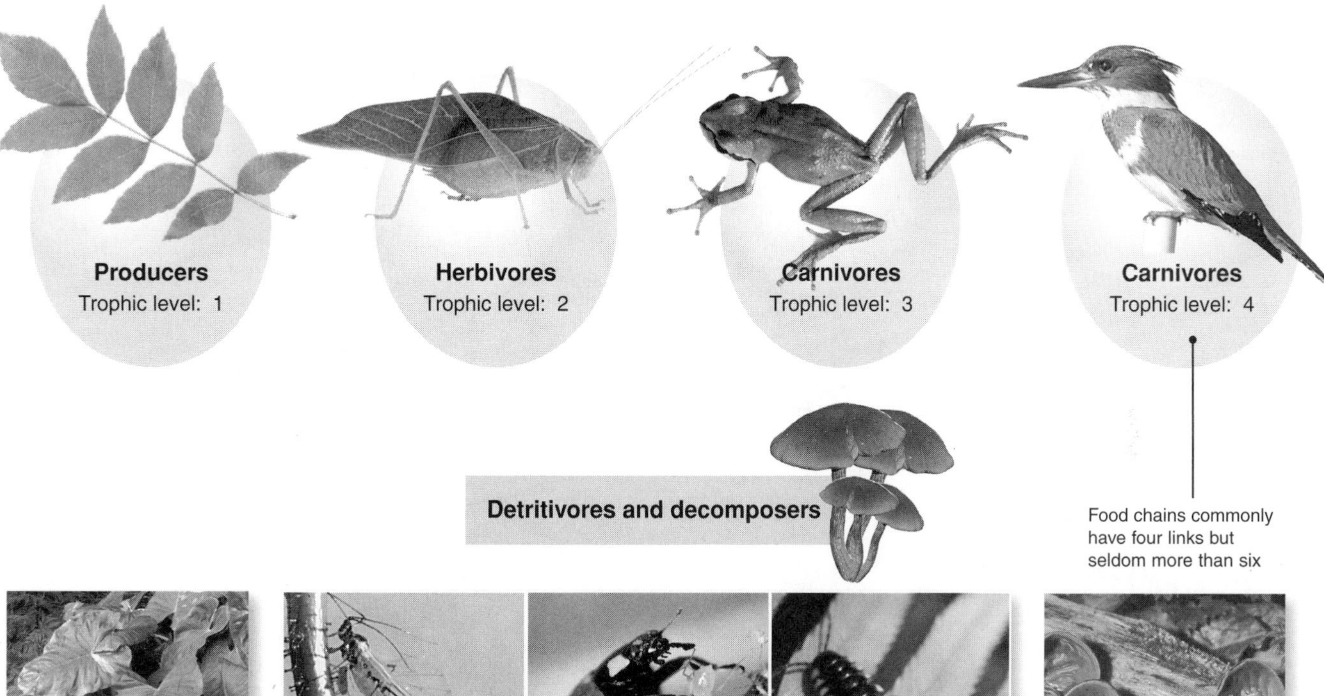

Green plants

Aphids

Ladybug eating aphid

Millipede

Wood-ear fungus

Producers (algae, green plants, and some bacteria) make their own food using simple inorganic carbon sources (e.g. CO_2). Sunlight is the most common energy source for this process.

Consumers (animals, non-photosynthetic protists, and some bacteria) rely on other living organisms or organic particulate matter for their energy and their source of carbon. First order consumers, such as aphids (left), feed on producers. Second (and higher) order consumers, such as ladybugs (center) eat other consumers. **Detritivores** consume (ingest and digest) detritus (decomposing organic material) from every trophic level. In doing so, they contribute to decomposition and the recycling of nutrients. Common detritivores include wood-lice, millipedes (right), and many terrestrial worms.

Decomposers (fungi and some bacteria) obtain their energy and carbon from the extracellular breakdown of dead organic matter (DOM). Decomposers play a central role in nutrient cycling.

Community Ecology

The diagram above represents the basic elements of a food chain. In the questions below, you are asked to add to the diagram the features that indicate the flow of energy through the community of organisms.

1. (a) State the original energy source for this food chain: __________

   (b) Draw arrows on the diagram above to show how the energy flows through the organisms in the food chain. Label each arrow with the process involved in the energy transfer. Draw arrows to show how energy is lost by respiration.

2. (a) Describe what happens to the **amount** of energy available to each successive trophic level in a food chain:

   ____

   (b) Explain why this is the case: __________

   ____

3. Explain what you could infer about the tropic level(s) of the kingfisher, if it was found to eat both katydids and frogs:

   ____

***Related activities**: Energy Flow in an Ecosystem, Food Webs, Ecological Pyramids*

Energy Inputs and Outputs

Light is the initial energy source for almost all ecosystems and photosynthesis is the main route by which energy enters most food chains (but see *Cave Food Webs*). Energy flows through ecosystems in the high energy chemical bonds within **organic matter** and, in accordance with the second law of thermodynamics, is dissipated as it is transferred through trophic levels. In contrast, nutrients move within and between ecosystems in **biogeochemical cycles** involving exchanges between the atmosphere, the Earth's crust, water, and living organisms. Energy flows through trophic levels rather inefficiently, with only 5-20% of usable energy being transferred to the subsequent level. Energy not used for metabolic processes is lost as heat.

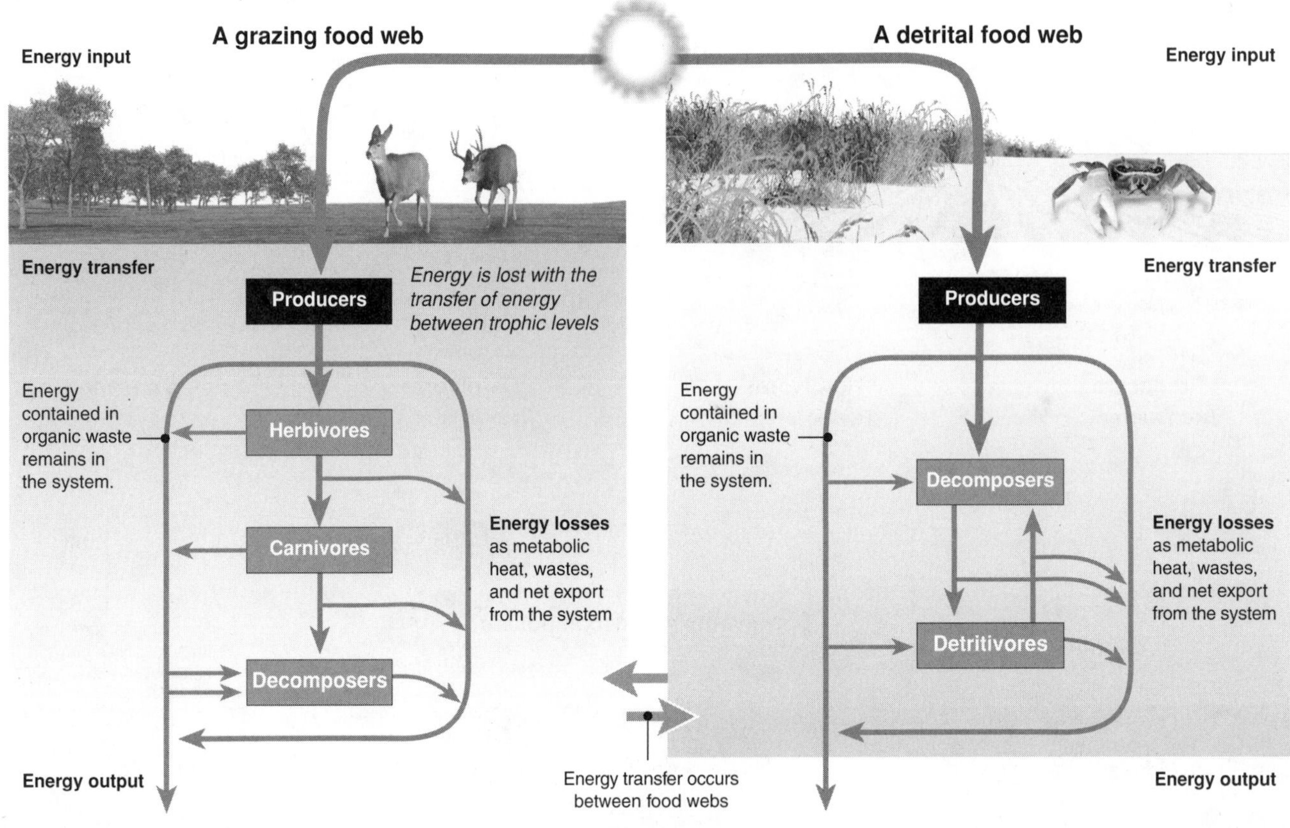

1. Discuss the nature of food chains. Include reference to trophic levels, producers, and first and second order consumers:

2. Describe the differences between **producers** and **consumers** with respect to their role in energy transfers:

3. With respect to energy flow, describe a major difference between a detrital and a grazing food web:

4. Distinguish between detritivores and decomposers with respect to how they contribute to nutrient cycling:

Related activities: *Food Chains, Food Webs, Energy Flow in an Ecosystem*

Food Webs

Every ecosystem has a **trophic structure**: a hierarchy of feeding relationships which determines the pathways for energy flow and nutrient cycling. Species are assigned to trophic levels on the basis of their sources of nutrition, with the first trophic level (the **producers**), ultimately supporting all other (consumer) levels. Consumers are ranked according to the trophic level they occupy, although some consumers may feed at several different trophic levels. The sequence of organisms, each of which is a source of food for the next, is called a **food chain**. The different food chains in an ecosystem are interconnected to form a complex web of feeding interactions called a **food web**. In the example of a lake ecosystem below, your task is assemble the organisms into a food web in a way that illustrates their trophic status and their relative trophic position(s).

Feeding Requirements of Lake Organisms

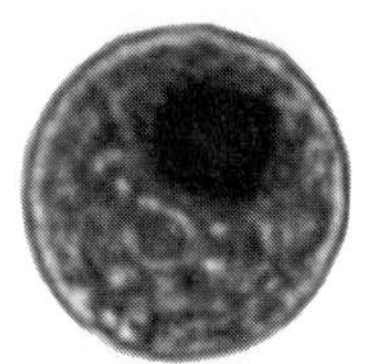

Autotrophic protists
Chlamydomonas (above), *Euglena*
Two of the many genera that form the phytoplankton.

Macrophytes (various species)
A variety of flowering aquatic plants are adapted for being submerged, free-floating, or growing at the lake margin.

Detritus
Decaying organic matter from within the lake itself or it may be washed in from the ake margins.

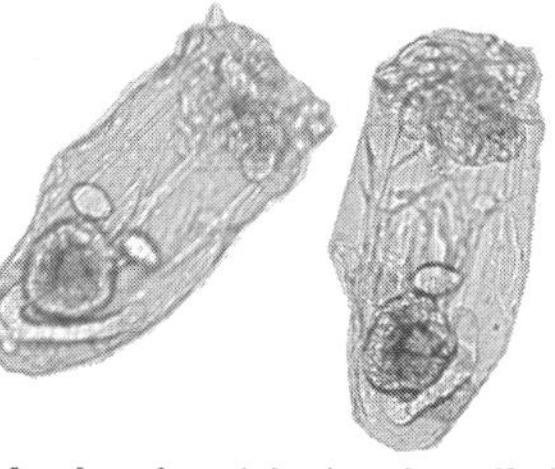

Asplanchna (planktonic rotifer)
A large, carnivorous rotifer that feeds on protozoa and young zooplankton (e.g. small *Daphnia*).

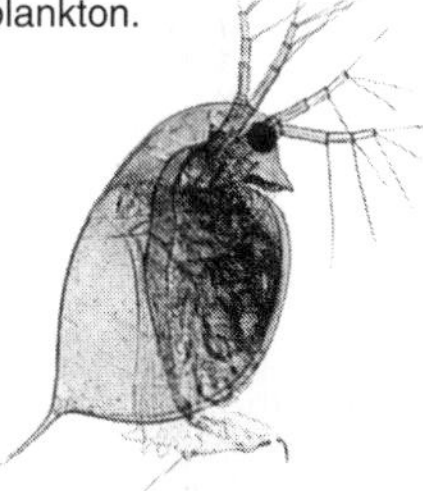

Daphnia
Small freshwater crustacean that forms part of the zooplankton. It feeds on planktonic algae by filtering them from the water with its limbs.

Leech (*Glossiphonia*)
Leeches are fluid feeding predators of smaller invertebrates, including rotifers, small pond snails and worms.

Three-spined stickleback (*Gasterosteus*)
A common fish of freshwater ponds and lakes. It feeds mainly on small invertebrates such as *Daphnia* and insect larvae.

Diving beetle (*Dytiscus*)
Diving beetles feed on aquatic insect larvae and adult insects blown into the lake community. The will also eat organic detritus collected from the bottom mud.

Carp (*Cyprinus*)
A heavy bodied freshwater fish that feeds mainly on bottom living insect larvae and snails, but will also take some plant material (not algae).

Dragonfly larva
Large aquatic insect larvae that are voracious predators of small invertebrates including *Hydra*, *Daphnia*, other insect larvae, and leeches.

Great pond snail (*Limnaea*)
Omnivorous pond snail, eating both plant and animal material, living or dead, although the main diet is aquatic macrophytes.

Herbivorous water beetles (*e.g. Hydrophilus*)
Feed on water plants, although the young beetle larvae are carnivorous, feeding primarily on small pond snails.

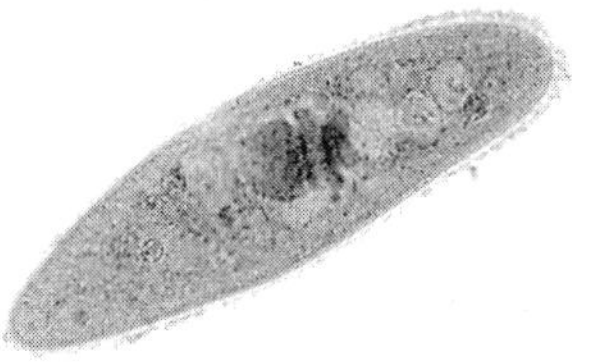

Protozan (*e.g. Paramecium*)
Ciliated protozoa such as *Paramecium* feed primarily on bacteria and microscopic green algae such as *Chlamydomonas*.

Pike (*Esox lucius*)
A top ambush predator of all smaller fish and amphibians. They are also opportunistic predators of rodents and small birds.

Mosquito larva (*Culex* spp.)
The larvae of most mosquito species, e.g. *Culex*, feed on planktonic algae before passing through a pupal stage and undergoing metamorphosis into adult mosquitoes.

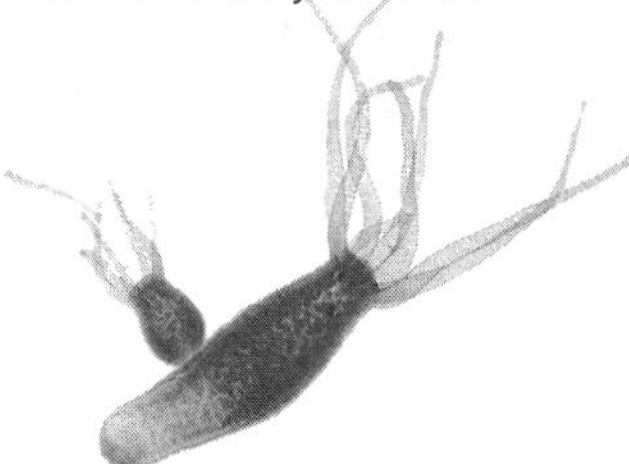

Hydra
A small carnivorous cnidarian that captures small prey items, e.g. small *Daphnia* and insect larvae, using its stinging cells on the tentacles.

Periodicals: *All life is here, The lake ecosystem*

Related activities: *Energy Inputs & Outputs, Energy Flow in an Ecosystem*
Web links: *Fitting Algae into the Food Web, Marine Food Webs*

1. From the information provided for the lake food web components on the previous page, construct **ten** different **food chains** to show the feeding relationships between the organisms. Some food chains may be shorter than others and most species will appear in more than one food chain. An example has been completed for you.

Example 1: Macrophyte ⟶ Herbivorous water beetle ⟶ Carp ⟶ Pike

(a) ______________________________

(b) ______________________________

(c) ______________________________

(d) ______________________________

(e) ______________________________

(f) ______________________________

(g) ______________________________

(h) ______________________________

(i) ______________________________

(j) ______________________________

2. (a) Use the food chains created above to help you to draw up a **food web** for this community. Use the information supplied to draw arrows showing the flow of **energy** between species (only energy **from** the detritus is required).

(b) Label each species to indicate its position in the food web, i.e. its trophic level (**T1, T2, T3, T4, T5**). Where a species occupies more than one trophic level, indicate this, e.g. **T2/3**:

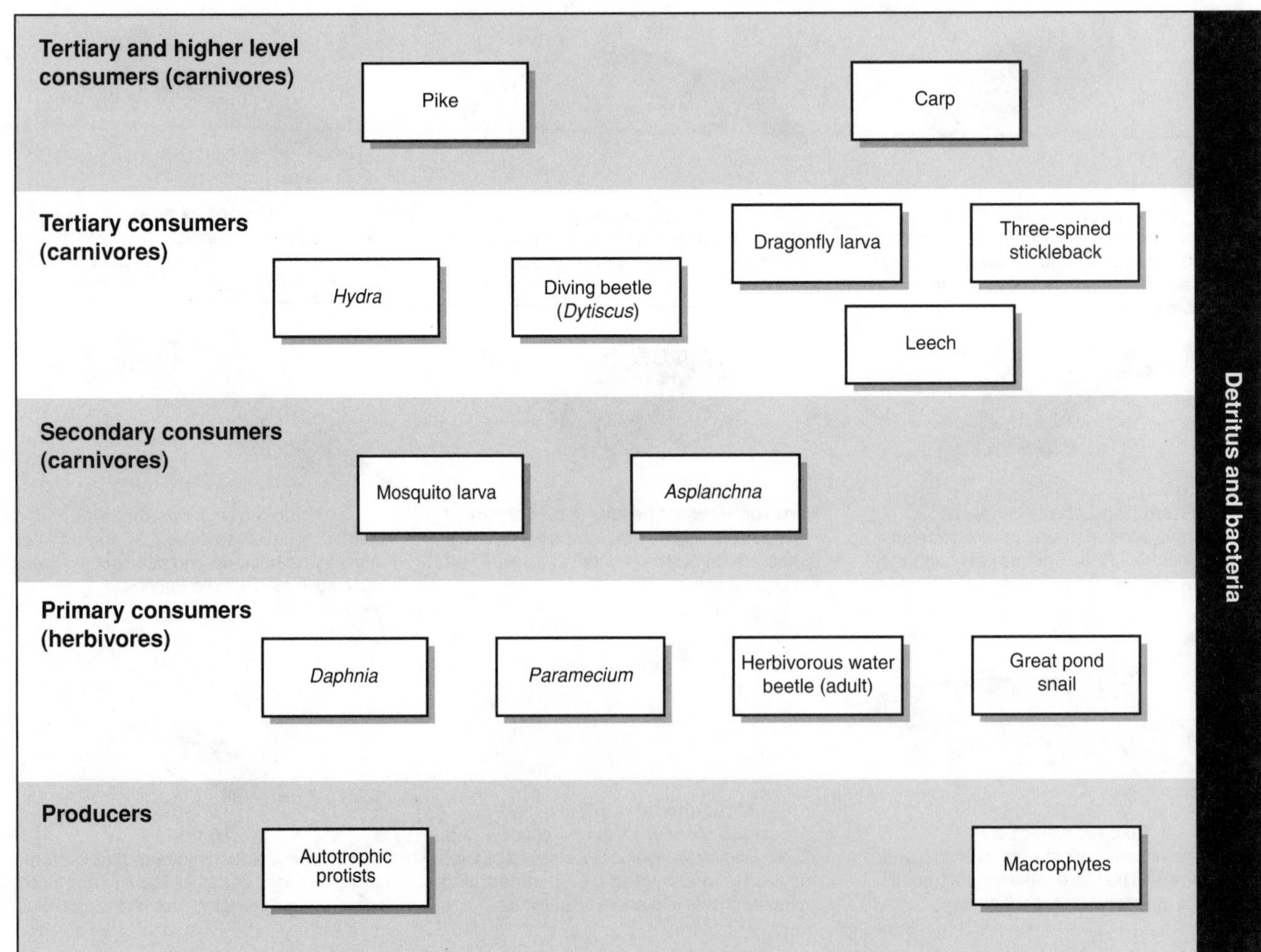

Cave Food Webs

A cave is a barren environment, without the light that normally sustains most ecosystems. Despite this, a wide range of animals inhabit caves and are specially adapted to that particular environment. Other animals, such as bats, do not live permanently in the cave but use it as a roosting or resting and breeding area, safe from many predators and other dangers. The food webs of caves are very fragile and based on few resources. **Around the entrance** of the cave, the owl (**1**) preys on the mouse (**2**) which itself feeds on the vegetation outside the cave. The owl and the mouse leave droppings that support the cave dung beetle (**3**) and the millipede (**4**). The cave cricket (**5**) scavenges dead birds and mammals near the cave entrance. The harvestman (**6**) is a predator of the dung beetle, the millipede, and the cricket. **Inside the cave**, the horseshoe bat (**7**) roosts and breeds in safety, leaving the cave to feed outside on slow flying insects. The bats produce vast quantities of guano (droppings). The guano is eaten by the blind cave beetle (**8**), the millipede (**4**) and the springtail (**9**). These invertebrates are hunted by the predatory cave spider (**10**). Occasionally, in tropical caves, snakes (not shown) may enter the cave and feed on bats. **In underground pools**, the bat guano supports the growth of bacteria (**11**). Flatworms (**12**) and isopods (**13**) feed on the bacteria and themselves are eaten by the blind cave shrimp (**14**). The blind cave fish (**15**) is the top predator, feeding on isopods and the blind cave shrimps.

1
Cave entrance
2
3
4
5
6
7
8
9
10
11
12
13
14
15
Pool
Sump
Guano and detritus layer

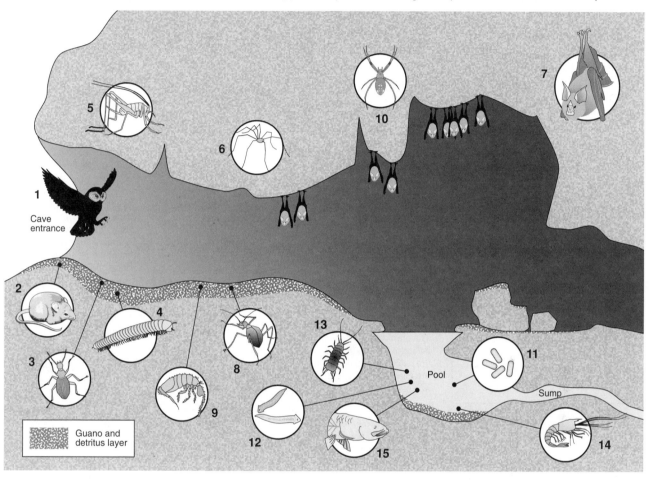

1. Using the previous lake food web activity as a guide, construct a food web for the cave ecosystem on a separate sheet of paper. For animals that feed outside the cave, do not include this outside source of food. As in the lake food web, label each species with the following codes to indicate its diet type (producer, herbivore, carnivore, omnivore) and its position in the food chain if it is a consumer (1st, 2nd, 3rd, 4th order consumer). Staple your finished web to this page.

2. Identify the major trophic of a usual food web that is missing from the cave food web: __________

3. Explain how energy is imported into the cave's food web: __________

4. Explain how energy from the cave ecosystem might be removed: __________

5. In many parts of the world, cave-dwelling bat species are endangered, often taken as food by humans or killed as pests. Explain how the cave food web would be affected if bat numbers were to fall substantially:

The Ecological Niche

The **ecological niche** describes the functional position of a species in its ecosystem; how it responds to the distribution of resources and how it, in turn, alters those resources for other species. The full range of environmental conditions under which an organism can exist describes its **fundamental niche**. As a result of direct and indirect interactions with other organisms, species are usually forced to occupy a niche that is narrower than this and to which they are best adapted. This is termed the **realized niche**. From this arose the idea that two species with exactly the same niche requirements could not coexist, because they would compete for the same resources, and one would exclude the other. This is known as **Gause's competitive exclusion principle**. The ecological niche is dynamic. Niche breadth increases when resources are abundant or intraspecific competition is strong. Similarly, niches contract when resources are limited or there is intense competition between species.

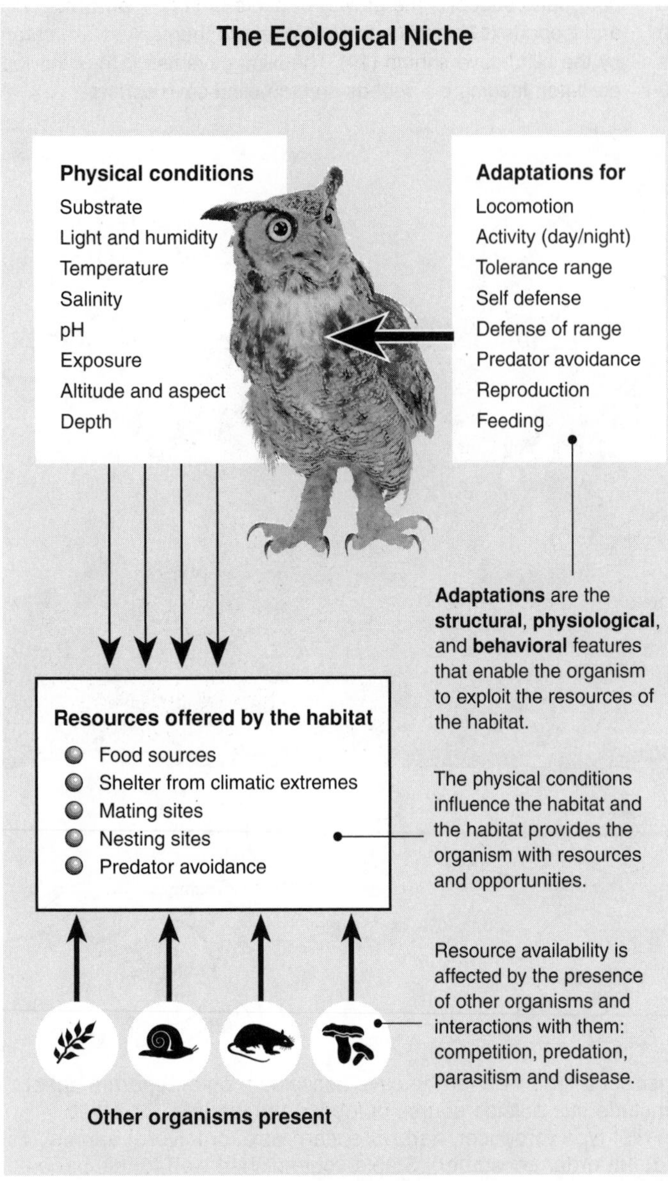

Competition and Niche Size

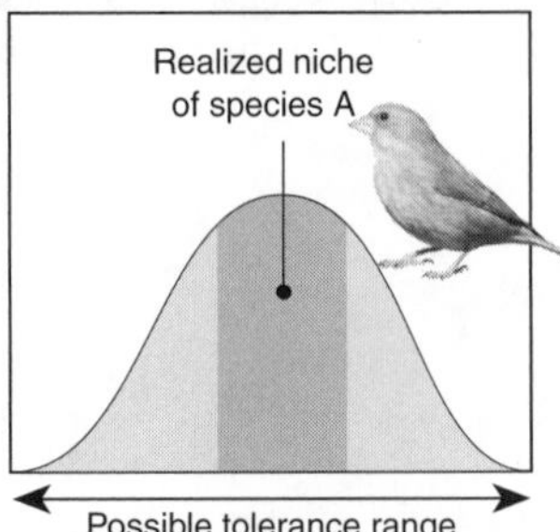

The realized niche

The tolerance range represents the potential (**fundamental**) niche a species could exploit. The actual or **realized** niche of a species is narrower than this because of competition with other species.

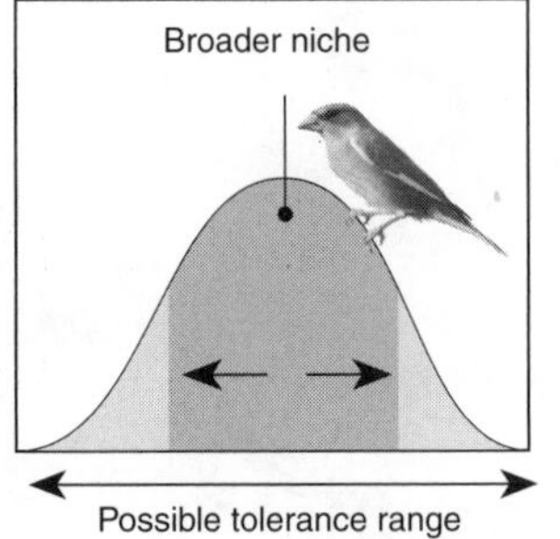

Intraspecific competition

Competition is strongest between individuals of the same species, because their resource needs exactly overlap. When intraspecific competition is intense, individuals are forced to exploit resources in the extremes of their tolerance range. This leads to expansion of the realized niche.

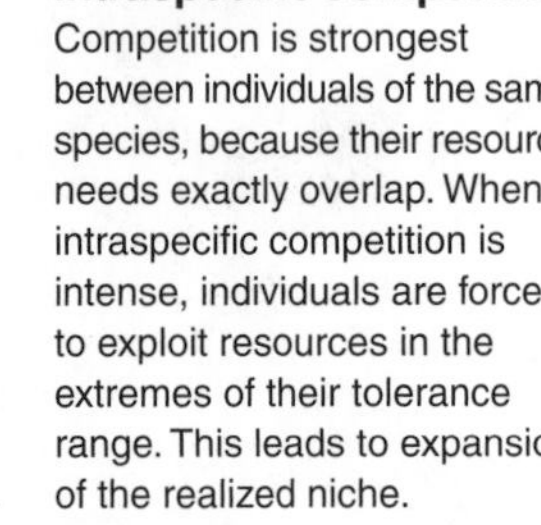

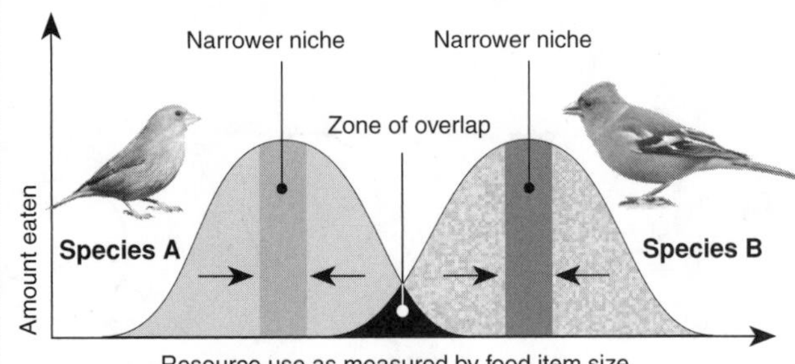

Interspecific competition

If two (or more) species compete for some of the same resources, their resource use curves will overlap. Within the zone of overlap, resource competition will be intense and selection will favor niche specialization so that one or both species occupy a narrower niche.

1. (a) Explain in what way the realized niche could be regarded as flexible: ______________________

(b) Describe factors that might constrain the extent of the realized niche: ______________________

2. Explain the contrasting effects of interspecific competition and intraspecific competition on niche breadth:

____

Related activities: Niche Differentiation, Interspecific Competition, Intraspecific Competition

Periodicals: The ecological niche

Adaptations to Niche

The adaptive features that evolve in species are the result of selection pressures on them through the course of their evolution. These features enable an organism to function most effectively in its niche, enhancing its exploitation of its environment and therefore its survival. The examples below illustrate some of the adaptations of two species: a British placental mammal and a migratory Arctic bird. Note that adaptations may be associated with an animal's structure (morphology), its internal physiology, or its behavior.

Northern or Common Mole
(Talpa europaea)

Head-body length: 113-159 mm, tail length: 25-40 mm, weight range: 70-130 g.

Moles (photos above) spend most of the time underground and are rarely seen at the surface. Mole hills are the piles of soil excavated from the tunnels and pushed to the surface. The cutaway view above shows a section of tunnels and a nest chamber. Nests are used for sleeping and raising young. They are dug out within the tunnel system and lined with dry plant material.

The northern (common) mole is a widespread insectivore found throughout most of Britain and Europe, apart from Ireland. They are found in most habitats but are less common in coniferous forest, moorland, and sand dunes, where their prey (earthworms and insect larvae) are rare. They are well adapted to life underground and burrow extensively, using enlarged forefeet for digging. Their small size, tubular body shape, and heavily buttressed head and neck are typical of burrowing species.

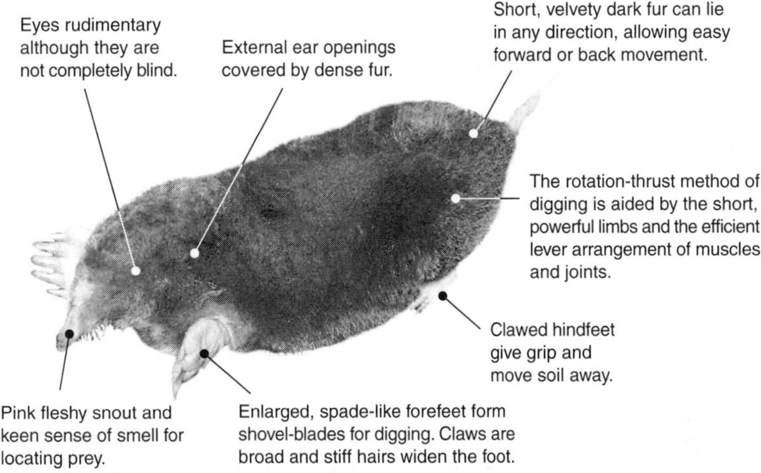

Eyes rudimentary although they are not completely blind.
External ear openings covered by dense fur.
Short, velvety dark fur can lie in any direction, allowing easy forward or back movement.
The rotation-thrust method of digging is aided by the short, powerful limbs and the efficient lever arrangement of muscles and joints.
Clawed hindfeet give grip and move soil away.
Pink fleshy snout and keen sense of smell for locating prey.
Enlarged, spade-like forefeet form shovel-blades for digging. Claws are broad and stiff hairs widen the foot.

Habitat and ecology: Moles spend most of their lives in underground tunnels. Surface tunnels occur where their prey is concentrated at the surface (e.g. land under cultivation). Deeper, permanent tunnels form a complex network used repeatedly for feeding and nesting, sometimes for several generations. **Senses and behavior**: Keen sense of smell but almost blind. Both sexes are solitary and territorial except during breeding. Life span about 3 years. Moles are prey for owls, buzzards, stoats, cats, and dogs. Their activities aerate the soil and they control many soil pests. Despite this, they are regularly trapped and poisoned as pests.

Snow Bunting
(Plectrophenax nivalis)

The snow bunting is a small ground feeding bird that lives and breeds in the Arctic and sub-Arctic islands. Although migratory, snow buntings do not move to traditional winter homes but prefer winter habitats that resemble their Arctic breeding grounds, such as bleak shores or open fields of northern Britain and the eastern United States. Snow buntings have the unique ability to molt very rapidly after breeding. During the warmer months, the buntings are a brown color, changing to white in winter (right). They must complete this color change quickly, so that they have a new set of feathers before the onset of winter and before migration. In order to achieve this, snow buntings lose as many as four or five of their main flight wing feathers at once, as opposed to most birds, which lose only one or two.

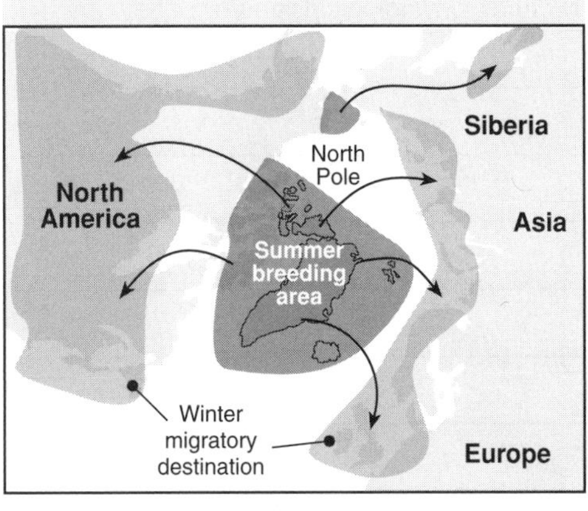

Very few small birds breed in the Arctic, because most small birds lose more heat than larger ones. In addition, birds that breed in the brief Arctic summer must migrate before the onset of winter, often traveling over large expanses of water. Large, long winged birds are better able to do this. However, the snow bunting is superbly adapted to survive in the extreme cold of the Arctic region.

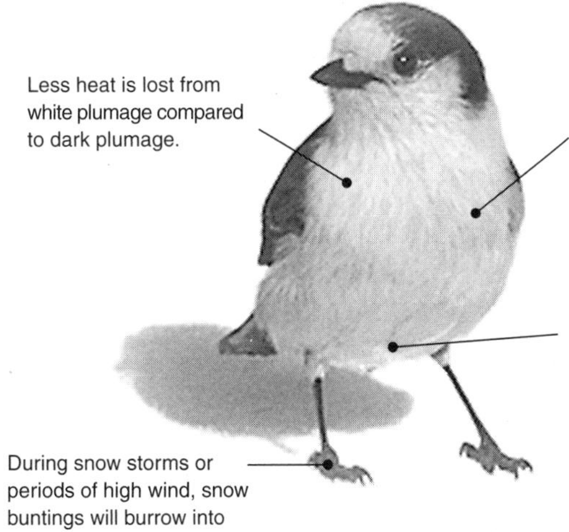

Less heat is lost from white plumage compared to dark plumage.
White feathers are hollow and filled with air, which acts as an insulator. In the dark colored feathers the internal spaces are filled with pigmented cells.
Snow buntings, on average, lay one or two more eggs than equivalent species further south. They are able to rear more young because the continuous daylight and the abundance of insects at high latitudes enables them to feed their chicks around the clock.
During snow storms or periods of high wind, snow buntings will burrow into snowdrifts for shelter.

Habitat and ecology: Widespread throughout Arctic and sub-Arctic Islands. Active throughout the day and night, resting for only 2-3 hours in any 24 hour period. Snow buntings may migrate up to 6000 km but are always found at high latitudes. **Reproduction and behavior**: The nest, which is concealed amongst stones, is made from dead grass, moss, and lichen. The male bird feeds his mate during the incubation period and helps to feed the young.

Community Ecology

1. Describe a structural, physiological, and behavioral adaptation of the **common mole**, explaining how each adaptation assists survival:

   (a) Structural adaptation: ______

   (b) Physiological adaptation: ______

   (c) Behavioral adaptation: ______

2. Describe a structural, physiological, and behavioral adaptation of the **snow bunting**, explaining how each adaptation assists survival:

   (a) Structural adaptation: ______

   (b) Physiological adaptation: ______

   (c) Behavioral adaptation: ______

3. The rabbit is a colonial mammal which lives underground in warrens (burrow systems) and feeds on grasses, cereal crops, roots, and young trees. Rabbits are a hugely successful species worldwide and often reach plague proportions. Through discussion, or your own knowledge and research, describe **six adaptations** of rabbits, identifying them as structural (S), physiological (P), or behavioral (B). The examples below are typical:

   Structural: *Widely spaced eyes gives wide field of vision for surveillance and detection of danger.*

   Physiological: *High reproductive rate. Short gestation and high fertility aids rapid population increases when food is available.*

   Behavioral: *Freeze behavior when startled reduces the possibility of detection by wandering predators.*

   (a) ______

   (b) ______

   (c) ______

   (d) ______

   (e) ______

   (f) ______

4. Examples of adaptations are listed below. Identify them as predominantly structural, physiological, and/or behavioral:

   (a) Relationship of body size and shape to latitude (tropical or Arctic): ______

   (b) The production of concentrated urine in desert dwelling mammals: ______

   (c) The summer and winter migratory patterns in birds and mammals: ______

   (d) The C_4 photosynthetic pathway and CAM metabolism of plants: ______

   (e) The thick leaves and sunken stomata of desert plants: ______

   (f) Hibernation or torpor in small mammals over winter: ______

   (g) Basking in lizards and snakes: ______

Species Interactions

No organism exists in isolation. Each takes part in many interactions, both with other organisms and with the non-living components of the environment. Species interactions may involve only occasional or indirect contact (predation or competition) or they may involve close association or **symbiosis**. Symbiosis is a term that encompasses a variety of interactions involving close species contact. There are three types of symbiosis: **parasitism** (a form of exploitation), **mutualism**, and **commensalism**. Species interactions affect population densities and are important in determining community structure and composition. Some interactions, such as **allelopathy**, may even determine species presence or absence in an area.

Examples of Species Interactions

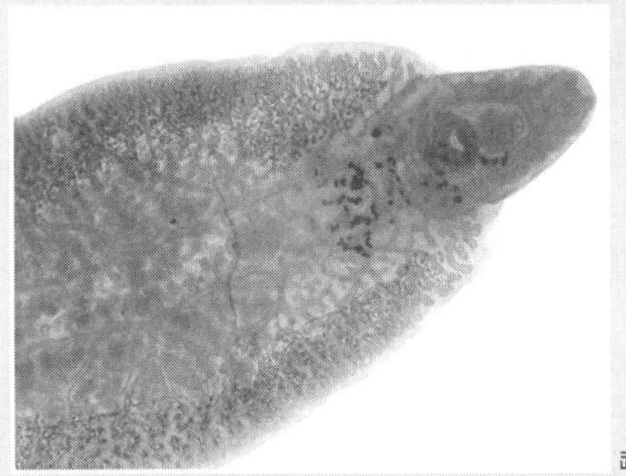

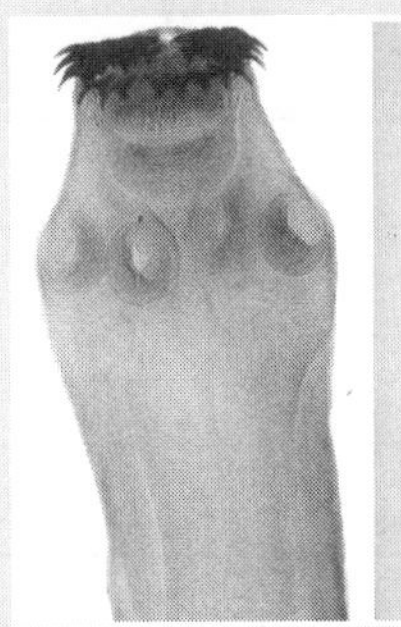

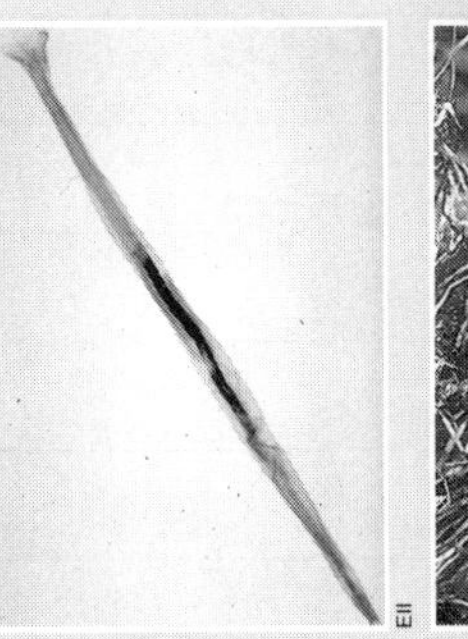

Parasitism is a common exploitative relationship in plants and animals. A parasite exploits the resources of its host (e.g. for food, shelter, warmth) to its own benefit. The host is harmed, but usually not killed. **Endoparasites**, such as liver flukes (left), tapeworms (center) and nematodes (right), are highly specialized to live inside their hosts, attached by hooks or suckers to the host's tissues.

Ectoparasites, such as ticks (above), mites, and fleas, live attached to the outside of the host, where they suck body fluids, cause irritation, and may act as vectors for disease causing microorganisms.

Mutualism involves an intimate association between two species that offers advantage to both. **Lichens** (above) are the result of a mutualism between a fungus and an alga (or cyanobacterium).

Termites have a mutualistic relationship with the cellulose digesting bacteria in their guts. A similar mutualistic relationship exists between ruminants and their gut microflora of bacteria and ciliates.

In **commensal** relationships, such as between this large grouper and a remora, two species form an association where one organism, the commensal, benefits and the other is neither harmed or helped.

Many species of decapod crustaceans, such as this anemone shrimp, are commensal with sea anemones. The shrimp gains by being protected from predators by the anemone's tentacles.

Interactions involving **competition** for the same food resources are dominated by the largest, most aggressive species. Here, hyaenas compete for a carcass with vultures and maribou storks.

Predation is an easily identified relationship, as one species kills and eats another (above). Herbivory is similar type of exploitation, except that the plant is usually not killed by the herbivore.

1. Discuss each of the following interspecific relationships, including reference to the species involved, their role in the interaction, and the specific characteristics of the relationship:

   (a) **Mutualism** between ruminant herbivores and their gut microflora: __________

Community Ecology

(b) **Commensalism** between a shark and a remora: ______

______

______

______

(c) **Parasitism** between a tapeworm and its human host: ______

______

______

______

(d) **Parasitism** between a cat flea and its host: ______

______

______

______

2. Summarize your knowledge of species interactions by completing the following, entering a (+), (–), or (0) for species B, and writing a brief description of each term. Codes: (+): species benefits, (–): species is harmed, (0): species is unaffected.

Interaction	Species A	Species B	Description of relationship
(a) Mutualism	+		
(b) Commensalism	+		
(c) Parasitism	–		
(d) Amensalism	0		
(e) Predation	–		
(f) Competition	–		
(g) Herbivory	+		
(h) Antibiosis	+ / 0		

3. For each of the interactions between two species described below, choose the correct term to describe the interaction and assign a +, – or 0 for each species involved in the space supplied. Use the completed table above to help you:

Description	Term	Species A	Species B
(a) A tiny cleaner fish picking decaying food from the teeth of a much larger fish (e.g. grouper).	Mutualism	Cleaner fish +	Grouper +
(b) Ringworm fungus growing on the skin of a young child.		Ringworm	Child
(c) Human effluent containing poisonous substances killing fish in a river downstream of discharge.		Humans	Fish
(d) Humans planting cabbages to eat only to find that the cabbages are being eaten by slugs.		Humans	Slugs
(e) A shrimp that gets food scraps and protection from sea anemones, which appear to be unaffected.		Shrimp	Anemone
(f) Birds follow a herd of antelopes to feed off disturbed insects, antelopes alerted to danger by the birds.		Birds	Antelope

Intraspecific Competition

Some of the most intense competition occurs between individuals of the same species (**intraspecific competition**). Most populations have the capacity to grow rapidly, but their numbers cannot increase indefinitely because environmental resources are finite. Every ecosystem has a **carrying capacity** (K), defined as the number of individuals in a population that the environment can support. Intraspecific competition for resources increases with increasing population size and, at carrying capacity, it reduces the per capita growth rate to zero. When the demand for a particular resource (e.g. food, water, nesting sites, nutrients, or light) exceeds supply, that resource becomes a **limiting factor**. Populations respond to resource limitation by reducing their population growth rate (e.g. through lower birth rates or higher mortality). The response of individuals to limited resources varies depending on the organism. In many invertebrates and some vertebrates such as frogs, individuals reduce their growth rate and mature at a smaller size. In some vertebrates, territoriality spaces individuals apart so that only those with adequate resources can breed. When resources are very limited, the number of available territories will decline.

Intraspecific Competition

Scramble competition in caterpillars

Direct competition for available food between members of the same species is called **scramble competition**. In some situations where scramble competition is intense, none of the competitors gets enough food to survive.

Contest competition in wolves

In some cases, competition is limited by hierarchies existing within a social group. Dominant individuals receive adequate food, but individuals low in the hierarchy must **contest** the remaining resources and may miss out.

Display of a male anole

Intraspecific competition may be for mates or breeding sites, as well as for food. In anole lizards (above), males have a bright red throat pouch and use much of their energy displaying to compete with other males for available mates.

Competition Between Tadpoles of *Rana tigrina*

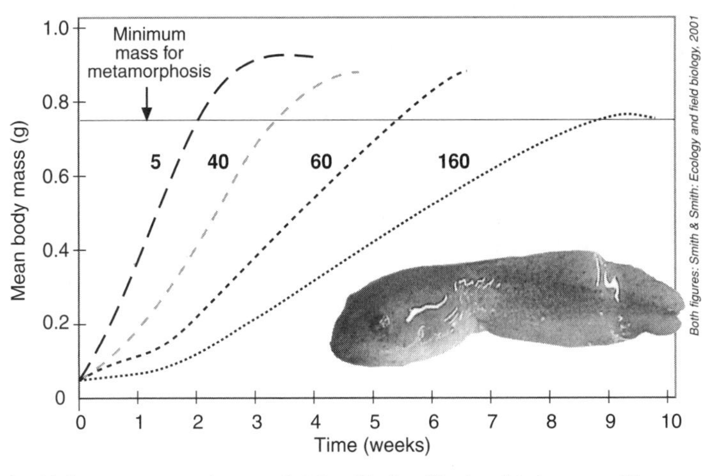

Food shortage reduces both individual growth rate and survival, and population growth. In some organisms, where there is a metamorphosis or a series of moults before adulthood (e.g. frogs, crustacean zooplankton, and butterflies), individuals may die before they mature.

The graph (left) shows how the growth rate of tadpoles (*Rana tigrina*) declines as the density increases from 5 to 160 individuals (in the same sized space).

- At high densities, tadpoles grow more slowly, taking longer to reach the minimum size for metamorphosis (0.75 g), and decreasing their chances of successfully metamorphosing from tadpoles into frogs.
- Tadpoles held at lower densities grow faster, to a larger size, metamorphosing at an average size of 0.889 g.
- In some species, such as frogs and butterflies, the adults and juveniles reduce the intensity of intraspecific competition by exploiting different food resources.

1. Using an example, predict the likely effects of **intraspecific competition** on each of the following:

(a) Individual growth rate: ______________________

(b) Population growth rate: ______________________

(c) Final population size: ______________________

Related activities: *Population Growth Curves*
Web links: *Intraspecific Relations*

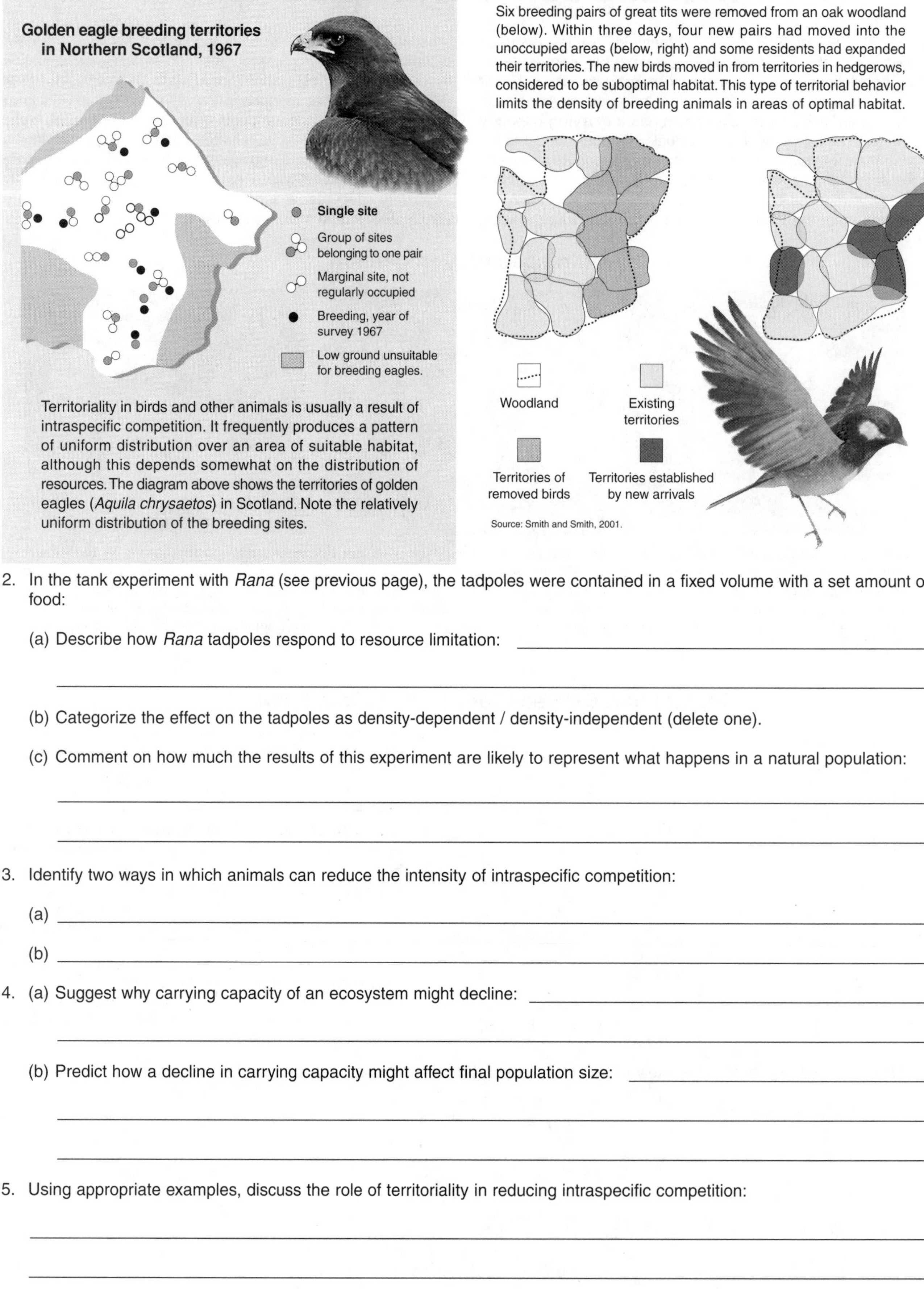

Territoriality in birds and other animals is usually a result of intraspecific competition. It frequently produces a pattern of uniform distribution over an area of suitable habitat, although this depends somewhat on the distribution of resources. The diagram above shows the territories of golden eagles (*Aquila chrysaetos*) in Scotland. Note the relatively uniform distribution of the breeding sites.

Territoriality in Great Tits (*Parus major*)

Six breeding pairs of great tits were removed from an oak woodland (below). Within three days, four new pairs had moved into the unoccupied areas (below, right) and some residents had expanded their territories. The new birds moved in from territories in hedgerows, considered to be suboptimal habitat. This type of territorial behavior limits the density of breeding animals in areas of optimal habitat.

Source: Smith and Smith, 2001.

2. In the tank experiment with *Rana* (see previous page), the tadpoles were contained in a fixed volume with a set amount of food:

   (a) Describe how *Rana* tadpoles respond to resource limitation: ______________________

   ____

   (b) Categorize the effect on the tadpoles as density-dependent / density-independent (delete one).

   (c) Comment on how much the results of this experiment are likely to represent what happens in a natural population:

   ____

   ____

3. Identify two ways in which animals can reduce the intensity of intraspecific competition:

   (a) ______________________

   (b) ______________________

4. (a) Suggest why carrying capacity of an ecosystem might decline: ______________________

   ____

   (b) Predict how a decline in carrying capacity might affect final population size: ______________________

   ____

   ____

5. Using appropriate examples, discuss the role of territoriality in reducing intraspecific competition:

   ____

   ____

   ____

   ____

   ____

   ____

Interspecific Competition

In naturally occurring populations, direct competition between different species (**interspecific competition**) is usually less intense than intraspecific competition because coexisting species have evolved slight differences in their realized niches, even though their fundamental niches may overlap (a phenomenon termed **niche differentiation**). However, when two species with very similar niche requirements are brought into direct competition through the introduction of a foreign species, one usually benefits at the expense of the other. The inability of two species with the same described niche to coexist is referred to as the **competitive exclusion principle**. In Britain, introduction of the larger, more aggressive, gray squirrel in 1876 has contributed to a contraction in range of the native red squirrel (below), and on the Scottish coast, this phenomenon has been well documented in barnacle species (see next page). The introduction of ecologically aggressive species is often implicated in the displacement or decline of native species, although there may be more than one contributing factor. Displacement of native species by introduced ones is more likely if the introduced competitor is also adaptable and hardy. It can be difficult to provide evidence of decline in a species as a direct result of competition, but it is often inferred if the range of the native species contracts and that of the introduced competitor shows a corresponding increase.

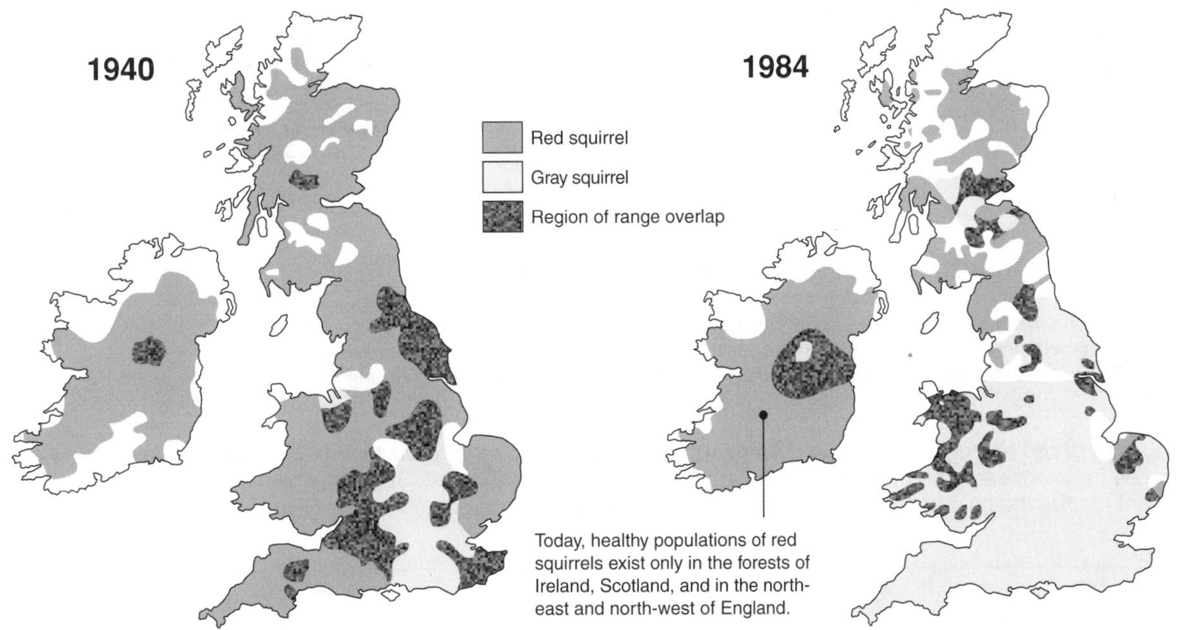

The **European red squirrel**, *Sciurus vulgaris*, was the only squirrel species in Britain until the introduction of the **American gray squirrel**, *Sciurus carolinesis*, in 1876. In 44 years since the 1940 distribution survey (above left), the more adaptable gray squirrel has displaced populations of the native red squirrels over much of the British Isles, particularly in the south (above right). Whereas the red squirrels once occupied both coniferous and broad leafed woodland, they are now almost solely restricted to coniferous forest and are completely absent from much of their former range.

1. Outline the evidence to support the view that the red-gray squirrel distributions in Britain are an example of the competitive exclusion principle:

2. Some biologists believe that competition with grey squirrels is only one of the factors contributing to the decline in the red squirrels in Britain. Explain the evidence from the 1984 distribution map that might support this view:

Periodicals:
The future of red squirrels in Britain, Black squirrels

Related activities: *The Ecological Niche*
Web links: *Red Squirrel, Gray Squirrel, Black Squirrel...*

Competitive Exclusion in Barnacles

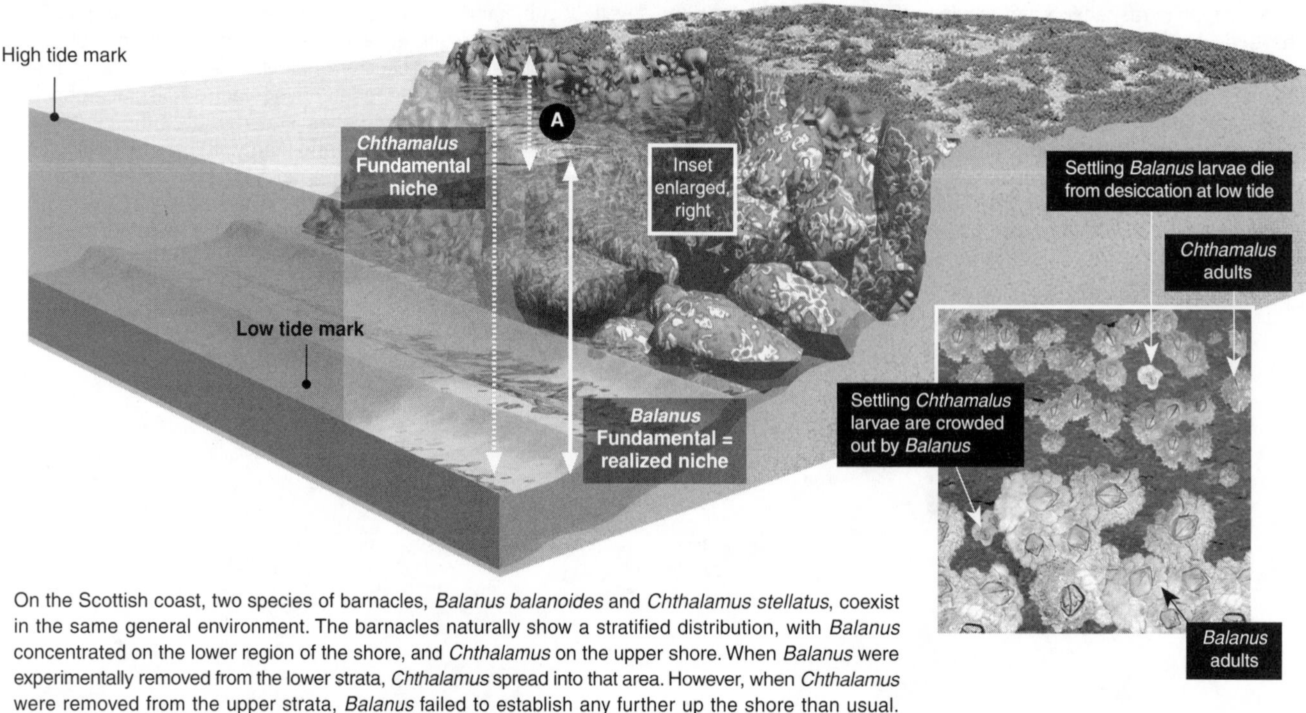

On the Scottish coast, two species of barnacles, *Balanus balanoides* and *Chthalamus stellatus*, coexist in the same general environment. The barnacles naturally show a stratified distribution, with *Balanus* concentrated on the lower region of the shore, and *Chthalamus* on the upper shore. When *Balanus* were experimentally removed from the lower strata, *Chthalamus* spread into that area. However, when *Chthalamus* were removed from the upper strata, *Balanus* failed to establish any further up the shore than usual.

3. The ability of red and gray squirrels to coexist appears to depend on the diversity of habitat type and availability of food sources (reds appear to be more successful in regions of coniferous forest). Suggest why careful habitat management is thought to offer the best hope for the long term survival of red squirrel populations in Britain:

4. Suggest other conservation methods that could possible aid the survival of viability of red squirrel populations:

5. (a) In the example of the barnacles (above), describe what is represented by the zone labeled with the arrow A:

   (b) Outline the evidence for the barnacle distribution being the result of competitive exclusion:

6. Describe two aspects of the biology of a named introduced species that have helped its success as an invading competitor:

   Species:

   (a)

   (b)

Niche Differentiation

Competition is most intense between members of the same species because their habitat and resource requirements are identical. In naturally occurring populations, **interspecific competition** (between different species) is usually less intense than intraspecific competition because coexisting species have developed (through evolution) slight differences in their realized niches. In fact, when the niches of naturally coexisting species are described, there is seldom much overlap. Species with similar ecological requirements may reduce competition by exploiting microhabitats within the ecosystem. In the eucalypt forest below, different bird species exploit tree trunks, leaf litter, different levels within the canopy, and air space. Competition may also be reduced by exploiting the same resources at a different time of the day or year.

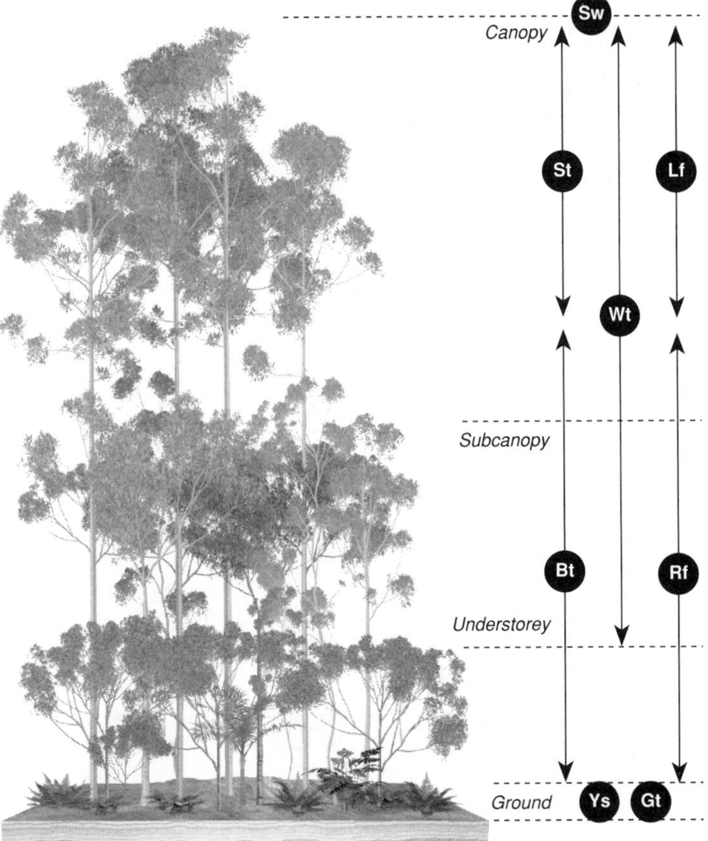

Reducing competition in a eucalypt forest

The diagram on the left shows the foraging heights of birds in an eastern Australian eucalypt forest. A wide variety of food resources are offered by the structure of the forest. Different layers of the forest allow birds to specialize in foraging at different heights. The ground-dwelling yellow-throated scrubwren and ground thrush have robust legs and feet, while the white-throated treecreeper has long toes and large curved claws and the swifts are extremely agile fliers capable of catching insects on the wing.

Key to bird species

Ys	Yellow-throated scrubwren	Lf	Leaden flycatcher
Bt	Brown thornbill	Gt	Ground thrush
Sw	Spine-tailed swift	Rf	Rufous fantail
St	Striated thornbill	Wt	White-throated treecreeper

Adapted from: Recher, Lunney & Dunn (1986): *A Natural Legacy. Ecology in Australia.* Maxwell Macmillan Publishing Australia.

Distribution of ecologically similar fish

The diagram below shows the distribution of ecologically similar damsel fish over a coral reef at Heron Island, Queensland, Australia. The habitat and resource requirements of these species overlap considerably.

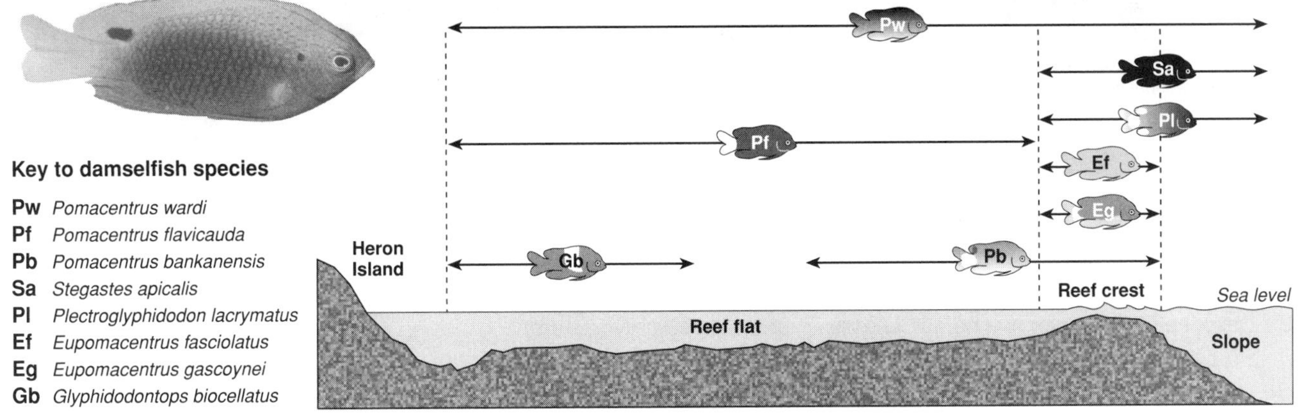

1. Describe two ways in which species can avoid directly competing for the same resources in their habitat:

   (a) ______________________________

   (b) ______________________________

2. Explain why **intraspecific** competition is more intense than **interspecific** competition: ______________________________

3. Suggest how the damsel fish on the reef at Heron Island (above) might reduce competition: ______________________________

Related activities: *The Ecological Niche, Interspecific Competition, Intraspecific Competition*

Predator-Prey Strategies

A predator with prey is one of the most conspicuous species interactions and in most, but not all, cases the predator and prey are different species. Predators have numerous adaptations for locating, identifying, and subduing prey. Prey can avoid being eaten by using passive defenses, such as hiding, or active ones, such as escaping or defending themselves against predators.

Predator Avoidance Strategies

Mimicry
Harmless prey gain immunity from attack by mimicking harmful animals. This is called **Batesian mimicry**.

Poisonous
Poisonous animals often advertise the fact that they are unpalatable by using brightly colored and gaudy markings.

Visual deception
Deceptive markings such as large, fake eyes can apparently deceive predators, allowing the prey to escape.

Chemical defense
Some animals can produce offensive smelling chemicals. American skunks squirt a nauseous fluid at attackers.

Offensive weapons
Offensive weapons are essential if prey are to actively fend off an attack by a predator.

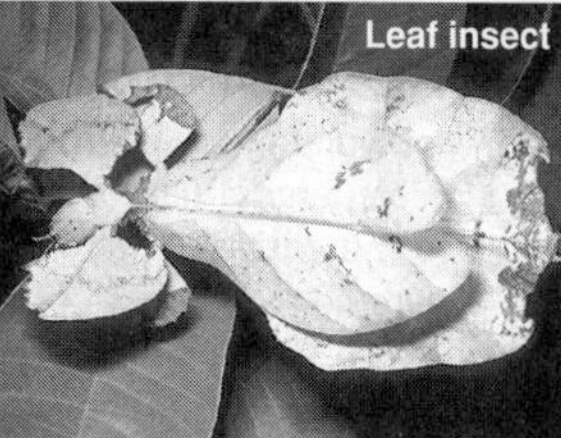

Camouflage
Cryptic shape and coloration allows some animals to blend into their background, like this insect above.

Prey Capturing Strategies

Concealment
Some animals camouflage themselves in their surroundings, striking when the prey comes within reach.

Filter feeding
Many marine animals (e.g. barnacles, baleen whales, sponges, manta rays) filter the water to extract tiny plankton.

Tool use
Some animals are gifted tool users. Chimpanzees use carefully prepared twigs to extract termites from mounds.

Stealth
The night hunting ability of some poisonous snakes is greatly helped by the presence of infrared senses.

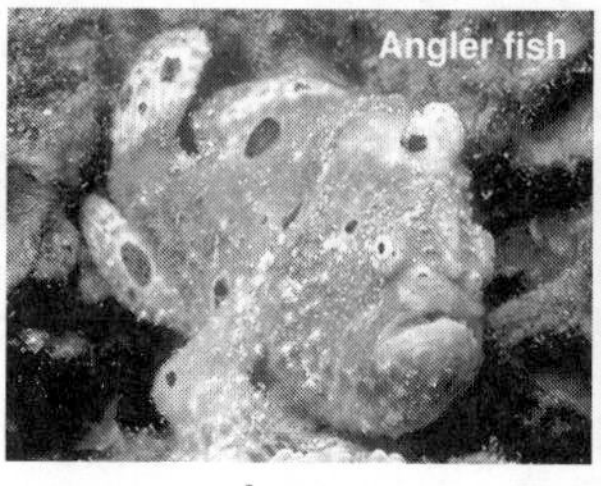

Lures
This angler fish, glow worms, and a type of spider all use lures to attract prey within striking range.

Traps
Spiders have developed a unique method of trapping their prey. Strong, sticky silk threads trap flying insects.

1. Describe a behavior of prey that is actively defensive: ____________________

____

2. Describe the behavior of a named predator that facilitates prey capture: ____________________

____

3. Explain why poisonous (unpalatable) animals are often brightly colored so that they are easily seen: ____________________

____

4. Describe the purpose of large, fake eyes on some butterflies and fish: ____________________

____

5. Explain how Batesian mimicry **benefits** the mimic (the one that is actually edible): ____________________

____

6. Describe a **behavior** typical of a (named) prey species that makes them difficult to detect by a predator:

____

Related activities: *Population Cycles*

Periodicals:
An uneasy Eden, Batesian mimicry in your backyard

Disturbance and Community Structure

Ecological theory suggests that all species in an ecosystem contribute in some way to ecosystem function. Therefore, species loss past a certain point is likely to have a detrimental effect on the functioning of the ecosystem and on its ability to resist change over time (its **stability**). Although many species still await discovery, we do know that the rate of species extinction is increasing. Scientists estimate that human destruction of natural habitats is implicated in the extinction of up to 100 000 species every year. This substantial loss of biodiversity has serious implications for the long term stability of many ecosystems.

The Concept of Ecosystem Stability

Ecosystem stability has various components, including **inertia** (the ability to resist disturbance) and **resilience** (ability to recover from external disturbances). Ecosystem stability is closely linked to the biodiversity of the system, although it is difficult to predict which factors will stress an ecosystem beyond its range of tolerance. It was once thought that the most stable ecosystems were those with the most species, because they had the greatest number of biotic interactions operating to buffer them against change. This assumption is supported by experimental evidence but there is uncertainty over what level of biodiversity provides an insurance against catastrophe.

Single species crops (monocultures), such as the soy bean crop (above, left), represent low diversity systems that can be vulnerable to disease, pests, and disturbance. In contrast, natural grasslands (above, right) may appear homogeneous, but contain many species which vary in their predominance seasonally. Although they may be easily disturbed (e.g. by burning) they are very resilient and usually recover quickly.

Tropical rainforests (above, left) represent the highest diversity systems on Earth. Whilst these ecosystems are generally resistant to disturbance, once degraded, (above, right) they have little ability to recover. The biodiversity of ecosystems at low latitudes is generally higher than that at high latitudes, where climates are harsher, niches are broader, and systems may be dependent on a small number of key species.

Community Response to Environmental Change

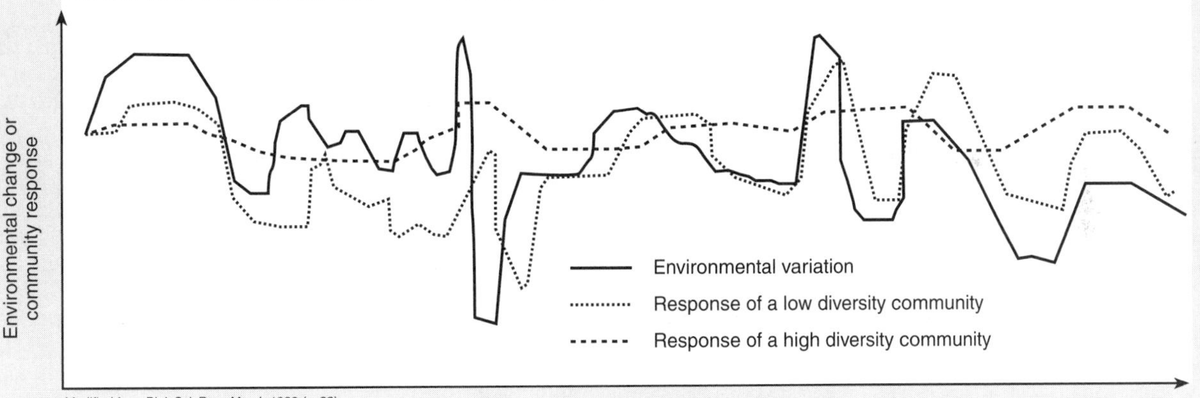

In models of ecosystem function, higher species diversity increases the stability of ecosystem functions such as productivity and nutrient cycling. In the graph above, note how the low diversity system varies more consistently with the environmental variation, whereas the high diversity system is buffered against major fluctuations. In any one ecosystem, some species may be more influential than others in the stability of the system. Such **keystone (key) species** have a disproportionate effect on ecosystem function due to their pivotal role in some ecosystem function such as nutrient recycling or production of plant biomass.

Elephants can change the entire vegetation structure of areas into which they migrate. Their pattern of grazing on taller plant species promotes a predominance of lower growing grasses with small leaves.

Termites are amongst the few larger soil organisms able to break down plant cellulose. They shift large quantities of soil and plant matter and have a profound effect on the rates of nutrient processing in tropical environments.

The starfish *Pisaster* is found along the coasts of North America where it feeds on mussels. If it is removed, the mussels dominate, crowding out most algae and leading to a decrease in the number of herbivore species.

Community Ecology

Periodicals: *Biodiversity and ecosystems*

Related activities: *Monitoring Aquatic Ecosystems, Loss of Biodiversity*

Keystone Species in North America

Gray wolf

Beaver, *Castor canadensis*

Sea otter, *Enhydra lutris*

istockphotos.com

Quaking aspen

JB-BU

Gray or **timber wolves** (*Canis lupus*) are a keystone predator and were once widespread in North American ecosystems. Historically, wolves were eliminated from Yellowstone National Park because of their perceived threat to humans and livestock. As a result, elk populations increased to the point that they adversely affected other flora and fauna. Wolves have since been reintroduced to the park and balance is returning to the ecosystem.

Two smaller mammals are also important keystone species in North America. **Beavers** (top) play a crucial role in biodiversity and many species, including 43% of North America's endangered species, depend partly or entirely on beaver ponds. **Sea otters** are also critical to ecosystem function. When their numbers were decimated by the fur trade, sea urchin populations exploded and the kelp forests, on which many species depend, were destroyed.

Quaking aspen (*Populus tremuloides*) is one of the most widely distributed tree species in North America, and aspen communities are among the most biologically diverse in the region, with a rich understorey flora supporting an abundance of wildlife. Moose, elk, deer, black bear, and snowshoe hare browse its bark, and aspen groves support up to 34 species of birds, including ruffed grouse, which depends heavily on aspen for its winter survival.

1. Suggest one probable reason why high biodiversity promotes greater ecosystem stability: ______

2. Explain why **keystone species** are so important to ecosystem function: ______

3. For each of the following species, discuss features of their biology that contribute to their position as keystone species:

   (a) Sea otter: ______

   (b) Beaver: ______

   (c) Gray wolf: ______

   (d) Quaking aspen: ______

4. Giving examples, explain how the actions of humans to remove a keystone species might result in ecosystem change: ______

Primary Succession

Ecological succession is the process by which communities in a particular area change over time. Succession takes place as a result of the complex interactions between biotic and abiotic factors. Early communities modify the physical environment causing it to change. This in turn alters the biotic community, which further alters the physical environment and so on. Each successive community makes the environment more favourable for the establishment of new species. An "idealized" succession (or sere) proceeds in seral stages, until the formation of a climax (old growth) community, which is generally stable until further disturbance. Early successional communities are characterised by a low species diversity, a simple structure, and broad niches. In contrast, climax communities are complex, with a large number of species interactions, narrow niches, and high species diversity.

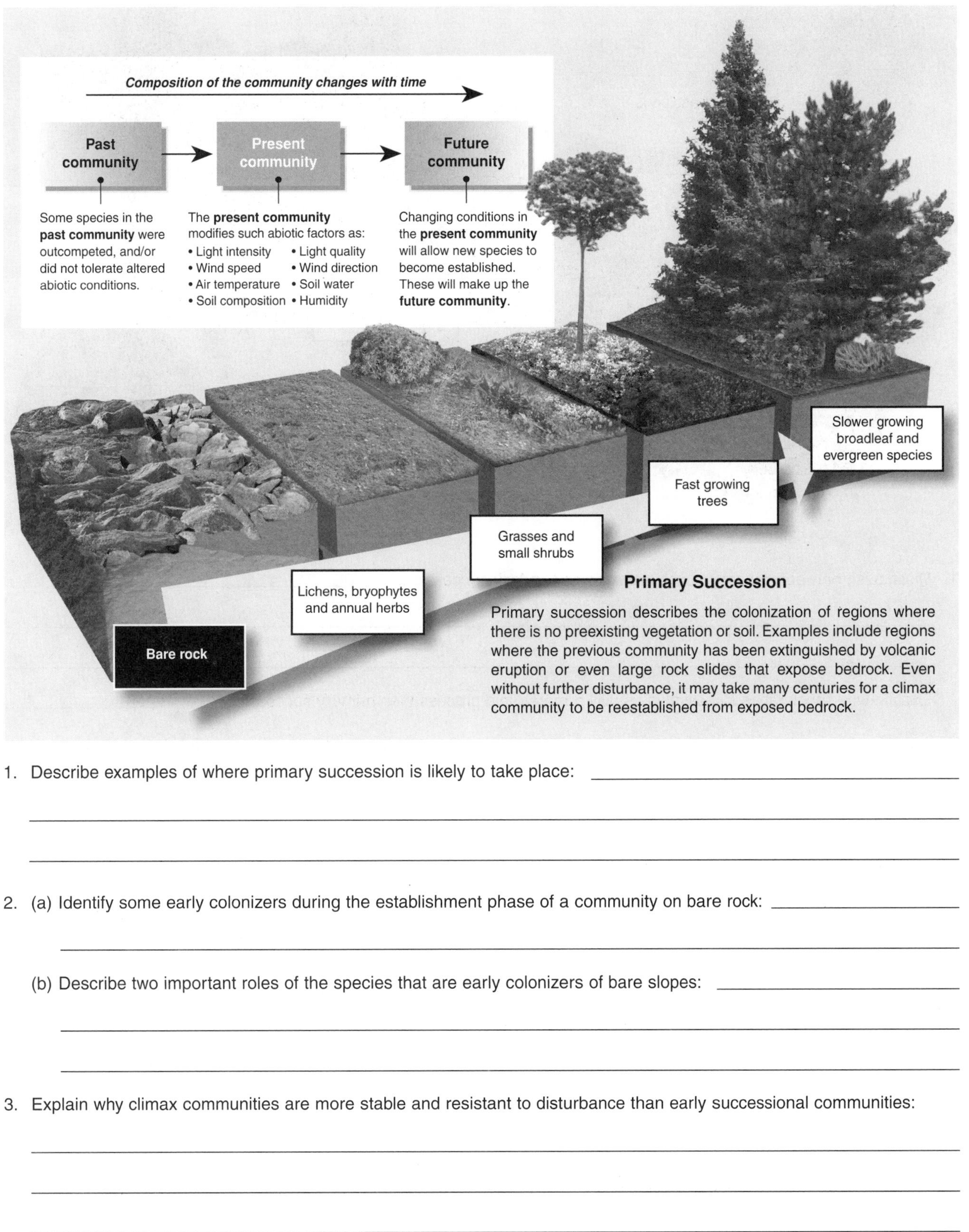
Composition of the community changes with time
Past community
Present community
Future community
Some species in the past community were outcompeted, and/or did not tolerate altered abiotic conditions.
The present community modifies such abiotic factors as:
• Light intensity • Light quality
• Wind speed • Wind direction
• Air temperature • Soil water
• Soil composition • Humidity
Changing conditions in the present community will allow new species to become established. These will make up the future community.
Bare rock
Lichens, bryophytes and annual herbs
Grasses and small shrubs
Fast growing trees
Slower growing broadleaf and evergreen species

Primary Succession

Primary succession describes the colonization of regions where there is no preexisting vegetation or soil. Examples include regions where the previous community has been extinguished by volcanic eruption or even large rock slides that expose bedrock. Even without further disturbance, it may take many centuries for a climax community to be reestablished from exposed bedrock.

1. Describe examples of where primary succession is likely to take place: ______

2. (a) Identify some early colonizers during the establishment phase of a community on bare rock: ______

(b) Describe two important roles of the species that are early colonizers of bare slopes: ______

3. Explain why climax communities are more stable and resistant to disturbance than early successional communities: ______

Community Ecology

Related activities: Secondary Succession, Disturbance and Community Structure
Web links: Mount St Helens, Forest Succession Animation, Primary Succession

Secondary Succession

A **secondary succession** takes place after a land clearance (e.g. following a fire or a landslide). Such events do not involve the loss of the soil and seed and root stocks are often undamaged. As a result, secondary succession tends to be more rapid than primary succession, although the time scale depends on the species involved and the climate and edaphic (soil) factors. Secondary succession events may occur over a wide area (such as after a forest fire), or in smaller areas where single trees have fallen.

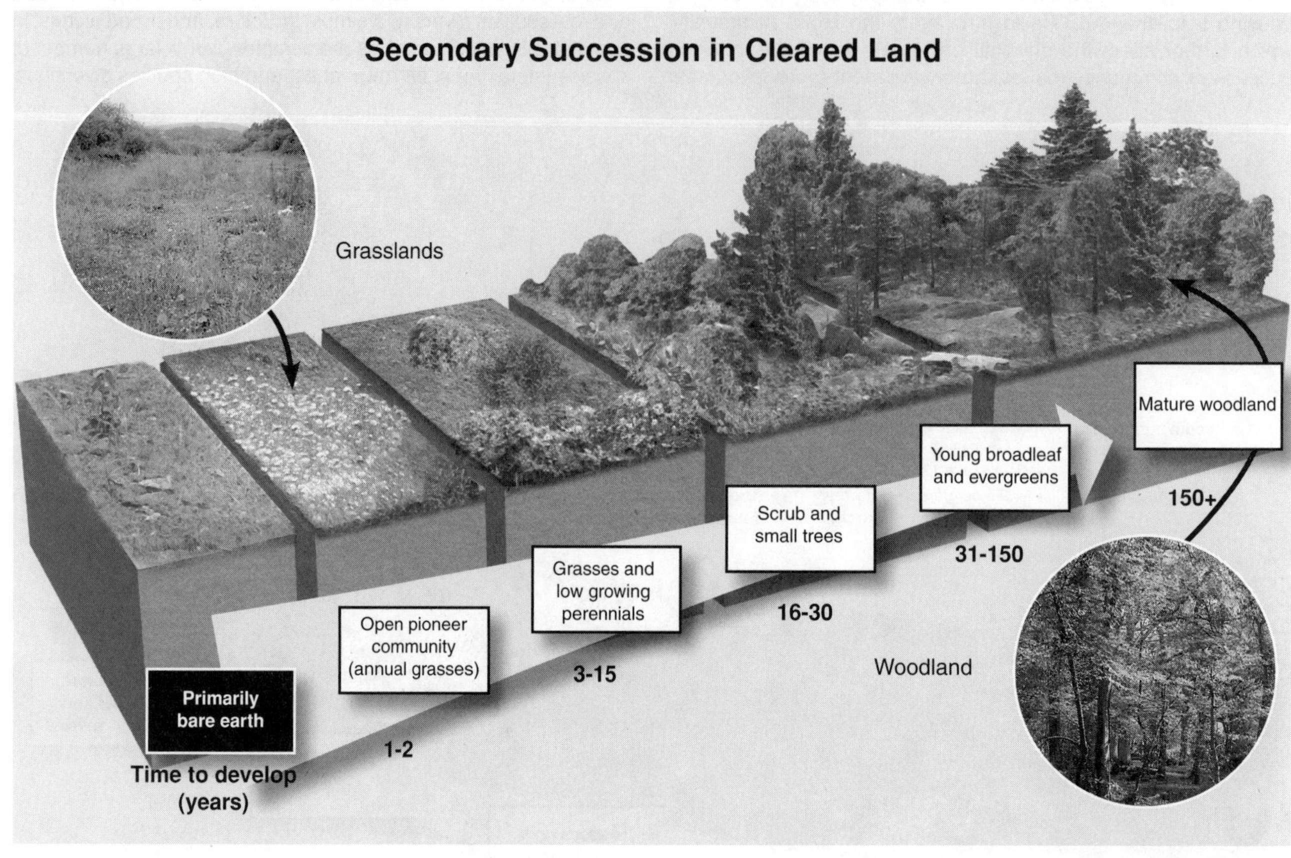

1. Distinguish between **primary** succession and **secondary** succession: ______

2. Explain why secondary succession is usually a more rapid process than primary succession: ______

3. Describe an event resulting in a secondary succession in a temperate ecosystem: ______

4. (a) Predict the likely effect of selective logging on the composition of a forest community: ______

(b) Suggest why selective logging could be considered preferable (for forest conservation) to clear felling of trees: ______

Wetland Succession

Wetland areas present a special case of ecological succession. Wetlands are constantly changing as plant invasion of open water leads to siltation and infilling. This process is accelerated by **eutrophication**. In well drained areas, pasture or **heath** may develop as a result of succession from freshwater to dry land. When the soil conditions remain non-acid and poorly drained, a swamp will eventually develop into a seasonally dry **fen**. In special circumstances (see below) an acid **peat bog** may develop. The domes of peat that develop produce a hummocky landscape with a unique biota. Wetland peat ecosystems may take more than 5000 years to form but are easily destroyed by excavation and lowering of the water table.

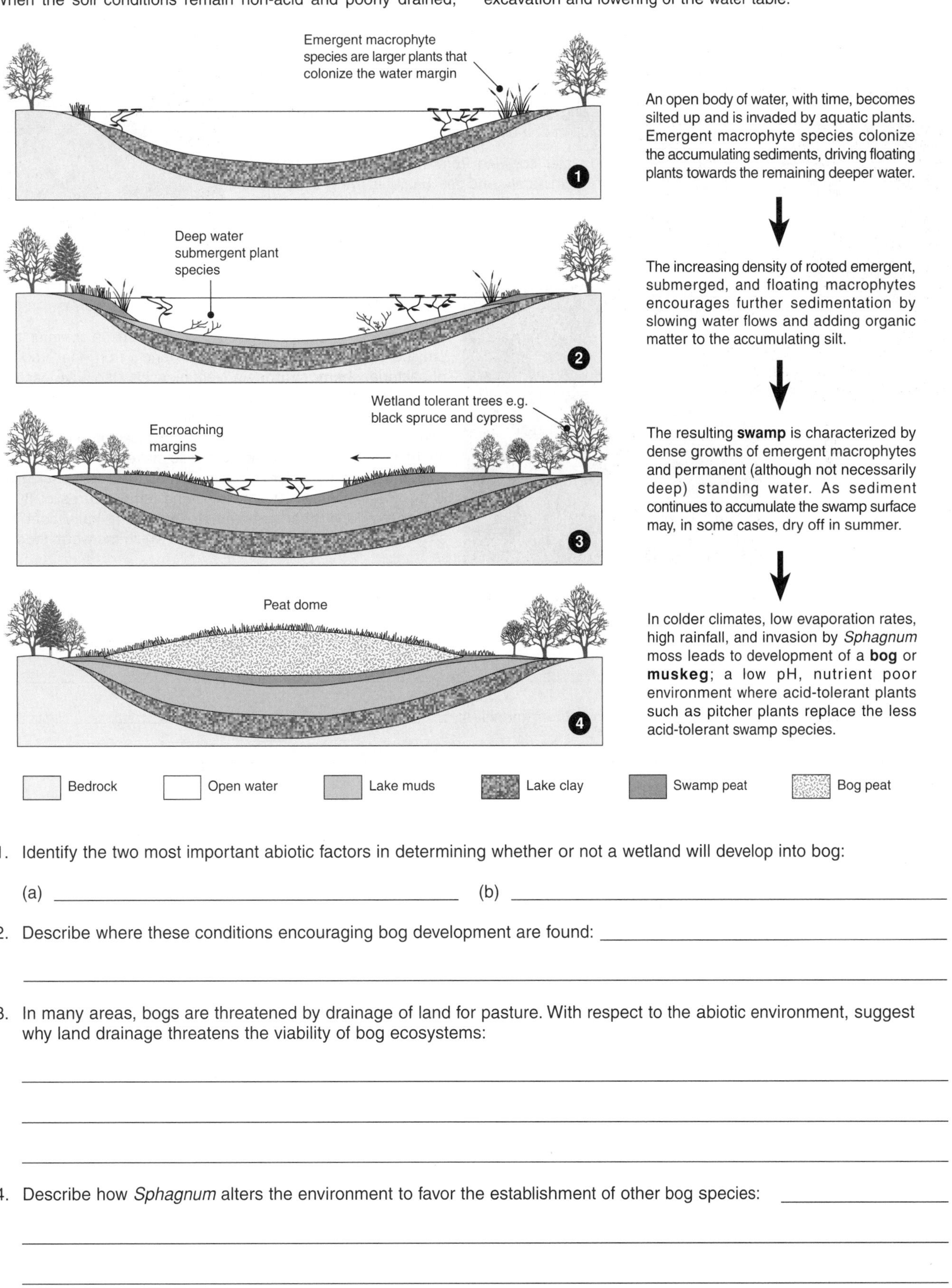

Emergent macrophyte species are larger plants that colonize the water margin
1
An open body of water, with time, becomes silted up and is invaded by aquatic plants. Emergent macrophyte species colonize the accumulating sediments, driving floating plants towards the remaining deeper water.
Deep water submergent plant species
2
The increasing density of rooted emergent, submerged, and floating macrophytes encourages further sedimentation by slowing water flows and adding organic matter to the accumulating silt.
Wetland tolerant trees e.g. black spruce and cypress
Encroaching margins
3
The resulting **swamp** is characterized by dense growths of emergent macrophytes and permanent (although not necessarily deep) standing water. As sediment continues to accumulate the swamp surface may, in some cases, dry off in summer.
Peat dome
4
In colder climates, low evaporation rates, high rainfall, and invasion by *Sphagnum* moss leads to development of a **bog** or **muskeg**; a low pH, nutrient poor environment where acid-tolerant plants such as pitcher plants replace the less acid-tolerant swamp species.
Bedrock
Open water
Lake muds
Lake clay
Swamp peat
Bog peat

1. Identify the two most important abiotic factors in determining whether or not a wetland will develop into bog:

   (a) ______________________ (b) ______________________

2. Describe where these conditions encouraging bog development are found: ______________________

   ____

3. In many areas, bogs are threatened by drainage of land for pasture. With respect to the abiotic environment, suggest why land drainage threatens the viability of bog ecosystems:

   ____

   ____

   ____

4. Describe how *Sphagnum* alters the environment to favor the establishment of other bog species: ______________________

   ____

   ____

   ____

Related activities: *Secondary Succession*
Web links: *Wetland Succession, A Hydrosere*

The Darkest Depths

Deep sea hydrothermal vents occur at around 2000 m depth and where tectonic plates meet. Buckling in the plates causes fault lines to form where water can move down into the crust before being heated and ejected at temperatures of up to 350°C. Temperatures this high are possible because the pressure of the ocean prevents the water from boiling into steam. The high water temperature dissolves minerals from crustal rocks and, when they reach the surface, they precipitate into formations called **black smokers** - chimneys of minerals that may reach 60m high.

Hydrothermal vents are the site of unique communities. At this depth, no light penetrates and the amount of organic debris falling from above is minimal because much of it has been used up by the time it reaches the bottom. The organisms living here are restricted in their movement; only a few tens of metres from the mouth of the vent, the water temperature plummets to barely above freezing. The degree of isolation has resulted in the evolution of a unique fauna.

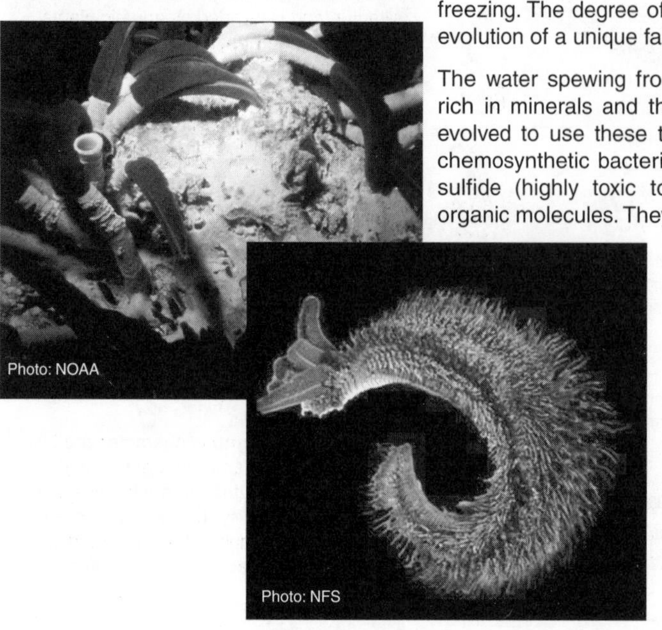

Photo: NOAA

Photo: NFS

The water spewing from the hydrothermal vents is rich in minerals and the bacteria living there have evolved to use these to manufacture food. These chemosynthetic bacteria use oxygen and hydrogen sulfide (highly toxic to most organisms) to build organic molecules. They are the producers on which the vent community is based. They form thick mats around the vents or float in aggregations resembling snow storms.

Tube worms, one of the larger organisms in these communities, provide shelter for the bacteria and benefit from the products of bacterial **chemosynthesis**. Vent mussels also have bacteria living within them and have abandoned a filter-feeding lifestyle to form a mutualistic relationship with bacteria in their tissues. Blind shrimps and crabs scavenge on decaying material and the bacterial mats. Octopi and fish also make up part of the food web, preying on smaller animals. Of most interest to scientists is the Pompeii worm. It can withstand temperatures of 80°C; higher than any other complex organism. The hairy coat that covers it is, in fact, mats of bacteria on which the worm feeds.

1. Describe the environmental conditions found around deep sea hydrothermal vents: ____________________

2. Explain reasons for the uniqueness of the vent communities: ____________________

3. Discuss the relationships between the organisms of the vent community and use the information to construct a basic food web in the space provided: ____________________

AP Topic 3C
IB SL/HL 5.1
5.2

Ecosystem Ecology & **Human Impact**

KEY CONCEPTS

- Energy transfers in communities can be quantified; this helps to describe ecosystem function.
- Nutrients move in and between ecosystems in dynamic biogeochemical (nutrient) cycles.
- Humans can interfere in nutrient cycles.
- Humans must manage their impact on the environment and cope with natural changes.

KEY TERMS

alien species
biodiversity
captive breeding
carbon cycle
carbon dioxide
conservation
denitrification
ecological pyramid
endangered species
energy flow diagram
eutrophication
gene bank
global warming
greenhouse effect
greenhouse gas
methane
nitrogen cycle
nitrogen fixation
nutrient cycle
pollution
precautionary principle
preservation
seed bank
sustainable management
threatened species

Periodicals: *listings for this chapter are on page 391*

Weblinks: *www.thebiozone.com/weblink/SB2-2603.html*

Teacher Resource CD-ROM: *Sustainable Futures*

OBJECTIVES

☐ 1. Use the **KEY TERMS** to help you understand and complete these objectives.

Quantifying Energy Transfers

pages 369-372

☐ 2. Recall that the efficiency of energy transfers between trophic levels is much less than 100% and explain the consequences of this.

☐ 3. Draw and interpret an **energy flow diagram** for a community.

☐ 4. Describe food chains quantitatively using **ecological pyramids** based on **numbers**, **biomass**, and **energy** at each trophic level. Describe communities for which pyramids of numbers and biomass may be inappropriate. Account for differences in the shapes of pyramids for different communities.

Humans and Nutrient Cycles

pages 368, 373-383

☐ 5. Describe the role of nutrient cycling in ecosystems, explaining how nutrients are exchanged within and between ecosystems, moving between the atmosphere, the Earth's crust, water, and organisms.

☐ 6. Describe the **carbon cycle**, using arrows to show the direction of nutrient flow and labels to identify the processes involved. Describe how human activity may intervene in various aspects of the carbon cycle.

☐ 7. Describe and explain the causes and consequences of the enhanced **greenhouse effect** (**global warming**). Include an analysis of the changes in concentration of atmospheric CO_2 as documented by historical records.

☐ 8. Discuss measures of reducing global warming or its effects. Include an evaluation of the **precautionary principle** (proof of no harm) to justify an immediate strong response to the threats posed by global warming.

☐ 9. Describe the **nitrogen cycle**, using arrows to show the direction of nutrient flow. Describe the processes involved and the role of microorganisms in the cycle. Describe how humans may intervene in the nitrogen cycle.

☐ 10. Describe and explain causes and consequences of nitrogen pollution and its particular relevance to agricultural regions.

Sustainable Futures

pages 384-386

☐ 11. Explain the importance of biodiversity. Describe regions of high biodiversity and explain their importance.

☐ 12. Describe the threats to biodiversity as a result of human activity. Examples (among many) include alien species, pollution, global warming, intensive agriculture, deforestation, and overfishing.

☐ 13. Describe measures to protect biodiversity and ensure sustainability of resources. Examples could include strategies for **conservation** and **preservation**, and legislative regulation (e.g. CITES, carbon trading, quota).

The Modern Atlantis?

Computer models of accelerated global warming show that the mean sea level may rise by between 100 and 900 mm over the next century. This is mainly due to the thermal expansion of the oceans as they increase in temperature, but also includes the melting of large ice sheets. This rise in sea level will have a significant effect on low lying islands and coral atolls as many are presently only a few metres above sea level. The following news article focuses on the island nation of Kiribati in the Pacific.

Vanishing Lands

The newspaper, and author name of the following article are fictitious, but the text is based on real events and information.

The Tribune
By Michael Anton: Saturday 7 June 2008

After many years of unanswered appeals for action on climate change, the tiny South Pacific nation of Kiribati has concluded that it is doomed. On Thursday its President, Anote Tong, used World Environment Day to request international help to evacuate his country before it disappears.

Kiribati consists of just 33 coral atolls scattered across 5 million square kilometres of the Pacific Ocean and it has limited scope for coping with impending global climate changes with most of the land being barely 2 metres above sea level.

Speaking from New Zealand, Mr Tong said his fellow countrymen, i-Kiribati, as they are known, had no alternative but to leave. "We may be beyond redemption," he said. "We may be at the point of no return, where the emissions in the atmosphere will carry on contributing to climate change, to produce a sea level change. So in time our small, low-lying islands will be submerged."

A London economics graduate, Mr Tong said the emigration of his people needed to start immediately. "We don't want to believe this, and our people don't want to believe this. It gives us a deep sense of frustration. What do we do?"

Kiribati is home to 97,000 people most of them living on the main atoll of Tarawa, a ring of islets surrounding a central lagoon. It is regarded as one of the places most vulnerable to climate change along with Vanauta, Tuvalu and the Marshall Islands.

Currently, the most serious problem Kiribati faces is erosion caused by flooding and storms. "We have to find the next highest spot," said Mr Tong. "At the moment there's only the coconut trees." But even the coconut trees are dying, caused by drought and a rising level of salt in ground water, which is also not being replaced due to the fact there has little rain at all for the last three years.

Mr Tong was in New Zealand – which, after committing itself to becoming carbon neutral, was chosen to host the UN's World Environment Day – for talks with Prime Minister, Helen Clark, whom he hopes to persuade to help resettle his people. But he also appealed to other countries for help relocating the i-Kiribati.

However, New Zealand, already with a large population of Pacific Islanders, would have immense trouble absorbing the 97,000 immigrants which would strain is generosity to the limit and total almost 2.5% of its current national population.

And that is just Kiribati. Talks have not yet begun with many of the other island nations which soon may also be submerged. In 2006, the Australian government issued a warning of a flood of environmental refugees across the Asian-Pacific region.

President Tong said he had heard many national leaders argue that measures to combat climate change could negatively affect their economic development, but pointed out that for the i-Kiribati it was not a matter of economics, it was a matter of survival. He said that while international scientists argue about the causes of climate change, the effects were already beginning to show on his nation. "I am not a scientist, but what I know is that things are happening we did not experience in the past... Every second week, when we get the high tides, there are always reports of erosion."

Villages, after occupying the same site for centuries, are having to be relocated due to the encroaching water. "We're doing it now... it's that urgent," he said. "Where they have been living over the past few decades is no longer there. It is being eroded."

Worse case scenarios suggest the i-Kiribati could be uninhabitable within 60 years, Mr Tong said. "I've appealed to the international community that we need to address this challenge. It's a challenge for the whole global community."

Leading industrialised nations last month pledged to cut their carbon emissions in half by 2050. But they stopped short of setting firm targets for 2020, which many scientists argue is crucial if the planet is to be saved. But for Kiribati, its saviour may have come too late.

1. From the information provided in the article, explain why Kiribati is vulnerable to the effects of global warming:

2. The article states that global warming is already adversely affecting the island nation of Kiribati. What are the physical effects it refers to, and what will be their impact on the inhabitants of Kiribati in the short term?

3. The inhabitants of Kiribati may be forced to relocate to other countries. What could this mean for the identity of the i-Kiribati and their culture and language?

4. Using other research tools, such as the library or internet, find and read other articles relating to sea level change in the Pacific. Do they all agree that there will be a rise in sea level if carbon emissions are not reduced? Explain why you think some articles take different viewpoints:

Related activities: *Global Warming*

Energy Flow in an Ecosystem

The flow of energy through an ecosystem can be measured and analyzed. It provides some idea as to the energy trapped and passed on at each trophic level. Each trophic level in a food chain or web contains a certain amount of biomass: the dry weight of all organic matter contained in its organisms. Energy stored in biomass is transferred from one trophic level to another (by eating, defecation etc.), with some being lost as low-grade heat energy to the environment in each transfer. Three definitions are useful:

- **Gross primary production**: The total of organic material produced by plants (including that lost to respiration).
- **Net primary production**: The amount of biomass that is available to consumers at subsequent trophic levels.
- **Secondary production**: The amount of biomass at higher trophic levels (consumer production). Production figures are sometimes expressed as rates (**productivity**).

The percentage of energy transferred from one trophic level to the next varies between 5% and 20% and is called the **ecological efficiency** (efficiency of energy transfer). An average figure of 10% is often used. The path of energy flow in an ecosystem depends on its characteristics. In a tropical forest ecosystem, most of the primary production enters the detrital and decomposer food chains. However, in an ocean ecosystem or an intensively grazed pasture more than half the primary production may enter the grazing food chain.

Energy Flow Through an Ecosystem

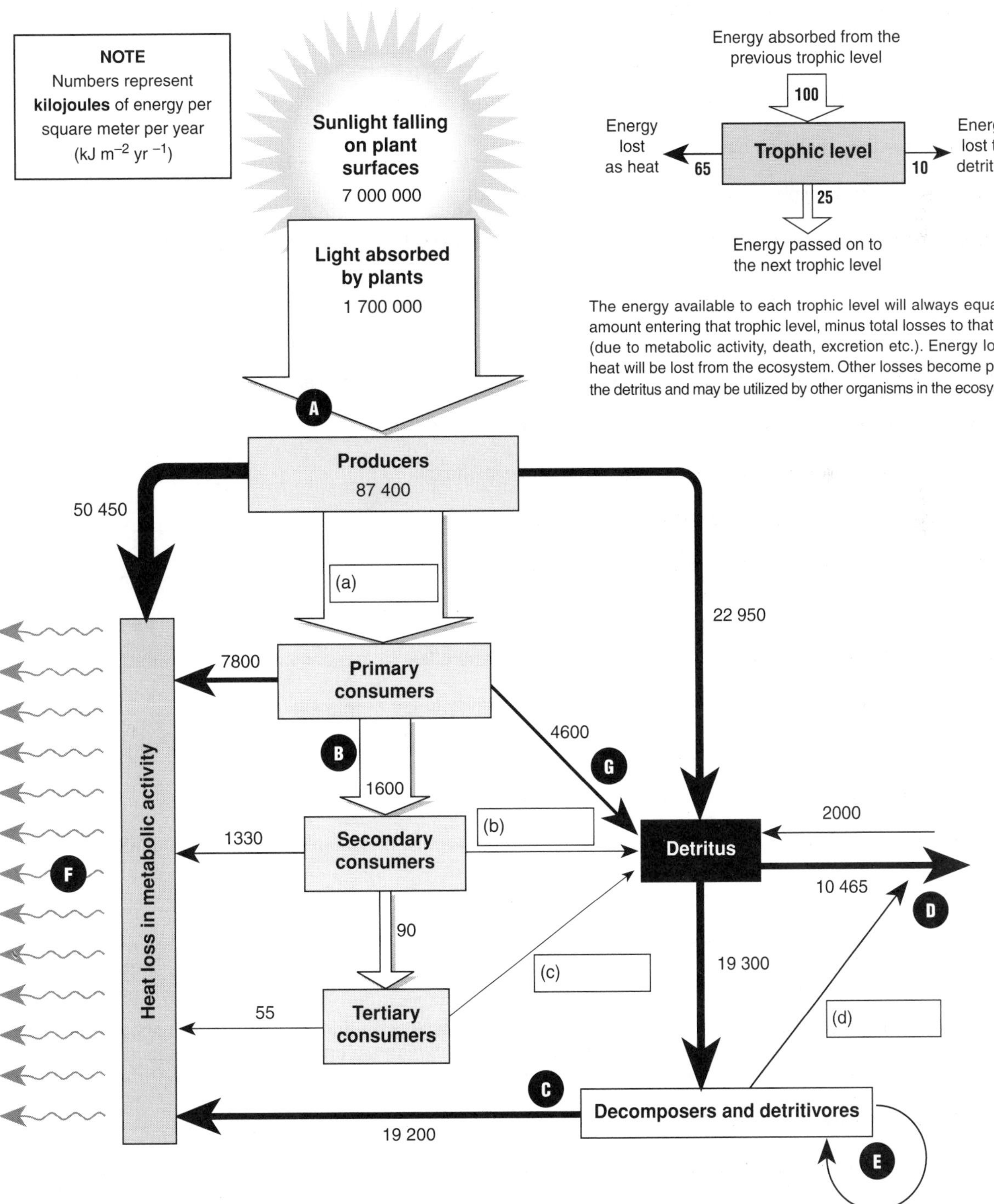

The energy available to each trophic level will always equal the amount entering that trophic level, minus total losses to that level (due to metabolic activity, death, excretion etc.). Energy lost as heat will be lost from the ecosystem. Other losses become part of the detritus and may be utilized by other organisms in the ecosystem.

1. Study the diagram on the previous page illustrating energy flow through a hypothetical ecosystem. Use the example at the top of the page as a guide to calculate the missing values (a)–(d) in the diagram. Note that the sum of the energy inputs always equals the sum of the energy outputs. Place your answers in the spaces provided on the diagram.

2. Describe the original source of energy that powers this ecosystem: ______________________

3. Identify the processes that are occurring at the points labeled **A – G** on the diagram:

   A. ______________________ E. ______________________

   B. ______________________ F. ______________________

   C. ______________________ G. ______________________

   D. ______________________

4. (a) Calculate the percentage of light energy falling on the plants that is absorbed at point A:

   Light absorbed by plants ÷ sunlight falling on plant surfaces x 100 = ______________________

   (b) Describe what happens to the light energy that is not absorbed: ______________________

   ____

5. (a) Calculate the percentage of light energy absorbed that is actually converted (fixed) into producer energy:

   Producers ÷ light absorbed by plants x 100 = ______________________

   (b) State the **amount** of light energy absorbed that is **not** fixed: ______________________

   (c) Account for the difference between the amount of energy absorbed and the amount actually fixed by producers:

   ____

   ____

6. Of the total amount of energy **fixed** by producers in this ecosystem (at point **A**) calculate:

   (a) The total amount that ended up as metabolic waste heat (in kJ): ______________________

   (b) The percentage of the energy fixed that ended up as waste heat: ______________________

7. (a) State the groups for which detritus is an energy source: ______________________

   (b) Describe by what means detritus could be removed or added to an ecosystem: ______________________

   ____

8. In certain conditions, detritus will build up in an environment where few (or no) decomposers can exist.

   (a) Describe the consequences of this lack of decomposer activity to the energy flow: ______________________

   ____

   (b) Add an additional arrow to the diagram on the previous page to illustrate your answer.

   (c) Describe three examples of materials that have resulted from a lack of decomposer activity on detrital material:

   ____

   ____

   ____

9. The **ten percent law** states that the total energy content of a trophic level in an ecosystem is only about one-tenth (or 10%) that of the preceding level. For each of the trophic levels in the diagram on the preceding page, determine the amount of energy passed on to the next trophic level as a percentage:

   (a) Producer to primary consumer: ______________________

   (b) Primary consumer to secondary consumer: ______________________

   (c) Secondary consumer to tertiary consumer: ______________________

Ecological Pyramids

The trophic levels of any ecosystem can be arranged in a pyramid shape. The first trophic level is placed at the bottom and subsequent trophic levels are stacked on top in their 'feeding sequence'. Ecological pyramids can illustrate changes in the **numbers**, **biomass** (weight), or **energy** content of organisms at each level. Each of these three kinds of pyramids tell us something different about the flow of energy and materials between one trophic level and the next. The type of pyramid you choose in order to express information about the ecosystem will depend on what particular features of the ecosystem you are interested in and, of course, the type of data you have collected.

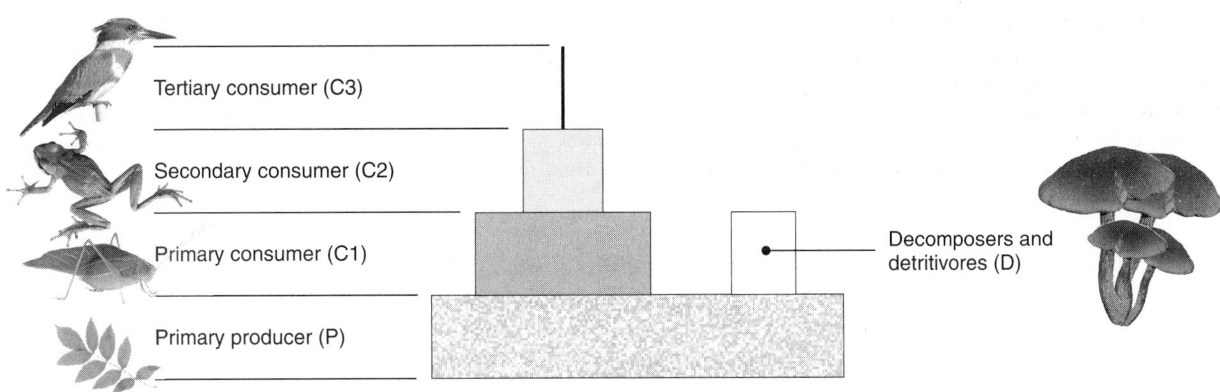

The generalized ecological pyramid pictured above shows a conventional pyramid shape, with a large number (or biomass) of producers forming the base for an increasingly small number (or biomass) of consumers. Decomposers are placed at the level of the primary consumers and off to the side. They may obtain energy from many different trophic levels and so do not fit into the conventional pyramid structure. For any particular ecosystem at any one time (e.g. the forest ecosystem below), the shape of this typical pyramid can vary greatly depending on whether the trophic relationships are expressed as numbers, biomass or energy.

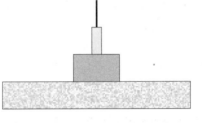

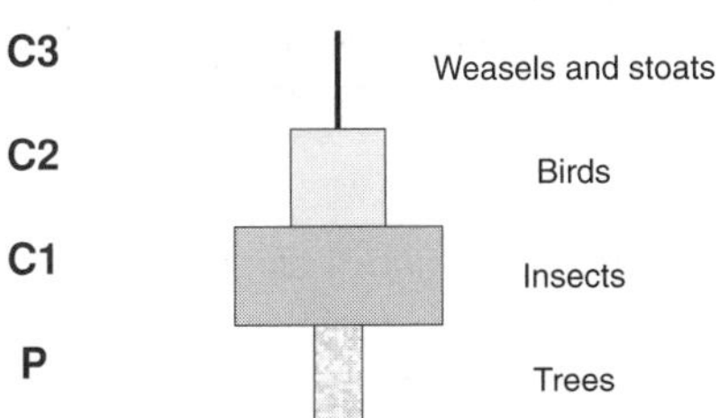

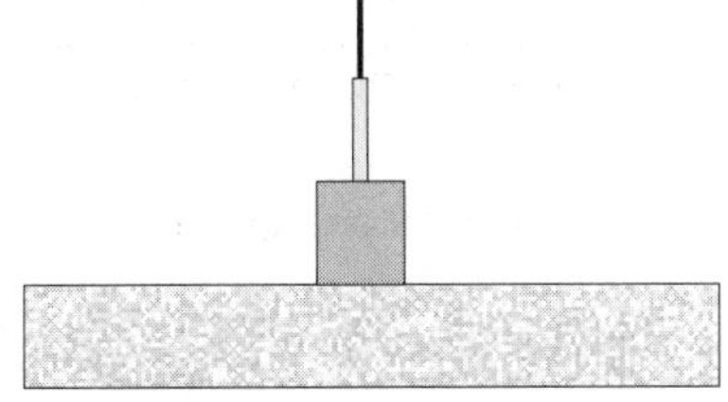

Numbers in a forest community

Pyramids of numbers display the number of individual organisms at each trophic level. The pyramid above has few producers, but they may be of a very large size (e.g. trees). This gives an 'inverted pyramid' although not all pyramids of numbers are like this.

Biomass in a forest community

Biomass pyramids measure the 'weight' of biological material at each trophic level. Water content of organisms varies, so 'dry weight' is often used. Organism size is taken into account, so meaningful comparisons of different trophic levels are possible.

Energy in a forest community

Pyramids of energy are often very similar to biomass pyramids. The energy content at each trophic level is generally comparable to the biomass (i.e. similar amounts of dry biomass tend to have about the same energy content).

1. Describe what the three types of ecological pyramids measure:

   (a) Number pyramid: ______________________________

   (b) Biomass pyramid: ______________________________

   (c) Energy pyramid: ______________________________

2. Explain the advantage of using a biomass or energy pyramid rather than a pyramid of numbers to express the relationship between different trophic levels:

   ____

   ____

   ____

3. Explain why it is possible for the forest ecosystem (on the next page) to have very few producers supporting a large number of consumers:

   ____

   ____

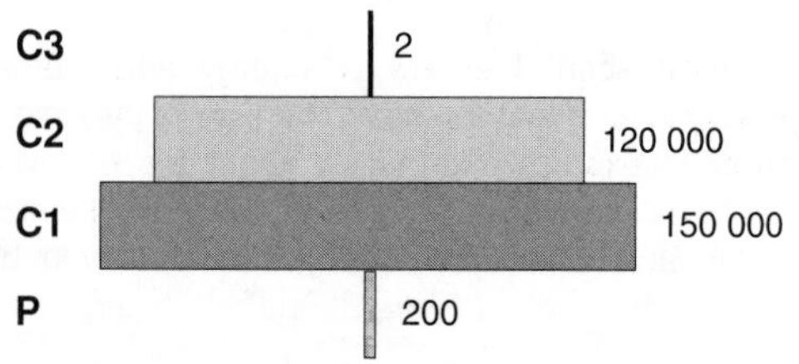

Pyramid of numbers: forest community

In a forest community a few producers may support a large number of consumers. This is due to the large size of the producers; large trees can support many individual consumer organisms. The example above shows the numbers at each trophic level for an oak forest in England, in an area of 10 m^2.

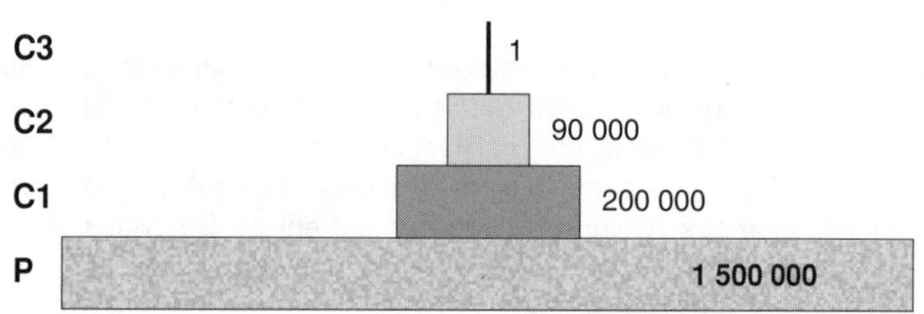

Pyramid of numbers: grassland community

In a grassland community a large number of producers are required to support a much smaller number of consumers. This is due to the small size of the producers. Grass plants can support only a few individual consumer organisms and take time to recover from grazing pressure. The example above shows the numbers at each trophic level for a derelict grassland area (10 m^2) in Michigan, United States.

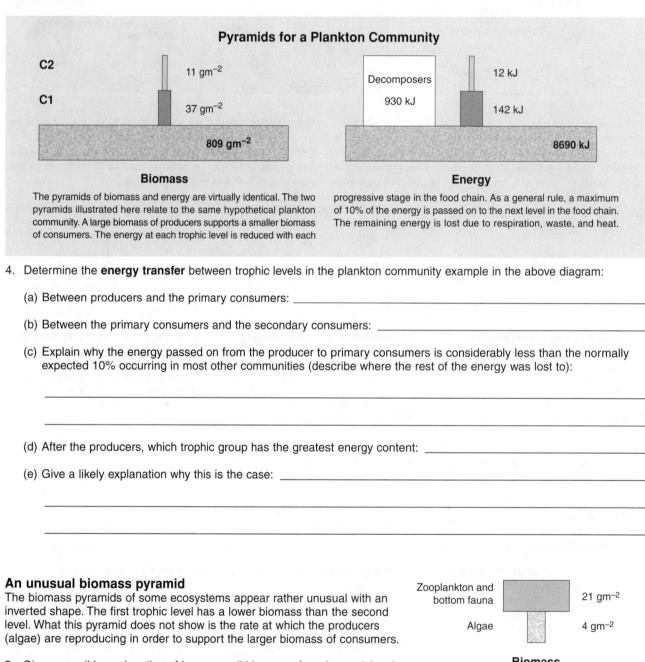

Pyramids for a Plankton Community

C2 — 11 gm^{-2}

C1 — 37 gm^{-2}

809 gm^{-2}

Biomass

Decomposers 930 kJ

12 kJ

142 kJ

8690 kJ

Energy

The pyramids of biomass and energy are virtually identical. The two pyramids illustrated here relate to the same hypothetical plankton community. A large biomass of producers supports a smaller biomass of consumers. The energy at each trophic level is reduced with each progressive stage in the food chain. As a general rule, a maximum of 10% of the energy is passed on to the next level in the food chain. The remaining energy is lost due to respiration, waste, and heat.

4. Determine the **energy transfer** between trophic levels in the plankton community example in the above diagram:

   (a) Between producers and the primary consumers:

   (b) Between the primary consumers and the secondary consumers:

   (c) Explain why the energy passed on from the producer to primary consumers is considerably less than the normally expected 10% occurring in most other communities (describe where the rest of the energy was lost to):

   (d) After the producers, which trophic group has the greatest energy content:

   (e) Give a likely explanation why this is the case:

An unusual biomass pyramid

The biomass pyramids of some ecosystems appear rather unusual with an inverted shape. The first trophic level has a lower biomass than the second level. What this pyramid does not show is the rate at which the producers (algae) are reproducing in order to support the larger biomass of consumers.

Zooplankton and bottom fauna — 21 gm^{-2}

Algae — 4 gm^{-2}

Biomass

5. Give a possible explanation of how a small biomass of producers (algae) can support a larger biomass of consumers (zooplankton):

The Carbon Cycle

Carbon is an essential element in living systems, providing the chemical framework to form the molecules that make up living organisms (e.g. proteins, carbohydrates, fats, and nucleic acids). Carbon also makes up approximately 0.03% of the atmosphere as the gas carbon dioxide (CO_2), and it is present in the ocean as carbonate and bicarbonate, and in rocks such as limestone. Carbon cycles between the living (biotic) and non-living (abiotic) environment: it is fixed in the process of photosynthesis and returned to the atmosphere in respiration. Carbon may remain locked up in biotic or abiotic systems for long periods of time as, for example, in the wood of trees or in fossil fuels such as coal or oil. Human activity has disturbed the balance of the carbon cycle (the global carbon budget) through activities such as combustion (e.g. the burning of wood and **fossil fuels**) and deforestation.

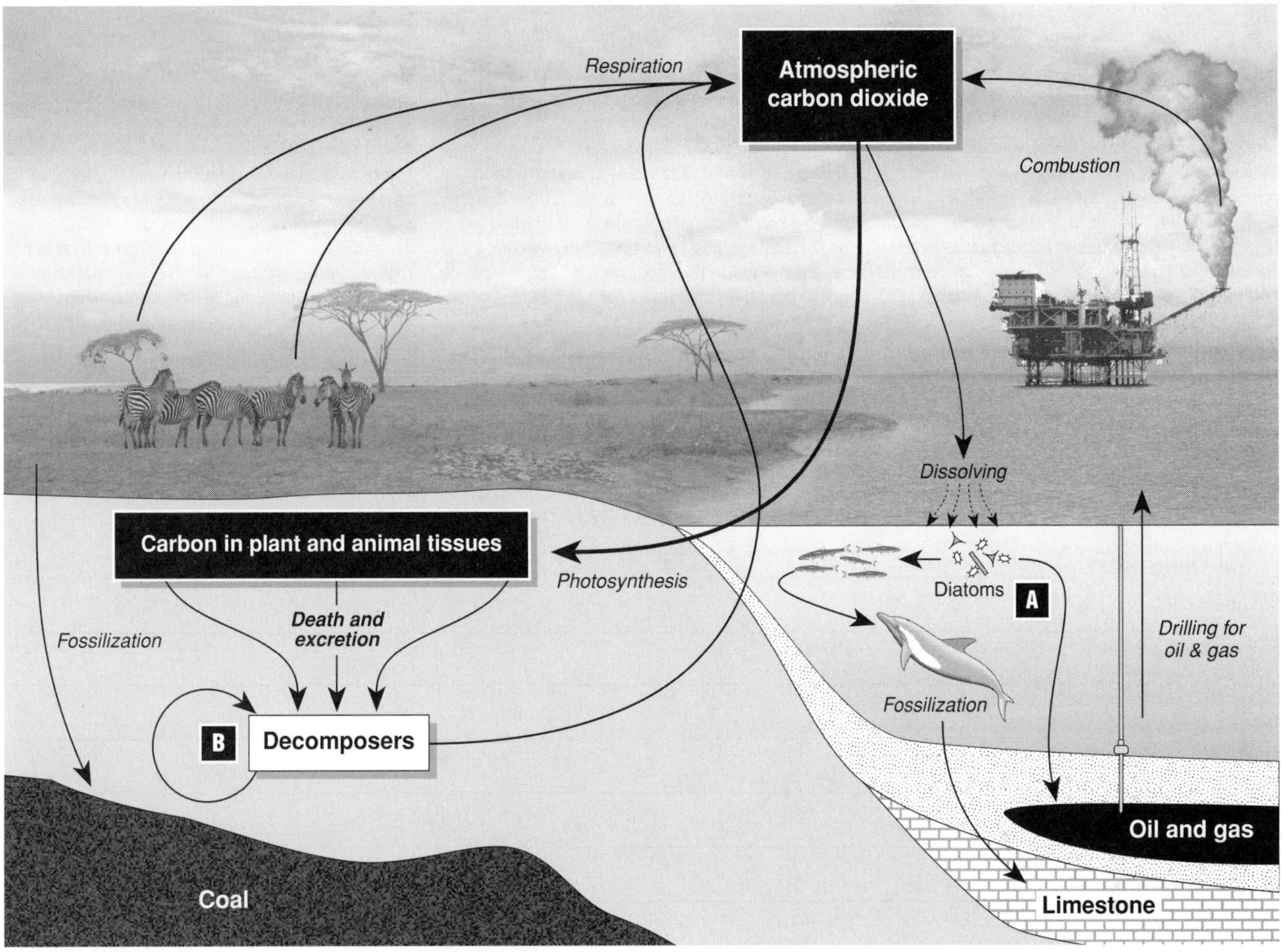

Respiration
Atmospheric carbon dioxide
Combustion
Dissolving
Carbon in plant and animal tissues
Photosynthesis
Diatoms
A
Drilling for oil & gas
Death and excretion
Fossilization
Fossilization
B
Decomposers
Oil and gas
Coal
Limestone

1. In the diagram above, add **arrows** and **labels** to show the following activities:

   (a) Dissolving of limestone by acid rain
   (b) Release of carbon from the marine food chain
   (c) Mining and burning of coal
   (d) Burning of plant material.

2. Describe the **biological origin** of the following geological deposits:

   (a) Coal: ______

   (b) Oil: ______

   (c) Limestone: ______

3. Describe the two processes that release carbon into the atmosphere: ______

4. Name the four geological reservoirs (sinks), in the diagram above, that can act as a source of carbon:

   (a) ______ (c) ______

   (b) ______ (d) ______

5. (a) Identify the process carried out by diatoms at point [**A**]: ______

   (b) Identify the process carried out by decomposers at [**B**]: ______

Termites: These insects play an important role in nutrient recycling. With the aid of symbiotic protozoans and bacteria in their guts, they can digest the tough cellulose of woody tissues in trees. Termites fulfill a vital function in breaking down the endless rain of debris in tropical rainforests.

Dung beetles: Beetles play a major role in the decomposition of animal dung. Some beetles merely eat the dung, but true dung beetles, such as the scarabs and *Geotrupes*, bury the dung and lay their eggs in it to provide food for the beetle grubs during their development.

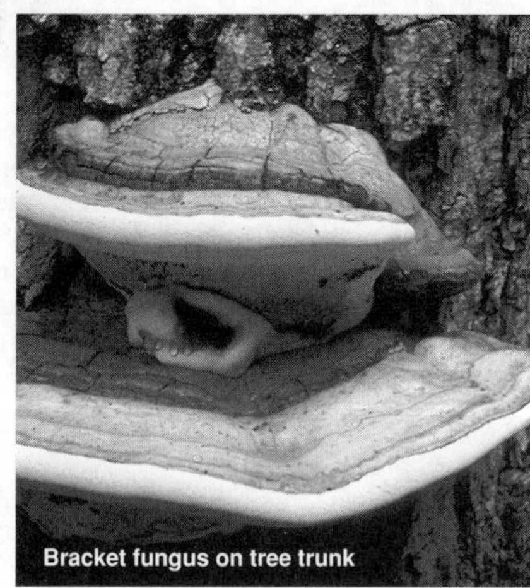

Fungi: Together with decomposing bacteria, fungi perform an important role in breaking down dead plant matter in the leaf litter of forests. Some mycorrhizal fungi have been found to link up to the root systems of trees where an exchange of nutrients occurs (a mutualistic relationship).

6. Predict the consequences to carbon cycling if there were no decomposers present in an ecosystem:

7. Explain how each of the three organisms listed below has a role to play in the carbon cycle:

   (a) Dung beetles:

   (b) Termites:

   (c) Fungi:

8. Using specific examples, explain the role of insects in carbon cycling:

9. In natural circumstances, accumulated reserves of carbon such as peat, coal and oil represent a **sink** or natural diversion from the cycle. Eventually the carbon in these sinks returns to the cycle through the action of geological processes which return deposits to the surface for oxidation.

   (a) Describe what effect human activity is having on the amount of carbon stored in sinks:

   (b) Describe two **global effects** resulting from this activity:

   (c) Suggest what could be done to prevent or alleviate these effects:

Global Warming

The Earth's atmosphere comprises a mixture of gases including nitrogen, oxygen, and water vapor. Also present are small quantities of carbon dioxide (CO_2), methane, and a number of other "trace" gases. In the past, our climate has shifted between periods of stable warm conditions to cycles of glacials and interglacials. The current period of warming climate is partly explained by the recovery after the most recent ice age that finished 10 000 years ago. Eight of the ten warmest years on record (records kept since the mid-1800s) were in the 1980s and 1990s. Global surface temperatures in 1998 set a new record by a wide margin, exceeding those of the previous record year, 1995. Many researchers believe the current warming trend has been compounded by human activity, in particular, the release of certain gases into the atmosphere. The term '**greenhouse effect**' describes a process of global climate warming caused by the release of 'greenhouse gases', which act as a thermal blanket in the atmosphere, letting in sunlight, but trapping the heat that would normally radiate back into space. About three-quarters of the natural greenhouse effect is due to water vapor. The next most significant agent is CO_2. Since the industrial revolution and expansion of agriculture about 200 years ago, additional CO_2 has been pumped into the atmosphere. The effect of global warming on agriculture, other human activities, and the biosphere in general, is likely to be considerable.

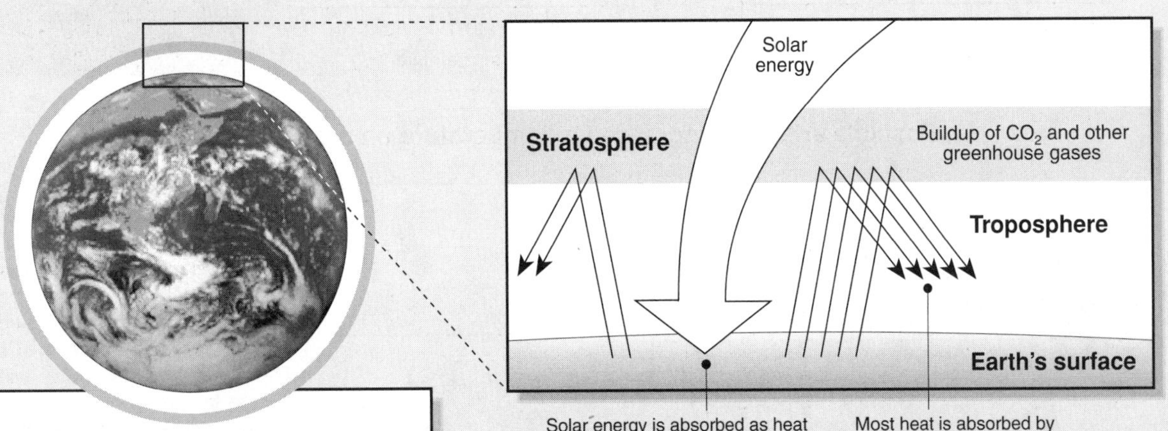

Sources of 'Greenhouse Gases'

Carbon dioxide
- Exhaust from cars
- Combustion of coal, wood, oil
- Burning rainforests

Methane
- Plant debris and growing vegetation
- Belching and flatus of cows

Chloro-fluoro-carbons (CFCs)
- Leaking coolant from refrigerators
- Leaking coolant from air conditioners

Nitrous oxide
- Car exhaust

Tropospheric ozone*
- Triggered by car exhaust (smog)

*Tropospheric ozone is found in the lower atmosphere (not to be confused with ozone in the stratosphere)

Greenhouse gas	Tropospheric conc.		Global warming potential *(compared to CO_2)¶*	Atmospheric lifetime *(years)§*
	Pre-industrial 1750	Present day (2008*)		
Carbon dioxide	280 ppm	383.9 ppm	1	120
Methane	700 ppb	1796 ppb	25	12
Nitrous oxide	270 ppb	320.5 ppb	310	120
CFCs	0 ppb	0.39 ppb	4000+	50-100
HFCs‡	0 ppb	0.045 ppb	1430	14
Tropospheric ozone	25 ppb	34 ppb	17	hours

ppm = parts per million; **ppb** = parts per billion; ‡Hydrofluorcarbons were introduced in the last decade to replace CFCs as refrigerants; * Data from July 2007-June 2008. ¶ Figures contrast the radiative effect of different greenhouse gases relative to CO_2 over 100 years, e.g. over 100 years, methane is 25 times more potent as a greenhouse gas than CO_2 § How long the gas persists in the atmosphere. *Source: CO_2 Information Analysis Centre, Oak Ridge National Laboratory, USA.*

The graph on the right shows how the mean temperature for each year from 1860 until 2008 (grey bars) compared with the average temperature between 1961 and 1990. The thick black line represents the mathematically fitted curve and shows the general trend indicated by the annual data. Most anomalies since 1977 have been above normal; warmer than the long term mean, indicating that global temperatures are tracking upwards. In 1998 the global temperature exceeded that of the previous record year, 1995, by about 0.2°C.

Source: Hadley Center for Prediction and Research

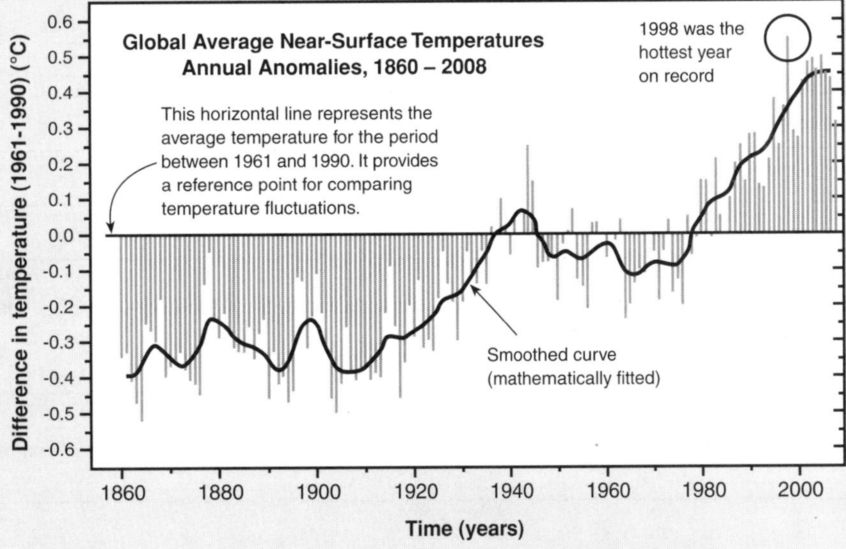

Effects of increases in temperature on crop yields

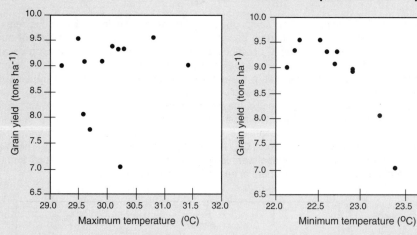

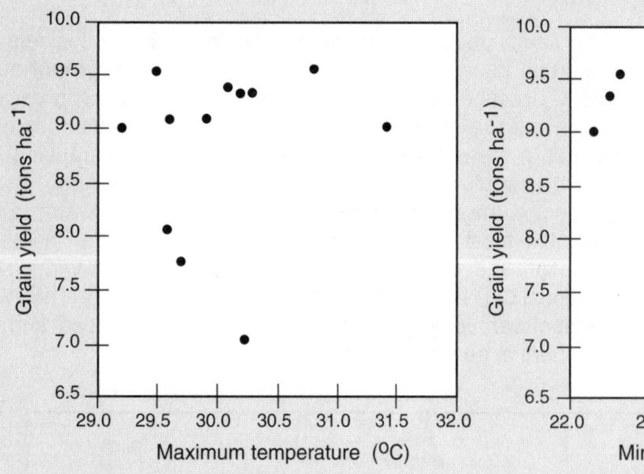

Studies on the grain production of rice have shown that maximum daytime temperatures have little effect on crop yield. However minimum night time temperatures lower crop yield by as much as 5% for every 0.5°C increase in temperature.

Possible effects of increases in temperature on crop damage

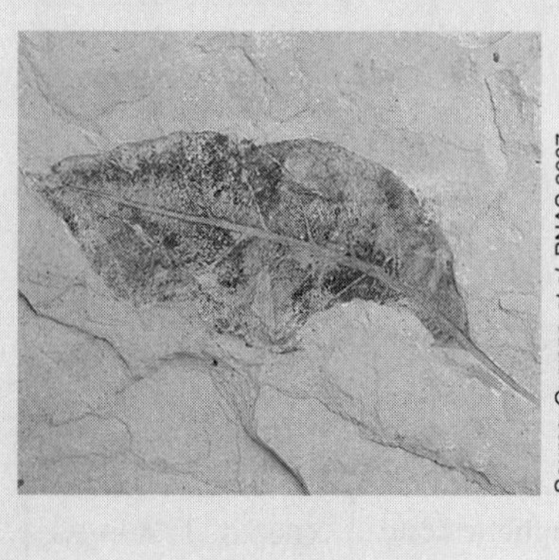

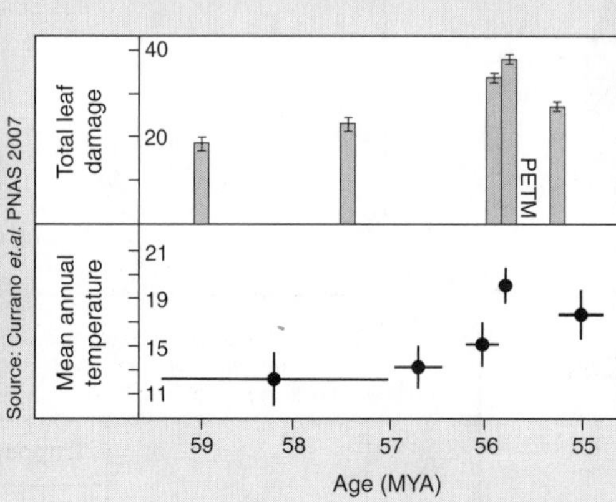

The fossil record shows that global temperatures rose sharply around 56 million years ago. Studies of fossil leaves with insect browse damage indicate that leaf damage peaked at the same time as the Paleocene Eocene Thermal Maximum (PETM). This gives some historical evidence that as temperatures increase, plant damage caused by insects also rises. This could have implications for agricultural crops.

The impacts of climate change on agriculture and horticulture in North America will vary because of the size and range of its geography. Examples of changes include citrus production shifting slightly north with reduced yields in Texas and Florida. Corn, soybean, and potato yields are all predicted to decline.

Many of North America's largest cities are located near to the coast. The rises in sea levels that are predicted could see these cities inundated. The devastating effects of disasters, such as hurricane Katrina, illustrate the vulnerability of low lying cities to sea level rises. Estimates vary, but rises may be between 30 and 50 cm by 2100.

1. Calculate the increase (as a %) in the 'greenhouse gases' between the pre-industrial era and the 2008 measurements (use the data from the table, see previous page). **HINT**: The calculation for carbon dioxide is: (383.9 - 280) ÷ 280 x 100 =

   (a) Carbon dioxide: ____________ (b) Methane: ____________ (c) Nitrous oxide: ____________

2. Describe the consequences of global temperature rise on low lying land: ____________

3. Explain the relationship between the rise in concentrations of atmospheric CO_2, methane, and oxides of nitrogen, and the enhanced greenhouse effect:

Biodiversity and Global Warming

Since the last significant period of climate change at the end of the ice age 10,000 years ago, plants and animals have adapted to survive in their current habitats. Accelerated global warming is again changing the habitats that plants and animals live in and this could have significant effects on the biodiversity of specific regions as well as on the planet overall. As temperatures rise, organisms will be forced to move to new areas where temperatures are similar to their current level. Those that cannot move face extinction, as temperatures move outside their limits of tolerance. Changes in precipitation as a result of climate change also affect where organisms can live. Long term changes in climate could see the contraction of many organisms' habitats while at the same time the expansion of others. Habitat migration, the movement of a habitat from its current region into another, will also become more frequent. Already there are a number of cases showing the effects of climate change on a range of organisms.

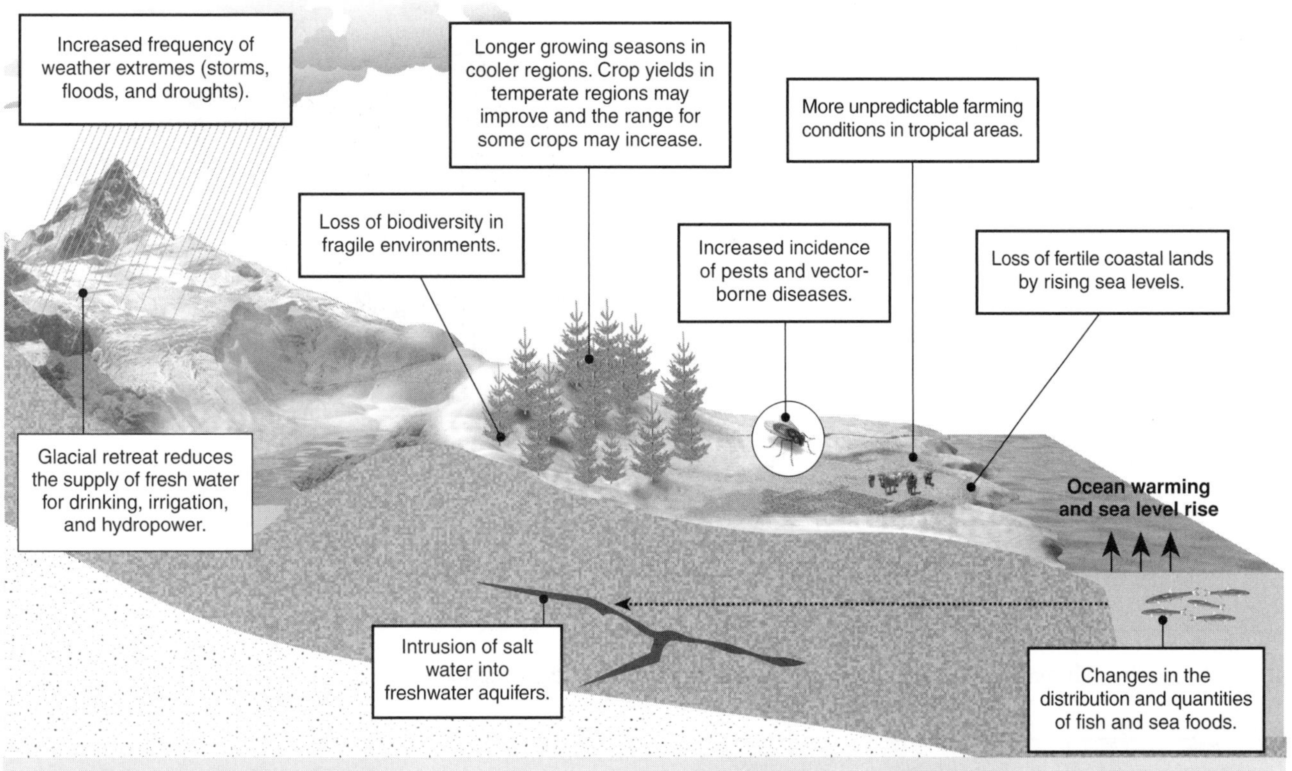

Increased frequency of weather extremes (storms, floods, and droughts).
Longer growing seasons in cooler regions. Crop yields in temperate regions may improve and the range for some crops may increase.
More unpredictable farming conditions in tropical areas.
Loss of biodiversity in fragile environments.
Increased incidence of pests and vector-borne diseases.
Loss of fertile coastal lands by rising sea levels.
Glacial retreat reduces the supply of fresh water for drinking, irrigation, and hydropower.
Ocean warming and sea level rise
Intrusion of salt water into freshwater aquifers.
Changes in the distribution and quantities of fish and sea foods.

Studies of forests in the United States have shown that although there will increases and decreases in the distribution ranges of various tree species, overall there will be an 11% decrease in forest cover, with an increase in savanna and arid woodland. Communities of oak/pine and oak/hickory are predicted to increase in range while spruce/fir and maple/beech/birch communities will decrease.

Photo: Walter Siegmund

Studies of the distributions of butterfly species in many countries show their populations are shifting. Surveys of Edith's checkerspot butterfly (*Euphydryas editha*) in western North America have shown it to be moving north and to higher altitudes.

Studies of sea life along the Californian coast have shown that between 1931 and 1996, shoreline ocean temperatures increased by 0.79ºC and populations of invertebrates including sea stars, limpets and snails moved northward in their distributions.

An Australian study in 2004 found the centre of distribution for the AdhS gene in *Drosophila*, which helps survival in hot and dry conditions, had shifted 400 kilometres south in the last twenty years.

A 2009 study of 200 million year old plant fossils from Greenland has provided evidence of a sudden collapse in biodiversity that is correlated with, and appears to be caused by, a very slight rise in CO_2 levels.

Periodicals:
Climate change and biodiversity

Related activities: *Global Warming*
Web links: *Climate Change, Global Warming Ecocentre*

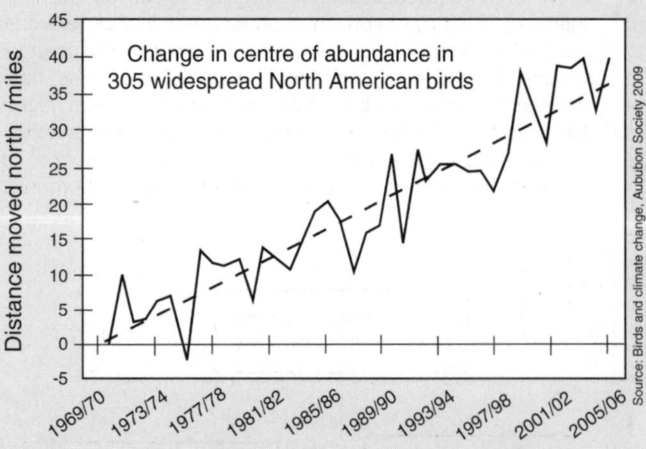

A number of studies indicate that animals are beginning to be affected by increases in global temperatures. Data sets from around the world show that birds are migrating up to two weeks earlier to summer feeding grounds and are often not migrating as far south in winter.

Animals living at altitude are also affected by warming climates and are being forced to shift their normal range. As temperatures increase, the snow line increases in altitude pushing alpine animals to higher altitudes. In some areas of North America this has resulting the local extinction of the North American pika (*Ochotona princeps*).

1. Describe some of the likely effects of global warming on physical aspects of the environment: ____________________

2. (a) Using the information on this and the previous activity, discuss the probable effects of global warming on plant crops:

(b) Suggest how farmers might be able to adjust to these changes: ____________________

3. Discuss the evidence that insect populations are affected by global temperature: ____________________

4. (a) Describe how increases in global temperatures have affected some migratory birds: ____________________

(b) Explain how these changes in migratory patterns might affect food availability for these populations: ____________________

5. Explain how global warming could lead to the local extinction of some alpine species: ____________________

The Nitrogen Cycle

Nitrogen is a crucial element for all living things, forming an essential part of the structure of proteins and nucleic acids. The Earth's atmosphere is about 80% nitrogen gas (N_2), but molecular nitrogen is so stable that it is only rarely available directly to organisms and is often in short supply in biological systems. Bacteria play an important role in transferring nitrogen between the biotic and abiotic environments. Some bacteria are able to fix atmospheric nitrogen, while others convert ammonia to nitrate and thus make it available for incorporation into plant and animal tissues. Nitrogen-fixing bacteria are found living freely in the soil *(Azotobacter)* and living symbiotically with some plants in root nodules *(Rhizobium)*. Lightning discharges also cause the oxidation of nitrogen gas to nitrate which ends up in the soil. Denitrifying bacteria reverse this activity and return fixed nitrogen to the atmosphere. Humans intervene in the nitrogen cycle by producing, and applying to the land, large amounts of nitrogen fertilizer. Some applied fertilizer is from organic sources (e.g. green crops and manures) but much is inorganic, produced from atmospheric nitrogen using an energy-expensive industrial process. Overuse of nitrogen fertilizers may lead to pollution of water supplies, particularly where land clearance increases the amount of leaching and runoff into ground and surface waters.

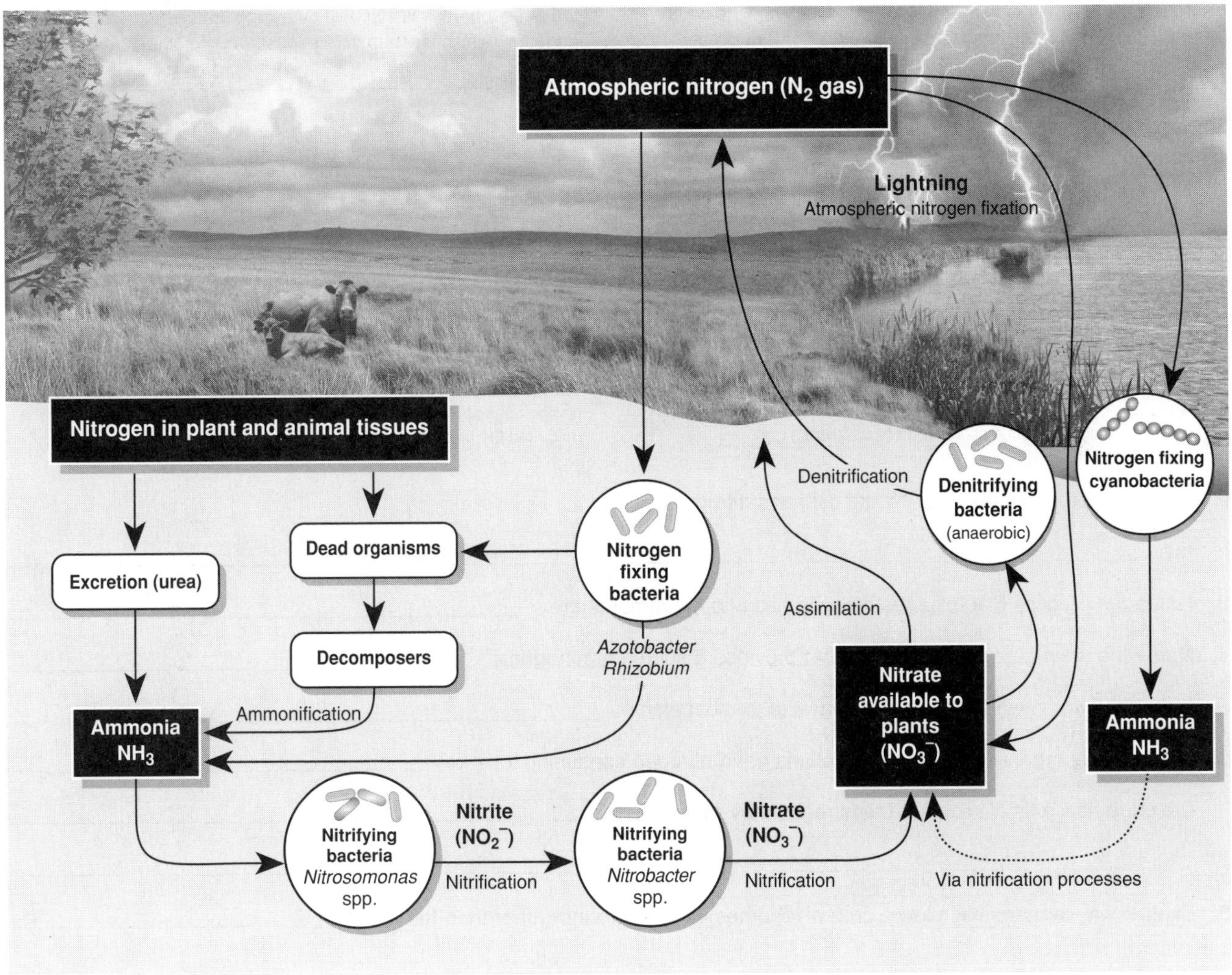

1. Describe five instances in the nitrogen cycle where **bacterial** action is important. Include the name of each of the processes and the changes to the form of nitrogen involved:

(a) ______________________________

____

(b) ______________________________

____

(c) ______________________________

____

(d) ______________________________

____

(e) ______________________________

____

Nitrogen Fixation in Root Nodules

Root nodules are a root **symbiosis** between a higher plant and a bacterium. The bacteria fix atmospheric nitrogen and are extremely important to the nutrition of many plants, including the economically important legume family. Root nodules are extensions of the root tissue caused by entry of a bacterium. In legumes, this bacterium is *Rhizobium*. Other bacterial genera are involved in the root nodule symbioses in non-legume species.

The bacteria in these symbioses live in the nodule where they fix atmospheric nitrogen and provide the plant with most, or all, of its nitrogen requirements. In return, they have access to a rich supply of carbohydrate. The fixation of atmospheric nitrogen to ammonia occurs within the nodule, using the enzyme **nitrogenase**. Nitrogenase is inhibited by oxygen and the nodule provides a low O_2 environment in which fixation can occur.

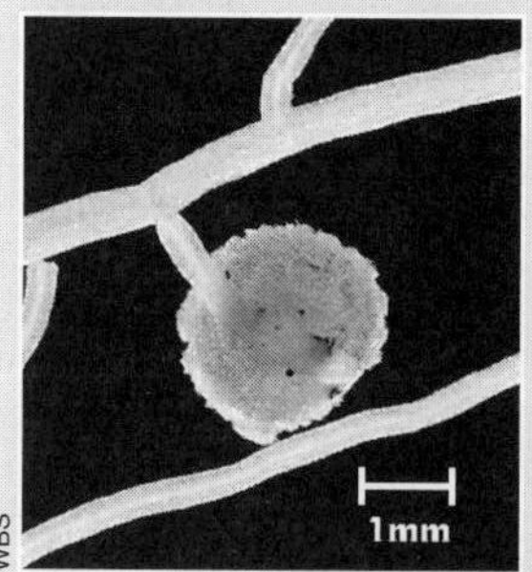

Two examples of legume nodules caused by *Rhizobium*. The photographs above show the size of a single nodule (left), and the nodules forming clusters around the roots of *Acacia* (right).

Human Intervention in the Nitrogen Cycle

Until about sixty years ago, microbial nitrogen fixation (left) was the only mechanism by which nitrogen could be made available to plants. However, during WW II, Fritz Haber developed the **Haber process** whereby nitrogen and hydrogen gas are combined to form gaseous ammonia. The ammonia is converted into ammonium salts and sold as inorganic fertilizer. Its application has revolutionized agriculture by increasing crop yields.

As well as adding nitrogen fertilizers to the land, humans use anaerobic bacteria to break down livestock wastes and release NH_3 into the soil. They also intervene in the nitrogen cycle by discharging **effluent** into waterways. Nitrogen is removed from the land through burning, which releases nitrogen oxides into the atmosphere. It is also lost by mining, harvesting crops, and irrigation, which leaches nitrate ions from the soil.

Two examples of human intervention in the nitrogen cycle. The photographs above show the aerial application of a commercial fertilizer (left), and the harvesting of an agricultural crop (right).

2. Identify three processes that **fix** atmospheric nitrogen:

   (a) ____________ (b) ____________ (c) ____________

3. Name the process that releases nitrogen gas into the atmosphere: ____________

4. Name the main geological reservoir that provides a source of nitrogen: ____________

5. State the form in which nitrogen is available to most plants: ____________

6. Name a vital organic compound that plants need nitrogen containing ions for: ____________

7. Describe how animals acquire the nitrogen they need: ____________

   ____

8. Explain why farmers may plow a crop of legumes into the ground rather than harvest it: ____________

   ____

9. Describe five ways in which humans may intervene in the nitrogen cycle and the effects of these interventions:

   (a) ____________

   ____

   (b) ____________

   ____

   (c) ____________

   ____

   (d) ____________

   ____

   (e) ____________

   ____

Nitrogen Pollution

The effect of excess nitrogen compounds on the environment is varied. Depending on the compound formed, nitrogen can cause smog in cities or algal blooms in lakes and seas. Nitrogen gas makes up almost 80% of the atmosphere but is unreactive at normal pressure and temperature. At the high pressures and temperatures reached in factories and combustion engines nitrogen gas forms nitric oxide along with other nitrogen oxides, most of which contribute to atmospheric pollution. Nitrates in fertilizers are washed into ground water by rain and slowly make their way to lakes and rivers and eventually out to sea. This process can take time to become noticeable as ground water can take many decades to reach a waterway. In many places where nitrate effects are only just becoming apparent, the immediate cessation of their use could take a long time to have any effect as it might take many years before the last of the ground water carrying the nitrates reaches a waterway.

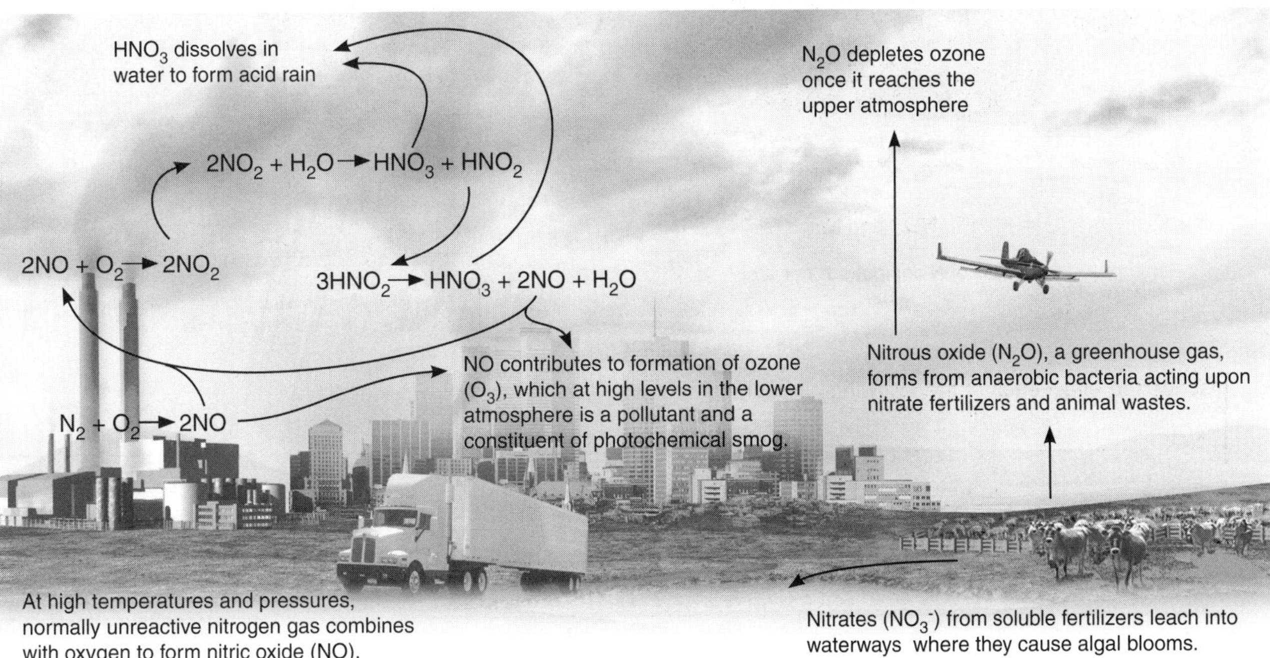

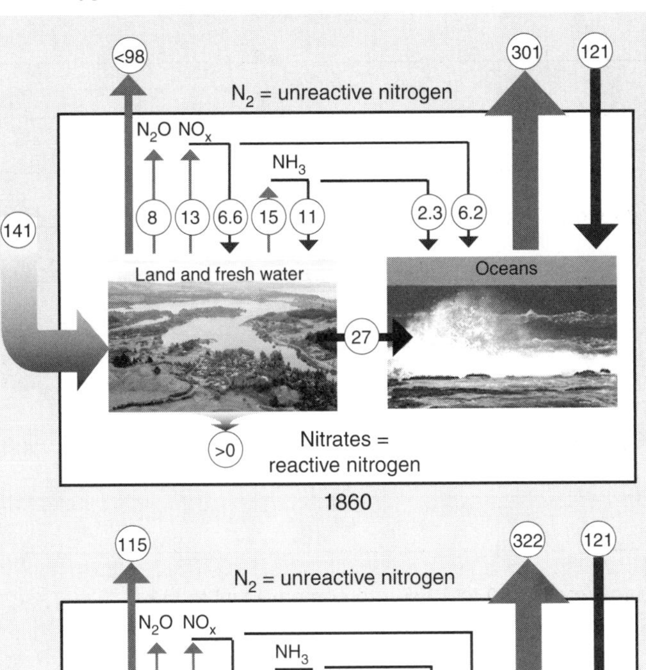

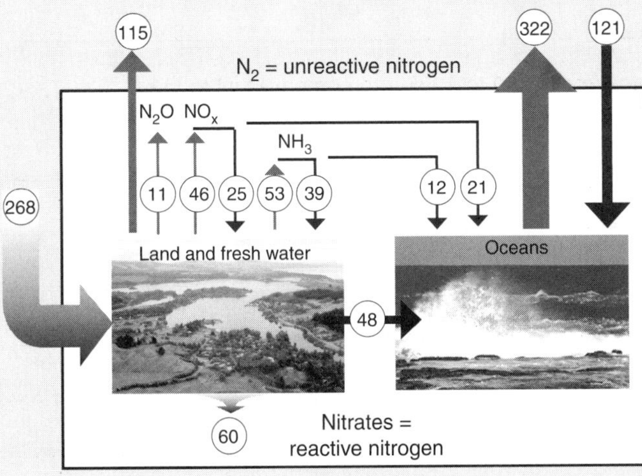

Changes in nitrogen inputs and outputs between 1860 and 1995 in million Tonne (modified from Galloway *et al* 2004)

Early last century, the Haber-Bosch process made nitrate fertilizers readily available for the first time. Since then, the use of nitrogen fertilizers has increased at an almost exponential rate. Importantly, this has led to an increase in the levels of nitrogen in land and water by up to 60 times those of 100 years ago. This extra nitrogen load is one of the causes of accelerated enrichment (**eutrophication**) of lakes and coastal waters. An increase in algal production also results in higher decomposer activity and, consequently, oxygen depletion, fish deaths, and depletion of aquatic biodiversity. Many aquatic microorganisms also produce toxins, which may accumulate in the water, fish, and shellfish. The diagrams (left) show the increase in nitrates in water sources from 1860 to 1995. The rate at which nitrates are added has increased faster than the rate at which nitrates are returned to the atmosphere as unreactive N_2 gas. This has led to the widespread accumulation of nitrogen.

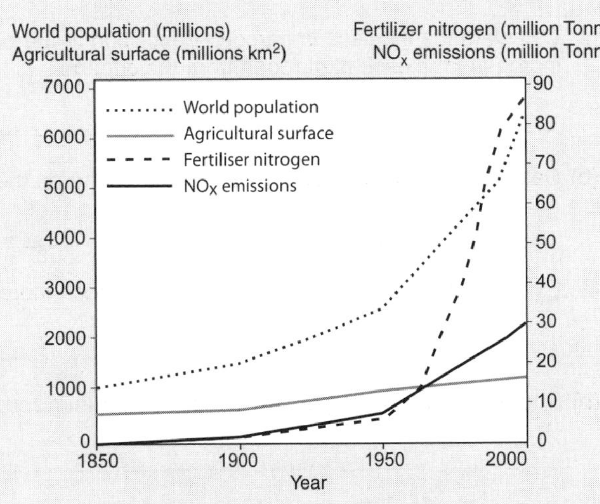

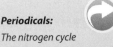

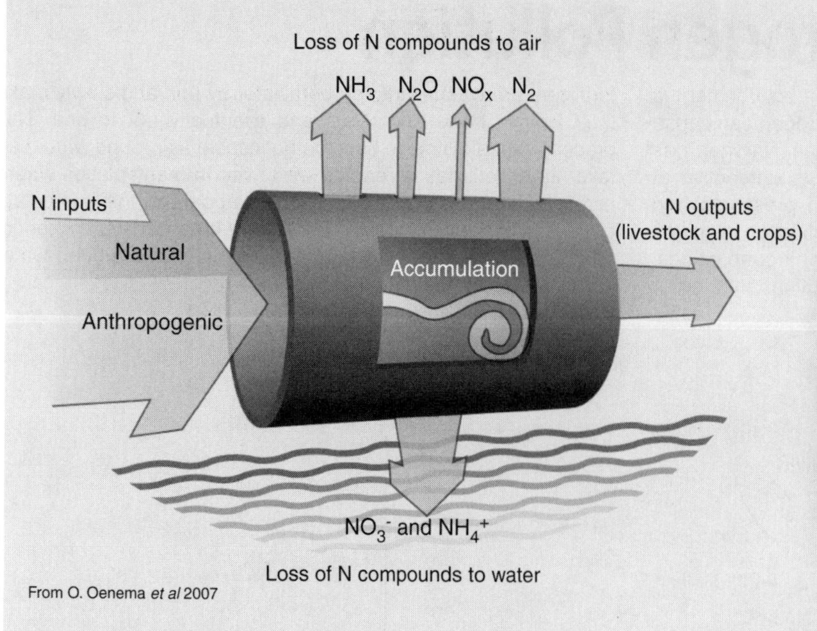

The "hole in the pipe" model (left) demonstrates inefficiencies in nitrogen fertilizer use. Nitrogen that is added to the soil and not immediately taken up by plants is washed into waterways or released into the air by bacterial action. These losses can be minimized to an extent by using slow release fertilizers during periods of wet weather and by careful irrigation practices.

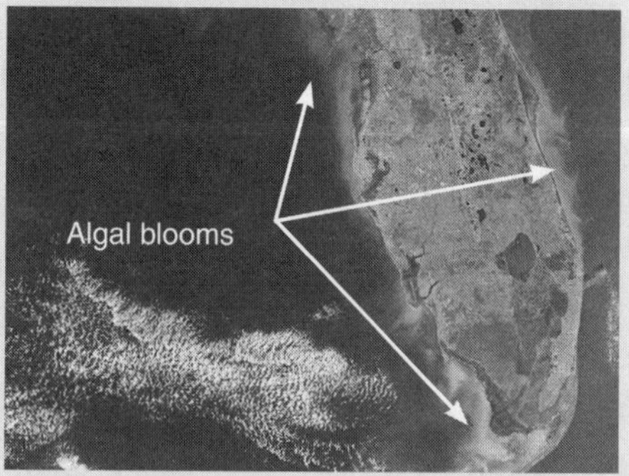

Satellite photo of algal blooms around Florida. Excessive nitrogen contributes to algal blooms in both coastal and inlands waters. *Image: NASA*

1. Describe the effect each of the following nitrogen compounds have on air and water quality:

   (a) NO: ______________________________

   (b) N_2O: ______________________________

   (c) NO_2: ______________________________

   (d) NO_3^-: ______________________________

2. Explain why the formation of NO can cause large scale and long term environmental problems: ______________________________

3. Explain why an immediate halt in the use of nitrogen fertilizers will not cause an immediate stop in their effects:

4. (a) Calculate the increase in nitrogen deposition in the oceans from 1860 to 1995 and compare this to the increase in release of nitrogen from the oceans.

   (b) Describe the effect these increases are having on the oceans: ______________________________

5. (a) Explain why nitrogen inputs tend to be so much more than outputs in livestock and crops: ______________________________

   (b) Suggest how the nitrogen losses could be minimized: ______________________________

Monitoring Aquatic Ecosystems

There is no single measure to objectively describe the quality of a stream, river, or lake. Rather it is defined in terms of various chemical, physical, and biological characteristics. Together, these factors define the 'health' of the aquatic ecosystem and its suitability for various desirable uses. It is normally not feasible to monitor for all contaminants potentially in water. For example, analysis for pesticides, dioxins, and other trace 'organics' can be a costly, ongoing expense. Water quality is determined by making measurements on-site or by taking samples back to a laboratory for physical, chemical, or microbiological analysis. Other methods, involving the use of **indicator species**, can also be used to biologically assess the health of a water body.

Some aspects of water quality, such as black disk clarity measurements (above), must be made in the field.

The collection of water samples allows many quality measurements to be carried out in the laboratory.

Telemetry stations relay continuous measurements of the water level of a water body to a control office.

Temperature and dissolved oxygen measurements must be carried out directly in the flowing water.

All Photos EW

Biological Indicators of Ecosystem Health

One of the best ways to determine the health of an ecosystem is to measure the variety of organisms living in it. Certain species, called **indicator species**, are typical of ecosystems in a particular state of health (e.g. polluted or pristine). An objective evaluation of an ecosystem's biodiversity can provide valuable insight into its health, particularly if the species assemblages have changed as a result of disturbance. Diversity can be quantified using a **diversity index**. Such indices are widely used in ecological work, particularly for monitoring ecosystem change or pollution.

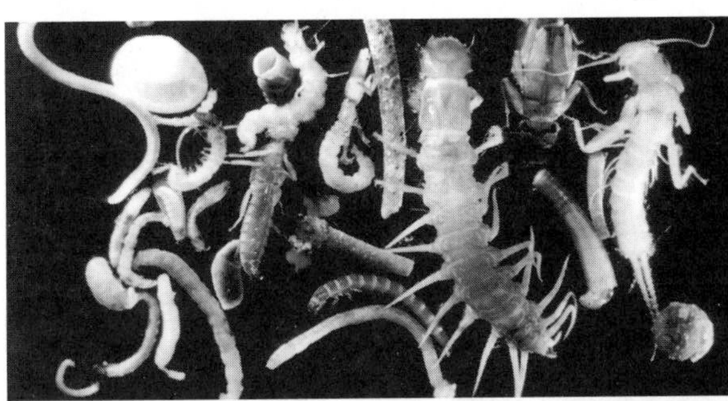

A stream community with a high diversity of macroinvertebrates (above) in contrast to a low diversity stream community (below).

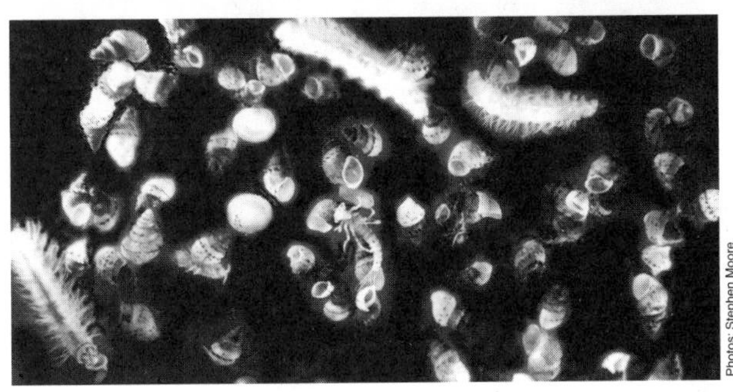

Photos: Stephen Moore

Water Quality Measurements

Measurement	Why measured
Dissolved oxygen	• A requirement for most aquatic life • Indicator of organic pollution • Indicator of photosynthesis (plant growth)
Temperature	• Organisms have specific temperature needs • Indicator of mixing processes • Computer modeling examining the uptake and release of nutrients
Conductivity	• Indicator of total salts dissolved in water • Indicator for geothermal input
pH (acidity)	• Aquatic life protection • Indicator of industrial discharges, mining
Clarity - turbidity - black disk	• Aesthetic appearance • Aquatic life protection • Indicator of catchment condition, land use
Color - light absorption	• Aesthetic appearance • Light availability for excessive plant growth • Indicator of presence of organic matter
Nutrients (Nitrogen and phosphorus)	• Enrichment, excessive plant growth • Limiting factor for plant and algal growth
Major ions (Mg^{2+}, Ca^{2+}, Na^+, K^+, Cl^-, HCO_3^-, SO_4^{2-})	• Baseline water quality characteristics • Indicator for catchment soil types, geology • Water hardness (magnesium/calcium) • Buffering capacity for pH change (HCO_3^-)
Organic carbon	• Indicator of organic pollution • Catchment characteristics
Fecal bacteria	• Indicator of pollution with fecal matter • Disease risk for swimming etc.

1. Explain why measurements of dissolved oxygen, temperature, and clarity are routinely made in the field:

2. Explain how you could use indicator species to detect pollution in a stream:

3. Discuss the link between water quality and land use:

Related activities: *Nitrogen Pollution, Sampling Populations*

Loss of Biodiversity

The species is the basic unit by which we measure biological diversity or **biodiversity**. Biodiversity is not distributed evenly on Earth, being consistently richer in the tropics and concentrated more in some areas than in others. Conservation International recognizes 25 **biodiversity hotspots**. These are biologically diverse and ecologically distinct regions under the greatest threat of destruction. They are identified on the basis of the number of species present, the amount of **endemism**, and the extent to which the species are threatened. More than a third of the planet's known terrestrial plant and animal species are found in these 25 regions, which cover only 1.4% of the Earth's land area. Unfortunately, biodiversity hotspots often occur near areas of dense human habitation and rapid human population growth. Most are located in the tropics and most are forests. Loss of biodiversity reduces the stability and resilience of natural ecosystems and decreases the ability of their communities to adapt to changing environmental conditions. With increasing pressure on natural areas from urbanization, roading, and other human encroachment, maintaining species diversity is paramount and should concern us all today.

Biodiversity Hotspots

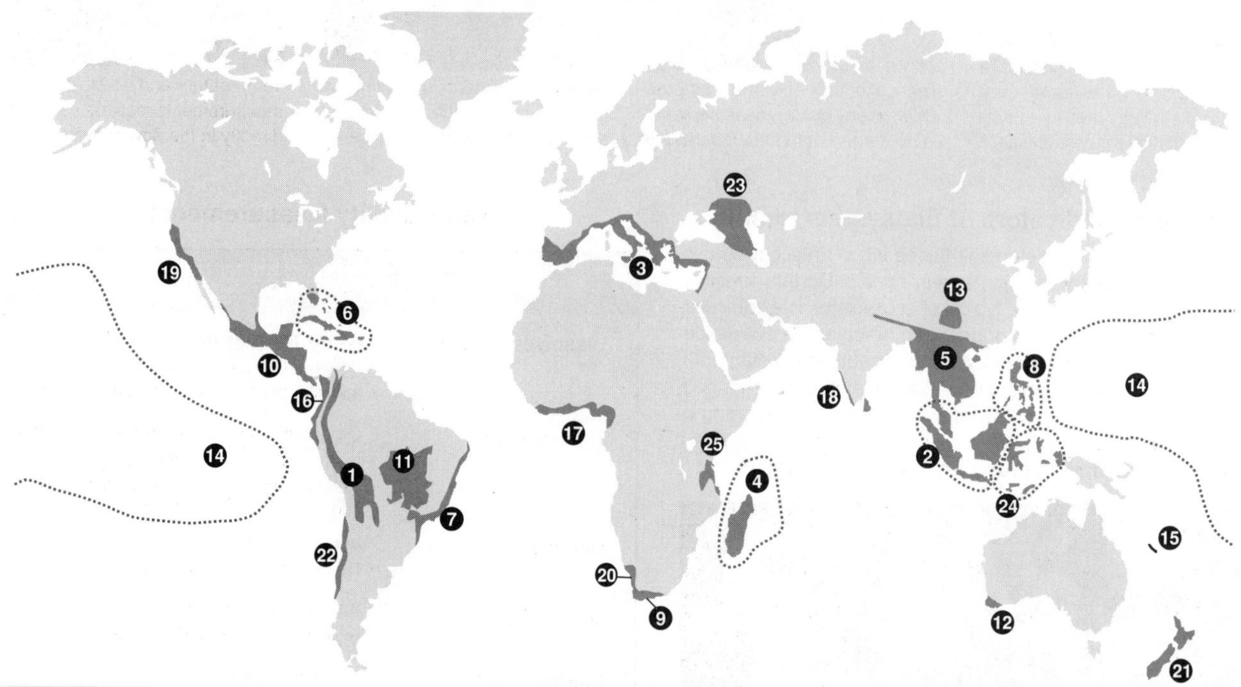

Threats to Biodiversity

Rainforests in some of the most species-rich regions of the world are being destroyed at an alarming rate as world demand for tropical hardwoods increases and land is cleared for the establishment of agriculture.

Illegal trade in species (for food, body parts, or for the exotic pet trade) is pushing some species to the brink of extinction. Despite international bans on trade, illegal trade in primates, parrots, reptiles, and big cats (among others) continues.

Pollution and the pressure of human populations on natural habitats threatens biodiversity in many regions. Environmental pollutants may accumulate through food chains or cause harm directly, as with this bird trapped in oil.

1. Use your research tools (e.g. textbook, internet, or encyclopaedia) to identify each of the 25 biodiversity hotspots illustrated in the diagram above. For each region, summarize the characteristics that have resulted in it being identified as a biodiversity hotspot. Present your summary as a short report and attach it to this page of your workbook.

2. Identify the threat to biodiversity that you perceive to be the most important and explain your choice:

Related activities: *Disturbance and Community Structure*
Web links: *Biodiversity Hotspots*

Periodicals:
Biodiversity: Taking stock of life, Earth's nine lives

Nature Reserves

Conservation on a national scale generally involves setting up reserves or protected areas to slow the loss of biodiversity. The different types of **nature reserves**, e.g. wildlife, scenic and scientific reserves, and National Parks, have varying levels of protection depending on country and local laws. Various management strategies (below) are used to protect species already at risk, and help those that are endangered return to sustainable population sizes. Internationally, there are a number of agencies concerned with monitoring and managing the loss of biodiversity. **The Nature Conservancy** is one such organization. The mission of the Conservancy is to preserve the natural communities that represent the diversity of life on Earth by protecting the lands and waters they need to survive. The Conservancy has purchased more than 12 million acres in the USA and a slightly smaller amount outside the USA. Larger nature reserves usually promote conservation of biodiversity more effectively than smaller ones, with **habitat corridors** for wildlife and **edge effects** also playing a part.

Ex-Situ Conservation Methods

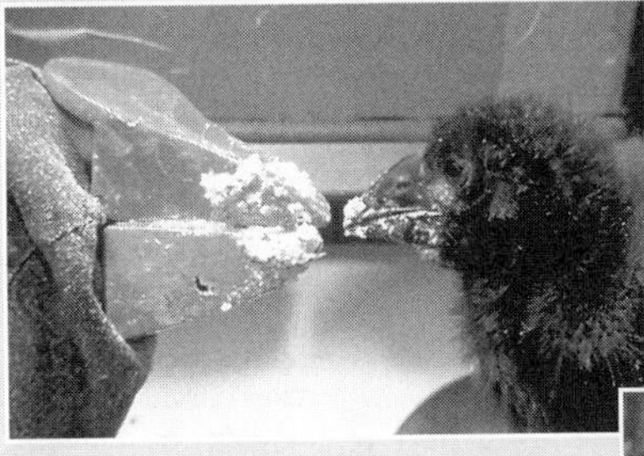

Captive Breeding and Relocation

Individuals are captured and bred under protected conditions. If breeding is successful and there is suitable habitat available, captive individuals are relocated to the wild where they can establish natural populations. Zoos now have an active role in captive breeding. *Photo left: A puppet 'mother' feeds a takahe chick.*

The Role of Zoos

Many zoos specialize in captive breeding programmes and have a major role in public education. Modern zoos tend to concentrate on particular species and are part of global programmes that work together to help retain genetic diversity in captive bred animals. *Photo right: Okapi; a rare forest antelope.*

The Role of Botanic Gardens

Botanic gardens have years of collective expertise and resources and play a critical role in plant conservation. They maintain seed banks, nuture rare species, maintain a living collection of plants, and help to conserve indigenous plant knowledge. They also have an important role in both research and education. *Photo left: The palm house at Kew Botanic Gardens.*

Seed and Gene Banks

Seed and **gene banks** around the world have a role in preserving the genetic diversity of species. A seed bank (right) stores seeds as a source for future planting in case seed reserves elsewhere are lost. The seeds may be from rare species whose genetic diversity is at risk, or they may be the seeds of crop plants, in some cases of ancient varieties no longer used in commercial production.

In-Situ Conservation Methods

Woodland-pond restoration (UK)

Habitat Protection and Restoration

Most countries have a system of parks and reserves focussed on **whole ecosystem conservation**. These areas aim to preserve habitats with special importance and they may be intensively managed through pest and weed control programs, revegetation, and reintroduction of threatened species. The most effective restoration programs are associated with ingoing monitoring and management of lfora and fauna to return threatened species to viable levels.

Orangutan (endangered species)

Ban on Trade in Endangered Species

The Convention on International Trade in Endangered Species (CITES) is an international agreement between governments which aims to ensure that international trade in species of wild animals and plants does not threaten their survival. Unfortunately, even under CITES, species are not guaranteed safety from illegal trade.

1. Explain how the following *ex situ* conservation measures are used in the restoration of endangered species:

   (a) Captive breeding of animals: ______________________________

   ____

   (b) Botanic gardens and gene banks: ______________________________

   ____

2. Describe one advantage of *in situ* (in place) conservation measures in conserving biodiversity: ______________________________

   ____

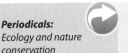

National parks are usually located in places which have been largely undeveloped, and they often feature areas with exceptional ecosystems such as those with endangered species, high biodiversity, or unusual geological features. Canada's National Parks are a country-wide system of representative natural areas of Canadian significance. They are protected by law for public understanding, appreciation, and enjoyment, while being maintained for future generations. National parks have existed in Canada for well over a century. Some 83 million acres (11% of public lands) of the USA are in National Parks and Preserves, which protect natural resources, while allowing restricted activities. National wildlife refuges form a network across the USA, with at least one in every state. They provide habitat for endangered species, migratory birds, and big game.

Parks and Reserves in North America

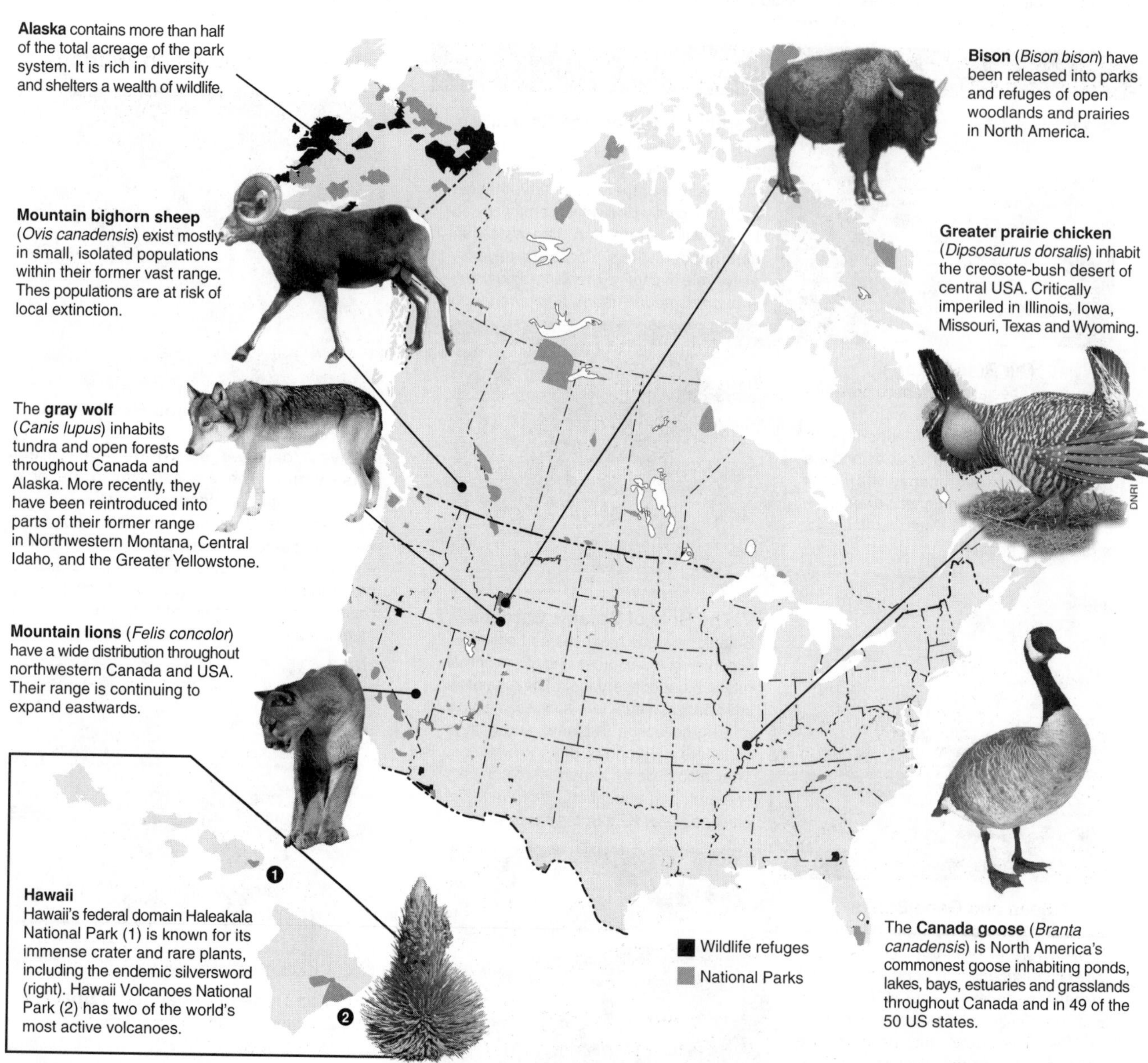

Alaska contains more than half of the total acreage of the park system. It is rich in diversity and shelters a wealth of wildlife.

Bison (*Bison bison*) have been released into parks and refuges of open woodlands and prairies in North America.

Mountain bighorn sheep (*Ovis canadensis*) exist mostly in small, isolated populations within their former vast range. Thes populations are at risk of local extinction.

Greater prairie chicken (*Dipsosaurus dorsalis*) inhabit the creosote-bush desert of central USA. Critically imperiled in Illinois, Iowa, Missouri, Texas and Wyoming.

The **gray wolf** (*Canis lupus*) inhabits tundra and open forests throughout Canada and Alaska. More recently, they have been reintroduced into parts of their former range in Northwestern Montana, Central Idaho, and the Greater Yellowstone.

Mountain lions (*Felis concolor*) have a wide distribution throughout northwestern Canada and USA. Their range is continuing to expand eastwards.

Hawaii
Hawaii's federal domain Haleakala National Park (1) is known for its immense crater and rare plants, including the endemic silversword (right). Hawaii Volcanoes National Park (2) has two of the world's most active volcanoes.

The **Canada goose** (*Branta canadensis*) is North America's commonest goose inhabiting ponds, lakes, bays, estuaries and grasslands throughout Canada and in 49 of the 50 US states.

3. Using a local example, discuss the role of **active management** strategies in conservation of an endangered species:

4. Use your research tools (e.g. textbook, internet, or encyclopaedia) to identify a National park or wildlife refuge. Summarize the features that have resulted in it being identified as a protected area:

KEY TERMS Mix and Match

INSTRUCTIONS: Test your vocab by matching each term to its correct definition, as identified by its preceding letter code.

ACID RAIN	**A** The harmful introduction of contaminants into an environment.
ALIEN SPECIES	**B** A graphical representation of the numbers, biomass, or productivity at each trophic level in a given ecosystem.
BIODIVERSITY	**C** The active management of a resource so that it is available for indefinite future use.
CAPTIVE BREEDING	**D** A greenhouse gas with a global warming potential that is 25 times that of carbon dioxide (over 100 years)
CARBON CYCLE	**E** The process of the Earth's surface steadily increasing in temperature. Usually attributed the rise in gases produced by fossil fuels and industrial processes.
CARBON DIOXIDE	**F** Any species that is vulnerable to extinction in the near future.
CONSERVATION	**G** The active management of natural populations in order to rebuild numbers and ensure species survival.
DENITRIFICATION	**H** The number of variety of species living within an area. The species richness.
ECOLOGICAL PYRAMID	**I** The act of retaining of environments in their natural state or in a state not greatly influenced by humans.
ENDANGERED SPECIES	**J** A population of organisms at risk of extinction because of low numbers or threats from habitat loss, hunting, pests, or diseases etc.
ENERGY FLOW DIAGRAM	**K** Low pH precipitation that can harm plants, aquatic animals, and infrastructure. It is caused by carbon, nitrogen, and sulphur compounds reacting with water molecules in the atmosphere to produce acids.
EUTROPHICATION	**L** The gain of excess nutrients into an environment, usually aquatic, often by farm runoff or pollution.
GENE BANK	**M** The natural biotic or abiotic processes, by which nitrogen (N_2) in the atmosphere is converted into ammonia.
GLOBAL WARMING	**N** The retention of solar energy in the Earth's atmosphere by gases that absorb heat and prevent it being released back into space.
GREENHOUSE EFFECT	**O** Any gas in the atmosphere that causes the retention of heat in the Earth's atmosphere. Major gases are water vapor and carbon dioxide.
GREENHOUSE GAS	**P** A species that evolved in one place and has been transported either intentionally or unintentionally by humans to a new habitat; a non-native.
METHANE	**Q** Breeding animals in human controlled environments with restricted settings, usually with the purpose of conservation management and species restoration.
NITROGEN CYCLE	**R** Any cycle which moves nutrients from one form or state to another through exchanges with the air, water, and organisms.
NITROGEN FIXATION	**S** A diagram quantifying the energy gains and energy losses in an ecosystem. It can be used to show the efficiency of a given system.
NUTRIENT CYCLE	**T** The expression of a need by decision-makers to anticipate harm and to mitigate against it before it occurs, even in the absence of scientific certainty.
POLLUTION	**U** Biogeochemical cycle by which carbon is exchanged among the biotic and abiotic components of the Earth.
PRECAUTIONARY PRINCIPLE	**V** The removal of nitrogen compounds from the soil by bacteria.
PRESERVATION	**W** A trace atmospheric gas used by plants in photosynthesis to make sugars.
SUSTAINABLE MANAGEMENT	**X** The collective biological and non-biological processes by which nitrogen is converted between its various chemical forms.
THREATENED SPECIES	**Y** A store of preserved genetic material.

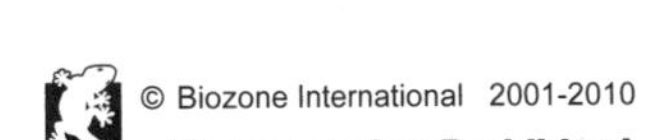

Appendix

PERIODICAL REFERENCES

CLASSIFICATION

- **What is a Species?**
Scientific American June 2008, pp. 48-55. *The science of classification; modern and traditional approaches, the value of each, and the importance of taxonomy to identifying and recognising diversity. Excellent.*

- **A Passion for Order**
National Geographic, 211(6) June 2007, pp. 73-87. *The history of Carl Linnaeus and the classification of plant species.*

- **The Loves of the Plants**
Scientific American, Feb. 1996, pp. 98-103. *The classification of plants and the development of keys to plant identification.*

- **World Flowers Bloom after Recount**
New Scientist, 29 June 2002, p. 11. *A systematic study of flowering plants indicates more species than expected, especially in regions of high biodiversity such as South American and Asia.*

- **The Family Line - The Human-Cat Connection**
National Geographic, 191(6) June 1997, pp. 77-85. *An examination of the genetic diversity and lineages within the felidae. A good context within which to study classification.*

KEEPING IN BALANCE

- **Homeostasis**
Biol. Sci. Rev., 12(5) May 2000, pp. 2-5. *Homeostasis: what it is, the role of negative feedback and the autonomic nervous system, and the adaptations of organisms for homeostasis in extreme environments (excellent).*

- **Growth Hormone**
Biol. Sci. Rev., 12 (4) March 2000, pp. 26-28. *The consequences of growth hormone deficiencies in humans.*

- **Metabolic Powerhouse**
New Scientist, 11 Nov. 2000 (Inside Science). *The myriad roles of the liver in metabolism, including discussion of amino acid and glucose metabolism.*

- **The Liver in Health and Disease**
Biol. Sci. Rev., 14(2) Nov. 2001, pp. 14-20. *The various roles of the liver, a major homeostatic organ.*

- **Temperature Regulation**
Biol. Sci. Rev., 17(2) Nov. 2004, pp. 2-7. *An account of thermoregulation in animals. It also discusses how animals cope with temperature extremes.*

- **Hair Growth in Mammals**
Biol. Sci. Rev., 17(4) April 2005, pp. 37-40. *How cycles of hair growth are timed so that the properties of the pelage meet the needs of the changing seasons.*

- **Glucose Center Stage**
Biol. Sci. Rev., 19(2), Nov. 2006, pp. 14-17. The homeostatic regulation of blood glucose, including the role of hormones.

- **Food for Thought**
Biol. Sci. Rev., 22(4, April 2010, pp. 22-25. *A clear, thorough account of how the body maintains its supply of glucose long after the nutrients absorbed from a meal have been exhausted.*

EATING TO LIVE

- **Insect Metamorphosis**
Biol. Sci. Rev., 12(4) March 2001, pp. 29-33. *The physiological and morphological changes associated with metamorphosis and associated dietary shifts in insects.*

- **Rumen Microbiology**
Biol. Sci. Rev., 14 (4) April 2002, pp. 14-17. *The pivotal role of microorganisms in ruminant digestion.*

- **The Anatomy of Digestion**
Biol. Sci. Rev., 22(3) Feb. 2010, pp. 18-21. *A well structured account of the stages food goes through as it moves through the human digestive tract.*

- **The Pancreas and Pancreatitis**
Biol. Sci. Rev., 13(5) May 2001, pp. 2-6. *The structure of the pancreas and its role in digestion.*

- **The Liver in Health and Disease**
Biol. Sci. Rev., 14(2) Nov. 2001, pp. 14-20. *The various roles of the liver, a major homeostatic organ.*

- **Diabetes**
Biol. Sci. Rev., 15(2), Nov. 2002, pp. 30-35. *The homeostatic imbalance that results in diabetes. The role of the pancreas in the hormonal regulation of blood glucose is discussed.*

THE BREATH OF LIFE

- **Getting in and Out**
Biol. Sci. Rev., 20(3), Feb. 2008, pp. 14-16. *Diffusion: some adaptations and some common misunderstandings*

- **Breathless**
New Scientist, 8 March 2003, pp. 46-49. *Adaptations for gas exchange in shark species able to withstand anoxia.*

- **Gas Exchange in the Lungs**
Bio. Sci. Rev. 16(1) Sept. 2003, pp. 36-38. *The structure and function of the alveoli of the lungs, with an account of respiratory problems and diseases.*

- **Humans with Altitude**
New Scientist, 2 Nov. 2002, pp. 36-39. *The short term adjustments and long term adaptations to life at altitude.*

LIFE BLOOD

- **Cunning Plumbing**
New Scientist, 6 Feb. 1999, pp. 32-37. *The arteries can actively respond to changes in blood flow, spreading the effects of mechanical stresses to avoid extremes.*

- **Blood Pressure**
Biol. Sci. Rev., 12(5) May 2000, pp. 9-12. *Blood pressure: its control, measurement, and significance to diagnosis.*

- **A Fair Exchange**
Biol. Sci. Rev., 13(1), Sept. 2000, pp. 2-5. *The role of tissue fluid in the body and how it is produced and reabsorbed.*

- **Red Blood Cells**
Bio. Sci. Rev. 11(2) Nov. 1998, pp. 2-4. *The structure and function of erythrocytes, including details of oxygen transport.*

- **The Heart**
Bio. Sci. Rev. 18(2) Nov. 2005, pp. 34-37. *The structure and physiology of the heart.*

- **The Search for Blood Substitutes**
Scientific American, Feb. 1998, pp. 60-65. *Finding a successful blood substitute depends on being able to replicate to the exact properties of blood. And it's not that easy.*

- **Venous Disease**
Biol. Sci. Rev., 19(3), Feb. 2007, pp. 15-17. *Valves in the deep veins of the legs assist venous return but when these are damaged, superficial veins are put under more pressure and circulation is compromised.*

- **Mending Broken Hearts**
National Geographic, 211(2), Feb. 2007, pp. 40-65. *Heart disease is becoming more prevalent. Assessing susceptibility.*

Appendix

PERIODICAL REFERENCES

to heart disease may be the key to treating the disease more effectively.

▶ **Keeping Pace - Cardiac Muscle & Heartbeat**
Biol. Sci. Rev., 19(3), Feb. 2007, pp. 21-24. *The structure and properties of cardiac muscle.*

▶ **Breaking Out of the Box**
The American Biology Teacher, 63(2), February 2001, pp. 101-115. *Investigating cardiovascular activity: a web-based activity on the cardiac cycle.*

DEFENDING AGAINST DISEASE

▶ **Koch's Postulates**
Biol. Sci. Rev., 15(3) February 2003, pp. 24-25. *Koch's postulates and the diagnosis of infectious disease.*

▶ **Mosquitoes**
Biol. Sci. Rev., 20(1) Sept. 2007, pp 34-37. *Life cycle of mosquitoes, and the disease they carry including malaria and dengue fever.*

▶ **Skin, Scabs and Scars**
Biol. Sci. Rev., 17(3) Feb. 2005, pp. 2-6. *The roles of skin, including its role in wound healing and the processes involved in its repair when damaged.*

▶ **Fight for Your Life!**
Biol. Sci. Rev., 18(1) Sept. 2005, pp. 2-6. *The mechanisms by which we recognize pathogens and defend ourselves against them (overview).*

▶ **What is the Human Microbiome?**
Biol. Sci. Rev., 22(2) Nov. 2009, pp. 38-41. *An informative acount of the nature and role of the microbes that inhabit our bodies.*

▶ **Looking Out for Danger: How White Blood Cells Protect Us**
Biol. Sci. Rev., 19 (4) April 2007, pp. 34-37. *The various types of lymphocytes (white blood cells) and they work together to protect the body against infection.*

▶ **Inflammation**
Biol. Sci. Rev., 17(1) Sept. 2004, pp. 18-20. *The role of this nonspecific defence response to tissue injury and infection. The processes involved in inflammation are discussed.*

▶ **Immunology**
Biol. Sci. Rev., 22(4) April 2010, pp. 20-21. *A pictorial but information-packed review of the basic of internal defense functions.*

▶ **Lymphocytes - The Heart of the Immune System**
Biol. Sci. Rev., 12 (1) Sept. 1999 pp. 32-35. *An excellent account of the role of lymphocytes in the immune response (includes the types and actions of different lymphocytes).*

▶ **Antibodies**
Biol. Sci. Rev., 11(3) January 1999, pp. 34-35. *The operation of the immune system and the production of antibodies (including procedures for producing monoclonal antibodies).*

▶ **Monoclonals as Medicines**
Biol. Sci. Rev., 18(4) April 2006, pp. 38-40. *The use of monoclonal antibodies in therapeutic and diagnostic medicine.*

▶ **AIDS**
Biol. Sci. Rev., 20(1) Sept. 2007, pp. 30-12. *The HIV virus can evade the immune system and acquire drug resistance. This has prevented effective cures from being developed.*

REGULATING FLUID AND REMOVING WASTES

▶ **Uric Acid-Life Saver and Liability**
Biol. Sci. Rev., 18(4) April 2006, pp. 7-9. *Nitrogen excretion and the situations (and taxa) in which uric acid is produced.*

▶ **Urea**
Biol. Sci. Rev., 17(4) April 2005, pp. 6-8. *An account of nitrogen balance and how and why urea is formed.*

▶ **Countercurrent Exchange Mechanisms**
Biol. Sci. Rev., 9(1) September 1996, pp. 2-6. *The diversity and roles of countercurrent multipliers in biological systems including the kidney nephron.*

▶ **The Kidney**
Bio. Sci. Rev. 16(2) Nov. 2003, pp. 2-7. *The structure and function of the human kidney, including countercurrent multiplication in the loop of Henlé.*

RESPONDING TO THE ENVIRONMENT

▶ **The Autonomic Nervous System**
Biol. Sci. Rev., 18(3) Feb. 2006, pp. 21-25. *Description of the structure and roles of the ANS.*

▶ **The Cerebral Cortex**
Biol. Sci. Rev., 21(1) Sept. 2008, pp. 2-6. *The structure, organization, and many roles of the cerebral cortex.*

▶ **Refractory Period**
Biol. Sci. Rev., 20(4) April 2008, pp. 7-9. *The nature and purpose of the refractory period in response stimuli. The biological principles involved are discussed with in the context of the refractory period of the human heart.*

▶ **Bridging the Gap**
Biol. Sci. Rev., 21(2) Nov. 2008, pp. 2-6. *Communication between nerve cells across synapses; the biology of chemical transmission.*

▶ **Infinite Sensation**
New Scientist, 11 August, 2001, pp. 24-28. *An examination of the nature of sensory perception and how we perceive the world. It has links to web sites where students explore their own responses to sensory inputs.*

▶ **What is a Pacinian Corpuscle?**
Biol. Sci. Rev., 21(3) Feb. 2009, pp. 11-13. *How these pressure receptors in the skin work to convert a mechanical stimulus into a nervous impulse.*

▶ **Remodelling the Eye**
Biol. Sci. Rev., 17(2) Nov. 2004, pp. 9-12. *An account of the structure of the human eye and disorders of vision.*

▶ **More than Meets the Eye**
Biol. Sci. Rev., 20(3) Feb. 2008, pp. 2-5. *The structure and function of vertebrate and invertebrate eyes. Visual adaptation and accommodation are also covered.*

▶ **From Genes to Color Vision**
Biol. Sci. Rev., 22(4) April 2010, pp. 2-5. *How do we distinguish color? This article describes the physiological basis of color vision and examines the case for why and how it may have evolved.*

▶ **Under Color of Darkness**
New Scientist, 6 Jan. 2007, pp. 36-39. *It has always been assumed that nocturnal animals see the world in black and white, but recent studies indicate that some nocturnal species can see color in dim light. This article includes a graph showing what animals see color at various points along the spectrum.*

▶ **Biological Clocks: Telling Time in the Arctic**
Biol. Sci. Rev. 21(2) Nov. 2008, pp. 14-17. *The nature of biological clocks and rhythms, adaptations to life in the Arctic light, and the role of rhythms in migration patterns.*

▶ **Times of our Lives**
Scientific American, Sept. 2002, pp. 40-47. *Biological clocks and their neurological basis; this article discusses how the brain synchronizes bodily functions.*

Appendix

PERIODICAL REFERENCES

- **The Hunger, The Horror?**
New Scientist, 30 May 2008, pp. 42-45. *The adaptive dispersal behaviour of locusts when they swarm.*

- **Flight of the Navigators**
New Scientist, 26 July 2008, pp. 36-39. *Better tracking technologies are solving the mysteries of bird migrations.*

- **Home Sweet Home**
New Scientist, 5 June 1999, pp. 34-38. *Navigation in honey bees: the waggle dance and its use in communicating information on foraging sites to fellow hive members.*

- **Honeybees: Workers Rule OK?**
Bio. Sci. Rev. 16(4) April 2004, pp. 23-26. *Queen bees rule through pheromones that convey information to others in the colony and influence their behaviour.*

- **All for One**
New Scientist, 13 June 1998, pp. 32-35. *Social insects such as ants and termites all show cooperative behaviour: workers are altruistic and the colony functions as a superorganism.*

- **Relative Distance**
Scientific American, Jan. 2008, pp. 15-16. *Cooperative breeding behaviour in hyaenas: spatial groups of striped hyenas comprise just one female defended by up to three males, which means some males don't breed.*

- **Animal Attraction**
National Geographic, July 2003, pp. 28-55. *An engaging and expansive account of mating in the animal world.*

MUSCLES & MOVEMENT

- **Energy Saving in Animal Movement**
Biol. Sci. Rev., 18(2) Nov. 2005, pp. 2-5. *Locomotion is costly in terms of energy; this account describes how structure, physiology, and behaviour are adapted to conserving energy during locomotion.*

- **How Skeletal Muscles Work**
Biol. Sci. Rev., 22(4) April 2010, pp. 10-15. *The structure and function of muscle in humans: contraction, sliding filament theory, and types of muscle.*

THE NEXT GENERATION

- **The Trouble with Sex**
New Scientist, 6 December 2003, pp. 44-47. *Sex must confer an advantage because it is so common. Just what does sex offer over asexual reproduction? Recent experiments point to the unpredictability of the environment being an important consideration.*

- **It's a Frog's Life**
Biol. Sci. Rev., 17(2) Nov. 2004, pp. 17-20. *Lifestyles and patterns of reproduction in leaf frogs.*

- **Why we don't Lay Eggs**
New Scientist, 12 June 1999, pp. 26-31. *Mammalian reproduction: the role of the placenta, the evolution of live birth, and exceptions to the usual mammalian pattern.*

- **Measuring Female Hormones in Saliva**
Biol. Sci. Rev., 13(3) Jan. 2001, pp. 37-39. *The female reproductive system, and the complex hormonal control of the female menstrual cycle.*

- **The Great Escape[1]**
New Scientist 29 Sept 2007, pp. 40-43. *Female reproductive physiology and hormonal control of the menstrual cycle.*

- **Male Contraception**
Biol. Sci. Rev., 13(2) Nov. 2000, pp. 6-9. *A new contraceptive technology involves the inhibition of spermatogenesis in males.*

- **Spermatogenesis**
Biol. Sci. Rev., 15(4) April 2003, pp. 10-14. *The process and control of sperm production in humans, with a discussion of the possible reasons for male infertility.*

- **The Great Escape[2]**
New Scientist, 15 Sept. 2001, (Inside Science). *How the foetus is accepted by the mother's immune system during pregnancy.*

- **The Placenta**
Biol. Sci. Rev., 12 (4) March 2000, pp. 2-5. *Placental function and the use of the placenta for prenatal diagnosis and gene therapy.*

- **Pregnancy Tests**
Scientific American, Nov. 2000, pp. 92-93. *Pregnancy tests: how they work and the role of HCG in signalling pregnancy.*

- **The Evolution of Human Birth**
Scientific American, Nov. 2001, pp. 60-65. *An examination of the various unique aspects of human reproduction and how they arose.*

- **The Biology of Milk**
Biol. Sci. Rev., 16(3) Feb. 2004, pp. 2-6. *The production and composition of milk, its role in mammalian biology, and the physiological processes controlling its release.*

- **Adolescence-Hormones Rule OK?**
Biol. Sci. Rev., 19(3) Feb. 2007, pp. 2-6. *The hormonal changes bringing about reproductive maturity.*

- **Menopause - Design Fault, or By Design?**
Biol. Sci. Rev., 14(1) Sept. 2001, pp. 2-6. *An excellent synopsis of the basic biology of menopause.*

- **Age - Old Story**
New Scientist, 23 Jan. 1999, (Inside Science). *The processes involved in ageing. An accessible, easy-to-read, but thorough account.*

PLANT STRUCTURE AND GROWTH RESPONSES

- **Cacti**
Biol. Sci. Rev., 20(1), Sept. 2007, pp. 26-30. *The growth forms and structural and physiological adaptations of cacti.*

- **Sending Plants Around the Bend**
Biol. Sci. Rev., 12(4) March 2000, pp. 14-17. *An account of how plants perceive and respond to stimuli around them. Tropisms are fully covered.*

- **How Plants Know Their Place**
Biol. Sci. Rev., 17(3) Feb. 2005, pp. 33-36. *Plant responses to light and the action of phytochromes.*

PLANT SUPPORT AND TRANSPORT

- **How Trees Lift Water**
Biol. Sci. Rev., 18(1), Sept. 2005, pp. 33-37. *Cohesion-tension theory and others on how trees lift water.*

- **High Tension**
Biol. Sci. Rev., 13(1), Sept. 2000, pp. 14-18. *Cell specialisation and transport in plants: an excellent account of the mechanisms by which plants transport water and solutes.*

PLANT REPRODUCTION

- **Hot Plants**
Biol. Sci. Rev. 20(4), April 2008, pp. 24-27. *Some plants attract pollinators by generating heat. these thermogenic plants have unusual, often spectacular flowering structures.*

- **Flower Power**
New Scientist, 9 January 1999, pp. 22-26. *Pollination, fertilization, and pollen competition in angiosperms.*

Appendix

INDEX OF LATIN & GREEK ROOTS

▶ **An Explosive Start for Plants**
New Scientist, 2 Jan. 1993, pp. 35-37. *Seed dispersal often involves the explosion of the seeds from the fruit.*

HABITAT & DISTRIBUTION

▶ **Grasslands**
Biol. Sci. Rev., 22(2), Nov. 2009, 2-5. *Grasslands are the dominant vegetation over vast regions of the world. The distribution of this immense biome is governed by rainfall and temperature patterns across the globe.*

POPULATION ECOLOGY

▶ **Fieldwork - Sampling Plants**
Biol. Sci. Rev., 10(5) May 1998, pp. 6-8. *Methodology for sampling plant communities. Includes thorough coverage of quadrat use.*

▶ **Population Bombshell**
New Scientist, 11 July 1998 (Inside Science). *Current and predicted growth rates in human populations, including analyses of population age distributions.*

▶ **Logarithms and Life**
Biol. Sci. Rev., 13(4) March 2001, pp. 13-15. *The basics of logarithmic growth and its application to real populations.*

▶ **Fieldwork - Sampling Animals**
Biol. Sci. Rev., 10(4) March 1998, pp. 23-25. *Appropriate methodology for collecting different types of animals in the field. Includes a synopsis of the mark and recapture technique.*

COMMUNITY ECOLOGY

▶ **Getting to Grips with Ecology**
Biol. Sci. Rev., 22(3) Feb. 2010, pp. 14-16. *Understanding the terms used in population and community ecology.*

▶ **Ecosystems**
Biol. Sci. Rev., 9(4) March 1997, pp. 9-14. *Ecosystems: food chains & webs, energy flows, nutrient cycles, and ecological pyramids.*

▶ **All Life is Here**
New Scientist, 15 March 1997, pp. 24-26. *Small water bodies provide ideal ecosystems in which to study ecology.*

▶ **The Lake Ecosystem**
Biol. Sci. Rev., 20(3) Feb. 2008, pp. 21-25. *An account of the components and functioning of lake ecosystems.*

▶ **Cave Dwellers: Living Without Light**
Biol. Sci. Rev., 17(1) Sept. 2004, pp. 38-41. *How do cave ecosystems thrive in the absence of light? An account of the functioning of a simple cave system.*

▶ **The Ecological Niche**
Biol. Sci. Rev., 12(4), March 2000, pp. 31-35. *An excellent account of the niche - an often misunderstood concept that is central to ecological theory.*

▶ **Inside Story**
New Scientist, 29 April 2000, pp. 36-39. *Ecological interactions between fungi and plants and animals.*

▶ **The Future of Red Squirrels in Britain**
Biol. Sci. Rev., 16(2) Nov. 2003, pp. 8-11. *An account of the impact of the grey squirrel on Britain's native red squirrel populations.*

▶ **Black Squirrels**
Biol. Sci. Rev., 21(2) Nov. 2008, pp. 39-41. *A recently recognized black morph of the gray squirrel is becoming the more common morph in Britain in some regions.*

▶ **An Uneasy Eden**
National Geographic, 214(1), July 2008, pp. 144-157. *Predator-prey relationships in marine reef ecosystems.*

▶ **Batesian Mimicry in Your Own Backyard**
Biol. Sci. Rev., 17(3) Feb. 2005, pp. 25-27. *Batesian mimicry is seen in even the most common species.*

▶ **Biodiversity and Ecosystems**
Biol. Sci. Rev., 11(4) March 1999, pp. 18-23. *Ecosystem diversity and its relationship to ecosystem stability.*

▶ **Plant Succession**
Biol. Sci. Rev., 14 (2) November 2001, pp. 2-6. *Primary and secondary succession, including the causes of different types of succession.*

MANAGING IN A CHANGING WORLD

▶ **The Case of the Missing Carbon**
National Geographic, 205(2), Feb. 2004, pp. 88-117. *The role of carbon sinks in the Earth's carbon cycling.*

▶ **Carbon Cowboys**
Sci. American, Special issue, Earth 3.0, 18(5) 2008, pp. 52-57. *An account of how carbon credits can operate to the benefit of the environment.*

▶ **Global Warming**
Time, special issue, 2007. *A. engaging special issue on global warming: the causes, perils, solutions, and actions.*

▶ **Climate Change and Biodiversity**
Biol. Sci. Rev., 16(1) Sept. 2003, pp. 10-14. *While the focus of this account is on climate change, it provides useful coverage of ecosystem structure and processes and how these are studied.*

▶ **Microbes and Nutrient Cycling**
Biol. Sci. Rev., 19(1) Sept. 2006, pp. 16-20. *The roles of microorganisms in nutrient cycling.*

▶ **The Nitrogen Cycle**
Biol. Sci. Rev., 13(2) Nov. 2000, pp. 25-27. *An excellent account of the nitrogen cycle: conversions, role in ecosystems, and the influence of human activity.*

▶ **Ecology & Nature Conservation**
Biol. Sci. Rev., 18(1) Sept. 2005, pp. 11-15. *The prnciples of conservation and restoration ecology.*

▶ **Biodiversity: Taking Stock**
National Geographic, 195(2) Feb. 1999 (entire issue). *Special issue exploring the Earth's biodiversity and what we can do to preserve it.*

▶ **Earth's Nine Lives**
New Scientist, 27 Feb. 2010, pp. 31-35. *How much can we push the Earth's support systems. This account examine the human interaction with nine critical global functions.*

▶ **Conflicted Conservation**
Scientific American, Sept. 2009, pp. 10-11. *Measures to protect biodiversity could force indigenous peoples off their land into poverty.*

INDEX OF LATIN & GREEK ROOTS, PREFIXES, & SUFFIXES

Many biological terms have a Latin or Greek origin. Understanding the meaning of these components in a word will help you to understand and remember its meaning and predict the probable meaning of new words. Recognizing some common roots, suffixes, and prefixes will make learning and understanding biological vocabulary easier.

The following terms are identified, together with an example illustrating their use in biology.

a(n)- without .... *anoxic*
ab- away from .... *abductor*
ad- towards .... *adductor*
affer- carrying to .... *afferent*
amphi- both .... *amphibian*

Appendix

INDEX OF LATIN & GREEK ROOTS

amyl- starch .... amylase
anemo- wind .... anemometer
ante- before .... antenatal
anthro- human...................... anthropology
anti- against, opposite ....antibiotic
apo- separate, from .... apoenzyme
aqua- water................................... aquatic
arach- spider .... arachnoid
arbor- tree ....arboreal
arch(ae/i)- ancient.......................Archaea
arthro- joint................................ arthropod
artic- jointed .... articulation
artio- even-numbered .... artiodactyl
auto- self .... autologous
avi- bird..avian
axi- axis...axillary

blast- germ............................... blastopore
brachy- short....................... brachycardia
brady- slow .... bradycardia
branch- gill................................ branchial
bronch- windpipe...................... bronchial
bucca- mouth cavity .... buccal

caec- blind.................................. caecum
cauda- tail .... caudal
centi- hundred .... centimorgan
ceph(al)- head .... cephalothorax
cerebro- brain .... cerebrospinal
cerv- neck ....cervix
chrom- color........................ chromoplast
chym- juice .... chyme
cili- eyelash ....cilia
cloaca- sewer ....cloacal
coel- hollow.............................. coelomate
contra- opposite................. contraception
cotyl- cup ....hypocotyl
crani- skull.................................... cranium
crypt- hidden .... crptic
cyan- blue .... cyanobacteria
cyt- cell ....cytoplasm

dactyl- finger.........................polydactylic
deci-(a) ten .... decibel, decapod
dendr- tree .... dendrogram
dent- tooth..................................edentate
derm- skin ....pachyderm
di- two .... dihybrid
dors- back.. dorsal
dur- hard .... dura mater

echino- spiny ....echinoderm
ecto- outside ....ectoderm
effer- carrying away .... efferent
endo- inside ....endoparasite
equi- horse, equal................... equilibrium
erythr- red .... erthyrocyte
eu- well, very............................. eukaryote
ex- out of.. explant
exo- outside .... exoskeleton
extra- beyond.................... extraperitoneal

foramen- opening ....foramen magnum

gast(e)r- stomach, pouch .... gastric
gymn- naked .... gymnosperm
hal- salty .... halophyte
haplo- single, simple .... hapolid
holo- complete, whole ....holozoic
hydr- water .... hydrophyte
hyper- above.............................. hypertoic
hypo- beneath.......................... hypotonic

inter- between ....interspecific
intra- within.............................intraspecific
iso- equal ....isotonic

kilo- thousand .... kilogram

leuc- white............................. leuc(k)ocyte
lip- fat .... lipoprotein
lith- stone ....Paleolithic
lumen- cavity.................................... lumen
lute- yellow .... corpus luteum
lymph- clear water ....lymphatic

magni- large.........................magnification
mamma- breast............................ mammal
mat(e)ri- mother .... maternal
mega- large .... megakaryocyte
melan- black ....melanocyte
meso- middle ....Mesolithic
meta- after....................... metamorphosis
micro- small .... microorganism
milli- thousand.............................millimetre
mono- one...............................monohybrid
morph- form ....morphology
motor- mover..........................motor nerve
multi- many .... multicellular
myo- muscle ....myofibril

necro- dead ....necrosis
neo- new .... Neolithic
nephr- kidney ....nephro
neur- nerve.. neural
notho- southern.......................Nothofagus
noto- back, south.......................notochord

oecious- house of ....monoecius
oed- swollen...................................o()edema
olfact- smelling............................ olfactory
os(s/t)- bone...............................osteocyte
ovo- egg............................. ovoviviparous

pachy- thick ....pachyderm
pal(a)e- old ....Pal(a)eocene
pect(or)- chest .... pectoral fin
ped- foot .... quadraped
pent- five .... pentose sugar
per(i)- through, beyond............. peristalsis
peri- around .... periosteum
perisso- odd-numbered...... perissodactyl
phag- eat .... phagosome
phyll- leaf .... sclerophyll
physio- nature ....physiology

phyto- plant......................... phytohormone
pisc- fish .... piscivorous
plagio- oblique.......................plagioclimax
pneu(mo/st)- air, lung............... pneumonia
pod- foot ....decapod
poly- many ....polydactyly
pre- before ....premolar
pro- in front of .... Prokaryote
prot- first.....................................protandry
pseud- false ....pseudopodia
pter- wing, fern.......................... Pterophyta
pulmo- lung .... pulmonary

radi- root .... radicle
ren- kidney .... renal
retic- network .... reticulated
rhin- nose, snout .... rhinoceros
rostr- beak, prow.......................... rostrum

sacchar- sugar................. polysaccharide
schizo- split..................... schizocoelomate
scler- hard .... sclerophyll
seba- tallow, wax .... sebaceous
semi- half .... semi-conservative
sept- seven, wall.......................... septum
soma- body......................................somatic
sperm- seed.......................spermatophyte
sphinct- closing .... sphincter
stereo- solid ....stereocilia
stom- mouth .... stoma
strat- layer..............................stratification
sub- below .... subtidal
sucr- sugar .... sucrase
sulc- furrow ....sulci
super- beyond............................... superior
supra- above.................supracoracoideus
sym- with.................................... symbiosis
syn- with.. synapsis

tact- touch ....tactile
tachy- fast .... tachycardia
trans- across .... transmembrane
tri- three...................................triploblastic
trich- hair .... trichome

ultra- above...................................ultraviolet
un- one ....unicellular
uro- tail .... urodele

vas- vessel ....vascular
ven- vein ....venous
ventr- belly .... ventral
vern- spring..vernal
visc- organs of body cavity .... viscera
vitr- glass .... in vitro
xanth- yellow .... xanthophyll
xen- stranger ....xenotransplant
xer- dry ....xerophyte
xyl- wood.. xylem

zo- animal .... zoological

Index

Index